# DIFFERENTIAL EQUATIONS AND LINEAR ALGEBRA

GILBERT STRANG

Department of Mathematics
Massachusetts Institute of Technology

**Differential Equations and Linear Algebra**

**ISBN 978-0-9802327-9-0**



**Typeset by Valutone** (www.valutone.co.in) Printed in the United States of America



| | |
|---|---|
| **Wellesley - Cambridge Press** | **diffeqla@gmail.com** |
| Box 812060 | **math.mit.edu/∼gs** |
| Wellesley MA 02482 USA | phone (781) 431-8488 |
| **www.wellesleycambridge.com** | fax (617) 253-4358 |

Our books are also distributed by SIAM (in North America) and by Cambridge University Press (in the rest of the world).

The website with solutions to problems in this textbook is **math.mit.edu/dela**
Linear Algebra and Differential Equations are on MIT's OpenCourseWare site **ocw.mit.edu**.
This provides video lectures of the full courses 18.03 and 18.06.
Course material is on the teaching website: **web.mit.edu/18.06**
Highlights of Calculus (17 lectures and text) are on **ocw.mit.edu**

The front cover shows the Lorenz attractor, drawn for this book by Gonçalo Morais. This is the first example of chaos, found by Edward Lorenz. The cover was designed by Lois Sellers and Gail Corbett.

# Table of Contents

# Preface

Differential equations and linear algebra are the two crucial courses in undergraduate mathematics. This new textbook develops those subjects separately and together. Separate is normal—these ideas are truly important. This book presents the basic course on differential equations, in full :

| | |
|---|---|
| Chapter 1 | First order equations |
| Chapter 2 | Second order equations |
| Chapter 3 | Graphical and numerical methods |
| Chapter 4 | Matrices and linear systems |
| Chapter 6 | Eigenvalues and eigenvectors |

I will write below about the highlights and the support for readers. Here I focus on the option to include more linear algebra. Many colleges and universities want to move in this direction, by connecting two essential subjects.

More than ever, the central place of linear algebra is recognized. Limiting a student to the mechanics of matrix operations is over. Without planning it or foreseeing it, my lifework has been the presentation of linear algebra in books and video lectures :

*Introduction to Linear Algebra* (Wellesley–Cambridge Press)
*MIT OpenCourseWare* (**ocw.mit.edu**, Mathematics 18.06 in 2000 and 2014).

Linear algebra courses keep growing because the need keeps growing. At the same time, a rethinking of the MIT differential equations course 18.03 led to a new syllabus. And independently, it led to this book.

The underlying reason is that time is short and precious. The curriculum for many students is just about full. Still these two topics cannot be missed—and linear differential equations go in parallel with linear matrix equations. The prerequisite is calculus, for a single variable only—the key functions in these pages are inputs $f(t)$ and outputs $y(t)$. For all linear equations, continuous and discrete, the complete solution has two parts :

$$\textbf{One particular solution } \boldsymbol{y_p} \qquad \boldsymbol{Ay_p = b}$$
$$\textbf{All null solutions } \boldsymbol{y_n} \qquad \boldsymbol{Ay_n = 0}$$

Those right hand sides add to $\boldsymbol{b} + \boldsymbol{0} = \boldsymbol{b}$. The crucial point is that the left hand sides add to $\boldsymbol{A}(\boldsymbol{y_p} + \boldsymbol{y_n})$. When the inputs add, and the equation is linear, the outputs add. The equality $\boldsymbol{A}(\boldsymbol{y_p} + \boldsymbol{y_n}) = \boldsymbol{b} + \boldsymbol{0}$ tells us all solutions to $\boldsymbol{Ay} = \boldsymbol{b}$ :

**The complete solution to a linear equation is $y$ = (one $y_p$) + (all $y_n$).**

The same steps give the complete solution to $dy/dt = f(t)$, for the same reason. We know the answer from calculus—it is the form of the answer that is important here :

$$\frac{dy_p}{dt} = f(t) \quad \text{is solved by} \quad \boldsymbol{y_p(t) = \int_0^t f(x)\,dx}$$

$$\frac{dy_n}{dt} = 0 \quad \text{is solved by} \quad \boldsymbol{y_n(t) = C} \text{ (any constant)}$$

$$\frac{dy}{dt} = f(t) \quad \text{is completely solved by} \quad \boldsymbol{y(t) = y_p(t) + C}$$

For every differential equation $dy/dt = Ay + f(t)$, our job is to find $y_p$ and $y_n$ : one particular solution and all homogeneous solutions. My deeper purpose is to build confidence, so the solution can be understood and used.

## Differential Equations

The whole point of learning calculus is to understand movement. An economy grows, currents flow, the moon rises, messages travel, your hand moves. The action is fast or slow depending on forces from inside and outside : competition, pressure, voltage, desire. Calculus explains the meaning of $dy/dt$, but to stop without putting it into an equation (a differential equation) is to miss the whole purpose.

That equation may describe growth (often exponential growth $e^{at}$). It may describe oscillation and rotation (with sines and cosines). Very frequently the motion approaches an equilibrium, where forces balance. That balance point is found by linear algebra, when the rate of change $dy/dt$ is zero.

The need is to explain what mathematics can do. I believe in looking partly outside mathematics, to include what scientists and engineers and economists actually remember and constantly use. My conclusion is that first place goes to linear equations. The essence of calculus is to linearize around a present position, to find the direction and the speed of movement.

Section 1.1 begins with the equations $dy/dt = y$ and $dy/dt = y^2$. It is simply wonderful that solving those two equations leads us here :

$$\boldsymbol{\frac{dy}{dt} = y \qquad y = 1 + t + \frac{1}{2}t^2 + \frac{1}{6}t^3 + \cdots \qquad y = e^t}$$

$$\boldsymbol{\frac{dy}{dt} = y^2 \qquad y = 1 + t + t^2 + t^3 + \cdots \qquad y = 1/(1-t)}$$

To meet the two most important series in mathematics, right at the start, that is pure pleasure. No better practice is possible as the course begins.

## Important Choices of $f(t)$

Let me emphasize that a textbook must do more than solve random problems. We could invent functions $f(t)$ forever, but that is not right. Much better to understand a small number of highly important functions:

| | | |
|---|---|---|
| $f(t) =$ | **sines and cosines** | (oscillating and rotating) |
| $f(t) =$ | **exponentials** | (growing and decaying) |
| $f(t) =$ | **1 for $t > 0$** | (a switch is turned on) |
| $f(t) =$ | **impulse** | (a sudden shock) |

The solution $y(t)$ is the response to those inputs—frequency response, exponential response, step response, impulse response. These particular functions and particular solutions are the best—the easiest to find and by far the most useful. All other solutions are built from these.

I know that an impulse (a delta function that acts in an instant) is new to most students. This idea deserves to be here! You will see how neatly it works. The response is like the inverse of a matrix—it gives a formula for *all* solutions. The book will be supplemented by video lectures on many topics like this, because a visual explanation can be so effective.

## Support for Readers

Readers should know all the support that comes with this book:

**math.mit.edu/dela** is the key website. The time has passed for printing solutions to odd-numbered problems in the back of the book. The website can provide more detailed solutions and serious help. This includes additional worked problems, and codes for numerical experiments, and much more. Please make use of everything and contribute.

**ocw.mit.edu** has complete sets of video lectures on both subjects (OpenCourseWare is also on YouTube). Many students know about the linear algebra lectures for 18.06 and 18.06 SC. I am so happy they are helpful. For differential equations, the 18.03 SC videos and notes and exams are extremely useful.

The new videos will be about special topics—possibly even the Tumbling Box.

## Linear Algebra

I must add more about linear algebra. My writing life has been an effort to present this subject clearly. Not abstractly, not with a minimum of words, but in a way that is helpful to the reader. It is such good fortune that the central ideas in matrix algebra (a basis for a vector space, factorization of matrices, the properties of symmetric and orthogonal matrices), are exactly the ideas that make this subject so useful. Chapter 5 emphasizes those ideas and Chapter 7 explains the applications of $A^{\mathrm{T}}A$.

Matrices are essential, not just optional. We are constantly acquiring and organizing and presenting data—the format we use most is a matrix. The goal is to see the relation between input and output. Often this relation is linear. In that case we can understand it.

The idea of a vector space is so central. Take *all* combinations of two vectors or two functions. I am always encouraging students to visualize that space—examples are really the best. When you see all solutions to $v_1 + v_2 + v_3 = 0$ and $d^2y/dt^2 + y = 0$, you have the idea of a vector space. This opens up the big questions of linear independence and basis and dimension—by example.

If $f(t)$ comes in continuous time, our model is a differential equation. If the input comes in discrete time steps, we use linear algebra. The model predicts the output $y(t)$ this is created by the input $f(t)$. But some inputs are simply more important than others—they are easier to understand and much more likely to appear. Those are the right equations to present in this course.

## Notes to Faculty (and All Readers)

One reason for publishing with Wellesley-Cambridge Press can be mentioned here. I work hard to keep book costs reasonable for students. This was just as important for *Introduction to Linear Algebra*. A comparison on Amazon shows that textbook prices from big publishers are more than double. Wellesley-Cambridge books are distributed by SIAM inside North America and Cambridge University Press outside, and from Wellesley, with the same motive. Certainly quality comes first.

I hope you will see what this book offers. The first chapters are a normal textbook on differential equations, for a new generation. The complete book is a year's course on differential equations and linear algebra, including Fourier and Laplace transforms—plus PDE's (Laplace equation, heat equation, wave equation) and the FFT and the SVD.

This is extremely useful mathematics! I cannot hope that you will read every word. But why should the reader be asked to look elsewhere, when the applications can come so naturally here?

A special note goes to engineering faculty who look for support from mathematics. I have the good fortune to teach hundreds of engineering students every year. My work with finite elements and signal processing and computational science helped me to know what students need—and to speak their language. I see texts that mention the impulse response (for example) in one paragraph or not at all. But this is the fundamental solution from which all particular solutions come. In the book it is computed in the time domain, starting with $e^{at}$, and again with Laplace transforms. The website goes further.

I know from experience that every first edition needs help. I hope you will tell me what should be explained more clearly. You are holding a book with a valuable goal—to become a textbook for a world of students and readers in a new generation and a new time, with limits and pressing demands on that time. The book won't be perfect. I will be so grateful if you contribute, in any way, to making it better.

## Acknowledgments

So many friends have helped this book. In first place is Ashley C. Fernandes, my early morning contact for 700 days. He leads the team at Valutone that prepared the LaTeX files. They gently allowed me to rewrite and rewrite, as the truly essential ideas of differential equations became clear. Working with friends is the happiest way to live.

The book began in discussions about the MIT course 18.03. Haynes Miller and David Jerison and Jerry Orloff wanted *change*—this is the lifeblood of a course. Think more about what we are doing ! Their starting point (I see it repeated all over the world) was to add more linear algebra. Matrix operations were already in 18.03, and computations of eigenvalues—they wanted bases and nullspaces and ideas.

I learned so much from their lectures. There is a wonderful moment when a class gets the point. Then the subject lives. The reader can feel this too, but only if the author does. I guess that is my philosophy of education.

Solutions to the Problem Sets were a gift from Bassel Khoury and Matt Ko. The example of a Tumbling Box came from Alar Toomre, it is the highlight of Section 3.3 (this was a famous experiment in his class, throwing a book in the air). Daniel Drucker watched over the text of Chapters 1-3, the best mathematics editor I know. My writing tries to be personal and direct—Dan tries to make it right.

The cover of this book was an amazing experience. Gonçalo Morais visited MIT from Portugal, and we talked. After he went home, he sent this very unusual picture of a strange attractor—a solution to the Lorenz equation. It became a way to honor that great and humble man, Ed Lorenz, who discovered chaos. Gail Corbett and Lois Sellers are the artists who created the cover—what they have done is beyond my thanks, it means everything.

At the last minute (every book has a crisis at the last minute) Shev MacNamara saved the day. Figures were missing. Big spaces were empty. The $S$-curve in Section 1.7, the direction fields in Section 3.1, the Euler and Runge-Kutta experiments, those and more came from Shev. He also encourages me to do an online course with new video lectures. I will think more about a MOOC when readers respond.

Thank you all, including every reader.

Gilbert Strang

## Outline of Chapter 1 : First Order Equations

| | | | |
|---|---|---|---|
| **1.3** | Solve | $dy/dt = ay$ | Construct the exponential $e^{at}$ |
| **1.4** | Solve | $dy/dt = ay + q(t)$ | Four special $q(t)$ and all $q(t)$ |
| **1.5** | Solve | $dy/dt = ay + e^{st}$ | Growth and oscillation : $s = a + i\omega$ |
| **1.6** | Solve | $dy/dt = a(t)y + q(t)$ | Integrating factor $=$ 1/growth factor |
| **1.7** | Solve | $dy/dt = ay - by^2$ | The equation for $z = 1/y$ is linear |
| **1.8** | Solve | $dy/dt = g(t)/f(y)$ | Separate $\int f(y)\,dy$ from $\int g(t)\,dt$ |

**The key formula in 1.4 gives the solution** $y(t) = e^{at}y(0) + \int_0^t e^{a(t-s)}q(s)ds$.

The website with solutions and codes and extra examples and videos is **math.mit.edu/dela**

Please contact **diffeqla@gmail.com** with questions and book orders and ideas.

# Chapter 1

# First Order Equations

## 1.1 Four Examples : Linear versus Nonlinear

A first order differential equation connects a function $y(t)$ to its derivative $dy/dt$. That rate of change in $y$ is decided by $y$ itself (and possibly also by the time $t$).

Here are four examples. Example **1** is the most important differential equation of all.

| | | | |
|---|---|---|---|
| 1) $\dfrac{dy}{dt} = y$ | 2) $\dfrac{dy}{dt} = -y$ | 3) $\dfrac{dy}{dt} = 2ty$ | 4) $\dfrac{dy}{dt} = y^2$ |

Those examples illustrate three **linear** differential equations (**1**, **2**, and **3**) and a **nonlinear** differential equation. The unknown function $y(t)$ is squared in Example **4**. The derivative $y$ or $-y$ or $2ty$ is proportional to the function $y$ in Examples **1**, **2**, **3**. The graph of $dy/dt$ versus $y$ becomes a parabola in Example **4**, because of $y^2$.

It is true that $t$ multiplies $y$ in Example **3**. That equation is still linear in $y$ and $dy/dt$. It has a *variable coefficient* $2t$, changing with time. Examples **1** and **2** have *constant coefficient* (the coefficients of $y$ are 1 and $-1$).

### Solutions to the Four Examples

We can write down a solution to each example. This will be one solution but it is not the *complete* solution, because each equation has a family of solutions. Eventually there will be a constant $C$ in the complete solution. This number $C$ is decided by the starting value of $y$ at $t = 0$, exactly as in ordinary integration. The integral of $f(t)$ solves the simplest differential equation of all, with $y(0) = C$ :

**5)** $\dfrac{dy}{dt} = f(t)$ The complete solution is $y(t) = \displaystyle\int_0^t f(s)\,ds + C$ .

For now we just write one solution to Examples **1 – 4**. They all start at $y(0) = 1$.

$$\mathbf{1} \quad \frac{dy}{dt} = y \quad \text{is solved by} \quad y(t) = \boldsymbol{e^t}$$

$$\mathbf{2} \quad \frac{dy}{dt} = -y \quad \text{is solved by} \quad y(t) = \boldsymbol{e^{-t}}$$

$$\mathbf{3} \quad \frac{dy}{dt} = 2ty \quad \text{is solved by} \quad y(t) = \boldsymbol{e^{t^2}}$$

$$\mathbf{4} \quad \frac{dy}{dt} = y^2 \quad \text{is solved by} \quad y(t) = \frac{\mathbf{1}}{\mathbf{1-t}}.$$

Notice: The three linear equations are solved by exponential functions (***powers of e***). The nonlinear equation **4** is solved by a different type of function; here it is $1/(1-t)$. Its derivative is $dy/dt = 1/(1-t)^2$, which agrees with $y^2$.

Our special interest now is in linear equations with *constant coefficients*, like **1** and **2**. In fact $dy/dt = y$ is the most important property of the great function $y = e^t$. Calculus had to create $e^t$, because a function from algebra (like $y = t^n$) cannot equal its derivative (the derivative of $t^n$ is $nt^{n-1}$). But a combination of all the powers $t^n$ can do it. That good combination is $e^t$ in Section 1.3.

The final example extends **1** and **2**, to allow **any constant coefficient $a$** :

$$\mathbf{6)} \qquad \frac{dy}{dt} = ay \quad \text{is solved by} \quad \boldsymbol{y = e^{at}} \quad \text{(and also} \quad \boldsymbol{y = Ce^{at}}).$$

If the constant growth rate $a$ is positive, the solution increases. If $a$ is negative, as in $dy/dt = -y$ with $a = -1$, the slope is negative and the solution $e^{-t}$ decays toward zero. Figure 1.1 shows three exponentials, with $dy/dt$ equal to $y$ and $2y$ and $-y$.

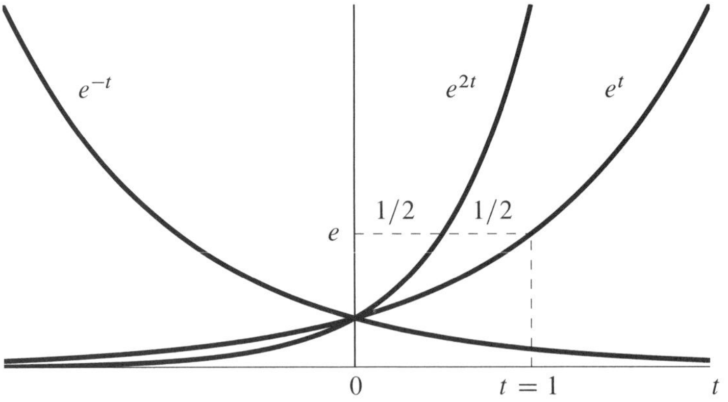


Figure 1.1: Growth, faster growth, and decay. The solutions are $e^t$ and $e^{2t}$ and $e^{-t}$.

When $a$ is larger than 1, the solution grows faster than $e^t$. That is natural. The neat thing is that we still follow the exponential curve—but $e^{at}$ climbs that curve faster. You could see the same result by *rescaling the time axis*. In Figure 1.1, the steepest curve (for $a = 2$) is the same as the first curve—but the time axis is compressed by 2.

Calculus sees this factor of 2 from the chain rule for $e^{2t}$. It sees the factor $2t$ from the chain rule for $e^{t^2}$. This exponent is $t^2$, the factor $2t$ is its derivative :

$$\frac{d}{dt}\left(e^u\right) = e^u\,\frac{du}{dt} \qquad \frac{d}{dt}\left(e^{2t}\right) = \left(e^{2t}\right) \text{ times } 2 \qquad \frac{d}{dt}\left(e^{t^2}\right) = \left(e^{t^2}\right) \text{ times } 2t$$

## Problem Set 1.1

**1** Draw the graph of $y = e^t$ by hand, for $-1 \le t \le 1$. What is its slope $dy/dt$ at $t = 0$ ? Add the straight line graph of $y = et$. Where do those two graphs cross ?

**2** Draw the graph of $y_1 = e^{2t}$ on top of $y_2 = 2e^t$. Which function is larger at $t = 0$ ? Which function is larger at $t = 1$ ?

**3** What is the slope of $y = e^{-t}$ at $t = 0$ ? Find the slope $dy/dt$ at $t = 1$.

**4** What "logarithm" do we use for the number $t$ (the exponent) when $e^t = 4$ ?

**5** State the chain rule for the derivative $dy/dt$ if $y(t) = f(u(t))$ (chain of $f$ and $u$).

**6** The *second* derivative of $e^t$ is again $e^t$. So $y = e^t$ solves $d^2y/dt^2 = y$. A second order differential equation should have another solution, different from $y = Ce^t$. What is that second solution ?

**7** Show that the nonlinear example $dy/dt = y^2$ is solved by $y = C/(1 - Ct)$ for every constant $C$. The choice $C = 1$ gave $y = 1/(1-t)$, starting from $y(0) = 1$.

**8** Why will the solution to $dy/dt = y^2$ grow faster than the solution to $dy/dt = y$ (if we start them both from $y = 1$ at $t = 0$) ? The first solution blows up at $t = 1$. The second solution $e^t$ grows exponentially fast but it never blows up.

**9** Find a solution to $dy/dt = -y^2$ starting from $y(0) = 1$. Integrate $dy/y^2$ and $-dt$. (Or work with $z = 1/y$. Then $\boldsymbol{dz/dt} = (dz/dy)\,(dy/dt) = (-1/y^2)(-y^2) = \mathbf{1}$. From $dz/dt = 1$ you will know $z(t)$ and $y = 1/z$.)

**10** Which of these differential equations are linear (in $y$) ?

(a) $y' + \sin y = t$ (b) $y' = t^2(y - t)$ (c) $y' + e^t y = t^{10}$.

**11** The product rule gives what derivative for $e^t e^{-t}$ ? This function is constant. At $t = 0$ this constant is 1. Then $e^t e^{-t} = 1$ for all $t$.

**12** $dy/dt = y + 1$ is not solved by $y = e^t + t$. Substitute that $y$ to show it fails. We can't just add the solutions to $y' = y$ and $y' = 1$. What number $c$ makes $y = e^t + c$ into a correct solution ?

## 1.2 The Calculus You Need

The prerequisite for differential equations is calculus. This may mean a year or more of ideas and homework problems and rules for computing derivatives and integrals. Some of those topics are essential, but others (as we all acknowledge) are not really of first importance. These pages have a positive purpose, to bring together essential facts of calculus. This section is to read and refer to—it doesn't end with a Problem Set.

I hope this outline may have value also at the end of a single-variable calculus course. Textbooks could include a summary of the crucial ideas, but usually they don't. Certainly the reader will not agree with every choice made here, and the best outcome would be a more perfect list. This one is a lot shorter than I expected.

At the end, a useful formula in differential equations is confirmed by the product rule, the derivative of $e^x$, and the Fundamental Theorem of Calculus.

**1. Derivatives of key functions : $x^n$ $\sin x$ $\cos x$ $e^x$ $\ln x$**

The derivatives of $x, x^2, x^3, \ldots$ come from first principles, as limits of $\Delta y/\Delta x$. The derivatives of $\sin x$ and $\cos x$ focus on the limit of $(\sin \Delta x)/\Delta x$. Then comes the great function $e^x$. It solves the differential equation $dy/dx = y$ starting from $y(0) = 1$. **This is the single most important fact needed from calculus : the knowledge of $e^x$.**

**2. Rules for derivatives : Sum rule Product rule Quotient rule Chain rule**

When we add, subtract, multiply, and divide the five original functions, these rules give the derivatives. The sum rule is the quiet one, applied all the time to *linear* differential equations. This equation is linear (*a crucial property*) :

$$\frac{dy}{dt} = ay + f(t) \text{ and } \frac{dz}{dt} = az + g(t) \text{ add to } \frac{d}{dt}(y+z) = a(y+z) + (f+g).$$

With $a = 0$ that is a straightforward sum rule for the derivative of $y + z$. We can always add equations as shown, because $a(t)y$ is linear in $y$. This confirms *superposition* of the separate solutions $y$ and $z$. Linear equations add and their solutions add.

The chain rule is the most prolific, in computing the derivatives of very remarkable functions. The chain $\boldsymbol{y = e^x}$ and $\boldsymbol{x = \sin t}$ produces $\boldsymbol{y = e^{\sin t}}$ (the composite of two functions). The chain rule gives $dy/dt$ by multiplying the derivatives $dy/dx$ and $dx/dt$ :

$$\textbf{Chain rule} \qquad \frac{dy}{dt} = \frac{dy}{dx}\frac{dx}{dt} = e^x \cos t = y \cos t.$$

Then $e^{\sin t}$ solves that differential equation $\dfrac{dy}{dt} = ay$ with varying growth rate $a = \cos t$.

### 3. The Fundamental Theorem of Calculus

**The derivative of the integral of $f(x)$ is $f(x)$.** The integral from 0 to $x$ of the derivative $df/dx$ is $f(x) - f(0)$. One operation inverts the other, when $f(0) = 0$. This is not so easy to prove, because both the derivative and the integral involve a limit step $\Delta x \to 0$.

One way to go forward starts with numbers $y_0, y_1, \ldots, y_n$. Their differences are like derivatives. Adding up those differences is like integrating the derivative:

$$\textbf{Sum of differences} \quad (y_1 - y_0) + (y_2 - y_1) + \cdots + (y_n - y_{n-1}) = y_n - y_0. \tag{1}$$

Only $y_n$ and $-y_0$ are left because all other numbers $y_1, y_2, \ldots$ come twice and cancel. To make that equation look like calculus, multiply every term by $\Delta x/\Delta x = 1$:

$$\left[\frac{y_1 - y_0}{\Delta x} + \frac{y_2 - y_1}{\Delta x} + \cdots + \frac{y_n - y_{n-1}}{\Delta x}\right]\Delta x = y_n - y_0. \tag{2}$$

Again, this is true for all numbers $y_0, y_1, \ldots, y_n$. Those can be heights of the graph of a function $y(x)$. The points $x_0, \ldots, x_n$ can be equally spaced between $x = a$ and $x = b$. Then each ratio $\Delta y/\Delta x$ is a *slope* between two points of the graph:

$$\frac{\Delta y}{\Delta x} = \frac{y_k - y_{k-1}}{x_k - x_{k-1}} = \frac{\text{distance up}}{\text{distance across}} = \text{slope}. \tag{3}$$

This slope is exactly correct if the graph is a straight line between the points $x_{k-1}$ and $x_k$. If the graph is a curve, the approximate slope $\Delta y/\Delta x$ becomes exact as $\Delta x \to 0$.

The delicate part is the requirement $n\Delta x = b - a$, to space the points evenly from $x_0 = a$ to $x_n = b$. Then $n$ will increase as $\Delta x$ decreases. Equation (2) remains correct at every step, with $y_0 = y(a)$ at the first point and $y_n = y(b)$ at the last point. As $\Delta x \to 0$ and $n \to \infty$, the slopes $\Delta y/\Delta x$ approach the derivative $dy/dx$. At the same time the sum approaches the integral of $dy/dx$. Equation (2) turns into equation (4):

$$\textbf{Fundamental Theorem of Calculus} \qquad \int_a^b \frac{dy}{dx}\,dx = y(b) - y(a) \qquad \frac{d}{dx}\int_a^x f(s)\,ds = f(x) \tag{4}$$

The limits of $\Delta y/\Delta x$ in (3) and the sum in (2) produce $dy/dx$ and its integral. Of course this presentation of the Fundamental Theorem needs more careful attention. But equation (1) holds a key idea: *a sum of differences*. This leads to *an integral of derivatives*.

### 4. The meaning of symbols and the operations of algebra

Mathematics is a language. The way to learn this language is to use it. So textbooks have thousands of exercises, to practice reading and writing symbols like $y(x)$ and $y(x + \Delta x)$. Here is a typical line of symbols:

$$\textbf{Derivative of } y \qquad \frac{dy}{dt}(t) = \lim_{\Delta t \to 0} \frac{y(t + \Delta t) - y(t)}{\Delta t}. \tag{5}$$

I am not very sure that this is clear. One function is $y$, the other function is its derivative $y'$.

*Could the symbol $y'$ be better than $dy/dt$*? Both are standard in this book. In calculus we know $y(t)$, in differential equations we don't. The whole point of the differential equation is to connect $y$ and $y'$. From that connection we have to discover what they are.

A first example is $y' = y$. That equation forces the unknown function $y$ to grow exponentially: $y(t) = Ce^t$. At the end of this section I want to propose a more complicated equation and its solution. But I could never find a more important example than $e^t$.

## 5. Three ways to use $dy/dx \approx \Delta y/\Delta x$

On the graph of a function $y(x)$, the exact slope is $dy/dx$ and the approximate slope (between nearby points) is $\Delta y/\Delta x$. If we know *any two* of the numbers $dy/dx$ and $\Delta y$ and $\Delta x$, then we have a good approximation to the third number. All three approximations are important, because $dy/dx$ is such a central idea in calculus.

**(A) When we know $\Delta x$ and $dy/dx$, we have $\Delta y \approx (\Delta x)(dy/dx)$.**

This is linear approximation. From a starting point $x_0$, we move a distance $\Delta x$. That produces a change $\Delta y$. The graph of $y(x)$ can go up or down, and the best information we have is the slope $dy/dx$ at $x_0$. (That number gives no way to account for *bending* of the graph, which appears in the next derivative $d^2y/dx^2$.)

Linear approximation is equivalent to following the tangent line —not the curve:

$$\boldsymbol{\Delta y \approx \Delta x \frac{dy}{dx}} \qquad y(x_0 + \Delta x) \approx y(x_0) + \Delta x \frac{dy}{dx}(x_0) \tag{6}$$

**(B) $\Delta y$ and $dy/dx$ lead to $\Delta x \approx (\Delta y)/(dy/dx)$. This is Newton's Method.**

Newton's Method is a way to solve $y(x) = 0$, starting at a point $x_0$. We want $y(x)$ to drop from $y(x_0)$ to zero at the new point $x_1$. *The desired change in $y$ is $\Delta y = 0 - y(x_0)$.* What we don't know is $\Delta x$, which locates $x_1$. The exact slope $dy/dx$ will be close to $\Delta y/\Delta x$, and that tells us a good $\Delta x$:

$$\textbf{Newton's Method} \qquad \boldsymbol{\Delta x \approx \frac{\Delta y}{dy/dx}} \qquad x_1 - x_0 = \frac{-y(x_0)}{dy/dx(x_0)} \tag{7}$$

Guess $x_0$, improve to $x_1$. This is an excellent way to solve nonlinear equations $y(x) = 0$.

**(C) Dividing $\Delta y$ by $\Delta x$ gives the approximation $dy/dx \approx \Delta y/\Delta x$.**

That is the point of equation (5), but something important often escapes our attention. *Are $x$ and $x + \Delta x$ the best two places to compute $y$*? Writing $\Delta y = y(x + \Delta x) - y(x)$ doesn't seem to offer other choices. If we notice that $\Delta x$ can be negative, this allows $x + \Delta x$ to be on the left side of $x$ (leading to a backward difference) . The best choice is not forward or backward but *centered around $x$ : a half step each way.*

$$\textbf{Centered difference} \qquad \frac{dy}{dx} \approx \frac{\Delta y}{\Delta x} = \boldsymbol{\frac{y(x + \frac{1}{2}\Delta x) - y(x - \frac{1}{2}\Delta x)}{\Delta x}} \tag{8}$$

Why is centering better? When $y = Cx + D$ has a straight line graph, all ratios $\Delta y/\Delta x$ give the correct slope $C$. But the parabola $y = x^2$ has the simplest possible bending, and **only this centered difference gives the correct slope $2x$** (varying with $x$).

**Exact slope for parabolas by centering** $$\frac{\Delta y}{\Delta x} = \frac{(x+\frac{1}{2}\Delta x)^2 - (x-\frac{1}{2}\Delta x)^2}{\Delta x} = \frac{x\,\Delta x - (-x\,\Delta x)}{\Delta x} = \mathbf{2x}$$

The key step in scientific computing is improving first order accuracy (forward differences) to second order accuracy (centered differences). For integrals, rectangle rules improve to trapezoidal rules. This is a big step to good algorithms.

## 6. Taylor series: Predicting $y(x)$ from all the derivatives at $x = x_0$

From the height $y_0$ and the slope $y_0'$ at $x_0$, we can predict the height $y(x)$ at nearby points. But the tangent line in equation (6) assumes that $y(x)$ has constant slope. That first order prediction becomes a second order prediction (*much more accurate*) when we use the second derivative $y_0''$ at $x_0$.

**Tangent parabola using $y_0''$** $$y(x_0 + \Delta x) \approx y_0 + (\Delta x)y_0' + \tfrac{1}{2}(\Delta x)^2 y_0''. \qquad (9)$$

Adding this $(\Delta x)^2$ term moves us from constant slope to constant bending. For the parabola $y = x^2$, equation (9) is exact: $(x_0 + \Delta x)^2 = (x_0^2) + (\Delta x)(2x_0) + \frac{1}{2}(\Delta x)^2(2)$.

Taylor added more terms—infinitely many. His formula gets *all derivatives correct* at $x_0$. The pattern is set by $\frac{1}{2}(\Delta x)^2 y_0''$. The $n^{\text{th}}$ derivative $y^{(n)}(x)$ contributes a new term $\frac{1}{n!}(\Delta x)^n y_0^{(n)}$. The complete Taylor series includes all derivatives at the point $x = x_0$:

**Taylor series**
**Stop at $y'$ for tangent line**
**Stop at $y''$ for parabola**

$$y(x_0 + \Delta x) = y_0 + (\Delta x)\,y_0' + \cdots + \frac{1}{n!}(\Delta x)^n\,y_0^{(n)} + \cdots = \sum_{n=0}^{\infty} \frac{(\Delta x)^n}{n!}\, y^{(n)}(x_0) \qquad (10)$$

Those equal signs are not always right. There is no way we can stop $y(x)$ from making a sudden change after $x$ moves away from $x_0$. Taylor's prediction of $y(x_0 + \Delta x)$ is exactly correct for $e^x$ and $\sin x$ and $\cos x$—good functions like those are "analytic" at all $x$.

Let me include here the two most important examples in all of mathematics. They are solutions to $dy/dx = y$ and $dy/dx = y^2$ — the most basic linear and nonlinear equations.

**Exponential series** with $y^{(n)}(0) = 1$ $$y = e^x = 1 + x + \frac{1}{2!}x^2 + \frac{1}{3!}x^3 + \cdots \qquad (11)$$

**Geometric series** with $y^{(n)}(0) = n!$ $$y = \frac{1}{1-x} = 1 + x + x^2 + x^3 + \cdots \qquad (12)$$

The center point is $x_0 = 0$. The series (11) gives $e^x$ for every $x$. The series (12) gives $1/(1-x)$ when $x$ is between $-1$ and $1$. Its derivative $1 + 2x + 3x^2 + \cdots$ is $1/(1-x)^2$.

For $x = 2$ that geometric series will certainly not produce $1/(1-2) = -1$. Notice that $1 + x + x^2 + \cdots$ becomes infinite at $x = 1$, exactly where $1/(1-x)$ becomes $1/0$.

The key point for $e^x$ is that its $n^{\text{th}}$ derivative is 1 at $x = 0$. The $n^{\text{th}}$ derivative of $1/(1-x)$ is $n!$ at $x = 0$. This pattern starts with $y, y', y'', y'''$ equal to $1, 1, 2, 6$ at $x = 0$:

$$y = (1-x)^{-1} \qquad y' = (1-x)^{-2} \qquad y'' = 2(1-x)^{-3} \qquad y''' = 6(1-x)^{-4}.$$

Taylor's formula combines the contributions of all derivatives at $x = 0$, to produce $y(x)$.

### 7. Application: An important differential equation

The linear differential equation $y' = ay + q(t)$ is a perfect multipurpose model. It includes the growth rate $a$ and the external source term $q(t)$. We want the particular solution that starts from $y(0) = 0$. Creating that solution uses the most essential idea behind integration. Verifying that the solution is correct uses the basic rules for derivatives. Many students in my graduate class had forgotten the derivative of the integral.

Here is the solution $y(t)$ followed by its interpretation, with $a = 1$ for simplicity:

$$\frac{dy}{dt} = y + q(t) \qquad \textbf{is solved by} \qquad y(t) = \int_0^t e^{t-s} q(s)\, ds. \tag{13}$$

*Key idea*: At each time $s$ between 0 and $t$, the input is a source of strength $q(s)$. That input grows or decays over the remaining time $t - s$. **The input $q(s)$ is multiplied by $e^{t-s}$ to give an output at time $t$.** Then the total output $y(t)$ is the *integral* of $e^{t-s}q(s)$.

We will reach $y(t)$ in other ways. Section 1.4 uses an "integrating factor." Section 1.6 explains "variation of parameters." The key is to see where the formula comes from. *Inputs lead to outputs, the equation is linear, and the principle of superposition applies.* The total output is the sum (in this case, the integral) of all those outputs.

We will confirm formula (13) by computing $dy/dt$. First, $e^{t-s}$ equals $e^t$ times $e^{-s}$. Then $e^t$ comes outside the integral of $e^{-s}q(s)$. Use the product rule on those two factors:

$$\textbf{Producing } y + q \qquad \frac{dy}{dt} = \left(\frac{d\,e^t}{dt}\right) \int_0^t e^{-s} q(s)\, ds + \left(e^t\right) \frac{d}{dt} \int_0^t e^{-s} q(s)\, ds. \tag{14}$$

The first term on the right side is exactly $y(t)$. How to recognize that last term as $q(t)$?

We don't need to know the function $q(t)$. What we do know (and need) is the *Fundamental Theorem of Calculus*. **The derivative of the integral of $e^{-t}q(t)$ is $e^{-t}q(t)$.** Then multiplying by $e^t$ gives the hoped-for result $q(t)$, because $e^t e^{-t} = 1$. The linear differential equation $y' = y + q$ with $y(0) = 0$ is solved by the integral of $e^{t-s}q(s)$.

## 1.3 The Exponentials $e^t$ and $e^{at}$

Here is the key message from this section: **The solutions to $dy/dt = ay$ are $y(t) = Ce^{at}$. That free constant $C$ matches the starting value $y(0)$. Then $y(t) = y(0)e^{at}$.**

I realize that you already know the function $y = e^t$. It is the star of precalculus and calculus. Now it becomes the key to linear differential equations. Here I focus on the two most important properties of this function $e^t$ :

**1.** *The slope $dy/dt$ equals the function $y$.* As $y$ grows, its graph gets steeper:

$$\frac{d}{dt}e^t = e^t. \tag{1}$$

**2.** $y(t) = e^t$ follows the *addition rule* for exponents:

$$e^t \text{ times } e^T \text{ equals } e^{t+T}. \tag{2}$$

How is this exponential function constructed? Only calculus can do it, because somewhere we must have a "limit step." Functions from ordinary algebra can get close to $e^t$, but they can't reach it. If we choose those functions to come closer and closer, then their limit is $e^t$.

This is like using fractions to approach the extraordinary number $\pi$. The fractions can start with 3/1 and 31/10 and 314/100. The neat fraction 22/7 is close to $\pi$. But "taking the limit" can't be avoided, because $\pi$ itself is not a fraction.

Similarly $e$ is not a fraction. On this book's home page **math.mit.edu/dela** is an article called *Introducing $e^x$*. It describes four popular ways to construct this function. The one chosen now is my favorite, because it is the most direct way.

$$\textbf{Construct } \mathbf{y = e^t} \textbf{ so that } \frac{\mathbf{dy}}{\mathbf{dt}} = \mathbf{y} \quad (\text{starting from } y = 1 \text{ at } t = 0)$$

To show how this construction works, here are ordinary polynomials $y$ and $dy/dt$ :

1. $y = 1 + t + \frac{1}{2}t^2$ The derivative is $dy/dt = 0 + 1 + t$
2. $y = 1 + t + \frac{1}{2}t^2 + \frac{1}{6}t^3$ The derivative is $dy/dt = 0 + 1 + t + \frac{1}{2}t^2$

You see that $dy/dt$ does not fully agree with $y$. It always falls one term short of $y$. We could get $t^3/6$ into the derivative by including $t^4/24$ in $y$. But now $dy/dt$ will be missing $t^4/24$.

You can see that $dy/dt$ won't catch up to $y$. *The way out is to have infinitely many terms: Don't stop.* Then you get $dy/dt = y$.

The limit step reaches an infinite series, adding new terms and never stopping. Every term has the form $t^n$ divided by $n\,!$ (*n factorial*). Its derivative is the previous term:

$$\textbf{The derivative of} \quad \frac{t^n}{(n)\dots(1)} = \frac{t^n}{n\,!} \quad \text{is} \quad \frac{t^{n-1}}{(n-1)\dots(1)} = \frac{t^{n-1}}{(n-1)\,!} \tag{3}$$

So if $t^n/n!$ is missing in $dy/dt$, we will capture it by including $t^{n+1}/(n+1)!$ in $y$.

Of course $dy/dt$ never completely catches up to $y$—*until we allow an infinite series. There is a term $t^n/n\,!$ for every $n$.* The term for $n=0$ is $t^0/0\,! = 1$.

$$\textbf{Construction of } e^t \qquad y = e^t = 1 + t + \frac{t^2}{2} + \frac{t^3}{6} + \frac{t^4}{24} + \cdots = \sum_{n=0}^{\infty} \frac{t^n}{n!} \tag{4}$$

**Taking the derivative of every term produces all the same terms. So $dy/dt = y$.** Notice: If you change every $t$ to $at$, **the derivative of $y = e^{at}$ becomes $a$ times $e^{at}$:**

$$\frac{d}{dt}\left(1 + at + \frac{a^2t^2}{2} + \frac{a^3t^3}{6} + \cdots\right) = a\left(1 + at + \frac{a^2t^2}{2} + \cdots\right) = \boldsymbol{ae^{at}} \tag{5}$$

This construction of $e^t$ brings up two questions, to be discussed in the Chapter 1 Notes. Does the infinite series add to a finite number (a different number for each choice of $t$)? Can we add the derivatives of each $t^n/n!$ and safely get the derivative of the sum $e^t$? Fortunately both answers are *yes*. The terms get very small, very fast, as $n$ increases. The limiting step is $n \to \infty$, producing the exact $e^t$.

When $t = 1$, we can watch the terms get small. We *must* do this, because $t = 1$ leads to the all-important number $e^1$ which is $e$:

$$\textbf{The series for } e \textbf{ at } t = 1 \qquad e = 1 + 1 + \frac{1}{2} + \frac{1}{6} + \frac{1}{24} + \cdots \approx 2.718$$

The first three terms add to 2.5. The first five terms almost reach 2.71. *We never reach* 2.72. With enough terms you can barely pass 2.71828. It is certain that the total sum $e$ is not a fraction. It never appears in algebra, but it is the key number for calculus.

## The Series for $e^t$ is a Taylor Series

The infinite series (4) for $e^t$ is the same as the Taylor series. Section 1.2 went from the tangent line $1 + t$ to the tangent parabola $1 + t + \frac{1}{2}t^2$. The next term will be $\frac{1}{6}t^3$, because that matches the third derivative $y''' = 1$ at $t = 0$. *All derivatives are equal to* 1 *at* $t = 0$, when we start from the basic equation $y' = y$. That equation gives $y'' = y' = y$ and the next derivative gives $y''' = y'' = y' = y$.

*Conclusion*: $t^n/n\,!$ has the correct $n^{\text{th}}$ derivative (which is 1) at the point $t = 0$. All these terms go into the Taylor series. The result is exactly the exponential series (4).

## Multiplying Powers by Adding Exponents

We write $3^2$ for 3 times 3. We write $e^2$ for $e$ times $e$. The question is, does $e = 2.718\ldots$ times $e = 2.718\ldots$ give the same answer as setting $t = 2$ in the infinite series to get $e^2$?

The answer is again *yes.* I could say "fortunately yes" but that might suggest a lucky accident. The amazing fact is that Property 1 ($y' = y$ is now confirmed) leads automatically to Property 2. The exponential starts from $y(0) = e^0 = 1$ at time $t = 0$.

**Property 2.** $\quad \boldsymbol{e^t}$ **times** $\boldsymbol{e^T}$ **equals** $\boldsymbol{e^{t+T}}$ $\qquad$ so $\;(e^1)\,(e^1) = e^2$

This is a differential equations course, so the proofs will use Property 1: $dy/dt = y$.

**First Proof.** We can solve $y' = (a+b)y$ two ways, starting from $y(0) = 1$. We know that $y(t) = \boldsymbol{e^{(a+b)t}}$. Another solution is $y(t) = \boldsymbol{e^{at}e^{bt}}$, as the product rule shows:

$$\frac{d}{dt}\left(e^{at}e^{bt}\right) = \left(ae^{at}\right)e^{bt} + e^{at}\left(be^{bt}\right) = (a+b)e^{at}e^{bt}. \tag{6}$$

This solution $e^{at}e^{bt}$ also starts at $e^0e^0 = 1$. It must be the same as the first solution $e^{(a+b)t}$. The equation $y' = (a+b)y$ only has one solution. At $t = 1$ this says that $e^{a+b} = e^ae^b$. QED.

**Second Proof.** Starting with $y = 1$ at $t = 0$, the solution out to time $t$ is $e^t$. The solution to time $t + T$ is $e^{t+T}$. The question is, do we also get that answer in two steps?

> Starting from $y = 1$ at $t = 0$, we go to $e^t$. Then start from $e^t$ at time $t$ and continue an additional time $T$. This would give $e^T$ starting from $y = 1$, but here the starting value is $e^t$. So $C = e^t$ multiplies $e^T$. At time $t + T$ we have perfect agreement:
>
> $e^t$ times $e^T$ (which is $C$ times $e^T$) agrees with one big step $e^{t+T}$.

## Negative Exponents

Remember the example $dy/dt = -y$ with solution $y = e^{-t}$. That exponent $-t$ is negative. The solution decays toward zero. The exponent rule $e^te^T = e^{t+T}$ still holds for negative exponents. In particular $e^t$ *times* $e^{-t}$ *is* $e^{t-t} = e^0 = 1$:

**Negative exponents** $\qquad \dfrac{1}{e^t} = e^{-t}$ and $\dfrac{1}{e} = e^{-1} = 1 - 1 + \dfrac{1}{2} - \dfrac{1}{6} + \dfrac{1}{24} - \cdots$

This number $1/e$ is about .36. The series always succeeds! The graph of $y = e^{-t}$ shows that $e^{-t}$ *stays positive*. It is very small for $t > 32$. Your computer might use 32 bit arithmetic and ignore numbers that are this small.

**Why does $\boldsymbol{e^t}$ grow so fast**? The slope is $y$ itself. So the slope increases when the function increases. That steep slope makes $y$ increase faster—and then the slope too.

## Interest Rates and Difference Equations

There is another approach to $e^t$ and $e^{at}$, which is not based on an infinite series. (At least, not at the start.) It connects to interest on bank accounts. For $e^t$ the rate is $a = 1 = 100\%$. For $e^{at}$ the differential equation is $dy/dt = ay$ and the interest rate is $a$.

*The different approach is to construct $e^t$ and $e^{at}$ as the limit of compound interest.*

$$\boxed{e^t = \lim_{N \to \infty} \left(1 + \frac{t}{N}\right)^N \qquad e^{at} = \lim_{N \to \infty} \left(1 + \frac{at}{N}\right)^N.} \tag{7}$$

The beauty of these formulas is that a bank does exactly what a computational scientist does. They both start with the differential equation $dy/dt = ay$ and the initial condition $y = 1$ at $t = 0$. Banks and scientists don't have computers that give exact solutions, when $y(t)$ changes continuously with time. Both take finite time steps $\Delta t$ instead of infinitesimal steps $dt$. **They reach time $t$ in $N$ steps of size $\Delta t = t/N$.** Their approximations are $Y_1, Y_2, \ldots, Y_N$ with $Y_0 = 1$. Compound interest produces a **difference equation**:

$$\frac{dy}{dt} = ay \quad \text{becomes} \quad \frac{Y_{n+1} - Y_n}{\Delta t} = a\,Y_n \quad \text{and} \quad \boldsymbol{Y_{n+1} = (1 + a\,\Delta t)Y_n}. \tag{8}$$

Each step multiplies the bank balance by $1 + a\Delta t$. The new balance is the old balance $Y_n$ plus $a\,\Delta t Y_n$ (the interest on $Y_n$ in the time interval $\Delta t$). This is ordinary compound interest that all banks offer, not continuous compounding as in $dy/dt$. The time step can be $\Delta t = 1$ year or 1 month. The balance at $t = 2$ years $= 24$ months is $Y_2$ or $Y_{24}$:

$$Y_2 = (1 + a)^2\, Y_0 \qquad Y_{24} = \left(1 + \frac{a}{12}\right)^{24} Y_0 \approx e^{2a} Y_0. \tag{9}$$

If the rate is $a = 3$ per cent per year $= .03$ per year, continuous compounding for 2 years would produce the exponential factor $e^{.06} \approx 1.06184$. Monthly compounding produces $(1.0025)^{24} \approx 1.06176$. We only lose a little, when the differential equation $y' = ay$ is approximated by the difference equation in (8).

The computational scientist is usually not willing to accept this loss of accuracy in $Y$. Equation (8) with a forward difference $Y_{n+1} - Y_n$ is called **Euler's method**. Its accuracy is not high and not hard to improve. It is the natural choice for a bank, because a backward difference costs them even more than continuous compounding:

$$\textbf{Backward difference} \quad \frac{Y_n - Y_{n-1}}{\Delta t} = aY_n \quad \text{or} \quad Y_n = \frac{1}{1 - a\Delta t}\, Y_{n-1}. \tag{10}$$

$Y_n$ connects backward to the earlier $Y_{n-1}$. Now each step divides by $1 - a\Delta t$. After $N$ steps of size $\Delta t = t/N$, we are again close to $e^{at}$. But with backward differences and $a > 0$, we overshoot the differential equation and the bank pays a little too much:

$$(1 + a\Delta t)^N \text{ is below } e^{at} \qquad \frac{1}{(1 - a\Delta t)^N} \textit{ is above } e^{at}.$$

## Complex Exponents

This isn't the time and place to study complex numbers in detail. It will be the pages about oscillations and $e^{i\omega t}$ that cannot go forward without the imaginary number $i$. Here we are solving $dy/dt = ay$, and all I want to do is to **choose $a = i$.**

I can think of two ways to solve the complex equation $\boldsymbol{dy/dt = iy}$. The fast way uses derivatives of sine and cosine, which we know well:

$$\begin{array}{llcl} \textbf{Proposed solution} & y & = & \mathbf{\cos t + i \sin t} \\ \textbf{Compare } \boldsymbol{dy/dt} & dy/dt & = & -\sin t + i \cos t \\ \textbf{with the right side } \boldsymbol{iy} & iy & = & i \cos t + i^2 \sin t \end{array} \qquad (11)$$

To check $dy/dt = iy$, compare the last two lines. **Use the rule $\mathbf{i^2 = -1}$.** (We had to imagine this number, because no real number has $x^2 = -1$.) Then $-\sin t$ is the same as $i^2 \sin t$. So $\boldsymbol{y = \cos t + i \sin t}$ solves the equation $dy/dt = iy$. This solution starts at $y = 1$ when $t = 0$, because $\cos 0 = 1$ and $\sin 0 = 0$.

The slower approach to $dy/dt = iy$ uses the infinite series. Since $a = i$, the solution $e^{at}$ becomes $e^{it}$. Formally, the series for $y = e^{it}$ certainly solves $dy/dt = iy$:

$$\textbf{Complex exponential} \qquad \boldsymbol{y = e^{it}} = 1 + (it) + \frac{1}{2}(it)^2 + \frac{1}{6}(it)^3 + \cdots \qquad (12)$$

The derivative of each term is $i$ times the previous term. Since the series never stops, the derivative $dy/dt$ perfectly matches $iy$. And we are still starting at $y = 1$ when we substitute $t = 0$. **This infinite series $\boldsymbol{e^{it}}$ equals the first solution $\boldsymbol{\cos t + i \sin t}$.**

*Now use the rule* $i^2 = -1$. For $(it)^2$ I will write $-t^2$. And $(it)^3$ equals $-it^3$. The fourth power of $i$ is $i^4 = i^2 i^2 = (-1)^2 = 1$. That sequence $\boldsymbol{i, -1, -i, 1}$ repeats forever.

$$i = i^5 \qquad i^2 = i^6 = -1 \qquad i^3 = i^7 = -i \qquad i^4 = i^8 = 1$$

The infinite series (12) includes those four numbers multiplying powers of $t$:

$$e^{it} = 1 + \left[it - 1\frac{t^2}{2!} - i\frac{t^3}{3!} + 1\frac{t^4}{4!}\right] + \left[i\frac{t^5}{5!} - 1\frac{t^6}{6!} - i\frac{t^7}{7!} + 1\frac{t^8}{8!}\right] + \cdots$$

This may be the first time a textbook has ever written out nine terms. You can see the full repeat of $i, -1, -i, 1$. That last coefficient divides by $8! = 8 \cdot 7 \cdot 6 \cdot 5 \cdot 4 \cdot 3 \cdot 2 \cdot 1$ which is 40320.

The main point is that the solution $y = \cos t + i \sin t$ in equation (11) must be the same as this series solution $e^{it}$. They both solve $dy/dt = iy$. They both start at $y = 1$ when $t = 0$. The equality between them is one of the greatest formulas in mathematics.

$$\boxed{\textbf{Euler's Formula is} \qquad \boldsymbol{e^{it} = \cos t + i \sin t}.} \qquad (13)$$

Then $e^{i\pi} = \cos \pi + i \sin \pi = \mathbf{-1}$. And $e^{i2\pi} = 1 + i2\pi + \frac{1}{2}(i2\pi)^2 + \cdots$ must add to **1**!

I cannot resist comparing $\cos t + i \sin t$ with the series for $e^{it}$. The **real part** of that series must be $\cos t$. The **imaginary part** (which multiplies $i$) must be $\sin t$. The even powers $1, t^2, t^4, \ldots$ give cosines. The odd powers $t, t^3, t^5, \ldots$ are multiplied by $i$:

$$\textbf{Cosine is even} \qquad \cos t = 1 - \frac{1}{2}t^2 + \frac{1}{24}t^4 - \frac{t^6}{6!} + \cdots \tag{14}$$

$$\textbf{Sine is odd} \qquad \sin t = t - \frac{1}{6}t^3 + \frac{1}{120}t^5 - \frac{t^7}{7!} + \cdots \tag{15}$$

These two pieces of the series for $e^{it}$ are famous functions on their own, and now we see their Taylor series . They are beautifully connected by Euler's Formula.

The derivative of the sine series is the cosine series:

$$\frac{\mathbf{d}}{\mathbf{dt}}\,\mathbf{\sin t = \cos t} \qquad \frac{d}{dt}\left(t - \frac{1}{6}t^3 + \cdots\right) = \mathbf{1 - \frac{1}{2}t^2 + \cdots = \text{cosine}}$$

The derivative of the cosine series is minus the sine series:

$$\frac{\mathbf{d}}{\mathbf{dt}}\,\mathbf{\cos t = -\sin t} \qquad \frac{d}{dt}\left(1 - \frac{1}{2}t^2 + \frac{1}{24}t^4 - \cdots\right) = \mathbf{-t + \frac{1}{6}t^3 \cdots = -\text{sine}}$$

All this important information came from allowing the exponent in $e^{it}$ to be imaginary. And $e^{it}$ times $e^{-it}$ is exactly $\cos^2 t + \sin^2 t = 1$.

## Matrix Exponents

One more thing, which you can safely ignore for now. The exponent in $e^{at}$ could become a **square matrix**. Instead of solving $dy/dt = ay$ by $e^{at}$, we can solve the matrix equation $d\mathbf{y}/dt = A\mathbf{y}$ by the matrix $e^{At}$. Start with the identity matrix $I$ instead of the number 1.

$$e^{At} \textbf{ is a matrix} \qquad \boldsymbol{e^{At} = I + At + \frac{1}{2}(At)^2 + \frac{1}{6}(At)^3 + \cdots} \tag{16}$$

The series has the usual form, with the matrix $A$ instead of the number $a$. Here I stop, because matrices come in Chapter 4: *Systems of Equations*. When the matrix $A$ is three by three, the equation $dy/dt = Ay$ represents three ordinary differential equations. Still first order linear, still constant coefficients, solved by $e^{At}$ in Section 6.4.

There is one big difference for matrices: $\boldsymbol{e^{At}e^{Bt} = e^{(A+B)t}}$ **is not true**. For numbers $a$ and $b$ this equation is correct. For matrices $A$ and $B$ something goes wrong in equation (6). When you look closely, you see that $b$ moved in front of $e^{at}$. But $e^{At}B = Be^{At}$ is false for matrices.

## ■ REVIEW OF THE KEY IDEAS ■

1. In the series for $e^t$, each term $t^n/n!$ is the derivative of the next term.
2. Then the derivative of $e^t$ is $e^t$, and the exponent rule holds: $e^t e^T = e^{t+T}$.
3. Another approach to $dy/dt = y$ is by finite differences $(Y_{n+1} - Y_n)/\Delta t = Y_n$. $Y_{n+1} = Y_n + \Delta t Y_n$ is the same as compound interest. Then $Y_n$ is close to $e^{n\Delta t} Y_0$.
4. $y = e^{at}$ solves $y' = ay$, and $a = i$ leads to $e^{it} = \cos t + i \sin t$ (Euler's Formula).
5. $\cos t = 1 - t^2/2 + \cdots$ and $\sin t = t - t^3/6 + \cdots$ are the even and odd parts of $e^{it}$.

# Problem Set 1.3

**1** Set $t = 2$ in the infinite series for $e^2$. The sum must be $e$ times $e$, close to 7.39. How many terms in the series to reach a sum of 7 ? How many terms to pass 7.3 ?

**2** Starting from $y(0) = 1$, find the solution to $dy/dt = y$ at time $t = 1$. Starting from that $y(1)$, solve $dy/dt = -y$ to time $t = 2$. Draw a rough graph of $y(t)$ from $t = 0$ to $t = 2$. What does this say about $e^{-1}$ times $e$ ?

**3** Start with $y(0) = \$5000$. If this grows by $dy/dt = .02y$ until $t = 5$ and then jumps to $a = .04$ per year until $t = 10$, what is the account balance at $t = 10$ ?

**4** Change Problem 3 to start with \$5000 growing at $dy/dt = .04y$ for the first five years. Then drop to $a = .02$ until $t = 10$. What is now the balance at $t = 10$ ?

**Problems 5–8 are about $y = e^{at}$ and its infinite series.**

**5** Replace $t$ by $at$ in the exponential series to find $e^{at}$ :

$$e^{at} = 1 + at + \frac{1}{2}(at)^2 + \cdots + \frac{1}{n!}(at)^n + \cdots$$

Take the derivative of every term (keep five terms). Factor out $a$ to show that *the derivative of $e^{at}$ equals $ae^{at}$*. At what time $T$ does $e^{at}$ reach 2 ?

**6** Start from $y' = ay$. Take the derivative of that equation. Take the $n^{\text{th}}$ derivative. Construct the Taylor series that matches all these derivatives at $t = 0$, starting from $1 + at + \frac{1}{2}(at)^2$. Confirm that this series for $y(t)$ is the series for $e^{at}$ in Problem 5.

**7** At what times $t$ do these events happen ?

(a) $e^{at} = e$ (b) $e^{at} = e^2$ (c) $e^{a(t+2)} = e^{at} e^{2a}$.

**8** If you multiply the series for $e^{at}$ in Problem 5 by itself you should get the series for $e^{2at}$. Multiply the first 3 terms by the same 3 terms to see the first 3 terms in $e^{2at}$.

**9** (recommended) Find $y(t)$ if $dy/dt = ay$ and $\boldsymbol{y(T) = 1}$ (instead of $y(0) = 1$).

**10** (a) If $dy/dt = (\ln 2)y$, explain why $y(1) = 2y(0)$.
(b) If $dy/dt = -(\ln 2)y$, how is $y(1)$ related to $y(0)$ ?

**11** In a one-year investment of $y(0) = \$100$, suppose the interest rate jumps from 6% to 10% after six months. Does the equivalent rate for a whole year equal 8%, or more than 8%, or less than 8% ?

**12** If you invest $y(0) = \$100$ at 4% interest compounded continuously, then $dy/dt = .04y$. Why do you have more than \$104 at the end of the year ?

**13** What linear differential equation $dy/dt = a(t)y$ is satisfied by $y(t) = e^{\cos t}$ ?

**14** If the interest rate is $a = 0.1$ per year in $y' = ay$, how many years does it take for your investment to be multiplied by $e$ ? How many years to be multiplied by $e^2$ ?

**15** Write the first four terms in the series for $y = e^{t^2}$. Check that $dy/dt = 2ty$.

**16** Find the derivative of $Y(t) = (1 + \frac{t}{n})^n$. If $n$ is large, this $dY/dt$ is close to $Y$ !

**17** Suppose the exponent in $y = e^{u(t)}$ is $u(t) =$ integral of $a(t)$. What equation $dy/dt =$ _____ $y$ does this solve ? If $u(0) = 0$ what is the starting value $y(0)$ ?

## Challenge Problems

**18** $e^{d/dx} = 1 + d/dx + \frac{1}{2}(d/dx)^2 + \cdots$ is a sum of higher and higher derivatives. Applying this series to $f(x)$ at $x = 0$ would give $f + f' + \frac{1}{2}f'' + \cdots$ at $x = 0$. The Taylor series says: This is equal to $f(x)$ at $x =$ _____ .

**19** (Computer or calculator, 2.xx is close enough) Find the time $t$ when $e^t = 10$. The initial $y(0)$ has increased by an order of magnitude—a factor of 10. The exact statement of the answer is $t =$ _____. At what time $t$ does $e^t$ reach 100?

**20** The most important curve in probability is the bell-shaped graph of $e^{-t^2/2}$. With a calculator or computer find this function at $t = -2, -1, 0, 1, 2$. Sketch the graph of $e^{-t^2/2}$ from $t = -\infty$ to $t = \infty$. *It never goes below zero.*

**21** Explain why $y_1 = e^{(a+b+c)t}$ is the same as $y_2 = e^{at}e^{bt}e^{ct}$. They both start at $y(0) = 1$. They both solve what differential equation?

**22** For $y' = y$ with $a = 1$, Euler's first step chooses $Y_1 = (1 + \Delta t)Y_0$. Backward Euler chooses $Y_1 = Y_0/(1 - \Delta t)$. Explain why $1 + \Delta t$ is smaller than the exact $e^{\Delta t}$ and $1/(1 - \Delta t)$ is larger than $e^{\Delta t}$. (Compare the series for $1/(1 - x)$ with $e^x$.)

**Note** Section 3.5 presents an accurate Runge-Kutta method that captures three more terms of $e^{a\Delta t}$ than Euler. For $dy/dt = ay$ here is the step to $Y_{n+1}$:

$$\textbf{Runge-Kutta for } \boldsymbol{y' = ay} \qquad Y_{n+1} = \left(1 + a\Delta t + \frac{a^2\Delta t^2}{2} + \frac{a^3\Delta t^3}{6} + \frac{a^4\Delta t^4}{24}\right)Y_n.$$

## 1.4 Four Particular Solutions

The equation $dy/dt = ay$ is solved by $y(t) = e^{at}y(0)$. All the input is in that starting value $y(0)$. The solution grows exponentially when $a > 0$ and it decays when $a < 0$. *This section allows new inputs $q(t)$ after the starting time $t = 0$.* That input $q$ is a "source" when we add to $y(t)$, and a "sink" when we subtract. If $y(t)$ is the balance in a bank account at time $t$, then $q(t)$ is the rate of new deposits and withdrawals.

The basic first order linear differential equation (1) is fundamental to this course. We must and will solve this equation. Please pay attention to this section. In every way, this Section 1.4 is important.

$$\frac{dy}{dt} = ay + q(t) \quad \textbf{starting from } y(0) \textbf{ at } t = 0. \tag{1}$$

**Important I will separate the solution $y(t)$ into two parts.** One part comes from the starting value $y(0)$. The other part comes from the source term $q(t)$. This separation is a crucial step for all linear equations, and I take this chance to give names to the two parts. The part $y_n = Ce^{at}$ is what we already know. The part $y_p$ from the source $q(t)$ is new.

**1 Homogeneous solution or null solution $y_n(t)$ with no source : $q = 0$**

This part $y_n(t) = Ce^{at}$ solves the equation $dy/dt = ay$. The source term $q$ is zero (null). We are really solving $y' - ay = 0$, an equation with zero on the right hand side. That equation is **homogeneous**—we can multiply a solution by any constant to get another solution $cy(t)$. This book will choose the simpler word *null* and the subscript $n$, because this connects differential equations to linear algebra.

**2 Particular solution $y_p(t)$ with source $q(t)$**

This part $y_p(t)$ comes from the source term $q(t)$. The previous section had no source and therefore no reason to mention $y_p(t)$. Now our whole task is to find a particular solution $y_p(t)$, because the null solutions $y_n(t) = Ce^{at}$ are already set.

**3 The complete solution is $y(t) = y_n(t) + y_p(t)$**

For linear equations—and only for linear equations—adding the two parts gives the complete solution $y = y_n + y_p$. This is also called the "general solution."

| | | | | | | |
|---|---|---|---|---|---|---|
| **Null** | $y_n'$ | $-$ | $ay_n$ | $=$ | $\mathbf{0}$ | $y_n$ can start from $y(0)$ |
| **Particular** | $y_p'$ | $-$ | $ay_p$ | $=$ | $q(t)$ | $y_p$ can start from $y = 0$ |
| $\boldsymbol{y = y_n + y_p}$ | $\boldsymbol{y'}$ | $-$ | $\boldsymbol{ay}$ | $=$ | $q(t)$ | $y$ must start from $y(0)$ |

A nonlinear equation could include a quadratic term $y^2$. In that case adding ${y_n}^2$ to ${y_p}^2$ would not give $(y_n + y_p)^2$. The null equation $y' - y^2 = 0$ would not be homogeneous, and we can't multiply $y$ by a constant $C$. This will happen for the "logistic equation" in Section 1.7. You will see that $y(0)$ enters the solution $y(t)$ in a more complicated way.

The back cover of this book shows one particular solution $y_p$ combining with all null solutions $y_n$. This important picture is repeated for matrix equations and linear algebra.

## Particular Solutions and the Complete Solution

We can draw the complete solution to $u + v = 6$. These points $(u, v)$ fill a straight line. We can also draw all the null solutions to $u + v = 0$. They fill a parallel straight line, going through the center point $(0, 0)$. Figure 1.2 shows how the null solutions combine with one particular solution $(3, 3)$ to give the line of complete solutions.

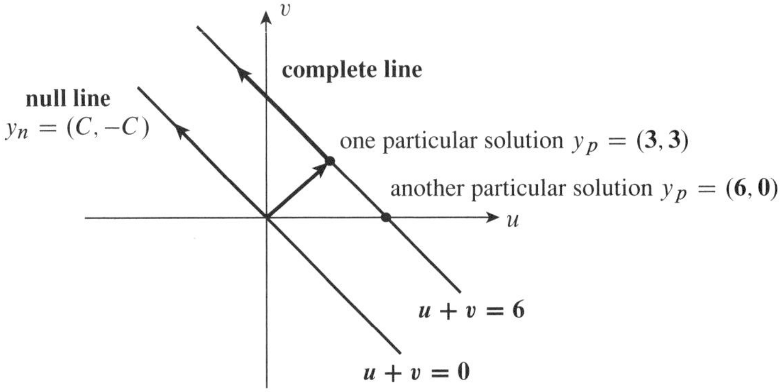


Figure 1.2: By adding all the null solutions to one particular solution, you get every solution (the complete line). You can start from *any* particular $y_p$ that solves $u + v = 6$.

Starting from $y_p = (3, 3)$, the complete solution has $u = 3 + C$ and $v = 3 - C$. This includes a **null** solution $C + (-C) = 0$, plus the **particular** solution $3 + 3 = 6$.

| | | | | | | | | | | |
|---|---|---|---|---|---|---|---|---|---|---|
| **Null** | $u_n$ | $+$ | $v_n$ | $=$ | $\mathbf{0}$ | $C$ | $+$ | $(-C)$ | $=$ | $0$ |
| **Particular** | $u_p$ | $+$ | $v_p$ | $=$ | $\mathbf{6}$ | $3$ | $+$ | $3$ | $=$ | $6$ |
| **Complete** | $u$ | $+$ | $v$ | $=$ | $\mathbf{6}$ | $\mathbf{(3+C)}$ | $+$ | $\mathbf{(3-C)}$ | $=$ | $\mathbf{6}$ |

The null solution $(C, -C)$ allows any constant $C$ (like $y(0)$). The particular solution could have any numbers $u_p$ and $v_p$ that add to 6. We made a special choice $u_p = 3$ and $v_p = 3$. In the equation $y' - ay = q$ we will often make the special choice $y_p(0) = 0$.

There are many particular solutions! You could say that we chose a *very particular* solution. In the differential equation we chose to start from $y_p(0) = 0$. For the equation $u + v = 6$ we chose $u = 3$ and $v = 3$. *We could equally well choose* $u = 6$ *and* $v = 0$. This particular solution is different, but we get the same complete solution line:

$$y_{\text{complete}} = (6 + c, 0 - c) \text{ is the same solution line as } y_{\text{complete}} = (3 + C, 3 - C).$$

If $c$ is 5, then $C$ is 8. From all $c$'s and all $C$'s, you get the same line.

I want to repeat this pattern of *null solution plus particular solution* by showing how it looks for an ordinary matrix equation $A\boldsymbol{v} = \boldsymbol{b}$ (Chapter 4 explains matrices):

**Null solution** $A\boldsymbol{v}_n = \mathbf{0}$ **Particular solution** $A\boldsymbol{v}_p = \boldsymbol{b}$ **Complete solution** $\boldsymbol{v} = \boldsymbol{v}_n + \boldsymbol{v}_p$

Always the key is *linearity*: $A\boldsymbol{v}$ *equals* $A\boldsymbol{v}_n + A\boldsymbol{v}_p$. Therefore $A\boldsymbol{v} = \mathbf{0} + \boldsymbol{b} = \boldsymbol{b}$.

Often the only solution to $A\boldsymbol{v}_n = \mathbf{0}$ is $\boldsymbol{v}_n = \mathbf{0}$. Then a particular solution $\boldsymbol{v}_p$ is also the complete solution. This will happen when $A$ is an "invertible matrix."

## Inputs $q(t)$ and Responses $y(t)$

For any input source $q(t)$, equation (4) will solve $dy/dt = ay + q(t)$. But when mathematics is applied to science and engineering and our society, problems don't involve "any $q(t)$." *Certain functions $q(t)$ are the most important.* Those functions are constantly met in applied mathematics. Here is a short list of special inputs:

1. **Constant source** $\quad q(t) = \boldsymbol{q}$
2. **Step function at $T$** $\quad q(t) = \boldsymbol{H(t-T)}$
3. **Delta function at $T$** $\quad q(t) = \boldsymbol{\delta(t-T)}$
4. **Exponential** $\quad q(t) = \boldsymbol{e^{ct}}$

This section will solve $dy/dt = ay + q(t)$ for the four functions on that short list. The next section adds one more source $q(t)$. It is a combination of sine and cosine. Or $q(t)$ can be a complex exponential (which has one term and is usually easier):

5. **Sinusoid** $\quad q(t) = \boldsymbol{A \cos \omega t + B \sin \omega t}$ or $\boldsymbol{R\, e^{i\omega t}}$

## Solving Linear Equations by an Integrating Factor

The equation $y' = ay + q$ is so important that I will solve it in different ways. The first way uses an integrating factor $M(t)$. Put both $y$ terms on the left. Keep $q(t)$ on the right.

**Problem** **Solve $\boldsymbol{y' - ay = q(t)}$ starting from any $\boldsymbol{y(0)}$**

**Method** ***Multiply both sides by the integrating factor $\boldsymbol{M(t) = e^{-at}}$.***

We chose that factor $e^{-at}$ so that $M$ times $y' - ay$ is exactly the derivative of $My$:

**Perfect derivative**
$$e^{-at}(y' - ay) \text{ agrees with } \frac{d}{dt}(e^{-at}y) = \frac{d}{dt}(My). \tag{2}$$

When both sides of $y' - ay = q$ are multiplied by $M = e^{-at}$, our equation is immediately ready to be integrated. The right side is $Mq$, the left side is the derivative of $My$.

$$\text{The integral of } \frac{d}{dt}(My) = Mq \quad \text{is} \quad M(t)\,y(t) - M(0)\,y(0) = \int_0^t M(s)\,q(s)\,ds \tag{3}$$

At $t = 0$ we know that $M(0) = e^0 = 1$. Multiply both sides of equation (3) by $e^{at}$ (which is $1/M$) to see $y(t) = y_n + y_p$. This solution comes many times in the book! To give meaning to formula (4), I will apply it to the most important inputs $q(t)$.

**The key formula**
**Solution to $\boldsymbol{y' = ay + q(t)}$**
$$\boldsymbol{y(t) = e^{at}\,y(0) + e^{at}\int_0^t e^{-as}\,q(s)\,ds.} \tag{4}$$

## Constant Source $q(t) = q$

When $q(t)$ is a constant, the integration for the particular solution in equation (4) is easy.

$$\int_0^t e^{-as}\, q\, ds = \left[\frac{qe^{-as}}{-a}\right]_{s=0}^{s=t} = \frac{q}{a}(1 - e^{-at}).$$

*Multiply by $e^{at}$ to find $y_p(t)$.* An important solution to an important equation.

| | | |
|---|---|---|
| **Solution for constant source $q$** | $\boldsymbol{y(t) = e^{at}\, y(0) + \frac{q}{a}(e^{at} - 1)}$ | (5) |

Example 1 has a positive growth rate $a > 0$. The solution will increase when $q > 0$. Example 2 will have a *negative* rate $a < 0$. In that case $y(t)$ approaches a *steady state.*

**Example 1** Solve $dy/dt - 5y = 3$ starting from $y(0) = 2$. Here $a = 5$ and $q = 3$. This fits perfectly with $y' - ay = q$. Equation (5) gives the solution $y(t)$:

*Solution* $y(t) = y_n + y_p = \boldsymbol{2e^{5t} + \frac{3}{5}(e^{5t} - 1)}$. Set $t = 0$ to check that $y(0) = 2$.

Looking at that solution, I have to admit that $y' - 5y = 3$ is not so obvious. This becomes much clearer when the two parts (null + particular) are separated:

$$\begin{aligned} y_n(t) &= 2e^{5t} \text{ certainly has } y_n' - 5y_n = 0 \text{ with } y_n(0) = 2 \\ y_p(t) &= \tfrac{3}{5}(e^{5t} - 1) \text{ has } y_p' = 3e^{5t}. \text{ This agrees with } 5y_p + 3. \end{aligned}$$

**Example 2** Solve $dy/dt = 3 - 6y$ starting from $y(0) = 2$.

Formula (5) still gives the answer, but this $y(t)$ is decreasing because $\boldsymbol{a = -6}$ **is negative**:

$$y(t) = 2e^{-6t} + \frac{3}{-6}(e^{-6t} - 1) = \boldsymbol{\frac{3}{2}e^{-6t} + \frac{1}{2}}.$$

When $t = 0$, that solution starts at $y(0) = 2$. The solution decreases because of $e^{-6t}$. **As $\boldsymbol{t \to \infty}$ the solution approaches $\boldsymbol{y_\infty = \frac{1}{2}}$.** This value $-q/a$ at $t = \infty$ is a *steady state.*

| |
|---|
| At $y = -\frac{q}{a} = \frac{1}{2}$ the equation $\frac{dy}{dt} = 3 - 6y$ becomes $\boldsymbol{\frac{dy}{dt} = 0}$. Nothing moves. |

Please notice that the steady state is $y_\infty = \frac{1}{2}$ for every initial value $y(0)$. That is because the null solution $y_n = y(0)e^{-6t}$ approaches zero. It is the particular solution that balances the source term $q = 3$ with the decay term $ay = -6y$ to approach $\boldsymbol{y_\infty = -q/a = 3/6}$.

*Question* If $y(0) = \frac{1}{2}$, what is $y(t)$? *Answer* $\boldsymbol{y(t) = \frac{1}{2}}$ **at all times**. $6y$ balances 3.

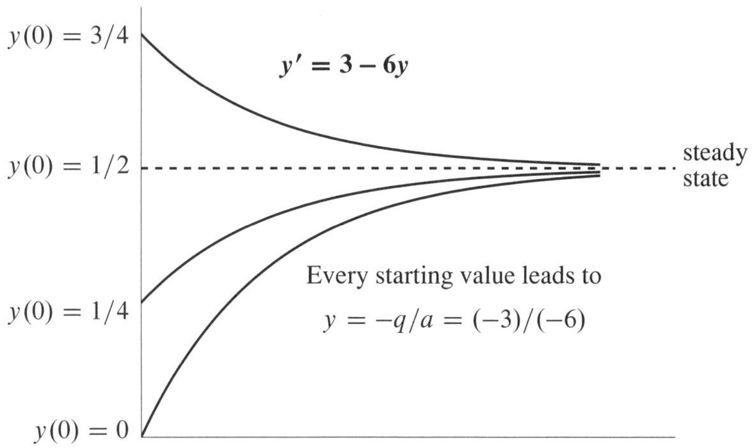


Figure 1.3: When $a$ is negative, $e^{at}$ approaches zero and $y(t)$ approaches $y_\infty = -q/a$.

Here is an important way to rewrite that basic equation $y' = ay + q$ when $a < 0$. The right hand side is the same as $a(y + \frac{q}{a})$. But $\boldsymbol{y + \frac{q}{a}}$ is exactly the distance $\boldsymbol{y - y_\infty}$. **Rewrite $\boldsymbol{y' = ay + q}$ as an easy equation $\boldsymbol{Y' = aY}$ by introducing $\boldsymbol{Y = y - y_\infty}$.**

**New unknown** $Y = y - y_\infty$ **New equation** $\boldsymbol{Y' = aY}$ **New start** $Y(0) = y(0) - y_\infty$

The solution to $Y' = aY$ is certainly $Y(t) = Y(0)e^{at}$. This approaches $Y_\infty = 0$ when $a < 0$. The original $y = Y + y_\infty$ still approaches $y_\infty$ which is $-q/a$ : see Figure 1.3.

$$(y - y_\infty)' = a(y - y_\infty) \quad \textbf{has solution} \quad y(t) - y_\infty = e^{at}(y(0) - y_\infty) \qquad (6)$$

Section 1.6 will present physical examples with $a < 0$: Newton's Law of Cooling, the level of messenger RNA, the decaying concentration of a drug in the bloodstream.

## Step Function

**The unit step function or "Heaviside step function" $\boldsymbol{H(t)}$ jumps from 0 to 1 at $\boldsymbol{t = 0}$.** Figure 1.4 shows its graph. The effect of $H(t)$ is like turning on a switch.

The second graph shows a *shifted step function* $H(t - T)$ which jumps from 0 to 1 at time $T$. This is the moment when $t - T = 0$, so $H$ jumps at that moment $T$.

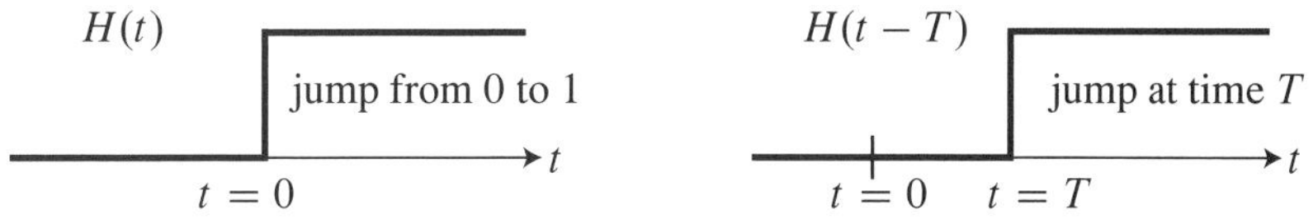


Figure 1.4: **The unit step function is $\boldsymbol{H(t)}$.** Its shift $H(t - T)$ jumps to 1 at $t = T$.

**When the step comes at $t = 0$, the solution to $y' - ay = H(t)$ is the step response.** That step response is easy to find because this equation is simply $y' - ay = 1$. The starting value is $y(0) = 0$. Put $q = 1$ into formula (5):

$$\textbf{Step response} \qquad y(t) = \frac{1}{a}(e^{at} - 1) \qquad (7)$$

**The interesting case is $a < 0$.** The solution starts at $y(0) = 0$. It grows to $y(\infty) = -1/a$. The system rises to that steady state after the switch is turned on. The graph of $y(t)$ is the bottom curve in Figure 1.3, except that $y_\infty$ is $1/6$ because the step function has $q = 1$.

*The step response is the output $y(t)$ when the step function is the input.* We are depositing at a constant rate $q = 1$. But when $a < 0$, we are losing $ay$ in real value because of inflation. Then growth stops at $y = -1/a$, where the deposits just balance the loss.

Now turn on the switch at time $T$ instead of time 0. The step function $H(t - T)$ is piecewise constant with two pieces: zero and one. If I multiply by any constant $q$, the source $q\, H(t - T)$ jumps from 0 to strength $q$ at time $T$.

The left side of our differential equation is still $y' - ay$, no change. The integrating factor $M = e^{-at}$ still makes that into a perfect derivative: $M(y' - ay)$ *equals* $(My)'$. The only change is on the right side, where the constant source doesn't start acting until the jump time $T$. At that time, the step function source $H(t - T)$ is turned on:

$$(e^{-at}y)' = e^{-at}\, H(t - T) \text{ now gives } e^{-at}y(t) - e^{0t}\, y(0) = \int_T^t e^{-as}\, ds. \qquad (8)$$

The only change for $t \geq T$ is to **start that integral at the turn-on time $T$**:

$$\int_T^t e^{-as}\, ds = \left[\frac{e^{-as}}{-a}\right]_{s=T}^{s=t} = \frac{1}{a}(e^{-aT} - e^{-at}). \qquad (9)$$

Multiply by $e^{at}$ to get the particular solution $y_p(t)$ beyond time $T$, and add $y_n = e^{at}\, y(0)$.

$$\textbf{Solution with unit step} \quad y(t) = e^{at}\, y(0) + \frac{1}{a}\left(e^{a(t-T)} - 1\right) \text{ for } t \geq T. \qquad (10)$$

As always, $y(0)$ grows or decays with $e^{at}$ in the null solution $y_n$. The step response is the particular solution, as soon as the input begins. But nothing enters until time $T$.

**Example 3** Suppose the input turns on at time $t = 0$ and turns off at $t = T$. Find $y(t)$.

**Solution** The input is $H(t) - H(t-T)$. The output is $y(t) = \frac{1}{a}\left(e^{at} - e^{a(t-T)}\right),\ t \geq T$.

## Delta Function

Now we meet a remarkable function $\delta(t)$. This "delta function" is everywhere zero, except at the instant $t = 0$. In that one moment it gives a unit input. Instead of a continuing source spread out over time, $\delta(t)$ is a **point source** completely concentrated at $t = 0$.

For a point source shifted to $\delta(t - T)$, **everything enters exactly at time $T$.** There is no source before that time or after that time. The delta function is zero except at one point. This "**impulse** " is by no means an ordinary function.

Here is one way to think about $\delta(t)$. **The delta function is the derivative of the unit step function $H(t)$.** But $H$ is constant and $dH/dt$ is zero except at $t = 0$. Take the integral of $\delta(t) = dH/dt$ from any negative number $N$ to any positive number $P$.

$$\textbf{Integral of } \boldsymbol{\delta(t)} \textbf{ is 1} \qquad \int_N^P \boldsymbol{\delta(t)}\, dt = \int_N^P \frac{\boldsymbol{dH}}{\boldsymbol{dt}}\, dt = H(P) - H(N) = \mathbf{1 - 0}. \tag{11}$$

"The area under the graph of $\delta(t)$ is 1. All that area is above the single point $t = 0$." Those words are in quotes because area at a point is impossible for ordinary functions. $\delta(t)$ may seem new and strange (it is useful!). Look at $dR/dt = H$ and $dH/dt = \delta$.

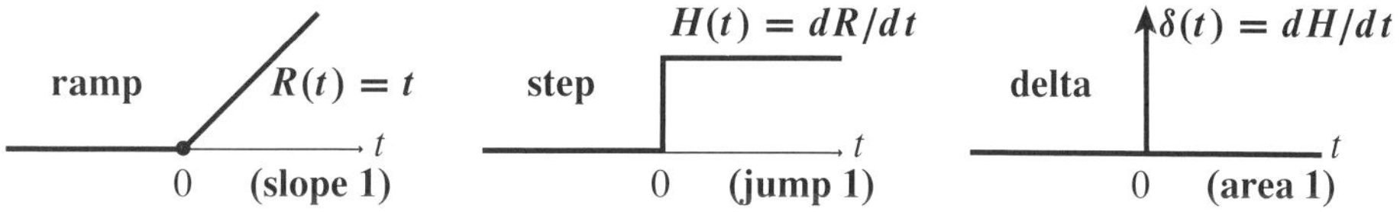


**Slope of the ramp jumps to 1. Slope of the step function is the delta function.**

The value of $\delta(0)$ is infinite. But that one word does not give full information. **The real way to understand delta functions is by their integrals.**

$$\int_{-\infty}^{\infty} \delta(t)\, dt = \mathbf{1} \qquad \int_{-\infty}^{\infty} \delta(t)\, F(t)\, dt = \boldsymbol{F(0)} \qquad \int_{-\infty}^{\infty} \delta(t - T)\, F(t)\, dt = \boldsymbol{F(T)} \tag{12}$$

Please visualize a tall thin box function—equal to $1/h$ between $t = 0$ and $t = h$. Now imagine $h$ going to zero. The width $h$ becomes zero and the height $1/h$ becomes infinite. *The area stays at* 1. All integrals of $\delta(t) F(t)$ are concentrated at $t = 0$: the "spike"

.

Here is a quick way to solve $y' - ay = \delta(t)$, and then we will do it more slowly. We know that the derivative of a step function $H(t)$ is the delta function $\delta(t)$. So the derivative of the step response must be the impulse response:

$$\frac{\boldsymbol{d}}{\boldsymbol{dt}}\textbf{(step)} = \textbf{delta} \qquad \frac{\boldsymbol{d}}{\boldsymbol{dt}}\left(\begin{array}{c}\textbf{step}\\ \textbf{response}\end{array}\right) = \frac{\boldsymbol{d}}{\boldsymbol{dt}}\left(\frac{\boldsymbol{e^{at} - 1}}{\boldsymbol{a}}\right) = \boldsymbol{e^{at}} = \begin{array}{c}\textbf{impulse}\\ \textbf{response}\end{array} \tag{13}$$

## The Impulse Response Solves $y' - ay = \delta(t)$

Start your bank account with one deposit. Start your heart with a sudden shock. Hit a golf ball. Fire a bullet. Many motions start with an "impulse" and then the source term is a delta function $\delta(t)$.

**The impulse response $y(t)$ jumps immediately to $y(0) = 1$.** You can see that by integrating every term in $dy/dt - ay = \delta(t)$. Integrating $\delta(t)$ from $t = -h$ to $h$ gives **1**. Integrating $dy/dt$ gives $y(h) - y(-h)$, which is $\boldsymbol{y(h)}$. The integral of $ay$ becomes zero as $h \to 0$. That limit step when $h \to 0$ leaves $\boldsymbol{y(0) = 1}$.

After the jump to $y(0) = 1$, the impulse $\delta(t)$ is immediately zero. So we just have the ordinary null solution to $y' = ay$ starting from $y(0) = 1$:

| **Impulse response** | $y' - ay = \delta(t)$ | $y(t) = e^{at}$ | (14) |
|---|---|---|---|

Notice the different responses to an impulse and a step function. The impulse deposits everything at $t = 0$. The step function goes on depositing forever. If $a < 0$ and inflation reduces our wealth, the impulse response dies out to $y_\infty = 0$. The step response increases from 0 to $y_\infty = -1/a$, where the deposits balance the loss from inflation.

I want to emphasize : **$e^{at}$ is the growth or decay factor $G(t)$ for all inputs**. When the input is $y(0)$, the output at time $t$ is $e^{at}y(0)$. When the input is $q(s)$ at time $s$, the output later at $t$ is $e^{a(t-s)}q(s)$. The growth is only over the remaining time $t - s$. **Our main formula (4) is adding up all the outputs that come from all the inputs.**

## Delayed Delta Function

The source $q(t) = \delta(t - T)$ turns on at time $T$. Then immediately it turns off. In that one instant of time, *the value of $y$ jumps by* 1. "We deposited \$1 at that moment." The integral of $dy/dt = \delta(t - T)$ is 1. This is the change in $y$, before $T$ to after $T$.

Coming up to time $T$, the solution is $y(t) = e^{at}y(0)$. At time $T$ we add 1. After time $T$, that input has the shorter period $t - T$ in which to grow. Multiply 1 by $e^{a(t-T)}$:

$$\textbf{Solution for } q = \delta(t - T) \qquad y(t) = y_n(t) + y_p(t) = e^{at}\, y(0) + e^{a(t-T)}. \qquad (15)$$

The solution $y$ jumps by $e^{a(T-T)} = e^0 = 1$, when that second term appears at $t = T$.

**Example 4** Solve the equation $\boldsymbol{y' - 5y = 3\delta(t - 4)}$ starting from $y(0) = 2$.
The null solution to $y' - 5y = 0$ starting at $y(0) = 2$ is $\boldsymbol{y_n(t) = 2e^{5t}}$. This we know. The particular solution is $y_p(t) = 0$ up to $t = 4$. At that moment $y$ jumps by 3, from $3\delta$. Its growth factor is $e^{5(t-4)}$. Then $\boldsymbol{y_p(t) = 3e^{5(t-4)}}$ after $t = 4$.

$$\textbf{Complete solution with jump of 3} \qquad y_n + y_p = 2e^{5t} + 3e^{5(t-4)}H(t - 4) \qquad (16)$$

The step function $H(t - 4)$ combines $y_p = 0$ before the jump and $y_p$ after the jump into one formula. At $t = 4$ the solution jumps by 3. Then this 3 grows to $3e^{5(t-4)}$.

***Remark* 1** This solution makes me realize that the initial value $y(0)$ is like having a delta function at time $t = 0$. *The solution "jumps" to* $y(0)$. I don't know if you agree with that.

***Remark* 2** $q(t) = -\delta(t - T)$ would be negative (a sink instead of a source). A bank account could be earning interest at the rate $a$, and suddenly you withdraw 1 at time $T$. The balance $y(T)$ had reached $e^{aT}y(0)$, and it drops by 1. From time $T$ onwards, the growth factor $e^{a(t-T)}$ multiplies the new balance, and $y(t) = e^{at}y(0) - e^{a(t-T)}$.

***Remark* 3** (a little mysterious) We could think of an ordinary continuous input $q(t)$ as a lot of delta functions—a delta function of strength $q(T)$ *at every time* $T$. Instead of "a lot" I need to say "an integral". Every continuous function $q(t)$ is an integral of delta functions $q(T)\,\delta(t - T)$ at all $T$. The integral picks out $q(t)$ at the spike point.

$$\textbf{Any } q(t) = \textbf{combination of delta functions} = \int q(T)\,\delta(t - T)\,dT. \qquad (17)$$

**Example 5** ($q = 1$) **The integral of all impulses for $T \geq 0$ is the step function $H(t)$.**

Then the integral of all impulse responses is the step response. The integral of $e^{at}$ from 0 to $t$ is $(e^{at} - 1)/a$. *Derivative of step response = impulse response* as in (13).

## Exponential Input $e^{ct}$

The source $q(t) = e^{ct}$ starts at time zero and continues forever. The particular solution $y_p(t)$ is easy to find, because **$y_p$ is a multiple $Ye^{ct}$ of this same exponential $e^{ct}$.** That is the beauty of exponentials . These are the most important functions and the best to work with. They allow growth or decay or oscillation from $c > 0$ and $c < 0$ and $c = i\omega$.

$$\textbf{Substitute } y_p = Ye^{ct} \textbf{ into } y' - ay = e^{ct} \qquad cYe^{ct} - aYe^{ct} = e^{ct}$$

When we cancel $e^{ct}$ this leaves a simple formula for the number $Y$ in $Ye^{ct}$ :

$$cY - aY = 1 \quad \textbf{gives} \quad \boxed{Y = \frac{1}{c - a}} \quad \textbf{and} \quad \boxed{y_p(t) = \frac{e^{ct}}{c - a}} \qquad (18)$$

**Example 6** Solve $y' - 5y = 3e^{4t}$ starting from $y(0) = 2$. Now $Y = \dfrac{3}{c - a} = \dfrac{3}{4 - 5}$.

The null solution still involves $e^{5t}$. The particular solution is $Y$ times $e^{4t}$ !

$$y_p(t) = Ye^{4t} \qquad y_p' - 5y_p = (4Y - 5Y)e^{4t} = 3e^{4t}. \quad \text{Then } \boldsymbol{Y = -3}.$$

This particular solution $-3e^{4t}$ starts at $-3$. Since $y(0) = 2$, the other part starts at $+5$.

$$\textbf{Complete solution} \qquad y(t) = 5e^{5t} - 3e^{4t}.$$

**The null solution grows at rate $a = 5$. One particular solution grows at rate $c = 4$.** The equation $y' - ay = e^{ct}$ is solved for $c \neq a$ but two final comments are needed.

1. This particular solution $y(t) = e^{ct}/(c - a)$ is not the "very particular" solution that starts from $y_p(0) = 0$. It is still perfectly good, except it starts at $1/(c - a)$. So the complete solution starting at $y(0)$ has to include the usual $y(0)e^{at}$ and **also a term to cancel $1/(c - a)$ at time zero**:

$$y' - ay = e^{ct} \qquad y_{\text{complete}} = y(0)\, e^{at} - \frac{e^{at}}{c-a} + \frac{e^{ct}}{c-a} \tag{19}$$

There you see a null solution $y_n$ (two terms) and our particular $y_p$ (the last term). Or the last two terms together are the very particular solution $(e^{ct} - e^{at})/(c - a)$.

2. For $c = a$ we are in serious trouble. The formulas fail because we can't divide by $c - a = 0$. This problem $y' - ay = e^{at}$ is a type of **resonance**, when the exponent $c$ in the source happens to equal the exponent $a$ in the natural growth from $y' = ay$. The integral in our main formula (4) becomes $\int e^{-as} e^{as}\, ds = \int 1\, ds = t$.

$$\textbf{Resonance} \quad c = a \qquad y' - ay = e^{at} \qquad y = y(0)e^{at} + te^{at} \tag{20}$$

That extra growth factor $t$ is because $y_n$ resonates with $y_p$. They both have $e^{at}$.

## ■ REVIEW OF THE KEY IDEAS ■

1. **Complete** solution to a linear equation = **null** solution(s) + **particular** solution.

2. The integrating factor $e^{-at}$ multiplies $y' - ay = q(t)$ to give $(e^{-at}y)' = e^{-at}q(t)$.
   Integrate and multiply by $e^{at}$: $y(t) = y_n + y_p = e^{at}y(0) + e^{at}\int e^{-as}q(s)\, ds$.

3. For $y' - ay = q =$ **constant**, the particular solution with $y_p(0) = 0$ is $q(e^{at} - 1)/a$.

4. $q(t) = H(t)$: the response to a unit step function is $y_p = (e^{at} - 1)/a$.

5. $q(t) = \delta(t)$: the impulse response to a unit delta function is $y_p = e^{at}$.

6. $q(t) = e^{ct}$ gives $y_p = (e^{ct} - e^{at})/(c - a)$. In case $c = a$, change to $y_p = te^{at}$.

## Problem Set 1.4

**1** All solutions to $dy/dt = -y + 2$ approach the steady state where $dy/dt$ is zero and $y = y_\infty =$ ___ . That constant $y = y_\infty$ is a particular solution $y_p$.

Which $y_n = Ce^{-t}$ combines with this steady state $y_p$ to start from $y(0) = 4$? This question chose $y_p + y_n$ to be $y_\infty +$ *transient* (decaying to zero).

**2** For the same equation $dy/dt = -y + 2$, choose the null solution $y_n$ that starts from $y(0) = 4$. Find the particular solution $y_p$ that starts from $y(0) = 0$.

This splitting chooses the two parts $e^{at}y(0)$ + integral of $e^{a(t-s)}q$ in equation (4).

**3** The equation $dy/dt = -2y + 8$ has two natural splittings $\boldsymbol{y_S + y_T = y_N + y_P}$ :

**1.** Steady ($\boldsymbol{y_S = y_\infty}$) + Transient ($\boldsymbol{y_T \to 0}$). What are those parts if $y(0) = 6$?

**2.** ($y_N' = -2y_N$ from $\boldsymbol{y_N(0) = 6}$) + ($y_P' = -2y_P + 8$ starting from $\boldsymbol{y_P(0) = 0}$).

**4** All null solutions to $u - 2v = 0$ have the form $(u, v) = (c,$ ____ $)$.

One particular solution to $u - 2v = 3$ has the form $(u, v) = (7,$ ____ $)$.

Every solution to $u - 2v = 3$ has the form $(7,$ ____ $) + c(1,$ ____ $)$.

But also every solution has the form $(3,$ ____ $) + C(1,$ ____ $)$ for $C = c + 4$.

**5** The equation $dy/dt = 5$ with $y(0) = 2$ is solved by $y =$ ____ . A natural splitting $y_n(t) =$ ___ and $y_p(t) =$ ___ comes from $y_n = e^{at}y(0)$ and $y_p = \int e^{a(t-s)}5\,ds$.

This small example has $\boldsymbol{a = 0}$ (so $ay$ is absent) and $\boldsymbol{c = 0}$ (the source is $q = 5e^{0t}$). When $a = c$ we have "resonance." A factor $t$ will appear in the solution $y$.

**Starting with Problem 6, choose the very particular $\boldsymbol{y_p}$ that starts from $\boldsymbol{y_p(0) = 0}$.**

**6** For these equations starting at $y(0) = 1$, find $y_n(t)$ and $y_p(t)$ and $y(t) = y_n + y_p$.

(a) $y' - 9y = 90$ (b) $y' + 9y = 90$

**7** Find a linear differential equation that produces $y_n(t) = e^{2t}$ and $y_p(t) = 5(e^{8t} - 1)$.

**8** Find a resonant equation ($a = c$) that produces $y_n(t) = e^{2t}$ and $y_p(t) = 3te^{2t}$.

**9** $y' = 3y + e^{3t}$ has $y_n = e^{3t}y(0)$. Find the resonant $y_p$ with $y_p(0) = 0$.

**Problems 10–13 are about $\boldsymbol{y' - ay}$ = constant source $\boldsymbol{q}$.**

**10** Solve these linear equations in the form $y = y_n + y_p$ with $y_n = y(0)e^{at}$.

(a) $y' - 4y = -8$ (b) $y' + 4y = 8$ Which one has a steady state?

**11** Find a formula for $y(t)$ with $y(0) = 1$ and draw its graph. What is $y_\infty$?

(a) $y' + 2y = 6$ (b) $y' + 2y = -6$

**12** Write the equations in Problem 11 as $Y' = -2Y$ with $Y = y - y_\infty$. What is $Y(0)$?

**13** If a drip feeds $q = 0.3$ grams per minute into your arm, and your body eliminates the drug at the rate $6y$ grams per minute, what is the steady state concentration $y_\infty$? Then *in* = *out* and $y_\infty$ is constant. Write a differential equation for $Y = y - y_\infty$.

**Problems 14–18 are about $y' - ay =$ step function $H(t - T)$:**

**14** Why is $y_\infty$ the same for $y' + y = H(t-2)$ and $y' + y = H(t-10)$?

**15** Draw the ramp function that solves $y' = H(t - T)$ with $y(0) = 2$.

**16** Find $y_n(t)$ and $y_p(t)$ as in equation (10), with step function inputs starting at $T = 4$.

(a) $y' - 5y = 3H(t-4)$ (b) $y' + y = 7H(t-4)$ (*What is* $y_\infty$?)

**17** Suppose the step function turns on at $T = 4$ and off at $T = 6$. Then $q(t) = H(t-4) - H(t-6)$. Starting from $y(0) = 0$, solve $y' + 2y = q(t)$. What is $y_\infty$?

**18** Suppose $y' = H(t-1) + H(t-2) + H(t-3)$, starting at $y(0) = 0$. Find $y(t)$.

**Problems 19–25 are about delta functions and solutions to $y' - ay = q\,\delta(t - T)$.**

**19** For all $t > 0$ find these integrals $a(t), b(t), c(t)$ of point sources and graph $b(t)$:

$$\text{(a)} \int_0^t \delta(T-2)\,dT \quad \text{(b)} \int_0^t (\delta(T-2) - \delta(T-3))\,dT \quad \text{(c)} \int_0^t \delta(T-2)\delta(T-3)\,dT$$

**20** Why are these answers reasonable? (They are all correct.)

$$\text{(a)} \int_{-\infty}^{\infty} e^t\delta(t)\,dt = 1 \quad \text{(b)} \int_{-\infty}^{\infty} (\delta(t))^2\,dt = \infty \quad \text{(c)} \int_{-\infty}^{\infty} e^T\delta(t-T)\,dT = e^t$$

**21** *The solution to* $y' = 2y + \delta(t-3)$ *jumps up by* 1 *at* $t = 3$. Before and after $t = 3$, the delta function is zero and $y$ grows like $e^{2t}$. Draw the graph of $y(t)$ when (a) $y(0) = 0$ and (b) $y(0) = 1$. Write formulas for $y(t)$ before and after $t = 3$.

**22** Solve these differential equations starting at $y(0) = 2$:

(a) $y' - y = \delta(t-2)$ (b) $y' + y = \delta(t-2)$. (*What is* $y_\infty$?)

**23** Solve $dy/dt = H(t-1) + \delta(t-1)$ starting from $y(0) = 0$: jump and ramp.

**24** (My small favorite) What is the steady state $y_\infty$ for $y' = -y + \delta(t-1) + H(t-3)$?

**25** Which $q$ and $y(0)$ in $y' - 3y = q(t)$ produce the step solution $y(t) = H(t-1)$?

**Problems 26–31 are about exponential sources $q(t) = Qe^{ct}$ and resonance.**

**26** Solve these equations $y' - ay = Qe^{ct}$ as in (19), starting from y(0) = 2:

(a) $y' - y = 8e^{3t}$ (b) $y' + y = 8e^{-3t}$ (*What is* $y_\infty$ ?)

**27** When $c = 2.01$ is very close to $a = 2$, solve $y' - 2y = e^{ct}$ starting from $y(0) = 1$. By hand or by computer, draw the graph of $y(t)$ : near resonance.

**28** When $c = 2$ is exactly equal to $a = 2$, solve $y' - 2y = e^{2t}$ starting from $y(0) = 1$. This is resonance as in equation (20). By hand or computer, draw the graph of $y(t)$.

**29** Solve $y' + 4y = 8e^{-4t} + 20$ starting from $y(0) = 0$. What is $y_\infty$ ?

**30** The solution to $y' - ay = e^{ct}$ didn't come from the main formula (4), but it could. Integrate $e^{-as}e^{cs}$ in (4) to reach the very particular solution $(e^{ct} - e^{at})/(c - a)$.

**31** *The easiest possible equation* $y' = 1$ *has resonance*! The solution $y = t$ shows the factor $t$. What number is the growth rate $a$ and also the exponent $c$ in the source?

**32** Suppose you know two solutions $y_1$ and $y_2$ to the equation $y' - a(t)y = q(t)$.

(a) Find a null solution to $y' - a(t)y = 0$.

(b) Find all null solutions $y_n$. Find all particular solutions $y_p$.

**33** Turn back to the first page of this Section 1.4. Without looking, can you write down a solution to $y' - ay = q(t)$ for all four source functions $\boldsymbol{q}, \boldsymbol{H(t)}, \boldsymbol{\delta(t)}, \boldsymbol{e^{ct}}$ ?

**34** Three of those sources in Problem 33 are actually the same, if you choose the right values for $q$ and $c$ and $y(0)$. What are those values?

**35** What differential equations $y' = ay + q(t)$ would be solved by $y_1(t)$ and $y_2(t)$? Jumps, ramps, corners—maybe harder than expected (math.mit.edu/dela/Pset1.4).

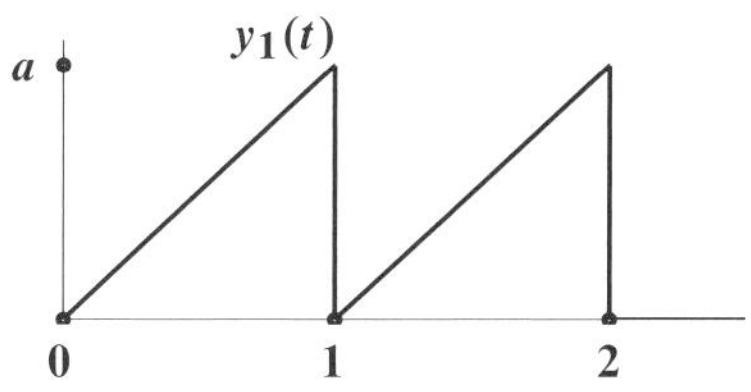


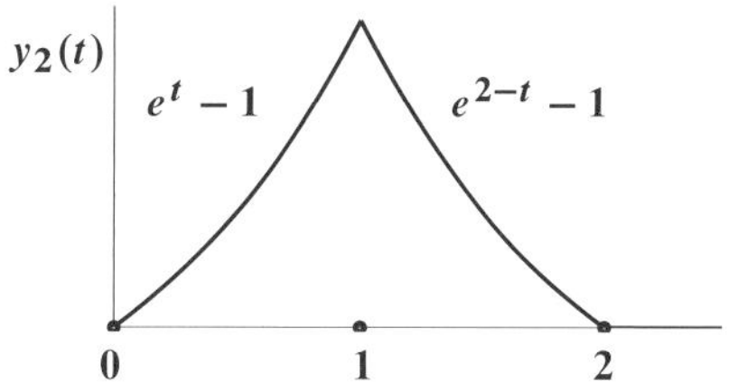

## 1.5 Real and Complex Sinusoids

Section 1.4 ended with the equation $y' - ay = e^{ct}$. A particular solution was easy to produce, because we kept $e^{ct}$. We simply chose the correct multiplier $Y = 1/(c-a)$ in $y_p(t) = Ye^{ct}$. **This section changes the real number $c$ to an imaginary number $i\omega$. The multiplier is now $Y = 1/(i\omega - a)$** . The solution formula $Ye^{i\omega t}$ will stay exactly the same, but we need complex numbers (with real part and imaginary part). The payoff is that we can solve all real problems $y' - ay = A\cos\omega t + B\sin\omega t$ at once.

Many scientific and engineering applications are driven by sources $q(t)$ that oscillate like $\cos\omega t$ and $\sin\omega t$ (**sinusoids** ). Pistons go up and down to drive a car, voltages go up and down to drive current (alternating current). The input frequency is $\omega$, and the output frequency is also $\omega$. The problem is to find the *amplitude* and the *phase* in the output (the response to the input). The real solution will be $\boldsymbol{y = M\cos\omega t + N\sin\omega t}$.

This $y(t)$ will be a particular solution (steady solution). It is not the transient solution $y_n(t)$ that decays to zero. *We solve $y' - ay = q(t)$ when the source $q(t)$ is a sinusoid.* For this section and the next, applications come from biology and chemistry and medicine and more. The number $a$ is often a rate constant. It tells the speed of a chemical reaction.

Note that RLC circuits (resistor-inductor-capacitor) produce equations with second derivatives. Those will go into Chapter 2, but RC and RL circuits (first order equations) belong here. Our plan for this section is straightforward: *Real then complex.*

**1** (**Real**) **Solve** $\boldsymbol{dy/dt - ay = q(t) = A\cos\omega t + B\sin\omega t}$.

This leads to two equations for the two coefficients $M, N$ in $y = M\cos\omega t + N\sin\omega t$.

**2** (**Complex**) **Solve** $\boldsymbol{dy/dt - ay = q(t) = R\,e^{i\omega t}}$.

This leads to one easy equation for the coefficient in $y = Ye^{i\omega t}$. But that number $Y$ is complex, so we still have two real numbers to find (real and imaginary parts of $Y$).

**3** (A key idea) **Write the complex number $\boldsymbol{1/(i\omega - a)}$ in its polar form $\boldsymbol{G\,e^{-i\alpha}}$.**

The positive number $G$ is the **gain** . The angle $\alpha$ is the **phase lag** . Those have important meanings and they are perfect to graph separately. In many problems (most problems) $G$ and $\alpha$ are more useful than the real and imaginary parts of $1/(i\omega - a)$.

So we need to explain and review complex numbers. They are worth knowing and not difficult. The next page will solve the real problem **1** and the complex problem **2**. We can't simplify the real problem by using cosines alone, because the term $dy/dt$ in the equation would unavoidably involve $\sin\omega t$.

The *Review of the Key Ideas* at the end organizes the important steps.

### Real Sinusoids

We want a particular real solution $y(t)$ when the source $q(t)$ oscillates with frequency $\omega$.

$$\boxed{\textbf{First order linear equation} \qquad \frac{dy}{dt} - ay = A\cos\omega t + B\sin\omega t.} \qquad (1)$$

The solution will have the same form $y = M\cos\omega t + N\sin\omega t$ as the source term. By matching the $\cos\omega t$ terms and separately the $\sin\omega t$ terms, you get two equations for $M$ and $N$. Just subtract $ay = aM\cos\omega t + aN\sin\omega t$ from $dy/dt = -\omega M\sin\omega t + \omega N\cos\omega t$.

$$\frac{dy}{dt} - ay = q \qquad \begin{array}{ll} \textbf{cos}\,\boldsymbol{\omega t}\ \textbf{terms} & -a\ M + \omega\ N = A \\ \textbf{sin}\,\boldsymbol{\omega t}\ \textbf{terms} & -\omega\ M - a\ N = B \end{array} \tag{2}$$

Those two equations tell us $M$ and $N$ in the real solution $y(t) = M\cos\omega t + N\sin\omega t$. I will write down the solution to equation (2), and then describe two ways to find it.

$$\begin{array}{l}\textbf{Source}\quad q = A\cos\omega t + B\sin\omega t \\ \textbf{Solution}\quad y = M\cos\omega t + N\sin\omega t\end{array} \qquad M = -\frac{aA + \omega B}{\omega^2 + a^2} \qquad N = \frac{\omega A - aB}{\omega^2 + a^2} \tag{3}$$

I would find $N$ by eliminating $M$ in equation (2). If you multiply the first equation by $\omega$ and the second equation by $a$, then *subtraction removes* $M$. The right side is $\omega A - aB$, the left side is $(\omega^2 + a^2)N$. Then $N$ is correct in equation (3). Similarly we find $M$.

For two equations it is also practical to find $M$ and $N$ from the 2 by 2 **inverse matrix** :

$$\begin{bmatrix} -a & \omega \\ -\omega & -a \end{bmatrix}\begin{bmatrix} M \\ N \end{bmatrix} = \begin{bmatrix} A \\ B \end{bmatrix} \quad \text{gives} \quad \begin{bmatrix} M \\ N \end{bmatrix} = \frac{1}{\omega^2 + a^2}\begin{bmatrix} -a & -\omega \\ \omega & -a \end{bmatrix}\begin{bmatrix} A \\ B \end{bmatrix}.$$

The matrix on the left times its inverse on the right gives the identity matrix $I$ in Chapter 4. That denominator $\omega^2 + a^2$ of the inverse matrix appears in $M$ and $N$, in the solution (3).

## Complex Sinusoid $e^{i\omega t}$

Now we come to the very important input $\boldsymbol{q(t) = R\,e^{i\omega t}}$. That input is oscillating with frequency $\omega$ radians per second. *The output $y(t)$ will oscillate with the same frequency $\omega$.* This is true because $a$ is constant in the differential equation. When $y(t) = Ye^{i\omega t}$ includes the same factor $e^{i\omega t}$, that factor cancels from every term in the equation:

$$\begin{array}{rcl} q(t) &=& Re^{i\omega t} \\ y(t) &=& Ye^{i\omega t} \end{array} \qquad y' - ay = q \ \text{ becomes } \ i\omega Ye^{i\omega t} - aYe^{i\omega t} = R\,e^{i\omega t}. \tag{4}$$

When we divide by $e^{i\omega t}$, this leaves an easy algebra problem for the complex number $Y$ :

$$\textbf{Response } Y(\omega) \qquad i\omega Y - aY = R \quad \text{gives} \quad \boldsymbol{Y = \frac{R}{i\omega - a}} \ \text{ and } \ \boldsymbol{y = Ye^{i\omega t}}. \tag{5}$$

The simplicity of the solution $y = Ye^{i\omega t}$ comes from one key fact: The derivative of $e^{i\omega t}$ is a multiple of $e^{i\omega t}$ (the multiplying factor is $i\omega$). This was not true for $\cos\omega t$. Its derivative brings in $\sin\omega t$. So we had to solve two real equations for $M$ and $N$, while (5) is one complex equation for $Y$.

## Complex Numbers : Rectangular and Polar

The complex number $z = x + iy$ has real part $x$ and imaginary part $y$. The basic ideas are explained here; more details are in Section 2.2. We plot all $z$ in the **complex plane** (the real-imaginary plane). Figure 1.5 shows the particular number $z = 4 + 3i$ with $x = \text{Re } z = 4$ and $y = \text{Im } z = 3$. No problem with the rectangular form $4 + 3i$, except that multiplying and dividing are not at all convenient in $x - y$ coordinates.

The first figure also shows the **polar form** of the same number $z$. The magnitude (or modulus) is $r$. The phase is the angle $\theta$. From $x$ and $y$ we can find $r$ and $\theta$.

*The magnitude is* $r = \sqrt{x^2 + y^2} = \sqrt{25} = 5$. *The angle* $\theta$ *has tangent* $y/x = 3/4$.

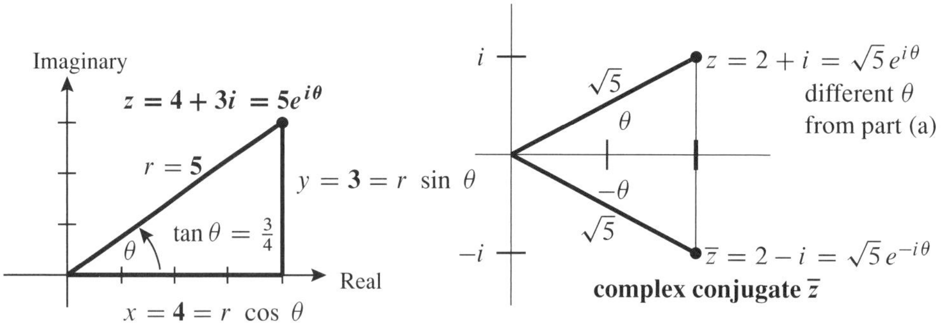


Figure 1.5: (a) $z = 4 + 3i$ is a point in the complex plane. Its polar form is $z = 5e^{i\theta}$.

The polar form is perfect for multiplication and division of complex numbers. To multiply $re^{i\theta}$ times $Re^{i\alpha}$, add the angles and multiply $r$ times $R$. To divide, subtract the angles and divide $r$ by $R$.

$$\textbf{Multiply} \quad (re^{i\theta})(Re^{i\alpha}) = \boldsymbol{rR\, e^{i(\theta+\alpha)}} \qquad \textbf{Divide} \quad \frac{re^{i\theta}}{Re^{i\alpha}} = \frac{\boldsymbol{r}}{\boldsymbol{R}}\,\boldsymbol{e^{i(\theta-\alpha)}} \tag{6}$$

The polar form is also perfect for squaring a complex number $re^{i\theta}$ and for $1/re^{i\theta}$ :

$$\textbf{Square} \quad z^2 = (re^{i\theta})(re^{i\theta}) = \boldsymbol{r^2 e^{2i\theta}} \qquad \textbf{Invert} \quad \frac{1}{z} = \frac{1}{re^{i\theta}} = \frac{\mathbf{1}}{\boldsymbol{r}}\,\boldsymbol{e^{-i\theta}} \tag{7}$$

Let me compare that polar form of $1/z$ with $1/(x+iy)$. Multiply by $(x-iy)/(x-iy) = 1$.

$$\frac{1}{z} = \frac{1}{x+iy} = \frac{1}{x+iy}\,\frac{x-iy}{x-iy} = \frac{\boldsymbol{x - iy}}{\boldsymbol{x^2 + y^2}} \qquad \frac{1}{4+3i} = \frac{4-3i}{4^2+3^2} = \frac{1}{5}e^{-i\theta}$$

This number $x - iy$ appears often. It is the **complex conjugate** $\bar{z}$ of the number $z = x + iy$.

Notice that $x + iy$ times $x - iy$ is $x^2 + y^2$. In other words $\boldsymbol{z}$ **times** $\boldsymbol{\bar{z}}$ **is** $\boldsymbol{|z|^2 = r^2}$.

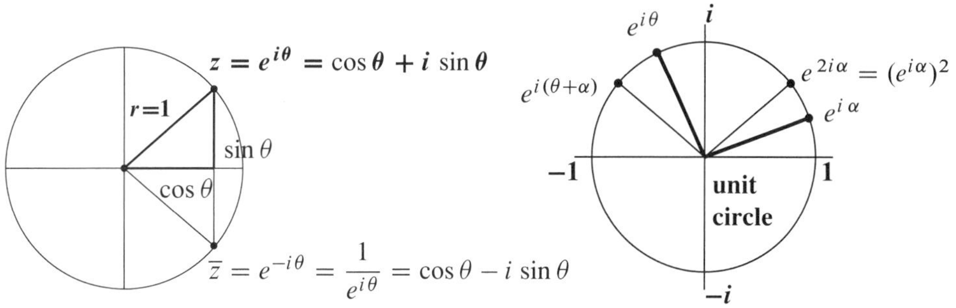


Figure 1.6: Points $e^{i\theta}$ on the unit circle have $r = 1$. When $e^{i\theta}$ multiplies $e^{i\alpha}$, angles add.

## The Unit Circle

Figure 1.6 shows the **unit circle** , where every radial distance is $r = 1$. Then we just add the angles to multiply, or double the angles to square, or subtract the angles to divide:

$$\textbf{On the circle} \qquad (e^{i\theta})(e^{i\alpha}) = e^{i(\theta+\alpha)} \qquad (e^{i\theta})(e^{-i\theta}) = 1 \qquad \frac{1}{e^{i\theta}} = e^{-i\theta}$$

$e^{-i\theta}$ is the complex conjugate of $e^{i\theta}$, the mirror image across the axis in Figure 1.6.

**Example 1** Describe the paths of the numbers $\boldsymbol{e^{st}}$ and $\boldsymbol{e^{i\omega t}}$ and $\boldsymbol{e^{(s+i\omega)t}}$ in the complex plane (real $s$ and real $\omega$). The time $t$ goes from 0 to $\infty$. Those paths start at 1.

*Solution* If $s > 0$, the number $e^{st}$ goes from 1 out the real axis to infinity. If $s < 0$, then $e^{st}$ goes from 1 in to zero. All real.

The path of $e^{i\omega t}$ goes around the unit circle with constant speed. At time $T = 2\pi/\omega$ (and also $2T$, $3T$, ...) it comes back to $e^{2\pi i} = 1$. The path goes clockwise if $\omega < 0$.

The path of $e^{(s+i\omega)t}$ **spirals outward** to infinity if $s > 0$. It spirals inward to zero if $s < 0$. At time $T = 2\pi/\omega$ it is a real number $e^{sT}$, because the factor $e^{i\omega T} = e^{2\pi i}$ is 1.

## The Gain $G$ and the Phase Lag $\alpha$

The complex number $1/(i\omega - a)$ multiplies the input $q(t) = Re^{i\omega t}$ to give the output $y(t) = Ye^{i\omega t}$. **What is the magnitude of $\boldsymbol{1/(i\omega - \alpha)}$ and what is its angle?** We need its polar form $\boldsymbol{1/(i\omega - a) = Ge^{-i\alpha}}$. Start with $i\omega - a = re^{i\alpha}$ and then invert:

$$\boldsymbol{i\omega - a = re^{i\alpha}} \qquad \boldsymbol{r = \sqrt{\omega^2 + a^2}} \quad \text{and} \quad \textbf{tan}\,\boldsymbol{\alpha} = \frac{\text{imaginary part}}{\text{real part}} = -\frac{\boldsymbol{\omega}}{\boldsymbol{a}}.$$

We want $1/(re^{i\alpha})$. This will be $Ge^{-i\alpha}$. The gain is $\boldsymbol{G = 1/r = 1/\sqrt{\omega^2 + \alpha^2}}$:

$$\textbf{Gain } G \quad \textbf{Phase angle } \alpha \qquad \frac{\mathbf{1}}{\boldsymbol{i\omega - a}} = \frac{1}{r}e^{-i\alpha} = \frac{1}{\sqrt{\omega^2 + a^2}}e^{-i\alpha} = \boldsymbol{Ge^{-i\alpha}}. \qquad (8)$$

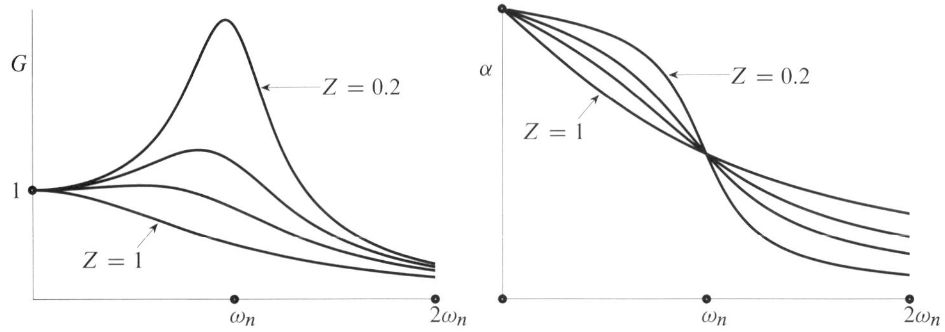


Figure 1.7: Dimensionless gain $G$ and phase angle $\phi$ as functions of frequency $\omega$.

The gain $G(\omega)$ and the angle $\alpha(\omega)$ are often graphed. The graphs below are variations of "*Bode plots*." The amplitude response $G(\omega)$ is especially important, and you are very likely to see that gain $G$ by itself—often including an extra factor $|a|$.

*Note* One common variation is to include the rate constant $a$ in the forcing term $q(t) = a\,R\,e^{i\omega t}$. We still think of $R\,e^{i\omega t}$ as the input, then $a$ gives $q$ the right physical units. That factor $a$ will appear in the output. So the gain $G = |\text{output}|\,/\,|\text{input}|$ will be increased by that factor $|a|$. *Then* $G = |a|/\sqrt{\omega^2 + a^2}$ is 1 *at the frequency* $\omega = 0$.

## Sinusoids $R\ \cos(\omega t - \phi)$

The next page will show that any combination of $\cos \omega t$ and $\sin \omega t$ is a *shifted cosine*. It has frequency $\omega$ and amplitude $R$ and phase lag $\phi$. If you know $\omega$ and $R$ and $\phi$, it is no problem to graph $y(t) = \boldsymbol{R\ \cos(\omega t - \phi)}$. To go the other way, and *read off those three numbers from the graph*, is much more interesting.

This mystery sinusoid came from lecture notes for MIT's course 18.03. The website **mathlets.org** has interactive experiments. The question here is : **Find $\boldsymbol{\omega}$, $\boldsymbol{R}$, and $\boldsymbol{\phi}$.**

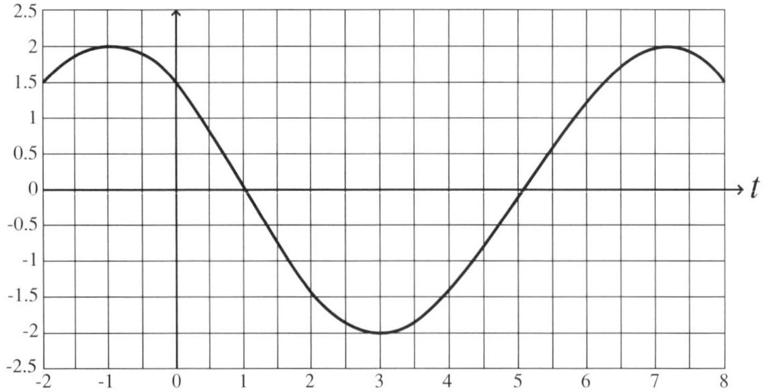

## The Sinusoidal Identity

We want to choose the magnitude $R$ and the angle $\phi$ so that $A \cos \omega t + B \sin \omega t$ is the *real part* of $Re^{i(\omega t - \phi)}$. We can and will solve $y' - ay = Re^{i(\omega t-\phi)}$ quickly. When we take the real part of all terms in this differential equation, the correct input $q(t) = R\cos(\omega t - \phi)$ will appear on the right side and the correct output $y(t)$ will appear on the left side. The real equation will be solved in one step.

So we want this identity for the "sinusoidal" input $q(t)$ :

**Sinusoidal identity** $$\boldsymbol{A \cos \omega t + B \sin \omega t = R \cos(\omega t - \phi)} \qquad (9)$$

The right side has the same period $2\pi/\omega$ as the left side—and only one term.

To find $R$ and $\phi$, expand $R\cos(\omega t - \phi)$ into $R\cos \omega t \cos \phi + R \sin \omega t \sin \phi$. Then match cosines to find $A$ and match sines to find $B$ :

$$A = R \cos \phi \quad \text{and} \quad B = R \sin \phi \qquad \boxed{\boldsymbol{A^2 + B^2 = R^2} \quad \text{and} \quad \boldsymbol{\tan \phi = \frac{B}{A}}.} \qquad (10)$$

So we know $R = \sqrt{A^2 + B^2}$ and $\phi = \tan^{-1}(B/A)$ in the sinusoidal identity. The beauty of $R$ and $\phi$ is that they match sinusoids to the polar form of complex numbers.

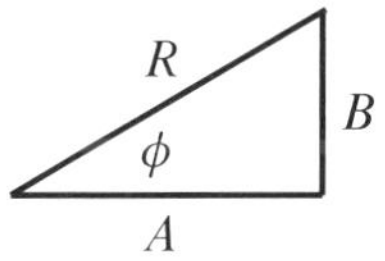


| | | |
|---|---|---|
| $A + iB$ | $= Re^{i\phi}$ | polar form of $A + iB$ |
| $R$ | $= \sqrt{A^2 + B^2}$ | produces $R$ and $\phi$ in the |
| $\tan \phi$ | $= B/A$ | sinusoidal identity (9) |

For practice with this important formula, Problem 1 will develop a slightly different proof.

**Example 2** Write $q(t) = \cos 3t + \sin 3t$ as $R\cos(3t - \phi)$ : the real part of $Re^{i(3t-\phi)}$.

*Solution* $A = 1$ and $B = 1$ so that $R = \sqrt{2}$. The angle $\phi = \frac{\pi}{4}$ has $\tan \phi = B/A = 1$. Then $\cos 3t + \sin 3t = \boldsymbol{\sqrt{2} \cos (3t - \frac{\pi}{4})}$.

**Example 3** Write the real part of $e^{i5t}/(\sqrt{3} + i)$ in the form $A \cos 5t + B \sin 5t$.

*Solution* $\sqrt{3} + i$ is $2e^{i\pi/6}$ (why?) Then $e^{i5t}/(\sqrt{3} + i)$ is $\frac{1}{2}e^{i(5t-\pi/6)}$. Its real part is

$$\frac{1}{2} \cos\left(5t - \frac{\pi}{6}\right) = \frac{1}{2}\left(\cos 5t \cos \frac{\pi}{6} + \sin 5t \sin \frac{\pi}{6}\right) = \frac{\sqrt{3}}{4} \cos 5t + \frac{1}{4} \sin 5t.$$

## Real Solution $y$ from Complex Solution $y_c$

The sinusoidal identity solves $y' - ay = A \cos \omega t + B \sin \omega t$ in three steps :

1. This equation is the real part of the complex equation $y_c' - ay_c = R\, e^{i(\omega t - \phi)}$.

2. The complex solution is $y_c = R\, e^{i(\omega t - \phi)}/(i\omega - a) = \boldsymbol{R\, G\, e^{i(\omega t - \phi - \alpha)}}$.

3. The real part of that complex solution $y_c$ is the desired real solution $y(t)$.

Those three steps are **1** (real to complex) **2** (solve complex) **3** (complex to real). This will succeed. The second step expresses $1/(i\omega - \alpha)$ as $Ge^{-i\alpha}$ to keep the polar form. The third step produces $y = M\cos\omega t + N\sin\omega t$ directly as $\boldsymbol{y = RG\cos(\omega t - \phi - \alpha)}$.

**Example 4** Take those three steps real-complex-real to solve $\boldsymbol{y' - y = \cos t - \sin t}$.

We have to find $R$, $\phi$, $G$, and $\alpha$ from the numbers $a = 1$, $\omega = 1$, $A = 1$, and $B = -1$. Notice that $\boldsymbol{RG = 1}$.

$$R = \sqrt{A^2 + B^2} = \sqrt{2} \qquad \tan\phi = \frac{B}{A} = -1 \text{ and } \boldsymbol{\phi = -\frac{\pi}{4}} \qquad G = \frac{1}{\sqrt{\omega^2 + a^2}} = \frac{\mathbf{1}}{\sqrt{\mathbf{2}}}$$

The angle for $i\omega - a = i - 1$ is $\boldsymbol{\alpha = \frac{3\pi}{4}}$. Its tangent is $-\frac{\omega}{a} = -1$.

1. The sinusoidal identity is $\cos t - \sin t = \sqrt{2}\cos(t - \phi) = \boldsymbol{\sqrt{2}\cos(t + \pi/4)}$.

2. $y_{\textbf{complex}} = \dfrac{\boldsymbol{\sqrt{2}\, e^{i(t+\pi/4)}}}{\boldsymbol{i - 1}}$. Here $\dfrac{1}{i\omega - a} = \dfrac{1}{i - 1} = Ge^{-i\alpha} = \dfrac{1}{\sqrt{2}}e^{-3\pi i/4}$.

3. $y_{\textbf{complex}} = \boldsymbol{RG\, e^{i(\omega t - \alpha - \phi)} = e^{i(t - \pi/2)}}$. Then $\boldsymbol{y_{\textbf{real}} = \cos\left(t - \frac{\pi}{2}\right) = \sin t}$.

That example was chosen so that $G = 1/\sqrt{2}$ cancelled $R = \sqrt{2}$. If we keep all the symbols $R$, $\phi$, $G$, $\alpha$ then the solution $y_{\text{real}} = \boldsymbol{RG\,\cos(\omega t - \phi - \alpha)}$ from Step **3** must agree with the solution $y = M\cos\omega t + N\sin\omega t$ at the start of this section.

The key point in many applications is not necessarily the numbers in the formula for $y(t)$. Very often the goal is to see from the formula how $y(t)$ depends on parameters like $a$ and $\omega$ in the differential equation. The gain $G = |\,\text{output}\,|/|\,\text{input}\,|$ is a convenient and very important guide.

The truth is that the complex solution is better. The sinusoidal identity shows how every combination $A\cos\omega t + B\sin\omega t$ is the real part $R\cos(\omega t - \phi)$ of a complex exponential $Re^{i(\omega t - \phi)}$. So we can convert real to complex and complex back to real.

In between, solve the complex form by using the ***frequency response*** $\boldsymbol{1/(i\omega - a)}$.

**Conclusion** When the input $q(t)$ is $Re^{i\omega t}$, the output $y(t)$ multiplies by $1/(i\omega - a)$. This multiplying factor is a complex number, and it changes with the frequency $\omega$. We absolutely need to understand that number $Y$ and graph its magnitude $G$ and its phase.

## ■ REVIEW OF THE KEY IDEAS ■

1. (*Real*) $y' - ay = A\cos\omega t + B\sin\omega t$ leads to $\boldsymbol{y_{\textbf{real}}} = M\cos\omega t + N\sin\omega t$.
2. (*Sinusoidal identity*) $A\cos\omega t + B\sin\omega t$ equals $R\cos(\omega t - \phi)$ with $R^2 = A^2 + B^2$.
3. (*Complex*) $y' - ay = Re^{i(\omega t - \phi)}$ leads to $\boldsymbol{y_{\textbf{complex}}} = Re^{i(\omega t - \phi)}/(i\omega - a)$.
4. (*Complex gain*) $1/(i\omega - a) = \boldsymbol{Ge^{-i\alpha}}$ with $G = 1/\sqrt{\omega^2 + a^2}$ and $\tan\alpha = -\omega/a$.
5. (*Real part of the complex solution*) $\boldsymbol{y_{\textbf{real}}} = \text{Re}(\boldsymbol{y_{\textbf{complex}}}) = \boldsymbol{RG\cos(\omega t - \alpha - \phi)}$.

## Problem Set 1.5

**Problems 1-6 are about the sinusoidal identity (9). It is stated again in Problem 1.**

**1** These steps lead again to the sinusoidal identity. This approach doesn't start with the usual formula $\cos(\omega t - \phi) = \cos \omega t \ \cos \phi + \sin \omega t \ \sin \phi$ from trigonometry. The identity says:

$$\textbf{If } A + iB = Re^{i\phi} \textbf{ then } A \cos \omega t + B \sin \omega t = R \cos(\omega t - \phi).$$

Here are the four steps to find that real part of $Re^{i(\omega t - \phi)}$. Explain $A - iB$ in Step 3.

$$R \cos(\omega t - \phi) = \text{Re}\left[R e^{i(\omega t - \phi)}\right] = \text{Re}\left[e^{i\omega t}(R e^{-i\phi})\right] = \ (\textit{what is } R e^{-i\phi}\,?)$$

$$= \text{Re}\left[(\cos \omega t + i \ \sin \omega t)\ (A - iB)\right] = A \cos \omega t + B \sin \omega t.$$

**2** To express $\sin 5t + \cos 5t$ as $R \ \cos(\omega t - \phi)$, what are $R$ and $\phi$?

**3** To express $6 \ \cos 2t + 8 \ \sin 2t$ as $R \ \cos(2t - \phi)$, what are $R$ and $\tan \phi$ and $\phi$?

**4** Integrate $\cos \omega t$ to find $(\sin \omega t)/\omega$ in this complex way.

(i) $dy_{\text{real}}/dt = \cos \omega t$ is the real part of $dy_{\text{complex}}/dt = e^{i\omega t}$.

(ii) Take the real part of the complex solution.

**5** The sinusoidal identity for $A = 0$ and $B = -1$ says that $-\sin \omega t = R \cos(\omega t - \phi)$. Find $R$ and $\phi$.

**6** Why is the sinusoidal identity useless for the source $q(t) = \cos t + \sin 2t$?

**7** Write $2 + 3i$ as $re^{i\phi}$, so that $\frac{1}{2+3i} = \frac{1}{r}e^{-i\phi}$. Then write $y = e^{i\omega t}/(2 + 3i)$ in polar form. Then find the real and imaginary parts of $y$. And also find those real and imaginary parts directly from $(2 - 3i)e^{i\omega t}/(2 - 3i)(2 + 3i)$.

**8** Write these functions $A \cos \omega t + B \sin \omega t$ in the form $R \cos(\omega t - \phi)$: Right triangle with sides $A$, $B$, $R$ and angle $\phi$.

1) $\cos 3t - \sin 3t$  2) $\sqrt{3} \cos \pi t - \sin \pi t$  3) $3 \cos(t - \phi) + 4 \sin(t - \phi)$

**Problems 9-15 solve real equations using the real formula (3) for $M$ and $N$.**

**9** Solve $dy/dt = 2y + 3 \cos t + 4 \sin t$ after recognizing $a$ and $\omega$. Null solutions $Ce^{2t}$.

**10** Find a particular solution to $dy/dt = -y - \cos 2t$.

**11** What equation $y' - ay = A \cos \omega t + B \sin \omega t$ is solved by $y = 3 \cos 2t + 4 \sin 2t$?

**12** The particular solution to $y' = y + \cos t$ in Section 1.4 is $y_p = e^t \int e^{-s} \cos s \, ds$. Look this up or integrate by parts, from $s = 0$ to $t$. Compare this $y_p$ to formula (3).

**13** Find a solution $y = M \cos \omega t + N \sin \omega t$ to $y' - 4y = \cos 3t + \sin 3t$.

**14** Find the solution to $y' - ay = A \cos \omega t + B \sin \omega t$ **starting from $y(0) = 0$**.

**15** If $a = 0$ show that $M$ and $N$ in equation (3) still solve $y' = A \cos \omega t + B \sin \omega t$.

**Problems 16-20 solve the complex equation $y' - ay = Re^{i(\omega t - \phi)}$.**

**16** Write down complex solutions $y_p = Ye^{i\omega t}$ to these three equations:

(a) $y' - 3y = 5e^{2it}$ (b) $y' = Re^{i(\omega t - \phi)}$ (c) $y' = 2y - e^{it}$

**17** Find complex solutions $z_p = Ze^{i\omega t}$ to these complex equations:

(a) $z' + 4z = e^{8it}$ (b) $z' + 4iz = e^{8it}$ (c) $z' + 4iz = e^{8t}$

**18** Start with the real equation $y' - ay = R \cos(\omega t - \phi)$. Change to the complex equation $z' - az = Re^{i(\omega t - \phi)}$. Solve for $z(t)$. Then take its real part $y_p = \operatorname{Re} z$.

**19** What is the initial value $y_p(0)$ of the particular solution $y_p$ from Problem 18? If the desired initial value is $y(0)$, how much of the null solution $y_n = Ce^{at}$ would you add to $y_p$?

**20** Find the real solution to $y' - 2y = \cos \omega t$ starting from $y(0) = 0$, in three steps: Solve the complex equation $z' - 2z = e^{i\omega t}$, take $y_p = \operatorname{Re} z$, and add the null solution $y_n = Ce^{2t}$ with the right $C$.

**Problems 21-27 solve real equations by making them complex. First a note on $\alpha$.**

Example 4 was $y' - y = \cos t - \sin t$, with growth rate $a = 1$ and frequency $\omega = 1$. The magnitude of $i\omega - a$ is $\sqrt{2}$ and the polar angle has $\tan \alpha = -\omega/a = -1$. Notice: *Both $\alpha = 3\pi/4$ and $\alpha = -\pi/4$ have that tangent*! How to choose the correct angle $\alpha$?

The complex number $i\omega - a = i - 1$ is in the *second quadrant*. Its angle is $\alpha = 3\pi/4$. **We had to look at the actual number and not just the tangent of its angle**.

**21** Find $r$ and $\alpha$ to write each $i\omega - a$ as $re^{i\alpha}$. Then write $1/re^{i\alpha}$ as $Ge^{-i\alpha}$.

(a) $\sqrt{3}i + 1$ (b) $\sqrt{3}i - 1$ (c) $i - \sqrt{3}$

**22** Use $G$ and $\alpha$ from Problem 21 to solve (a)-(b)-(c). Then take the real part of each equation and the real part of each solution.

(a) $y' + y = e^{i\sqrt{3}t}$ (b) $y' - y = e^{i\sqrt{3}t}$ (c) $y' - \sqrt{3}y = e^{it}$

**23** Solve $y' - y = \cos\omega t + \sin\omega t$ in three steps: real to complex, solve complex, take real part. This is an important example.

(1) Find $R$ and $\phi$ in the sinusoidal identity to write $\cos\omega t + \sin\omega t$ as the real part of $Re^{i(\omega t-\phi)}$.

(2) Solve $y' - y = e^{i\omega t}$ by $y = Ge^{-i\alpha}e^{i\omega t}$. Multiply by $Re^{-i\phi}$ to solve $z' - z = Re^{i(\omega t-\phi)}$.

(3) Take the real part $y(t) = \text{Re}\, z(t)$. Check that $y' - y = \cos\omega t + \sin\omega t$.

**24** Solve $y' - \sqrt{3}y = \cos t + \sin t$ by the same three steps with $a = \sqrt{3}$ and $\omega = 1$.

**25** (**Challenge**) Solve $y' - ay = A\cos\omega t + B\sin\omega t$ in two ways. First, find $R$ and $\phi$ on the right and $G$ and $\alpha$ on the left. Show that the final real solution $RG\cos(\omega t - \phi - \alpha)$ agrees with $M\cos\omega t + N\sin\omega t$ in equation (2).

**26** We don't have resonance for $y' - ay = Re^{i\omega t}$ when $a$ and $\omega \neq 0$ are real. *Why not*? (Resonance appears when $y_n = Ce^{at}$ and $y_p = Ye^{ct}$ share the exponent $a = c$.)

**27** If you took the imaginary part $y = \text{Im}\, z$ of the complex solution to $z' - az = Re^{i(\omega t-\phi)}$, what equation would $y(t)$ solve? Answer first with $\phi = 0$.

**Problems 28-31 solve first order circuit equations: not RLC but RL and RC.**

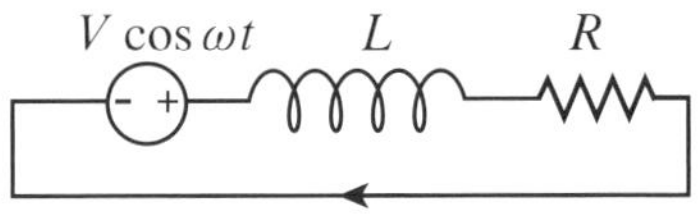


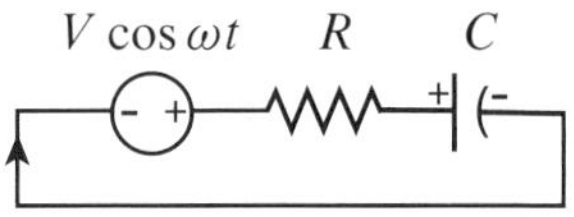


**28** Solve $\boldsymbol{L\, dI/dt + RI(t) = V\cos\omega t}$ for the current $I(t) = I_n + I_p$ in the RL loop.

**29** With $L = 0$ and $\omega = 0$, that equation is Ohm's Law $V = IR$ for direct current. The **complex impedance** $Z = R + i\omega L$ replaces $R$ when $L \neq 0$ and $I(t) = Ie^{i\omega t}$.

$$L\,dI/dt + RI(t) = \boldsymbol{(i\omega L + R)Ie^{i\omega t}} = \boldsymbol{Ve^{i\omega t}} \quad \text{gives} \quad \boldsymbol{Z\,I = V}.$$

What is the magnitude $|Z| = |R + i\omega L|$? What is the phase angle in $Z = |Z|e^{i\theta}$? Is the current $|I|$ larger or smaller because of $L$?

**30** Solve $\boldsymbol{R\dfrac{dq}{dt} + \dfrac{1}{C}q(t) = V\cos\omega t}$ for the charge $q(t) = q_n + q_p$ in the RC loop.

**31** Why is the complex impedance now $Z = R + \frac{1}{i\omega C}$? Find its magnitude $|Z|$. **Note that mathematics prefers $\boldsymbol{i = \sqrt{-1}}$, we are not conceding yet to $\boldsymbol{j = \sqrt{-1}}$!**

## 1.6 Models of Growth and Decay

This is an important section. It combines formulas with their applications. The formulas solve the key linear equation $y' - a(t)y = q(t)$—we are very close to the solution. Now $a$ can vary with $t$. The final step is to see the *purpose* of those formulas.

The point of this subject and this course is to understand change. **Calculus is about change**. A differential equation is a model of change. It connects $dy/dt$ to the current value of $y$ and to inputs/outputs that produce change. We see this as a math equation and solve it by a formula. If we stop there, we miss the whole reason for differential equations.

I will select five models of growth or decay, and five equations to describe them. Often the hardest part is to get the right equation. (Definitely harder than the right solution formula.) This section presents both steps of applied mathematics:

1. From the **model** to the **equation** 2. From the **equation** to the **solution**.

Our plan is to take the second step (the easier step) first: *Solve the equation*. Find the output $y(t)$ from inputs $a(t)$ and $q(t)$ and $y(0)$. Then come the models.

Here is the differential equation for $y(t)$. We want a formula to solve it—and we want to understand where that formula comes from. The solution $y(t)$ must use the three inputs $a(t)$ and $q(t)$ and $y(0)$, because they define the problem. **Sometimes $a(t)$ changes with time**. This possibility was not allowed in Sections 1.4 and 1.5.

**Differential equation** $$\boxed{\frac{dy}{dt} = a(t)y + q(t)} \quad \text{starting from } y(0) \text{ at } t = 0. \tag{1}$$

Up to now, our models had limited options for those inputs (and $a$ was constant):

| | |
|---|---|
| **Growth rate** $a(t)$ | The classic exponential $y(t) = e^t$ had $\mathbf{a = 1}$ |
| **Source term** $q(t)$ | Sections 1.4 and 1.5 had five particular inputs like $\mathbf{e}^{ct}$ and $\mathbf{e}^{i\omega t}$ |
| **Initial value** $y(0)$ | The starting value for $y(t) = e^t$ was $\mathbf{y(0) = 1}$ |

The "initial value" $y(0)$ is like a deposit to open a bank account. The source or sink $q(t)$ comes from **saving or spending** as time goes on. The solution $y(t)$ is the balance in the account at time $t$. I will reveal the final formula now, so you know where we are going.

**Growth factor $G(s,t)$ from time $s$ to time $t$**
$$\boxed{y(t) = G(0,t)\,y(0) + \int_0^t G(s,t)\,q(s)\,ds.} \tag{2}$$

Formula (2) has two parts. The first part $y_n = G(0,t)y(0)$ has $q = 0$: no source. The second part $y_p$ introduces the source $q(t)$, which adds fresh growth $G$ times $q$ (or subtracts when $q(t)$ is negative). *Go forward 2 pages to see the factor $G(s,t)$.*

$y =$ ( Null solution with $q = 0$) + (Particular solution from the input $q$).

### Particular Solution from $q(t)$

*On this page $a$ is constant.* The particular solution $y_p(t)$ is so important that we will reach it in three ways. Of course those three approaches will be closely related—but they are different enough and valuable enough to be presented separately :

**1. Integrating factor 2. Variation of parameters 3. Combine all outputs.**

1. The *integrating factor* $M(t) = e^{-at}$ was seen in Section 1.4. It solves $M' = -aM$. For constant growth rate $a$, multiplying the equation $y' - ay = q(t)$ by $M = e^{-at}$ turns the left side into an exact derivative of $My$ :

$$\frac{d}{dt}(e^{-at}y) = e^{-at}(y' - ay) = e^{-at}q(t). \tag{3}$$

Then we integrate the left and right hand sides to find $y = y_p(t)$ with $y_p(0) = 0$ :

$$e^{-at}\,y(t) = \int_0^t e^{-as}\,q(s)\,ds \quad \text{and} \quad y(t) = \int_0^t \boldsymbol{e^{a(t-s)}\,q(s)\,ds}. \tag{4}$$

2. *Variation of parameters* starts with the solutions $y_n = Ce^{at}$ to the null equation $y' - ay = 0$. **The new idea is to let $C$ vary with time in the particular solution.** Substitute $y = C(t)e^{at}$ into the equation $y' - ay = q(t)$ to find $C'e^{at} = q(t)$ :

$$(Ce^{at})' - a\,Ce^{at} = C'e^{at} + a\,Ce^{at} - a\,Ce^{at} = \boldsymbol{C'e^{at} = q(t)}. \tag{5}$$

Then $C' = e^{-at}q(t)$. Integrate to find $C$ and the solution formula we want :

$$C(t) = \int_0^t e^{-as}\,q(s)\,ds \qquad y(t) = C(t)e^{at} = \int_0^t \boldsymbol{e^{a(t-s)}\,q(s)\,ds}. \tag{6}$$

The integrating factor $M$ changes the equation. Varying $C(t)$ changes the solution. $C(t)$ will stay important for *systems* of $n$ equations ; integrating factors lose out.

3. **Each input $q(s)$ grows to $e^{a(t-s)}q(s)$ in the time between $s$ and $t$.** Then the solution $y(t)$ comes from these inputs $q(t)$ and growth factor $G = e^{a(t-s)}$. Add up (integrate) all those outputs :

$$\textbf{Growing time for } q(s) \textbf{ is } t-s \qquad \textbf{Output } y(t) = \int_0^t \boldsymbol{e^{a(t-s)}\,q(s)\,ds}. \tag{7}$$

To me, this third approach captures the meaning of the formulas (4) = (6) = (7). I like to think of each input $q(s)$ growing by the factor $G(s,t) = e^{a(t-s)}$ in the time $t - s$.

## Changing Growth Rate $a(t)$

The next step is to let $a(t)$ change in time. For example $a(t)$ could be $1+\cos t$, varying between 2 and 0. Certainly interest rates do change. The growth rate $a$ of your bank balance often slows down or speeds up. **Then the growth factor $G(0,t)$ is not just $e^{at}$.**

The null solution to $y_n' = a(t)y_n$ shows this clearly—the growth from time 0 to time $t$ :

**Integrate $a$ from 0 to $t$**
**Take the exponential**

$$\boxed{G(0,t) = e^{\int_0^t a(s)\,ds}} \qquad y_n(t) = G(0,t)\,y(0). \tag{8}$$

**The key point is that $dG/dt = a(t)\,G$.** First, the derivative of the integral of $a(t)$ is $a(t)$—by the Fundamental Theorem of Calculus. Second, the chain rule produces the derivative of $G$, when that integral goes into the exponent. Here is $dG/dt$ :

$$\frac{d}{dt}\left(e^{\text{integral of } a}\right) = \left(e^{\text{integral of } a}\right)\frac{d}{dt}(\text{integral of } a) \qquad \frac{dG}{dt} = (G)(a(t)) \tag{9}$$

When $a$ is constant, that integral is just $at$. This leads to the usual growth $G = e^{at}$. When $a$ varies, the exponent is messier than $at$ but the idea is the same: $\boldsymbol{dG/dt = aG}$.

Our example is $a(t) = 1+\cos t$. The integral of $a(t)$ is $t+\sin t$. This is the exponent:

**Growth factor $G(0,t) = e^{t+\sin t}$** **Null solution $y_n(t) = e^{t+\sin t}\,y(0)$**

Now we tackle the particular solution that comes from the inputs $q(t)$ when they grow. Again this $y_p(t)$ can come from an *integrating factor* or *variation of parameters* or an *integral of all outputs from all inputs*.

1. **The integrating factor is $M(t) = 1/G(t) = e^{-\int_0^t a(s)\,ds}$**. This has $M' = -a(t)M$.

Then the derivative of $My$ is exactly $Mq$, when we use $M' = -aM$.

**Product rule**
**Chain rule**

$$\frac{d}{dt}(My) = My' + M'y = M(y' - a(t)y) = Mq(t). \tag{10}$$

Integrate both sides of $(My)' = Mq$ starting from $y_p(0) = 0$. Then divide by $M$ :

$$M(t)y_p(t) = \int_0^t M(s)\,q(s)\,ds \qquad y_p(t) = e^{\int_0^t a(s)\,ds}\int_0^t e^{-\int_0^s a(s)\,ds}\,q(s)\,ds \tag{11}$$

When you multiply those exponentials, the exponents combine. The integral from 0 to $t$, minus the integral from 0 to $s$, equals the integral from $s$ to $t$. Each $q(s)$ enters at $s$. **The exponential of the integral of $a$ from $s$ to $t$ is the growth factor $G(s,t)$**:

$$\textbf{Growth factor } G(s,t) = e^{\int_s^t a(T)\,dT} \qquad \textbf{Solution } y_p(t) = \int_0^t G(s,t)\,q(s)\,ds \tag{12}$$

2. **Variation of parameters** . I will save this method to use in Chapter 2 for second order equations (with $y''$). Then all three methods get an equal chance—variation of parameters can solve equations that go beyond $y' = a(t)y + q(t)$.

3. **Integral of outputs** (my own choice). The input $q(s)$ enters at time $s$. It grows or decays until time $t$. The growth factor multiplying $q$ over that time is $G(s,t)$. Since $a(t)$ changes, the growth factor needs the integral of $a$. **The inputs are $q(s)$, the outputs are $G(s,t)\,q(s)$, and the total output $y_p(t)$ agrees with (12)**:

$$\boxed{G(s,t) = e^{\int_s^t a(T)\,dT} \qquad y_p(t) = \int_0^t G(s,t)\,q(s)\,ds} \tag{13}$$

When $q$ is a delta function at time $s$ (an impulse), the response is $y_p = G(s,t)$ at time $t$.

**Example 1** The growth rate $a(t) = 2t$ puts the economy into serious inflation. The integral of $a(t)$ is $\int_s^t 2T\,dT = t^2 - s^2$. Then $G$ is the growth from $s$ to $t$:

$$G(s,t) = e^{t^2-s^2} \qquad y' = 2t\,y + q(t) \text{ has } y_p(t) = \int_0^t e^{t^2-s^2}\,q(s)\,ds.$$

**Example 2** Here is an interesting case for investors. **Suppose the interest rate $a$ goes to zero**. What happens to the solution formula? The first term $y_n$ becomes $y(0)$. This deposit doesn't grow or disappear, it stays fixed. The growth factor is $G = 1$ and we just add up all the inputs (they didn't grow):

$$a = 0 \qquad y' = q(t) \text{ has the particular solution } y_p(t) = \int_0^t q(s)\,ds.$$

The problem comes when we start with the formula to solve $y' = ay + q$ (*constant* $q$):

$$y(t) = e^{at}y(0) + \int_0^t e^{a(t-s)}q\,ds = e^{at}y(0) + q\frac{e^{at}-1}{a}.$$

That looks bad at $a = 0$ because of dividing by $a$. But the factor $e^{at} - 1$ is also zero. *This is a case for l'Hôpital's Rule. Wonderful*! We can make sense of $0/0$:

$$\lim_{a \to 0} \frac{e^{at}-1}{a} = \frac{\text{Derivative with respect to } a}{\text{Derivative with respect to } a} = \frac{t}{1} = t.$$

The particular solution from $y' = q$ reduces to $q$ times $t$. That is the total savings during the time from 0 to $t$. With $a = 0$ it doesn't grow. Like putting money under a mattress, $a = 0$ means no risk and no gain. Then $dy/dt = q$ has $y(t) = y(0) + qt$.

Now the solution formula can be applied to real problems.

## Models of Growth and Decay

The whole point of a differential equation is to give a mathematical model of a practical problem. It is my duty to show you examples. This section will offer growth equations ($a > 0$), decay equations ($a < 0$), and the balance equation that controls the temperature of the Earth. That balance equation is not linear.

Please understand that a linear equation is only an approximation to reality. The approximation can be very good over an important range of values. Newton's Law $F = ma$ is linear and we live by it every day. But Einstein showed that the mass $m$ is not a constant, it increases with the velocity. We don't notice this until we are near the speed of light.

Similarly the stretch in a spring is proportional to the force—for a while. A really large force will stretch the spring way out of shape. That takes us to nonlinear elasticity. Eventually the spring breaks.

The same for analysis of a car crash. Linear at very slow speed, nonlinear at normal speeds, total wreck at high speeds. A crash is a very difficult problem in computational mechanics. So is the effect of dropping a cell phone. This has been studied in great detail.

Back to linear equations, starting with constant $a$ and $y(0)$ and $q$.

**Model 1** $\quad$ $\boldsymbol{y(t)}$ **= money in a savings account**

This is the example we already started. We have a formula for the answer, now we use it. That formula is based on a *continuous* savings rate $q(t)$ (deposits every instant, not every month). It also has *continuous* interest $ay$ (computed every instant, not every month or every year). Continuous compounding does not bring instant riches. Just a little more income, by computing interest day and night.

Suppose we get 3% interest. This number is $a = .03$, but what are the "units" of $a$? The rate is 3% **per year**. There is a time dimension. If we change to months, the same rate is now $a = \frac{3}{12}\% = .0025$ **per month**.

**Units of $\boldsymbol{a}$ are** $\dfrac{\mathbf{1}}{\textbf{time}}$ $\quad$ To change from years to months, divide $a$ by 12.

You can see this in the equation $dy/dt = ay$. Both sides have $y$. So $a$ on the right agrees dimensionally with $1/t$ on the left. Frequency is also $1/$ time; $i\omega - a$ is good!

The savings rate $q$ has the same dimension as $ay$. The dimension of $q$ is **money / time**. We see that in the words too: $q = 100$ **dollars per month**.

**Question : Does $\boldsymbol{y(t)}$ grow or decay ?** $\quad$ This depends on $y(0)$ and $a$ and $q$.

So far $a$ and $q$ have been positive; we were saving. If we spend money constantly, then $q$ changes to *negative*. Interest is still entering because $a$ is positive. Does $q$ win or does $a$ win? Do we spend all our deposit and drop to $y = 0$, or does the interest $ay(t)$ allow us to keep up the spending level $q$ forever?

**Answer : If we start with $\boldsymbol{ay(0) + q > 0}$, then $\boldsymbol{y(t)}$ will grow even if $\boldsymbol{q < 0}$.**

The reason is in the differential equation $dy/dt = ay(t) + q$. If the right side is positive at time $t = 0$, then $y$ starts growing. So the right side stays positive, and $y$ keeps growing.

Common sense gives the same answer: If $ay + q > 0$, the interest $ay$ coming in stays ahead of the spending going out.

*A question for you.* Suppose $a < 0$ but $q > 0$. Your investment is going down at rate $a$. You are adding new investments at rate $q$. Overall, does your account go up or down?

You won't actually hit zero, because $e^{at}$ stays positive forever, even if $a < 0$. You approach the steady state $y_\infty = -q/a$. In reality, the end of prosperity has come.

Now I will compare continuous compounding (expressed by a differential equation) with ordinary compounding (a difference equation). The difference equation starts with the same $Y_0 = y(0)$. This changes to $Y_1$ and then $Y_2$ and $Y_3$, taking a finite step each year. When the time step $\Delta t$ is one year, the interest rate is $A$ **per year** and the saving rate is $Q$ **dollars per year**:

$$\frac{dy}{dt} = ay + q \qquad \text{changes to} \qquad \frac{Y_{n+1} - Y_n}{\Delta t} = AY_n + Q \tag{14}$$

We don't need calculus for difference equations. The derivative enters when the time step $\Delta t$ approaches zero. The model looks simpler if I multiply equation (14) by $\Delta t$:

$$\textbf{One step, n to n + 1} \qquad Y_{n+1} = (1 + A\,\Delta t)Y_n + Q\,\Delta t \tag{15}$$

At the end of year $n$, the bank adds interest $A\Delta t Y_n$ to the balance $Y_n$ you already have. You also put in new savings (or you spend if $Q < 0$). The new year starts with $Y_{n+1}$.

In case $A\,\Delta t = at/N$ and $Q = 0$, we are back to $Y_{n+1} = (1 + at/N)Y_n$:

$$\boldsymbol{N} \textbf{ steps from 0 to } \boldsymbol{N} \qquad \boldsymbol{Y_N = \left(1 + \frac{at}{N}\right)^N Y_0 \;\to\; e^{at}\,y(0)} \quad \text{as } N \to \infty.$$

### Model 2 Radioactive Decay

The next models will deal with decay. The growth rate $a$ is *negative*. The solution $y$ is decreasing. Decay is an expected and natural result when $a < 0$. In fact the differential equation is called **stable** when all solutions approach zero. In many applications this is highly desired.

Exponential growth with $a > 0$ may be good for bank accounts, but not for a drug in our bloodstream. Here are examples where any starting amount $y(0)$ decays exponentially:

A radioactive isotope like Carbon 14
Newton's Law of Cooling
The concentration of a drug in our bloodstream

I will emphasize the **half-life**—the time for half of the Carbon 14 to decay, or half of the drug to disappear. This is decided by the decay rate $a < 0$ in the equation $y' = ay$.

The half-life $H$ is the opposite of the **doubling time** $\boldsymbol{D}$, when $a > 0$ and $e^{aD} = 2$.

### Half-life and Doubling Time

How long does it take for $y(t)$ to be reduced to half of $y(0)$? The equation $y' = ay$ has the solution $e^{at}y(0)$, and we know that $a < 0$.

$$\textbf{Half-life } \boldsymbol{H} \qquad e^{aH} = \frac{1}{2} \qquad aH = \ln\frac{1}{2} = -\ln 2 \qquad \boldsymbol{H = \frac{-\ln 2}{a}}$$

That answer $H$ is positive because $a < 0$. For Carbon 14 the half-life $H$ is 5730 years.

It has just taken 150 hours on a Cray XT5 supercomputer to find 8 eigenvalues of a matrix of size 1 billion—to explain that long half-life. Other carbon isotopes have $H = 20$ minutes. Going in reverse, $H$ tells us the decay rate:

$$\textbf{Decay rate } \boldsymbol{a} \qquad a = \frac{-\ln 2}{5730} \approx 1.216 \times 10^{-4} \text{ per year.}$$

The "quarter-life" would be $2H$, twice as long as the half-life. The time to divide by $e$ is

$$\textbf{Relaxation time } \boldsymbol{\tau} \qquad e^{a\tau} = e^{-1} \approx 0.368 \qquad a\tau = -1 \qquad \boldsymbol{\tau = \frac{-1}{a}}$$

*Question.* Suppose we find a sample where 60 % of the Carbon 14 is gone. *How old is the sample*? If the carbon came from a tree, its decay started at the moment when the tree died.

*Answer.* The age $T$ is the time when $e^{aT} = 0.6$. At that time

$$aT = \ln(0.6) \qquad T = \frac{-0.51}{a} = 4200 \text{ years.}$$

The doubling time $D$ uses the same ideas but now the growth rate is $\boldsymbol{a > 0}$:

$$\textbf{Doubling time} \qquad e^{aD} = 2 \qquad aD = \ln 2 \qquad \boldsymbol{D = \frac{\ln 2}{a}}$$

At 5% interest ($a = .05$/year) the doubling time is less than 14 years. Not 20 years.

**Model 3 Newton's Law of Cooling**

When you put water in a freezer, it cools down. So does a cup of hot coffee on a table. The rate of cooling is proportional to the temperature difference.

| | | |
|---|---|---|
| **Newton's Law** | $\boldsymbol{\frac{dT}{dt} = k(T_\infty - T)}$ | $\boldsymbol{T_\infty}$ **= surrounding temperature** |

This is a linear constant coefficient equation. The solution approaches $T_\infty$. Include that constant on the left side, to make the equation and the solution clear:

$$\frac{d(T - T_\infty)}{dt} = k(T_\infty - T) \qquad \boldsymbol{T - T_\infty = e^{-kt}(T - T_0)}$$

*Question.* Suppose the starting temperature difference $T_0 - T_\infty$ is 80°. After 90 minutes the difference $T_1 - T_\infty$ has dropped to 20°. At what time will the difference be 10° ? When will the temperature reach $T_\infty$ ?

*Answer.* The starting difference 80° is divided by 4 in 90 minutes. To divide again by 2 takes 45 minutes from 20° to 10°. There you see a fundamental rule for exponentials:

$$\text{If } e^{90k} = 1/4 \text{ then } e^{45k} = \sqrt{1/4} = 1/2. \text{ It is not necessary to know } k.$$

The temperature never reaches $T_\infty$ exactly. The exponential $e^{-kt}$ never reaches 0 exactly.

### Model 4 Drug Elimination

The concentration $C(t)$ of a drug in the bloodstream drops at a rate proportional to $C(t)$ itself. Then $dC/dt = -kC$. The elimination constant $k > 0$ is carefully measured, and $C(t) = e^{-kt}C(0)$.

Suppose you want to maintain at least $G$ grams in your body. If you are taking the drug every 8 hours, what dose should you take?

$$t = 8 \text{ hours} \qquad k = \text{ decay rate per hour} \qquad \text{Take } e^{8k}G \text{ grams.}$$

### Model 5 Population growth

Certainly the world population is increasing. Its growth rate $a$ is the birth rate minus the death rate. A reasonable estimate for $a$ right now is 1.3% a year, or $a = .013/\text{year}$ (the dimension of $a$ is 1/time). A first model assumes this growth rate to be constant, continuing forever: Now we ask for the *doubling time*, a number that is independent of the starting value $y(0)$:

**Doubling time $D$** $\qquad e^{aD} = 2 \quad \text{or} \quad D = \dfrac{\ln 2}{.013} \text{ years} = 53 \text{ years.}$

**World population** $\qquad \dfrac{dy}{dt} = .013\,y \quad \text{and} \quad y(t) = e^{.013t}y(0).$

The "forever" part is unrealistic. After 1000 years, it produces $e^{13}y(0)$. That number $e^{13}$ is enormous. If we start today (so that $t = 0$ is the year we are living in) then eventually we will have about one atom each. Ridiculous. But it is quite possible that the pure growth equation $y' = ay$ does describe the real population for a short time.

Eventually the equation has to be corrected. We need a **nonlinear term** like $-by^2$, to model the effect of competition ($y$ against $y$). As $y$ gets large, $y^2$ gets much larger. Then $-by^2$ subtracts from $dy/dt$ and eventually competition stops growth.

This is the famous **"logistic equation"** $\boldsymbol{dy/dt = ay - by^2}$. It is solved in Section 1.7. Here I want to end with a problem of scientific importance—the changing temperature of the Earth. The equations are nonlinear. The data is incomplete. There is no solution formula. This is the reality of science.

## Energy Balance Equations

The Earth gets practically all its energy from the Sun. A lot of that energy goes back out into space. This is radiation in and radiation out. The energy that doesn't go back is responsible for changing the Earth's temperature $T$.

This energy balance is crucial to our lives. It won't permit life on Mercury (too hot), and certainly not on Pluto (too cold). We are extremely fortunate to live on Earth. The form of the temperature equation is completely typical of balance equations in applied mathematics:

**Energy in minus energy out**
**This raises the temperature $T$**

$$C\frac{dT}{dt} = E_{\text{in}} - E_{\text{out}} \tag{16}$$

There is a coefficient $C$ in every equation like this. Let me show you another balance equation, to emphasize how the problem can change but the form stays the same.

**Flow into a bathtub minus flow out**
**This raises the water height $H$**

$$A\,\frac{dH}{dt} = F_{\text{in}} - F_{\text{out}} \tag{17}$$

The tap controls the incoming flow $F_{\text{in}}$. The drain controls the outgoing flow $F_{\text{out}}$. The volume of water changes according to $dV/dt = F_{\text{in}} - F_{\text{out}}$. That volume change $\boldsymbol{dV/dt}$ is a height change $\boldsymbol{dH/dt}$ multiplied by $\boldsymbol{A} =$ area of the water surface. Check units:

$H =$ **meters** $A =$ **(meters)$^2$** $V =$ **(meters)$^3$** $t =$ **seconds** $F =$ **(meters)$^3$/second**

I include this bathtub example because it makes the balance clear:

1. Flow rate in minus flow rate out equals fill rate $dV/dt$.

2. Volume change $dV/dt$ splits into $(A)\,(dH/dt) =$ area times height change.

In a curved bathtub, the water area $A$ changes with the height $H$. Then equation (17) is nonlinear. Every scientist looks immediately at the balance equation: Can it be linear? Can its coefficients be constant? The true answer is no, the practical answer is often yes. (Numerical methods are slowed by nonlinearity. Analytical methods are usually destroyed.)

## Energy Balance for the Earth

The energy balance equation $CT' = E_{\text{in}} - E_{\text{out}}$ is the start. Temperature is in Kelvin (degrees Celsius are also used). The *heat capacity* $C$ is the energy needed to raise the temperature by 1 degree (just as the area $A$ was the volume of water that raises the height of water by 1 meter). That heat capacity $C$ truly changes between ice and ocean and land. Exactly as predicted, the starting simplification is $C =$ *constant*.

On the right side of the equation, the energy $E_{\text{in}}$ is coming from the Sun. A serious fraction $\alpha$ of the arriving energy bounces back and is never absorbed. This fraction $\alpha$ is the **albedo**. It can vary from .80 for snow to .08 for ocean. On a global scale, we have to simplify the albedo formula to a constant, and then improve it:

$$\text{Constant } \alpha = .30 \text{ for all } T \qquad \text{Piecewise linear } \alpha = \begin{cases} .60 & \text{if} \quad T \leq 255\,K \\ .20 & \text{if} \quad T \geq 290\,K \end{cases}$$

The main point is that $\boldsymbol{E_{\text{in}} = (1-\alpha)Q}$, where $Q$ measures energy flow from the Sun to a unit area of the Earth. Now we turn to $\boldsymbol{E_{\text{out}}}$.

Radiation of energy is theoretically proportional to $T^4$ (the Stefan-Boltzmann law). There is an ideal constant $\sigma$ from quantum theory, but the Earth is not ideal. The "greenhouse effect" of particles in the atmosphere reduces $\sigma$ by an emission factor close to $\epsilon = .62$. For a unit area, the radiation $E_{\text{out}}$ is $\epsilon\,\sigma\,T^4$ and the radiation $E_{\text{in}}$ is $(1-\alpha)Q$:

$$\textbf{Energy balance } \boldsymbol{E_{\text{in}} = E_{\text{out}}} \qquad \boldsymbol{(1-\alpha)Q = \epsilon\,\sigma\,T^4} \qquad \boldsymbol{T = \left(\frac{(1-\alpha)Q}{\epsilon\sigma}\right)^{1/4}}$$

You understand that these are not fixed laws like Einstein's $e = mc^2$. Satellites measure the actual radiation, sensors measure the actual temperature. That nonlinear $T^4$ formula is often replaced by a linear $A + BT$. This gives the most basic model of a steady state.

## Multiple Steady States

I will take one more step with that model—we are on the edge of real science. You know that the albedo $\alpha$ (the bounceback of solar energy) depends on the temperature $T$. The coefficients $A$ and $B$ and $\epsilon$ also depend on $T$. The temperature balance equation $C\,dT/dt = E_{\text{in}} - E_{\text{out}}$ and the steady equilibrium equation $E_{\text{in}} = E_{\text{out}}$ are **not linear**. From a nonlinear model, what can we learn?

**Point 1** $E_{\text{in}}(T) = E_{\text{out}}(T)$ can easily have **more than one solution $\boldsymbol{T}$**.

**Point 2** Those steady states when $dT/dt = 0$ can be **stable or unstable**.

**Point 3** You can see $T_1$ and $T_3$ (**stable**) and $T_2$ (**unstable**) in this graph of $E_{\text{in}}$ and $E_{\text{out}}$.

*Why is $T_2$ unstable?* If $T$ is just above $T_2$, then $E_{\text{in}} > E_{\text{out}}$. Therefore $dT/dt > 0$ and the temperature climbs further away from $T_2$. If $T$ is just below $T_2$, then $E_{\text{in}} < E_{\text{out}}$. Therefore $dT/dt < 0$ and $T$ falls further below $T_2$.

The next section 1.7 shows how to decide stability or instability for any equation $dT/dt = f(T)$ or $dy/dt = f(y)$. Just as here, each steady state has $f(T) = 0$. **Stable steady states also have $\boldsymbol{df/dT < 0}$ or $\boldsymbol{df/dy < 0}$**. Simple and important.

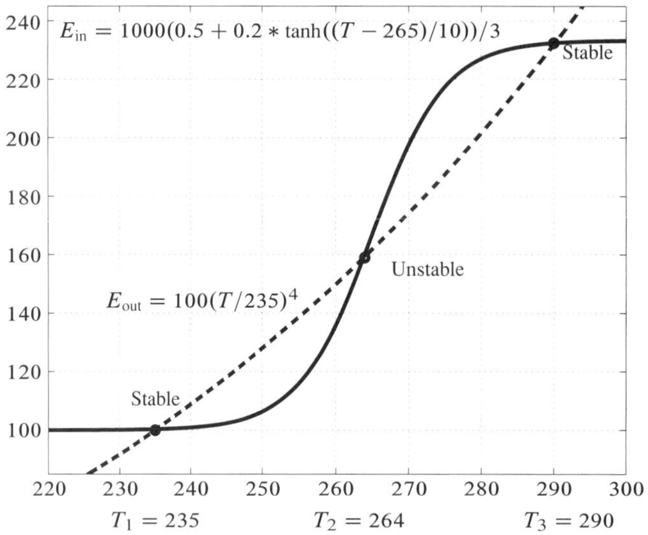


Figure 1.8: The analysis and the graph are from *Mathematics and Climate* by Hans Kaper and Hans Engler (SIAM, 2013). $E_{in} - E_{out}$ has slope $< 0$ at two stable steady states.

## Problem Set 1.6

**1** Solve the equation $dy/dt = y + 1$ up to time $t$, starting from $y(0) = 4$.

**2** You have \$1000 to invest at rate $a = 1 = 100\,\%$. Compare after one year the result of depositing $y(0) = 1000$ immediately with $q = 0$, or choosing $y(0) = 0$ and $q = 1000/\text{year}$ to deposit continually during the year. In both cases $dy/dt = y + q$.

**3** If $dy/dt = y - 1$, when does your original deposit $y(0) = \frac{1}{2}$ drop to zero?

**4** Solve $\dfrac{dy}{dt} = y + t^2$ from $y(0) = 1$ with increasing source term $t^2$.

**5** Solve $\dfrac{dy}{dt} = y + e^t$ (resonance $a = c$ !) from $y(0) = 1$ with exponential source $e^t$.

**6** Solve $\dfrac{dy}{dt} = y - t^2$ from an initial deposit $y(0) = 1$. The spending $q(t) = -t^2$ is growing. When (if ever) does $y(t)$ drop to zero ?

**7** Solve $\dfrac{dy}{dt} = y - e^t$ from an initial deposit $y(0) = 1$. This spending term $-e^t$ grows at the same $e^t$ rate as the initial deposit. When (if ever) does $y$ drop to zero ?

**8** Solve $\dfrac{dy}{dt} = y - e^{2t}$ from $y(0) = 1$. At what time $T$ is $y(T) = 0$ ?

**9** Which solution ($y$ or $Y$) is eventually larger if $y(0) = 0$ and $Y(0) = 0$ ?

$$\frac{dy}{dt} = y + 2t \quad \text{or} \quad \frac{dY}{dt} = 2Y + t.$$

**10** Compare the linear equation $y' = y$ to the separable equation $y' = y^2$ starting from $y(0) = 1$. Which solution $y(t)$ must grow faster ? It grows so fast that it blows up to $y(T) = \infty$ at what time $T$ ?

**11** $Y' = 2Y$ has a larger growth factor (because $a = 2$) than $y' = y + q(t)$. What source $q(t)$ would be needed to keep $y(t) = Y(t)$ for all time ?

**12** Starting from $y(0) = Y(0) = 1$, does $y(t)$ or $Y(t)$ eventually become larger ?

$$\frac{dy}{dt} = 2y + e^t \qquad \frac{dY}{dt} = Y + e^{2t}.$$

**Questions 13-18 are about the growth factor $G(s,t)$ from time $s$ to time $t$.**

**13** What is the factor $G(s,s)$ in zero time ? Find $G(s,\infty)$ if $a = -1$ and if $a = 1$.

**14** Explain the important statement after equation (13): *The growth factor $G(s,t)$ is the solution to $y' = a(t)y + \delta(t-s)$.* The source $\delta(t-s)$ deposits \$1 at time $s$.

**15** Now explain this meaning of $G(s,t)$ when $t$ *is less than* $s$. We go backwards in time. *For $t < s$, $G(s,t)$ is the value at time $t$ that will grow to equal* 1 *at time $s$.*

When $t = 0$, $G(s,0)$ is the "present value" of a promise to pay \$1 at time $s$. If the interest rate is $a = 0.1 = 10\,\%$ per year, what is the present value $G(s,0)$ of a million dollar inheritance promised in $s = 10$ years ?

**16** (a) What is the growth factor $G(s,t)$ for the equation $y' = (\sin t)y + Q \sin t$ ?

(b) What is the null solution $y_n = G(0,t)$ to $y' = (\sin t)y$ when $y(0) = 1$ ?

(c) What is the particular solution $y_p = \int_0^t G(s,t)\, Q \sin s\, ds$ ?

**17** (a) What is the growth factor $G(s,t)$ for the equation $y' = y/(t+1) + 10$ ?

(b) What is the null solution $y_n = G(0,t)$ to $y' = y/(t+1)$ with $y(0) = 1$ ?

(c) What is the particular solution $y_p = 10\int_0^t G(s,t)\, ds$ ?

**18** Why is $G(t,s) = 1/G(s,t)$ ? Why is $G(s,t) = G(s,S)G(S,t)$ ?

**Problems 19–22 are about the "units" or "dimensions" in differential equations.**

**19** (recommended) If $dy/dt = ay + qe^{i\omega t}$, with $t$ in seconds and $y$ in meters, what are the units for $a$ and $q$ and $\omega$ ?

**20** The logistic equation $dy/dt = ay - by^2$ often measures the time $t$ in years (and $y$ counts people). What are the units of $a$ and $b$ ?

**21** Newton's Law is $m\,d^2y/dt^2 + ky = F$. If the mass $m$ is in grams, $y$ is in meters, and $t$ is in seconds, what are the units of the stiffness $k$ and the force $F$ ?

**22** Why is our favorite example $y' = y + 1$ very unsatisfactory dimensionally ? Solve it anyway starting from $y(0) = -1$ and from $y(0) = 0$.

**23** The difference equation $Y_{n+1} = cY_n + Q_n$ produces $Y_1 = cY_0 + Q_0$. Show that the next step produces $Y_2 = c^2Y_0 + cQ_0 + Q_1$. After $N$ steps, the solution formula for $Y_N$ is like the solution formula for $y' = ay + q(t)$. Exponentials of $a$ change to powers of $c$, the null solution $e^{at}y(0)$ becomes $c^N Y_0$. The particular solution

$$Y_N = c^{N-1}Q_0 + \cdots + Q_{N-1} \quad \text{is like} \quad y(t) = \int_0^t e^{a(t-s)}q(s)ds.$$

**24** Suppose a fungus doubles in size every day, and it weighs a pound after 10 days. If another fungus was twice as large at the start, would it weigh a pound in 5 days ?

## 1.7 The Logistic Equation

This section presents one particular nonlinear differential equation—the *logistic equation*. It is a model of growth *slowed down by competition*. In later chapters, one group $y_1$ will compete against another group $y_2$. Here the competition is inside one group. The growth comes from $ay$ as usual. The competition ($y$ against $y$) comes from $-by^2$.

**Logistic equation / nonlinear** $$\frac{dy}{dt} = ay - by^2 \qquad (1)$$

We will discuss the meaning of this equation, and its solution $y(t)$.

One key idea comes right away: the **steady state**. Any time we have $dy/dt = f(y)$, it is important to know when $f(y)$ is zero. Growth stops at that point because $dy/dt$ is zero. If the number $Y$ solves $f(Y) = 0$, the constant function $y(t) = Y$ solves the equation $dy/dt = f(y)$: both sides are zero. For the special starting value $y(0) = Y$, the solution would stay at $Y$. It is a steady solution, not changing with time.

**The logistic equation has two steady states with $f(Y) = 0$:**

$$\frac{dy}{dt} = ay - by^2 = 0 \text{ when } aY = bY^2. \text{ Then } \boldsymbol{Y = 0} \text{ or } \boldsymbol{Y = a/b}. \qquad (2)$$

That point $a/b$ is where competition balances growth. It is the top of the "$S$-curve" in Figure 1.9, where the curve goes flat. It is the end of growth. The solution $y(t)$ cannot get past the value $a/b$. At the start of the $S$-curve, the other steady state $Y = 0$ is **unstable**. The curve goes *away* from $Y = 0$ and *toward* $Y = a/b$.

In some applications, this number $a/b$ is the **carrying capacity** ($K$) of the system. If $a/b = K$ then $b = a/K$. So the logistic equation can be written in terms of $a$ and $K$:

$$\frac{dy}{dt} = ay - by^2 = ay - \frac{a}{K}y^2 = ay\left(1 - \frac{y}{K}\right). \qquad (3)$$

Mathematically, we have done nothing interesting. But the number $K$ may be easier to work with than $b$. We might have an estimate like $K = 12$ billion people for the maximum population that the world can deal with. Rewriting the equation doesn't change the solution, but it can help our understanding.

### Solution of the Logistic Equation

What is $y(t)$ ? The logistic equation is nonlinear because of $y^2$, and most nonlinear equations have no solution formula. ($y = Ce^{at}$ is extremely unlikely.) But the particular equation $dy/dt = ay - by^2$ can be solved, and I want to present two ways to do it:

**1** (by magic) The equation for $z = 1/y$ happens to be linear: $\boldsymbol{dz/dt = -az + b}$. We can solve that equation and then we know $y$.

**2** (by partial fractions) This systematic approach takes longer. In principle, partial fractions can be used any time $dy/dt$ is a ratio of polynomials in $y$.

You will appreciate method **1** (only two steps A and B) after you see method **2**.

(**A**) If $z = \dfrac{1}{y}$, the chain rule gives $\dfrac{dz}{dt} = \dfrac{-1}{y^2}\dfrac{dy}{dt}$. Substitute $ay - by^2$ for $\dfrac{dy}{dt}$ :

$$\frac{\boldsymbol{dz}}{\boldsymbol{dt}} = \frac{1}{y^2}\left(-ay + by^2\right) = -\frac{a}{y} + b = \boldsymbol{-az+b}. \tag{4}$$

(**B**) This is the linear equation $z' + az = b$ that was solved in the previous sections. Change $a$ to $-a$ in the solution formula. Change $y$ and $q$ to $z$ and $b$ :

$$\textbf{Solution} \qquad z(t) = e^{-at}z(0) - \frac{b}{a}\left(e^{-at} - 1\right) = \frac{de^{-at} + b}{a} \tag{5}$$

The number $d$ collects all the constants $a, y(0), b$ in one place :

$$\frac{d}{a} = z(0) - \frac{b}{a} \quad \text{and} \quad z(0) = \frac{1}{y(0)} \quad \text{produce} \quad \boldsymbol{d = \frac{a}{y(0)} - b}. \tag{6}$$

Now turn equation (5) upside down to find $y = 1/z$ :

$$\textbf{Solution to the logistic equation} \qquad \boldsymbol{y(t) = \frac{a}{de^{-at} + b}} \tag{7}$$

This is a beautiful solution. Look at its value for large positive $t$ and large negative $t$ :

| | | | |
|---|---|---|---|
| Approaching $t = +\infty$ | $e^{-at} \to 0$ | and | $\boldsymbol{y(t) \to \dfrac{a}{b}}$ |
| Approaching $t = -\infty$ | $e^{-at} \to \infty$ | and | $\boldsymbol{y(t) \to 0}$ |

Far back in time, the population was near $Y = 0$. Far forward in time, the population will approach $Y = a/b$. Those are the two steady states, the points where $ay - by^2$ is zero and the curve becomes flat. Then $dy/dt$ is zero and $y$ never changes.

In between, the population $y(t)$ is following an **S-curve**, climbing toward $a/b$. It is symmetric around the halfway point $y = a/2b$. The world is near that point right now.

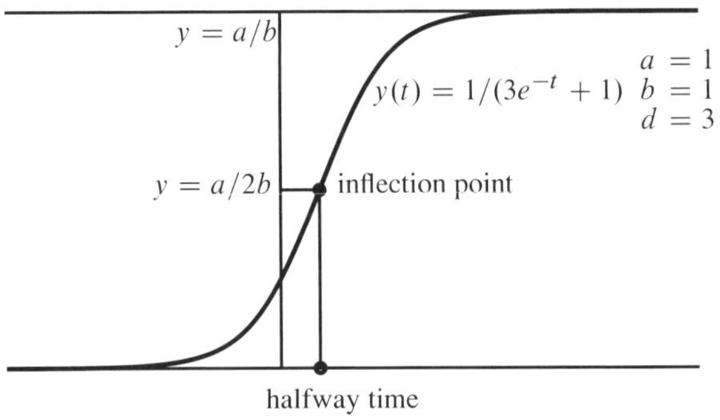


Figure 1.9: The $S$-curve solves the logistic equation. The inflection point is halfway.

## Simplest Example of the $S$ - curve

The best example has $a = b = 1$. The top of the $S$-curve is $Y = a/b = 1$. The bottom is $Y = 0$. The halfway time is $t = 0$, where $y(0) = \frac{1}{2}$. Then the logistic equation and its solution are as simple as possible:

$$\frac{dy}{dt} = y - y^2 \text{ has the solution } y(t) = \frac{1}{1+e^{-t}} \text{ starting from } y(0) = \frac{1}{2}. \quad (8)$$

That solution $1/(1+e^{-t})$ approaches 1 when $t \to \infty$. It approaches 0 when $t \to -\infty$. Let me review the "$z = 1/y$ method" to solve the logistic equation $y' = y - y^2$.

$$\frac{dz}{dt} = \frac{-1}{y^2}\frac{dy}{dt} = \frac{-y+y^2}{y^2} = -z + 1.$$

Then $z(t) = 1 + Ce^{-t}$. Take $C = 1$ to match $y(0) = \frac{1}{2}$ and $z(0) = 2$. Now $y = \dfrac{1}{1+e^{-t}}$.

## World Population and the Carrying Capacity $K$

What are the numbers $a$ and $b$ for human population ? Ecologists estimate the natural growth rate at $a = .029$ per year. This is not the actual rate, because of $b$. About 1930, the world population was near $y = 3$ billion. The $ay$ term predicts a one-year increase of (.029) (3 billion) = 87 million. The actual growth was more like $dy/dt = 60$ million/year. In this simple model, that difference of 27 million/year was caused by $by^2$:

$$27 \text{ million/year} = b\ (3 \text{ billion})^2 \quad \text{leads to} \quad b = 3 \text{ times } 10^{-12}/\text{year}.$$

When we know $b$, we know the steady state $y(\infty) = K = a/b$. At that point the loss $by^2$ from competition balances the gain $ay$ from growth:

$$\textbf{Estimated capacity} \quad K = \frac{a}{b} = \frac{.029}{3}10^{12} \approx 9.7 \text{ billion people.}$$

This number is low, and $y$ is growing faster. The estimates I see now are closer to

$$y(\infty) > 10 \text{ billion} \quad \text{and} \quad y(2014) \approx 7.2 \text{ billion.}$$

Our world is beyond the halfway point $y = a/2b$ on the curve. That looks like an inflection point (by symmetry of the graph), and the test $d^2y/dt^2 = 0$ confirms that it is.

**The inflection point with $y'' = 0$ is halfway up the curve in Figure 1.9**

$$\frac{d}{dt}\left(\frac{dy}{dt}\right) = \frac{d}{dt}(ay - by^2) = (a - 2by)\frac{dy}{dt} = 0 \text{ when } y = \frac{a}{2b} \quad (9)$$

After this halfway point, the $S$-curve bends downward. The population $y$ is still increasing, but its growth rate $dy/dt$ is decreasing. (Notice the difference.) The inflection point separates "bending up" from "bending down" and the *rate* of growth is a maximum at that point. You will understand that this simple model must be and has been improved.

## Partial Fractions

The logistic equation is nonlinear but it is ***separable*** . We can separate $y$ from $t$ as follows :

$$\frac{dy}{dt} = ay - by^2 = a\left(y - \frac{b}{a}y^2\right) \quad \text{leads to} \quad \frac{\boldsymbol{dy}}{\boldsymbol{y - \frac{b}{a}y^2}} = \boldsymbol{a\ dt}. \tag{10}$$

In this separated form, the problem is reduced to two ordinary integrations ($y$-integration on the left side, $t$-integration on the right side). The integral of $a\,dt$ on the right side is certainly $at + C$. The left side can be looked up in a table of integrals or produced by software like *Mathematica* or discovered by ourselves.

I will explain the idea of **partial fractions** that produces this integral. You may know it as a "Technique of Integration" from first-year calculus (it is really just algebra). The plan is to split the fraction in two pieces so the integration becomes easy :

$$\textbf{Partial fractions} \qquad \frac{1}{y - \frac{b}{a}y^2} \quad \text{separates into} \quad \frac{A}{y} + \frac{B}{1 - \frac{b}{a}y} \tag{11}$$

I factored $y - \frac{b}{a}y^2$ into $y$ times $1 - \frac{b}{a}y$. I put those two denominators on the right side. *We need to know A and B*. To compare with the left side, combine those two fractions:

$$\textbf{Common denominator} \qquad \frac{A}{y} + \frac{B}{1 - \frac{b}{a}Y} = \frac{A\left(1 - \frac{b}{a}y\right) + By}{y\left(1 - \frac{b}{a}y\right)}. \tag{12}$$

The correct $A$ and $B$ must produce 1 in the numerator, to match the 1 in equation (11):

$$A\left(1 - \frac{b}{a}y\right) + By = 1 \quad \text{when} \quad \boldsymbol{A = 1} \quad \text{and} \quad \boldsymbol{B = \frac{b}{a}}. \tag{13}$$

This completes the algebra of partial fractions, by finding $A$ and $B$ in equation (11):

$$\textbf{Two fractions} \qquad \frac{1}{y - \frac{b}{a}y^2} = \frac{1}{y(1 - \frac{b}{a}y)} = \frac{\mathbf{1}}{\boldsymbol{y}} + \frac{\boldsymbol{b/a}}{\boldsymbol{1 - \frac{b}{a}y}}. \tag{14}$$

## Integrate the Partial Fractions

With $A = 1$ and $B = b/a$, integrate the two partial fractions separately :

$$\int \frac{1\,dy}{y} + \int \frac{(b/a)dy}{1 - (b/a)y} = \ln y - \ln\left(1 - \frac{b}{a}y\right). \tag{15}$$

This is the calculus part (the integration) in solving the logistic equation. After the integration, use algebra to write the answer $y(t)$ in a good form.

Actually that good form of $y(t)$ was already found by our first method. The magic of $z = 1/y$ produced a linear equation $dz/dt = -az + b$. Then returning to $y = 1/z$ put the crucial factor $e^{-at}$ into the denominator of (7), and we repeat that solution here:

$$\textbf{Solution in (7)} \qquad y(t) = \frac{a}{de^{-at} + b} \quad \text{with} \quad d = \frac{a}{y(0)} - b. \tag{16}$$

This same answer must come from the integral (15) that used partial fractions. The integral has the form $\ln y - \ln x$, which is the same as $\ln(y/x)$ (and $x$ is $1 - (b/a)y$).

$$\int \frac{dy}{y - \frac{b}{a}y^2} = \int a\,dt \quad \text{gives} \quad \ln \frac{y}{1 - \frac{b}{a}y} = at + C = at + \ln \frac{y(0)}{1 - \frac{b}{a}y(0)}. \tag{17}$$

I chose the integration constant $C$ to make (17) correct at $t = 0$. Now take exponentials of both sides:

$$\frac{y}{1 - \frac{b}{a}y} = e^{at} \frac{y(0)}{1 - \frac{b}{a}y(0)}. \tag{18}$$

The final algebra part is to solve this equation for $y$. Let me move that into Problem 3. Then we recover the good formula (16) that came so much faster from $y = 1/z$.

Looking ahead, partial fractions will appear again in Section 2.7. They simplify the Laplace transform so you can recognize the inverse transform. That section gives a formula **PF2** for the numbers $A$ and $B$ in the fractions—it is previewed here in Problem 14.

Again, we solved $dy/dt = f(y)$ by separating $\int dy/f(y)$ from $\int dt$.

## Autonomous Equations $dy/dt = f(y)$

*The logistic equation is autonomous.* This means that $f$ depends only on $y$, and not on $t$: $\boldsymbol{dy/dt = f(y)}$. A linear example is $y' = y$. The big advantage of an autonomous equation is that the solution curve can stay the same, when the starting value $y(0)$ is changed. "We just climb onto the curve at height $y(0)$ and keep going."

You saw how Figure 1.9 had the same $S$-curve for every $y(0)$ between 0 and $a/b$. The equation $dy/dt = y$ has the same exponential curve $y = e^t$ for every $y(0) > 0$. Just mark the $t = 0$ point wherever the height is $y(0)$.

This means that time $t$ is not essential in the graphs. **The graph of $\boldsymbol{f(y)}$ against $\boldsymbol{y}$ is the key**. For the logistic equation, the parabola $f(y) = ay - by^2$ tells you everything (except the time for each $y$). $y(t)$ increases when this parabola $f(y)$ is above the axis (because $dy/dt > 0$ when $f > 0$). So I only drew one $S$-curve.

There is also a decreasing curve starting from $y(0) > a/b$. **It approaches the steady state $\boldsymbol{Y = a/b}$ from above**. Another curve starts below $Y = 0$ and drops to $-\infty$. The upgoing $S$-curve is sandwiched between two downgoing curves, because in Figure 1.10 the positive piece of $ay - by^2$ is sandwiched between two negative pieces.

## Stability of Steady States

The steady states of $dy/dt = f(y)$ are solutions of $f(Y) = 0$. The differential equation becomes $0 = 0$ when $y(t) = Y$ is constant (steady). Here is the stability question:

**Starting close to $Y$, does $y(t)$ approach $Y$ (stable) or does it leave $Y$ (unstable)?**

We had a formula for the $S$-curve. So we could answer this stability question. One $Y$ is stable (that is $Y = a/b$ at the end). The steady state $Y = 0$ is unstable. It is important (and not hard) to be able to decide stability *without a formula for* $y(t)$.

Everything depends on the derivative $df/dy$ at the steady value $y = Y$. That slope of $f(y)$ will be called $c$. Here is the test for stability, followed by a reason and examples.

**Stable if $c < 0$  The steady state $Y$ is stable if $df/dy < 0$ at $y = Y$.**

**Reason**: Near the steady state, $f(y)$ is close to $c(y - Y)$. Then $y' = f(y)$ is close to $(y - Y)' = c(y - Y)$. Then **$y - Y$ is like $e^{ct}$**, and **$y \to Y$ when $c < 0$ and $e^{ct} \to 0$.**

Let me explain in detail for any autonomous equation $dy/dt = f(y)$. Suppose that $Y = 0$ is a steady state. This means that $f(0) = 0$. Calculus gives the linear approximation $f(y) \approx cy$, where $c$ is the slope of the tangent line. That number is $c = df/dy$ at $Y = 0$. **If $c$ is negative then $y(t)$ will move toward $Y = 0$ (stability)**:

| | | |
|---|---|---|
| For small $y(0) > 0$ | $dy/dt = f(y) \approx cy < 0$ | $y(t)$ decreases toward 0 |
| For small $y(0) < 0$ | $dy/dt = f(y) \approx cy > 0$ | $y(t)$ increases toward 0 |

For any other steady state $Y$, calculus gives the linear approximation $f(y) \approx c(y-Y)$. Now that number is $c = df/dy$, the slope of the tangent line **at $y = Y$**.

| | | |
|---|---|---|
| For $y(0)$ just above $Y$ | $dy/dt = f(y) \approx c(y - Y) < 0$ | **$y(t)$ decreases toward $Y$** |
| For $y(0)$ just below $Y$ | $dy/dt = f(y) \approx c(y - Y) > 0$ | **$y(t)$ increases toward $Y$** |

**Example 1** (logistic) The derivative of $ay - by^2$ is $\boldsymbol{df/dy = a - 2by}$.

At the steady state $Y = 0$, $df/dy$ is $a > 0$: **$Y = 0$ is unstable**.

At $Y = a/b$, this derivative is $a - 2b(a/b) = -a$. **$Y = a/b$ is stable**.

For $dy/dt = ay - by^2$ this **stability line** shows which way $y(t)$ moves from any $y(0)$.

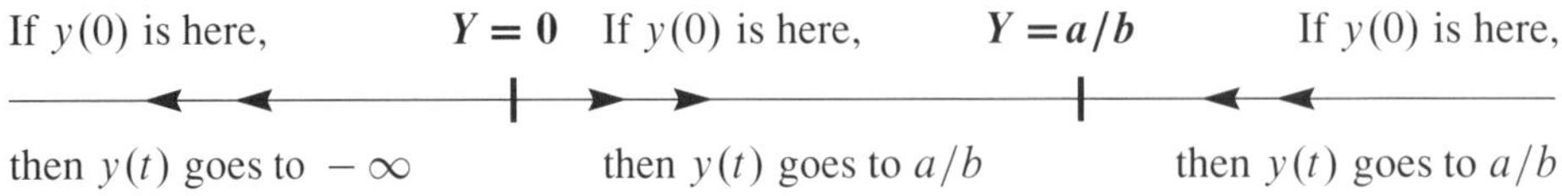
If $y(0)$ is here,
$Y = 0$
If $y(0)$ is here,
$Y = a/b$
If $y(0)$ is here,
then $y(t)$ goes to $-\infty$
then $y(t)$ goes to $a/b$
then $y(t)$ goes to $a/b$

The steady states have to alternate between stable and unstable, because $df/dy$ will alternate between negative and positive. I am excluding the undecided cases when $f(Y) = 0$ and also $df/dy(Y) = 0$. This is a borderline case for critical harvesting.

## The Harvesting Equation

Suppose the logistic equation also includes a constant harvesting rate $-h$. This will reduce the growth rate $dy/dt$. Let me start with the logistic equation $dy/dt = 4y - y^2$, where the $S$-curve rises from $Y = 0$ to the other steady state $Y = a/b = 4/1$. If the new harvesting term is $-h = -3$, the steady states change from 0 and 4 to 1 and 3 :

$$\frac{\boldsymbol{dy}}{\boldsymbol{dt}} = \mathbf{4}\boldsymbol{y} - \boldsymbol{y}^{\mathbf{2}} - \mathbf{3} \quad \text{has new steady states} \quad Y = 1 \quad \text{and} \quad Y = 3. \tag{19}$$

I found 1 and 3 by factoring $4Y - Y^2 - 3$ into $-(Y-1)(Y-3)$. Those populations $Y = 1$ and $Y = 3$ are the points where the equation is $dy/dt = 0$. Then $y = Y$ stays steady.

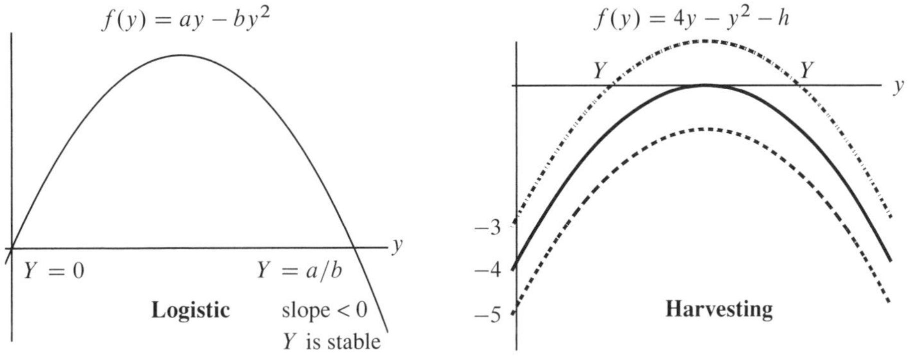


Figure 1.10: Harvesting lowers the parabola $f(y) = ay - by^2 - h$. Steady $Y$'s disappear.

This figure shows the stability or instability of the steady states. $Y = 0$ in the logistic graph and $Y = 1$ in the harvesting graph are **unstable**. At those points $f(y)$ climbs from negative to positive. Above $Y$, the graph shows $dy/dt = f(y)$ as positive. So $y(t)$ will increase, and it moves away from $Y$.

$Y = a/b$ in the logistic graph and $Y = 3$ in the harvesting graph are **stable**. Beyond those points $f(y)$ is negative. This is $dy/dt$. So $y(t)$ decreases back toward $Y$. The graphs are a little tricky to read, because they don't show $y(t)$. **They show the phase plane with** $\boldsymbol{y' = f(y)}$ **against** $\boldsymbol{y}$ : Velocity versus position, not position versus time !

Looking again at the figure, $h = 4$ gives critical harvesting: **One double stationary point** $\boldsymbol{Y = 2}$. That curve shows $dy/dt = f(y)$ as always negative, so $y(t)$ will decrease. If $y(0)$ is greater than 2, then $y(t)$ must come back toward $Y = 2$. But this is *one-sided stability*, because if $y(0)$ is smaller then 2, then $y(t)$ will decrease and go far away from 2.

The lowest curve has $h = 5$ and **no steady states**. At all points $dy/dt = f(y)$ is negative. All solutions $y(t)$ are decreasing. If we can find a formula for $y(t)$, we can watch this happen: $\boldsymbol{y(t) \to -\infty}$. The logistic and harvesting equations are terrific nonlinear examples, because we can actually find $y(t)$.

## Solving the Harvesting Equation

We have three types of harvesting equations, with 2 or 1 or 0 steady states:

**$h < 4$** $\quad y' = 4y - y^2 - h$ will reduce to a logistic equation: **underharvesting**

**$h = 4$** $\quad y' = -(y-2)^2$ has a double steady state: **critical harvesting**

**$h > 4$** $\quad y'$ stays below zero and $y(t)$ approaches $-\infty$: **overharvesting**.

All these equations are autonomous, so they separate into $dy/f(y) = dt$. *Integrate* $1/f(y)$.

**Small $h = 3$** $\qquad$ Factor $f(y)$ into $-(y-1)(y-3)$ $\qquad$ Then $\boldsymbol{Y = 1}$ and $\boldsymbol{Y = 3}$

Let me shift those steady states down to $V = 0$ and $V = 2$, by shifting $y(t)$ to $\boldsymbol{v(t) = y(t) - 1}$. The equation for $v(t)$ is logistic, and its $S$-curve climbs from 0 to 2:

$$(1+v)' = -(v)(v-2) \text{ is } \boldsymbol{v' = 2v - v^2} \tag{20}$$

When you add back the 1 to get $y = 1 + v$, its $S$-curve climbs from 1 to 3.

**Critical $h = 4$** $\qquad$ Factor $f(y) = 4y - y^2 - 4 = -(y-2)^2$ $\qquad$ Then $\boldsymbol{Y = 2}$ and **2**

The equation is $y' = -(y-2)^2$. Shifting to $v(t) = y(t) - 2$ gives $\boldsymbol{dv/dt = -v^2}$. Page 1 of this book had the equation $dy/dt = +y^2$ (with time going the other way). The solution looks so innocent:

$$v(t) = \frac{v(0)}{1 + tv(0)} \qquad \begin{array}{l} \text{goes gently to } v = 0 \text{ as } t \to \infty \text{ provided } v(0) > 0 \\ \text{goes suddenly to } v = -\infty \text{ when } 1 + tv(0) = 0 \end{array}$$

This shows (one-sided) stability if $y(0) > 2$ and $v(0) > 0$.

When harvesting is more than critical, the population dies out from every $y(0)$.

**Overharvesting $h = 5$** Write $y' = 4y - y^2 - 5 = -1 - (y-2)^2$. Always $\boldsymbol{y' < 0}$.

Now $v = y - 2$ simplifies the equation to $\boldsymbol{v' = -1 - v^2}$. Integrate $dv/(1+v^2) = -dt$ to get $\tan^{-1} v = -t + C$. If $v(0) = 0$ then $C = 0$. Now go back to $y = v + 2$:

$$\frac{dv}{dt} = -1 - v^2 \text{ with } v(0) = 0 \text{ gives } v(t) = \tan(-t). \text{ Then } \boldsymbol{y(t) = 2 - \tan t}. \tag{21}$$

When the tangent reaches 2, the population $y = 0$ is all gone. If the solution continues to $t = \pi/2$, then $\tan t$ is infinite. The model loses meaning and $y(\pi/2) = -\infty$.

Overall, I hope you see how a simple stability test tells so much about $y' = f(y)$:

| |
|---|
| **1** Find all solutions to $f(y) = 0$ $\qquad$ **2** If $df/dy < 0$ at $y = Y$, that state is stable. |

## ■ REVIEW OF THE KEY IDEAS ■

1. The logistic equation $dy/dt = ay - by^2$ has steady states at $Y = 0$ and $Y = a/b$.

2. The $S$-curve $y(t) = a/(de^{-at} + b)$ approaches the carrying capacity $y(\infty) = a/b$.

3. The equation for $z = \dfrac{1}{y}$ is linear ! Or we can separate into $dy/\left(y - \dfrac{b}{a}y^2\right) = a\,dt$.

4. The stability test $df/dy = a - 2by < 0$ is passed at $Y = a/b$ and failed at $Y = 0$.

5. This stability test applies to all equations $y' = f(y)$ including $y' = ay - by^2 - h$.

# Problem Set 1.7

**1** If $y(0) = a/2b$, the halfway point on the $S$-curve is at $t = 0$. Show that $d = b$ and $y(t) = \dfrac{a}{d\,e^{-at} + b} = \dfrac{a}{b}\dfrac{1}{e^{-at} + 1}$. Sketch the curve from $y_{-\infty} = 0$ to $y_{\infty} = \dfrac{a}{b}$.

**2** If the carrying capacity of the Earth is $K = a/b = 14$ billion people, what will be the population at the inflection point ? What is $dy/dt$ at that point ? The actual population was 7.14 billion on January 1, 2014.

**3** Equation (18) must give the same formula for the solution $y(t)$ as equation (16). If the right side of (18) is called $R$, we can solve that equation for $y$ :

$$y = R\left(1 - \frac{b}{a}y\right) \quad \rightarrow \quad \left(1 + R\frac{b}{a}\right)y = R \quad \rightarrow \quad y = \frac{R}{\left(1 + R\frac{b}{a}\right)}.$$

Simplify that answer by algebra to recover equation (16) for $y(t)$.

**4** Change the logistic equation to $y' = y + y^2$. Now the nonlinear term is positive, and *cooperation of $y$ with $y$* promotes growth. Use $z = 1/y$ to find and solve a linear equation for $z$, starting from $z(0) = y(0) = 1$. Show that $y(T) = \infty$ when $e^{-T} = 1/2$. Cooperation looks bad, the population will explode at $t = T$.

**5** The US population grew from 313, 873, 685 in 2012 to 316, 128, 839 in 2014. If it were following a logistic $S$-curve, what equations would give you $a, b, d$ in the formula (4) ? Is the logistic equation reasonable and how to account for immigration ?

**6** The **Bernoulli equation** $y' = ay - by^n$ has competition term $by^n$. Introduce $z = y^{1-n}$ which matches the logistic case when $n = 2$. Follow equation (4) to show that $z' = (n-1)(-az + b)$. Write $z(t)$ as in (5)-(6). Then you have $y(t)$.

**Problem 7–13 develop better pictures of the logistic and harvesting equations.**

7 $y' = y - y^2$ is solved by $y(t) = 1/(de^{-t}+1)$. This is an $S$-curve when $y(0) = 1/2$ and $d = 1$. But show that $y(t)$ is very different if $y(0) > 1$ or if $y(0) < 0$.

If $y(0) = 2$ then $d = \frac{1}{2} - 1 = -\frac{1}{2}$. Show that $y(t) \to 1$ from above.

If $y(0) = -1$ then $d = \frac{1}{-1} - 1 = -2$. At what time $T$ is $y(T) = -\infty$?

8 (recommended) Show those 3 solutions to $y' = y - y^2$ in one graph! They start from $y(0) = 1/2$ and 2 and $-1$. The $S$-curve climbs from $\frac{1}{2}$ to 1. Above that, $y(t)$ descends from 2 to 1. Below the $S$-curve, $y(t)$ drops from $-1$ to $-\infty$.

Can you see 3 regions in the picture? **Dropin curves above $y = 1$ and $S$-curves sandwiched between 0 and 1 and dropoff curves below $y = 0$.**

9 Graph $f(y) = y - y^2$ to see the unstable steady state $Y = 0$ and the stable $Y = 1$. Then graph $f(y) = y - y^2 - 2/9$ with harvesting $h = 2/9$. What are the steady states $Y_1$ and $Y_2$? The 3 regions in Problem 8 now have $Z$-curves above $y = 2/3$, $S$-curve sandwiched between $1/3$ and $2/3$, dropoff curves below $y = 1/3$.

10 What equation produces an $S$-curve climbing to $y_\infty = K$ from $y_{-\infty} = L$?

11 $y' = y - y^2 - \frac{1}{4} = -(y - \frac{1}{2})^2$ shows *critical harvesting* with a double steady state at $y = Y = \frac{1}{2}$. The layer of $S$-curves shrinks to that single line. Sketch a dropin curve that starts above $y(0) = \frac{1}{2}$ and a dropoff curve that starts below $y(0) = \frac{1}{2}$.

12 Solve the equation $y' = -(y-\frac{1}{2})^2$ by substituting $v = y-\frac{1}{2}$ and solving $v' = -v^2$.

13 With overharvesting, every curve $y(t)$ drops to $-\infty$. There are no steady states. Solve $Y - Y^2 - h = 0$ (quadratic formula) to find only complex roots if $4h > 1$.

The solutions for $h = \frac{5}{4}$ are $y(t) = \frac{1}{2} - \tan(t + C)$. Sketch that dropoff if $C = 0$. Animal populations don't normally collapse like this from overharvesting.

14 With **two partial fractions**, this is my preferred way to find $A = \dfrac{1}{r-s}$, $B = \dfrac{1}{s-r}$

**PF2** $$\frac{1}{(y-r)(y-s)} = \frac{1}{(y-r)(r-s)} + \frac{1}{(y-s)(s-r)}$$

Check that equation: The common denominator on the right is $(\boldsymbol{y}-\boldsymbol{r})(\boldsymbol{y}-\boldsymbol{s})(\boldsymbol{r}-\boldsymbol{s})$. The numerator should cancel the $\boldsymbol{r}-\boldsymbol{s}$ when you combine the two fractions.

$$\text{Separate } \frac{1}{y^2-1} \text{ and } \frac{1}{y^2-y} \text{ into two fractions } \frac{A}{y-r} + \frac{B}{y-s}.$$

*Note* When $y$ approaches $r$, the left side of **PF2** has a blowup factor $1/(y-r)$. The other factor $1/(y-s)$ correctly approaches $A = 1/(r-s)$. So the right side of **PF2** needs the same blowup at $y = r$. The first term $A/(y-r)$ fits the bill.

**15** The **threshold equation** is the logistic equation backward in time:

$$-\frac{dy}{dt} = ay - by^2 \quad \text{is the same as} \quad \frac{dy}{dt} = -ay + by^2.$$

Now $Y = 0$ is the stable steady state. $Y = a/b$ is the unstable state (why?). If $y(0)$ is below the threshold $a/b$ then $y(t) \to 0$ and the species will die out.

Graph $y(t)$ with $y(0) < a/b$ (reverse $S$-curve). Then graph $y(t)$ with $y(0) > a/b$.

**16** (Cubic nonlinearity) The equation $y' = y(1-y)(2-y)$ has **three steady states**: $Y = 0, 1, 2$. By computing the derivative $df/dy$ at $y = 0, 1, 2$, decide whether each of these states is stable or unstable.

Draw the *stability line* for this equation, to show $y(t)$ leaving the unstable $Y$'s. Sketch a graph that shows $y(t)$ starting from $y(0) = \frac{1}{2}$ and $\frac{3}{2}$ and $\frac{5}{2}$.

**17** (a) Find the steady states of the **Gompertz equation** $dy/dt = y(1 - \ln y)$.

(b) Show that $z = \ln y$ satisfies the linear equation $dz/dt = 1 - z$.

(c) The solution $z(t) = 1 + e^{-t}(z(0) - 1)$ gives what formula for $y(t)$ from $y(0)$?

**18** Decide stability or instability for the steady states of

(a) $dy/dt = 2(1-y)(1-e^y)$ (b) $dy/dt = (1-y^2)(4-y^2)$

**19** Stefan's Law of Radiation is $dy/dt = K(M^4 - y^4)$. It is unusual to see fourth powers. Find all real steady states and their stability. Starting from $y(0) = M/2$, sketch a graph of $y(t)$.

**20** $dy/dt = ay - y^3$ has how many steady states $Y$ for $a < 0$ and then $a > 0$? Graph those values $Y(a)$ to see a *pitchfork bifurcation*—new steady states suddenly appear as $a$ passes zero. The graph of $Y(a)$ looks like a pitchfork.

**21** (Recommended) The equation $dy/dt = \sin y$ has **infinitely many steady states**. What are they and which ones are stable? Draw the stability line to show whether $y(t)$ increases or decreases when $y(0)$ is between two of the steady states.

**22** Change Problem 21 to $dy/dt = (\sin y)^2$. The steady states are the same, but now the derivative of $f(y) = (\sin y)^2$ is zero at all those states (because $\sin y$ is zero). What will the solution actually do if $y(0)$ is between two steady states?

**23** (*Research project*) Find actual data on the US population in the years 1950, 1980, and 2010. What values of $a, b, d$ in the solution formula (7) will fit these values? Is the formula accurate at 2000, and what population does it predict for 2020 and 2100?

You could reset $t = 0$ to the year 1950 and rescale time so that $t = 3$ is 1980.

**24** If $dy/dt = f(y)$, what is the limit $y(\infty)$ starting from each point $y(0)$ ?

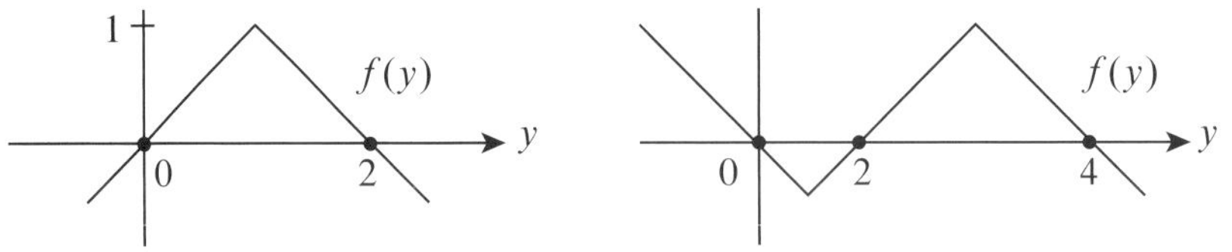


**25** (a) Draw a function $f(y)$ so that $y(t)$ approaches $y(\infty) = 3$ from every $y(0)$.

(b) Draw $f(y)$ so that $y(\infty) = 4$ if $y(0) > 0$ and $y(\infty) = -2$ if $y(0) < 0$.

**26** Which exponents $n$ in $dy/dt = y^n$ produce blowup $y(T) = \infty$ in a finite time? You could separate the equation into $dy/y^n = dt$ and integrate from $y(0) = 1$.

**27** Find the steady states of $dy/dt = y^2 - y^4$ and decide whether they are stable, unstable, or one-sided stable. Draw a stability line to show the final value $y(\infty)$ from each initial value $y(0)$.

**28** For an autonomous equation $y' = f(y)$, why is it impossible for $y(t)$ to be increasing at one time $t_1$ and decreasing at another time $t_2$ ?

**The website math.mit.edu/dela has more graph questions for autonomous $y' = f(y)$.**

**Notes on feedback** The $S$-curve represents a good response from an elevator. The transient response in the middle of the $S$ is the fast movement between floors. The elevator slows down as it approaches steady state (the floor it is going to). There is a *feedback loop* to tell the elevator how far it is from its destination, and control its speed.

An **open-loop** system has no feedback. A simple toaster will keep going and burn your toast. The end time is entirely controlled by the input setting. A **closed-loop** system feeds back the difference between the state $y(t)$ and the desired steady state $y_\infty$. A toaster oven can avoid burning by feeding back the temperature.

The logistic equation is nonlinear because of its feedback term $-by^2$. This is so common in other examples of movement and growth. Our brain controls arm movement and brings it to a stop. Your car has thousands of computer chips and controllers that measure position and speed, to slow down and stop before disaster.

I admit that I don't use cruise control because the car might keep cruising—I am not too sure it will stop. But it does have a feedback loop to keep the car below a set speed.

## 1.8 Separable Equations and Exact Equations

This section presents two special types of first order nonlinear differential equations. They are a bridge between $y' = ay$ and the very general form $y' = f(t, y)$. These pages explain how to solve the two types in between, by ordinary integration. *Separable* equations are the simplest. For *exact* equations, see formulas (12) and (15).

| Separable | Exact |
|---|---|
| $\dfrac{\mathbf{dy}}{\mathbf{dt}} = \dfrac{\mathbf{g(t)}}{\mathbf{f(y)}}$ | $\dfrac{\mathbf{dy}}{\mathbf{dt}} = \dfrac{\mathbf{g(y,t)}}{\mathbf{f(y,t)}}$ when $\dfrac{\partial f}{\partial t} = -\dfrac{\partial g}{\partial y}$ |

### 1. Separable Equations $f(y)dy = g(t)dt$

With $f(y)$ on one side and $g(t)$ on the other side, you see the meaning of *separable*. The ordinary way to write this equation would be

$$\frac{dy}{dt} = \frac{g(t)}{f(y)} \quad \text{starting from } y(0) \text{ at time } t = 0. \tag{1}$$

When $dy/dt$ has this separable form, we combine $f(y)$ with $dy$ and $g(t)$ with $dt$. Those functions $f$ and $g$ need to be integrated. **The integrals $F(y)$ and $G(t)$ start at** $y = y(0)$ **and** $t = 0$:

$$F(y) = \int_{y(0)}^{y} f(u)\,du \qquad G(t) = \int_{x=0}^{t} g(x)\,dx \tag{2}$$

The dummy variables $u$ and $x$ were chosen because $y$ and $t$ are needed in the upper limits of integration. Every author faces this question, to select variables. To show that the letters $u$ and $x$ don't matter, I could change them to $Y$ and $T$.

After integrating $f$ and $g$, we have *implicitly* solved the differential equation:

$$\textbf{Solution} \quad \frac{dy}{dt} = \frac{g(t)}{f(y)} \quad \text{integrates to} \quad \boldsymbol{F(y) = G(t)}. \tag{3}$$

To get an *explicit* solution $y = \ldots$ we have to solve this equation $F(y) = G(t)$ to find $y$.

**Example 1** $\dfrac{dy}{dt} = \dfrac{t}{y}$ is $\boldsymbol{y\,dy = t\,dt}$. Integrate to find $\dfrac{1}{2}\left(y(t)^2 - y(0)^2\right) = \dfrac{1}{2}t^2$.

Solve this implicit equation to find $y(t)$ explicitly:

**Solution** $\boldsymbol{y(t) = \sqrt{y(0)^2 + t^2}}$. Then $\dfrac{dy}{dt} = \dfrac{t}{\sqrt{y(0)^2 + t^2}} = \dfrac{t}{y}$.

**Example 2** $\boldsymbol{dy/dt = 2ty}$ has $\boldsymbol{g(t) = 2t}$ divided by $\boldsymbol{f(y) = 1/y}$.

**Solution** Separate $1/y$ from $2t$ and integrate to get $F = \ln y - \ln y(0)$ and $G = t^2$:

$$\frac{dy}{y} = 2t\,dt \quad \text{leads to} \quad \int_{y(0)}^{y} \frac{du}{u} = \ln y - \ln y(0) \quad \text{and} \quad \int_0^t 2x\,dx = t^2$$

In this example, $F(y) = G(t)$ produces $\ln y = \ln y(0) + t^2$. Take exponentials of both sides to find the solution $y$:

$$\mathbf{y = e^{\ln y(0)} e^{t^2} = y(0)\, e^{t^2}}. \tag{4}$$

I always check the derivative $dy/dt$ and the starting value $y(0)$:

$$\frac{d}{dt}\left(y(0)\,e^{t^2}\right) = 2t\left(y(0)\,e^{t^2}\right) = 2ty \qquad y(0)\,e^{t^2} = y(0) \ \text{ at } \ t = 0. \tag{5}$$

**Example 3** Our favorite equation $\frac{dy}{dt} = ay + q$ is separable when $a$ and $q$ are constant. Move $y + \frac{q}{a}$ to the left side below $dy$. Keep $a\,dt$ on the right side. Then integrate both sides, and you have solved this equation once more!

$$\frac{dy}{y + \frac{q}{a}} = a\,dt \quad \text{gives} \quad \ln\left(y + \frac{q}{a}\right) = at + C. \tag{6}$$

Take exponentials to find $y$, and set $t = 0$ to find $C$:

$$\textbf{Exponential growth} \qquad y(t) + \frac{q}{a} = e^{at} e^C \quad \text{and} \quad y(0) + \frac{q}{a} = e^C. \tag{7}$$

Substitute for $e^C$ in the left equation, to get the answer we know:

$$\boxed{y(t) + \frac{q}{a} = e^{at}\left(y(0) + \frac{q}{a}\right) \quad \text{and then} \quad \boldsymbol{y(t) = e^{at}\,y(0) + \frac{q}{a}(e^{at} - 1)}.} \tag{8}$$

This answer was the key to Section 1.4. Here the formulas came faster (the first one in that box looks attractive). But I like the old way: *Follow each input as it grows.*

**Example 4** **(Logistic equation )**

$$\frac{dy}{dt} = ay - by^2 \qquad\qquad \int_{y(0)}^{y} \frac{du}{au - bu^2} = \int_{t(0)}^{t} dx \tag{9}$$

The right side is certainly $G(t) = t - t(0)$. I am including $t(0)$ to show how the system allows any starting value for $t$ as well as $y$. We don't know a perfect starting time for the Earth's population, so we pick a year like $t(0) = 2000$ and work from there. The key point is that two integrals $F(y)$ and $G(t)$ give the answer.

Section 1.7 computed those integrals and solved the logistic equation.

## 2. Exact Equations $f(y,t)dy = g(y,t)dt$

A separable equation has $dy/dt = g(t)/f(y)$. We wrote this as $f(y)dy = g(t)dt$. We integrated the two sides separately to get $F(y) = G(t)$. This solved the equation.

*Exact equations are not required to be separable.* The functions $f$ and $g$ can depend on both variables $t$ and $y$. The equation does not split into a pure $y$-integration and a pure $t$-integration. We now have $\mathbf{f(y,t)\,dy = g(y,t)\,dt}$. But it sometimes succeeds to integrate the left side $f(y,t)$ with respect to $y$, as if $t$ were a constant which it is not.

**Step 1 Integrate $f$ with respect to $y$** $\quad \int f(y,t)\,dy = \boldsymbol{F(y,t) + C(t)}. \quad (10)$

Normally, any constant $C$ can be added to an integral. The answer stays correct, because the derivative of $C$ is zero. Here, *any function of $t$ can be added to the integral*, because the $y$ derivative of any $C(t)$ is zero. Now $F(y,t) + C(t)$ has more flexibility.

**Step 2 (if possible) Choose** $C(t)$ so that $\boldsymbol{\frac{\partial}{\partial t}(F(y,t) + C(t)) = -g(y,t)}. \quad (11)$

If that choice of $C(t)$ is possible, our original equation involving $g$ and $f$ is solved:

**Step 3** $\quad \frac{dy}{dt} = \frac{g(y,t)}{f(y,t)}$ is solved by $\boldsymbol{F(y,t) + C(t) = \text{any constant}}. \quad (12)$

Before I show when and why this works, here is an example of success.

**Example 5** The equation $\frac{dy}{dt} = \boldsymbol{\frac{2yt - 1}{y^2 - t^2}}$ has $g = 2yt - 1$ and $f = y^2 - t^2$.

**Step 1** Integrate $f dy = (y^2 - t^2)dy$ to find $F(y,t) = \frac{1}{3}y^3 - yt^2$. Then $\frac{\partial F}{\partial t} = -2ty$.

**Step 2** Solve equation (11) for $C(t)$. For our particular $f$ and $g$, this is possible :

$$-2ty + \frac{dC}{dt} = -(2yt - 1) \text{ gives } \frac{dC}{dt} = 1 \text{ and } C(t) = t.$$

**Step 3** The original $\frac{dy}{dt} = \frac{g}{f}$ is solved by $F(y,t) + C(t) = \text{constant}$:

*Solution from $F + C$*
*Constant is set by $y(0)$*

$$\boldsymbol{\frac{1}{3}y^3 - yt^2 + t = \frac{1}{3}y(0)^3}.$$

To check this answer, take its time derivative implicitly (which means : just do it).

$$\textbf{Implicit differentiation} \quad y^2\frac{dy}{dt} - t^2\frac{dy}{dt} - 2yt + 1 = 0.$$

This is our equation $dy/dt = \boldsymbol{(2yt - 1)/(y^2 - t^2)}$ as we hoped. Now to explain why.

## The Exactness Condition

*When is Step 2 possible ?* Sometimes there is $C(t)$ to solve equation (11), but usually not. To find the condition for exactness, take the $y$-derivative of both sides in Step 2:

$$\frac{\partial}{\partial y}\frac{\partial}{\partial t}\left(F(y,t)+C(t)\right) = -\frac{\partial}{\partial y}\left(g(y,t)\right). \tag{13}$$

The order of $\dfrac{\partial}{\partial y}$ and $\dfrac{\partial}{\partial t}$ can always be reversed. Certainly $\dfrac{\partial}{\partial y}C(t)=0$ and $\dfrac{\partial}{\partial y}F=f$.

**The left side of (13) is** $\dfrac{\partial}{\partial y}\dfrac{\partial}{\partial t}F(y,t)=\dfrac{\partial}{\partial t}\dfrac{\partial}{\partial y}F(y,t)$ which is $\dfrac{\partial}{\partial t}f(y,t)$. (14)

Comparing (14) with (13), Step 2 is only possible when our original differential equation $dy/dt = g/f$ is exact:

$$\textbf{Exactness condition} \qquad \frac{\partial}{\partial t}f(y,t) = -\frac{\partial}{\partial y}g(y,t). \tag{15}$$

When the equation is exact, Step 2 will produce $C(t)$. The final question is about Step 3. Why is $F(y,t)+C(t) = \text{constant}$ for the original differential equation $dy/dt = g/f$ ? To see this, take the time derivative of $F(y,t)+C(t)$ using the (implicit) chain rule:

$$\frac{\partial F}{\partial y}\frac{dy}{dt}+\frac{\partial F}{\partial t}+\frac{\partial C}{\partial t}=0. \tag{16}$$

Step 1 produced $\dfrac{\partial F}{\partial y}=f$. Step 2 produced $\dfrac{\partial F}{\partial t}+\dfrac{\partial C}{\partial t}=-g$. We have success:

*Equation* (16) *is* $f\dfrac{dy}{dt}-g=0$. *This is our original problem* $\dfrac{dy}{dt}=\dfrac{g}{f}$.

Example 5 was exact because $g=2yt-1$ and $f=y^2-t^2$ agree on $\dfrac{\partial f}{\partial t}=-\dfrac{\partial g}{\partial y}=-2t$.

**Example 6** Steps 1, 2, 3 must be possible because this non-separable equation is exact :

$$\frac{dy}{dt}=\frac{t-y}{t+y}=\frac{g(y,t)}{f(y,t)} \quad \text{has} \quad \frac{\partial f}{\partial t}=-\frac{\partial g}{\partial y}=1. \tag{17}$$

**Step 1** Integrate $\int f dy = \int (t+y)dy$ to find $F = ty+\frac{1}{2}y^2$.

**Step 2** Write out $\dfrac{\partial}{\partial t}(F+C)=-g=y-t$ to find $C(t)=-\frac{1}{2}t^2$

**Step 3** The example is solved by $F+C = ty+\dfrac{1}{2}y^2-\dfrac{1}{2}t^2 = \text{constant} = \frac{1}{2}\,y(0)^2$.

To check that solution, find the total time derivative of $F+C$ by the chain rule :

$$t\frac{dy}{dt}+y+y\frac{dy}{dt}-t=0. \text{ This is } \frac{dy}{dt}=\frac{t-y}{t+y} \text{ as desired.}$$

### Final Note : Separable is Exact

Notice that **a separable equation $dy/dt = g(t)/f(y)$ is always exact**:

$$\textbf{(15) is satisfied} \qquad \frac{\partial}{\partial t}\, f(y) = -\frac{\partial}{\partial y}\, g(t) \text{ becomes } \mathbf{0} = \mathbf{0}.$$

No problem with integrating $\int f(y)\,dy$ and $\int g(t)\,dt$ to find $F(y)$ and $G(t) = -C(t)$.

## ■ REVIEW OF THE KEY IDEAS ■

1. A **separable** equation $\dfrac{dy}{dt} = \dfrac{g(t)}{f(y)}$ is solved by $\int f(y)dy = \int g(t)dt +$ any constant.
2. That solution gives $y$ implicitly. Solve to find $y$ explicitly as a function of $t$.
3. An **exact** equation $\dfrac{dy}{dt} = \dfrac{g(y,t)}{f(y,t)}$ has $\boldsymbol{\dfrac{\partial g}{\partial y} = -\dfrac{\partial f}{\partial t}}$. Then $F(y,t) + C(t) =$ constant.
4. The solution has $F(y,t) = \int f(y,t)dy$ for each $t$, and $C(t) = -\displaystyle\int \left(\frac{\partial F}{\partial t} + g\right) dt$.
5. The exactness condition in **3** removes $y$ from that integral for $C(t)$ in **4**.

## Problem Set 1.8

**1** Finally we can solve the example $dy/dt = y^2$ in Section 1.1 of this book.

Start from $y(0) = 1$. Then $\displaystyle\int_1^y \frac{dy}{y^2} = \int_0^t dt$. Notice the limits on $y$ and $t$. Find $y(t)$.

**2** Start the same equation $dy/dt = y^2$ from any value $y(0)$. At what time $t$ does the solution blow up? For which starting values $y(0)$ does it never blow up?

**3** Solve $dy/dt = a(t)y$ as a separable equation starting from $y(0) = 1$, by choosing $f(y) = 1/y$. This equation gave the growth factor $G(0,t)$ in Section 1.6.

**4** Solve these separable equations starting from $y(0) = 0$:

(a) $\dfrac{dy}{dt} = ty$ (b) $\dfrac{dy}{dt} = t^m\, y^n$

**5** Solve $\dfrac{dy}{dt} = a(t)y^2 = \dfrac{a(t)}{1/y^2}$ as a separable equation starting from $y(0) = 1$.

**6** The equation $\dfrac{dy}{dt} = y + t$ is not separable or exact. But it is linear and $y =$ _____.

**7** The equation $\dfrac{dy}{dt} = \dfrac{y}{t}$ has the solution $y = At$ for every constant $A$. Find this solution by separating $f = 1/y$ from $g = 1/t$. Then integrate $dy/y = dt/t$. Where does the constant $A$ come from ?

**8** For which number $A$ is $\dfrac{dy}{dt} = \dfrac{ct - ay}{At + by}$ an exact equation ? For this $A$, solve the equation by finding a suitable function $F(y,t) + C(t)$.

**9** Find a function $y(t)$ different from $y = t$ that has $dy/dt = y^2/t^2$.

**10** These equations are separable after factoring the right hand sides :

$$\text{Solve} \quad \frac{dy}{dt} = e^{y+t} \quad \text{and} \quad \frac{dy}{dt} = yt + y + t + 1.$$

**11** These equations are linear and separable : Solve $\dfrac{dy}{dt} = (y+4)\cos t$ and $\dfrac{dy}{dt} = ye^t$.

**12** Solve these three separable equations starting from $y(0) = 1$ :

$$\text{(a)}\ \frac{dy}{dt} = -4ty \qquad \text{(b)}\ \frac{dy}{dt} = ty^3 \qquad \text{(c)}\ (1+t)\frac{dy}{dt} = 4y$$

**Test the exactness condition $\partial g/\partial y = -\partial f/\partial t$ and solve Problems 13-14.**

**13** (a) $\dfrac{dy}{dt} = \dfrac{-3t^2 - 2y^2}{4ty + 6y^2}$ (b) $\dfrac{dy}{dt} = -\dfrac{1 + ye^{ty}}{2y + te^{ty}}$

**14** (a) $\dfrac{dy}{dt} = \dfrac{4t - y}{t - 6y}$ (b) $\dfrac{dy}{dt} = -\dfrac{3t^2 + 2y^2}{4ty + 6y^2}$

**15** Show that $\dfrac{dy}{dt} = -\dfrac{y^2}{2ty}$ is exact but the same equation $\dfrac{dy}{dt} = -\dfrac{y}{2t}$ is not exact. Solve both equations. (This problem suggests that many equations become exact when multiplied by an integrating factor.)

**16** Exactness is really the condition to solve two equations with the same function $H(t,y)$ :

$$\frac{\partial H}{\partial y} = f(t,y) \ \text{ and } \ \frac{\partial H}{\partial t} = -g(t,y) \ \text{ can be solved if } \ \frac{\partial f}{\partial t} = -\frac{\partial g}{\partial y}.$$

Take the $t$ derivative of $\partial H/\partial y$ and the $y$ derivative of $\partial H/\partial t$ to show that exactness is *necessary*. It is also *sufficient* to guarantee that a solution $H$ will exist.

**17** The linear equation $\dfrac{dy}{dt} = aty + q$ is not exact or separable. Multiply by the integrating factor $e^{-\int at\,dt}$ and solve the equation starting from $y(0)$.

**Second order equations $F(t, y, y', y'') = 0$ involve the second derivative $y''$. This reduces to a first order equation for $y'$ (not $y$) in two important cases:**

| |
|---|
| **I.** When $y$ is missing in $F$, set $y' = v$ and $y'' = v'$. Then $\boldsymbol{F(t, v, v') = 0}$. |
| **II.** When $t$ is missing in $F$, set $y'' = \dfrac{dv}{dt} = \dfrac{dv}{dy}\dfrac{dy}{dt} = v\dfrac{dv}{dy}$. Then $\boldsymbol{F\left(y, v, v\dfrac{dv}{dy}\right) = 0}$. |

See the website for **reduction of order** when one solution $y(t)$ is known.

**18** ($y$ is missing) Solve these differential equations for $v = y'$ with $v(0) = 1$. Then solve for $y$ with $y(0) = 0$.

(a) $y'' + y' = 0$ (b) $2ty'' - y' = 0$.

**19** Both $y$ and $t$ are missing in $\boldsymbol{y'' = (y')^2}$. Set $v = y'$ and go two ways:

I. ($y$ missing) Solve $\dfrac{dv}{dt} = v^2$ for $v(t)$ and then $\dfrac{dy}{dt} = v(t)$ with $y(0) = 0$, $y'(0) = 1$.

II. ($t$ missing) Solve $v\dfrac{dv}{dy} = v^2$ for $v(y)$ and then $\dfrac{dy}{dt} = v(y)$ with $y(0) = 0$, $y'(0) = 1$.

**20** An **autonomous equation $\boldsymbol{y' = f(y)}$** has no terms that contain $t$ ($t$ is missing).

Explain why every autonomous equation is separable. A non-autonomous equation could be separable or not. For a linear equation we usually say LTI (**linear time-invariant** ) when it is autonomous: coefficients are constant, not varying with $t$.

**21** $my'' + ky = 0$ is a highly important LTI equation. Two solutions are $\cos\omega t$ and $\sin\omega t$ when $\omega^2 = k/m$. Solve differently by reducing to a first order equation for $y' = dy/dt = v$ with $y'' = v\,dv/dy$ as above:

$$mv\frac{dv}{dy} + ky = 0 \text{ integrates to } \frac{1}{2}mv^2 + \frac{1}{2}ky^2 = \text{ constant } E.$$

For a mass on a spring, kinetic energy $\frac{1}{2}mv^2$ plus potential energy $\frac{1}{2}ky^2$ is a constant energy $E$. What is $E$ when $y = \cos\omega t$ ? What integral solves the separable $m(y')^2 = 2E - ky^2$ ? I would not solve the linear oscillation equation this way.

**22** $my'' + k\sin y = 0$ is the *nonlinear* oscillation equation: not so simple. Reduce to a first order equation as in Problem 21 :

$$mv\frac{dv}{dy} + k\sin y = 0 \text{ integrates to } \frac{1}{2}mv^2 - k\cos y = \text{ constant } E.$$

With $v = dy/dt$ what impossible integral is needed for this first order separable equation? Actually that integral gives the period of a nonlinear pendulum—this integral is extremely important and well studied even if impossible.

## ■ CHAPTER 1 NOTES ■

**The great function of calculus is $e^t$. How best to define this exponential function?**

Section 1.3 constructed $y = e^t$ from its infinite series $1 + t + \frac{1}{2}t^2 + \frac{1}{6}t^3 + \cdots$. Euler would approve! Taking the derivative of each term brings back $e^t$. This property $dy/dt = y$ is the most important tool we have—it is the foundation of our subject.

I like this approach to $e^t$ for at least two reasons:

1. It is based on the derivatives of $t$ and $t^2$ and $t^n$ : well known.
2. The Chapter 3 Notes solve nonlinear equations in exactly the same way.

The limiting step required here is to add up an infinite series. We don't expect a simple answer like $1 + \frac{1}{2} + \frac{1}{4} + \frac{1}{8} + \cdots = 2$. But the numbers $1/n!$ in $e^t$ are (*much smaller*) than these numbers $1/2^n$.

This is really the key point, to see that the terms $t^n/n!$ approach zero quickly.

**The infinite series $1 + t + t^2/2 + \cdots + t^n/n! + \cdots$ converges for every $t$.**

*Proof.* Each term $t^n/n!$ multiplies the previous term $t^{n-1}/(n-1)!$ by $t/n$. At some point $n = N$, that number $t/N$ goes below $\frac{1}{2}$. From this point on, we know that

$$\frac{t^n}{N!} + \frac{t^{N+1}}{(N+1)!} + \frac{t^{N+2}}{(N+2)!} + \cdots \quad \text{is less than} \quad \frac{t^N}{N!}\left(1 + \frac{1}{2} + \frac{1}{4} + \cdots\right)$$

The right side is $t^N/N!$ times 2. The left side is smaller. The first $N$ terms that come before $t^N/N!$ have no effect on convergence of the series (they just enter the final sum). *So the series for $e^t$ always converges.*

If $t$ is negative, use its absolute value $|t|$ and the proof still succeeds. The series for the derivative of $e^t$ is the same as the series for $e^t$. So we know: This series is absolutely convergent. We can safely say that $y' = y$.

**Four approaches to $e^t$** Looking back at my own teaching and writing, I really missed the importance of this big step in calculus. Just another function? *Not at all.* Textbooks offer four main ways to construct $y = e^t$ :

1. Add all the terms $t^n/n!$. The derivative of each term is the previous $t^{n-1}/(n-1)!$
2. Take the $n$th power of $(1 + t/n)$ as in compound interest. Let $n$ approach infinity.
3. The slope of $b^t$ is $C$ times $b^t$. Choose $e$ as the value of $b$ that makes $C = 1$.
4. Integrate $1/y$ to construct $t = \ln y$. Invert this function to find $y = e^t$.

I believe that **3** and **4** are too tricky. Explicit constructions are the winners. You want to say, "*Here is the function.*" In method **2** you are working with $(1 + t/n)^n$ : not too bad. In **1** you see step by step and term by term that $dy/dt = y$.

# Chapter 2

# Second Order Equations

## 2.1 Second Derivatives in Science and Engineering

Second order equations involve the second derivative $d^2y/dt^2$. Often this is shortened to $y''$, and then the first derivative is $y'$. In physical problems, $y'$ can represent velocity $v$ and the second derivative $y'' = a$ is **acceleration**: the rate $dy'/dt$ that velocity is changing.

The most important equation in dynamics is Newton's Second Law $\boldsymbol{F = ma}$. Compare a second order equation to a first order equation, and allow them to be nonlinear:

$$\textbf{First order} \quad y' = f(t, y) \qquad \textbf{Second order} \quad y'' = F(t, y, y') \tag{1}$$

The second order equation needs **two initial conditions**, normally $y(0)$ and $y'(0)$— the initial velocity as well as the initial position. Then the equation tells us $y''(0)$ and the movement begins.

When you press the gas pedal, that produces acceleration. The brake pedal also brings acceleration but it is *negative* (the velocity decreases). The steering wheel produces acceleration too! Steering changes the direction of velocity, not the speed.

Right now we stay with straight line motion and one–dimensional problems:

$$\frac{d^2y}{dt^2} > 0 \quad \text{(speeding up)} \qquad\qquad \frac{d^2y}{dt^2} < 0 \quad \text{(slowing down)}.$$

The graph of $y(t)$ bends upwards for $y'' > 0$ (the right word is *convex*). Then the velocity $y'$ (slope of the graph) is increasing. The graph bends downwards for $y'' < 0$ (*concave*). Figure 2.1 shows the graph of $\boldsymbol{y = \sin t}$, when the acceleration is $a = d^2y/dt^2 = -\sin t$. The important equation $\boldsymbol{y'' = -y}$ leads to $\sin t$ and $\cos t$.

Notice how the velocity $dy/dt$ (slope of the graph) changes sign in between zeros of $y$.

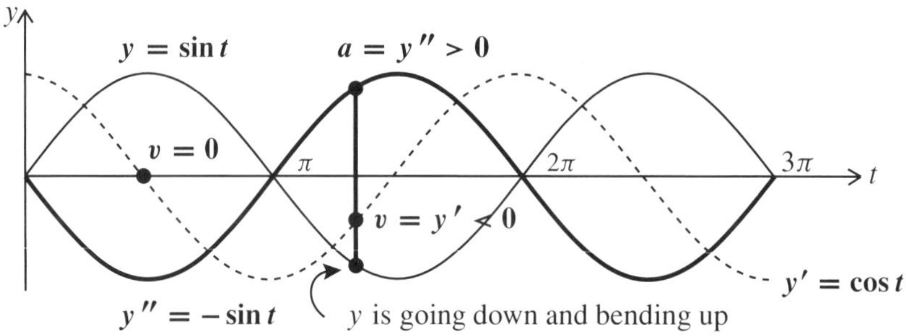


Figure 2.1: $y'' > 0$ means that velocity $y'$ (or slope) increases. The curve bends upward.

The best examples of $F = ma$ come when the force $F$ is $-ky$, a constant $k$ times the "position" or "displacement" $y(t)$. This produces the oscillation equation.

**Fundamental equation of mechanics** $$m\frac{d^2y}{dt^2} + ky = 0 \tag{2}$$

Think of a mass hanging at the bottom of a spring (Figure 2.2). The top of the spring is fixed, and the spring will stretch. Now stretch it a little more (move the mass downward by $y(0)$) and let go. The spring pulls back on the mass. Hooke's Law says that the force is $F = -ky$, proportional to the stretching distance $y$. Hooke's constant is $k$.

The mass will oscillate up and down. The oscillation goes on forever, because equation (2) does not include any friction (damping term $b\,dy/dt$). The oscillation is a perfect cosine, with $y = \cos\omega t$ and $\omega = \sqrt{k/m}$, because the second derivative has to produce $k/m$ to match $y'' = -(k/m)y$.

**Oscillation at frequency** $\omega = \sqrt{\dfrac{k}{m}}$ $$y = y(0)\cos\left(\sqrt{\frac{k}{m}}\,t\right). \tag{3}$$

At time $t = 0$, this shows the extra stretching $y(0)$. The derivative of $\cos\omega t$ has a factor $\omega = \sqrt{k/m}$. The second derivative $y''$ has the required $\omega^2 = k/m$, so $my'' = -ky$.

The movement of one spring and one mass is especially simple. There is only one frequency $\omega$. When we connect $N$ masses by a line of springs there will be $N$ frequencies—then Chapter 6 has to study the eigenvalues of $N$ by $N$ matrices.

$$m\frac{d^2y}{dt^2} = -ky$$

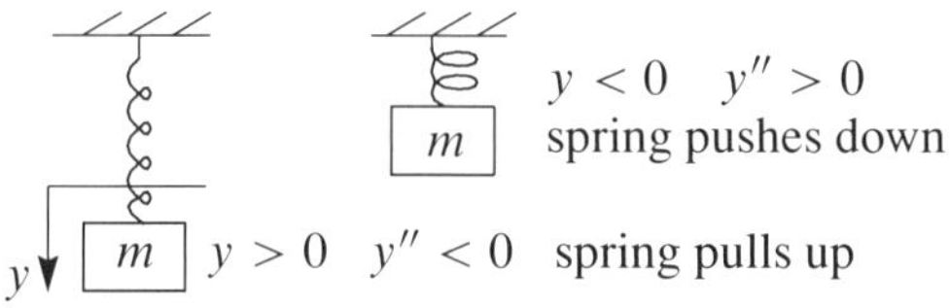


Figure 2.2: Larger $k$ = stiffer spring = *faster* $\omega$. Larger $m$ = heavier mass = *slower* $\omega$.

### Initial Velocity $y'(0)$

Second order equations have *two* initial conditions. The motion starts in an initial position $y(0)$, and its initial velocity is $y'(0)$. We need both $y(0)$ and $y'(0)$ to determine the two constants $c_1$ and $c_2$ in the complete solution to $my'' + ky = 0$:

$$\textbf{"Simple harmonic motion"} \qquad y = c_1 \cos\left(\sqrt{\frac{k}{m}}\,t\right) + c_2 \sin\left(\sqrt{\frac{k}{m}}\,t\right). \qquad (4)$$

Up to now the motion has started from rest ($y'(0) = 0$, no initial velocity). Then $c_1$ is $y(0)$ and $c_2$ is zero: only cosines. As soon as we allow an initial velocity, the sine solution $y = c_2 \sin \omega t$ must be included. But its coefficient $c_2$ is not just $y'(0)$.

$$\text{At} \quad t = 0, \quad \frac{dy}{dt} = c_2\,\omega \cos \omega t \quad \text{matches} \quad y'(0) \quad \text{when} \quad \boldsymbol{c_2 = \frac{y'(0)}{\omega}}. \qquad (5)$$

The original solution $y = y(0) \cos \omega t$ matched $y(0)$, with zero velocity at $t = 0$. The new solution $y = (y'(0)/\omega) \sin \omega t$ has the right initial velocity and it starts from zero. When we combine those two solutions, $y(t)$ matches both conditions $y(0)$ and $y'(0)$:

$$\textbf{Unforced oscillation} \qquad \boldsymbol{y(t) = y(0)\cos \omega t + \frac{y'(0)}{\omega} \sin \omega t \;\; \text{with} \;\; \omega = \sqrt{\frac{k}{m}}}. \qquad (6)$$

With a trigonometric identity, I can combine those two terms (cosine and sine) into one.

### Cosine with Phase Shift

We want to rewrite the solution (6) as $\boldsymbol{y(t) = R\cos(\omega t - \alpha)}$. The amplitude of $y(t)$ will be the positive number $R$. The phase shift or lag in this solution will be the angle $\alpha$. By using the right identity for the cosine of $\omega t - \alpha$, we match both $\cos \omega t$ and $\sin \omega t$:

$$R \cos(\omega t - \alpha) = R \cos \omega t \, \cos \alpha + R \sin \omega t \, \sin \alpha. \qquad (7)$$

This combination of $\cos \omega t$ and $\sin \omega t$ agrees with the solution (6) if

$$R \cos \alpha = y(0) \quad \text{and} \quad R \sin \alpha = \frac{y'(0)}{\omega}. \qquad (8)$$

Squaring those equations and adding will produce $R^2$:

$$\textbf{Amplitude } \boldsymbol{R} \qquad R^2 = R^2(\cos^2 \alpha + \sin^2 \alpha) = (y(0))^2 + \left(\frac{y'(0)}{\omega}\right)^2. \qquad (9)$$

The ratio of the equations (8) will produce the tangent of $\alpha$:

$$\textbf{Phase lag } \boldsymbol{\alpha} \qquad \tan \alpha = \frac{R \sin \alpha}{R \cos \alpha} = \frac{y'(0)}{\omega\, y(0)}. \qquad (10)$$

Problem 14 will discuss the angle $\alpha$ we should choose, since different angles can have the same tangent. The tangent is the same if $\alpha$ is increased by $\pi$ or any multiple of $\pi$.

The pure cosine solution that started from $y'(0) = 0$ has *no phase shift*: $\alpha = 0$. Then the new form $y(t) = R \cos(\omega t - \alpha)$ is the same as the old form $y(0) \cos \omega t$.

### Frequency $\omega$ or $f$

If the time $t$ is measured in *seconds*, the frequency $\omega$ is in *radians per second*. Then $\omega t$ is in radians—it is an angle and $\cos \omega t$ is its cosine. But not everyone thinks naturally about radians. Complete cycles are easier to visualize. So frequency is also measured in *cycles per second*. A typical frequency in your home is $f = 60$ cycles per second. One cycle per second is usually shortened to $\boldsymbol{f = 1}$ **Hertz**. A complete cycle is $2\pi$ radians, so $f = 60$ *Hertz is the same frequency as* $\omega = 120\pi$ *radians per second*.

The **period** is the time $T$ for one complete cycle. Thus $T = 1/f$. This is the only page where $f$ is a frequency—on all other pages $f(t)$ is the driving function.

**Frequency** $\boldsymbol{\omega = 2\pi f}$ **Period** $\boldsymbol{T = \dfrac{1}{f} = \dfrac{2\pi}{\omega}}$.

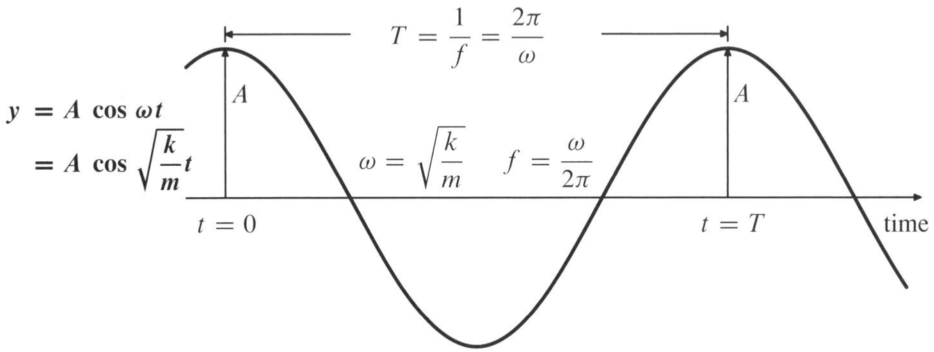


Figure 2.3: Simple harmonic motion $y = A \cos \omega t$ : amplitude $A$ and frequency $\omega$.

### Harmonic Motion and Circular Motion

Harmonic motion is up and down (or side to side). **When a point is in circular motion, its projections on the $x$ and $y$ axes are in harmonic motion**. Those motions are closely related, which is why a piston going up and down can produce circular motion of a flywheel. The harmonic motion "speeds up in the middle and slows down at the ends" while the point moves with constant speed around the circle.

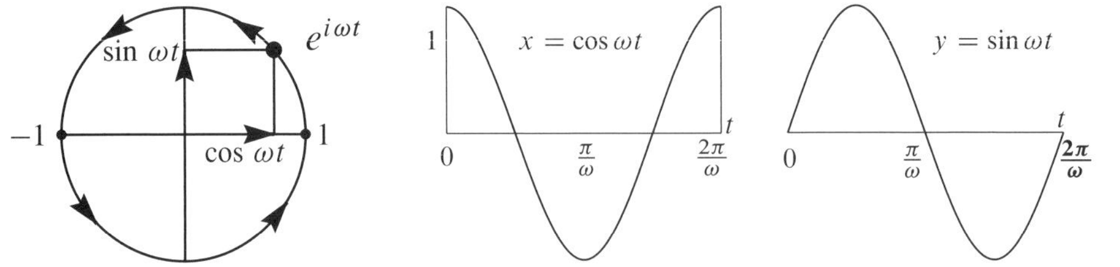


Figure 2.4: Steady motion around a circle produces cosine and sine motion along the axes.

## Response Functions

I want to introduce some important words. The **response** is the output $y(t)$. Up to now the only inputs were the initial values $y(0)$ and $y'(0)$. In this case $y(t)$ would be the *initial value response* (but I have never seen those words). When we only see a few cycles of the motion, initial values make a big difference. In the long run, what counts is the response to a *forcing function* like $f = \cos \omega t$.

Now $\omega$ is the **driving frequency** on the right hand side, where the **natural frequency** $\omega_n = \sqrt{k/m}$ is decided by the left hand side: $\omega$ comes from $y_p$, $\omega_n$ comes from $y_n$. When the motion is driven by $\cos \omega t$, a particular solution is $y_p = Y \cos \omega t$:

**Forced motion $y_p(t)$ at frequency $\omega$**
$$my'' + ky = \cos \omega t \qquad y_p(t) = \frac{1}{k - m\omega^2} \cos \omega t. \tag{11}$$

To find $y_p(t)$, I put $Y \cos \omega t$ into $my'' + ky$ and the result was $(k - m\omega^2) Y \cos \omega t$. This matches the driving function $\cos \omega t$ when $Y = 1/(k - m\omega^2)$.

The initial conditions are nowhere in equation (11). Those conditions contribute the null solution $y_n$, which oscillates at the natural frequency $\omega_n = \sqrt{k/m}$. Then $k = m\omega_n^2$.

If I replace $k$ by $m\omega_n^2$ in the response $y_p(t)$, I see $\omega_n^2 - \omega^2$ in the denominator:

**Response to $\cos \omega t$**
$$y_p(t) = \frac{1}{m\left(\omega_n^2 - \omega^2\right)} \cos \omega t. \tag{12}$$

Our equation $my'' + ky = \cos \omega t$ has no damping term. That will come in Section 2.3. It will produce a phase shift $\alpha$. Damping will also reduce the amplitude $|Y(\omega)|$. The amplitude is all we are seeing here in $Y(\omega) \cos \omega t$:

**Frequency response**
$$Y(\omega) = \frac{1}{k - m\omega^2} = \frac{1}{m\left(\omega_n^2 - \omega^2\right)}. \tag{13}$$

The mass and spring, or the inductance and capacitance, decide the natural frequency $\omega_n$. The response to a driving term $\cos \omega t$ (or $e^{i\omega t}$) is multiplication by the frequency response $Y(\omega)$. *The formula changes when $\omega = \omega_n$—we will study resonance!*

With damping in Section 2.3, the frequency response $Y(\omega)$ will be a complex number. We can't escape complex arithmetic and we don't want to. The magnitude $|Y(\omega)|$ will give the **magnitude response** (or amplitude response). The angle $\theta$ in the complex plane will decide the **phase response** (then $\alpha = -\theta$ because we measure the phase lag).

The response is $Y(\omega)e^{i\omega t}$ to $f(t) = e^{i\omega t}$ and the response is $g(t)$ to $f(t) = \delta(t)$. These show the frequency response $Y$ from equation (13) and the impulse response $g$ from equation (15). *$Ye^{i\omega t}$ and $g(t)$ are the two key solutions to $my'' + ky = f(t)$.*

## Impulse Response = Fundamental Solution

The most important solution to a linear differential equation will be called $g(t)$. In mathematics $g$ is the *fundamental solution*. In engineering $g$ is the *impulse response*. It is a particular solution when the right side $f(t) = \delta(t)$ is an impulse (a delta function).

The same $g(t)$ solves $mg'' + kg = 0$ when the initial velocity is $g'(0) = 1/m$.

| | | | |
|---|---|---|---|
| **Fundamental solution** | $mg'' + kg = \delta(t)$ | **with zero initial conditions** | (14) |
| **Null solution also** | $g(t) = \dfrac{\sin \omega_n t}{m\omega_n}$ | **has $g(0)=0$ and $g'(0)=\dfrac{1}{m}$.** | (15) |

To find that null solution, I just put its initial values 0 and $1/m$ into equation (6). The cosine term disappeared because $g(0) = 0$.

I will show that those two problems give the same answer. Then this whole chapter will show why $g(t)$ is so important. For first order equations $y' = ay + q$ in Chapter 1, the fundamental solution (impulse response, growth factor) was $g(t) = e^{at}$. The first two names were not used, but you saw how $e^{at}$ dominated that whole chapter.

I will first explain the response $g(t)$ in physical language. *We strike the mass and it starts to move.* All our force is acting at one instant of time: *an impulse*. A finite force within one moment is impossible for an ordinary function, only possible for a delta function. Remember that the integral of $\delta(t)$ jumps to 1 when we pass the point $t = 0$.

If we integrate $mg'' = \delta(t)$, nothing happens before $t = 0$. In that instant, the integral jumps to 1. The integral of the left side $mg''$ is $mg'$. Then $mg' = 1$ instantly at $t = 0$. This gives $g'(0) = 1/m$. You see that computing with an impulse $\delta(t)$ needs some faith.

**The point of $g(t)$ is that it solves the equation for any forcing function $f(t)$:**

$$\boldsymbol{my'' + ky = f(t)} \text{ has the particular solution } \boldsymbol{y(t) = \int_0^t g(t-s)\, f(s)\, ds}. \tag{16}$$

That was the key formula of Chapter 1, when $g(t - s)$ was $e^{a(t-s)}$ and the equation was first order. Section 2.3 will find $g(t)$ when the differential equation includes damping. The coefficients in the equation will stay constant, to allow a neat formula for $g(t)$.

You may feel uncertain about working with delta functions—a means to an end. We will verify this final solution $y(t)$ in three different ways:

**1** Substitute $y(t)$ from (16) directly into the differential equation (Problem 21)

**2** Solve for $y(t)$ by variation of parameters (Section 2.6)

**3** Solve again by using the Laplace transform $Y(s)$ (Section 2.7).

## ■ REVIEW OF THE KEY IDEAS ■

1. $my'' + ky = 0$: A mass on a spring oscillates at the natural frequency $\omega_n = \sqrt{k/m}$.
2. $my'' + ky = \cos\omega t$: This driving force produces $y_p = (\cos\omega t)/m\left(\omega_n^2 - \omega^2\right)$.
3. There is resonance when $\omega_n = \omega$. The solution $y_p = t\sin\omega t$ includes a new factor $t$.
4. $mg'' + kg = \delta(t)$ gives $\boldsymbol{g(t) = (\sin\omega_n t)/m\omega_n}$ = null solution with $g'(0) = 1/m$.
5. Fundamental solution $g$: Every driving function $f$ gives $y(t) = \int_0^t g(t-s)f(s)\,ds$.
6. Frequency: $\omega$ radians per second or $f$ cycles per second ($f$ Hertz). Period $T = 1/f$.

## Problem Set 2.1

**1** Find a cosine and a sine that solve $d^2y/dt^2 = -9y$. This is a second order equation so we expect *two constants* $C$ and $D$ (from integrating twice):

$$\textbf{Simple harmonic motion} \qquad y(t) = C\cos\omega t + D\sin\omega t. \quad \text{What is } \omega\,?$$

If the system starts from rest (this means $dy/dt = 0$ at $t = 0$), which constant $C$ or $D$ will be zero?

**2** In Problem 1, which $C$ and $D$ will give the starting values $y(0) = 0$ and $y'(0) = 1$?

**3** Draw Figure 2.3 to show simple harmonic motion $y = A\cos(\omega t - \alpha)$ with phases $\alpha = \pi/3$ and $\alpha = -\pi/2$.

**4** Suppose the circle in Figure 2.4 has radius 3 and circular frequency $f = 60$ Hertz. If the moving point starts at the angle $-45°$, find its $x$-coordinate $A\cos(\omega t - \alpha)$. The phase lag is $\alpha = 45°$. When does the point first hit the $x$ axis?

**5** If you drive at 60 miles per hour on a circular track with radius $R = 3$ miles, what is the time $T$ for one complete circuit? Your circular frequency is $f =$ _____ and your angular frequency is $\omega =$ _____ (with what units?). The period is $T$.

**6** The total energy $E$ in the oscillating spring-mass system is

$$E = \textbf{kinetic} \text{ energy in mass} + \textbf{potential} \text{ energy in spring} = \frac{m}{2}\left(\frac{dy}{dt}\right)^2 + \frac{k}{2}y^2.$$

Compute $E$ when $y = C\cos\omega t + D\sin\omega t$. The energy is constant!

**7** Another way to show that the total energy $E$ is constant:

$$\text{Multiply } \boldsymbol{my'' + ky = 0} \text{ by } \boldsymbol{y'}. \text{ Then integrate } my'y'' \text{ and } kyy'.$$

**8** A **forced oscillation** has another term in the equation and in the solution:

$$\frac{d^2y}{dt^2} + 4y = F\cos\omega t \quad \text{has} \quad y = C\cos 2t + D\sin 2t + A\cos\omega t.$$

(a) Substitute $y$ into the equation to see how $C$ and $D$ disappear (they give $y_n$). Find the forced amplitude $A$ in the particular solution $y_p = A\cos\omega t$.

(b) In case $\omega = 2$ (forcing frequency = natural frequency), what answer does your formula give for $A$? The solution formula for $y$ breaks down in this case.

**9** Following Problem 8, write down the complete solution $y_n + y_p$ to the equation

$$m\frac{d^2y}{dt^2} + ky = F\cos\omega t \quad \text{with } \omega \neq \omega_n = \sqrt{k/m} \text{ (no resonance).}$$

The answer $y$ has free constants $C$ and $D$ to match $y(0)$ and $y'(0)$ ($A$ *is fixed* by $F$).

**10** Suppose Newton's Law $F = ma$ has the force $F$ in the *same* direction as $a$:

$$my'' = +ky \qquad \text{including} \qquad y'' = 4y.$$

Find two possible choices of $s$ in the exponential solutions $y = e^{st}$. The solution is not sinusoidal and $s$ is real and the oscillations are gone. Now $y$ is unstable.

**11** Here is a *fourth order* equation: $d^4y/dt^4 = 16y$. Find *four* values of $s$ that give exponential solutions $y = e^{st}$. You could expect four initial conditions on $y$: $y(0)$ is given along with what three other conditions?

**12** To find a particular solution to $y'' + 9y = e^{ct}$, I would look for a multiple $y_p(t) = Ye^{ct}$ of the forcing function. What is that number $Y$? When does your formula give $Y = \infty$? (Resonance needs a new formula for $Y$.)

**13** In a particular solution $y = Ae^{i\omega t}$ to $y'' + 9y = e^{i\omega t}$, what is the amplitude $A$? The formula blows up when the forcing frequency $\omega$ = what natural frequency?

**14** Equation (10) says that the tangent of the phase angle is $\tan\alpha = y'(0)/\omega y(0)$. First, check that $\tan\alpha$ is dimensionless when $y$ is in meters and time is in seconds. Next, if that ratio is $\tan\alpha = 1$, should you choose $\alpha = \pi/4$ or $\alpha = 5\pi/4$? Answer:

$$\textbf{Separately you want } R\cos\alpha = y(0) \textbf{ and } R\sin\alpha = y'(0)/\omega.$$

If those right hand sides are positive, choose the angle $\alpha$ between 0 and $\pi/2$.
If those right hand sides are negative, add $\pi$ and choose $\alpha = 5\pi/4$.

*Question*: If $y(0) > 0$ and $y'(0) < 0$, does $\alpha$ fall between $\pi/2$ and $\pi$ or between $3\pi/2$ and $2\pi$? If you plot the vector from $(0,0)$ to $(y(0), y'(0)/\omega)$, its angle is $\alpha$.

**15** Find a point on the sine curve in Figure 2.1 where $y > 0$ but $v = y' < 0$ and also $a = y'' < 0$. The curve is sloping down and bending down.

Find a point where $y < 0$ but $y' > 0$ and $y'' > 0$. The point is below the $x$-axis but the curve is sloping _____ and bending _____.

**16** (a) Solve $y'' + 100y = 0$ starting from $y(0) = 1$ and $y'(0) = 10$. (**This is $y_n$.**)

(b) Solve $y'' + 100y = \cos \omega t$ with $y(0) = 0$ and $y'(0) = 0$. (**This can be $y_p$.**)

**17** Find a particular solution $y_p = R\cos(\omega t - \alpha)$ to $y'' + 100y = \cos \omega t - \sin \omega t$.

**18** Simple harmonic motion also comes from a linear pendulum (like a grandfather clock). At time $t$, the height is $A \cos \omega t$. What is the frequency $\omega$ if the pendulum comes back to the start after 1 second? The period does not depend on the amplitude (a large clock or a small metronome or the movement in a watch can all have $T = 1$).

**19** If the phase lag is $\alpha$, what is the time lag in graphing $\cos(\omega t - \alpha)$?

**20** What is the response $y(t)$ to a delayed impulse if $my'' + ky = \delta(t - T)$?

**21** (Good challenge) Show that $y = \int_0^t g(t-s) f(s)\, ds$ has $my'' + ky = f(t)$.

**1** Why is $y' = \int_0^t g'(t-s) f(s)\, ds + g(0) f(t)$ ? Notice the two $t$'s in $y$.

**2** Using $g(0) = 0$, explain why $y'' = \int_0^t g''(t-s) f(s)\, ds + g'(0) f(t)$.

**3** Now use $g'(0) = 1/m$ and $mg'' + kg = 0$ to confirm $my'' + ky = f(t)$.

**22** With $f = 1$ (direct current has $\omega = 0$) verify that $my'' + ky = 1$ for this $y$:

$$\textbf{Step response} \quad y(t) = \int_0^t \frac{\sin \omega_n (t-s)}{m\omega_n}\, 1\, ds = y_p + y_n \text{ equals } \frac{\mathbf{1}}{\boldsymbol{k}} - \frac{\mathbf{1}}{\boldsymbol{k}} \cos \boldsymbol{\omega_n t}.$$

**23** (Recommended) For the equation $d^2y/dt^2 = 0$ find the null solution. Then for $d^2g/dt^2 = \delta(t)$ find the fundamental solution (start the null solution with $g(0) = 0$ and $g'(0) = 1$). For $y'' = f(t)$ find the particular solution using formula (16).

**24** For the equation $d^2y/dt^2 = e^{i\omega t}$ find a particular solution $y = Y(\omega)e^{i\omega t}$. Then $Y(\omega)$ is the frequency response. Note the "resonance" when $\omega = 0$ with the null solution $y_n = 1$.

**25** Find a particular solution $Ye^{i\omega t}$ to $my'' - ky = e^{i\omega t}$. The equation has $-ky$ instead of $ky$. What is the frequency response $Y(\omega)$? For which $\omega$ is $Y$ infinite?

## 2.2 Key Facts About Complex Numbers

The solutions to differential equations involve *real* numbers $\boldsymbol{a}$ and *imaginary* numbers $\boldsymbol{i\omega}$. They combine into *complex* numbers $\boldsymbol{s = a + i\omega}$ (real plus imaginary). Here are three equations and their solutions:

$$\frac{dy}{dt} = ay \qquad \frac{d^2y}{dt^2} + \omega^2 y = 0 \qquad \frac{d^2y}{dt^2} - 2a\,\frac{dy}{dt} + (\omega^2 + a^2)y = 0$$

$$y = \boldsymbol{Ce^{at}} \qquad y = \boldsymbol{c_1 e^{i\omega t} + c_2 e^{-i\omega t}} \qquad y = \boldsymbol{c_1 e^{(a+i\omega)t} + c_2 e^{(a-i\omega)t}}$$

Chapter 1 solved $y' = ay$. Section 2.1 solved $y'' + \omega^2 y = 0$. Section 2.3 will solve the last equation $Ay'' + By' + Cy = 0$. The balance between real and imaginary (between $a$ and $i\omega$) will come down to a competition between $B^2$ and $4AC$.

This course cannot go forward without complex numbers. You see their rectangular form in $s = a + i\omega$ (real part and imaginary part). What you must also see is their **polar form**. It is $e^{st}$, more than $s$ by itself, that demands to be seen in polar form:

$$\boldsymbol{e^{st} = e^{(a+i\omega)t} = e^{at}\, e^{i\omega t}}$$

$\boldsymbol{e^{at}}$ **gives growth or decay** $\quad \boldsymbol{e^{i\omega t}}$ **gives oscillation and rotation**

The real part $a$ is the rate of growth. The imaginary part $\omega$ is the frequency of oscillation. The addition $a + i\omega$ turns into the multiplication $e^{at}e^{i\omega t}$ because of the rule for exponentials. We will surely see exponentials everywhere, because they solve all constant coefficient equations: *The solution to* $y' = sy$ *is* $y = Ce^{st}$. With a forcing function $e^{i\omega t}$, a particular solution to $y' - sy = e^{i\omega t}$ is $y_p = e^{i\omega t}/(i\omega - s)$: a complex function.

Euler's formula $e^{i\omega t} = \cos \omega t + i \sin \omega t$ brings back two real functions (cosine and sine). Real equations have real solutions. When the forcing function on the right side is $f = A \cos \omega t + B \sin \omega t$, a good particular solution is $y_p = M \cos \omega t + N \sin \omega t$.

**In this real world, the amplitudes $\sqrt{A^2 + B^2}$ and $\sqrt{M^2 + N^2}$ are all-important.** The amplitude is what we see (in light) and hear (in sound) and feel (in vibration).

The null solutions $y_n$ and the particular solution $y_p$ need complex numbers. The form of $y_n$ is $Ce^{st}$. The form of $y_p$ is $Ye^{i\omega t}$. The complex gain is $Y$. Notice that the $\omega$ in $s = a + i\omega$ is the *natural frequency* in the null solution $y_n$. The $\omega$ in the right hand side $e^{i\omega t}$ is the *driving frequency* in the particular solution $y_p$.

If $\omega_{natural} = \omega_{driving}$, we will see "resonance" and we will need new formulas.

*Here is the plan for this section.*

**1** Multiply complex numbers $s_1$ and $s_2$ (review).

**2** Use the polar form $s = re^{i\theta}$ to find the powers $s^n = r^n e^{in\theta}$ (review).

**3** Look especially at the equation $s^n = 1$. It has $n$ roots, all on the **unit circle**.

**4** Find the **exponential** $\boldsymbol{e^{st}}$ and watch it move in the complex plane.

## Complex Numbers : Rectangular and Polar

A complex number $a+i\omega$ has a real part $a$ and an imaginary part $\omega$. Two complex numbers are easy to add: real part $a_1 + a_2$, imaginary part $\omega_1 + \omega_2$. It is multiplication that looks messy in equation (1). The good way is in equation (5).

**Multiplication** $\quad (a_1 + i\omega_1)(a_2 + i\omega_2) = (a_1a_2 - \omega_1\omega_2) + i(a_1\omega_2 + a_2\omega_1)$. (1)

Just multiply each part $a_1$ and $i\omega_1$ by each part $a_2$ and $i\omega_2$.

**Important case $s$ times $\overline{s}$** $\quad (a + i\omega)(a - i\omega) = a^2 + \omega^2$ : Real number. (2)

$\overline{s} = a - i\omega$ is the **complex conjugate** of $s = a + i\omega$. Equation (2) says that $\boldsymbol{s\overline{s} = |s|^2}$.

$\boldsymbol{|s| = \sqrt{a^2 + \omega^2}}$ is the *absolute value* or *magnitude* or *modulus* of $s = a + i\omega$.

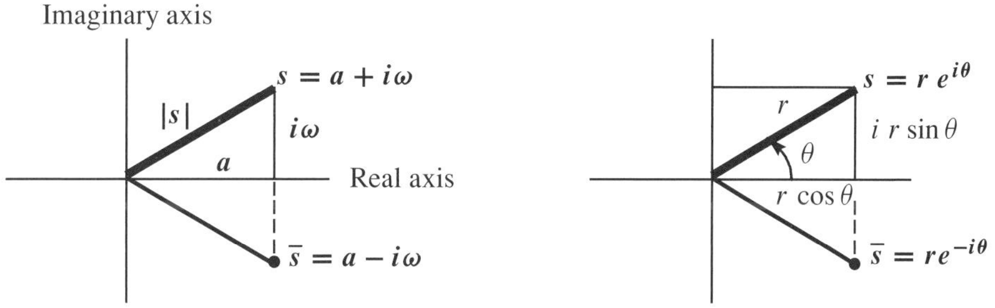


Figure 2.5: (i) The rectangular form $s = a + i\omega$. (ii) The polar form $s = re^{i\theta}$ with absolute value $r = |s| = \sqrt{a^2 + \omega^2}$. The complex conjugate of $s$ is $\overline{s} = a - i\omega = re^{-i\theta}$.

The polar form of $s$ uses that distance $r = |s|$ to the center point $(0,0)$. The real numbers $a$ and $\omega$ (rectangular) are connected to $r$ and $\theta$ (polar) by

$$\boxed{\boldsymbol{a = r\cos\theta} \qquad \boldsymbol{\omega = r\sin\theta} \qquad s = a + i\omega = r(\cos\theta + i\sin\theta) = \boldsymbol{re^{i\theta}}.} \quad (3)$$

At that moment you see Euler's Formula $e^{i\theta} = \cos\theta + i\sin\theta$. I could regard this as the complex *definition* of the exponential. Or I can separate the infinite series for $e^{i\theta}$ into its real part (the series for $\cos\theta$) and imaginary part (the series for $\sin\theta$).

Euler's Formula is used all the time, to express $e^{i\theta}$ in terms of $\cos\theta$ and $\sin\theta$. It is useful to go the other way, and express the cosine and sine in terms of $e^{i\theta}$ and $e^{-i\theta}$:

**Cosines from exponentials** $\quad \cos\theta = \dfrac{e^{i\theta} + e^{-i\theta}}{2} \qquad \sin\theta = \dfrac{e^{i\theta} - e^{-i\theta}}{2i}$ (4)

The sine comes from subtraction. Cancel $\cos\theta$ to get $2i\sin\theta$. We need to divide by $2i$.

## The Polar Form of $s^n$ and $1/s$

The polar form is perfect for multiplication and for powers $s^n$. We just multiply absolute values of $s_1$ and $s_2$, and *add* their angles. Multiply $r_1 r_2$ and add $\theta_1 + \theta_2$.

$$\textbf{Multiplication } s_1 s_2 \qquad \left(r_1\, e^{i\theta_1}\right)\left(r_2\, e^{i\theta_2}\right) = \boldsymbol{r_1 r_2\, e^{i(\theta_1+\theta_2)}} \tag{5}$$

$$\textbf{Powers of } s = re^{i\theta} \qquad s^n = \left(r\, e^{i\theta}\right)^n = \boldsymbol{r^n\, e^{in\theta}} \tag{6}$$

If $n = 2$, we are multiplying $re^{i\theta}$ times $re^{i\theta}$ to get $r^2e^{i2\theta}$. ($\theta$ is added to $\theta$.) If $n = -1$, we are dividing. The rectangular form of $1/(a + i\omega)$ matches the polar form of $1/(re^{i\theta})$ :

$$\frac{\mathbf{1}}{\boldsymbol{a+i\omega}} = \frac{1}{a+i\omega}\,\frac{a-i\omega}{a-i\omega} = \frac{\boldsymbol{a-i\omega}}{\boldsymbol{a^2+\omega^2}} \qquad \frac{\mathbf{1}}{\boldsymbol{re^{i\theta}}} = \frac{1}{r}\,\frac{1}{e^{i\theta}} = \frac{\mathbf{1}}{\boldsymbol{r}}\boldsymbol{e^{-i\theta}}. \tag{7}$$

That magnitude is $r = |a + i\omega| = \sqrt{a^2 + \omega^2}$. Equation (7) says that $1/s$ equals $\overline{s}/|s|^2$. In solving $y' - ay = e^{i\omega t}$, what we meet is $y = e^{i\omega t}/(i\omega - a)$ :

$$\textbf{Gain } \boldsymbol{G} \textbf{ and Phase } \boldsymbol{\alpha} \qquad i\omega - a = re^{i\alpha} \qquad \frac{1}{i\omega - a} = \frac{1}{r}e^{-i\alpha} = Ge^{-i\alpha} \tag{8}$$

I prefer this polar form. When $s = re^{i\theta}$, the absolute value of $1/s$ is $1/r$. The angle is $-\theta$.

**Examples** The polar form of $1 + i$ is $\sqrt{2}e^{i\pi/4}$ : absolute value $r = \sqrt{1+1} = \boldsymbol{\sqrt{2}}$.
The polar form of its conjugate $1 - i$ is $\boldsymbol{\sqrt{2}}e^{-\pi i/4}$.
The polar form of its reciprocal $1/(1 + i)$ is $(\mathbf{1}/\boldsymbol{\sqrt{2}})e^{-\pi i/4}$.

Notice that *we can add $2\pi$ to the angle $\theta$*. That brings us around a circle and back to the same point. Then $e^{i\theta} = e^{i(\theta+2\pi)}$ and $e^{-i\pi/4} = e^{7\pi i/4}$.

## The Unit Circle

The polar form brings out the importance of the unit circle in the complex plane. That circle contains all complex numbers with absolute value $r = |s| = 1$. The numbers on the unit circle are exactly $\boldsymbol{s = e^{i\theta} = \cos\theta + i\ \sin\theta}$.

Since $r = 1$, every $r^n$ is also 1. All powers like $s^2$ and $s^{-1}$ stay on the unit circle. The angles in Figure 2.6 become $2\theta$ and $-\theta$. The $n^{\text{th}}$ power $s^n$ has angle $n\theta$.

Here is a nice application of complex numbers to trigonometry. The "double angle" formulas for $\cos 2\theta$ and $\sin 2\theta$ are not so easy to remember. The "triple angle" formulas for $\cos 3\theta$ and $\sin 3\theta$ are even harder. But all these formulas come from one simple fact:

$$\boldsymbol{(e^{i\theta})^n = e^{in\theta}} \qquad (\cos\theta + i\ \sin\theta)^{\boldsymbol{n}} = \cos n\theta + i\ \sin n\theta. \tag{9}$$

If you take $n = 2$, you are squaring $e^{i\theta} = \cos\theta + i\ \sin\theta$ to get $e^{i2\theta}$ :

$$(\cos\theta + i\sin\theta)^2 = \cos^2\theta - \sin^2\theta + 2i\ \cos\theta\ \sin\theta = \cos 2\theta + i\ \sin 2\theta. \tag{10}$$

The real part $\cos^2\theta - \sin^2\theta$ is $\boldsymbol{\cos 2\theta}$. The imaginary part $2\sin\theta\cos\theta$ is $\boldsymbol{\sin 2\theta}$. For triple angles, multiply again by $\cos\theta + i\ \sin\theta$ (in Problem 4).

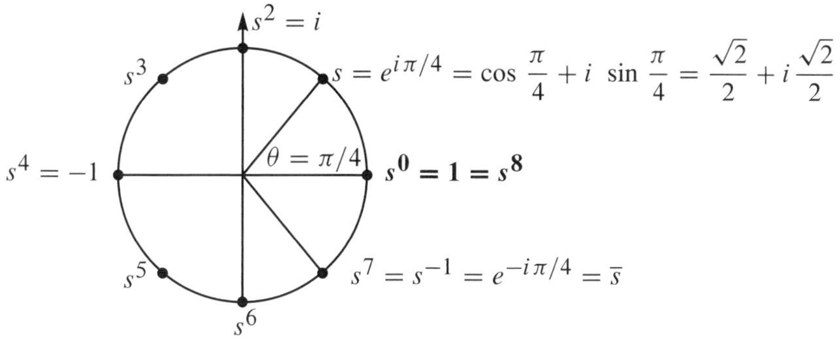


Figure 2.6: The number $s = e^{i\theta}$ has $s^2 = e^{i2\theta}$ and $s^{-1} = e^{-i\theta}$, all on the circle with $r = 1$. Here $\theta = 45°$ which is $\pi/4$ radians. So $2\theta = 90°$ and $\mathbf{s^2 = i}$. **Then** $\mathbf{s^8 = 1}$.

## The Equation $s^n = 1$

There are two numbers with $s^2 = 1$ (they are $s = 1$ and $-1$). There are four numbers with $s^4 = 1$ (they are 1 and $-1$ and $i$ and $-i$). Those four numbers are *equally spaced around the unit circle.* This is the pattern for every equation $s^n = 1$ : $n$ numbers equally spaced around the unit circle, starting with $s = 1$. The Fundamental Theorem of Algebra says that $n^{\text{th}}$ degree equations have $n$ (possibly complex) solutions. The equation $s^n = 1$ is no exception, and all its roots are on the unit circle.

$$\textbf{\textit{n} roots of } \mathbf{s^n = 1} \qquad s = e^{2\pi i/n}, s = e^{4\pi i/n}, \ldots, s = e^{2n\pi i/n} = e^{2\pi i} = 1.$$

These are the powers $s, s^2, \ldots, s^n$ of the special complex number $s = e^{2\pi i/n}$. This number $s = e^{2\pi i/8}$ is the first of the 8 solutions to $s^8 = 1$, going around the circle in Figure 2.6.

Here is a remarkable fact about the solutions to $s^n = 1$. **Those $n$ numbers add to zero.** In Figure 2.6, you can see that $s^5 = -s$ and $s^6 = -s^2$ and $s^7 = -s^3$ and $s^8 = -s^4$. The roots pair off. Each pair adds to zero. So the 8 roots add to zero.

For $n = 3$ or 5 or 7, this pairing off will not work. The three solutions to $s^3 = 1$ are at 120° angles. ($s$ and $s^2$ are $e^{2\pi i/3}$ and $e^{4\pi i/3}$, at angles 120° and 240°. Then comes 360°.) To show that those three numbers add to zero, I will factor $s^3 - 1 = 0$:

$$0 = s^3 - 1 = (s-1)(s^2 + s + 1) \quad \text{leads to} \quad \mathbf{s^2 + s + 1 = 0}. \tag{11}$$

The $n$ numbers on the unit circle go into the Fourier matrix. They are the key to the overwhelming success of the Fast Fourier Transform in Section 8.2.

## The Exponentials $e^{i\omega t}$ and $e^{ist}$

We use complex numbers to solve differential equations. For $dy/dt = ay$ the solution $y = Ce^{at}$ is real. But second order equations can bring oscillations $e^{i\omega t}$ together with growth/decay from $e^{at}$. Now $y$ has sines and cosines, or complex exponentials.

$$\boxed{y = c_1 e^{(a+i\omega)t} + c_2 e^{(a-i\omega)t} \quad \text{or} \quad y = C_1 e^{at}\cos\omega t + C_2 e^{at}\sin\omega t.} \tag{12}$$

Our goal is to follow those pieces of the complete solution to $Ay'' + By' + Cy = 0$. *Where does the point $e^{(a+i\omega)t}$ travel in the complex plane ?* The next section connects $a$ and $\omega$ to the numbers $A, B, C$ and solves the differential equation.

The best way to track the path of $e^{(a+i\omega)t}$ is to separate $a$ from $i\omega$. The path of $e^{i\omega t}$ is a circle. The factor $e^{at}$ turns the circle into a spiral.

**Rule for exponentials** $\qquad e^{(a+i\omega)t} = e^{at}e^{i\omega t}. \qquad (13)$

This is the polar form! The factor $e^{at}$ is the absolute value $r$. The angle $\omega t$ is the phase angle $\theta$. As the time $t$ increases, we follow those two parts:

| | | |
|---|---|---|
| **Absolute value** | $e^{at}$ grows with $t$ if $a > 0$ | $e^{at}$ decays if $a < 0$ |
| **Phase angle** | $e^{i\omega t}$ goes around the unit circle when $t$ increases by $2\pi/\omega$ | |

The real part $a$ decides stability. This is just like Chapter 1. We will see that damping produces $a < 0$ which is stability. In that case $B > 0$ in $y'' + By' + Cy = 0$.

This section is about the $i\omega$ part of the exponent $s$. That produces the $e^{i\omega t}$ part of the solution $y = e^{st}$. The pure oscillations in Section 2.1 came from $my'' + ky = 0$ with no damping. They had only this $e^{i\omega t}$ part (along with $e^{-i\omega t}$, which travels in the opposite direction around the unit circle). The frequency is $\omega = \sqrt{k/m}$.

Watch $e^{i\omega t}$ as it goes around the circle. If you follow its horizontal motion (its shadow on the $x$ axis) you will see $\cos\omega t$. If you follow its height on the $y$ axis, you will see $\sin\omega t$. The circle is complete when $\omega t = 2\pi$. So the period is $T = 2\pi/\omega$.

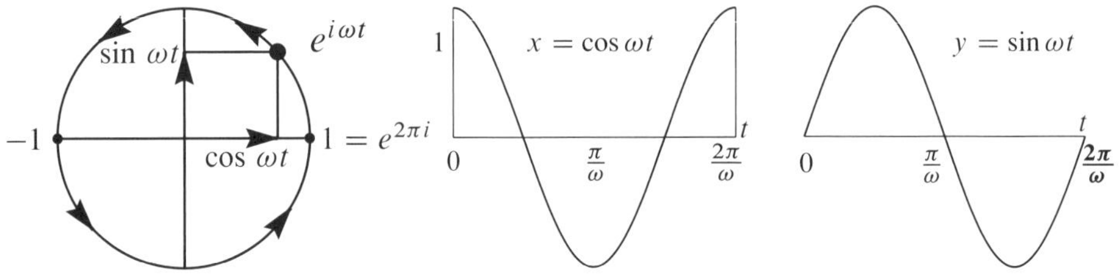


Figure 2.7: $y'' + \omega^2 y = 0$: One complex solution $e^{i\omega t}$ produces two real solutions.

When we multiply $e^{i\omega t}$ by $e^{at}$, **their product $e^{st}$ gives a spiral.** The spiral goes in to the center if $a$ is negative. The spiral goes outward $a > 0$. You are seeing the benefit of complex numbers, to merge oscillation and decay into one function. The real functions are $e^{at}\cos\omega t$ and $e^{at}\sin\omega t$. The complex function is $e^{at}e^{i\omega t} = e^{st}$.

*Question* What will be the time $T$ and the crossing point $X$, when the spiral completes one loop and returns to the positive $x$–axis ?

*Answer* The time $T$ will be $2\pi/\omega$, to complete each loop of the spiral. The crossing point on the $x$–axis will be $X = e^{aT}$. At time $2T$, the crossing will be at $X^2$.

## Problem Set 2.2

**1** Mark the numbers $s_1 = 2 + i$ and $s_2 = 1 - 2i$ as points in the complex plane. (The plane has a real axis and an imaginary axis.) Then mark the sum $s_1 + s_2$ and the difference $s_1 - s_2$.

**2** Multiply $s_1 = 2 + i$ times $s_2 = 1 - 2i$. Check absolute values: $|s_1||s_2| = |s_1 s_2|$.

**3** Find the real and imaginary parts of $1/(2 + i)$. Multiply by $(2 - i)/(2 - i)$:

$$\frac{1}{2+i}\,\frac{2-i}{2-i} = \frac{2-i}{|2+i|^2} = ?$$

**4** *Triple angles* Multiply equation (10) by another $e^{i\theta} = \cos\theta + i\sin\theta$ to find formulas for $\cos 3\theta$ and $\sin 3\theta$.

**5** *Addition formulas* Multiply $e^{i\theta} = \cos\theta + i\sin\theta$ times $e^{i\phi} = \cos\phi + i\sin\phi$ to get $e^{i(\theta+\phi)}$. Its real part is $\cos(\theta + \phi) = \cos\theta\cos\phi - \sin\theta\sin\phi$. What is its imaginary part $\sin(\theta + \phi)$?

**6** Find the real part and the imaginary part of each cube root of 1. Show directly that the three roots add to zero, as equation (11) predicts.

**7** The three cube roots of 1 are $z$ and $z^2$ and 1, when $z = e^{2\pi i/3}$. What are the three cube roots of 8 and the three cube roots of $i$? (The angle for $i$ is $90°$ or $\pi/2$, so the angle for one of its cube roots will be ____. The roots are spaced by $120°$.)

**8** (a) The number $i$ is equal to $e^{\pi i/2}$. Then its $i^{\text{th}}$ power $\boldsymbol{i^i}$ comes out equal to a real number, using the fact that $(e^s)^t = e^{st}$. What is that real number $\boldsymbol{i^i}$?

(b) $e^{i\pi/2}$ is also equal to $e^{5\pi i/2}$. Increasing the angle by $2\pi$ does not change $e^{i\theta}$ — it comes around a full circle and back to $i$. Then $i^i$ has another real value $(e^{5\pi i/2})^i = e^{-5\pi/2}$. What are all the possible values of $i^i$?

**9** The numbers $s = 3 + i$ and $\bar{s} = 3 - i$ are complex conjugates. Find their sum $s + \bar{s} = -B$ and their product $(s)(\bar{s}) = C$. Then show that $s^2 + Bs + C = 0$ and also $\bar{s}^2 + B\bar{s} + C = 0$. Those numbers $s$ and $\bar{s}$ are the two roots of the quadratic equation $x^2 + Bx + C = 0$.

**10** The numbers $s = a + i\omega$ and $\bar{s} = a - i\omega$ are complex conjugates. Find their sum $s + \bar{s} = -B$ and their product $(s)(\bar{s}) = C$. Then show that $s^2 + Bs + C = 0$. The two solutions of $x^2 + Bx + C = 0$ are $s$ and $\bar{s}$.

**11** (a) Find the numbers $(1 + i)^4$ and $(1 + i)^8$.

(b) Find the polar form $re^{i\theta}$ of $(1 + i\sqrt{3})/(\sqrt{3} + i)$.

**12** The number $z = e^{2\pi i/n}$ solves $z^n = 1$. The number $Z = e^{2\pi i/2n}$ solves $Z^{2n} = 1$. How is $z$ related to $Z$? (This plays a big part in the Fast Fourier Transform.)

**13** (a) If you know $e^{i\theta}$ and $e^{-i\theta}$, how can you find $\sin\theta$?

(b) Find all angles $\theta$ with $e^{i\theta} = -1$, and all angles $\phi$ with $e^{i\phi} = i$.

**14** Locate all these points on one complex plane:

(a) $2 + i$ (b) $(2+i)^2$ (c) $\dfrac{1}{2+i}$ (d) $|2+i|$

**15** Find the absolute values $r = |z|$ of these four numbers. If $\theta$ is the angle for $6 + 8i$, what are the angles for these four numbers?

(a) $6 - 8i$ (b) $(6-8i)^2$ (c) $\dfrac{1}{6-8i}$ (d) $8i + 6$

**16** What are the real and imaginary parts of $e^{a + i\pi}$ and $e^{a + i\omega}$?

**17** (a) If $|s| = 2$ and $|z| = 3$, what are the absolute values of $sz$ and $s/z$?

(b) Find upper and lower bounds in $L \le |s + z| \le U$. When does $|s + z| = U$?

**18** (a) Where is the product $(\sin\theta + i\cos\theta)(\cos\theta + i\sin\theta)$ in the complex plane?

(b) Find the absolute value $|S|$ and the polar angle $\phi$ for $S = \boldsymbol{\sin\theta + i\cos\theta}$.

This is my favorite problem, because $S$ combines $\cos\theta$ and $\sin\theta$ in a new way. To find $\phi$, you could plot $S$ or add angles in the multiplication of part $(a)$.

**19** Draw the spirals $e^{(1-i)t}$ and $e^{(2-2i)t}$. Do those follow the same curves? Do they go clockwise or anticlockwise? When the first one reaches the negative $x$-axis, what is the time $T$? What point has the second one reached at that time?

**20** The solution to $d^2y/dt^2 = -y$ is $y = \cos t$ if the initial conditions are $y(0) =$ _____ and $y'(0) =$ _____. The solution is $y = \sin t$ when $y(0) =$ _____ and $y'(0) =$ _____. Write each of those solutions in the form $c_1 e^{it} + c_2 e^{-it}$, to see that real solutions can come from complex $c_1$ and $c_2$.

**21** Suppose $y(t) = e^{-t} e^{it}$ solves $y'' + By' + Cy = 0$. What are $B$ and $C$? If this equation is solved by $y = e^{3it}$, what are $B$ and $C$?

**22** From the multiplication $e^{iA}\, e^{-iB} = e^{i(A-B)}$, find the "subtraction formulas" for $\cos(A - B)$ and $\sin(A - B)$.

**23** (a) If $r$ and $R$ are the absolute values of $s$ and $S$, show that $rR$ is the absolute value of $sS$. (Hint: Polar form!)

(b) If $\overline{s}$ and $\overline{S}$ are the complex conjugates of $s$ and $S$, show that $\overline{s}\,\overline{S}$ is the complex conjugate of $sS$. (Polar form!)

**24** Suppose a complex number $s$ solves a real equation $s^3 + As^2 + Bs + C = 0$ (with $A$, $B$, $C$ real). Why does the complex conjugate $\overline{s}$ also solve this equation? "*Complex solutions to real equations come in conjugate pairs $s$ and $\overline{s}$.*"

**25** (a) If two complex numbers add to $s + S = 6$ and multiply to $sS = 10$, what are $s$ and $S$? (They are complex conjugates.)

(b) If two numbers add to $s + S = 6$ and multiply to $sS = -16$, what are $s$ and $S$? (Now they are real.)

**26** If two numbers $s$ and $S$ add to $s + S = -B$ and multiply to $sS = C$, show that $s$ and $S$ solve the quadratic equation $s^2 + Bs + C = 0$.

**27** Find three solutions to $s^3 = -8i$ and plot the three points in the complex plane. What is the sum of the three solutions?

**28** (a) For which complex numbers $s = a + i\omega$ does $e^{st}$ approach 0 as $t \to \infty$? Those numbers $s$ fill which "half–plane" in the complex plane?

(b) For which complex numbers $s = a + i\omega$ does $s^n$ approach 0 as $n \to \infty$? Those numbers $s$ fill which part of the complex plane? Not a half-plane!

## 2.3 Constant Coefficients $A$, $B$, $C$

Section 2.1 presented the important equation $my'' + ky = 0$. That is a special case of this second order constant coefficient equation. We still need two initial conditions:

$$\boxed{A\frac{d^2y}{dt^2} + B\frac{dy}{dt} + Cy = 0} \quad \text{starting from } y(0) \text{ and } y'(0). \tag{1}$$

The coefficients $A, B, C$ can be *any constants*. For pure oscillation, $A$ was the mass $m$ and $C$ was the spring constant $k$, both positive. **$B > 0$ introduces damping**. In this section the numbers $A, B, C$ can be positive or negative or zero, so we may have exponential growth or decay or (damped) oscillation. With zero on the right hand side of equation (1), this section is finding null solutions $y_n$ : *unforced motion*.

Our first job is to solve equation (1). When the coefficients are constant, we always look for exponentials $e^{st}$. That number $s$ can be positive ($y$ will grow) or negative ($y$ decays) or pure imaginary ($y$ oscillates). If $s$ is a complex number $a + i\omega$, then its real part $a$ controls growth or decay. The imaginary part $\omega$ controls oscillation.

We will see the solutions clearly, because $A, B, C$ are constant. The right choice of $y(0)$ and $y'(0)$ will produce the growth factor $g(t)$ that multiplies all inputs to give $y_p$.

**The key step is to find the rate $s$ in $y = e^{st}$.** A second order equation normally has two possible rates $s_1$ and $s_2$. To find those numbers, substitute $y = e^{st}$ into equation (1):

$$As^2e^{st} + B\,se^{st} + Ce^{st} = 0. \tag{2}$$

The factor $e^{st}$ can be divided out because it is never zero. This leaves an all-important equation to determine $s$ :

**Characteristic equation** $\quad As^2 + Bs + C = 0. \quad (3)$

This is an ordinary quadratic equation for $s$. Every quadratic has two roots $s_1$ and $s_2$. They could be real, they could be complex, they could be equal. The two roots come from the quadratic formula:

**Two values for s**
$$s_1 = \frac{-B + \sqrt{B^2 - 4AC}}{2A} \qquad s_2 = \frac{-B - \sqrt{B^2 - 4AC}}{2A}. \tag{4}$$

Those roots add up to $s_1 + s_2 = -B/A$. The roots multiply to give $s_1 s_2 = C/A$. The question of real roots or complex roots is highly important, and it has a direct answer:

| Real roots $B^2 > 4AC$ | Equal roots $B^2 = 4AC$ | Complex roots $B^2 < 4AC$ |
|---|---|---|

When $B^2 - 4AC$ is positive, its square root is real. Then we have real roots $s_1 > s_2$. When $B^2 - 4AC = 0$, its square root is zero and $s_1 = s_2$ (borderline case: equal roots). When $B^2 - 4AC$ is negative, its square root is *imaginary*. The quadratic formula (4) produces two complex numbers $a + i\omega$ and $a - i\omega$ with the same real part $a = -B/2A$.

Let me look at all three cases, starting with examples.

## Two Real Roots, One Double Root, No Real Roots

A picture will show you how $B^2 - 4AC$ decides real vs. complex. The three parabolas in Figure 2.8 have $C = 0$ and $C = 1$ and $C = 2$. **By increasing $C$ we lift the parabolas.** The critical value is $C = 1$, when the middle parabola barely touches $y = 0$ at $s = 1$. $C = 1$ gives a double root and in this case $B^2 = 4AC = 4$.

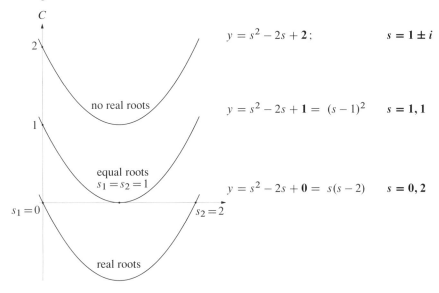

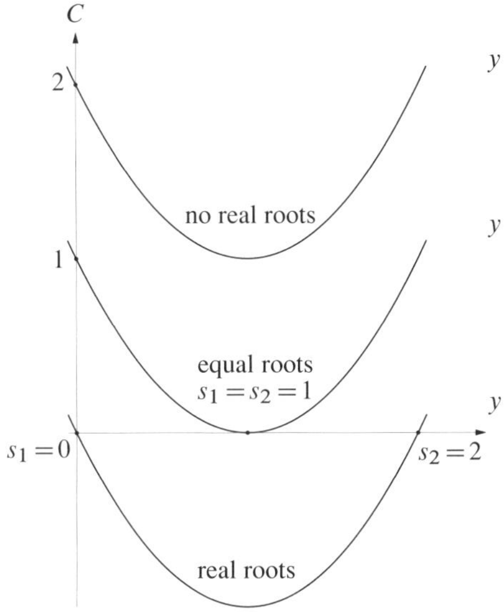


$y = s^2 - 2s + \mathbf{2}$; $\quad$ $\mathbf{s = 1 \pm i}$

$y = s^2 - 2s + \mathbf{1} = (s-1)^2$ $\quad$ $\mathbf{s = 1, 1}$

$y = s^2 - 2s + \mathbf{0} = s(s-2)$ $\quad$ $\mathbf{s = 0, 2}$

Figure 2.8: Lowest curve: Two roots for $C = 0$. Middle curve: Double root for $C = 1$. Highest curve misses the axis: No real roots for $C = 2 \rightarrow$ *complex roots* $a + i\omega$.

All three parabolas have $A = 1$ and $B = -2$ and $B^2 = 4$. The test that compares $B^2$ to $4AC$ is comparing 4 to $4C$. This shows again that $C = 1$ is at the critical borderline $B^2 = 4AC$. Any value $C > 1$ will lift the parabola above the $y = 0$ axis. The roots of $s^2 - 2s + C = 0$ will be complex, and $y'' - 2y' + Cy = 0$ will give damped oscillation.

For $C = 2$ that equation becomes $(s-1)^2 = -1$. Then $s - 1 = i$ or $s - 1 = -i$. The two complex roots are $s = 1 + i$ and $s = 1 - i$. The quadratic formula (4) agrees.

## Real Roots $s_1 > s_2$

**Example 1** $\quad y'' + 3y' + 2y = 0$ with $\boldsymbol{y = e^{st}}$ $\qquad$ Substitute $A, B, C = 1, 3, 2$ to find $s$.

$$As^2 + Bs + C = \mathbf{s^2 + 3s + 2} = 0 \text{ factors into } \mathbf{(s+1)(s+2)} = 0. \tag{5}$$

The roots are both negative: $\mathbf{s_1 = -1}$ and $\mathbf{s_2 = -2}$. Those numbers come from the quadratic formula (4) and they come faster from the factors in (5): The first factor $s + 1$ is zero when $s_1 = -1$, and $s + 2 = 0$ when $s_2 = -2$. **Damping $\rightarrow$ negative $s$ $\rightarrow$ stability.**

The complete solution to our linear differential equation is any combination of the two pure exponential solutions. These are null solutions (homogeneous solutions).

**Null solutions** $\quad y(t) = c_1 e^{s_1 t} + c_2 e^{s_2 t} = c_1 e^{-t} + c_2 e^{-2t} \quad (6)$

The numbers $c_1$ and $c_2$ are chosen to make $y(0)$ and $y'(0)$ correct when $t = 0$:

$$\textbf{Set } t = 0 \qquad y(0) = c_1 + c_2 \qquad \text{and} \qquad y'(0) = -c_1 - 2c_2. \qquad (7)$$

Those two equations safely determine $c_1 = 2y(0) + y'(0)$ and $c_2 = -y(0) - y'(0)$:

$$\textbf{Final solution} \quad y(t) = c_1 e^{-t} + c_2 e^{-2t} = y(0)(2e^{-t} - e^{-2t}) + y'(0)(e^{-t} - e^{-2t}).$$

**Example 2** Solve $y'' - 3y' + 2y = 0$. The coefficient $B$ has changed from 3 to $-3$.
*Solution* Substitute $y = e^{st}$ as before. *Negative damping gives positive* $s$.

$$s^2 - 3s + 2 = 0 \qquad (s-1)(s-2) = 0 \qquad s_1 = 2 \text{ and } s_2 = 1.$$

The complete solution is now $\boldsymbol{y(t) = c_1 e^{2t} + c_2 e^{t}}$. Exponential growth = instability.

## Equal Roots $s_1 = s_2$

The roots of $As^2 + Bs + C$ will be equal when $B^2 = 4AC$. When you factor the quadratic, you see $(s - s_1)^2$ times $A$. The factor $s - s_1$ appears twice: **$s = s_1$ is now a double root.**

Our $e^{st}$ method has a problem when it finds one double root $s = s_1$. After $y = e^{s_1 t}$, what is a *second solution* to our second order equation?

We will show that **$y = te^{s_1 t}$ is also a solution** when $s_2 = s_1$.

**Example 3** Solve $y'' - 2y' + y = 0$. Those coefficients $1, -2, 1$ have $B^2 = 4AC$.
*Solution* Substitute $y = e^{st}$ as usual. The root $s = 1$ is repeated: *two equal roots.*

$$s^2 - 2s + 1 = 0 \qquad (s-1)^2 = 0 \qquad \boldsymbol{s_1 = 1 = s_2}$$

With that root, $y = e^t$ solves the equation: easy to check. A second solution is needed! We now confirm that $\boldsymbol{y = te^{st} = te^{t}}$ is also a solution of $y'' - 2y' + y = 0$:

$$y' = (te^t)' = te^t + e^t \qquad y'' - 2y' + y = (te^t + 2e^t) - 2(te^t + e^t) + (te^t) = 0$$

**A double root of $As^2 + Bs + C = 0$ must be $s_1 = -B/2A$.**
**Then $y_1 = e^{s_1 t}$ and also $y_2 = te^{s_1 t}$ solve $Ay'' + By' + Cy = 0$.**

*Proof* With simple roots, the lowest parabola in Figure 2.8 cuts across $Y = 0$. The middle parabola $Y = (s-1)^2$ is *tangent* to the $Y = 0$ axis at the double root 1, 1. "*The graph touches twice at the same point* $s = s_1$." The root is $s_1 = s_2 = -B/2A$.

**Height zero**
**Slope zero** $\quad Y = As_1^2 + Bs_1 + C = 0$ and also $\dfrac{dY}{ds} = 2As_1 + B = 0. \quad (8)$

To confirm that $Ay'' + By' + Cy$ is zero for $y = te^{s_1 t}$, look at $y$ and $y'$ and $y''$:

$$\begin{aligned} \boldsymbol{y'} &= s_1 t e^{s_1 t} + e^{s_1 t} = \boldsymbol{s_1 y + e^{s_1 t}} \\ \boldsymbol{y''} &= s_1 y' + s_1 e^{s_1 t} = s_1(s_1 y + e^{s_1 t}) + s_1 e^{s_1 t} = \boldsymbol{s_1^2 y + 2 s_1 e^{s_1 t}} \end{aligned}$$

Substituting $y''$ and $y'$ and $y$ into $Ay'' + By' + Cy$, we get $0 + 0$ from equation (8):

$$A(s_1^2 y + 2s_1 e^{s_1 t}) + B(s_1 y + e^{s_1 t}) + Cy = (As_1^2 + Bs_1 + C)y + (2As_1 + B)e^{s_1 t} = 0 + 0.$$

The quadratic formula agrees with $s_1 = -B/2A = s_2$, because $B^2 - 4AC = 0$. The square root disappears, leaving $-B/2A$ for both solutions. Here is the simplest example of a double root $s_1 = s_2$ and a factor $t$ in the second solution.

**Example 4** **Solve $\boldsymbol{y'' = 0}$.** The coefficients $1, 0, 0$ have $B^2 = 4AC$.

*Solution* Substitute $y = e^{st}$ to find $s^2 e^{st} = 0$ and $\boldsymbol{s^2 = 0}$. *The double root is* $s = 0$. The usual solution $y = e^{st} = e^{0t} = 1$ does have $y'' = 0$. We need a second solution.

The rule $y = te^{st}$ still applies when $s = 0$. That second solution is $\boldsymbol{y = te^{0t} = t}$. We know this already: $y = 1$ *and* $y = t$ *solve* $y'' = 0$.

## Higher Order Equations

Problem 18 will extend these ideas to $n^{\text{th}}$ order equations (still constant coefficients!). Substitute $y = e^{st}$ to get an $n^{\text{th}}$ degree polynomial in $s$. *Now there are $n$ roots.* If those roots $s_1, s_2, \ldots, s_n$ are all different, they give $n$ independent solutions $y = e^{st}$. But if a root $s_1$ is repeated two or three or $m$ times, we need $m$ different solutions for $s = s_1$:

**Multiplicity $\boldsymbol{m}$** $\quad$ The $m$ solutions are $y = \boldsymbol{e^{s_1 t}}$, $y = \boldsymbol{t\, e^{s_1 t}}, \ldots,$ $y = \boldsymbol{t^{m-1}\, e^{s_1 t}}$. $\quad (9)$

A simple example would be the equation $y'''' = 0$. Substituting $y = e^{st}$ leads to $s^4 = 0$. This equation has four zero roots (multiplicity $m = 4$). The four solutions predicted by equation (9) are $y = 1, t, t^2, t^3$. No surprise that those all satisfy the equation $y'''' = 0$: their fourth derivatives are zero.

Here is a fourth order equation that produces two real roots and two complex roots:

$$\boldsymbol{y'''' - y = 0} \qquad y = e^{st} \text{ leads to } \boldsymbol{s^4 - 1 = 0} \qquad (10)$$

The four roots are $s_1 = \boldsymbol{1}$ and $s_2 = \boldsymbol{-1}$ and $s_3 = \boldsymbol{i}$ and $s_4 = \boldsymbol{-i}$. Then the complete solution to $y'''' = y$ is $y = c_1 e^t + c_2 e^{-t} + c_3 e^{it} + c_4 e^{-it}$.

### Complex Roots $s_1 = a + i\omega$ and $s_2 = a - i\omega$

The formula for the roots of a quadratic includes the square root of $B^2 - 4AC$. When that number is negative, the square root is *imaginary*. The example $y'' + y = 0$ has $A, B, C$ equal to $1, 0, 1$, so $B^2 - 4AC = -4$. The quadratic is $As^2 + Bs + C = \mathbf{s^2 + 1}$.

The solutions to $s^2 + 1 = 0$ are $s = i$ and $s = -i$. The solutions to $s^2 + 4 = 0$ are $s = 2i$ and $s = -2i$. The oscillations from $y'' + 4y = 0$ can be written in two ways:

$$\mathbf{B = 0: \text{No damping}} \qquad \boxed{y = c_1 e^{2it} + c_2 e^{-2it} = C_1 \cos 2t + C_2 \sin 2t.} \tag{11}$$

The real part of $s$ is zero when $B = 0$: pure oscillation.

**Now bring in damping**: $\mathbf{y'' + y' + y = 0}$. For the solutions to $s^2 + s + 1 = 0$, go to the quadratic formula: $A, B, C$ are $1, 1, 1$ and $B^2 - 4AC$ is $-3$:

$$\mathbf{s^2 + s + 1 = 0} \qquad s_1 = \frac{-1 + \sqrt{-3}}{2} = -\frac{\mathbf{1}}{\mathbf{2}} + \frac{\sqrt{\mathbf{3}}}{\mathbf{2}}\,i \qquad s_2 = -\frac{\mathbf{1}}{\mathbf{2}} - \frac{\sqrt{\mathbf{3}}}{\mathbf{2}}\,i.$$

The two complex roots $s_1$ and $s_2$ have the same real part $a = -1/2$. Their imaginary parts $\omega$ and $-\omega$ have opposite signs (as in $\sqrt{3}/2$ and $-\sqrt{3}/2$). Those are the plus and minus signs on the square root of $B^2 - 4AC$. Assuming that $A, B, C$ are real numbers, the two roots of $As^2 + Bs + C = 0$ are *complex conjugates*. If I place $s_1$ and $s_2$ onto the complex plane, they are symmetric mirror images across the real axis.

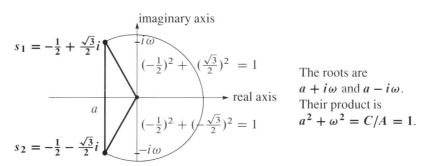

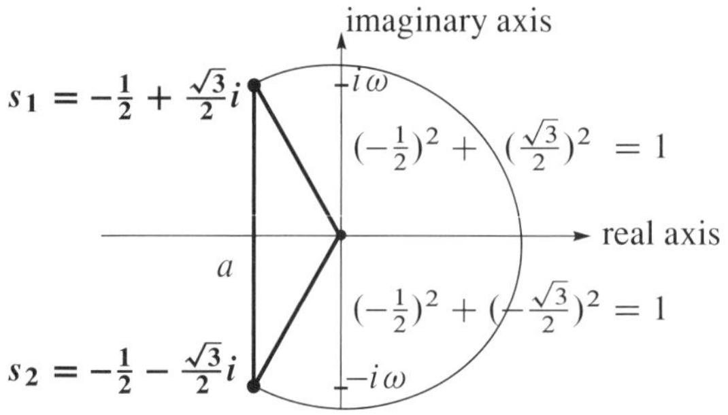


The roots are
$\boldsymbol{a + i\omega}$ and $\boldsymbol{a - i\omega}$.
Their product is
$\mathbf{a^2 + \omega^2 = C/A = 1}$.

The **conjugate** of $s = a + i\omega$ is $\bar{s} = a - i\omega$. The magnitude is $|s| = \sqrt{a^2 + \omega^2}$.

In the example with $a = -1/2$ and $\omega = \sqrt{3}/2$, the magnitude is exactly $|s| = 1$. This is because $(-1/2)^2 + (\sqrt{3}/2)^2 = 1$. The circle in the picture has radius 1. The unit circle is extremely important to recognize. The complex numbers on that circle have the form $s = \cos\theta + i\sin\theta$, because $(\text{cosine})^2 + (\text{sine})^2 = 1$. The angle $\theta$ is measured clockwise from the positive real axis. In the figure this angle is $120°$ or $\pi/3$.

**The points on the unit circle are given by Euler's Formula $\boldsymbol{e^{i\theta} = \cos\theta + i\sin\theta}$.**

We can switch between the complex form for $y(t)$ and its equivalent real form.

$$\textbf{Complex } y(t) = e^{at}(c_1 e^{i\omega t} + c_2 e^{-i\omega t}) \qquad \textbf{Real } y(t) = e^{at}(C_1 \cos\omega t + C_2 \sin\omega t)$$

Euler's formula for $e^{i\omega t}$ and $e^{-i\omega t}$ shows that $C_1 = c_1 + c_2$ and $C_2 = ic_1 - ic_2$.

With those key facts about complex numbers $a + i\omega$, we come back to the example $s^2 + s + 1 = 0$ and the differential equation it comes from:

$$\frac{d^2y}{dt^2} + \frac{dy}{dt} + y = 0 \qquad y_1 = e^{s_1 t} = e^{(a+i\omega)t} \qquad y_2 = e^{s_2 t} = e^{(a-i\omega)t}$$

This number $e^{(a+i\omega)t}$ is *not* on the unit circle. The real part $a = -1/2$ is responsible. When $a = 0$, $e^{i\omega t}$ goes around the circle. When $a < 0$, $e^{(a+i\omega)t}$ spirals to zero: **damped**.

The magnitude of $e^{i\omega t}$ is 1, but $e^{at}$ grows large or small depending on the sign of $a$:

| | | | |
|---|---|---|---|
| **Growth** | $a > 0$ | Magnitude | $\lvert e^{(a+i\omega)t}\rvert = \boldsymbol{e^{at}} \to \infty$ |
| **Decay** | $a < 0$ | Magnitude | $\lvert e^{(a+i\omega)t}\rvert = \boldsymbol{e^{at}} \to \mathbf{0}$ |

That real part is always $\boldsymbol{a = -B/2A}$. Every equation $Ay'' + By' + Cy = 0$ will have damping and decay if $A$ and $B$ are positive. Here is an example with $B = -1$:

**Negative damping → growth** $\quad y'' - y' + y = 0 \quad s^2 - s + 1 = 0$.

That changes $a$ to $+\frac{1}{2}$. The roots $a \pm i\omega$ are now coming from $s^2 - s + 1 = 0$:

$$s_1 = a + i\omega = +\frac{1}{2} + \frac{\sqrt{3}}{2}i \quad \text{has magnitude} \quad |s_1| = \sqrt{a^2 + \omega^2} = 1.$$

This point $s_1$ is on the unit circle, because $|s_1| = 1$. Its real part $a$ is $+\frac{1}{2}$, so $s_1$ is on the right side (not left side) of the imaginary axis. The angle in $s_1 = e^{i\theta}$ changes to $\theta = 60°$. Now $s_1$ and $s_2$ are on the *right half* of the unit circle (the unstable half: $e^{st}$ grows).

**"Anti-damping"** $B = -1$ $\quad$ **Growth rate** $a = \dfrac{1}{2}$ $\quad$ **Magnitude** $|e^{st}| = e^{at} = e^{t/2}$

In most physical problems we expect positive damping $B > 0$ and negative growth rate $a < 0$. Then the differential equation is stable and its null solutions die out as $t \to \infty$.

## Overdamping versus Underdamping

This section emphasizes the difference between $B^2 > 4AC$ and $B^2 < 4AC$. That is the difference between real roots and complex roots. This is a difference you can see—with your own eyes and not just with formulas. For damping coefficients $B = 1, 2, 3$ the solutions to $y'' + By' + y = 0$ will approach zero in different ways (Figure 2.9).

At this time I want to vary the damping $B$ instead of the stiffness $C$.

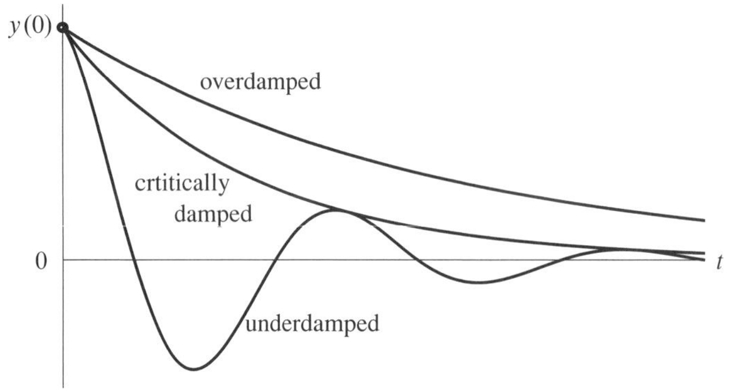


Figure 2.9: $y(t)$ goes directly to zero (overdamped) or it oscillates (underdamped).

The four damping possibilities match the four possibilities for roots of $As^2 + Bs + C = 0$. This table brings the whole section together:

| | | | |
|---|---|---|---|
| **Overdamping** | $B^2 > 4AC$ | **Real roots** | $e^{s_1 t}$ **and** $e^{s_2 t}$ |
| **Critical damping** | $B^2 = 4AC$ | **Double root** | $e^{s_1 t}$ **and** $te^{s_1 t}$ |
| **Underdamping** | $B^2 < 4AC$ | **Complex roots** | $e^{at}\cos\omega t,\ e^{at}\sin\omega t$ |
| **No damping** | $B = 0$ | **Imaginary roots** | $\cos\omega t$ **and** $\sin\omega t$ |

Figure 2.9 shows how *the graph crosses zero and comes back*, for underdamping. This is like a child's swing that is settling to zero (so the child can get off the swing). When $B = 0$ we have $a = 0$ and imaginary roots $\pm i\omega$ and pure spring–mass oscillation.

Figure 2.10 shows four parabolas all with $A = C = 1$. The damping coefficients are $B = 0, 1, 2, 3$. When $B = 3$ the damping is strong and $s^2 - 3s + 1 = 0$ has *real roots*. When $B = 2$ the damping is critical and $s^2 - 2s + 1 = 0$ has a *double root* $s = 1, 1$. When $B = 1$ the damping is weak and the roots are *complex*. The solutions $y = e^{at}\cos\omega t$ and $y = e^{at}\sin\omega t$ oscillate as the $e^{at}$ term goes to zero. When $B = 0$ there is no decay.

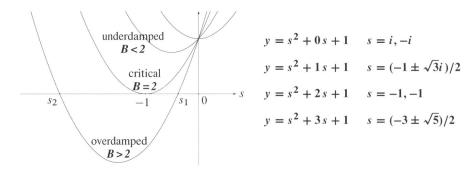

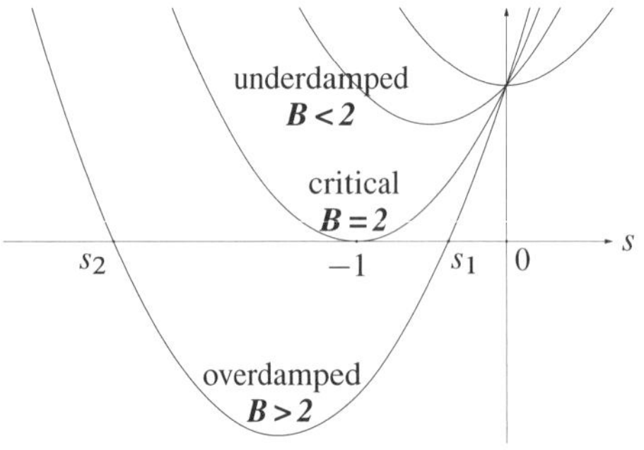


$y = s^2 + 0s + 1 \qquad s = i, -i$

$y = s^2 + 1s + 1 \qquad s = (-1 \pm \sqrt{3}i)/2$

$y = s^2 + 2s + 1 \qquad s = -1, -1$

$y = s^2 + 3s + 1 \qquad s = (-3 \pm \sqrt{5})/2$

Figure 2.10: As $B$ increases, the lowest point on the parabola moves left and down.

### Fundamental Solution = Growth Factor = Impulse Response

One special choice of initial conditions is all-important: $\boldsymbol{g(0) = 0}$ **and** $\boldsymbol{g'(0) = 1/A}$. The letter $g$ instead of $y$ picks out this fundamental solution. This is a null solution with the jump start $g'(0)$. It is also a particular solution to $Ag'' + Bg' + Cg = \delta(t)$. This fundamental solution from the delta function will lead us to *all* solutions.

Review: The roots of $As^2 + Bs + C = 0$ are $s_1$ and $s_2$. They give two solutions $e^{s_1 t}$ and $e^{s_2 t}$ to the null equation, if $s_1 \neq s_2$. We want the combination $g = c_1 e^{s_1 t} + c_2 e^{s_2 t}$ that matches $g(0) = 0$ and $g'(0) = 1/A$. Choose the right $c_1$ and $c_2$:

| | | |
|---|---|---|
| $g(0) = \quad c_1 + \quad c_2 = 0$ | *Multiply by* $s_2$ | $s_2 c_1 + s_2 c_2 = 0$ |
| $\boldsymbol{g'(0) = s_1 c_1 + s_2 c_2 = 1/A}$ | *Then subtract* | $\boldsymbol{(s_1 - s_2) c_1 = 1/A}$ |

**The fundamental solution** $$\boldsymbol{g(t) = \frac{e^{s_1 t} - e^{s_2 t}}{A(s_1 - s_2)}} \quad \textbf{has } \boldsymbol{c_1 = \frac{1}{A(s_1 - s_2)} = -c_2} \qquad (12)$$

**No damping** For the oscillation equation $my'' + ky = 0$, the roots of $ms^2 + k = 0$ are imaginary: $s_1 = i\sqrt{k/m} = i\omega$ and $s_2 = -i\sqrt{k/m} = -i\omega$. Then the fundamental solution has a simple form with $A = m$:

$$\boldsymbol{g(t)} = \frac{e^{s_1 t} - e^{s_2 t}}{m(s_1 - s_2)} = \frac{e^{i\omega t} - e^{-i\omega t}}{m(2i\omega)} = \frac{2i \sin \omega t}{2i\, m\omega} = \boldsymbol{\frac{\sin \omega t}{A\omega}}. \qquad (13)$$

This is exactly the impulse response from Section 2.1. Clearly $g(0) = 0$ and $g'(0) = 1/A$.

**Underdamping** Now $s_1 = a + i\omega$ and $s_2 = a - i\omega$. There is decay from $a = -B/2A$ and oscillation from $\omega$. Soon we will write $p$ for $B/2A$ and $\omega_d$ for $\omega$.

$$\boldsymbol{g(t)} = \frac{e^{(a+i\omega)t} - e^{(a-i\omega)t}}{A(2i\omega)} = \boldsymbol{e^{at} \frac{\sin \omega t}{A\omega}} = \boldsymbol{e^{-pt} \frac{\sin \omega_d t}{A\omega_d}}. \qquad (14)$$

**Critical damping** Now $B^2 = 4AC$ and the roots are equal: $\boldsymbol{s_1 = s_2 = -B/2A}$. The second solution to the differential equation (after $e^{s_1 t}$) is $g(t) = te^{s_1 t}$. Dividing by $A$, this is exactly the solution that has $g(0) = 0$ and $g'(0) = 1/A$.

$$\boldsymbol{g(t)} = \frac{te^{s_1 t}}{A} = \boldsymbol{\frac{t\, e^{-Bt/2A}}{A}}. \qquad (15)$$

**Overdamping** When $B^2 > 4AC$, the roots $s_1$ and $s_2$ are real. Formula (12) is best.

The real purpose of $g(t)$ is to solve $Ay'' + By' + Cy = f(t)$ with any right side $f(t)$. This impulse response $g$ is the fundamental solution that gives all other solutions:

**Solution for any** $\boldsymbol{f(t)}$ $$\boldsymbol{y_p(t) = \int_0^t g(t-s) f(s)\, ds} \qquad (16)$$

*The* **step response** *to* $f(t) = 1$ *is* $y_p =$ *integral of* $g(t)$. This comes in Section 2.5.

## Delta Function and Impulse Response

In this section $g(t)$ is a **null solution** with initial velocity $g'(0) = 1/A$. The same $g(t)$ is a **particular solution** in the next section, with initial velocity zero but driven by an impulse $f(t) = \delta(t)$. Only a delta function could make this possible: $g(t)$ is $y_n$ for one problem and $y_p$ for another problem.

The informal explanation is to integrate all terms in $Ag'' + Bg' + Cg = \delta(t)$. On the right side the integral is 1. The integration is over a *very short interval* 0 to $\Delta$. On the left side the integral of $Ag''$ is $Ag'(\Delta)$, plus terms of order $\Delta$ going to 0. To match 1 on the right side, the impulse response $g(t)$ starts immediately with $g' = 1/A$.

**Example 5** **The best example is $g''(t) = \delta(t)$ with ramp function $g(t) = t$.**

The derivative of the ramp is a step function. You see the sudden jump to $g' = 1$. The ramp $g(t) = t$ agrees with formula (15) in this case with $A = 1$ and $B = C = 0$. The null equation $g'' = 0$ starting from $g(0) = 0$ and $g'(0) = 1$ is solved by $g(t) = t$. Everything is zero for $t < 0$. Then we see the ramp $g(t)$ and the step $g'(t)$ and $g'' = \delta(t)$. This is the limiting case of equation (12) when $B$ and $C$ and $s_1$ and $s_2$ approach zero.

*A personal note* Thank you for accepting the slightly illegal input $\delta(t)$ and its response $g(t)$. I could have left those out of the book. But I couldn't have lived with myself. They are truly the key to theory and applications.

## Shift Invariance from Constant Coefficients

For a constant coefficient equation, the growth from time $s$ to time $t$ is exactly equal to the growth from 0 to $t - s$. The problem is **shift invariant**. We can start the time interval anywhere we want. For all intervals of the same length, we will see the same growth factor $g(t - s)$. This is the growth of input

**Inputs $f(s)$ at times $s$** **Total output** $$y(t) = \int_0^t g(t-s)\, f(s)\, ds. \qquad (17)$$

This is exactly like the main formula $y(t) = \int e^{a(t-s)} q(s)\, ds$ in Chapter 1. There the growth factor was $g(t) = e^{at}$. The equation $dy/dt - ay = q(t)$ had *constant* $a$.

Shift invariance is lost if any of the coefficients $A, B, C$ change with time. The growth factor becomes $g(s,t)$, depending on the specific start $s$ and end $t$ (*not just on the elapsed time* $t - s$). In this harder case the solution is $y(t) = \int g(s,t)\, f(s)\, ds$.

For a first order equation, Section 1.6 found $g(s,t)$. But second order equations with time-varying coefficients are usually impossible to solve with familiar functions. We often have no formula for $g(s,t)$—the response at time $t$ to an impulse at time $s$. *Shift invariance* (constant coefficients) is the key to successful solution formulas.

## Better Formulas for $s_1$ and $s_2$

The solutions to $As^2 + Bs + C = 0$ are $s_1$ and $s_2$. The formula for those two roots involves $B^2 - 4AC$. We have seen that $B^2 > 4AC$ is very different from $B^2 < 4AC$. Overdamping leads to real roots, underdamping leads to complex roots and oscillations. The formulas are so important that the whole world of science and engineering has tried to make them simpler.

Here is the natural way to start. Assign letters to the ratios $B/2A$ and $C/A$. We know $C/A$ as $\omega_n^2$. This is $k/m$ in mechanics. It gives the "natural frequency" with no damping. *For the ratio $B/2A$ I will use the letter $p$.* The main point is to simplify $s_1$ and $s_2$:

$$s_1, s_2 = \frac{-B \pm \sqrt{B^2 - 4AC}}{2A} = -p \pm \sqrt{p^2 - \omega_n^2} \tag{18}$$

A big improvement! Two symbols instead of three, which makes sense because we can divide $As^2 + Bs + C = 0$ by $A$. By introducing $p = B/2A$ we remove the 2 and the 4 in equation (18).

The comparison of $B^2$ to $4AC$ is now the comparison of $p^2$ to $\omega_n^2$. When $p^2 > \omega_n^2$, the roots are real (overdamping). When $p^2 - \omega_n^2$ is negative, $s_1$ and $s_2$ will be complex. **We have oscillation at a damped frequency $\omega_d$, lower than the natural frequency $\omega_n$:**

$$\omega_d^2 = \omega_n^2 - p^2 \qquad s_1 \text{ and } s_2 = -p \pm i\sqrt{\omega_n^2 - p^2} = -p \pm i\omega_d \tag{19}$$

## The Damping Ratio $Z$

The presentation could stop there. We see that the ratio of $p$ to $\omega_n$ is highly important. This fact suggests one final step, that we take now: **$Z = p/\omega_n$ is the damping ratio $Z$.** In engineering this ratio is called zeta (the Greek letter is $\zeta$). To make it easier to write, allow me to use $Z$ (capital zeta in Greek = capital $Z$ in Roman.) *Then we can replace $p$ by $Z\omega_n$.* Now the formula $s = -p \pm i\omega_d$ uses $\omega_n$ and $Z$:

$$\textbf{Damping ratio } Z = \frac{p}{\omega_n} \qquad s = -Z\omega_n \pm i\omega_d = -Z\omega_n \pm i\omega_n\sqrt{1 - Z^2} \tag{20}$$

The damped $\omega_d^2$ is $\omega_n^2 - p^2 = \omega_n^2(1 - Z^2)$. Its square root $\omega_d$ is the damped frequency. The null solutions are $y_n(t) = e^{-Z\omega_n t}(c_1 \cos \omega_d t + c_2 \sin \omega_d t)$.

Underdamping is $Z < 1$, critical damping is $Z = 1$, and overdamping is $Z > 1$. The key points become clear because *this ratio $Z$ is dimensionless*:

$$\textbf{Damping ratio} \quad Z = \frac{p}{\omega_n} = \frac{B/2A}{\sqrt{C/A}} = \frac{\mathbf{B}}{\sqrt{\mathbf{4AC}}} = \frac{b}{\sqrt{4mk}}. \tag{21}$$

If time is measured in minutes instead of seconds, the numbers $A, B, C$ are changed by $60^2$ and 60 and 1. **The ratio of $B$ to $\sqrt{4AC}$ is not changed**: a factor of 60 for both. This confirms that $B^2 - 4AC$ is a suitable quantity to appear in the quadratic formula, because $B^2$ and $4AC$ have the same units.

One last point is a good approximation when $Z$ is small. The square root of $1 - Z^2$ is close to $1 - \frac{1}{2}Z^2$. This comes from calculus (linear approximation using the tangent line). The good way to confirm it is to square both sides. Then $Z^4/4$ is very small.

$$\sqrt{\mathbf{1 - Z^2}} \approx \mathbf{1 - \frac{1}{2}Z^2} \quad \text{becomes} \quad 1 - Z^2 \approx 1 - Z^2 + \frac{1}{4}Z^4. \tag{22}$$

The good measure of damping is the **ratio** $Z = B/\sqrt{4AC}$. This key dimensionless number decides everything:

| | |
|---|---|
| $Z > 1$ | $B^2 > 4AC$ and real roots: *Overdamping and no oscillation.* |
| $Z < 1$ | $B^2 < 4AC$ and complex roots: *Underdamping and slow oscillation.* |
| $Z = 1$ | $B^2 = 4AC$ and a double root $-B/2A$: *critical damping.* |

Here is a curious fact. For very large $B$, the roots are approximately $s_1 = -1/B$ and $s_2 = -B$. That root $s_2$ gives fast decay. But the actual decay of $y(t)$ is controlled by $s_1$, which approaches zero! So increasing $B$ actually slows down this dominant decay mode.

Note that many authors refer to $s_1$ and $s_2$ as **poles**. They are poles of the transfer function $Y(s) = 1/(As^2 + Bs + C)$, where $Y$ becomes $1/0$. We will come back to transfer functions! Some authors emphasize **time constants** rather than exponents. The exponential $e^{-pt}$ has time constant $\tau = 1/p$. In that time $\tau$, $e^{-pt}$ decays by a factor $e$.

## ■ REVIEW OF THE KEY IDEAS ■

1. The equation $Ay'' + By' + Cy = 0$ is solved by $y = e^{st}$ when $As^2 + Bs + C = 0$.

2. The roots $s_1, s_2$ are *real* if $B^2 > 4AC$, *equal* if $B^2 = 4AC$, *complex* if $B^2 < 4AC$.

3. Negative real roots give stability and overdamping: $y(t) = \boldsymbol{c_1 e^{s_1 t} + c_2 e^{s_2 t}} \to 0$.

4. Equal roots $s = -B/2A$ when $B^2 = 4AC$. Change the second solution to $\boldsymbol{y_2 = t\, e^{st}}$.

5. Complex roots $a \pm i\omega$ give underdamped oscillations: $e^{at}(C_1 \cos \omega t + C_2 \sin \omega t)$.

6. The initial values $g(0) = 0$ and $g'(0) = 1/A$ give $g(t) = \boldsymbol{\left(e^{s_1 t} - e^{s_2 t}\right)/A(s_1 - s_2)}$. The same $g(t)$ solves $Ag'' + Bg' + Cg = \delta(t)$. This is the fundamental solution.

7. $s_1$ and $s_2$ become $\boldsymbol{-p \pm i\omega_d}$ with $p = B/2A$ and $\omega_d^2 = \omega_n^2 - p^2$. With damping ratio $Z = B/\sqrt{4AC} < 1$, those complex $s_1$ and $s_2$ are $\boldsymbol{-Z\omega_n \pm i\omega_n\sqrt{1 - Z^2}}$.

# Problem Set 2.3

**1** Substitute $y = e^{st}$ and solve the characteristic equation for $s$:

(a) $2y'' + 8y' + 6y = 0$ (b) $y'''' - 2y'' + y = 0$.

**2** Substitute $y = e^{st}$ and solve the characteristic equation for $s = a + i\omega$:

(a) $y'' + 2y' + 5y = 0$ (b) $y'''' + 2y'' + y = 0$

**3** Which second order equation is solved by $y = c_1 e^{-2t} + c_2 e^{-4t}$? Or $y = te^{5t}$?

**4** Which second order equation has solutions $y = c_1 e^{-2t} \cos 3t + c_2 e^{-2t} \sin 3t$?

**5** Which numbers $B$ give (under) (critical) (over) damping in $4y'' + By' + 16y = 0$?

**6** If you want oscillation from $my'' + by' + ky = 0$, then $b$ must stay below ____ .

**Problems 7–16 are about the equation $As^2 + Bs + C = 0$ and the roots $s_1, s_2$.**

**7** The roots $s_1$ and $s_2$ satisfy $s_1 + s_2 = -2p = -B/2A$ and $s_1 s_2 = \omega_n^2 = C/A$. Show this two ways:

(a) Start from $As^2 + Bs + C = A(s - s_1)(s - s_2)$. Multiply to see $s_1 s_2$ and $s_1 + s_2$.

(b) Start from $s_1 = -p + i\omega_d$, $s_2 = -p - i\omega_d$

**8** Find $s$ and $y$ at the bottom point of the graph of $y = As^2 + Bs + C$. At that minimum point $s = s_{\min}$ and $y = y_{\min}$, the slope is $dy/ds = 0$.

**9** The parabolas in Figure 2.10 show how the graph of $y = As^2 + Bs + C$ is raised by increasing $B$. Using Problem 8, show that the bottom point of the graph moves left (change in $s_{\min}$) and down (change in $y_{\min}$) when $B$ is increased by $\Delta B$.

**10** (recommended) Draw a picture to show the paths of $s_1$ and $s_2$ when $s^2 + Bs + 1 = 0$ and the damping increases from $B = 0$ to $B = \infty$. At $B = 0$, the roots are on the ____ axis. As $B$ increases, the roots travel on a circle (why?). At $B = 2$, the roots meet on the real axis. For $B > 2$ the roots separate to approach 0 and $-\infty$. *Why is their product $s_1 s_2$ always equal to* 1?

**11** (this too if possible) Draw the paths of $s_1$ and $s_2$ when $s^2 + 2s + k = 0$ and the stiffness increases from $k = 0$ to $k = \infty$. When $k = 0$, the roots are ____ . At $k = 1$, the roots meet at $s =$ ____. For $k \to \infty$ the two roots travel up/down on a ____ in the complex plane. *Why is their sum $s_1 + s_2$ always equal to* $-2$?

**12** If a polynomial $P(s)$ has a double root at $s = s_1$, then $(s - s_1)$ is a double factor and $P(s) = (s - s_1)^2 Q(s)$. Certainly $P = 0$ at $s = s_1$. Show that also $dP/ds = 0$ at $s = s_1$. Use the product rule to find $dP/ds$.

**13** Show that $y'' = 2ay' - (a^2 + \omega^2)y$ leads to $s = a \pm i\omega$. Solve $y'' - 2y' + 10y = 0$.

**14** The undamped *natural frequency* is $\omega_n = \sqrt{k/m}$. The two roots of $ms^2 + k = 0$ are $s = \pm i\omega_n$ (pure imaginary). With $p = b/2m$, the roots of $ms^2 + bs + k = 0$ are $\boldsymbol{s_1, s_2 = -p \pm \sqrt{p^2 - \omega_n^2}}$. The coefficient $p = b/2m$ has the units of $1/\text{time}$. Solve $s^2 + 0.1s + 1 = 0$ and $s^2 + 10s + 1 = 0$ with numbers correct to two decimals.

**15** With large overdamping $\boldsymbol{p >> \omega_n}$, the square root $\sqrt{p^2 - \omega_n^2}$ is close to $p - \omega_n^2/2p$. Show that the roots of $ms^2 + bs + k$ are $s_1 \approx \boldsymbol{-\omega_n^2/2p}$ = (small) and $s_2 \approx -2p = \boldsymbol{-b/m}$ (large).

**16** With small underdamping $\boldsymbol{p << \omega_n}$, the square root of $p^2 - \omega_n^2$ is approximately $i\omega_n - ip^2/2\omega_n$. Square that to come close to $p^2 - \omega_n^2$. Then the frequency for small underdamping is reduced to $\omega_d \approx \omega_n - p^2/2\omega_n$.

**17** Here is an 8th order equation with eight choices for solutions $y = e^{st}$:

$$\frac{d^8 y}{dt^8} = y \quad \text{becomes} \quad s^8 e^{st} = e^{st} \quad \text{and} \quad \boldsymbol{s^8 = 1}: \quad \text{Eight roots in Figure 2.6.}$$

Find two solutions $e^{st}$ that don't oscillate ($s$ is real). Find two solutions that only oscillate ($s$ is imaginary). Find two that spiral in to zero and two that spiral out.

**18** $A_n \dfrac{d^n y}{dt^n} + \cdots + A_1 \dfrac{dy}{dt} + A_0 y = 0$ leads to $\boldsymbol{A_n s^n + \cdots + A_1 s + A_0 = 0}$.

The $n$ roots $s_1, \ldots, s_n$ produce $n$ solutions $y(t) = e^{st}$ (if those roots are distinct). Write down $n$ equations for the constants $c_1$ to $c_n$ in $y = c_1 e^{s_1 t} + \cdots + c_n e^{s_n t}$ by matching the $n$ initial conditions for $y(0)$, $y'(0)$, ..., $D^{n-1} y(0)$.

**19** **Find two solutions to $\boldsymbol{d^{2015} y/dt^{2015} = dy/dt}$.** Describe all solutions to $s^{2015} = s$.

**20** The solution to $y'' = 1$ starting from $y(0) = y'(0) = 0$ is $y(t) = t^2/2$. The fundamental solution to $g'' = \delta(t)$ is $g(t) = t$ by Example 5. Does the integral $\int g(t - s) f(s) ds = \int (t - s) ds$ from 0 to $t$ give the correct solution $y = t^2/2$?

**21** The solution to $y'' + y = 1$ starting from $y(0) = y'(0) = 0$ is $y = 1 - \cos t$. The solution to $g'' + g = \delta(t)$ is $\boldsymbol{g(t) = \sin t}$ by equation (13) with $\omega = 1$ and $A = 1$. Show that $1 - \cos t$ agrees with the integral $\int g(t - s) f(s) ds = \int \sin(t - s) ds$.

**22** The step function $H(t) = 1$ for $t \geq 0$ is the integral of the delta function. **So the step response $\boldsymbol{r(t)}$ is the integral of the impulse response.** This fact must also come from our basic solution formula:

$$Ar'' + Br' + Cr = 1 \quad \text{with} \quad r(0) = r'(0) = 0 \quad \text{has} \quad \boldsymbol{r(t) = \int_0^t g(t - s)\, 1\, ds}$$

Change $t - s$ to $\tau$ and change $ds$ to $-d\tau$ to confirm that $r(t) = \int_0^t g(\tau) d\tau$.

Section 2.5 will find two good formulas for the step response $r(t)$.

## 2.4 Forced Oscillations and Exponential Response

The equation $Ay'' + By' + Cy = 0$ has no forcing term. Its right side is zero. This equation is *homogeneous*. The null solution $y_n(t) = c_1 e^{s_1 t} + c_2 e^{s_2 t}$ is controlled by the initial conditions $y(0)$ and $y'(0)$. If those are zero, the system never moves.

The equation $Ay'' + By' + Cy = f(t)$ is **forced** or **driven** by that new term $f(t)$. Previously $y = 0$ was a possible solution. Now we can expect a particular solution $y_p$.

This section is about driving forces $f = e^{st}$ and $e^{i\omega t}$ and $\cos \omega t$ and $\sin \omega t$. For $f = e^{st}$, the next example will show you how to find $y_p$.

### Exponential Driving Force

In this example, one particular solution $y_p(t) = Ye^{st}$ is a multiple of the input $e^{4t}$. All we have to do is find that number $Y$, by substituting into the differential equation.

**Example 1** Solve $\boldsymbol{y'' + 5y' + 6y = e^{4t}}$. One particular solution will be $\boldsymbol{y_p = Ye^{4t}}$.

When $Ye^{4t}$ is substituted into the equation, all terms contain $e^{4t}$ :

$$y'' + 5y' + 6y = 16Ye^{4t} + 20Ye^{4t} + 6Ye^{4t} = e^{4t}. \tag{1}$$

The left side is $42\,Ye^{4t}$. This matches the right side $e^{4t}$ when $Y = 1/42$:

$$\textbf{Particular } y_p \qquad 42\,Ye^{4t} = e^{4t} \text{ gives } 42\,Y = 1 \qquad y_p(t) = e^{4t}/42 \tag{2}$$

The complete solution has the form $y = y_p + y_n$. There are two arbitrary constants $c_1$ and $c_2$ in the solution $y_n(t)$ to the homogeneous equation (the null equation with forcing term = zero). Look for the two exponents $s_1$ and $s_2$ that solve the quadratic equation $As^2 + Bs + C = 0$. We know how to find the null solution $y_n$.

**Substitute $y = e^{st}$ into $y'' + 5y' + 6y = 0$. Cancel $e^{st}$ to find $s^2 + 5s + 6 = 0$.**

That quadratic factors into $(s + 2)(s + 3)$. This is zero for $\boldsymbol{s = -2}$ and $\boldsymbol{s = -3}$. Those roots of the "characteristic equation" are the exponents in the null solution $y_n(t)$. This is the homogeneous solution = complementary solution = **transient solution**, which decays to zero at $t = \infty$ when there is damping.

$$\textbf{Null solution} \qquad y_n(t) = c_1 e^{-2t} + c_2 e^{-3t}.$$

The final step is to choose $c_1$ and $c_2$ so that $y = y_p + y_n = \frac{1}{42}e^{4t} + y_n$ satisfies the initial conditions. This will complete Example 1, by getting it right at $t = 0$.

$$\textbf{Initial position} \qquad y(0) = \frac{1}{42} + c_1 + c_2$$

$$\textbf{Initial velocity} \qquad y'(0) = \frac{4}{42} - 2c_1 - 3c_2$$

Those two equations tell us the correct values $c_1$ and $c_2$, when $y(0)$ and $y'(0)$ are given.

## Exponential Response Formula

We can turn that example into a formula for $Y$ that almost always succeeds. Put $y = Ye^{st}$ into the equation. Each derivative multiplies $y$ by $s$. So $Ay'' + By' + Cy$ will multiply $y = Ye^{st}$ by the number $As^2 + Bs + C$. Divide by that number to see $Y$ :

$$\boldsymbol{Ay'' + By' + Cy = e^{st}} \quad \textbf{is solved by} \quad y = Ye^{st} = \frac{1}{As^2 + Bs + C}\, e^{st} \tag{3}$$

That fraction $Y$ is called the **transfer function**. It 'transfers' the exponential input $e^{st}$ into the exponential output $y_p = Ye^{st}$. The formula allows $s$ to be an imaginary $i\omega$ or any complex number $s = a + i\omega$. Use the exponent $s$ that is in the driving force $f$ :

$$Ay'' + By' + Cy = e^{i\omega t} \quad \text{leads to} \quad y_p(t) = \frac{1}{A(i\omega)^2 + B(i\omega) + C}\, e^{i\omega t}. \tag{4}$$

**Example 2** $\boldsymbol{y'' + y' = e^{it}}$ **has** $\boldsymbol{s = i\omega = i}$. Substitute $y = Ye^{it}$ and solve for $Y$ :

$$i^2\, Ye^{it} + i\, Ye^{it} = e^{it} \qquad (i^2 + i)\,Y = 1 \qquad \boldsymbol{y_p(t) = \frac{1}{-1+i}\, e^{it}}. \tag{5}$$

**Example 3** (important) **Solve** $\boldsymbol{y'' + y' = \cos t}$. The cosine is the real part of $e^{it}$.

Warning: The solution will *not* have the form $y = Y \cos t$. The derivative $-Y \sin t$ would appear in the differential equation, with no other term to cancel it. The correct solution involves *both* $\cos t$ *and* $\sin t$. Damping from $y'$ delays the cosine.

Here $y_p(t)$ in Example 3 is the real part of $y_p(t)$ in Example 2. Please use this idea: **The real part of the input $\boldsymbol{e^{i\omega t}}$ produces the real part of the output $\boldsymbol{Ye^{i\omega t}}$.**

Step 1 Write $Y = \dfrac{1}{-1+i} = \dfrac{1}{-1+i}\left(\dfrac{-1-i}{-1-i}\right) = \dfrac{-1-i}{2}$.

Step 2 The real part of $Ye^{it} = \dfrac{-1-i}{2}(\cos t + i \sin t)$ is $\boldsymbol{y_p = \frac{1}{2}(-\cos t + \sin t)}$.

The exponential response formulas are (3) and (4). The only time they fail is when the denominator in the fraction is zero. The formula would then contain **1/0**. That happens when the exponent $s$ in the driving term equals one of the exponents $s_1$ and $s_2$ in the null solution $y_n = c_1 e^{s_1 t} + c_2 e^{s_2 t}$. This is called **resonance**: $\boldsymbol{s = s_1}$ **or** $\boldsymbol{s = s_2}$.

You see that we cannot allow $y_p$ to be included among the null solutions $y_n$. If the right side is $f \neq 0$ for $y_p$, it cannot also be $f = 0$ as required by $y_n$. We will see that the correct form for a resonant solution $y_p$ includes an extra factor $\boldsymbol{t}$ in $Y\boldsymbol{t}e^{st}$.

A special effort goes into the oscillating case $s = i\omega$. Null solutions $y_n = e^{st}$ depend only on $A, B, C$. That part comes from the roots of $As^2 + Bs + C = 0$. The new part is the forced oscillation $\boldsymbol{y_p(t)}$, a particular solution that is driven by $\cos \omega t$. It will be $\boldsymbol{y_p(t) = G\cos(\omega t - \alpha)}$ **with a phase shift $\boldsymbol{\alpha}$ and a gain $\boldsymbol{G}$ in the amplitude**.

## Equations of Order $N$ and Order 2

I would like to outline the work ahead, because this section is important. It started with a specific example $y'' + 5y' + 6y = e^{4t}$. Those numbers $1, 5, 6, 4$ changed to letters $A, B, C, s$. We solved the second order equation $Ay'' + By' + Cy = e^{st}$. The solution $Ye^{st}$ introduced the transfer function $Y = 1/(As^2 + Bs + C)$.

Now we have two ways to go, both essential. One is to see the same formula $y = Ye^{st}$ for **every constant coefficient equation**. $Y$ comes from the "exponential response formula" because $Ye^{st}$ is the response to the exponential $f(t) = e^{st}$. One formula covers almost all equations (but resonance is special and $Y$ has to change).

The other crucial step is to focus on **second order equations driven by** $\boldsymbol{f = e^{i\omega t}}$. Yes, this is covered by the formula. But if we are serious, we won't stop with $Y(i\omega)$. We truly need the rectangular and polar forms of that complex number:

$$Y(i\omega) = \frac{1}{A(i\omega)^2 + B(i\omega) + C} = M - iN = Ge^{-i\alpha}. \qquad (6)$$

$M, N, G, \alpha$ will be in equations (23) to (27). The solution driven by $f = \cos\omega t$ becomes $y = M\cos\omega t + N\sin\omega t$. Damped motion ($B > 0$) can be compared with undamped. And the big applications in Section 2.5 need the better notation using $Z$:

**Natural frequency** $\omega_n^2 = \dfrac{C}{A}$ **Damping ratio** $Z = \dfrac{B}{\sqrt{4AC}}$ **Damped frequency** $\omega_d^2 = \omega_n^2(1 - Z^2)$ (7)

The damping ratio $Z$ and those frequencies $\omega_n$ and $\omega_d$ give meaning to the solution $y(t)$.

## Complete Solution $y_p + y_n$

Let me summarize the case of **undamped forced oscillation** (driving force $F\cos\omega t$). If $B = 0$, the complete solution to $Ay'' + Cy = F\cos\omega t$ is one particular solution $y_p$ plus any null solution $y_n$ at the natural frequency $\omega_n = \sqrt{C/A}$. Notice the two $\omega$'s:

**Particular solution** $(\omega)$
**Unforced solution** $(\omega_n)$

$$y = \frac{F}{C - A\omega^2}\cos\omega t + c_1 e^{i\omega_n t} + c_2 e^{-i\omega_n t} \qquad (8)$$

To repeat: Any time we have a linear equation $\boldsymbol{Ly = f}$, the complete solution has the form $y = y_p + y_n$. The particular solution solves $Ly_p = f$. The null solution solves $Ly_n = 0$. Linearity of $L$ guarantees that $y = y_p + y_n$ solves $Ly = f$:

**Complete solution** $\boldsymbol{y = y_p + y_n}$ If $Ly_p = f$ and $Ly_n = 0$ then $Ly = f$. (9)

This book emphasizes linear equations. You will see $y_p + y_n$ again, always with the rule of linearity $Ly = Ly_p + Ly_n$. This applies to linear differential equations and matrix equations. In differential equations, $L$ is called a *linear operator*.

**Linear operator** $\quad Ly = Ay'' + By' + Cy \;$ or $\; Ly = A_N \dfrac{d^N y}{dt^N} + \cdots + A_1 \dfrac{dy}{dt} + A_0 y$

For an operator $L$, the inputs $y$ and the outputs $Ly$ are functions.

*Every solution to* $Ly = f$ *has the form* $y_p + y_n$. Suppose we start with one particular solution $y_p$. If $y$ is any other solution, then $L(y - y_p) = 0$:

$$\boldsymbol{y_n = y - y_p} \textbf{ is a null solution} \qquad \boldsymbol{Ly_n} = Ly - Ly_p = f - f = \mathbf{0}. \qquad (10)$$

**Example 4** Suppose the linear equation is just $\boldsymbol{Ly} = x_1 - x_2 = 1$: one equation in two unknowns $x_1$ and $x_2$. The solutions are vectors $\boldsymbol{y} = (x_1, x_2)$. The right side $f = 1$ is not zero. The bold line in Figure 2.11 is the graph of all solutions.

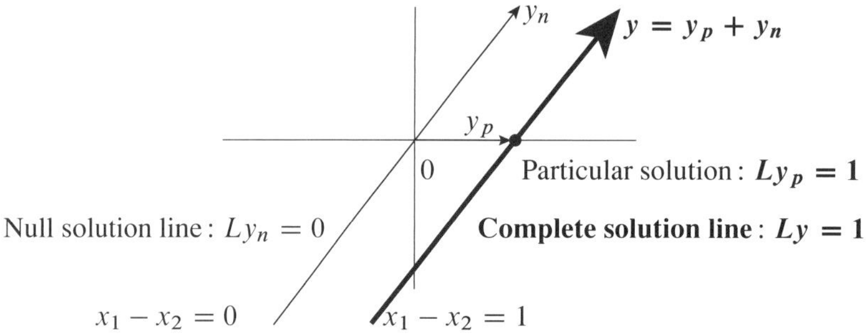


Figure 2.11: Complete solution = *one* particular solution + *all* null solutions.

Every point on that bold line is a particular solution to $x_1 - x_2 = 1$. We marked only one $y_p$. **Null solutions lie on a parallel line $\boldsymbol{x_1 - x_2 = 0}$ through the center $\mathbf{(0, 0)}$.**

**Example 5** Second order equations $Ay'' + By' + Cy = e^{st}$ or $e^{i\omega t}$ have complete solutions $y = y_p + y_n$. The particular solution $y_p = Ye^{st}$ is a multiple of $e^{st}$. The null solutions are $y_n = c_1 e^{s_1 t} + c_2 e^{s_2 t}$. If $s_2 = s_1$, replace $e^{s_2 t}$ by $te^{s_1 t}$.

**Example 6** The complete solution to the impressive equation $5y = 10$ is $y = 2$. This is our only choice for the particular solution, $y_p = 2$. The null solutions solve $5y_n = 0$, and the only possibility is $y_n = 0$. *The one and only solution is* $y = y_p + y_n = 2 + 0$.

That seems boring, when $y_n = 0$ is the only null solution. But this is what we want (and usually get) for matrix equations. If $A$ is an invertible matrix, the only solution to $Ay = b$ is $y = y_p = A^{-1}b$. Then the only null solution to $Ay_n = 0$ is $y_n = 0$.

## Higher Order Equations

Up to this moment, third derivatives have not been seen. They don't arise often in physical problems. But exponential solutions $Ye^{st}$ and $Ye^{i\omega t}$ still appear. The one essential requirement is that the equation must have *constant coefficients*.

**Equation of order $N$** $$A_N\frac{d^Ny}{dt^N}+\cdots+A_1\frac{dy}{dt}+A_0y=f(t) \qquad (11)$$

When $f=0$, the best solutions of the null equation are still exponentials $y_n=e^{st}$. Substitute $e^{st}$ into the equation to find $N$ possible exponents $s_1, s_2, \ldots, s_N$.

$$\boldsymbol{f=0 \text{ and } y_n=e^{st}} \qquad \left(A_Ns^N+\cdots+A_1s+A_0\right)e^{st}=0. \qquad (12)$$

The exponents $s$ in $y_n$ are the $N$ roots of that polynomial. So we (usually) have $N$ independent solutions $e^{s_1t}, \ldots, e^{s_Nt}$. All their combinations are still solutions. If the polynomial in (12) happens to have a double root at $s$, our two solutions are $e^{st}$ and $te^{st}$.

**Example 7** **Solve the third order equation** $y'''+2y''+y'=e^{3t}$.

**Solution** To find the null solutions $y_n$, substitute $y_n=e^{st}$ with right hand side zero:

$$\boldsymbol{s^3+2s^2+s=0} \qquad s(s^2+2s+1)=0 \qquad \boldsymbol{s(s+1)^2=0}.$$

The exponents are $s=0,-1,-1$. The null solutions are $c_1e^{0t}$ and $c_2e^{-t}$ and $c_3te^{-t}$ (the extra $t$ comes from the double root). A particular solution $y_p$ is $Ye^{3t}$ (since 3 is not one of the exponents 0 and $-1$ in $y_n$). Substitute $Ye^{3t}$ to find $\boldsymbol{Y=1/48}$:

$$27\,Ye^{3t}+18Ye^{3t}+3Ye^{3t}=e^{3t} \text{ and } 48Y=1 \text{ and } \boldsymbol{y_p=e^{3t}/48}.$$

**The transfer function is $Y(s)=1/(s^3+2s^2+s)$. For $e^{3t}$ put $s=3$. Then $Y=1/48$.**

Here is the plan for this section on constant coefficient equations with forced oscillations.

1 Find the **exponential response $y(t)=Y(s)e^{st}$** to the driving function $f(t)=e^{st}$.

2 Adjust that formula when $Y(s)=\infty$ because of **resonance**.

3 Solve the **real equation** $Ay''+By'+Cy=\cos\omega t$ to see the effect of damping.

This is the key example for applications: $y$ is the real part of $Y(s)e^{st}$ when $s=i\omega$. The solution in equation (23) is $\boldsymbol{y(t)=M\cos\omega t+N\sin\omega t=G\cos(\omega t-\alpha)}$.

## Exponential Response Function = Transfer Function

This book concentrates on first and second order equations. When the coefficients are constant and the right side is an exponential, we have solved three important problems:

| | | |
|---|---|---|
| **First order** | $y'-ay=\boldsymbol{e^{ct}}$ | $y_p=e^{ct}/\boldsymbol{(c-a)}$ |
| **Oscillation** | $my''+ky=\boldsymbol{e^{i\omega t}}$ | $y_p=e^{i\omega t}/\boldsymbol{(k-m\omega^2)}$ |
| **Second order** | $Ay''+By'+Cy=e^{st}$ | $y_p=\boldsymbol{e^{st}/(As^2+Bs+C)}$ |

It is natural (natural to a mathematician) to try to solve all constant coefficient equations of all orders by one formula. We can almost do it, but resonance gets in the way.

Let me write $D$ for each derivative $d/dt$. Then $D^2$ is $d^2/dt^2$. All our equations involve powers of $D$, and equations of order $N$ involve $D^N$. Here $N = 2$.

$$\textbf{Polynomial } \boldsymbol{P(D)} \qquad Ay'' + By' + Cy = (\boldsymbol{AD^2 + BD + C})\, y = \boldsymbol{P(D)}\, y. \tag{13}$$

The null solutions and the particular solution all come from this polynomial $P(D)$.

$$\textbf{Find } \boldsymbol{N} \textbf{ null solutions } \boldsymbol{y_n = e^{st}} \qquad As^2 + Bs + C = 0 \text{ is exactly } \boldsymbol{P(s) = 0} \tag{14}$$

$$\textbf{Find a particular } \boldsymbol{y_p = Ye^{ct}} \qquad P(D)y = e^{ct} \text{ gives the number } \boldsymbol{Y = 1/P(c)} \tag{15}$$

**The value $Y$ of the transfer function gives the exponential response** $\quad \boldsymbol{y_p = e^{ct}/P(c)}$.

Please understand: In the null solutions, $s$ has $N$ specific values $s_1, \ldots, s_N$. Those are the roots of the $N$th degree characteristic equation $P(s) = 0$. In the particular solution $e^{ct}/P(c)$, the specific value $s = c$ is the exponent in the right hand side $f = e^{ct}$.

The exponents $c$ and $s$ are completely allowed to be imaginary or complex.

$$\boldsymbol{P(D)y = e^{ct}} \qquad \boldsymbol{y = y_p + y_n = \frac{e^{ct}}{P(c)} + c_1 e^{s_1 t} + \cdots + c_N e^{s_N t}} \tag{16}$$

That fraction $Y = 1/P(c)$ "transfers" the input $f = e^{ct}$ into the output $y = Ye^{ct}$. You often see it as $1/P(s)$ with the variable $s$. It is sometimes called the **system function**.

There is only one exception to this simple and beautiful exponential response formula. The forcing exponent $c$ might be one of the exponents $s_1, \ldots, s_N$ in the null solution. **In this case $P(c)$ is zero**. We cannot divide by $P(c)$ when it is zero.

**Exception** $\qquad$ If $P(c) = 0$ then $y = e^{ct}/P(c)$ cannot solve $P(D)y = e^{ct}$.

$P(c) = 0$ is the exceptional case of **resonance**. The formula $e^{ct}/P(c)$ has to change.

## Resonance

We may be pushing a swing at its natural frequency. Then $c = i\omega_n = i\sqrt{k/m}$. The polynomial $P(D)$ from $my'' + ky$ is $mD^2 + k$, and we have $P(c) = 0$ at this natural frequency. Here is the exponential response formula adjusted for resonance.

$$\textbf{Resonant response} \qquad \textbf{If } \boldsymbol{P(c) = 0} \textbf{ then } \boldsymbol{y_p = \frac{t}{P'(c)}\, e^{ct}} \tag{17}$$

That extra factor $t$ enters the solution when $P(c) = 0$. We replace $1/P(c)$ by $\boldsymbol{t/P'(c)}$. This succeeds unless there is "double resonance" and $P'(c)$ is also zero. Then the formula moves on to the second derivative of $P$, and $y_p(t) = t^2 e^{ct}/P''(c)$.

The odds against double resonance are pretty high. The point is that the equation $\boldsymbol{P(D)y = e^{ct}}$ has a neat solution in terms of the polynomial $P$ : usually $\boldsymbol{y = e^{ct}/P(c)}$.

I can explain that resonant solution $y = te^{ct}/P'(c)$ when $P(c) = 0$ and $P'(c) \neq 0$. We have seen this happen in Section 1.5 for the first order equation $y' - ay = e^{ct}$. That equation has $P(D) = D - a$ and $P(c) = c - a$ and resonance when $c = a$:

$$y' - ay = e^{ct} \quad \text{has the very particular solution} \quad \boldsymbol{y_{vp} = \frac{e^{ct} - e^{at}}{c - a}}$$

$$\text{As } c \text{ approaches } a, \; y_{vp} \text{ approaches } \frac{\text{derivative of top}}{\text{derivative of bottom}} = \boldsymbol{\frac{te^{at}}{1}}$$

That is l'Hôpital's Rule! The only unusual thing is that we have $c$ in place of $x$, and $c$-derivatives in place of $x$-derivatives. The very particular solution is the one starting from $y_{vp} = 0$ at $t = 0$. The resonant solution $te^{at}$ fits our formula $te^{ct}/P'(c)$ because $c = a$ and $P(c) = c - a$ and $P'(c) = 1$.

When the equation has order $N$, the polynomial $P$ has degree $N$. Suppose the exponent $c$ is *close* to $a$—which is one of the exponents $s_1, \ldots, s_N$ in the null solution. Then $P(a) = 0$ and $e^{at}$ is a null solution and $e^{ct}/P(c)$ is one particular solution:

$$\text{A very particular solution to } P(D)y = e^{ct} \text{ is } \boldsymbol{y_{vp} = \frac{e^{ct} - e^{at}}{P(c) - P(a)}}. \tag{18}$$

To emphasize: $c$ close to $a$ is fine. But $c = a$ is *not* fine. Formula (16) changes at $c = a$:

**Resonance** **If $c = a$ then l'Hôpital's limit in (16) is** $\boldsymbol{y_{vp} = \dfrac{te^{at}}{P'(a)}}$. (19)

Take the $c$-derivatives of $e^{ct} - e^{at}$ and $P(c) - P(a)$ at $c = a$, to get $te^{at}$ and $P'(a)$.

**Summary** The transfer function is $Y(s) = 1/P(s)$. It has "poles" at the $N$ roots of $P(s) = 0$. Those are the exponents in the null solutions $y_n(t)$. The particular solution $y_p = Ye^{ct}$ has the same exponent $c$ as the driving term $f = e^{ct}$. The transfer function $Y(c) = 1/P(c)$ decides the amplitude of $y_p(t)$. If $c$ is a pole of $Y$, we have resonance.

**Example 8** The 4th degree equation $\boldsymbol{D^4y = d^4y/dt^4 = 1}$ has 4-*way resonance.*

What are the null solutions to $y'''' = 0$? By trying $y = e^{st}$ we get $s^4 = 0$. This has *all four roots at* $s = 0$. Then one null solution is $y = e^{0t}$, which is $y = 1$. The other null solutions have factors $t, t^2, t^3$ because of the four-way zero. Altogether:

**The null solutions to $y'''' = 0$ have the form $y_n(t) = c_1 + c_2t + c_3t^2 + c_4t^3$.**

Now find a particular solution to $y'''' = e^{ct}$. For most exponents $c$ we get $\boldsymbol{y_p = e^{ct}/c^4}$. This is exactly $e^{ct}/P(c)$. *But $c = 0$ gives* **quadruple resonance**: $c^4 = 0$ has a 4-way root. A quadruple l'Hôpital rule gives the fourth derivative $P''''$ and the very particular solution to $y'''' = 1$ that you knew before taking this course and seeing this book:

$$\boldsymbol{y'''' = 1 = e^{0t} \quad \text{has} \quad c = a = 0 \quad \text{and} \quad P = s^4 \qquad y_p(t) = \frac{t^4e^{0t}}{P''''(0)} = \frac{t^4}{24}}.$$

## Real Second Order Equations with Damping

Now we focus on the key equation: **second order**. The left side is $Ay'' + By' + Cy$. The transfer function is $Y(s) = 1/(As^2 + Bs + C)$. When the right side is $f(t) = e^{i\omega t}$, the exponent is $s = i\omega$. When $A, B, C$ are nonzero, we won't have resonance:

$$\textbf{No resonance} \quad A(i\omega)^2 + B(i\omega) + C = (C - A\omega^2) + i(B\omega) \neq 0.$$

We know that the response to $f(t) = e^{i\omega t}$ is $y_p(t) = Y(i\omega)e^{i\omega t}$. This is a perfect example, except that those functions are not real.

In applications to real life (and this equation has many), we want $f(t) = \cos\omega t$. We *must* solve this problem. You will say, just solve for $e^{i\omega t}$ and $e^{-i\omega t}$, and take half of each solution. Even faster than that, **solve for $e^{i\omega t}$ and take the real part of $y_p(t)$.** Or you could stay entirely real and look for a solution $y(t) = M \cos \omega t + N \sin \omega t$.

All those ideas will succeed. They all give the same answer (in different forms). The best form has to bring out the most important number in the answer $y(t)$. That number is the **amplitude $G$ of the forced oscillation**. So first place goes to the **polar form $y(t) = G \cos(\omega t - \alpha)$**, because this shows the gain $G$.

The null solutions decay because the solutions $s_1$ and $s_2$ to $As^2 + Bs + C = 0$ have negative real parts $-B/2A$. The particular solution $G \cos(\omega t - \alpha)$ does not decay, because it is driven by a forcing function $f = \cos \omega t$ that never stops.

The next pages will find $G$ and $\alpha$. This is algebra put to good use. We are working with letters $A, B, C$ that represent physical quantities. In Section 2.5 they will be mass-damping-stiffness or inductance-resistance-inverse capacitance. Those are not the only possible examples! Biology and chemistry and management and the economics of a whole country also see damped oscillations. I hope you will find those models.

## Damped Oscillations in Rectangular Form

I will start with the rectangular form $y(t) = M \cos\omega t + N \sin\omega t$. It is not as useful as the polar form, but it is easier to compute. Substitute this $y(t)$ into the differential equation $Ay'' + By' + Cy = \cos \omega t$. Match the cosine terms and the sine terms:

| | | |
|---|---|---|
| **Cosines on both sides** | $-A\omega^2 M + B\omega N + CM = 1$ | (20) |
| **Sines on the left side** | $-A\omega^2 N - B\omega M + CN = 0$ | (21) |

To solve for $M$, multiply equation (20) by $C - A\omega^2$. Then multiply equation (21) by $B\omega$ and subtract from (20). The coefficient of $N$ will be zero. So $N$ is eliminated and we have an equation for $M$ alone. $M$ is multiplied by the important number $D$:

**$C - A\omega^2$ times (20)**
**minus $B\omega$ times (21)**

$$[(C - A\omega^2)^2 + (B\omega)^2]M = \boldsymbol{D}M = C - A\omega^2. \quad (22)$$

We divide by $D$ to find $M = (C - A\omega^2)/D$. Then equation (21) tells us $N = B\omega/D$. And equation (27) will tell us that $\boldsymbol{M^2 + N^2 = 1/D}$.

$$\textbf{Real solution } y_p \textbf{ is } M\cos\omega t + N\sin\omega t \qquad M = \frac{C - A\omega^2}{D} \qquad N = \frac{B\omega M}{C - A\omega^2} = \frac{B\omega}{D} \tag{23}$$

Let me say right away: **The complex number $Y(i\omega)$ is just $M - iN$.** This calculation will connect real to complex and rectangular to polar. When I multiply and divide by $Y(-i\omega)$, you will see that the denominator of $Y(i\omega)$ is $D = (C - A\omega^2)^2 + (B\omega)^2$:

$$\frac{1}{(C - A\omega^2) + iB\omega} \times \frac{(C - A\omega^2) - iB\omega}{(C - A\omega^2) - iB\omega} = \frac{(C - A\omega^2) - iB\omega}{D} = \boldsymbol{M - iN}. \tag{24}$$

$Y = M - iN$ is exactly what we want and need. The input $f = \cos\omega t$ is the real part of $e^{i\omega t}$, so the output $y$ is the real part of $Ye^{i\omega t}$. That real part is the rectangular form $y = M\cos\omega t + N\sin\omega t$:

$$\mathbf{Re}\,(\boldsymbol{Ye^{i\omega t}}) = \mathrm{Re}\,[(M - iN)(\cos\omega t + i\sin\omega t)] = \boldsymbol{M\cos\omega t + N\sin\omega t} \tag{25}$$

## Damped Oscillations in Polar Form

The solution we want is the real part of $Y(i\omega)e^{i\omega t}$. Equation (25) computed that solution in its rectangular from. To compute $y(t)$ in polar form, the first step (almost the only step) is to put $Y(i\omega)$ in polar form. This number is the complex gain:

$$\textbf{Complex gain} \quad \boldsymbol{Y(i\omega) = M - iN = Ge^{i\alpha}} \text{ with } \boldsymbol{G = \frac{1}{\sqrt{D}}} \text{ and } \boldsymbol{\tan\alpha = \frac{N}{M}}. \tag{26}$$

That amplitude $G$ is simply called the "gain". It is the most important quantity in all these pages of calculations. The input cos $\omega t$ had amplitude 1, the output $y(t)$ has amplitude $G$. Of course that output is not $y = G\cos\omega t$! Damping produces a phase lag $\alpha$. At the same time damping reduces the amplitude of the output.

**The undamped amplitude $|Y| = 1/|C - A\omega^2|$ is reduced to $G = 1/\sqrt{D}$:**

$$G = \sqrt{M^2 + N^2} = \left(\frac{(C - A\omega^2)^2}{D^2} + \frac{(B\omega)^2}{D^2}\right)^{1/2} = \left(\frac{D}{D^2}\right)^{1/2} = \frac{1}{\sqrt{D}}. \tag{27}$$

I will collect all these beautiful (?) important (!) formulas after one example.

**Example 9** Solve $y'' + y' + 2y = \cos t$ in rectangular form and also in polar form.

*Solution* The equation has $A = 1$, $B = 1$, $C = 2$, and $\omega = 1$. We are finding a particular solution. Let me use the formulas directly and then comment briefly. The numbers give $C - A\omega^2 = 1$ and $B\omega = 1$, so $D = 1^2 + 1^2 = 2$.

Therefore the solution has $G = \sqrt{1/2}$ and $M = N = \frac{1}{2}$ and $\tan\alpha = 1$ and $\alpha = \pi/4$:

**Rectangular** $\quad y(t) = M \cos\omega t + N \sin\omega t = \boldsymbol{\frac{1}{2}(\cos t + \sin t)}$

**Polar** $\quad y(t) = \text{Re}\left(Ge^{-i\alpha}e^{i\omega t}\right) = G\cos(\omega t - \alpha) = \boldsymbol{\frac{1}{\sqrt{2}}\cos\left(t - \frac{\pi}{4}\right)}.$

For this example we verify directly that polar = rectangular:

$$\boldsymbol{G\cos\left(t - \frac{\pi}{4}\right)} = \frac{1}{\sqrt{2}}\left(\cos t\ \cos\frac{\pi}{4} + \sin t\ \sin\frac{\pi}{4}\right) = \boldsymbol{\frac{1}{2}(\cos t + \sin t)}.$$

The rectangular form has simpler numbers. But the polar form has the most important number $G = 1/\sqrt{2}$. That gain $G$ is less than the undamped gain $|Y|$ by a factor $\cos\alpha$.

$$\text{Undamped}\quad |Y| = \frac{1}{|C - A\omega^2|} = 1 \qquad \text{Damped}\quad G = \frac{1}{\sqrt{D}} = \frac{1}{\sqrt{2}} = \cos\alpha.$$

## Undamped versus Damped

The undamped equation $Ay'' + Cy = \cos\omega t$ has $B = 0$ and $Y = 1/(C - A\omega^2)$. Compare that amplitude of $y(t) = Y\cos\omega t$ from Section 2.1 with the harder problem we just solved. The comparison lets you see how the damping contributes $Bs = Bi\omega$ in the transfer function that multiplies the input $e^{i\omega t}$. Damping causes a phase lag $\alpha$. Damping also reduces the amplitude to $G = Y\cos\alpha$. Here are the key formulas:

| | **Undamped** | **Damped** |
|---|---|---|
| **Equation** | $Ay'' + Cy = \cos\omega t$ | $Ay'' + By' + Cy = \cos\omega t$ |
| **Solution** | $y = Y\cos\omega t$ | $y = G\cos(\omega t - \alpha)$ |
| **Magnitude** | $\lvert Y\rvert = \dfrac{1}{\lvert C - A\omega^2\rvert}$ | $G = \dfrac{1}{\sqrt{D}} = Y\cos\alpha$ |
| **Phase lag** | zero | $\tan\alpha = \dfrac{N}{M} = \dfrac{B\omega}{C - A\omega^2}$ |

When the driving function is $F\cos\omega t$, the solutions include that extra factor $F$. When the driving function is $\sin\omega t$, that is the same as $\cos\left(\omega t - \frac{\pi}{2}\right)$. So the solutions have $\phi = \pi/2$ as an additional phase lag: $y = G\cos(\omega t - \alpha - \pi/2) = G\sin(\omega t - \alpha)$.

When the driving function is $A\cos\omega t + B\sin\omega t$, that equals $R\cos(\omega t - \phi)$. This is the sinusoidal identity from Section 1.5. Then the solution is $\boldsymbol{RG\cos(\omega t - \alpha - \phi)}$. This is the particular solution $y_p$ that oscillates with the same frequency $\omega$ as the input.

Let me show why the gain is reduced to $G = Y\cos\alpha$ from its undamped value $|Y| = 1/|C - A\omega^2|$. We know from (27) that $G = \sqrt{M^2 + N^2} = 1/\sqrt{D}$. And we know from (23) that $YM = 1/D$:

$$\textbf{Damped gain} \qquad Y\cos\alpha = \frac{YM}{\sqrt{M^2+N^2}} = \frac{1/D}{1/\sqrt{D}} = G. \tag{28}$$

## Better Notation

A good plan is to divide $my'' + by' + ky = kF(t)$ by the mass $m$, for several reasons:

$$y'' + \frac{b}{m}y' + \frac{k}{m}y = \frac{k}{m}F(t). \tag{29}$$

First, the coefficient of $y''$ becomes 1. Second, replacing $k/m$ by $\omega_n^2$ gives it meaning. Third, the input $F$ has the same units as the output $y$. So now the gain $G = |y|/|F|$ is dimensionless. This happened because the original $f(t)$ with unsuitable units was replaced by $kF(t)$—which is now divided by $m$.

Most valuable of all is a new way to write the damping term $b/m$, which is $B/A$. The key point is that **$b^2$ and $mk$ have the same dimensions**. From the equation, $my''$ and $by'$ and $ky$ have the same dimensions. Then so do $(by')^2$ and $(my'')(ky)$. And also $(y')^2$ and $(y'')(y)$—they both contain $1/(\text{time})^2$. *This leaves $b^2$ and $mk$.*

This quantity $Z = b/\sqrt{4mk}$ is highly useful. Overdamping is $Z > 1$. Underdamping is $Z < 1$. The coefficient $b/m$ in equation (29) has a better form $2Z\omega_n$ in (30).

$$\frac{b}{m} = \frac{2b}{\sqrt{4mk}}\sqrt{\frac{k}{m}} = 2Z\omega_n \qquad \boldsymbol{y'' + 2Z\omega_n y' + \omega_n^2 y = \omega_n^2 F(t)} \tag{30}$$

***Z* is the damping ratio. The correct symbol is a Greek zeta** ($\zeta$). But a capital zeta $= Z$ is so much easier to read and write. (The MATLAB command is also named zeta.) Watch how this ratio of $B$ to $\sqrt{4AC}$ brings out the important parts of every formula. If $Z < 1$, the natural frequency $\omega_n$ is reduced to the **damped frequency $\boldsymbol{\omega_d = \omega_n\sqrt{1-Z^2}}$**.

| | | |
|---|---|---|
| **Roots $s_1$ and $s_2$** | $s^2 + 2Z\omega_n s + \omega_n{}^2 = 0$ gives $s = -Z\omega_n \pm \omega_n\sqrt{Z^2-1}$ | (31) |
| **Underdamping** | $Z^2 = \dfrac{b^2}{4mk} < 1$ and $\boldsymbol{s = -Z\omega_n \pm i\omega_d}$ | (32) |
| **Null solutions** | $y_n(t) = \boldsymbol{e^{-Z\omega_n t}(c_1\cos\omega_d t + c_2\sin\omega_d t)}$ | (33) |

The null solutions are not pure oscillations. They include the exponential $e^{-Z\omega_n t}$. Their frequency changes to $\omega_d$. The graph of $y(t)$ oscillates as it approaches zero, and the peak times when $y = y_{\max}$ are spaced by $2\pi/\omega_d$.

**The page after Problem Set 2.4 collects our solution formulas in one place.**

## ■ REVIEW OF THE KEY IDEAS ■

1. A particular solution to $Ay'' + By' + Cy = e^{st}$ is $e^{st}/(As^2 + Bs + C)$.

2. This is a constant coefficient equation $P(D)y = e^{ct}$ with solution $\boldsymbol{y_p = e^{ct}/P(c)}$.

3. Resonance occurs if $e^{ct}$ is a null solution of $P(D)y = 0$. This means that $P(c) = 0$.

4. Resonance leads to an extra $t$ : $\boldsymbol{y_p(t) = te^{ct}/P'(c)}$ when $P(c) = 0$ and $P'(c) \neq 0$.

5. For second order equations with $f = \cos \omega t$ the gain is $G = 1/|P(i\omega)| = 1/\sqrt{D}$.

6. The real solution is $M \cos \omega t + N \sin \omega t = G \cos(\omega t - \alpha)$ with $\tan \alpha = N/M$.

7. With damping ratio $Z = B/\sqrt{4AC}$, the equation is $y'' + 2\omega_n Z y' + \omega_n^2 y = \omega_n^2 F(t)$.

8. If $Z < 1$, the damped frequency is $\omega_d = \omega_n \sqrt{1 - Z^2}$. Then $s_1, s_2$ are $-Z\omega_n \pm i\omega_d$.

# Problem Set 2.4

**Problems 1-4 use the exponential response $\boldsymbol{y_p = e^{ct}/P(c)}$ to solve $\boldsymbol{P(D)y = e^{ct}}$.**

**1** Solve these constant coefficient equations with exponential driving force:

(a) $y_p'' + 3y_p' + 5y_p = e^t$ (b) $2y_p'' + 4y_p = e^{it}$ (c) $y'''' = e^t$

**2** These equations $P(D)y = e^{ct}$ use the symbol $D$ for $d/dt$. Solve for $y_p(t)$:

(a) $(D^2 + 1)y_p(t) = 10e^{-3t}$ (b) $(D^2 + 2D + 1)y_p(t) = e^{i\omega t}$

(c) $(D^4 + D^2 + 1)y_p(t) = e^{i\omega t}$

**3** How could $y_p = e^{ct}/P(c)$ solve $y'' + y = e^t e^{it}$ and then $y'' + y = e^t \cos t$ ?

**4** (a) What are the roots $s_1$ to $s_3$ and the null solutions to $y_n''' - y_n = 0$ ?

(b) Find particular solutions to $y_p''' - y_p = e^{it}$ and to $y_p''' - y_p = e^t - e^{i\omega t}$.

**Problems 5-6 involve repeated roots $\boldsymbol{s}$ in $\boldsymbol{y_n}$ and resonance $\boldsymbol{P(c) = 0}$ in $\boldsymbol{y_p}$.**

**5** Which value of $C$ gives resonance in $y'' + Cy = e^{i\omega t}$ ? Why do we never get resonance in $y'' + 5y' + Cy = e^{i\omega t}$ ?

**6** Suppose the third order equation $P(D)y_n = 0$ has solutions $y = c_1 e^t + c_2 e^{2t} + c_3 e^{3t}$. What are the null solutions to the sixth order equation $P(D)P(D)y_n = 0$ ?

**7** Complete this table with equations for $s_1$ and $s_2$ $y_n$ and $y_p$ :

| | | |
|---|---|---|
| **Undamped free oscillation** | $my'' + ky = 0$ | $y_n =$ ____ |
| **Undamped forced oscillation** | $my'' + ky = e^{i\omega t}$ | $y_p =$ ____ |
| **Damped free motion** | $my'' + by' + ky = 0$ | $y_n =$ ____ |
| **Damped forced motion** | $my'' + by' + ky = e^{ct}$ | $y_p =$ ____ |

**8** Complete the same table when the coefficients are 1 and $2Z\omega_n$ and $\omega_n^2$ with $Z < 1$.

| | | |
|---|---|---|
| **Undamped and free** | $y'' + \omega_n^2 y = 0$ | $y_n =$ ____ |
| **Undamped and forced** | $y'' + \omega_n^2 y = e^{i\omega t}$ | $y_p =$ ____ |
| **Underdamped and free** | $y'' + 2Z\omega_n y' + \omega_n^2 y = 0$ | $y_n =$ ____ |
| **Underdamped and forced** | $y'' + 2Z\omega_n y' + \omega_n^2 y = e^{ct}$ | $y_p =$ ____ |

**9** What equations $y'' + By' + Cy = f$ have these solutions ?

(a) $y = c_1 \cos 2t + c_2 \sin 2t + \cos 3t$

(b) $y = c_1 e^{-t} \cos 4t + c_2 e^{-t} \sin 4t + \cos 5t$

(c) $y = c_1 e^{-t} + c_2 t e^{-t} + e^{i\omega t}$

**10** If $y_p = te^{-6t} \cos 7t$ solves a second order equation $Ay'' + By' + Cy = f$, what does that tell you about $A$, $B$, $C$, and $f$ ?

**11** (a) Find the steady oscillation $y_p(t)$ that solves $y'' + 4y' + 3y = 5 \cos \omega t$.

(b) Find the amplitude $A$ of $y_p(t)$ and its phase lag $\alpha$.

(c) Which frequency $\omega$ gives maximum amplitude (maximum gain)?

**12** Solve $y'' + y = \sin \omega t$ starting from $y(0) = 0$ and $y'(0) = 0$. Find the limit of $y(t)$ as $\omega$ approaches 1, and the problem approaches resonance.

**13** Does critical damping and a double root $s = 1$ in $y'' + 2y' + y = e^{ct}$ produce an extra factor $t$ in the null solution $y_n$ or in the particular $y_p$ (proportional to $e^{ct}$)? What is $y_n$ with constants $c_1, c_2$ ? What is $y_p = Ye^{ct}$ ?

**14** If $c = i\omega$ in Problem 13, the solution $y_p$ to $y'' + 2y' + y = e^{i\omega t}$ is ____. That fraction $Y$ is the transfer function at $i\omega$. What are the magnitude and phase in $Y = Ge^{-i\alpha}$ ?

**By rescaling both $t$ and $y$, we can reach $A = C = 1$. Then $\omega_n = 1$ and $B = 2Z$. The model problem is $y'' + 2Zy' + y = f(t)$.**

**15** What are the roots of $s^2 + 2Zs + 1 = 0$? Find two roots for $Z = 0, \frac{1}{2}, 1, 2$ and identify each type of damping. The natural frequency is now $\omega_n = 1$.

**16** Find two solutions to $y'' + 2Zy' + y = 0$ for every $Z$ except $Z = 1$ and $-1$. Which solution $g(t)$ starts from $g(0) = 0$ and $g'(0) = 1$? What is different about $Z = 1$?

**17** The equation $my'' + ky = \cos \omega_n t$ is exactly at resonance. The driving frequency on the right side equals the natural frequency $\omega_n = \sqrt{k/m}$ on the left side. Substitute $y = Rt \sin(\sqrt{k/m}\, t)$ to find $R$. This resonant solution grows in time because of the factor $t$.

**18** Comparing the equations $Ay'' + By' + Cy = f(t)$ and $4Az'' + Bz' + (C/4)z = f(t)$, what is the difference in their solutions?

**19** Find the fundamental solution to the equation $g'' - 3g' + 2g = \delta(t)$.

**20** (Challenge problem) Find the solution to $y'' + By' + y = \cos t$ that starts from $y(0) = 0$ and $y'(0) = 0$. Then let the damping constant $B$ approach zero, to reach the resonant equation $y'' + y = \cos t$ in Problem 17, with $m = k = 1$.

Show that your solution $y(t)$ is approaching the resonant solution $\frac{1}{2}t \sin t$.

**21** Suppose you know three solutions $y_1, y_2, y_3$ to $y'' + B(t)y' + C(t)y = f(t)$. How could you find $B(t)$ and $C(t)$ and $f(t)$?

## Solution Page    Linear Constant Coefficient Equations

**First order** $\dfrac{dy}{dt} = ay + f(t)$    **Second order** $A\dfrac{d^2y}{dt^2} + B\dfrac{dy}{dt} + Cy = f(t)$

**Nth order** $A_N\dfrac{d^Ny}{dt^N} + \cdots + A_1\dfrac{dy}{dt} + A_0 y = (A_N D^N + \cdots + A_0)y = P(D)y = f(t)$

**Null solutions $y_n$ have $f(t) = 0$    Substitute $y = e^{st}$ to find the $N$ exponents $s$**

| | | |
|---|---|---|
| First order | $\dfrac{d}{dt}(e^{st}) = ae^{st}$ | $s = a$ and $y_n = ce^{at}$ |
| Second order | $As^2 + Bs + C = 0$ | $y_n = c_1 e^{s_1 t} + c_2 e^{s_2 t}$ |
| Nth order | $P(s) = 0$ | $y_n = c_1 e^{s_1 t} + \cdots + c_N e^{s_N t}$ |

**Exponential response to $f(t) = e^{ct}$    Step response for $c = 0$    Look for $y = Ye^{ct}$**

| | | |
|---|---|---|
| First order | $\dfrac{d}{dt}\left(Ye^{ct}\right) - aYe^{ct} = e^{ct}$ | $y_p = \dfrac{e^{ct}}{c-a}$ has $Y = \dfrac{1}{c-a}$ |
| Second order | $Y(Ac^2 + Bc + C)e^{ct} = e^{ct}$ | $y_p = \dfrac{e^{ct}}{Ac^2 + Bc + C} = Ye^{ct}$ |
| Nth order | $YP(c)e^{ct} = e^{ct}$ | $y_p = \dfrac{e^{ct}}{P(c)}$ or $\dfrac{te^{ct}}{P'(c)}$ when $P(c) = 0$ |

**Fundamental solution $g(t)$ = Impulse response when $f(t) = \delta(t)$**

| | | |
|---|---|---|
| First order | $g(t) = e^{at}$ | starting from $g(0) = 1$ |
| Second order | $g(t) = \dfrac{e^{s_1 t} - e^{s_2 t}}{A(s_1 - s_2)}$ | starting from $g(0) = 0$ and $g'(0) = 1/A$ |
| Undamped | $g(t) = \dfrac{\sin \omega_n t}{A\omega_n}$ | underdamped $g(t) = e^{-Z\omega_n t}\dfrac{\sin \omega_d t}{A\omega_d}$ |
| Nth order | $g(t) = y_n(t)$ | $g(0) = g'(0) = \ldots = 0, g^{(N-1)}(0) = 1/A_N$ |

**Very particular solution for each driving function $f(t)$**: zero initial conditions on $y_{vp}$

**Multiply input at every time $s$**
**by the growth factor over $t - s$**

$$y(t) = \int_0^t g(t-s)\, f(s)\, ds$$

| | |
|---|---|
| **Undetermined coefficients** | Direct solution for special $f(t)$ in Section 2.6 |
| **Variation of parameters** | $y_p(t)$ comes from $y_n(t)$ in Section 2.6 |
| **Solution by Laplace transform** | Transfer function = transform of $g(t)$ in Section 2.7 |
| **Solution by convolution** | $y(t) = g(t) * f(t)$ in Section 8.6 |

## 2.5 Electrical Networks and Mechanical Systems

Section 2.4 solved the equation $Ay'' + By' + Cy = \cos \omega t$. Now we want to understand the meaning of $A, B, C$ in real applications. This is the fundamental equation of engineering for a one-unknown system, when the forcing function is a sinusoid. It is a perfect opportunity to use the **transfer function**. This connects the input to the response.

For mechanical engineers the unknown $y$ gives the position of one mass—oscillating or rotating or vibrating. For electrical engineers the unknown $y$ is the voltage $V(t)$ or the current $I(t)$ in a one-loop RLC circuit. Those letters R, L, C represent a resistor, an inductor, and a capacitor. For a chemical engineer or a scientist or an economist the equation is a model of ..... I have to stop or this presentation will go out of control.

The great differential equations of applied mathematics are *first order or second order*. The equations we understand best are *linear with constant coefficients*.

In later chapters the single unknown becomes a vector. Its coefficients become square matrices in $\boldsymbol{dy/dt = Ay}$ and $\boldsymbol{d^2y/dt^2 = -Sy}$. We have a system of $n$ equations for voltages at nodes or currents along edges or positions of $n$ masses. Linear algebra will organize the equations and their solutions. *Matrix differential equations give us the right language to express applied mathematics.*

Our goals are to find and solve the equations for $y(t)$ in real applications. These are **balance equations**: balance of forces and balance of currents. **Flow in equals flow out.**

### Spring-Mass-Dashpot Equation and Loop Equation

In mechanics, $y$ and $y'$ and $y''$ are the position, the velocity, and the acceleration. The numbers $A, B, C$ represent the *mass* $m$, the *damping* $b$, and the *stiffness* $k$ :

**Newton's Law $F = ma$** $\qquad \boldsymbol{my'' + by' + ky =}$ **applied force.** (1)

The picture in Figure 2.12 shows the mass $m$ attached to a spring and also a dashpot. Those two are responsible for the forces $-ky$ and $-by'$. The stretched spring pulls back on the mass. By Hooke's Law that force is $-ky$. The damping force comes from a dashpot (old-fashioned word, key idea). You could visualize the mass moving in a heavy liquid like oil. The friction force is $-by'$, proportional to velocity and in the opposite direction.

For an electrical network, it was Kirchhoff and not Newton who provided the balance equations. **Kirchhoff's Voltage Law says that the sum of voltage drops around any closed loop is zero**. The current is $I(t)$ and we start with one loop:

**Voltage law KVL :** $$L\frac{dI}{dt} + RI + \frac{1}{C}\int I\,dt = \text{applied voltage.} \qquad (2)$$

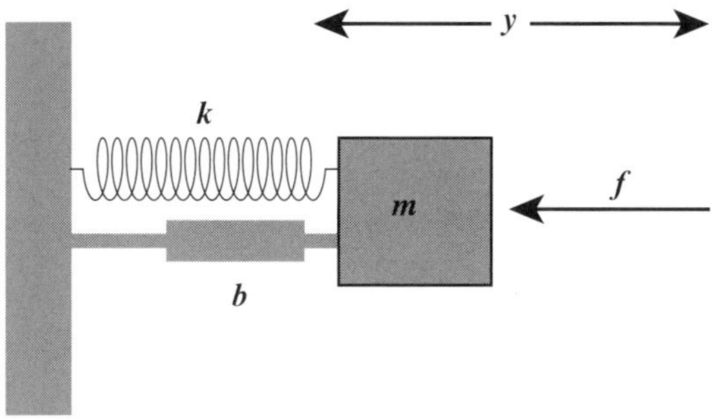


Figure 2.12: Three forces enter $F = my''$: spring force $ky$, friction $by'$, driving force $f$.

The numbers $L, R, C$ are the inductance, the resistance, and the capacitance. (Unfortunately we divide by the capacitance $C$. In the end the equation has constant coefficients and regardless of the letters we solve it.) To produce a second order differential equation for $I(t)$, and to remove the integration in equation (2), take the derivative of every term:

**Loop equation for the current $I(t)$** $$LI'' + RI' + \frac{1}{C}I = F\cos\omega t. \quad (3)$$

That force $F\cos\omega t$ comes from a battery or a generator, when we close the switch. We will be looking for a **particular solution $I_p(t)$**. That solution is produced by the applied force. We are *not* looking at initial conditions and $y_n(t)$. Those null solutions $y_n$ are transient, with $f = 0$. They die out exponentially fast.

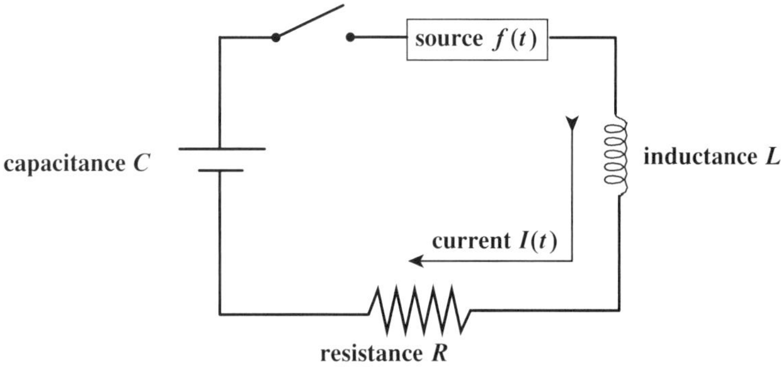


Figure 2.13: A one-loop RLC circuit with a source and a switch.

## The Mechanical-Electrical Analogy

Both applications produce second order equations $Ay'' + By' + Cy = f(t)$. This means we can solve both problems at once—not only mathematically but also physically. We can predict the behavior of a mechanical system by testing an electrical analog, when simple circuit elements are more convenient to work with. The basic idea is to match the three numbers $m, b, k$ with the numbers $L$, $R$, and $1/C$.

| **Mechanical System** | | **Electrical System** |
|---|---|---|
| Mass $m$ | $\longleftrightarrow$ | Inductance $L$ |
| Damping constant $\boldsymbol{b}$ | $\longleftrightarrow$ | Resistance $R$ |
| Spring constant $k$ | $\longleftrightarrow$ | Reciprocal capacitance $1/C$ |
| Natural frequency $\omega_n^2 = k/m$ | $\longleftrightarrow$ | Natural frequency $\omega_n^2 = 1/LC$ |

Before solving for the loop current $I(t)$, let me outline three solution methods—our past method, our present method, and our future method.

## $\cos\omega t$ to $e^{i\omega t}$ to $Y(\omega)$

**Past method** Section 2.4 solved $Ay'' + By' + Cy = F\cos\omega t$. The equation was real and the solution was real. That solution had a sine-cosine form and also an amplitude-phase form:

$$y(t) = M\cos\omega t + N\sin\omega t = G\cos(\omega t - \alpha). \tag{4}$$

The connections between inputs $F$ and outputs $M, N$ came by substituting $y(t)$ into the differential equation and matching terms. Then $G^2 = M^2 + N^2$ and $M = G\cos\alpha$.

**Present method** Instead of working with $\cos\omega t$ and $\sin\omega t$, it is much cleaner to work with a **complex input** $\boldsymbol{Ve^{i\omega t}}$. Then the output (the current) is *a multiple of* $Ve^{i\omega t}$. That multiple $Y$ is a complex number. It tells us amplitudes and also phase shifts.

This is the right way to see the response of a one-loop RLC circuit. When the input frequency is $\omega$, the output frequency is also $\omega$.

**Equation** $$L\frac{dI}{dt} + RI + \frac{1}{C}\int I\,dt = \text{applied voltage} = Ve^{i\omega t} \tag{5}$$

**Solution** $$I(t) = \frac{Ve^{i\omega t}}{i\omega L + R + 1/i\omega C} = \frac{\textbf{input}}{\textbf{impedance}} \tag{6}$$

We will study that complex impedance in detail.

**Future method** Once we see the advantages of a complex $e^{i\omega t}$, we won't go back. What we are really doing is *to change a differential equation for $y$ in the* **time domain** *into an algebraic equation for $Y$ in the* **frequency domain**:

**Set $y = Ye^{i\omega t}$** $\quad Ay'' + By' + Cy = e^{i\omega t}$ becomes $(i^2\omega^2 A + i\omega B + C)Y = 1.$

Derivatives of $y(t)$ become multiplications by $i\omega$. We are talking here about the most important and useful simplification in applied mathematics. It requires constant coefficients $A, B, C$. This allows us to factor out $e^{i\omega t}$.

The **transfer function** $\boldsymbol{Y(s)}$ takes two more steps from derivatives to algebra. First, it changes $e^{i\omega t}$ to $e^{st}$. That exponent $s$ can be pure imaginary ($s = i\omega$). It can also be any complex number ($s = a + i\omega$). We recover the freedom of Chapter 1, to allow growth or decay from $a > 0$ or $a < 0$. We are interested in *all* $s$ and not just the special $s_1$ and $s_2$ that came from solving $As^2 + Bs + C = 0$.

The exponentials $e^{s_1 t}$ and $e^{s_2 t}$ went into the transient solution $y_n(t)$. Now we are working with the long-time solution $y_p(t)$ coming from an applied force $Fe^{st}$.

The second contribution of the transfer function is to give a name to the all-important multiplier in the system. It multiplies the input to give the output.

$$\textbf{The transfer function is } Y(s) = \frac{1}{As^2 + Bs + C}. \quad \textbf{The output is } \boldsymbol{Y(s)} \textbf{ times } \boldsymbol{e^{st}}.$$

Derivatives and integrals become multiplications and divisions (by $s$). One more name is needed. $Y(s)$ is the **Laplace transform** of the impulse response $g(t)$.

| | | |
|---|---|---|
| Input $f = \boldsymbol{\delta(t)}$ | Output $y = g(t) =$ **impulse response** | Transform $\boldsymbol{Y(s)}$ |
| Input $f =$ **step** | Output $y = r(t) =$ **step response** | Transform $\boldsymbol{Y(s)/s}$ |

The step function is the integral of the impulse $\delta(t)$. The step response is the integral of the impulse response $g(t)$. For their Laplace transforms, integration becomes division by $s$. *Calculus in the time domain becomes algebra in the frequency domain.*

The rules for the transforms of $dy/dt$ and $\int y(t)\,dt$, and also a table of inverse Laplace transforms to recover $y(t)$ from $Y(s)$, will come in Section 2.7.

## Complex Impedance

The present method uses $Ve^{i\omega t}$ for the alternating current input. The output divides that input by the impedance $Z$. This is like Ohm's Law $I = E/R$, but the resistance $R$ changes to the impedance $Z$ for this RLC loop:

$$\textbf{Current} \qquad I(t) = \frac{Ve^{i\omega t}}{i\omega L + R + 1/i\omega C} = \frac{Ve^{i\omega t}}{Z} = \frac{\textbf{input}}{\textbf{impedance}}. \tag{7}$$

The complex impedance $Z$ depends on $\omega$. The real part of $Z$ is the resistance $R$. The imaginary part of $Z$ is the "reactance" $\omega L - 1/\omega C$. From those rectangular coordinates Re $Z$ and Im $Z$, we know the polar form $|Z|e^{i\alpha}$ of this complex number:

$$\textbf{Magnitude} \qquad |Z| = \sqrt{R^2 + (\omega L - 1/\omega C)^2} \tag{8}$$

$$\textbf{Phase angle} \qquad \tan\alpha = \frac{\operatorname{Im} Z}{\operatorname{Re} Z} = \frac{\omega L - 1/\omega C}{R} \tag{9}$$

$$\textbf{Loop current} \qquad I(t) = \frac{Ve^{i\omega t}}{Z} = \frac{V}{|Z|}\,e^{i(\omega t - \alpha)} \tag{10}$$

The phase angle $\alpha$ tells us the time lag of the current behind the voltage.

Remember that $R$ is the damping constant, like the coefficient $B$ in $Ay'' + By' + Cy$. In the language of Section 2.4, we have *forced damped motion*. The damping keeps us away from exact resonance with the natural frequency of free undamped motion—which has $\omega L = 1/\omega C$ and $\omega = 1/\sqrt{LC}$. The magnitude $|Z|$ is smallest and $V/|Z|$ is largest at that natural frequency. We tune a radio to this $\omega$ to get a loud clear signal.

**Example 1** Suppose the RLC circuit has resistance $R = 10$ ohms and inductance $L = 0.1$ henry and capacitance $C = 10^{-4}$ farad. The units of $R$ and $\omega L$ and $1/\omega C$ must agree. Since frequency $\omega$ is measured in inverse seconds, all three units can be given in terms of $V =$ volts and $A =$ amps (for current) and seconds:

| | | | |
|---|---|---|---|
| **R** | Ohm $\Omega$ | $= V/A$ | $= 1$ volt per amp |
| **L** | Henry $H$ | $= V \cdot \text{sec}/A$ | $= 1$ volt-second per amp |
| **C** | Farad $F$ | $= A \cdot \text{sec}/V$ | $= 1$ amp-second per volt |

**Example 2** Find the impedance $Z$, its magnitude $|Z|$, and the phase angle $\alpha$ for an RLC loop when the frequency is $\omega = 60$ cycles/second $= 60$ Hz $= 120\pi$ radians/second.

The impedance of this loop is $\quad Z = R + i\left(\omega L - \dfrac{1}{\omega C}\right) = |Z|e^{-i\alpha}$.

The magnitude of the impedance is $\quad |Z| = \ldots$

The phase angle producing time delay is $\quad \alpha = \ldots$

**Example 3** To tune a radio to a station with frequency $\omega$, what should be the capacitance $C$ (which you adjust)? Suppose $R$ and $L$ are fixed and known.

*Solution* The goal of tuning is to achieve $\omega L = 1/\omega C$. Then the imaginary part of $Z$ is zero: *inductance cancels capacitance*. Tuning achieves $Z = R$, that real part $R$ is fixed.

$$\omega L = \frac{1}{\omega C} \qquad \omega^2 = \frac{1}{LC} \qquad C = \frac{1}{L\omega^2}$$

**Example 4** Suppose the network contains **two RLC branches in parallel**. Find the total impedance $Z_{12}$ from the impedances $Z_1$ and $Z_2$ of the two separate branches.

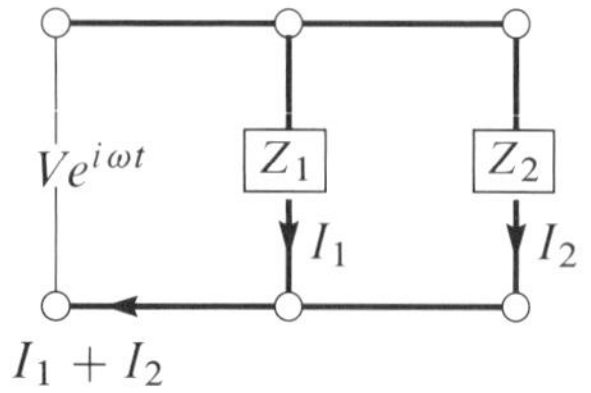


$$\frac{1}{Z_{12}} = \frac{1}{Z_1} + \frac{1}{Z_2} = \frac{Z_1 + Z_2}{Z_1 Z_2}$$

$$I_{12} = I_1 + I_2 = \frac{Z_1 Z_2}{Z_1 + Z_2}\, Ve^{i\omega t}$$

## Loop Equations Versus Node Equations : KVL or KCL

Equation (2) expressed Kirchhoff's Voltage Law. **The sum of voltage drops around a closed loop is zero**. In principle, we could find a set of independent loops in any larger electrical network. Then the Voltage Law will give an equation like (2) around each of the independent loops. Those loop currents determine the currents on all the edges of the network and the voltages at all the nodes.

Most codes to solve problems on large networks *do not use the voltage law*! The preferred approach is **Kirchhoff's Current Law**: **The net current into each node is zero**.

The balance equations of KCL say that "current in = current out" at every node.

Let me illustrate nodal analysis using the network in Figure 2.14. The unknowns are the voltages $V_1$ and $V_2$. The currents are easy to find once those voltages are known.

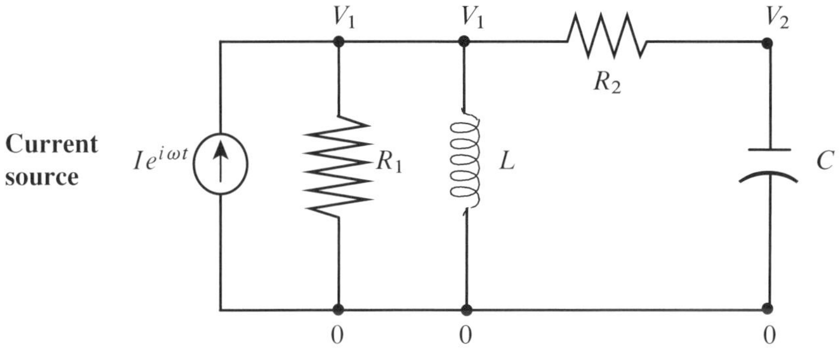


Figure 2.14: Four currents in and out of Node 1. *Node* 2 : Current in, current out.

A problem of this size can be solved symbolically or numerically :

**Symbolically** Work in the $s$-domain and find the transfer function. Since $R_1$ is in parallel with $L$, and $R_2$ is in series with $C$, we can find the currents on all the edges in terms of $V_1$ and $V_2$. Here is Kirchhoff's Current Law at those nodes :

$$\frac{V_1}{R_1} + \frac{V_1}{Ls} + \frac{V_1 - V_2}{R_2} = I \qquad \text{and} \qquad \frac{V_2 - V_1}{R_2} + sCV_2 = 0 \qquad (11)$$

**Numerically** Assign values to $R_1$, $L$, $R_2$, $C$ and $\omega$. Compute $V_1$ and $V_2$ from current balance at the nodes. Compute the currents from $V_1/R_1$ and $V_2/iL\omega$.

For a larger network, the algebra in the $s$-domain ($i\omega$ domain) becomes humanly impossible. A symbolic package could go further but in the end (and for nonlinear networks) the numerical approach will win. Widely known codes developed from the original SPICE code created at UC Berkeley. The SPICE codes use nodal analysis instead of loop analysis, for realistic networks.

Computational mechanics faced the same choice between nodal analysis and loop analysis. It reached the same conclusion. A complicated structure is broken up into *finite elements*—small pieces in which linear or quadratic approximation is adequate. The choice is between displacements at nodes or stresses inside the elements, as the primary unknowns. The finite element community has made the same decision as the circuit simulation community: *Work with displacements* (and work with voltages) *at the nodes*.

A network produces a large system of equations—linear equations with simple RLC elements and nonlinear equations for circuit elements like transistors. **The nodes connected by the edges form a graph**. To organize the equations, you need the basic concepts of graph theory in Section 5.5:

An **incidence matrix** $A$ tells which pairs of nodes are connected by which edges.

A **conductivity matrix** $C$ expresses the physical properties along each edge.

Then the overall conductance matrix is $\boldsymbol{K = A^{\mathrm{T}}CA}$. The system we solve, for linear problems in circuit simulation and in structural mechanics, has the matrix form $\boldsymbol{Ky = f}$.

Chapter 4 will explain matrices and Section 5.5 will focus on the incidence matrix $A$ of a graph. Those are necessary preparations for Kirchhoff's Current Law at all the nodes. Then Sections 7.4 and 7.5 create the stiffness matrix $K$ (for mechanics) and the graph Laplacian matrix (for networks): basic ideas in applied mathematics.

## Step Response

This book has emphasized the two fundamental problems for differential equations. One is the response to a delta function. The other is the response to a step function. For second order equations the impulse response $g(t)$ was computed in Section 2.3. This is our chance to find the step response, and we have to take it.

The two responses are closely related because the two inputs are related. The delta function is the derivative of the step function $H(t)$. The step function is the integral of the delta function. For constant coefficient equations, we can integrate every term. **The integral of the impulse response $\boldsymbol{g(t)}$ is the step response $\boldsymbol{r(t)}$.**

| | | |
|---|---|---|
| **Impulse response** $\boldsymbol{g(t)}$ | $Ag'' + Bg' + Cg = C\delta(t)$ | (12) |
| **Step response** $\boldsymbol{r(t)}$ | $Ar'' + Br' + Cr = CH(t)$ | (13) |

We are following the "better notation" convention that includes the coefficient $C$ on the right hand side. Its purpose is to give the output $y$ or $g$ or $r$ the same units as the forcing term. Then the gain $G = |\text{output/input}|$ is dimensionless. For the step function with input $H(t) = 1$, **the steady state of the step response will be** $\boldsymbol{r(\infty) = 1}$.

I see two ways to compute that step response. One is to integrate the impulse response. The other is to solve equation (13) directly. The particular solution is $\boldsymbol{r_p(t)} = 1$. The null solution is a combination of $e^{s_1 t}$ and $e^{s_2 t}$, using the two roots of $As^2 + Bs + C = 0$.

To be safe, it seems reasonable to find $r(t)$ both ways.

$$\textbf{Method 1} \quad \text{Integrate the impulse response } g(t) = \frac{C}{A}\frac{e^{s_1 t} - e^{s_2 t}}{s_1 - s_2} \tag{14}$$

$$\textbf{Method 2} \quad \text{Solve } Ar'' + Br' + Cr = C \text{ with } r(0) = r'(0) = 0. \tag{15}$$

## Computing the Step Response

Method 2 is the normal way to solve differential equations. Substitute $e^{st}$ to find $s_1$

$$\textbf{Null solutions } e^{st} \qquad As^2 + Bs + C = 0 \text{ has roots } s_1 \text{ and } s_2.$$

The complete solution to $Ar'' + Br' + Cr = C$ is *particular* + *null*:

$$r(t) = 1 + c_1 e^{s_1 t} + c_2 e^{s_2 t}. \tag{16}$$

The step response starts from $r(0) = 0$ and $r'(0) = 0$. A switch is turned on at $t = 0$, and the solution rises to $r(\infty) = 1$. The conditions at $t = 0$ determine $c_1$ and $c_2$:

$$r(0) = 1 + c_1 + c_2 = 0 \qquad r'(0) = c_1 s_1 + c_2 s_2 = 0. \tag{17}$$

Those coefficients are $c_1 = s_2/(s_1 - s_2)$ and $c_2 = -s_1/(s_1 - s_2)$. Then we know $r(t)$:

$$\textbf{Step response} \qquad r(t) = 1 + \frac{1}{s_1 - s_2}\left(s_2 e^{s_1 t} - s_1 e^{s_2 t}\right). \tag{18}$$

The same answer must come from integrating $g(t)$ in equation (14) from 0 to $t$. Remember that the roots of any quadratic multiply to give $s_1 s_2 = C/A$.

$$\textbf{Step response} = \textbf{integral of } g(t) \qquad r(t) = \frac{s_1 s_2}{s_1 - s_2}\left[\frac{e^{s_1 t} - 1}{s_1} - \frac{e^{s_2 t} - 1}{s_2}\right]. \tag{19}$$

The coefficient of $e^{s_1 t}$ is the same $s_2/(s_1 - s_2)$ as in (18). Similarly for the coefficient of $e^{s_2 t}$. The constant term equals 1, so (18) and (19) are the same:

$$\frac{s_1 s_2}{s_1 - s_2}\left[-\frac{1}{s_1} + \frac{1}{s_2}\right] = \frac{s_1 s_2}{s_1 - s_2}\left[\frac{s_1 - s_2}{s_1 s_2}\right] = 1.$$

## Better Notation

Our formula for the step response $r(t)$ can't stop with equation (18). Those roots $s_1$ and $s_2$ will depend on the physical parameters $A$, $B$, $C$. In mechanics these numbers are $m, b, k$. For a one-loop network the numbers are $L$, $R$, $1/C$. We need to express $r(t)$ with numbers we know, instead of $s_1$ and $s_2$.

Remember that *combinations* of $A$, $B$, $C$ are especially useful. The simplest choices are $p = B/2A$ and $\omega_n^2$:

$$r'' + \frac{B}{A}r' + \frac{C}{A}r = \frac{C}{A} \quad \text{becomes} \quad \boldsymbol{r'' + 2pr' + \omega_n^2 r = \omega_n^2}. \tag{20}$$

The same exponents $s_1$ and $s_2$ are now roots of $s^2 + 2ps + \omega_n^2 = 0$. Suppose $p < \omega_n$:

$$\textbf{Null solutions } \boldsymbol{e^{st}} \qquad \boldsymbol{s_1, s_2} = -p \pm \sqrt{p^2 - \omega_n^2} = \boldsymbol{-p \pm i\omega_d}. \tag{21}$$

Substituting for $s_1$ and $s_2$ in equation (18) gives a beautiful expression for $r(t)$:

$$\textbf{Step response} \qquad \boldsymbol{r(t) = 1 - \frac{\omega_n}{\omega_d} e^{-pt} \sin(\omega_d t + \phi)}. \tag{22}$$

That angle $\phi$ is in the right triangle that connects $\omega_n$ to $p$ and $\omega_d$:

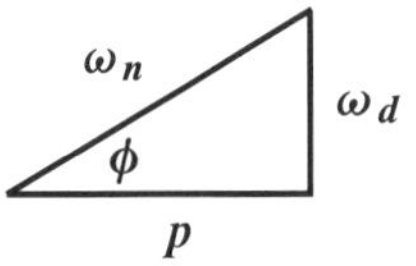


$$\omega_d^2 + p^2 = \omega_n^2 \qquad \sin\phi = \frac{\omega_d}{\omega_n} \qquad \cos\phi = \frac{p}{\omega_n}$$

Now we check that $r(0) = 0$ and $r'(0) = 0$—then formula (22) must be correct:

$$r(0) = 1 - \frac{\omega_n}{\omega_d}\sin\phi = 0 \qquad r'(0) = \frac{\omega_n}{\omega_d}(p\sin\phi - \omega_d\cos\phi) = 0.$$

That final solution (22) combines $e^{-pt}\sin\omega_d t$ and $e^{-pt}\cos\omega_d t$. This null solution is a combination of $e^{s_1 t}$ and $e^{s_2 t}$ with $s = -p \pm i\omega_d$, as required. The particular solution is $r(\infty) = 1$. We see this steady state appear when the transients decay to zero with $e^{-pt}$. *The step response rises to* 1.

The number $p = B/2A$ can be replaced by $\omega_n$ times the damping ratio, if preferred.

## Practical Resonance: Minimum $D$, Maximum Gain

The gain is $1/\sqrt{D}$. *If $D$ is small then the gain is large.* That is how you tune a radio, by choosing the frequency $\omega_{res}$ that minimizes $D$ and maximizes $G$. Then you can hear the signal. It is not perfect resonance—the gain does not become infinite—but it is resonance in practice.

$$\begin{array}{lll} \textbf{Practical resonance} & \text{Minimize} & D = \left(C - A\omega^2\right)^2 + (B\omega)^2 \\ \textbf{Derivative of } \boldsymbol{D} \textbf{ is zero} & & -4A\omega(C - A\omega^2) + 2B^2\omega = 0. \end{array}$$

When you cancel $\omega$ and solve $2B^2 = 4A(C - A\omega^2)$, that gives the frequency $\omega_{res}$ with largest gain. When $B = 0$ this is the natural frequency $\omega_n$ with infinite gain: $A\omega_n^2 = C$.

For $2Z^2 < 1$ there is practical resonance when $2B^2 = 4A(C - A\omega^2)$ at $\omega_{res}$ :

**Largest gain** $\quad \omega_{res}^2 = \dfrac{C}{A} - \dfrac{B^2}{2A^2} = \dfrac{C}{A}\left(1 - \dfrac{B^2}{2AC}\right) = \omega_n^2(1 - 2Z^2).$

## ■ REVIEW OF THE KEY IDEAS ■

1. $L, R, C$ in $LI'' + RI' + \frac{1}{C}I = e^{i\omega t}$ are the inductance, resistance, capacitance.
2. For networks, node equations replace that loop equation: KCL instead of KVL.
3. The response to a step function rises from $r(0) = 0$ to a steady value $r(\infty) = 1$.
4. Practical resonance (the maximum gain) is at the frequency $\omega_{res} = \omega_n\sqrt{1 - 2\zeta^2}$.

**Important note** We computed the step response $r(t)$ in the time domain. Using the Laplace transform in Section 2.7, this computation can be moved to the $s$-domain. The transform of a unit step is $1/s$. *Derivatives in $t$ become multiplications by $s$:*

**The state equation $Ar'' + Br' + Cr = C$ transforms to $(As^2 + Bs + C)R(s) = \dfrac{C}{s}$.**

The problem is to find the inverse Laplace transform $r(t)$ of this function $R(s)$. There are excellent control engineering textbooks that leave this as an exercise in partial fractions. The time domain (state space) solution in this section reached $r(t)$ successfully.

## Problem Set 2.5

**1** (Resistors in parallel) Two parallel resistors $R_1$ and $R_2$ connect a node at voltage $V$ to a node at voltage zero. The currents are $V/R_1$ and $V/R_2$. What is the total current $I$ between the nodes? Writing $R_{12}$ for the ratio $V/I$, what is $R_{12}$ in terms of $R_1$ and $R_2$?

**2** (Inductor and capacitor in parallel) Those elements connect a node at voltage $Ve^{i\omega t}$ to a node at voltage zero (grounded node). The currents are $(V/i\omega L)e^{i\omega t}$ and $V(i\omega C)e^{i\omega t}$. The total current $Ie^{i\omega t}$ between the nodes is their sum. Writing $Z_{12}e^{i\omega t}$ for the ratio $Ve^{i\omega t}/Ie^{i\omega t}$, what is $Z_{12}$ in terms of $i\omega L$ and $i\omega C$?

**3** The impedance of an RLC loop is $Z = i\omega L + R + 1/i\omega C$. This impedance $Z$ is real when $\omega =$ _____. This impedance is pure imaginary when _____. This impedance is zero when _____.

**4** What is the impedance $Z$ of an RLC loop when $R = L = C = 1$? Draw a graph that shows the magnitude $|Z|$ as a function of $\omega$.

**5** Why does an LC loop with no resistor produce a $90°$ phase shift between current and voltage? Current goes around the loop from a battery of voltage $V$ in the loop.

**6** The mechanical equivalent of zero resistance is zero damping: $my'' + ky = \cos \omega t$. Find $c_1$ and $Y$ starting from $y(0) = 0$ and $y'(0) = 0$ with $\omega_n^2 = k/m$.

$$y(t) = c_1 \cos \omega_n t + Y \cos \omega t.$$

That answer can be written in two equivalent ways:

$$y = Y(\cos \omega t - \cos \omega_n t) = 2Y \sin \frac{(\omega_n - \omega)t}{2} \sin \frac{(\omega_n + \omega)t}{2}.$$

**7** Suppose the driving frequency $\omega$ is close to $\omega_n$ in Problem 6. A fast oscillation $\sin[(\omega_n + \omega)t/2]$ is multiplying a very slow oscillation $2Y \sin[(\omega_n - \omega)t/2]$. By hand or by computer, draw the graph of $y = (\sin t)(\sin 9t)$ from $0$ to $2\pi$.

You should see a fast sine curve inside a slow sine curve. This is a **beat**.

**8** What $m, b, k, F$ equation for a mass-dashpot-spring-force corresponds to Kirchhoff's Voltage Law around a loop? What force balance equation on a mass corresponds to Kirchhoff's Current Law?

**9** If you only know the natural frequency $\omega_n$ and the damping coefficient $b$ for one mass and one spring, why is that *not enough* to find the damped frequency $\omega_d$? If you know all of $m, b, k$ what is $\omega_d$?

**10** Varying the number $a$ in a first order equation $y' - ay = 1$ changes the *speed* of the response. Varying $B$ and $C$ in a second order equation $y'' + By' + Cy = 1$ changes the *form* of the response. Explain the difference.

**11** Find the step response $r(t) = y_p + y_n$ for this overdamped system:

$$r'' + 2.5r' + r = 1 \quad \text{with} \quad r(0) = 0 \quad \text{and} \quad r'(0) = 0.$$

**12** Find the step response $r(t) = y_p + y_n$ for this critically damped system. The double root $s = -1$ produces what form for the null solution?

$$r'' + 2r' + r = 1 \quad \text{with} \quad r(0) = 0 \quad \text{and} \quad r'(0) = 0.$$

**13** Find the step response $r(t)$ for this underdamped system using equation (22):

$$r'' + r' + r = 1 \quad \text{with} \quad r(0) = 0 \quad \text{and} \quad r'(0) = 0.$$

**14** Find the step response $r(t)$ for this undamped system and compare with (22):

$$r'' + r = 1 \quad \text{with} \quad r(0) = 0 \quad \text{and} \quad r'(0) = 0.$$

**15** For $b^2 < 4mk$ (underdamping), what parameter decides the speed at which the step response $r(t)$ rises to $r(\infty) = 1$? Show that the **peak time** is $T = \pi/\omega_d$ when $r(t)$ reaches its maximum before settling back to $r = 1$. At peak time $r'(T) = 0$.

**16** If the voltage source $V(t)$ in an RLC loop is a unit step function, what resistance $R$ will produce an overshoot to $r_{\max} = 1.2$ if $C = 10^{-6}$ Farads and $L = 1$ Henry? (Problem 15 found the peak time $T$ when $r(T) = r_{\max}$).

Sketch two graphs of $r(t)$ for $p_1 < p_2$. Sketch two graphs as $\omega_d$ increases.

**17** What values of $m, b, k$ will give the step response $r(t) = 1 - \sqrt{2}e^{-t}\sin(t + \frac{\pi}{4})$?

**18** What happens to the $p - \omega_d - \omega_n$ right triangle as the damping ratio $\omega_n/p$ increases to 1 (critical damping)? At that point the damped frequency $\omega_d$ becomes _____. The step response becomes $r(t) =$ _____.

**19** **The roots $s_1, s_2 = -p \pm i\omega_d$ are poles of the transfer function $1/(As^2 + Bs + C)$**

Show directly that the product of the roots $s_1 = -p + i\omega_d$ and $s_2 = -p - i\omega_d$ is $s_1 s_2 = \omega_n^2$. The sum of the roots is $-2p$. The quadratic equation with those roots is $s^2 + 2ps + \omega_n^2 = 0$.

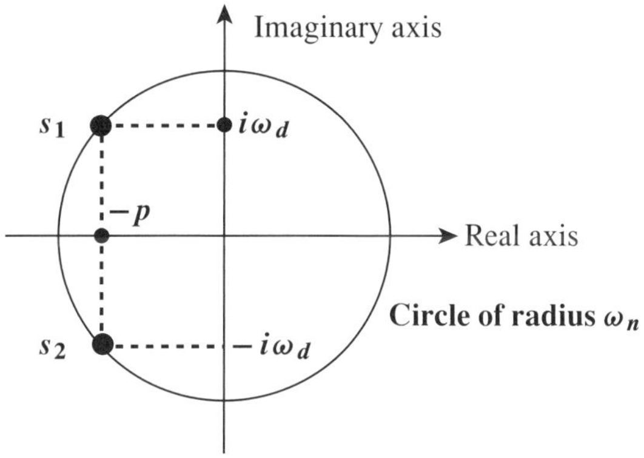


**20** Suppose $p$ is increased while $\omega_n$ is held constant. How do the roots $s_1$ and $s_2$ move?

**21** Suppose the mass $m$ is increased while the coefficients $b$ and $k$ are unchanged. What happens to the roots $s_1$ and $s_2$?

**22** **Ramp response** How could you find $y(t)$ when $F = t$ is a ramp function?

$$y'' + 2py' + \omega_n^2 y = \omega_n^2 t \quad \text{starting from} \quad y(0) = 0 \text{ and } y'(0) = 0.$$

A particular solution (straight line) is $y_p =$ _____. The null solution still has the form $y_n =$ _____. Find the coefficients $c_1$ and $c_2$ in the null solution from the two conditions at $t = 0$.

This ramp response $y(t)$ can also be seen as the integral of _____.

## 2.6 Solutions to Second Order Equations

Up to now, all forcing terms $f(t)$ for second order equations have been $e^{st}$ or $\cos\omega t$. How can you find a particular solution when $f(t)$ is not a sinusoid or exponential? This section gives one answer for constant coefficients $A, B, C$ and then a general answer ***VP*** :

***UC*** If $f(t)$ is a polynomial in $t$, then $y_p(t)$ is also a polynomial in $t$.

***VP*** Suppose we know the null solutions $y_n = c_1 y_1(t) + c_2 y_2(t)$. Then a particular solution has the form $y_p = c_1(t) y_1(t) + c_2(t) y_2(t)$.

Those methods are called "**undetermined coefficients**" and "**variation of parameters**".

The special method is simple to execute (you will like it). When $f(t)$ is a quadratic, then one solution is also a quadratic: $y_p(t) = at^2 + bt + c$. Those numbers $a, b, c$ are the **undetermined coefficients**. The differential equation will determine them. This succeeds for any constant coefficient differential equation—always limited to special $f(t)$.

That method ***UC*** can be pushed further. If $f(t)$ is a polynomial times an exponential, then $y_p(t)$ has the same form. The highest power of $t$ allowed in $y_p$ is the same as in $f$. Those polynomials normally have the same degree.

Only in the case of resonance must we allow an extra factor $t$ in the solution. This is like the exponential response to $f(t) = e^{ct}$ in Section 2.4. That presented a perfect example of an undetermined coefficient $Y$ in $y_p(t) = Ye^{st}$. The coefficient $Y = 1/(As^2 + Bs + C)$ was determined by the equation. This is $Y = 1/P(s)$ for all equations $P(D)y = e^{st}$. With resonance we move to $y_p = te^{st}/P'(s)$.

**Variation of parameters** is a more powerful method. It applies to all $f(t)$. It even applies when the equation $A(t)y'' + B(t)y' + C(t)y = f(t)$ has variable coefficients. But it starts with a big assumption: **We have to know the null solutions** $y_1(t)$ **and** $y_2(t)$.

The method will succeed completely when the coefficients $A, B, C$ are constant. This important case gives formula (17). Variation of parameters also succeeded in Chapter 1, for first order equations $y' - a(t)y = q(t)$. In that case we could solve the null equation $y' = a(t)y$. For second order equations with variable coefficients, like Airy's equation $y'' = ty$, the null equation is a difficult obstacle.

I guess we have to realize that not all problems lead to simple formulas.

### The Method of Undetermined Coefficients

This direct approach finds a particular solution $y_p$, when the forcing term $f(t)$ has a special form. I can explain the method of undetermined coefficients by four examples.

**Example 1** $\boldsymbol{y'' + y = t^2}$ **has a solution of the form** $\boldsymbol{y = at^2 + bt + c}$.

The reason for this choice of $y$ is that $y'$ and $y''$ will have a similar form. They will also be combinations of $t^2$ and $t$ and 1. *All the terms in* $y'' + y = t^2$ *will have this special form.*

Choose the numbers $a, b, c$ to satisfy that equation:

$$\boldsymbol{y'' + y = (at^2 + bt + c)'' + (at^2 + bt + c) = t^2.} \qquad (1)$$

Key idea: **We can separately match the coefficients of $t^2$ and $t$ and 1 in equation (1)**:

$$(\boldsymbol{t^2}) \quad a = 1 \qquad (\boldsymbol{t}) \quad b = 0 \qquad (\mathbf{1}) \quad 2a + c = 0 \qquad (2)$$

Then $c = -2a = -2$ and the answer is $y = at^2 + c = \boldsymbol{t^2 - 2}$. This solves $y'' + y = t^2$.

**Example 2** Find the complete solution to $\boldsymbol{y'' + 4y' + 3y = e^{-t} + t}$.

**Answer** First find the null solution to $y_n'' + 4y_n' + 3y_n = 0$, by substituting $y_n = e^{st}$:

$$(s^2 + 4s + 3)e^{st} = 0 \text{ leads to } s^2 + 4s + 3 = (s+1)(s+3) = 0.$$

The roots are $s_1 = -1$ and $s_2 = -3$. The null solutions are $y_n = c_1 e^{-t} + c_2 e^{-3t}$.

Now find one particular solution. With $f = e^{-t} + t$, the usual form with undetermined coefficients would be $y_p = ae^{-t} + bt + c$ (notice $c$ in the polynomial). *But $e^{-t}$ is a null solution.* Therefore the assumed form for $y$ needs an extra factor $t$ multiplying $e^{-t}$.

**Substitute $\boldsymbol{y = ate^{-t} + bt + c}$** into the differential equation, so $y' = ae^{-t} - ate^{-t} + b$:

$$y'' + 4y' + 3y = (-2ae^{-t} + ate^{-t}) + 4(ae^{-t} - ate^{-t} + b) + 3(ate^{-t} + bt + c) = e^{-t} + t.$$

The coefficients of $te^{-t}$ are $a - 4a + 3a = 0$. No problem with this $te^{-t}$ term. We must balance the coefficients of $e^{-t}$ and $t$ and 1:

$$\textbf{Find } \boldsymbol{a, b, c} \qquad \boldsymbol{-2a + 4a = 1} \qquad \boldsymbol{3b = 1} \qquad \boldsymbol{4b + 3c = 0}$$

Then $a = \frac{1}{2}$ and $b = \frac{1}{3}$ and $c = -\frac{4}{9}$ produce the particular $y_p = \boldsymbol{\frac{1}{2}}te^{-t} + \boldsymbol{\frac{1}{3}}t - \boldsymbol{\frac{4}{9}}$. The null solution is $c_1 e^{-t} + c_2 e^{-3t}$. The complete solution is always $y = y_p + y_n$.

The method only applies to very special forcing functions, but when it succeeds it is as fast and simple as possible. Let me list special inputs $f(t)$ and the form of a solution $y(t)$ when the differential equation $Ay'' + By + Cy = f(t)$ has **constant coefficients**.

| | |
|---|---|
| **1. $\boldsymbol{f(t)}$ = polynomial in $\boldsymbol{t}$** | **$\boldsymbol{y(t)}$ = polynomial in $\boldsymbol{t}$ (same degree)** |
| **2. $\boldsymbol{f(t) = A\cos\omega t + B\sin\omega t}$** | **$\boldsymbol{y(t) = M\cos\omega t + N\sin\omega t}$** |
| **3. $\boldsymbol{f(t)}$ = exponential $\boldsymbol{e^{st}}$** | **$\boldsymbol{y(t) = Ye^{st}}$** |
| **4. $\boldsymbol{f(t)}$ = product $\boldsymbol{t^2 e^{st}}$** | **$\boldsymbol{y(t) = (at^2 + bt + c)e^{st}}$** |

$t^2 e^{st}$ is included in **4** by multiplying possibilities **1** and **3**. The good form for $y(t)$ multiplies the solutions to **1** and **3**. The coefficients $M, N, Y, a, b, c$ are "undetermined" until you substitute $y(t)$ into the differential equation. *That equation determines $a$, $b$, $c$.*

*Note to professors* It seems to me that a polynomial times $e^{t^2}$ shares the key property. Its derivatives have the same form. But their polynomial degree goes up. Not good.

**Example 3** Find a particular solution to $\boldsymbol{y'' + y = t\,e^{st}}$ = **polynomial times** $\boldsymbol{e^{st}}$.
The good form to assume for $y(t)$ is $(at + b)\,e^{st}$. *Please notice that $b\,e^{st}$ is included.* Even though $f$ doesn't have $e^{st}$ by itself, that will appear in the derivatives of $t\,e^{st}$. To be sure we capture every derivative, $at + b$ must include that constant $b$.
I need to find the second derivative of the undetermined $\boldsymbol{y(t) = (at+b)\,e^{st}}$.

$$y' = s(at+b)\,e^{st} + a\,e^{st} \qquad y'' = s^2(at+b)\,e^{st} + 2as\,e^{st}.$$

Substitute $y$ and $y''$ into the equation $y'' + y = t\,e^{st}$ and match terms to find $a$ and $b$:

$$\begin{array}{lrcl} \text{Coefficient of } t\,e^{st} & as^2 + a & = & 1 \\ \text{Coefficient of } e^{st} & bs^2 + 2as + b & = & 0 \end{array}$$

Those two equations produce $\quad a = \dfrac{1}{1+s^2} \quad$ and $\quad b = \dfrac{-2as}{1+s^2} = \dfrac{-2s}{(1+s^2)^2}. \qquad (3)$

Now $y(t) = (at+b)\,e^{st}$ is a particular solution of $y'' + y = t\,e^{st}$.

**Possible difficulty of the method** **Suppose $\boldsymbol{s = i}$ or $\boldsymbol{-i}$ in the forcing term $\boldsymbol{f = t\,e^{st}}$**

Those exponents $s = i$ and $s = -i$ have $1 + s^2 = 0$. Our answer in (3) for $a$ and $b$ is dividing by zero. The result is useless. What went wrong?

**Explanation** If $s = i$, the assumed form $y = (at+b)e^{it}$ includes a solution $be^{it}$ of $y'' + y = 0$. We have accidentally included a null solution $y_n = be^{it}$. There is no hope of determining $b$. That coefficient is truly undetermined and it stays that way.

We are seeing a problem of resonance, when the hoped-for $y_p$ is already a part of $y_n$. The result in Section 2.4 was that *resonant solutions have and need an extra factor $t$*. The same is true here. When $s = i$ or $s = -i$, the good form to assume is $y_p = \boldsymbol{t}(at+b)\,e^{st}$.

When you substitute this $y_p$ into $y'' + y = t\,e^{st}$, the coefficients $a$ and $b$ will be properly determined. If $s = i$, could you verify that $a = -1/4$ and $b = i/4$?

**Example 4** Let me apply "undetermined coefficients" to an equation you already know:

$$Ay'' + By' + Cy = \cos\omega t. \qquad (4)$$

*Solution by undetermined coefficients* Look for $y(t) = M\cos\omega t + N\sin\omega t$. Those coefficients $M$ and $N$ are also in equation (21) of Section 2.4.

$$M = \frac{C - A\,\omega^2}{D} \qquad N = \frac{B\,\omega}{D} \qquad D = (C - A\omega^2)^2 + B^2\omega^2.$$

Is this perfect? Not quite. In case the denominator is $D = 0$, the method will fail. That is exactly the case of resonance, when $A\,\omega^2 = C$ and $B = 0$. The coefficients $M$ and $N$ become $0/0$. The equation becomes $A\,(y'' + \omega^2\,y) = \cos\omega t$. The particular $y_p$ cannot be $M\cos\omega t + N\sin\omega t$ ***because*** $\boldsymbol{\cos\omega t}$ ***and*** $\boldsymbol{\sin\omega t}$ ***are null solutions*** $\boldsymbol{y_n}$. They have $y'' + \omega^2\,y = 0$. The same $\omega$ is on both sides of the equation.

**Resonant solutions** In case $D = 0$, the particular solution again has an extra factor $t$.

Then put $y_p = Mt\cos\omega t + Nt\sin\omega t$ into equation (4) to find $M = 0$ and $N = 1/2$.

### Summary of the Method of Undetermined Coefficients

When the forcing term $f(t)$ is a polynomial or a sinusoid or an exponential, look for a particular solution $y_p(t)$ of the same form. Derivatives of polynomials are polynomials, derivatives of sinusoids are sinusoids, derivatives of exponentials are exponentials. Then all terms in $Ay'' + By' + Cy = f$ will share the same form.

When $f(t) =$ sum of exponentials, look for $y(t) =$ sum of exponentials. When $f$ is a polynomial times a sinusoid or an exponential, $y(t)$ has the same form. When a sinusoid or an exponential in $f$ happens to be a null solution (*resonance*), include an extra $t$ in $y_p$.

**Question** What form would you assume for $y(t)$ when $f(t) = 4e^t + 5\cos 2t + t$ ?

**Answer** Look for $y(t) = Ye^t + M\cos 2t + N\sin 2t + at + b$. The coefficients in the differential equation need to be constants. Then $Ay''$, $By'$, $Cy$ and $f$ all look like $y$.

### Variation of Parameters

Now we want to allow any forcing function $f(t)$. The equation might even have variable coefficients. If we know the null solutions, the method called "variation of parameters" can find a particular solution.

Suppose the null solution with $f = 0$ is $y_n(t) = c_1 y_1(t) + c_2 y_2(t)$. We know $y_1$ and $y_2$. *For a particular solution when* $f(t) \neq 0$, *allow* $c_1$ *and* $c_2$ *to vary with time*:

**Variation of parameters** $\quad \boldsymbol{y_p(t) = c_1(t)y_1(t) + c_2(t)y_2(t)} \quad (5)$

This idea applies to any second order linear differential equation like

$$\frac{d^2y}{dt^2} + B(t)\frac{dy}{dt} + C(t)y = f(t). \tag{6}$$

Substituting $y_p(t)$ from (5) gives a first equation for $c_1{}'$ and $c_2{}'$. Those are the parameters varying with $t$. To recognize a convenient second equation for $c_1{}'$ and $c_2{}'$, compute the derivative of $y_p$ by the product rule :

$$y_p{}' = (c_1(t)y_1{}' + c_2(t)y_2{}') + (c_1{}'(t)y_1 + c_2{}'(t)y_2). \tag{7}$$

A good choice is to require that the second sum be zero :

**Second equation for** $c_1{}', c_2{}'$ $\quad \boldsymbol{c_1{}'(t)y_1(t) + c_2{}'(t)y_2(t) = 0}. \quad (8)$

Now the second sum in (7) drops out and we compute $y_p{}''$ (product rule again) :

$$y_p{}'' = (c_1(t)y_1{}'' + c_2(t)y_2{}'') + (c_1{}'(t)y_1{}' + c_2{}'(t)y_2{}'). \tag{9}$$

Put $y_p$, $y_p{}'$, $y_p{}''$ from (5), (7), (9) into the differential equation to get a wonderful result :

**First equation for** $c_1{}', c_2{}'$ $\quad \boldsymbol{c_1{}'(t)y_1{}'(t) + c_2{}'(t)y_2{}'(t) = f(t)}. \quad (10)$

That became simple because the null solutions $y_1$ and $y_2$ satisfy $y'' + B(t)y' + C(t)y = 0$.

We now have two equations (8) and (10) for two unknowns $c_1{}'(t)$ and $c_2{}'(t)$. At each time $t$, the four coefficients $P, Q, R, S$ in the two equations are the numbers $y_1(t)$, $y_2(t)$, $y_1{}'(t)$, $y_2{}'(t)$. Solve those two equations, first using $P, Q, R, S$:

$$\begin{aligned} Pc_1{}' + Qc_2{}' &= 0 \\ Rc_1{}' + Sc_2{}' &= f \end{aligned} \quad \text{lead to} \quad c_1{}' = \frac{-Qf}{PS - QR} \quad \text{and} \quad c_2{}' = \frac{Pf}{PS - QR}. \tag{11}$$

When you multiply those fractions by $P$ and $Q$, they cancel. When you multiply the fractions by $R$ and $S$ and add, the result is the second equation $Rc_1{}' + Sc_2{}' = f(t)$.

Linear equations come at the beginning of linear algebra in Chapter 4. Here we have a separate problem for each time $t$, and the solution (11) becomes (12) when $P, Q, R, S$ are $y_1(t)$, $y_2(t)$, $y_1{}'(t)$, $y_2{}'(t)$. I will write $W$ for $PS - QR$:

$$\boldsymbol{c_1{}'(t) = \frac{-y_2(t)f(t)}{W(t)} \qquad c_2{}'(t) = \frac{y_1(t)f(t)}{W(t)} \qquad W(t) = y_1y_2{}' - y_2y_1{}'} \tag{12}$$

This denominator $W(t)$ is the **Wronskian** of the two null solutions $y_1(t)$ and $y_2(t)$. It was introduced in Section 2.1, and the independence of $y_1(t)$ and $y_2(t)$ guarantees that $W(t) \neq 0$. The divisions by $W(t)$ in (12) are safe. **The varying parameters $c_1(t)$ and $c_2(t)$ are the integrals of $c_1{}'(t)$ and $c_2{}'(t)$ in** (12).

We have found a particular solution $c_1y_1 + c_2y_2$ to the differential equation (6):

If $y_1$ and $y_2$ are independent null solutions to $y'' + B(t)y' + C(t)y = 0$, then a particular solution $y_p(t)$ with right side $f(t)$ is $c_1(t)y_1(t) + c_2(t)y_2(t)$:

**Variation of Parameters**

$$\boldsymbol{y_p(t) = -y_1(t)\int \frac{y_2(t)f(t)}{W(t)}\,dt + y_2(t)\int \frac{y_1(t)f(t)}{W(t)}\,dt.} \tag{13}$$

**Example 5 Variation of parameters: Find a particular solution for $y'' + y = t$.**

The right side $f(t) = t$ is not a sinusoid. No problem to find the independent solutions $y_1(t) = \cos t$ and $y_2(t) = \sin t$ to the null equation $y'' + y = 0$. The Wronskian is 1:

$$W(t) = y_1y_2{}' - y_2y_1{}' = \cos^2 t + \sin^2 t = 1 \quad \text{(never zero as predicted).}$$

The particular solution $y_p(t) = c_1(t)\ \cos t + c_2(t)\ \sin t$ needs integrals of $c_1{}'$ and $c_2{}'$:

$$c_1(t) = \int \frac{(-\sin t)t\,dt}{1} = \boldsymbol{t\cos t - \sin t} \qquad c_2(t) = \int \frac{(\cos t)t\,dt}{1} = \boldsymbol{t\sin t + \cos t}.$$

Variation of parameters has found a particular solution $c_1y_1 + c_2y_2$, and it simplifies:

$$y_p = (\boldsymbol{t\ \cos t - \sin t})\cos t + (\boldsymbol{t\ \sin t + \cos t})\sin t = \boldsymbol{t}. \tag{14}$$

Apologies! We could have seen by ourselves that $y = t$ solves $y'' + y = t$. And the method of undetermined coefficients would find $y = t$ much faster: no integrations.

**Example 6** Solve $y'' + y = \delta(t)$ by variation of parameters. The null solutions $\cos t$ and $\sin t$ still give $W(t) = 1$. The delta function $f$ goes into the integrals for $c_1$ and $c_2$:

$$c_1 = \int \frac{(\sin t)\,\delta(t)\,dt}{1} = \sin 0 = \mathbf{0} \qquad c_2 = \int \frac{(\cos t)\,\delta(t)\,dt}{1} = \cos 0 = \mathbf{1}$$

Then $y_p(t) = (1)y_2(t) = \mathbf{\sin t}$. With $f = \delta(t)$, this is the fundamental solution $g(t)$ (the impulse response). Then $\sin t$ is also the solution to $y'' + y = 0$ that starts from $y(0) = 0$ and $y'(0) = 1$. We will find this growth factor again in (17) with $s_1 = -s_2 = i$.

## Constant Coefficients and the Solution Formula

The one time we are sure to know the null solutions $y_1$ and $y_2$ is when the differential equation has constant coefficients. Substituting $y = e^{st}$ into $Ay'' + By' + Cy = 0$ leads to $As^2 + Bs + C = 0$. The roots are $s_1$ and $s_2$. The null solutions are $e^{s_1 t}$ and $e^{s_2 t}$. Notice that we are free to assume that $\boldsymbol{A = 1}$. (If not, divide the equation by $A$.)

Variation of parameters gives the solution (13). All we need is the Wronskian $W(t)$, and for these null solutions it is beautiful:

$$\boldsymbol{W(t)} = y_1 y_2{}' - y_2 y_1{}' = (e^{s_1 t})(s_2 e^{s_2 t}) - (e^{s_2 t})(s_1 e^{s_1 t}) = \boldsymbol{(s_2 - s_1)e^{s_1 t}e^{s_2 t}}. \quad (15)$$

Immediately we know that $W(t) \neq 0$ unless $s_1 = s_2$. With equal roots we expect to need the special null solution $y_2 = te^{st}$. Even in that case the Wronskian looks terrific:

$$\boldsymbol{W(t)} = (e^{st})(te^{st})' - (te^{st})(e^{st})' = (e^{st})(ste^{st} + e^{st}) - (te^{st})(se^{st}) = \boldsymbol{e^{2st}}. \quad (16)$$

When you substitute $y_1$ and $y_2$ and $W$ into (13), that "*VP* formula" produces $y_p(t)$.

**Unequal roots $\boldsymbol{s_1 \neq s_2}$.** The first integral has $y_2/W = e^{-s_1 t}/(s_2 - s_1)$. The second integral has $y_1/W = e^{-s_2 t}/(s_2 - s_1)$. Put those into (13):

**Particular solution**
**Constant coefficients**
$$y_p(t) = \frac{-e^{s_1 t}}{s_2 - s_1}\int_0^t e^{-s_1 T} f(T)\,dT + \frac{e^{s_2 t}}{s_2 - s_1}\int_0^t e^{-s_2 T} f(T)\,dT$$

To me, a growth factor $g(t - T)$ is multiplying the inputs $f(T)$. *The integrals just sum up the outputs.* Here is the same formula for $y_p(t)$ written so it uses $g(t)$:

**Growth factor** $\boldsymbol{g(t) = \dfrac{e^{s_1 t} - e^{s_2 t}}{s_1 - s_2}}$ **Solution** $\boldsymbol{y_p(t) = \displaystyle\int_0^t g(t - T)\,f(T)\,dT}$ (17)

That might be the nicest formula in the book. Probably I am writing those words because I didn't see this formula coming. Section 2.3 discovered the same response $g(t)$!

Forgive me for that personal note. I will go on to the other case, with $s_1 = s_2$.

**Equal roots $s_1 = s_2 = s$ with $W = e^{2st}$.** The first integral in (13) still has $y_1 = e^{st}$ and now $y_2/W = te^{-st}$. The second integral has $y_2 = te^{st}$ and $y_1/W = e^{-st}$:

**Particular solution $y_p$**
**Null solutions $e^{st}, te^{st}$**

$$y_p(t) = -e^{st}\int_0^t Te^{-sT} f(T)\,dT + te^{st}\int_0^t e^{-sT} f(T)\,dT.$$

This also has a perfect form when you identify the factor $g(t-T)$ that is multiplying $f$:

$$\textbf{Growth factor}\quad g(t) = te^{st} \qquad \textbf{Solution}\quad y_p(t) = \int_0^t g(t-T)f(T)\,dT \tag{18}$$

Formulas that good never happen by accident, $g(t)$ must mean something important:

**The growth factor $g(t)$ is the impulse response:** $\quad$ **$y_p(t)$ is $g(t)$ when $f(t)$ is $\delta(t)$.**

Let me close Section 2.6 on that high note. Then Section 2.7 will take the Laplace transform of the growth factors $g(t)$ to get the **transfer function $Y(s)$**:

$$\text{The transform of } g(t) = \frac{e^{s_1 t} - e^{s_2 t}}{s_1 - s_2} \text{ is } \frac{1}{(s-s_1)(s-s_2)} = \frac{1}{s^2 + Bs + C} = Y(s).$$

$$\text{The transform of } g(t) = te^{s_1 t} \text{ is } \frac{1}{(s-s_1)^2} = \frac{1}{s^2 + Bs + C} \text{ when } s_1 = s_2.$$

$Y(s)$ comes from $B$ and $C$. **The solution $y(t)$ comes from $g(t)$ = "Green's function."** The last pages of the book will see the integral of $g(t-T)f(T)$ as a convolution.

## ■ REVIEW OF THE KEY IDEAS ■

1. **Undetermined coefficients** in $y_p$ apply when $f(t)$ has only $e^{st}$, $\cos\omega t$, $\sin\omega t$, $t^n$.
2. Set $y_p$ = exponential/sinusoid/polynomial. Find coefficients $a, b, \ldots$ to match $f(t)$.
3. **Variation of parameters**: $c_1$ and $c_2$ vary with $t$ in $y_p = c_1(t)\,y_1(t) + c_2(t)\,y_2(t)$.
4. Two equations for $c_1{}'$ and $c_2{}'$ lead to $c_1$ and $c_2$ = integrals of $-y_2\,f/W$ and $y_1\,f/W$.
5. For constant coefficients $c_1$ and $c_2$ those are integrals of $e^{-s_1 t} f(t)$ and $e^{-s_2 t} f(t)$.
6. Then $y_p = \int g(t-s)f(s)\,ds$ when $g(t)$ = response to the impulse $f = \delta(t)$.

## Problem Set 2.6

**Find a particular solution by inspection** (or the method of undetermined coefficients)

**1** (a) $y'' + y = 4$ (b) $y'' + y' = 4$ (c) $y'' = 4$

**2** (a) $y'' + y' + y = e^t$ (b) $y'' + y' + y = e^{ct}$

**3** (a) $y'' - y = \cos t$ (b) $y'' + y = \cos 2t$ (c) $y'' + y = t + e^t$

**4** For these $f(t)$, predict the form of $y(t)$ with undetermined coefficients:

(a) $f(t) = t^3$ (b) $f(t) = \cos 2t$ (c) $f(t) = t \cos t$

**5** Predict the form for $y(t)$ when the right hand side is

(a) $f(t) = e^{ct}$ (b) $f(t) = te^{ct}$ (c) $f(t) = e^t \cos t$

**6** For $f(t) = e^{ct}$ when is the prediction for $y(t)$ different from $Ye^{ct}$?

**Use the method of undetermined coefficients to find a solution $y_p(t)$.**

**7** (a) $y'' + 9y = e^{2t}$ (b) $y'' + 9y = te^{2t}$

**8** (a) $y'' + y' = t + 1$ (b) $y'' + y' = t^2 + 1$

**9** (a) $y'' + 3y = \cos t$ (b) $y'' + 3y = t \cos t$

**10** (a) $y'' + y' + y = t^2$ (b) $y'' + y' + y = t^3$

**11** (a) $y'' + y' + y = \cos t$ (b) $y'' + y' + y = t \sin t$

**Problems 12–14 involve resonance. Multiply the usual form of $y_p$ by $t$.**

**12** (a) $y'' + y = e^{it}$ (b) $y'' + y = \cos t$

**13** (a) $y'' - 4y' + 3y = e^t$ (b) $y'' - 4y' + 3y = e^{3t}$

**14** (a) $y' - y = e^t$ (b) $y' - y = te^t$ (c) $y' - y = e^t \cos t$

**15** For $y'' + 4y = e^t \sin t$ (exponential times sinusoidal) we have two choices:

1 (Real) Substitute $y_p = Me^t \cos t + Ne^t \sin t$: determine $M$ and $N$
2 (Complex) Solve $z'' + 4z = e^{(1+i)t}$. Then $y$ is the imaginary part of $z$.

Use both methods to find the same $y(t)$—which do you prefer?

**16** (a) Which values of $c$ give resonance for $y'' + 3y' - 4y = te^{ct}$?

(b) What form would you substitute for $y(t)$ if there is no resonance?

(c) What form would you use when $c$ produces resonance?

**17** This is the rule for equations $P(D)y = e^{ct}$ with resonance $P(c) = 0$:

If $P(c) = 0$ and $P'(c) \neq 0$, look for a solution $y_p = Cte^{ct}$ $(m = 1)$
If $c$ is a root of multiplicity $m$, then $y_p$ has the form ______.

**18** (a) To solve $d^4y/dt^4 - y = t^3e^{5t}$, what form do you expect for $y(t)$?

(b) If the right side becomes $t^3 \cos 5t$, which 8 coefficients are undetermined?

**19** For $y' - ay = f(t)$, the method of undetermined coefficients is looking for all $f(t)$ so that the usual formula $y_p = e^{at} \int e^{-as} f(s)ds$ is easy to integrate. Find these integrals for $f = e^{ct}$, $f = e^{i\omega t}$, and $f = t$:

$$\int e^{-as}e^{cs}ds \qquad \int e^{-as}e^{i\omega s}ds \qquad \int e^{-as}s\,ds$$

**Problems 20–27 develop the method of variation of parameters.**

**20** Find two solutions $y_1, y_2$ to $y'' + 3y' + 2y = 0$. Use those in formula (13) to solve

(a) $y'' + 3y' + 2y = e^t$ (b) $y'' + 3y' + 2y = e^{-t}$

**21** Find two solutions to $y'' + 4y' = 0$ and use variation of parameters for

(a) $y'' + 4y' = e^{2t}$ (b) $y'' + 4y' = e^{-4t}$

**22** Find an equation $y'' + By' + Cy = 0$ that is solved by $y_1 = e^t$ and $y_2 = te^t$. If the right side is $f(t) = 1$, what solution comes from the *VP* formula (13)?

**23** $y'' - 5y' + 6y = 0$ is solved by $y_1 = e^{2t}$ and $y_2 = e^{3t}$, because $s = 2$ and $s = 3$ lead to $s^2 - 5s + 6 = 0$. Now solve $y'' - 5y' + 6y = 12$ in two ways:

**1.** Undetermined coefficients (or inspection) **2.** Variation of parameters using (13)

The answers are different. Are the initial conditions different?

**24** What are the initial conditions $y(0)$ and $y'(0)$ for the solution (13) coming from variation of parameters, starting from any $y_1$ and $y_2$?

**25** The equation $y'' = 0$ is solved by $y_1 = 1$ and $y_2 = t$. Use variation of parameters to solve $y'' = t$ and also $y'' = t^2$.

**26** Solve $y_s{}'' + y_s = 1$ for the step response using variation of parameters, starting from the null solutions $y_1 = \cos t$ and $y_2 = \sin t$.

**27** Solve $y_s{}'' + 3y_s{}' + 2y_s = 1$ for the step response starting from the null solutions $y_1 = e^{-t}$ and $y_2 = e^{-2t}$.

**28** Solve $Ay'' + Cy = \cos \omega t$ when $A\omega^2 = C$ (the case of resonance). Example 4 suggests to substitute $y = Mt \cos \omega t + Nt \sin \omega t$. Find $M$ and $N$.

**29** Put $g(t)$ into the great formulas (17)-(18) to see the equations above them.

## 2.7 Laplace Transforms $Y(s)$ and $F(s)$

If you think about the functions that have dominated this book, the list is not very long. They are the right hand sides of linear differential equations and also the solutions $y(t)$ :

1. Exponentials $\boldsymbol{e^{at}}$
2. Sinusoids $\boldsymbol{\cos \omega t}$ and $\boldsymbol{\sin \omega t}$
3. Polynomials starting with $\mathbf{1}$ and $\boldsymbol{t}$ and $\boldsymbol{t^2}$
4. Step functions $\boldsymbol{H(t-T)}$
5. Delta functions $\boldsymbol{\delta(t-T)}$
6. Products of **1** to **5**

Why are these functions special? I believe this is an important question.

The answer that strikes me first is something I had not thought about:

**The derivatives and integrals of these functions are also on the list** *(almost)*.

That was true from the very start of Chapter 1. Example 1 on page 1 was $y = e^t$. Its fundamental property is $dy/dt = y$. The derivative leaves it unchanged, which puts it on the list. And the product of two exponentials is another exponential. In fact exponentials could be a short list by themselves.

Cosines and sines were written separately, but those are combinations of $e^{i\omega t}$ and $e^{-i\omega t}$. They just move us to complex numbers. The constant polynomial is $e^{0t} = 1$. Integrals and derivatives of polynomials are polynomials. The product rule for derivatives (and the reverse rule which is integration by parts) keep the list self-contained: no new functions.

There is one flaw but it is easily fixed. The delta function $\delta(t)$ is the derivative of the step function $H(t)$, but we need all derivatives and integrals. Include them on the list! Solving $dy/dt =$ step function gives $\boldsymbol{y(t) =}$ *ramp function*. This is zero for $t \leq 0$, and $y(t) = t$ for $t \geq 0$. Its graph has a corner and its slope has a jump. The integral of that linear ramp is a *parabolic ramp*. The next integral leads toward a *cubic spline*. The derivative of a delta function is a very singular object (see Problem 25).

In the end, all these ideal functions can go on the list which is now complete.

### The Algebra of Differential Equations

With those special functions, solving a constant coefficient linear differential equation is not so difficult. It reduces to an algebra problem. The null solution $y_n$ is a combination of exponentials (possibly times powers of $t$). The particular solution $y_p$ has a known form like $Ye^{i\omega t}$—the differential equation will decide the undetermined coefficient $Y$. For functions **1** to **6**, the integrals using variation of parameters are already on the list.

**The Laplace transform gives a systematic way to do the algebra**. *Functions of $t$ become functions of $s$.* Instead of derivatives $dy/dt$, we have multiplications $sY(s)$. Then differential equations in $t$ become algebra equations in $s$. Start with these examples:

**Left side** $\boldsymbol{y(t) \to Y(s)}$ | $y'(t) \to \boldsymbol{sY(s)}$ and $y''(t) \to \boldsymbol{s^2Y(s)}$ when $y(0) = y'(0) = 0$

**Right side** $\boldsymbol{f(t) \to F(s)}$ | $f = e^{at} \to F = \boldsymbol{1/(s-a)}$ and impulse $f = \delta(t) \to F = \mathbf{1}$.

Solving a differential equation by using the Laplace transform involves three steps:

**1** Transform every term **2** Solve for $Y(s)$ **3** Find $y(t)$ whose transform is $Y(s)$.

You will see how initial values for $y(0)$ and $y'(0)$ go into the $s$-equation for $Y(s)$. And most important, you will see how the **zeros** of the polynomial $s^2 + Bs + C$ become "**poles**" of $Y(s)$. Those exponents $s_1$ and $s_2$ give us the null solution $y_n(t)$. Dividing by that polynomial gives the transfer function $1/(s^2 + Bs + C)$. Now we see all of this as a natural part of the Laplace transform.

**Example 1** Start from $y(0) = 0$ and $y'(0) = 0$. With those initial conditions, the transform of $y'$ is $sY$ and the transform of $y''$ is $s^2Y$. We can transform a whole equation:

*Step* 1 $\boldsymbol{y'' - 4y' + 3y = e^{at}}$ transforms to $\boldsymbol{(s^2 - 4s + 3)\, Y(s) = \dfrac{1}{s-a}}$

*Step* 2 The transform of $y(t)$ is $Y(s) = \dfrac{1}{(s^2 - 4s + 3)(s-a)} = \boldsymbol{\dfrac{1}{(s-3)(s-1)(s-a)}}$

*Step* 3 The inverse Laplace transform of $Y(s)$ is $\boldsymbol{y(t) = C_1e^{3t} + C_2e^t + Ge^{at}}$.

$C_1$ and $C_2$ come from matching the initial conditions $y(0) = 0$ and $y'(0) = 0$. The gain $G = 1/(a^2 - 4a + 3)$ is the transfer function at $s = a$. The inverse transform of $Y(s)$ is computed in equations (12) and (14). Step 2 revealed the poles of $Y(s)$:

$$\frac{1}{(s-3)(s-1)(s-a)} \quad \textbf{has poles at } s = 3 \textbf{ and } s = 1 \textbf{ and } s = a.$$

Those three numbers are the all-important exponents in $y(t) = C_1e^{3t} + C_2e^t + Ge^{at}$. Now they are seen as the poles **3**, **1**, $\boldsymbol{a}$ where $\boldsymbol{Y(s)}$ becomes infinite.

**Example 2** Change from $f = e^{at}$ to $\boldsymbol{f = \delta(t)}$ = **impulse**. Keep $y(0) = y'(0) = 0$.

*Step* 1 $\boldsymbol{y'' + By' + Cy = \delta(t)}$ transforms to $\boldsymbol{(s^2 + Bs + C)\, Y(s) = 1}$.

*Step* 2 The transform of $y(t)$ is $Y(s) = \boldsymbol{\dfrac{1}{s^2 + Bs + C}}$ = **transfer function**.

*Step* 3 **The inverse transform is** $\boldsymbol{y(t) = g(t) = \dfrac{e^{s_1t} - e^{s_2t}}{s_1 - s_2}}$ = **impulse response**.

Those roots $s_1, s_2$ of $s^2 + Bs + C = (s - s_1)(s - s_2)$ give poles in $Y(s)$ and exponentials in $y(t)$. You have to be impressed by how quickly steps 1-2-3 led to this central fact.

**When $f = \delta(t)$, the transform of the impulse response $g$ is the transfer function $Y$.**

## The Laplace Transform

Our first Table of Transforms will include the most essential functions and no more. A more complete presentation of this transform will be saved for the last sections of the book. We will define $Y(s)$ here, but the shift rule for transforms will be developed there. All step functions $H(t - T)$ are left for Chapter 8, except for one comment below.

Especially we point to the final Section 8.6 on "*convolutions*". These are the inverse transforms of products $Y(s) = F(s)G(s)$. Convolution is exactly what we need when $f(t)$ is not a simple function like $e^{at}$ and $F(s)$ is not a simple function like $1/(s - a)$.

To create the Table of Transforms we start with the integral that defines $F(s)$:

$$\textbf{The Laplace transform of } f(t) \textbf{ is } F(s) = \int_0^\infty f(t)\, e^{-st}\, dt. \tag{1}$$

The first function to transform is certainly $f(t) = e^{at}$. Then $F(s) = 1/(s-a)$ as expected:

$$F(s) = \int_0^\infty e^{at} e^{-st}\, dt = \left[\frac{e^{(a-s)t}}{a - s}\right]_{t=0}^{t=\infty} = 0 - \frac{1}{a - s} = \frac{\mathbf{1}}{\boldsymbol{s - a}}. \tag{2}$$

That integral would be infinite if $a \geq s$. It is typical of Laplace transforms to require $s > a$. Then the factor $e^{-st}$ in the integral brings us safely to zero at $t = \infty$. The following rule is natural for all functions $f(t)$, when you look at the integral (1) from $t = 0$ to $t = \infty$:

**By definition $f(t) = 0$ for all $t < 0$. Functions don't start until $t = 0$.**

Then the step function $H(t)$ and the constant function $f = 1$ have the same transform!

$$\textbf{The transform of } f(t) = 1 \textbf{ is } F(s) = \int_0^\infty 1e^{-st}\, dt = \frac{1}{s}. \tag{3}$$

This is the transform of $e^{at}$ when the exponent $a$ goes to 0 and $1/(s - a)$ goes to $1/s$.

## Transform of the Derivative

Now comes the most important rule—the whole basis for solving differential equations. If the transform of $y(t)$ is $Y(s)$, what is the transform of the derivative $dy/dt$?

**Derivative Rule**   The transform of $dy/dt$ is $\boldsymbol{sY(s) - y(0)}$.

The derivative rule shows how the initial conditions enter the transformed problem—not as separate side conditions, but directly into the equation for $Y(s)$. The proof uses integration by parts. The integral of $dy/dt$ is $y(t)$ and the derivative of $e^{-st}$ is $-se^{-st}$ :

$$\int_0^\infty \frac{dy}{dt} e^{-st}\,dt = -\int_0^\infty y(t)(-se^{-st})\,ds + \left[y(t)e^{-st}\right]_0^\infty$$

**Transform of $dy/dt$** $\quad = \mathbf{sY(s) - y(0)}$ (4)

Again $s$ must be large enough—or more exactly, the real part of $s$ must be large enough—to assure that $y(t)e^{-st}$ drops to zero at $t = \infty$.

We can immediately solve the model problem of Chapter 1: A first order linear equation. The solution steps 1, 2, 3 produce $Y(s)$ with poles (blowup values for $s$) at the two key exponents $s = a$ and $s = c$ :

**Example 3** Solve $\boldsymbol{\dfrac{dy}{dt} - ay = e^{ct}}$ starting from any $y(0)$.

*Step* 1 Transform the equation to $sY(s) - y(0) - aY(s) = \dfrac{1}{s-c}$. (5)

*Step* 2 $(s-a)Y(s) = y(0) + \dfrac{1}{s-c}$ gives $Y(s) = \dfrac{y(0)}{s-a} + \dfrac{1}{(s-a)(s-c)}$. (6)

*Step* 3 The inverse transform of $\dfrac{y(0)}{s-a}$ is the null solution $\boldsymbol{y_n(t) = y(0)e^{at}}$. (7)

The inverse transform of $\dfrac{1}{(s-a)(s-c)}$ is the very particular solution $\boldsymbol{\dfrac{e^{ct} - e^{at}}{c-a}}$. (8)

I have to say, this is beautiful. The effort we made in Chapter 1 has been reduced to its bare minimum. All that is left is the derivative rule, the transform of exponentials, and "partial fractions." Those partial fractions were the algebra from Step 2 to Step 3: separating $1/(s-a)(s-c)$ with two poles $a$ and $c$ into **two fractions with one pole each**.

**PF2** $$\frac{1}{(\boldsymbol{s}-\boldsymbol{a})(\boldsymbol{s}-\boldsymbol{c})} = \frac{1}{(\boldsymbol{s}-\boldsymbol{a})(a-c)} + \frac{1}{(c-a)(\boldsymbol{s}-\boldsymbol{c})} \qquad (9)$$

PF2 was used in Example 2 to find the impulse response. In that case $a$ and $c$ were $s_1$ and $s_2$. Partial fractions were also used in Example 1, with $f = e^{at}$ and *three* poles 3, 1 $a$.

## Partial Fractions

Example 1 reached $Y(s) = 1/(s+3)(s+1)(s-a)$. We didn't immediately know its inverse transform $y(t)$. But finding $y(t)$ becomes simple when $Y(s)$ is separated into **three terms with one pole each. Those three pieces are the Partial Fractions in PF3**:

$$\frac{1}{(\boldsymbol{s-3})(\boldsymbol{s-1})(\boldsymbol{s-a})} = \frac{1}{(\boldsymbol{s-3})(3-1)(3-a)} + \frac{1}{(1-3)(\boldsymbol{s-1})(1-a)} + \frac{1}{(a-3)(a-1)(\boldsymbol{s-a})}$$

Usually I would show you where this PF3 formula comes from. In this case I would rather show you that it is correct. Above all, you must see the main point: The three separate terms with one pole each lead immediately to the three parts $\boldsymbol{C_1 e^{3t}}$ and $\boldsymbol{C_2 e^t}$ and $\boldsymbol{Ye^{at}}$.

Officially, correctness can be proved by multiplying PF3 by $(s-3)(s-1)(s-a)$.

$$1 = \frac{(s-1)(s-a)}{(3-1)(3-a)} + \frac{(s-3)(s-a)}{(1-3)(1-a)} + \frac{(s-3)(s-1)}{(a-3)(a-1)}. \tag{10}$$

At $s = 3$, the last two terms disappear and we have $1 = 1$ (as desired). At $s = 1$, the second term equals 1. At $s = a$, the third term equals 1. Thus (10) is an equation of the form $1 = As^2 + Bs + C$, and the equation is correct at three values $s = 3, 1, a$. Therefore the equation must be always correct, and PF3 is shown to be true.

*Remark* The theory of partial fractions usually computes $C_1$ and $C_2$ and $Y$ so that

$$\frac{\mathbf{1}}{(\boldsymbol{s-3})(\boldsymbol{s-1})(\boldsymbol{s-a})} = \frac{\boldsymbol{C_1}}{\boldsymbol{s-3}} + \frac{\boldsymbol{C_2}}{\boldsymbol{s-1}} + \frac{\boldsymbol{Y}}{\boldsymbol{s-a}}. \tag{11}$$

The idea is to put the right side over a common denominator, which is on the left side. Matching the coefficients of $s^2$ and $s$ and 1 gives three equations for $C_1$ and $C_2$ and $Y$. My shortcut was to go directly to the answers $C_1, C_2, G$ that you see in PF3:

$$C_1 = \frac{1}{(3-1)(3-a)} \qquad C_2 = \frac{1}{(1-3)(1-a)} \qquad Y = \frac{1}{(a-3)(a-1)}. \tag{12}$$

I think it is easier to remember this pattern than to solve for a new $C_1$ and $C_2$ and $Y$, every time you change the poles 3 and 1 and $a$. *To repeat, from the three partial fractions in* PF3 *we read off the coefficients* $C_1, C_2, Y$ *in equation* (12).

## Very Particular Solution

Look at what we have in those three parts. The last part $Ye^{at}$ is a particular solution—the one that comes from the transfer function and the exponential response formula. The equation was $y'' - 4y' + 3y = e^{at}$, and the response to $e^{at}$ is

$$y_p(t) = Ye^{at} = \frac{1}{a^2 - 4a + 3}\, e^{at} = \frac{1}{(a-3)(a-1)}\, e^{at}. \tag{13}$$

That is old news. This is not the very particular solution, it doesn't start at $y(0) = 0$ and $y'(0) = 0$. The solution with that particular start is the one from the Laplace transform:

$$\textbf{The very particular solution is all of } y_{vp}(t) = C_1 e^{3t} + C_2 e^t + Y e^{at}. \tag{14}$$

Remember, any null solution $y_n$ can be added to one particular $y_p$. That gives another $y_p$. The very particular solution $y_{vp}$ starts from rest.

The complete solution adjusts the free constants $c_1$ and $c_2$ (note the small $c$) to match any starting values $y(0)$ and $y'(0)$:

$$y_{\text{complete}} = c_1 e^{3t} + c_2 e^t + Y e^{at}. \tag{15}$$

You could solve for $c_1$ and $c_2$ as usual, by setting $t = 0$ in $y$ and $y'$. Then you are working in the time domain. Or you could use $y(0)$ and $y'(0)$ in finding $Y(s)$, when you transform the equation in the first place. Let me show you that way, compared to the usual way.

.

## Including $y(0)$ and $y'(0)$ in the Transform

We know that the transform of $y'$ is $sY(s) - y(0)$. To find the transform of $y''$, use that first derivative rule twice. This brings in $y'(0)$ along with $y(0)$.

$$\begin{aligned} \textbf{transform of } y'' &= s(\text{transform of } y') - y'(0) \\ &= s(sY(s) - y(0)) - y'(0) \\ &= \boldsymbol{s^2 Y(s) - s y(0) - y'(0)}. \end{aligned} \tag{16}$$

Now we can solve the equation $y'' - 4y' + 3y = e^{at}$ entirely by Laplace transform:

*Step* 1 Transform to $(s^2 Y(s) - sy(0) - y'(0)) - 4(sY(s) - y(0)) + 3Y(s) = \dfrac{1}{s-a}$

*Step* 2 Rewrite as $(s^2 - 4s + 3)Y(s) = (s-4)y(0) + y'(0) + 1/(s-a)$.

$$\text{Solve for } Y(s): \quad Y(s) = \frac{(s-4)y(0) + y'(0)}{s^2 - 4s + 3} + \frac{1}{(s^2 - 4s + 3)(s-a)}. \tag{17}$$

*Step* 3 Invert both pieces of $Y(s)$ to find $y_n(t) + y_p(t)$.

This looks more painful to me! The last part of $Y(s)$ is fine—that is what we already worked with to find $y_p$. Its inverse transform is the very particular solution in (14). The first part of $Y(s)$ involves $y(0)$ and $y'(0)$. We have to do partial fractions again: *not good.*

The denominator $s^2 - 4s + 3$ has two factors $(s-3)(s-1)$ and not three factors. But I would prefer to find $c_1$ and $c_2$ in the complete solution (15), by setting $t = 0$ and solving these two equations:

$$\begin{aligned} c_1 + c_2 + \phantom{a}Y &= y(0) \\ 3c_1 + c_2 + aY &= y'(0) \end{aligned} \tag{18}$$

When $y(0)$ and $y'(0)$ are zero, that's when $c_1$ and $c_2$ and $y$ equal $C_1$ and $C_2$ and $y_{vp}$.

## Transforms at Resonance

The reader will remember that when two exponents come together, and two solutions become one solution like $e^{at}$, another solution is born. It is like atomic fission or fusion. The new solution has the form $\boldsymbol{te^{at}}$. We want to find its Laplace transform.

Equal exponents can happen in two different ways for $y'' + By' + Cy = f(t)$.

**1** (*Null solution*) Two roots $s_1$ and $s_2$ of the characteristic polynomial become equal.

**2** (*Particular solution*) The exponent in $f = e^{at}$ equals $s_1$ or $s_2$ in the null solution.

In a truly extreme case we might have $s_1 = s_2 = a$, three equal exponents. Then the null solution is $c_1e^{at} + c_2te^{at}$, and a particular solution is $Gt^2e^{at}$.

We are seeing these possibilities in the "time domain" and we can see them in the "frequency domain". **Double roots in the $t$-domain become double poles in $Y(s)$.**

$$\textbf{The Laplace transform of } \boldsymbol{te^{at}} \textbf{ is } \frac{\mathbf{1}}{\boldsymbol{(s-a)^2}} \textbf{ with a double pole.} \tag{19}$$

A nice proof starts with a simple pole in the transform. The transform of $e^{at}$ is $1/(s-a)$. Now take derivatives of both sides *with respect to $a$*:

$$\int_0^\infty e^{at}e^{-st}\,dt = \frac{1}{s-a} \qquad \int_0^\infty te^{at}e^{-st}\,dt = \frac{d}{da}\left(\frac{1}{s-a}\right) = \frac{1}{(s-a)^2}$$

If we take another $a$-derivative, the transform of $t^2e^{at}$ is seen as $\boldsymbol{2(s-a)^{-3}}$ with a triple pole. The simplest example of this extreme case would be the equation $y'' = 2$.

$$y'' = 2 \text{ has exponents } 0 \text{ and } 0 \text{ in } y_n(t) = c_1 + c_2t \quad \text{and } a = 0 \text{ in } y_p(t) = t^2e^{0t} = t^2.$$

The initial conditions give $c_1 = y(0)$ and $c_2 = y'(0)$. The solution is easy to check:

$$\boldsymbol{y = y(0) + ty'(0) + t^2} \quad \textbf{solves} \quad \boldsymbol{y''} = 2. \tag{20}$$

To find this solution by Laplace transform, start by transforming $y''$ and $2$:

$$s^2Y(s) - y(0)s - y'(0) = \frac{2}{s} \qquad \text{gives} \quad Y(s) = \frac{y(0)}{s} + \frac{y'(0)}{s^2} + \frac{2}{s^3}. \tag{21}$$

The inverse transforms of $1/s$ and $1/s^2$ are $1$ and $t$. The inverse transform of $2/s^3$ is $t^2$. So the inverse transform of $Y(s)$ is the correct $y = y(0) + ty'(0) + t^2$ in (20).

Those are really $e^{0t}$ and $te^{0t}$ and $t^2e^{0t}$ : three zero exponents, a truly extreme case.

The inverse of equation (19) tells us the fundamental solution $g(t)$ when the transfer function $1/(s^2 + Bs + C)$ has a double pole and $s^2 + Bs + C = 0$ has $s_1 = s_2$:

$$\textbf{If } \boldsymbol{s^2 + Bs + C = (s - s_1)^2} \textbf{ then the fundamental solution is } \boldsymbol{g(t) = te^{s_1t}}.$$

### The Transforms of $\cos \omega t$ and $\sin \omega t$

In all of this section on Laplace transforms, there is no requirement that $a$ must be real. That exponent can be $i\omega$ or $-i\omega$ or any complex number $a + i\omega$. From the identity $\cos \omega t = \frac{1}{2}(e^{i\omega t} + e^{-i\omega t})$, and from the linearity of the formula for $F(s) = \int f(t)e^{-st}\,dt$, we can combine the known transforms of $e^{i\omega t}$ and $e^{-i\omega t}$:

The transform of $f(t) = \mathbf{cos}\ \boldsymbol{\omega t}$ is $F(s) = \dfrac{1}{2}\left(\dfrac{1}{s - i\omega} + \dfrac{1}{s + i\omega}\right) = \dfrac{\boldsymbol{s}}{\boldsymbol{s^2 + \omega^2}}$ (22)

The twin identity $\sin \omega t = \dfrac{1}{2i}(e^{i\omega t} - e^{-i\omega t})$ also comes from Euler's formula.

The transform of $f(t) = \mathbf{sin}\ \boldsymbol{\omega t}$ is $F(s) = \dfrac{1}{2i}\left(\dfrac{1}{s - i\omega} - \dfrac{1}{s + i\omega}\right) = \dfrac{\boldsymbol{\omega}}{\boldsymbol{s^2 + \omega^2}}.$ (23)

Those transforms appear in the fundamental example of a mass hanging from a spring:

Step 1 $\quad my'' + ky = \cos \omega t$ transforms to $m(s^2Y(s) - sy(0) - y'(0)) + kY(s) = \dfrac{s}{s^2 + \omega^2}.$

The transform $Y(s)$ is multiplied by $ms^2 + k$. **The transfer function is $\boldsymbol{1/(ms^2 + k)}$.**

*The transfer function multiplies the input to give the output.* The input is on the right hand side, the output is the solution. Both of those are now in transform space!

Step 2 $\quad$ Solve for $Y(s) = \dfrac{1}{ms^2 + k}\left(sy(0) + y'(0) + \dfrac{s}{s^2 + \omega^2}\right).$ (24)

We are ready for Step 3, but it doesn't look so easy. It requires the inverse transform of this $Y(s)$. Our simple mass-spring problem has led us to a *fourth degree* denominator $(ms^2 + k)(s^2 + \omega^2)$. We need partial fractions to separate $Y(s)$ into two pieces with *second degree* denominators. That algebra is not so bad, and it can be left for Problem 26.

The result is that $y(t)$ has a term in $\cos \omega t$ and another term in $\cos \omega_n t$. The driving frequency is $\omega$, the natural frequency $\omega_n = \sqrt{k/m}$ comes from the zeros of $ms^2 + k$.

**The frequencies in the solution $\boldsymbol{y(t)}$ are the poles $\boldsymbol{\pm i\omega}$ and $\boldsymbol{\pm i\omega_n}$ in its transform $\boldsymbol{Y(s)}$.**

That bold statement is really the important message from a Laplace transform. We engineer the system or the network by moving those poles. Often we keep them well separated to avoid instability. And we add damping to push the zeros of $ms^2 + bs + k$ (poles of $Y(s)$) off the imaginary axis and into the stable left halfplane where Re $s < 0$.

| $f(t)$ | $1, t, t^2$ | $e^{at}, te^{at}, t^2e^{at}$ | $\cos \omega t, \sin \omega t$ | $y, y', y''$ |
|---|---|---|---|---|
| $F(s)$ | $\dfrac{1}{s}, \dfrac{1}{s^2}, \dfrac{2}{s^3}$ | $\dfrac{1}{s-a}, \dfrac{1}{(s-a)^2}, \dfrac{2}{(s-a)^3}$ | $\dfrac{s}{s^2+\omega^2}, \dfrac{\omega}{s^2+\omega^2}$ | $Y,\ sY - y(0),$ $s^2Y - sy(0) - y'(0)$ |

## Complex Roots $a \pm i\omega$

Finally we come to the most typical case for physical systems. It has damping, and it has oscillation. *The roots of $s^2 + 2s + 5$ are complex.* Their real parts are $a = -2/2 = -1$. Their imaginary parts $\pm\sqrt{B^2 - 4AC}/2$ are $\pm i\omega = \pm\sqrt{-16}/2 = \pm 2i$. We are in the underdamped case and the solutions to $y'' + 2y' + 5y = 0$ can be written two ways:

$$y = c_1 e^{(-1+2i)t} + c_2 e^{(-1-2i)t} \quad \text{or} \quad y = e^{-t}(C_1 \cos 2t + C_2 \sin 2t). \tag{25}$$

What does this problem look like in the $s$-domain, after a Laplace transform?

$$y'' + 2y' + 5y = 0 \quad \text{transforms to} \quad (s^2 + 2s + 5)\,Y(s) - (s+2)\,y(0) - y'(0) = 0. \tag{26}$$

That quadratic $s^2 + 2s + 5$ will go into the denominator of $Y(s)$, as always. **This part of $Y(s)$ is the transfer function $1/(s^2 + 2s + 5)$.** The numerator is $(s + 2)y(0) + y'(0)$ from the initial conditions. The right hand side of our null equation (26) is zero and the transfer function is connecting the inputs $y(0)$ and $y'(0)$ to the solution:

$$\text{The transform of } y(t) \quad \text{is} \quad Y(s) = \frac{(s+2)\,y(0) + y'(0)}{s^2 + 2s + 5}. \tag{27}$$

This is the point where partial fractions can enter, if we choose. We can separate $s^2 + 2s + 5$ into its linear factors $(s - s_1)(s - s_2)$. *I suggest not to do it.* Those roots $s_1$ and $s_2$ are complex numbers, and it is easier to stay with one real quadratic.

We are close to the transforms of $\cos \omega t$ and $\sin \omega t$, already in the Table above. The new factor is $e^{at} = e^{-t}$ from the real part, and it gives decay.

$$\boldsymbol{e^{at} \cos \omega t \text{ and } e^{at} \sin \omega t \text{ transform to } \frac{s-a}{(s-a)^2 + \omega^2} \text{ and } \frac{\omega}{(s-a)^2 + \omega^2}}. \tag{28}$$

For (27), the key is to separate $s^2 + 2s + 5$ into $(s + 1)^2 + 4$. From this we recognize $a = -1$ and $\omega = 2$ as expected. Then the inverse transform combines $e^{-t} \cos 2t$ and $e^{-t} \sin 2t$. The numerator in (27) is linear, call it $Hs + K$. To fit perfectly with the numerator $s - a$ in (28), we can split any $Hs + K$ into $H(s - a) + (K + Ha)$:

$$\text{The inverse transform of } \frac{Hs + K}{(s-a)^2 + \omega^2} \quad \text{is} \quad He^{at} \cos \omega t + (K + Ha)\, e^{at} \frac{\sin \omega t}{\omega} \tag{29}$$

For higher order equations, and for equations with exponential driving functions $f(t)$, the transform $Y(s)$ involves polynomials of higher degree. In principle, partial fractions can reduce to degree 1 and degree 2. Those produce the real poles and complex poles of $Y(s)$—the real and complex exponentials $e^{st}$ in $y(t)$. I would certainly turn first to the method of undetermined coefficients in Section 2.6.

The best contribution of Laplace transforms is to focus attention on transfer functions like $1/(As^2 + Bs + C)$ and their poles.

## ■ REVIEW OF THE KEY IDEAS ■

1. The Laplace transform of $f(t)$ is $F(s) = \int_0^\infty f(t)e^{-st}\,dt$. $\boldsymbol{f = e^{at} \to F = \frac{1}{s-a}}$.
2. $Ay'' + By' + Cy$ transforms to $(As^2 + Bs + C)\,Y(s) - (As + B)y(0) - Ay'(0)$.
3. Step 1 transforms the equation, Step 2 solves for $Y(s)$, Step 3 inverts $Y(s)$ to $y(t)$.
4. The exponents in the solutions $y_n(t)$ and $y_p(t)$ are the poles in $Y(s)$.
5. Partial fractions can simplify $Y(s)$ using **PF**2 and **PF**3, to help invert to $y(t)$.

# Problem Set 2.7

**1** Take the Laplace transform of each term in these equations and solve for $Y(s)$, with $y(0) = 0$ and $y'(0) = 1$. Find the roots $s_1$ and $s_2$ — the poles of $Y(s)$:

| | |
|---|---|
| *Undamped* | $y'' + 0y' + 16y = 0$ |
| *Underdamped* | $y'' + 2y' + 16y = 0$ |
| *Critically damped* | $y'' + 8y' + 16y = 0$ |
| *Overdamped* | $y'' + 10y' + 16y = 0$ |

For the overdamped case use PF2 to write $Y(s) = A/(s - s_1) + B/(s - s_2)$.

**2** Invert the four transforms $Y(s)$ in Problem 1 to find $y(t)$.

**3** (a) Find the Laplace transform $Y(s)$ from the equation $y' = e^{at}$ with $y(0) = A$.

(b) Use PF2 to break $Y(s)$ into two fractions $C_1/(s - a) + C_2/s$.

(c) Invert $Y(s)$ to find $y(t)$ and check that $y' = e^{at}$ and $y(0) = A$.

**4** (a) Find the transform $Y(s)$ when $y'' = e^{at}$ with $y(0) = A$ and $y'(0) = B$.

(b) Split $Y(s)$ into $C_1/(s - a) + C_2/(s - a)^2 + C_3/s$.

(c) Invert $Y(s)$ to find $y(t)$. Check $y'' = e^{at}$ and $y(0) = A$ and $y'(0) = B$.

**5** Transform these differential equations to find $Y(s)$:

(a) $y'' - y' = 1$ with $y(0) = 4$ and $y'(0) = 0$

(b) $y'' + y = \cos \omega t$ with $y(0) = y'(0) = 0$ and $\omega \neq 1$

(c) $y'' + y = \cos t$ with $y(0) = y'(0) = 0$. What changed for $\omega = 1$?

**6** Find the Laplace transforms $F_1, F_2, F_3$ of these functions $f_1, f_2, f_3$:

$$f_1(t) = e^{at} - e^{bt} \qquad f_2(t) = e^{at} + e^{-at} \qquad f_3(t) = t\cos t$$

**7** For any real or complex $a$, the transform of $f = te^{at}$ is ____. By writing $\cos \omega t$ as $(e^{i\omega t} + e^{-i\omega t})/2$, transform $g(t) = t \cos \omega t$ and $h(t) = te^t \cos \omega t$. *(Notice that the transform of h is new.)*

**8** Invert the transforms $F_1, F_2, F_3$ using PF2 and PF3 to discover $f_1, f_2, f_3$:

$$F_1(s) = \frac{1}{(s-a)(s-b)} \qquad F_2(s) = \frac{s}{(s-a)(s-b)} \qquad F_3(s) = \frac{1}{s^3 - s}$$

**9** Step 1 transforms these equations and initial conditions. Step 2 solves for $Y(s)$. Step 3 inverts to find $y(t)$:

(a) $y' - ay = t$ with $y(0) = 0$

(b) $y'' + a^2 y = 1$ with $y(0) = 1$ and $y'(0) = 2$

(c) $y'' + 3y' + 2y = 1$ with $y(0) = 4$ and $y'(0) = 5$.

What particular solution $y_p$ to (c) comes from using "undetermined coefficients"?

**Questions 10-16 are about partial fractions.**

**10** Show that PF2 in equation (9) is correct. Multiply both sides by $(s-a)(s-b)$:

$$(*) \qquad 1 = \underline{\quad\quad} + \underline{\quad\quad}.$$

(a) What do those two fractions in $(*)$ equal at the points $s = a$ and $s = b$?

(b) The equation $(*)$ is correct at those two points $a$ and $b$. It is the equation of a straight ____. So why is it correct for every $s$?

**11** Here is the PF2 formula with numerators. Formula $(*)$ had $K = 1$ and $H = 0$:

$$\textbf{PF2}' \qquad \frac{Hs + K}{(s-a)(s-b)} = \frac{Ha + K}{(s-a)(a-b)} + \frac{Hb + K}{(b-a)(s-b)}$$

To show that PF2′ is correct, multiply both sides by $(s-a)(s-b)$. You are left with the equation of a straight ____. Check your equation at $s = a$ and at $s = b$. Now it must be correct for all $s$, and PF2′ is proved.

**12** Break these functions into two partial fractions using PF2 and PF2′:

(a) $\dfrac{1}{s^2 - 4}$ (b) $\dfrac{s}{s^2 - 4}$ (c) $\dfrac{Hs + K}{s^2 - 5s + 6}$

**13** Find the integrals of $(a)(b)(c)$ in Problem 12 by integrating each partial fraction. The integrals of $C/(s-a)$ and $D/(s-b)$ are logarithms.

**14** Extend PF3 to PF3′ in the same way that PF2 extended to PF2′:

$$\textbf{PF3}' \qquad \frac{Gs^2 + Hs + K}{(s-a)(s-b)(s-c)} = \frac{Ga^2 + Ha + K}{(s-a)(a-b)(a-c)} + \frac{?}{?} + \frac{?}{?}.$$

**15** The linear polynomial $(s-b)/(a-b)$ equals 1 at $s=a$ and 0 at $s=b$. Write down a quadratic polynomial that equals 1 at $s=a$ and 0 at $s=b$ and $s=c$.

**16** What is the number $C$ so that $C(s-b)(s-c)(s-d)$ equals 1 at $s=a$?

*Note* A complete theory of partial fractions must allow double roots (when $b=a$). The formula can be discovered from l'Hôpital's Rule (in PF3 for example) when $b$ approaches $a$. Multiple roots lose the beauty of PF3 and PF3′—we are happy to stay with simple roots $a, b, c$.

**Questions 17-21 involve the transform $F(s) = 1$ of the delta function $f(t) = \delta(t)$.**

**17** Find $F(s)$ from its definition $\int_0^\infty f(t)e^{-st}\,dt$ when $f(t) = \delta(t-T)$, $T \geq 0$.

**18** Transform $y''-2y'+y=\delta(t)$. The **impulse response** $y(t)$ transforms into $Y(s) =$ **transfer function**. The double root $s_1 = s_2 = 1$ gives a double pole and a new $y(t)$.

**19** Find the inverse transforms $y(t)$ of these transfer functions $Y(s)$:

(a) $\dfrac{s}{s-a}$ (b) $\dfrac{s}{s^2-a^2}$ (c) $\dfrac{s^2}{s^2-a^2}$

**20** Solve $y''+y=\delta(t)$ by Laplace transform, with $y(0)=y'(0)=0$. If you found $y(t)=\sin t$ as I did, this involves a serious mystery: *That sine solves* $y''+y=0$, and it doesn't have $y'(0)=0$. *Where does* $\delta(t)$ *come from*? In other words, what is the derivative of $y'=\cos t$ if all functions are zero for $t<0$?

**If $y = \sin t$, explain why $y'' = -\sin t + \delta(t)$. Remember that $y=0$ for $t<0$.**

Problem (20) connects to a remarkable fact. The same impulse response $y=g(t)$ solves both of these equations: **An impulse at $t=0$ makes the velocity $y'(0)$ jump by 1**. Both equations start from $y(0)=0$.

$$y''+By'+Cy=\boldsymbol{\delta(t)} \text{ with } y'(0)=\mathbf{0} \qquad y''+By'+Cy=\mathbf{0} \text{ with } y'(0)=\mathbf{1}.$$

**21** (Similar mystery) These two problems give the same $Y(s)=s/(s^2+1)$ and the same impulse response $y(t)=g(t)=\cos t$. How can this be?

$$y'=-\sin t \text{ with } y(0)=\mathbf{1} \qquad y'=-\sin t+\boldsymbol{\delta(t)} \text{ with ``}y(0)=0\text{''}$$

**Problems 22-24 involve the Laplace transform of the integral of $y(t)$.**

**22** If $f(t)$ transforms to $F(s)$, what is the transform of the integral $h(t)=\int_0^t f(T)\,dT$? Answer by transforming the equation $dh/dt=f(t)$ with $h(0)=0$.

**23** Transform and solve the integro-differential equation $y' + \int_0^t y\,dt = 1,\ y(0) = 0.$

A mystery like Problem 20: $y = \cos t$ seems to solve $y' + \int_0^t y\,dt = 0,\ y(0) = 1.$

**24** Transform and solve the amazing equation $dy/dt + \int_0^t y\,dt = \delta(t).$

**25** The derivative of the delta function is not easy to imagine—it is called a "doublet" because it jumps up to $+\infty$ and back down to $-\infty$. Find the Laplace transform of the doublet $d\delta/dt$ from the rule for the transform of a derivative.

A doublet $\delta'(t)$ is known by its integral: $\int \delta'(t)F(t)dt = -\int \delta(t)F'(t)dt = \boldsymbol{-F'(0)}$.

**26** (Challenge) What function $y(t)$ has the transform $Y(s) = 1/(s^2+\omega^2)(s^2+a^2)$? First use partial fractions to find $H$ and $K$:

$$Y(s) = \frac{H}{s^2+\omega^2} + \frac{K}{s^2+a^2}$$

**27** Why is the Laplace transform of a unit step function $H(t)$ the same as the Laplace transform of a constant function $f(t) = 1$?

# Chapter 3

# Graphical and Numerical Methods

The world of differential equations is large (very large). This page aims to see what is already done and what remains to do.

Chapters 1 and 2 concentrated on *equations we can solve*. Compared to digging for coal or drilling for oil, this was the equivalent of picking up gold. Solutions were waiting for us. Looking back honestly, we just wrote them down (not so easy in Chapter 2).

Above all I am thinking of $e^{at}$ in Chapter 1 and $e^{st}$ in Chapter 2 and $e^{\lambda t}\boldsymbol{x}$ coming in Chapter 6 (with eigenvalues and eigenvectors). When the equation is linear, and its coefficients are constant, then its solutions are exponentials.

**Chapter 1** First order equations (linear or separable or exact or special)

**Chapter 2** Second order equations $Ay'' + By' + Cy = f(t)$

**Chapter 6** First order systems $\boldsymbol{y}' = A\boldsymbol{y} + \boldsymbol{f}(t)$ with matrices $A$ and vectors $\boldsymbol{y}$.

Chapter 3 will be different. Instead of $f(t)$ we have $f(t, y)$. Most nonlinear problems don't allow a formula for $y(t)$. "A solution exists but it has no formula." This is the hard reality of differential equations $y' = f(t, y)$. The equations are important but they don't have exponential answers. This chapter **pictures** the solution, **computes** the solution, and decides if the solution is **stable**.

**Section 3.1** Pictures for nonlinear equations $y' = f(t, y)$ : Stability decided by $\partial f/\partial y$.

**Section 3.2** Pictures for linear second order equations and 2 by 2 systems : Stable or not.

**Section 3.3** Test for stability at critical points by linearizing systems of equations.

**Section 3.4** Euler methods (safe but slow) for computing approximations to $y$.

**Section 3.5** Fast and accurate computations, by methods more efficient than Euler. Science and engineering and finance constantly use Runge-Kutta.

After this chapter, the book will move into high dimensions : **the world of linear algebra**. One particle and one resistor and one spring and one of anything : that was only a start. The reality is a network of connections : a brain, a living body, a modern machine, a web of processors. Every network leads to a matrix. *You will learn how to read a matrix.*

In my opinion, linear algebra is pure gold.

## 3.1 Nonlinear Equations $y' = f(t, y)$

This section aims to get a picture of $y(t)$, not a formula. The pictures will be graphs in the $t - y$ plane ($t$ across and $y(t)$ up). The differential equation is $dy/dt = f(t, y)$ and everything depends on that function $f$. I can start with a linear equation $y' = 2y$.

> The solutions to $y' = 2y$ are $y(t) = Ce^{2t}$. For every number $C$ this gives a solution curve from $t = -\infty$ to $t = \infty$. Those curves cover every point in the $t - y$ plane. This is the "solution picture" we want for nonlinear equations $y' = f(t, y)$.

That solution $y = Ce^{2t}$ has a graph. The plane is filled with those graphs. Every point $t, y$ has one of those curves going through it (choose the right $C$). A different equation $y' = \sin ty$ won't have a formula. Its picture starts with just this one fact:

$dy/dt = \sin ty$ **The solution curve through the point $t$, $y$ has the slope $\sin ty$.**

From that *point* picture we have to build a *curve* picture. This section tries to connect small arrows at points into solution curves through those points. The arrow at the point $t, y$ has the right slope $f(t, y)$. Connecting with other arrows is the hard part.

I will separate this section into facts about $y(t)$ and pictures of $y(t)$.

### Facts About $y(t)$

The facts will be answers to these questions, and the Chapter 3 Notes add more:

1. Starting from $y(0)$ at $t = 0$, **does $dy/dt = f(t, y)$ have a solution**?
2. **Could there be two or more solutions** that start from the same $y(0)$?

Question **1** is about *existence* of $y(t)$. Is there a solution curve through $t = 0$, $y = y(0)$?
Question **2** is about *uniqueness* of $y(t)$. Could two solution curves go through one point?

When $f(t, y)$ is reasonable, we expect exactly one curve through every point $t, y$: *existence and also uniqueness*. Which functions are reasonable? Here are answers:

1. A solution exists if $f(t, y)$ is a continuous function for $t$ near 0 and $y$ near $y(0)$.
2. There can't be two solutions with the same $y(0)$ when $\partial f/\partial y$ is also continuous.

The word "continuous" has a precise technical meaning. Let me be imprecise and nontechnical. Continuity at a point rules out jumps and infinities in a small neighborhood of that point. The particular function $f = y/t$ is certainly ruled out at points where $t = 0$:

$$\frac{dy}{dt} = \frac{y}{t} \quad \text{with } y(0) = 0 \text{ has infinitely many solutions } \boldsymbol{y = Ct}.$$

The particular function $f = t/y$ is also ruled out when $y(0) = 0$ (no division by 0):

$$\frac{dy}{dt} = \frac{t}{y} \quad \text{with } y(0) = 0 \text{ has two solutions } \boldsymbol{y(t) = t} \textbf{ and } \boldsymbol{y(t) = -t}.$$

In those examples, $y/t$ and $t/y$ are starting from $0/0$. Solutions do exist (that fact wasn't guaranteed). Solutions are not unique (no surprise). We ask more from $f(t, y)$.

There is one important point that we emphasize here, because it could easily be missed.

**Continuity of $f$ and $\dfrac{\partial f}{\partial y}$ at all points does not guarantee that solutions reach $t = \infty$.**

Yes, there will be a solution starting from $y(0)$. That solution will be unique. But $y(t)$ could blow up at some finite time $t$. The first nonlinear equation in the book (Section 1.1) was an example of early explosion:

**Blow-up at $t = 1$** The solution to $\dfrac{dy}{dt} = y^2$ with $y(0) = 1$ is $y(t) = \dfrac{1}{1-t}$.

That function $f = y^2$ is certainly continuous. Its derivative $\partial f/\partial y = 2y$ is also continuous. But the derivative $2y$ grows when the solution grows. To be sure there is no explosion at a finite time $t$, we ask for an upper bound $L$ on the continuous function $\partial f/\partial y$:

**If $\left|\dfrac{\partial f}{\partial y}\right| \leq L$ for all $t$ and $y$ there is a unique solution through $y(0)$ reaching all $t$.**

For a linear differential equation $y' = a(t)y + q(t)$, the derivative $\partial f/\partial y$ of the right hand side is just $a(t)$. Then if $|a(t)| \leq L$ and $q(t)$ is continuous for all time, solution curves go from $t = -\infty$ to $t = \infty$. Chapter 1 found a formula for $y(t)$ in this linear case.

I will end with one final nonlinear fact. The condition $|\partial f/\partial y| \leq L$ is pushed to its limit when $\partial f/\partial y = L$ exactly. Then $y' = Ly + q(t)$. A comparison with this linear equation gives information about the nonlinear equation, when $|\partial f/\partial y| \leq L$:

$$\text{If } y' = f(t, y) \text{ and } z' = f(t, z), \text{ then } |y(t) - z(t)| \leq e^{Lt}|y(0) - z(0)|. \quad (1)$$

*If $y(t)$ and $z(t)$ start very close, they stay close.* This is the opposite of what you see on the cover of this book. The cover shows a famous example of **chaos**: solutions go wild. A slight change in $y(0)$ will send the solution on a completely different (and distant) path. We now know that Pluto's orbit is chaotic: very very unpredictable. The equations allow it, because they don't have $|\partial f/\partial y| \leq L$. Pluto is not a planet.

## Pictures of the Solution

**Example 1** $dy/dt = 2 - y$ **Solution** $y(t) = 2 + Ce^{-t}$ $y(\infty) = 2$

The perfect picture of $y' = 2 - y$ would show a small arrow at every point $t, y$. **The arrow would have slope $s = 2 - y$.** Along the all-important "steady state line" $y = 2$, this slope would be *zero*. The arrows are flat ($s = 0$) along that line: a constant solution.

Above that steady line, the slope $2 - y$ is negative. The vectors have components $dt$ across and $dy = (2 - y)dt$ down. We don't have space for an arrow at every point, but Figure 3.1 gives the idea. MATLAB calls the field of arrows a "quiver".

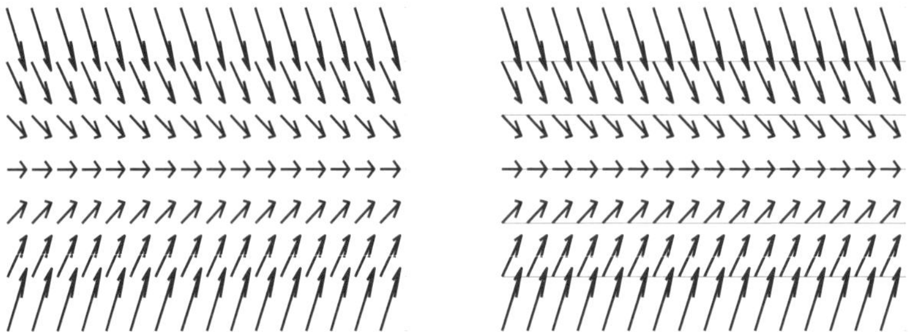

Figure 3.1: (a) Arrows with slopes $f(t, y)$ show the direction of the solution curves $y(t)$. (b) **Along an isocline $f(t, y) = s$, all arrows have the same slope $s$.** Here $s = 2 - y$.

Notice that all arrows point **toward** the line $y = 2$. That steady state solution is **stable**. The formula $y(t) = 2 + Ce^{-t}$ confirms that the solutions approach $y = 2$.

*First key idea*: **The solution curves $y(t) = 2 + Ce^{-t}$ are tangent to the arrows.** Tangent means: The curves have the same slope $s = 2 - y$ as the arrows! The curves solve the equation, the equation specifies the slopes, the arrows have correct slopes.

*Second key idea*: **Put your arrows along isoclines.** An isocline (meaning "same slope") is a curve $f(t, y) =$ constant. This idea makes the arrows much easier to draw. All the isoclines $2 - y = s$ are horizontal lines for this equation $y' = 2 - y$. When the differential equation is $dy/dt = f(t, y)$, **each choice of slope $s$ produces an isocline $f(t, y) = s$.**

In our example, those isoclines $2 - y = s$ are flat because $f(t, y) = 2 - y$ does not depend on $t$ (autonomous equation). I start the picture by drawing a few isoclines. I always draw the isocline $f(t, y) = 0$ (here $2 - y = 0$ is the steady state line $y = 2$). For this equation, that "nullcline" or "zerocline" with $s = 0$ **is also a solution curve**. The arrows have slope zero when $y = 2$, so they point along the flat line.

How to understand these pictures? **The arrows are pointing along the solution curves.** The curves cross over isoclines. But they don't cross over the zero isocline $y = 2$.

All arrows are pointing toward the line $y = 2$. Those arrows will eventually take us across every other isocline. The pictures say that the solution curves $y(t)$ are asymptotic to that line $y = 2$. For this equation $dy/dt = 2 - y$ we know the solutions $y = 2 + Ce^{-t}$.

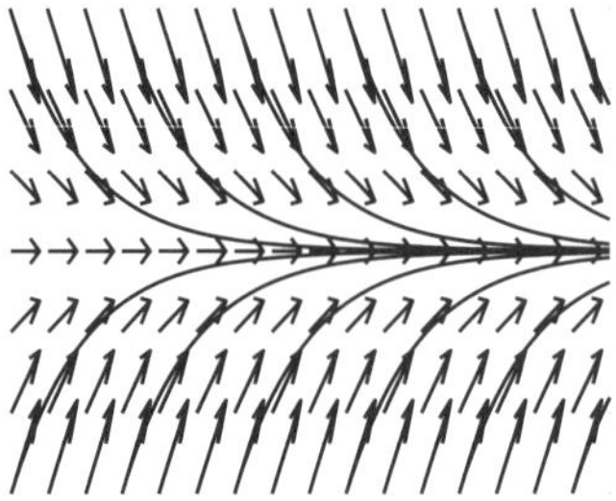

Figure 3.2: Solution curves (tangent to arrows) go through isoclines: $y' = 2 - y$.

**Example 2** $\dfrac{dy}{dt} = y - y^2$ **Solutions** $y(t) = \dfrac{1}{1 + Ce^{-t}}$ $y(t) \to 1$ or $-\infty$

The slope of every small arrow is $y - y^2$. In the range $0 < y < 1$, $y$ will be larger than $y^2$. The arrows have positive slope $y - y^2$ in this range (small slope near $y = 0$, small slope near $y = 1$, all up and to the right). The other two ranges are above $y = 1$ and below $y = 0$. There the slopes $y - y^2$ are negative—arrows go down and right. *The solution curves are steep when* $y$ *is large*, because $y^2 >> y$.

Figure 3.3 shows the isoclines $f(t, y) = y - y^2 = s =$ constant. Again $f$ does not depend on $t$ ! The equation is autonomous, the isoclines are flat lines. There are **two zeroclines $y = 1$ and $y = 0$** (where $dy/dt = 0$ and $y$ is constant). Those arrows have zero slope and the graph of $y(t)$ runs along each zerocline : a steady state.

The question is about all the other solution curves : What do they do ? We happen to have a formula for $y(t)$, but the point is that *we don't need it*. Figure 3.3 shows the three possibilities for the solution curves to the *logistic equation* $y' = y - y^2$ :

1. Curves above $y = 1$ go from $+\infty$ down toward the line $y = 1$ (**dropin curves**)
2. Curves between $y = 0$ and $y = 1$ go up toward that line $y = 1$ (**$S$-curves**)
3. Curves below $y = 0$ go down (fast) toward $y = -\infty$ (**dropoff curves**).

The solution curves go across all isoclines except the two zeroclines where $y - y^2 = 0$.

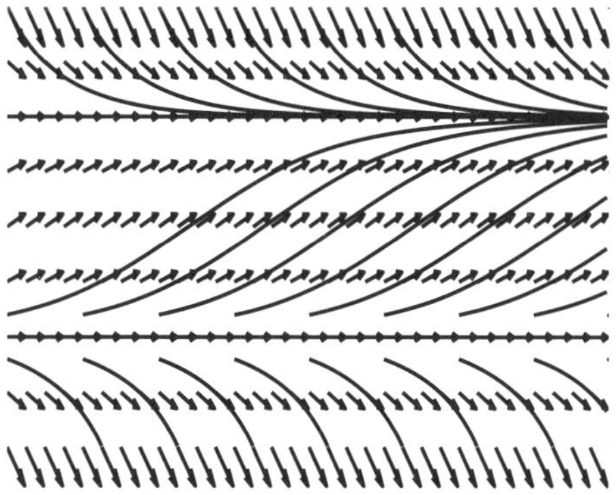

Figure 3.3: The arrows form a "direction field". Isoclines $y - y^2 = s$ attract or repel.

You see the $S$-curves between 0 and 1. The arrows are flat as they leave $y = 0$, steepest at $y = \frac{1}{2}$, flat again as they approach $y = 1$. The dropoff curves are below $y = 0$. Those arrows get very steep and the curves never reach $t = \infty$: $y = 1/(1 - e^{-t})$ gives $1/0 =$ *minus infinity* when $t = 0$. That dropoff curve never gets out of the third quadrant.

**Important** Solution curves have a special feature for autonomous equations $y' = f(y)$. Suppose the curve $y(t)$ is shifted right or left to the curve $Y(t) = y(t + C)$. Then $Y(t)$ solves the same equation $Y' = f(Y)$—both sides are just shifted in the same way.

Conclusion: The solution curves for autonomous equations $y' = f(y)$ just shift along *with no change in shape*. You can also see this by integrating $dy/f(y) = dt$ (separable equation). The right side integrates to $t + C$. We get all solutions by allowing all $C$.

In the logistic example, all $S$-curves and dropin curves and dropoff curves come from shifting *one* $S$-curve and *one* dropin curve and *one* dropoff curve.

## Solution Curves Don't Meet

Is there a solution curve through every point $(t, y)$? Could two solution curves meet at that point? Could a solution curve suddenly end at a point? These "picture questions" are already answered by the facts.

At the start of this section, the functions $f$ and $\partial f/\partial y$ were required to be continuous near $t = 0$, $y = y(0)$. Then there is a unique solution to $y' = f(t, y)$ with that start. In the picture this means: **There is exactly one solution curve going through the point.** The curve doesn't stop. By requiring $f$ and $\partial f/\partial y$ to be continuous at and near *all* points, we guarantee one non-stopping solution curve through every point.

Example 3 will fail! The solution curves for $dy/dt = -t/y$ are half-circles and not whole circles. **They start and stop and meet on the line $y = 0$** (where $f = -t/y$ is **not continuous**). Exactly one semicircular curve passes through every point with $y \neq 0$.

**Example 3** $dy/dt = -t/y$ is separable. Then $y\,dy = -t\,dt$ leads to $y^2 + t^2 = C$.

Start again with pictures. The isocline $f(t, y) = -t/y = s$ is the line $y = (-1/s)t$. All those isoclines go through $(0, 0)$ which is a very singular point. In this example the direction arrows with slope $s$ are perpendicular to the isoclines with slope $dy/dt = -1/s$.

The isoclines are rays out from $(0, 0)$. The arrow directions are perpendicular to those rays and tangent to the solution curves. **The curves are half-circles $y^2 + t^2 = C$.** (There is another half-circle on the opposite side of the axis. So two solutions start from $y = 0$ at time $-T$ and go forward to $y = 0$ at time $T$.) The solution curves stop at $y = 0$, where the function $f = -t/y$ loses its continuity and the solution loses its life.

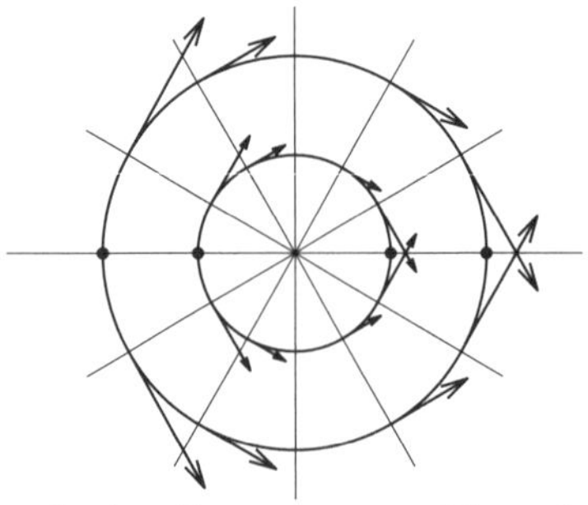

Figure 3.4: For $y' = -t/y$ the isoclines are rays. The solution curves are half-circles.

**Example 4** $\;y' = 1 + t - y$ is linear but not separable. The isoclines trap the solution.

Trapping between isoclines is a neat part of the picture. It is based on the arrows. **All arrows go one way across an isocline, so all solution curves go that way**. Solutions that cross the isocline can't cross back. The zero isocline $f(t, y) = 1 + t - y = 0$ in Figure 3.5 is the line $y = t + 1$. Along that isocline the arrows have slope 0. The solution curves must cross from above to below.

The central isocline $1 + t - y = 1$ in Figure 3.5 is the 45° line $y = t$. This solves the differential equation! The arrow directions are exactly along the line: slope $s = 1$. Other solution curves could never touch this one.

The picture shows solution curves in a "lobster trap" between the lines: the curves can't escape. They are trapped between the line $y = t$ and every isocline $1 + t - y = s$ above or below it. The trap gets tighter and tighter as $s$ increases from 0 to 1, and the isocline gets closer to $y = t$. *Conclusion from the picture*: **The solution $y(t)$ must approach $t$.**

This is a linear equation $y' + y = 1 + t$. The null solutions to $y' + y = 0$ are $Ce^{-t}$. The forcing term $1 + t$ is a polynomial. A particular solution comes by substituting $y_p(t) = at + b$ into the equation and solving for those undetermined coefficients $a$ and $b$:

$$(at + b)' = 1 + t - (at + b) \qquad \mathbf{a = 1 \text{ and } b = 0} \qquad \mathbf{y = y_n + y_p = Ce^{-t} + t} \qquad (2)$$

The solution curves $y = Ce^{-t} + t$ do approach the line $y = t$ asymptotically as $t \to \infty$.

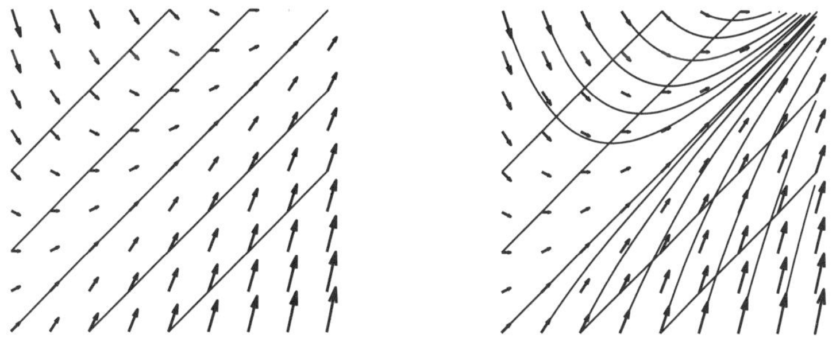

Figure 3.5: The solution curves for $y' = 1 + t - y$ get trapped between the 45° isoclines.

## ■ REVIEW OF THE KEY IDEAS ■

1. The direction field for $y' = f(t, y)$ has an arrow with slope $f$ at each point $t, y$.
2. Along the isocline $f(t, y) = s$, all arrows have the same slope $s$.
3. The solution curves $y(t)$ are tangent to the arrows. One way through isoclines!
4. Fact: When $f$ and $\partial f/\partial y$ are continuous, the curves cover the plane and don't meet.
5. The solution curves for autonomous $y' = f(y)$ shift left - right to $Y(t) = y(t - T)$.

## Problem Set 3.1

**1** (a) Why do two isoclines $f(t, y) = s_1$ and $f(t, y) = s_2$ never meet ?

(b) Along the isocline $f(t, y) = s$, what is the slope of all the arrows ?

(c) Then all solution curves go only one way across an _____ .

**2** (a) Are isoclines $f(t, y) = s_1$ and $f(t, y) = s_2$ always parallel ? Always straight ?

(b) An isocline $f(t, y) = s$ is a solution curve when its slope equals _____ .

(c) The zerocline $f(t, y) = 0$ is a solution curve only when $y$ is _____ : slope 0.

**3** If $y_1(0) < y_2(0)$, what continuity of $f(t, y)$ assures that $y_1(t) < y_2(t)$ for all $t$ ?

**4** The equation $dy/dt = t/y$ is completely safe if $y(0) \neq 0$. Write the equation as $y\,dy = t\,dt$ and find its unique solution starting from $y(0) = -1$. The solution curves are hyperbolas—can you draw two on the same graph ?

**5** The equation $dy/dt = y/t$ has many solutions $y = Ct$ in case $y(0) = 0$. It has no solution if $y(0) \neq 0$. When you look at all solution curves $y = Ct$, which points in the $t$, $y$ plane have no curve passing through ?

**6** For $y' = ty$ draw the isoclines $ty = 1$ and $ty = 2$ (those will be hyperbolas). On each isocline draw four arrows (they have slopes 1 and 2). Sketch pieces of solution curves that fit your picture between the isoclines.

**7** The solutions to $y' = y$ are $y = Ce^t$. Changing $C$ gives a higher or lower curve. But $y' = y$ is autonomous, its solution curves should be shifting right and left!

Draw $y = 2e^t$ and $y = -2e^t$ to show that they really are *right-left shifts* of $y = e^t$ and $y = -e^t$. The shifted solutions to $y' = y$ are $e^{t+C}$ and $-e^{t+C}$.

**8** For $y' = 1 - y^2$ the flat lines $y =$ constant are isoclines $1 - y^2 = s$. Draw the lines $y = 0$ and $y = 1$ and $y = -1$. On each line draw arrows with slope $1 - y^2$. The picture says that $y =$ _____ and $y =$ _____ are steady state solutions. From the arrows on $y = 0$, guess a shape for the solution curve $y = (e^t - e^{-t})/(e^t + e^{-t})$.

**9** The parabola $y = t^2/4$ and the line $y = 0$ are both solution curves for $\boldsymbol{y' = \sqrt{|y|}}$. Those curves meet at the point $t = 0$, $y = 0$. What continuity requirement is failed by $f(y) = \sqrt{|y|}$, to allow more than one solution through that point ?

**10** Suppose $y = 0$ up to time $T$ is followed by the curve $y = (t - T)^2/4$. Does this solve $\boldsymbol{y' = \sqrt{|y|}}$? Draw this $y(t)$ going through flat isoclines $\sqrt{|y|} = 1$ and 2.

**11** The equation $\boldsymbol{y' = y^2 - t}$ is often a favorite in MIT's course 18.03 : not too easy. Why do solutions $y(t)$ rise to their maximum on $y^2 = t$ and then descend ?

**12** Construct $f(t, y)$ with two isoclines so solution curves go *up* through the higher isocline and other solution curves go *down* through the lower isocline. *True or false* : Some solution curve will stay between those isoclines : **A continental divide**.

## 3.2 Sources, Sinks, Saddles, and Spirals

The pictures in this section show solutions to $Ay'' + By' + Cy = 0$. These are linear equations with constant coefficients $A, B$, and $C$. The graphs show solutions $y$ on the horizontal axis and their slopes $y' = dy/dt$ on the vertical axis. These pairs $(y(t), y'(t))$ depend on time, *but time is not in the pictures.* The paths show *where* the solution goes, but they don't show when.

Each specific solution starts at a particular point $(y(0), y'(0))$ given by the initial conditions. The point moves along its path as the time $t$ moves forward from $t = 0$. We know that the solutions to $Ay'' + By' + Cy = 0$ depend on the two solutions to $As^2 + Bs + C = 0$ (an ordinary quadratic equation for $s$). When we find the roots $s_1$ and $s_2$, we have found all possible solutions:

$$y = c_1 e^{s_1 t} + c_2 e^{s_2 t} \qquad y' = c_1 s_1 e^{s_1 t} + c_2 s_2 e^{s_2 t} \tag{1}$$

The numbers $s_1$ and $s_2$ tell us which picture we are in. Then the numbers $c_1$ and $c_2$ tell us which path we are on.

Since $s_1$ and $s_2$ determine the picture for each equation, it is essential to see the six possibilities. We write all six here in one place, to compare them. Later they will appear in six different places, one with each figure. The first three have real solutions $s_1$ and $s_2$. The last three have complex pairs $\boldsymbol{s = a \pm i\omega}$.

| **Sources** | **Sinks** | **Saddles** | **Spiral out** | **Spiral in** | **Center** |
|---|---|---|---|---|---|
| $s_1 > s_2 > 0$ | $s_1 < s_2 < 0$ | $s_2 < 0 < s_1$ | $a = \text{Re}\, s > 0$ | $a = \text{Re}\, s < 0$ | $a = \text{Re}\, s = 0$ |

In addition to those six, there will be limiting cases $s = 0$ and $s_1 = s_2$ (as in resonance).

**Stability** This word is important for differential equations. *Do solutions decay to zero?* The solutions are controlled by $e^{s_1 t}$ and $e^{s_2 t}$ (and in Chapter 6 by $e^{\lambda_1 t}$ and $e^{\lambda_2 t}$). We can identify the two pictures (out of six) that are displaying full stability: the sinks. **A center $s = \pm i\omega$ is at the edge of stability** ($e^{i\omega t}$ is neither decaying or growing).

| | | |
|---|---|---|
| **2. Sinks are stable** | $\boldsymbol{s_1 < s_2 < 0}$ | **Then $y(t) \to 0$** |
| **5. Spiral sinks are stable** | $\text{Re}\, s_1 = \text{Re}\, s_2 < 0$ | **Then $y(t) \to 0$** |

**Special note.** May I mention here that the same six pictures also apply to a system of *two first order equations.* Instead of $y$ and $y'$, the equations have unknowns $y_1$ and $y_2$. Instead of the constant coefficients $A, B, C$, the equations will have a 2 by 2 matrix. Instead of the roots $s_1$ and $s_2$, that matrix will have eigenvalues $\lambda_1$ and $\lambda_2$. **Those eigenvalues are the roots of an equation $\boldsymbol{A\lambda^2 + B\lambda + C = 0}$**, just like $s_1$ and $s_2$.

We will see the same six possibilities for the $\lambda$'s, and the same six pictures. The eigenvalues of the 2 by 2 matrix give the growth rates or decay rates, in place of $s_1$ and $s_2$.

$$\begin{bmatrix} y_1' \\ y_2' \end{bmatrix} = \begin{bmatrix} a & b \\ c & d \end{bmatrix} \begin{bmatrix} y_1 \\ y_2 \end{bmatrix} \text{ has solutions } \begin{bmatrix} y_1(t) \\ y_2(t) \end{bmatrix} = \begin{bmatrix} v_1 \\ v_2 \end{bmatrix} e^{\lambda t}.$$

The eigenvalue is $\lambda$ and the eigenvector is $\boldsymbol{v} = (v_1, v_2)$. The solution is $\boldsymbol{y(t) = ve^{\lambda t}}$.

## The First Three Pictures

We are starting with the case of *real roots* $s_1$ and $s_2$. In the equation $Ay'' + By' + Cy = 0$, this means that $\boldsymbol{B^2 \geq 4AC}$. Then $B$ is relatively large. The square root in the quadratic formula produces a real number $\sqrt{B^2 - 4AC}$. If $A, B, C$ have the same sign, we have overdamping and **negative roots** and stability. The solutions decay to $(0, 0)$ : a **sink**.

If $A$ and $C$ have opposite sign to $B$ as in $y'' - 3y' + 2y = 0$, we have negative damping and **positive roots** $s_1, s_2$. The solutions grow (this is instability : a **source** at $(0, 0)$).

Suppose $A$ and $C$ have different signs, as in $y'' - 3y' - 2y = 0$. Then $s_1$ and $s_2$ also have **different signs** and the picture shows a **saddle**. The moving point $(y(t), y'(t))$ can start in toward $(0, 0)$ before it turns out to infinity. The positive $s$ gives $e^{st} \to \infty$. *Second example for a saddle*: $y'' - 4y = 0$ leads to $s^2 - 4 = (s - 2)(s + 2) = 0$. The roots $s_1 = \mathbf{2}$ and $s_2 = \mathbf{-2}$ have opposite signs. Solutions $c_1 e^{2t} + c_2 e^{-2t}$ grow unless $c_1 = 0$. Only that one line with $c_1 = 0$ has arrows inward.

In every case with $B^2 \geq 4AC$, the roots are real. The solutions $y(t)$ have growing exponentials or decaying exponentials. We don't see sines and cosines and oscillation.

The first figure shows growth: $0 < s_2 < s_1$. Since $e^{s_1 t}$ grows faster than $e^{s_2 t}$, the larger number $s_1$ will dominate. The solution path for $(y, y')$ will approach the straight line of slope $s_1$. That is because the ratio of $y' = c_1 s_1 e^{s_1 t}$ to $y = c_1 e^{s_1 t}$ is exactly $s_1$.

If the initial condition is on the "$s_1$ line" then the solution $(y, y')$ stays on that line: $c_2 = 0$. If the initial condition is exactly on the "$s_2$ line" then the solution stays on that secondary line : $c_1 = 0$. You can see that if $c_1 \neq 0$, the $c_1 e^{s_1 t}$ part takes over as $t \to \infty$.

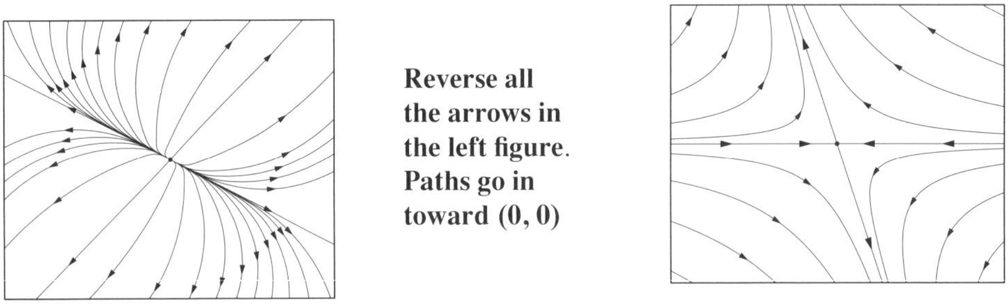
**Reverse all the arrows in the left figure. Paths go in toward (0, 0)**

| $\mathbf{0 < s_2 < s_1}$ | $\mathbf{s_1 < s_2 < 0}$ | $\mathbf{s_2 < 0 < s_1}$ |
|---|---|---|
| **Source : Unstable** | **Sink : Stable** | **Saddle : Unstable** |

Figure 3.6: **Real roots $s_1$ and $s_2$.** The paths of the point $(y(t), y'(t))$ lead out when roots are positive and lead in when roots are negative. With $s_2 < 0 < s_1$, the $s_2$-line leads in but all other paths eventually go out near the $s_1$-line : *The picture shows a saddle point.*

*Example for a source*: $y'' - 3y' + 2y = 0$ leads to $s^2 - 3s + 2 = (s-2)(s-1) = 0$. The roots **1** and **2** are positive. The solutions grow and $e^{2t}$ dominates.

*Example for a sink*: $y'' + 3y' + 2y = 0$ leads to $s^2 + 3s + 2 = (s+2)(s+1) = 0$. The roots **−2** and **−1** are negative. The solutions decay and $e^{-t}$ dominates.

## The Second Three Pictures

We move to the case of **complex roots** $s_1$ and $s_2$. In the equation $Ay'' + By' + Cy = 0$, this means that $B^2 < 4AC$. Then $A$ and $C$ have the same signs and $B$ is relatively small (underdamping). The square root in the quadratic formula (2) is an imaginary number. *The exponents $s_1$ and $s_2$ are now a complex pair $a \pm i\omega$*:

**Complex roots of** **$As^2 + Bs + C = 0$**

$$s_1, s_2 = -\frac{B}{2A} \pm \frac{\sqrt{B^2 - 4AC}}{2A} = a \pm i\omega. \qquad (2)$$

The path of $(y, y')$ **spirals around the center**. Because of $e^{at}$, the spiral goes out if $a > 0$: **spiral source**. Solutions spiral in if $a < 0$: **spiral sink**. The frequency $\omega$ controls how fast the solutions oscillate and how quickly the spirals go around $(0, 0)$.

In case $a = -B/2A$ is zero (no damping), we have a **center** at $(0, 0)$. The only terms left in $y$ are $e^{i\omega t}$ and $e^{-i\omega t}$, in other words $\cos \omega t$ and $\sin \omega t$. Those paths are ellipses in the last part of Figure 3.7. The solutions $y(t)$ are **periodic**, because increasing $t$ by $2\pi/\omega$ will not change $\cos\,\omega t$ and $\sin\,\omega t$. That circling time $2\pi/\omega$ is the **period**.

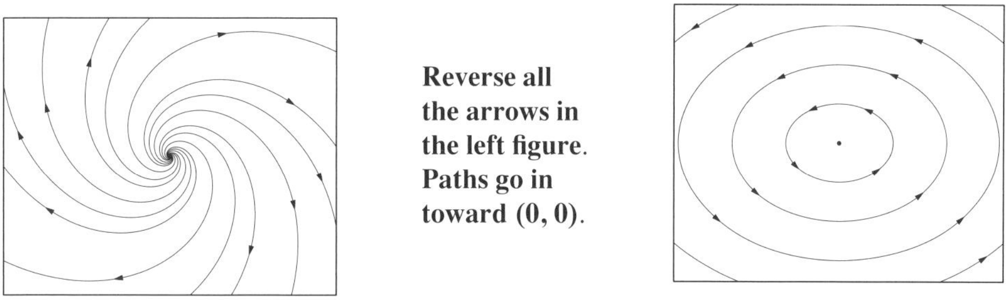
Reverse all
the arrows in
the left figure.
Paths go in
toward (0, 0).

**$a = \mathrm{Re}\, s > 0$**
**Spiral source : Unstable**

**$a = \mathrm{Re}\, s < 0$**
**Spiral sink : Stable**

**$a = \mathrm{Re}\, s = 0$**
**Center : Neutrally stable**

Figure 3.7: **Complex roots $s_1$ and $s_2$**. The paths go once around $(0, 0)$ when $t$ increases by $2\pi/\omega$. The paths spiral in when $A$ and $B$ have the same signs and $a = -B/2A$ is negative. They spiral out when $a$ is positive. If $B = 0$ (no damping) and $4AC > 0$, we have a center. The simplest center is $y = \sin\, t$, $y' = \cos\, t$ (circle) from $y'' + y = 0$.

## First Order Equations for $y_1$ and $y_2$

On the first page of this section, a "Special Note" mentioned another application of the same pictures. Instead of graphing the path of $(y(t), y'(t))$ for one second order equation, we could follow the path of $(y_1(t), y_2(t))$ for **two first order equations**. The two equations look like this:

| | | |
|---|---|---|
| **First order system $\boldsymbol{y}' = A\boldsymbol{y}$** | $$\begin{aligned} dy_1/dt &= ay_1 + by_2 \\ dy_2/dt &= cy_1 + dy_2 \end{aligned}$$ | (3) |

The starting values $y_1(0)$ and $y_2(0)$ are given. The point $(y_1, y_2)$ will move along a path in one of the six figures, depending on the numbers $a, b, c, d$.

Looking ahead, those four numbers will go into a 2 by 2 matrix $A$. Equation (3) will become $d\boldsymbol{y}/dt = A\boldsymbol{y}$. The symbol $\boldsymbol{y}$ in boldface stands for the vector $\boldsymbol{y} = (y_1, y_2)$. And most important for the six figures, *the exponents $s_1$ and $s_2$ in the solution $\boldsymbol{y}(t)$ will be the* **eigenvalues $\lambda_1$ and $\lambda_2$** of the matrix $A$.

## Companion Matrices

Here is the connection between a second order equation and two first order equations. All equations on this page are linear and all coefficients are constant. I just want you to see the special "*companion matrix*" that appears in the first order equations $\boldsymbol{y}' = A\boldsymbol{y}$.

Notice that $\boldsymbol{y}$ is printed in **boldface type** because it is a **vector**. It has two components $y_1$ and $y_2$ (those are in lightface type). The first $y_1$ is the same as the unknown $y$ in the second order equation. The second component $y_2$ is the velocity $dy/dt$:

$$\begin{matrix} y_1 = y \\ y_2 = y' \end{matrix} \qquad \boldsymbol{y'' + 4y' + 3y} = 0 \quad \text{becomes} \quad \boldsymbol{y_2' + 4y_2 + 3y_1} = 0. \tag{4}$$

On the right you see one of the first order equations connecting $y_1$ and $y_2$. We need a second equation (two equations for two unknowns). *It is hiding at the far left!* There you see that $\boldsymbol{y_1' = y_2}$. In the original second order problem this is the trivial statement $y' = y'$. In the vector form $\boldsymbol{y}' = A\boldsymbol{y}$ it gives the first equation in our system. The first row of our matrix is **0 1**. When $y$ and $y'$ become $y_1$ and $y_2$,

$$y'' + 4y' + 3y = 0 \quad \text{becomes} \quad \begin{aligned} y_1' &= \qquad\quad y_2 \\ y_2' &= -3y_1 - 4y_2 \end{aligned} = \begin{bmatrix} \mathbf{0} & \mathbf{1} \\ \mathbf{-3} & \mathbf{-4} \end{bmatrix} \begin{bmatrix} y_1 \\ y_2 \end{bmatrix} \tag{5}$$

That first row **0 1** makes this a 2 by 2 **companion matrix**. It is the companion to the second order equation. The key point is that the first order and second order problems have the same "characteristic equation" because they are the same problem.

The equation $s^2 + 4s + 3 = 0$ gives the exponents $s_1 = \mathbf{-3}$ and $s_2 = \mathbf{-1}$

The equation $\lambda^2 + 4\lambda + 3 = 0$ gives the eigenvalues $\lambda_1 = \mathbf{-3}$ and $\lambda_2 = \mathbf{-1}$

The problems are the same, the exponents $-3$ and $-1$ are the same, the figures will be the same. Those figures show a **sink** because $-3$ and $-1$ are real and both negative. Solutions approach $(0,0)$. These equations are **stable**.

$$\textbf{The companion matrix for } y'' + By' + Cy = 0 \textbf{ is } A = \begin{bmatrix} 0 & 1 \\ -C & -B \end{bmatrix}.$$

Row 1 of $y' = Ay$ is $y_1' = y_2$. Row 2 is $y_2' = -Cy_1 - By_2$. When you replace $y_2$ by $y_1'$, this means that $y_1'' + By_1' + Cy_1 = 0$: *correct*.

## Stability for 2 by 2 Matrices

I can explain when a 2 by 2 system $\boldsymbol{y}' = A\boldsymbol{y}$ is stable. This requires that all solutions $\boldsymbol{y}(t) = (y_1(t), y_2(t))$ approach zero as $t \to \infty$. When the matrix $A$ is a companion matrix, this 2 by 2 system comes from one second order equation $y'' + By' + Cy = 0$. In that case we know that stability depends on the roots of $s^2 + Bs + C = 0$. **Companion matrices are stable when $B > 0$ and $C > 0$.**

From the quadratic formula, the roots have $s_1 + s_2 = -B$ and $s_1 s_2 = C$.

If $s_1$ and $s_2$ are negative, this means that $B > 0$ and $C > 0$.

If $s_1 = a + i\omega$ and $s_2 = a - i\omega$ and $a < 0$, this again means $B > 0$ and $C > 0$

Those complex roots add to $s_1 + s_2 = 2a$. Negative $a$ (stability) means positive $B$, since $s_1 + s_2 = -B$. Those roots multiply to $s_1 s_2 = a^2 + \omega^2$. This means that $C$ is positive, since $s_1 s_2 = C$.

For companion matrices, stability is decided by $B > 0$ and $C > 0$. **What is the stability test for any 2 by 2 matrix**? This is the key question, and Chapter 6 will answer it properly. We will find the equation for the eigenvalues of any matrix (Section 6.1). We will test those eigenvalues for stability (Section 6.4). Eigenvalues and eigenvectors are a major topic, the most important link between differential equations and linear algebra. Fortunately, the eigenvalues of 2 by 2 matrices are especially simple.

$$\text{The eigenvalues of the matrix } A = \begin{bmatrix} a & b \\ c & d \end{bmatrix} \text{ have } \lambda^2 - T\lambda + D = 0.$$

The number $T$ is $a + d$. The number $D$ is $ad - bc$.

Companion matrices have $a = 0$ and $b = 1$ and $c = -C$ and $d = -B$. Then the characteristic equation $\lambda^2 - T\lambda + D = 0$ is exactly $s^2 + Bs + C = 0$.

$$\text{Companion matrices have } \begin{bmatrix} 0 & 1 \\ -C & -B \end{bmatrix} \quad T = a+d = -B \text{ and } D = ad-bc = C.$$

**The stability test $B > 0$ and $C > 0$ is turning into the stability test $T < 0$ and $D > 0$.**

This is the test for any 2 by 2 matrix. Stability requires $T < 0$ and $D > 0$. Let me give four examples and then collect together the main facts about stability.

$$A_1 = \begin{bmatrix} 0 & 1 \\ -2 & 3 \end{bmatrix} \text{ is } \textit{unstable} \text{ because } T = 0 + 3 \text{ is positive}$$

$$A_2 = \begin{bmatrix} 0 & 1 \\ 2 & -3 \end{bmatrix} \text{ is } \textit{unstable} \text{ because } D = -(1)(2) \text{ is negative}$$

$$A_3 = \begin{bmatrix} 0 & 1 \\ -2 & -3 \end{bmatrix} \text{ is } \textit{stable} \text{ because } T = -3 \text{ and } D = +2$$

$$A_4 = \begin{bmatrix} -1 & 1 \\ -1 & -1 \end{bmatrix} \text{ is } \textit{stable} \text{ because } T = -1 - 1 \text{ is negative and } D = 1 + 1 \text{ is positive}$$

The eigenvalues always come from $\lambda^2 - T\lambda + D = 0$. For that last matrix $A_4$, this eigenvalue equation is $\lambda^2 + 2\lambda + 2 = 0$. The eigenvalues are $\lambda_1 = \mathbf{-1+i}$ and $\lambda_2 = \mathbf{-1-i}$. They add to $T = -2$ and they multiply to $D = +2$. **This is a spiral sink and it is stable.**

| **Stability for 2 by 2 matrices** | $A = \begin{bmatrix} a & b \\ c & d \end{bmatrix}$ is stable if $\begin{matrix} T = a + d < 0 \\ D = ad - bc > 0 \end{matrix}$ |
|---|---|

The six pictures for $(y, y')$ become six pictures for $(y_1, y_2)$. The first three pictures have real eigenvalues from $T^2 \geq 4D$. The second three pictures have complex eigenvalues from $T^2 < 4D$. This corresponds perfectly to the tests for $y'' + By' + Cy = 0$ and its companion matrix:

| | | | |
|---|---|---|---|
| Real eigenvalues | $T^2 \geq 4D$ | $B^2 \geq 4C$ | Overdamping |
| Complex eigenvalues | $T^2 < 4D$ | $B^2 < 4C$ | Underdamping |

That gives one picture of eigenvalues $\lambda$ : *Real or complex.* The second picture is different: *Stable or unstable.* Both of those splittings are decided by $T$ and $D$ (or $-B$ and $C$).

| | | | |
|---|---|---|---|
| 1. | **Source** | $T > 0,\ D > 0,\ T^2 \geq 4D$ | *Ustable* |
| 2. | **Sink** | $T < 0,\ D > 0,\ T^2 \geq 4D$ | *Stable* |
| 3. | **Saddle** | $D < 0$ and $T^2 \geq 4D$ | *Unstable* |
| 4. | **Spiral source** | $T > 0,\ D > 0,\ T^2 < 4D$ | *Unstable* |
| 5. | **Spiral Sink** | $T < 0,\ D > 0,\ T^2 < 4D$ | *Stable* |
| 6. | **Center** | $T = 0,\ D > 0,\ T^2 < 4D$ | *Neutral* |

That neutrally stable center has eigenvalues $\lambda_1 = i\omega$ and $\lambda_2 = -i\omega$ and undamped oscillation.

Section 3.3 will use this information to decide the stability of *nonlinear* equations.

## Eigenvectors of Companion Matrices

Eigenvalues of $A$ come with eigenvectors. If we stay a little longer with a companion matrix, we can see its eigenvectors. Chapter 6 will develop these ideas for any matrix, and we need more linear algebra to understand them properly. But our vectors $(y_1, y_2)$ come from $(y, y')$ in a differential equation, and that connection makes the eigenvectors of a companion matrix especially simple.

The fundamental idea for constant coefficient linear equations is always the same: **Look for exponential solutions**. For a second order equation those solutions are $y = e^{st}$. For a system of two first order equations those solutions are $\boldsymbol{y} = \boldsymbol{v}e^{\lambda t}$. **The vector $\boldsymbol{v} = (v_1, v_2)$ is the eigenvector that goes with the eigenvalue $\boldsymbol{\lambda}$.**

Substitute $\begin{aligned} y_1 &= v_1 e^{\lambda t} \\ y_2 &= v_2 e^{\lambda t} \end{aligned}$ into the equations $\begin{aligned} y_1' &= a y_1 + b y_2 \\ y_2' &= c y_1 + d y_2 \end{aligned}$ and factor out $e^{\lambda t}$.

Because $e^{\lambda t}$ is the same for both $y_1$ and $y_2$, it will appear in every term. When all factors $e^{\lambda t}$ are removed, we will see the equations for $v_1$ and $v_2$. That vector $\boldsymbol{v} = (v_1, v_2)$ will satisfy the eigenvector equation $A\boldsymbol{v} = \lambda\boldsymbol{v}$. This is the key to Chapter 6.

Here I only look at eigenvectors for companion matrices, because $v$ has a specially nice form. The equations are $y_1' = y_2$ and $y_2' = -Cy_1 - By_2$.

Substitute $\begin{aligned} y_1 &= v_1 e^{\lambda t} \\ y_2 &= v_2 e^{\lambda t} \end{aligned}$ Then $\begin{aligned} \lambda v_1 e^{\lambda t} &= v_2 e^{\lambda t} \\ \lambda v_2 e^{\lambda t} &= -C v_1 e^{\lambda t} - B v_2 e^{\lambda t}. \end{aligned}$

Cancel every $e^{\lambda t}$. The first equation becomes $\lambda v_1 = v_2$. This is our answer:

$$\text{Eigenvectors of companion matrices are multiples of the vector } \boldsymbol{v} = \begin{bmatrix} 1 \\ \lambda \end{bmatrix}.$$

## ■ REVIEW OF THE KEY IDEAS ■

1. If $B^2 \neq 4AC \neq 0$, six pictures show the paths of $(y, y')$ for $Ay'' + By' + Cy = 0$.

2. Real solutions to $As^2 + Bs + C = 0$ lead to sources and sinks and saddles at $(0, 0)$.

3. Complex roots $s = a \pm i\omega$ give spirals around $(0, 0)$ (or closed loops if $a = 0$).

4. Roots $s$ become eigenvalues $\lambda$ for $\begin{bmatrix} y \\ y' \end{bmatrix}' = \begin{bmatrix} 0 & 1 \\ -C & -B \end{bmatrix} \begin{bmatrix} y \\ y' \end{bmatrix}$. Same six pictures.

## Problem Set 3.2

**1** Draw Figure 3.6 for a sink (the missing middle figure) with $y = c_1 e^{-2t} + c_2 e^{-t}$. Which term dominates as $t \to \infty$? The paths approach the dominating line as they go in toward zero. **The slopes of the lines are −2 and −1** (the numbers $s_1$ and $s_2$).

**2** Draw Figure 3.7 for a spiral sink (the missing middle figure) with roots $s = -1 \pm i$. The solutions are $y = C_1 e^{-t} \cos t + C_2 e^{-t} \sin t$. They approach zero because of the factor $e^{-t}$. They spiral around the origin because of $\cos t$ and $\sin t$.

**3** Which path does the solution take in Figure 3.6 if $y = e^t + e^{t/2}$? Draw the curve $(y(t), y'(t))$ more carefully starting at $t = 0$ where $(y, y') = (2, 1.5)$.

**4** Which path does the solution take around the saddle in Figure 3.6 if $y = e^{t/2} + e^{-t}$? Draw the curve more carefully starting at $t = 0$ where $(y, y') = (2, -\frac{1}{2})$.

**5** Redraw the first part of Figure 3.6 when the roots are equal: $s_1 = s_2 = 1$ and $y = c_1 e^t + c_2 t e^t$. *There is no $s_2$-line.* Sketch the path for $y = e^t + te^t$.

**6** The solution $y = e^{2t} - 4e^t$ gives a source (Figure 3.6), with $y' = 2e^{2t} - 4e^t$. Starting at $t = 0$ with $(y, y') = (-3, -2)$, where is $(y, y')$ when $e^t = 1.1$ and $e^t = .25$ and $e^t = 2$?

**7** The solution $y = e^t(\cos t + \sin t)$ has $y' = 2e^t \cos t$. This spirals out because of $e^t$. Plot the points $(y, y')$ at $t = 0$ and $t = \pi/2$ and $t = \pi$, and try to connect them with a spiral. Note that $e^{\pi/2} \approx 4.8$ and $e^{\pi} \approx 23$.

**8** The roots $s_1$ and $s_2$ are $\pm 2i$ when the differential equation is ______. Starting from $y(0) = 1$ and $y'(0) = 0$, draw the path of $(y(t), y'(t))$ around the center. Mark the points when $t = \pi/2, \pi, 3\pi/2, 2\pi$. Does the path go clockwise?

**9** The equation $y'' + By' + y = 0$ leads to $s^2 + Bs + 1 = 0$. For $B = -3, -2, -1, 0, 1, 2, 3$ decide which of the six figures is involved. For $B = -2$ and $2$, why do we not have a perfect match with the source and sink figures?

**10** For $y'' + y' + Cy = 0$ with damping $B = 1$, the characteristic equation will be $s^2 + s + C = 0$. Which $C$ gives the changeover from a *sink* (overdamping) to a spiral *sink* (underdamping)? Which figure has $C < 0$?

**Problems 11–18 are about $dy/dt = Ay$ with companion matrices $\begin{bmatrix} 0 & 1 \\ -C & -B \end{bmatrix}$.**

**11** The eigenvalue equation is $\boldsymbol{\lambda^2 + B\lambda + C = 0}$. Which values of $B$ and $C$ give complex eigenvalues? Which values of $B$ and $C$ give $\lambda_1 = \lambda_2$?

**12** Find $\lambda_1$ and $\lambda_2$ if $B = 8$ and $C = 7$. Which eigenvalue is more important as $t \to \infty$? Is this a sink or a saddle?

**13** Why do the eigenvalues have $\lambda_1 + \lambda_2 = -B$? Why is $\lambda_1 \lambda_2 = C$?

**14** Which second order equations did these matrices come from?

$$A_1 = \begin{bmatrix} 0 & 1 \\ 1 & 0 \end{bmatrix} \text{ (saddle)} \qquad A_2 = \begin{bmatrix} 0 & 1 \\ -1 & 0 \end{bmatrix} \text{ (center)}$$

**15** The equation $y'' = 4y$ produces a saddle point at $(0,0)$. Find $s_1 > 0$ and $s_2 < 0$ in the solution $y = c_1 e^{s_1 t} + c_2 e^{s_2 t}$. If $c_1 c_2 \neq 0$, this solution will be (large) (small) as $t \to \infty$ and also as $t \to -\infty$.

The only way to go toward the saddle $(y, y') = (0,0)$ as $t \to \infty$ is $c_1 = 0$.

**16** If $B = 5$ and $C = 6$ the eigenvalues are $\lambda_1 = 3$ and $\lambda_2 = 2$. The vectors $\boldsymbol{v} = (1,3)$ and $\boldsymbol{v} = (1,2)$ are *eigenvectors* of the matrix $A$: Multiply $A\boldsymbol{v}$ to get $3\boldsymbol{v}$ and $2\boldsymbol{v}$.

**17** In Problem 16, write the two solutions $\boldsymbol{y} = \boldsymbol{v}e^{\lambda t}$ to the equations $\boldsymbol{y}' = A\boldsymbol{y}$. Write the complete solution as a combination of those two solutions.

**18** The eigenvectors of a companion matrix have the form $v = (1, \lambda)$. Multiply by $A$ to show that $Av = \lambda v$ gives one trivial equation and the characteristic equation $\lambda^2 + B\lambda + C = 0$.

$$\begin{bmatrix} 0 & 1 \\ -C & -B \end{bmatrix} \begin{bmatrix} 1 \\ \lambda \end{bmatrix} = \lambda \begin{bmatrix} 1 \\ \lambda \end{bmatrix} \quad \text{is} \quad \begin{aligned} \lambda &= \lambda \\ -C - B\lambda &= \lambda^2 \end{aligned}$$

Find the eigenvalues and eigenvectors of $A = \begin{bmatrix} 3 & 1 \\ 1 & 3 \end{bmatrix}$.

**19** An equation is stable and all its solutions $y = c_1 e^{s_1 t} + c_2 e^{s_2 t}$ go to $y(\infty) = 0$ exactly when

($s_1 < 0$ **or** $s_2 < 0$)  ($s_1 < 0$ **and** $s_2 < 0$)  (Re $s_1 < 0$ **and** Re $s_2 < 0$)?

**20** If $Ay'' + By' + Cy = D$ is stable, what is $y(\infty)$?

## 3.3 Linearization and Stability in 2D and 3D

The logistic equation $y' = y - y^2$ has two steady states $Y = 0$ and $Y = 1$. Those are **critical points**, where the function $f(y) = y - y^2$ is zero. Along the lines $Y = 0$ and $Y = 1$ the equation $y' = f(y)$ becomes $0 = 0$. We have those two steady solutions, and their stability or instability is important. Do nearby solutions approach $Y$ or not?

**The stability test requires $df/dy < 0$ at $Y$.** This is the slope of the tangent to $f(y)$:

$$\boldsymbol{f(y-Y)} \approx f(Y) + \left(\frac{df}{dy}\right)(y-Y) = \boldsymbol{0 + A(y-Y)}. \tag{1}$$

The linearization of $y' = f(y)$ at the critical point $y = Y$ comes from $f \approx A(y - Y)$. Replace $f$ by this linear part and include the constant $Y$ on the left side too:

**Linearized equation near a critical point $Y$** $\quad \boldsymbol{(y-Y)' = A(y-Y)}.$ (2)

The solution $y - Y = Ce^{At}$ grows if $A > 0$ (instability). The solution decays if $A < 0$. The logistic equation has $f(y) = y - y^2$ with derivative $A = 1 - 2y$. At the steady state $Y = 0$ this shows instability ($A = +1$). The other critical point $Y = 1$ is stable ($A = -1$).

The **stability line** or **phase line** in Section 1.7 showed $Y = 1$ as the attractor:

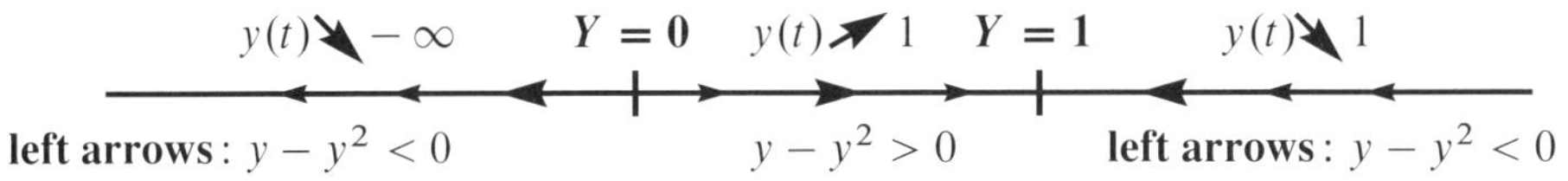


The arrows in Section 3.1 had slopes $f(t, y)$. Stability is decided by the slope $df/dy$.

*Note* The most basic example is $y' = y$. The only steady state solution is $Y = 0$. That must be unstable, because $f = y$ has $A = df/dy = 1$. All other solutions $y(t) = Ce^t$ travel far away from $Y = 0$, even when $C = y(0)$ is close to zero.

*Opposite case*: $y' = 6 - y$ is stable ($A = -1$). Solutions approach $Y = y_\infty = 6$.

### Solution Curves in the $yz$ Plane

Those paragraphs were review for one unknown $y(t)$. Section 3.2 had two unknowns $y$ and $z$ in two linear first order equations (or $y$ and $y'$ in a linear second order equation).

**Move now to nonlinear**. The equations will be **autonomous**, the same at all times $t$:

$$\boldsymbol{\frac{dy}{dt} = f(y,z)} \ \textbf{ and } \ \boldsymbol{\frac{dz}{dt} = g(y,z)} \quad \textbf{starting from } \boldsymbol{y(0)} \textbf{ and } \boldsymbol{z(0)}. \tag{3}$$

A **critical point** $Y, Z$ solves $f(Y, Z) = 0$ and $g(Y, Z) = 0$. It is a steady solution: constant $y = Y$ and constant $z = Z$.

| | | |
|---|---|---|
| **Critical point** | $f(Y, Z) = 0$ and $g(Y, Z) = 0$ | (4) |

For every critical point $Y, Z$ we must decide: **stable or unstable or neutral**?

To graph the solutions, there is a problem with $y$ and $z$ and $t$. Three variables won't fit into a 2D picture. Our solution curves for autonomous equations will omit $t$. The curves $y(t), z(t)$ show the paths of solutions in the $y, z$ plane but not the times along those paths.

*Those pictures do not show the time* $t$, as the solution moves. Different equations $dy/dt = cf(y,z)$ and $dz/dt = cg(y,z)$ will produce the same picture for all $c \neq 0$. That constant $c$ just rescales the time and the speed along the same path $y(ct), z(ct)$. Time and speed are not shown by the pictures.

Each steady state $y(t) = Y, z(t) = Z$ will be one point in the picture! The stability question is whether paths near that point (those are nearby solutions) go in toward $Y, Z$ or away from $Y, Z$ or around $Y, Z$: stable or unstable or neutrally stable.

That stability question is answered by the eigenvalues of a 2 by 2 matrix $A$.

## Solutions Near a Critical Point

Here is the key to this section. **Very close to a critical point where $f(Y, Z) = 0$ and $g(Y, Z) = 0$, solution curves have the same six possibilities that we already know**:

| | | | |
|---|---|---|---|
| **Stable** | Sink | **Unstable** | Source |
| | Spiral sink | | Spiral source |
| **Neutral** | Center | | Saddle point |

The pictures for linear equations were in Section 3.2. They came from six possibilities for the roots of $As^2 + Bs + C = 0$, and from six types of 2 by 2 matrices $A$:

| | | | |
|---|---|---|---|
| **Linear equations**<br>**Constant coefficients** | $y' = ay + bz$<br>$z' = cy + dz$ | $\begin{bmatrix} y \\ z \end{bmatrix}' = \begin{bmatrix} a & b \\ c & d \end{bmatrix} \begin{bmatrix} y \\ z \end{bmatrix}$ | (5) |

Those model problems in 2D have the critical point $Y = 0, Z = 0$. That is the point where $f(y,z) = ay + bz = 0$ and $g(y,z) = cy + dz = 0$. There is one critical point $(0,0)$ at the center of each picture in Section 3.2. Now we are saying that **nonlinear equations look like linear equations when you look near each critical point**.

This is the 2D equivalent of one equation $(y - Y)' = A(y - Y)$. That number $A$ was $df/dy$. Now we have two unknowns $y$ and $z$, and two functions $f(y,z)$ and $g(y,z)$. **There are four partial derivatives of $f$ and $g$, and they go into the 2 by 2 matrix $A$**:

| | | |
|---|---|---|
| **First derivative matrix**<br>**"Jacobian matrix"** | $A = \begin{bmatrix} \partial f/\partial y & \partial f/\partial z \\ \partial g/\partial y & \partial g/\partial z \end{bmatrix}$ | (6) |

## Linearization of a Nonlinear Equation

For one equation, linearization was based on the tangent line. The beginning of the Taylor series around $Y$ is $f(Y) + (df/dy)(y - Y)$. Critical points have $f(Y) = 0$, removing the constant term. Two variables $y$ and $z$ lead to the same idea, but now it is a tangent plane:

$$\begin{aligned} f(y,z) &\approx f(Y,Z) + \left(\frac{\partial f}{\partial y}\right)(y-Y) + \left(\frac{\partial f}{\partial z}\right)(z-Z) \\ g(y,z) &\approx g(Y,Z) + \left(\frac{\partial g}{\partial y}\right)(y-Y) + \left(\frac{\partial g}{\partial z}\right)(z-Z) \end{aligned} \tag{7}$$

A critical point has $f(Y,Z) = g(Y,Z) = 0$. The four linear terms take over:

$$\boxed{\begin{bmatrix} (y-Y)' \\ (z-Z)' \end{bmatrix} \approx \begin{bmatrix} \partial f/\partial y & \partial f/\partial z \\ \partial g/\partial y & \partial g/\partial z \end{bmatrix} \begin{bmatrix} y-Y \\ z-Z \end{bmatrix} = A \begin{bmatrix} y-Y \\ z-Z \end{bmatrix}.} \tag{8}$$

There stands the linearized equation. It is centered and linearized around the special point $(Y, Z)$. If we reset by shifting $(Y, Z)$ to $(0, 0)$, equation (8) is one of our model problems:

$$\begin{bmatrix} y' \\ z' \end{bmatrix} = A \begin{bmatrix} y \\ z \end{bmatrix} = \begin{bmatrix} a & b \\ c & d \end{bmatrix} \begin{bmatrix} y \\ z \end{bmatrix}. \tag{9}$$

**Example 1** Linearize $y' = \sin(ay + bz)$ and $z' = \sin(cy + dz)$ at $Y = 0, Z = 0$.

**Solution** Check first: $f = \sin(ay + bz)$ and $g = \sin(cy + dz)$ are zero at $(Y, Z) = (0, 0)$. *This is a critical point.* The first derivatives of $f$ and $g$ at that point go into $A$.

$$\partial f/\partial y = a\cos(ay + bz) = a\cos 0 = a \text{ when } (y, z) = (0, 0)$$

The other three partial derivatives give $b$ and $c$ and $d$. They enter the matrix $A$:

$$\begin{aligned} y' &= \sin(ay+bz) \\ z' &= \sin(cy+dz) \end{aligned} \quad \text{linearizes to} \quad \begin{aligned} y' &= ay + bz \\ z' &= cy + dz \end{aligned} = \begin{bmatrix} a & b \\ c & d \end{bmatrix} \begin{bmatrix} y \\ z \end{bmatrix}. \tag{10}$$

That example just moved the simple linearization $\sin x \approx x$ into two variables.

**Example 2 (Predator-Prey) Linearize** $\begin{aligned} y' &= y - yz \\ z' &= yz - z \end{aligned}$ **at all critical points.**

**Meaning of these predator-prey equations** The prey $y$ is like rabbits, the predator $z$ is like foxes. On their own with no foxes, the rabbits grow by nibbling grass: $y' = y$. On their own with no rabbits, the foxes don't eat well and $z' = -z$. Then the multiplication $yz$ accounts for the interactions between $y$ rabbits and $z$ foxes. Those interactions end up in more foxes and fewer rabbits.

This example has simplified coefficients 1 and $-1$ multiplying $y$ and $z$ and $yz$. The predator-prey model is a great example and we will develop it further.

## Linearize Predator – Prey at Critical Points

Set $f = Y - YZ = 0$ and also $g = YZ - Z = 0$. Solve for all critical points $Y, Z$.

$$\boldsymbol{Y - YZ = Y(1 - Z) = 0} \qquad \text{and} \qquad \boldsymbol{YZ - Z = (Y - 1)Z = 0}.$$

The critical points $Y, Z$ are $0, 0$ and $1, 1$. Track their stability using the matrix $A$.

$$\textbf{At } \boldsymbol{Y, Z = 0, 0} \qquad A = \begin{bmatrix} \partial f/\partial y & \partial f/\partial z \\ \partial g/\partial y & \partial g/\partial z \end{bmatrix} = \begin{bmatrix} \mathbf{1 - Z} & \mathbf{-Y} \\ \boldsymbol{Z} & \mathbf{Y - 1} \end{bmatrix} = \begin{bmatrix} \mathbf{1} & 0 \\ 0 & \mathbf{-1} \end{bmatrix}.$$

This is a **saddle point**: **unstable**. Starting near $0, 0$ the rabbit population $y(t)$ will grow. The eigenvalues are 1 (for the rabbits) and $-1$ (for the foxes) from $\boldsymbol{y' = y}$ and $\boldsymbol{z' = -z}$. An all-fox population would decay (this is the only path in to the saddle point).

$$\textbf{At } \boldsymbol{Y, Z = 1, 1} \qquad A = \begin{bmatrix} 1 - Z & -Y \\ Z & Y - 1 \end{bmatrix} = \begin{bmatrix} 0 & \mathbf{-1} \\ \mathbf{1} & 0 \end{bmatrix}.$$

This matrix has imaginary eigenvalues $\lambda_1 = \boldsymbol{i}$ and $\lambda_2 = \boldsymbol{-i}$. Their real parts are **zero**. The stability is **neutral**. **The critical point $\boldsymbol{Y = 1, Z = 1}$ is a center**. A solution that starts near that point will go around $1, 1$ and return where it started:

**Extra rabbits** → Foxes increase → Rabbits decrease → Foxes decrease → **Extra rabbits**

We can see without eigenvalues that the solution to the linearized equations makes a perfect circle around $(1, 1)$. The matrix $A$ has $-1$ in row 1 and $+1$ in row 2.

$$\begin{aligned} (y-1)' &= -\,(z-1) \\ (z-1)' &= +\,(y-1) \end{aligned} \qquad \text{is solved by} \qquad \begin{aligned} y - 1 &= r\cos t \\ z - 1 &= r\sin t \end{aligned} \tag{11}$$

The actual nonlinear solution $y(t)$, $z(t)$ won't make a perfect circle. Usually we can't find its exact path, but in this case we can. The $y - z$ equation is separable and solvable:

$$\frac{dy}{dz} = \frac{dy/dt}{dz/dt} = \frac{f}{g} = \frac{y(1-z)}{(y-1)z} \quad \text{separates into} \quad \boldsymbol{\frac{y-1}{y}\,dy = \frac{1-z}{z}\,dz}. \tag{12}$$

Integration of 1 and $1/y$ and $1/z$ gives $\boldsymbol{y - \ln y = \ln z - z + C}$. That constant is $C = 2$ when $y = z = 1$ (critical). These solution curves are drawn in Figure 3.8 for $C = 2.1, 2.2, 2.3, 2.4$. They are nearly circular near $C = 2$. That is linearization!

As $C$ increases, $y$ and $z$ move further away from 1 and the circles are lost. But the nonlinear solution is still **periodic**. The rabbit-fox population comes back to its starting point and goes around again. Populations can be close to cyclic.

Equation (12) took time out of the picture. A numerical solution (Euler or Runge-Kutta) puts time back. This famous model came from Lotka and Volterra in 1925.

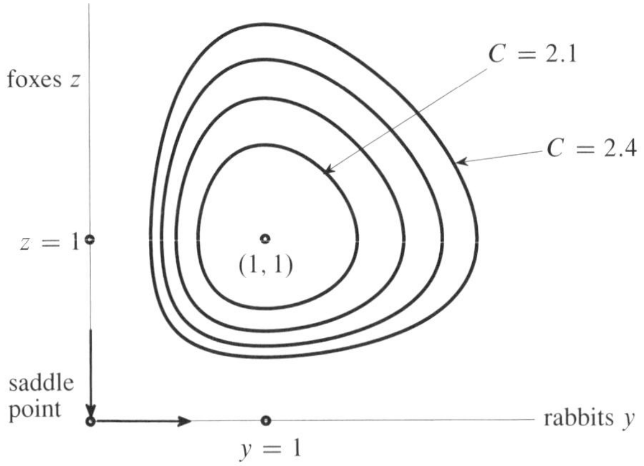


Figure 3.8: Solution paths $y + z - \ln y - \ln z = C$ around the critical point: a *center*.

## Predator – Prey – Logistic Equation

When Example 2 has no foxes ($z = 0$), the rabbit equation is $y' = y$. There is no control of rabbits and $y = Ce^t$. When we add a logistic term like $-qy^2$ (rabbits eventually competing with rabbits for available lettuce) this makes the equations more realistic.

We also allow different coefficients $p, r, s, t$ (not all 1 or $-1$) in the other terms:

**Rabbits** $\quad y' = y\,(p - qy - rz)$

**Foxes** $\quad z' = z\,(-s + wy)$

**First critical point $(Y, Z) = (0, 0)$**

**Second point $(Y, Z) = (p/q, 0)$**

**Third $s = wY$ and $p = qY + rZ$**

At those critical points, $y'$ and $z'$ are zero. The solutions are steady states $y = Y$, $z = Z$.

Near those points we linearize the equation to decide stability. The derivatives of $f(y, z)$ and $g(y, z)$ are in control, because $f = g = 0$ at the critical points:

**First derivatives**
**Jacobian at 0, 0**
$$\begin{bmatrix} \partial f/\partial y & \partial f/\partial z \\ \partial g/\partial y & \partial g/\partial z \end{bmatrix} = \begin{bmatrix} p - 2qy - rz & -ry \\ wz & -s + wy \end{bmatrix} = \begin{bmatrix} \boldsymbol{p} & 0 \\ 0 & \boldsymbol{-s} \end{bmatrix}.$$

$(0, 0)$ is a saddle point: unstable. Small populations have $y' \approx py$ and $z' \approx -sz$. Rabbits increase and foxes decrease. One eigenvalue $p$ is positive, the other eigenvalue $-s$ is negative. Near this $(0, 0)$ point, the competition terms $-qy^2$ and $-ryz$ and $wyz$ are higher order. Those terms disappear in the linearization.

The second critical point has $Y = p/q$ and $Z = 0$. This point is a sink or a saddle:

**Linearization around $(p/q, 0)$** $\quad \begin{bmatrix} y - Y \\ z - Z \end{bmatrix}' = A \begin{bmatrix} y - Y \\ z - Z \end{bmatrix}$ **with** $A = \begin{bmatrix} -q & -rp/q \\ 0 & -s + wp/q \end{bmatrix}$

If $s > wp/q$, that last entry is negative. So is $-q$, and we have a sink: two negative eigenvalues.
If $s < wp/q$, that last entry is positive. In this case we have a saddle.

The third critical point $(Y, Z)$ is different. At this point $p = qY + rZ$ and $s = wY$. This leaves only three simple terms in the first derivative matrix above:

**Linearization around $(Y, Z)$** $\quad \begin{bmatrix} y - Y \\ z - Z \end{bmatrix}' = A \begin{bmatrix} y - Y \\ z - Z \end{bmatrix}$ with $A = \begin{bmatrix} -qY & -rY \\ wZ & 0 \end{bmatrix}$

The new term $-qy^2$ in the rabbit equation has produced $-qY = -qs/w$ in the matrix $A$. This is a negative number, it stabilizes the equation. It pulls both of the eigenvalues (previously imaginary) to negative real parts. **Neutral stability changes to full stability.**

2 by 2 matrices are special (with only two eigenvalues $\lambda_1$ and $\lambda_2$). I can reveal the two facts that produce those two eigenvalues of $A$: Add the $\lambda$'s and multiply the $\lambda$'s.

| | | |
|---|---|---|
| **Sum** | $\lambda_1 + \lambda_2$ equals the sum $\boldsymbol{T}$ of diagonal entries | $\boldsymbol{T = -qY}$ |
| **Product** | $\lambda_1\lambda_2$ equals the determinant $\boldsymbol{D}$ of the matrix | $\boldsymbol{D = rYwZ}$ |

**Our matrix has $\boldsymbol{\lambda_1 + \lambda_2 < 0}$ and $\boldsymbol{\lambda_1\lambda_2 > 0}$.** This suggests two negative eigenvalues $\lambda_1$ and $\lambda_2$ (a sink). It also allows $\lambda_1 = a + ib$ and $\lambda_2 = a - ib$ ($a < 0$, a spiral sink). Our conclusion is: *The third critical point $Y, Z$ is stable.*

## Final Tests for Stability: Trace and Determinant

We can bring this whole section together. It started with finding the critical points $Y, Z$ and linearizing the differential equations. Now we can give simple tests on the 2 by 2 linearized matrix $A$. We don't need to compute the eigenvalues before testing them—because the matrix immediately tells us their sum $\lambda_1 + \lambda_2$ and their product $\lambda_1\lambda_2$. *That sum and product* (**the trace and determinant of $A$**) *are all we need.*

**Step 1** Find all critical points (steady states) of $y' = f(y, z)$ and $z' = g(y, z)$ by solving $f(Y, Z) = 0$ and $g(Y, Z) = 0$.

**Step 2** At each critical point find the matrix $A$ from derivatives of $f$ and $g$

$$A = \begin{bmatrix} a & b \\ c & d \end{bmatrix} = \begin{bmatrix} \partial f/\partial y & \partial f/\partial z \\ \partial g/\partial y & \partial g/\partial z \end{bmatrix} \text{ at the point } Y, Z$$

**Step 3** Decide stability from the **trace $\boldsymbol{T = a + d}$ and determinant $\boldsymbol{D = ad - bc}$**

| | |
|---|---|
| **Unstable** | $T > 0$ or $D < 0$ or both |
| **Neutral** | $T = 0$ and $D \geq 0$ |
| **Stable** | $T < 0$ and $D > 0$ |

If $T^2 \geq 4D > 0$, the stable critical point is a **sink**: real eigenvalues less than zero.
If $T^2 < 4D$, the stable critical point is a **spiral sink**: complex eigenvalues with Re $\lambda < 0$.
Section 6.4 will explain these rules and draw the stable region $T < 0$, $D > 0$.

The solution curves $y(t)$, $z(t)$ are paths in the $yz$ plane. Near each critical point $Y$, $Z$, the paths are close to one of the six possibilities in Section 3.2. **Source**, **sink**, or **saddle** for real eigenvalues; **Spiral source**, **spiral sink**, or **center** for complex eigenvalues.

## A Special 3 by 3 System: A Tumbling Box

You understand that 3 by 3 systems will be more complicated. The pictures don't stay in a plane. There are 9 partial derivatives of $f$, $g$, $h$ with respect to $x$, $y$, $z$. The matrix $A$ with those entries is 3 by 3. Its three eigenvalues decide stability ($T$ and $D$ are not enough).

But we live in three dimensions. The most ordinary motions will follow a space curve and not a plane curve. We can imagine the whole of three-dimensional space filled with those curves—that picture is hard to draw. Still there are important special motions that we can understand (and even test for ourselves). Here is a beautiful example.

**Throw a closed box up in the air**. **Throw a cell phone**. **Throw this book**. Those all have unequal sides $s_1 < s_2 < s_3$. Gravity will bring the book or the box back down, but that is not the interesting part. The key is *how it turns in space*.

There are three special ways to throw the box. It can rotate around the short side $s_1$. It can rotate around the longest side $s_3$. The box can try to rotate around its middle side $s_2$. Those three motions will be critical points. Your throwing experiment will quickly find that **two of the rotations are stable and one is unstable**. In this book on differential equations, we want to understand why. Please put a rubber band around the book.

Since the up and down motion from gravity is not important, we will remove it. Keep the origin $(0, 0, 0)$ at the center of the box. The box turns around that center point. At every moment in time, a 3D rotation is around an **axis**. If the box tumbles around in the air, that rotation axis is changing with time.

After writing about boxes I thought of another important example. **Throw a football**. If you throw it the right way, spinning around its long axis, it flies smoothly. Any quarterback does that automatically. But if your arm is hit while throwing, the ball wobbles. A football has one long axis and two equal short axes, $s_1 = s_2 < s_3$.

One more: A well-thrown frisbee spins around its short axis (very short). Its long axes go out to the edges of the frisbee, so $s_1 < s_2 = s_3$. A bad throw will make it tumble.

**Tumbling indicates an unstable critical point for the equations of motion**.

## Equations of Motion: Simplest Form

For a box of the right shape, Euler found these three equations. The unknowns $x, y, z$ give the angular momentum around axes 1, 2, 3 (short, medium, long).

| | | |
|---|---|---|
| $f(x,y,z)$ | $dx/dt = yz$ | **Critical points $X, Y, Z$ have $f = g = h = 0$** |
| $g(x,y,z)$ | $dy/dt = -2xz$ | **There are 6 critical points on a sphere** |
| $h(x,y,z)$ | $dz/dt = xy$ | **$(X, Y, Z) = (\pm 1, 0, 0)$ $(0, \pm 1, 0)$ $(0, 0, \pm 1)$** |

Multiply the three equations by $x, y, z$ and add them together, to see the sphere :

$$x\frac{dx}{dt} + y\frac{dy}{dt} + z\frac{dz}{dt} = xyz - 2xyz + xyz = 0 \qquad \boldsymbol{x^2 + y^2 + z^2 = \textbf{constant}}.$$

The point $x, y, z$ travels on a sphere. There are six critical points $X, Y, Z$ (steady rotations). The question is, which steady states are stable ? Try the experiment. Toss up a book.

## Linearize at Each Critical Point

When you take 9 partial derivatives of $f = yz$ and $g = -2xz$ and $h = xy$, you get the 3 by 3 Jacobian matrix $J$. Its first row $\mathbf{0} \;\; \boldsymbol{z} \;\; \boldsymbol{y}$ contains the partial derivatives of $f = yz$. At each critical point, substitute $X, Y, Z$ into $J$ to see the matrix $A$ in the linearized equations. The six critical points $(X, Y, Z)$ are $(\pm 1, 0, 0)$ and $(0, \pm 1, 0)$ and $(0, 0, \pm 1)$.

$$\boldsymbol{J} = \begin{bmatrix} 0 & z & y \\ -2z & 0 & -2x \\ y & x & 0 \end{bmatrix} \qquad \pm\boldsymbol{A} = \begin{bmatrix} 0 & 0 & 0 \\ 0 & 0 & \mathbf{-2} \\ 0 & \mathbf{1} & 0 \end{bmatrix} \begin{bmatrix} 0 & 0 & \mathbf{1} \\ 0 & 0 & 0 \\ \mathbf{1} & 0 & 0 \end{bmatrix} \begin{bmatrix} 0 & \mathbf{1} & 0 \\ \mathbf{-2} & 0 & 0 \\ 0 & 0 & 0 \end{bmatrix}$$

That middle matrix $A$ with two ones gives instability around the point $(0, 1, 0)$. Start the linearized equations from the nearby point $(c, 1, c)$.

$$\begin{bmatrix} x' \\ y' \\ z' \end{bmatrix} = \begin{bmatrix} 0 & 0 & \mathbf{1} \\ 0 & 0 & 0 \\ \mathbf{1} & 0 & 0 \end{bmatrix} \begin{bmatrix} x \\ y \\ z \end{bmatrix} \quad \text{is} \quad \begin{aligned} \boldsymbol{x' = z} \\ \boldsymbol{y' = 0} \\ \boldsymbol{z' = x} \end{aligned} \quad \text{Then} \quad \begin{aligned} \boldsymbol{x = ce^t} \\ \boldsymbol{y = 1} \\ \boldsymbol{z = ce^t} \end{aligned} \tag{13}$$

Those solutions with $e^t$ are leaving the critical point. You are seeing the eigenvalue $\lambda = 1$. The other eigenvalues are 0 and $-1$ : a **saddle point**. When you try to spin a box around its middle axis, the wobble quickly gets worse. *It is humanly impossible to spin the box perfectly because that axis is unstable.*

The other two axes are neutrally stable. Their matrices $A$ have $-2$ and $+1$. Their eigenvalues are $\sqrt{2}\,i$ and $-\sqrt{2}\,i$ and 0. Around the short axis $(1, 0, 0)$, the essential part of $A$ is 2 by 2. We see sines and cosines (not $e^t$ and instability) :

$$\begin{bmatrix} \boldsymbol{x'} \\ \boldsymbol{y'} \\ \boldsymbol{z'} \end{bmatrix} = \begin{bmatrix} 0 & 0 & 0 \\ 0 & 0 & \mathbf{-2} \\ 0 & \mathbf{1} & 0 \end{bmatrix} \begin{bmatrix} \boldsymbol{x} \\ \boldsymbol{y} \\ \boldsymbol{z} \end{bmatrix} = \begin{bmatrix} \mathbf{0} \\ \mathbf{-2}\boldsymbol{z} \\ \boldsymbol{y} \end{bmatrix}. \quad \text{Then} \quad \begin{aligned} x &= 1 \\ y &= \sqrt{2}c \cos(\sqrt{2}\,t) \\ z &= c \sin(\sqrt{2}\,t) \end{aligned}$$

The turning axis $(x, y, z)$ travels in an ellipse around $(1, 0, 0)$. This indicates a *center*. Let me go back to the nonlinear equations to see that elliptical cylinder $y^2 + 2z^2 = C$.

Multiply $\; x' = yz, y' = -2xz, z' = xy \;$ by $\; 0, y, 2z$. Add to get $\boldsymbol{yy' + 2zz' = 0}$.

The derivative of $y^2 + 2z^2$ is zero. Every path $x(t), y(t), z(t)$ is an ellipse on the sphere.

## Alar Toomre's Picture of the Solutions

At this point we know a lot about every solution to $x' = yz$ and $y' = -2xz$ and $z' = xy$.

| | | |
|---|---|---|
| Stays on a sphere | $x^2 + y^2 + z^2 = C_1$ | Multiply the equations by $x, y, z$. |
| Stays on an elliptical cylinder | $2x^2 + y^2 = C_2$ | Multiply by $2x, y, 0$ and add. |
| Stays on an elliptical cylinder | $y^2 + 2z^2 = C_3$ | Multiply by $0, y, 2z$ and add. |
| Stays on a hyperbolic cylinder | $x^2 - z^2 = C_4$ | Multiply by $x, 0, -z$ and add. |

Professor Alar Toomre made the tumbling box famous among MIT students. The year when I went to his 18.03 lecture, he tossed up a book several times (in all three ways). The book turned or tumbled around its short and middle and long axes: *stable*, *unstable*, and *stable*. Actually the stability is only neutral, and wobbles don't grow or disappear.

Maybe you can see those ellipses around two critical points: cylinders intersect a sphere. The website will show one of those cylinders going around $(1, 0, 0)$ : a neutrally stable case. It is harder to visualize the hyperbolas $x^2 - z^2 = C_4$ around the unstable point $(0, 1, 0)$.

This figure shows the value of *seeing* a solution—not just its formula. With good fortune a video of this experiment will go onto the book's website **math.mit.edu/dela**.

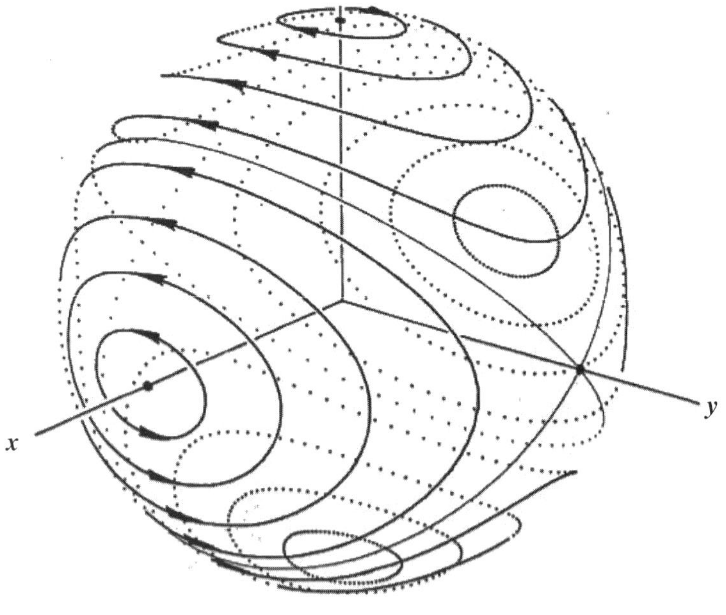


Figure 3.9: Toomre's picture of solution paths $x(t), y(t), z(t)$ from Euler's three equations.

I will end this example with a square box: two equal axes. The symmetry of a football also produces two equal axes. The Earth itself is flatter near the North Pole and South Pole, and symmetric around that short axis. *Fortunately for us this case is neutrally stable.*

The Earth's wobble doesn't go away, at the same time it doesn't get worse. The spin axis passes about five meters from the North Pole.

| | | |
|---|---|---|
| **Flattened sphere** | $dx/dt = 0$ | **Critical points $(\pm 1, 0, 0)$ at Poles** |
| **Square book** | $dy/dt = -xz$ | **Critical plane $(0, y, z)$** |
| **Two equal axes** | $dz/dt = xy$ | **(the plane of the Equator)** |

The partial derivatives of $-xz$ and $xy$ are quick to compute at $(X, Y, Z) = (1, 0, 0)$:

$$A = \begin{bmatrix} 0 & 0 & 0 \\ 0 & 0 & -1 \\ 0 & 1 & 0 \end{bmatrix} \quad \text{has eigenvalues } \lambda = i \text{ and } \lambda = -i \text{ and } \lambda = 0$$

The path of $x, y, z$ is a circle around the North Pole (for the nonlinear equations too). The Earth wobbles as it spins, but it stays stable. Not like a tumbling box.

## Epidemics and the SIR Model

An epidemic can spread until a serious fraction of the population gets sick—or the epidemic can die out early. Unstable or stable: always the important question. Suppose it is a flu epidemic on a closed campus (with no flu shots). The population divides into three groups:

$S$ = Susceptible (may catch the flu)
$I$ = Infected (sick with the flu)
$R$ = Recovered (after having the flu)

The equations for $S(t)$, $I(t)$, $R(t)$ will involve an infection constant $\beta$ and a recovery constant $\alpha$. The infection rate is $\beta SI$, proportional to the susceptible fraction $S$ times the infected (and infectious) fraction $I$. The recovery rate is simply $\alpha I$. This simple model has been improved in many ways—$SIR$ is now a highly developed technique. Epidemiology has major importance, and we want to present this small model:

$$\begin{aligned} dS/dt &= -\beta SI = f(S, I) \\ dI/dt &= \beta SI - \alpha I = g(S, I) \\ dR/dt &= \alpha I \end{aligned}$$

We work with fractions of the total population, so $S + I + R = 1$. Adding the equations confirms that $S + I + R$ is constant (their derivatives add to zero). It is enough to study $S$ and $I$. We are ignoring births and deaths—our system is closed and the epidemic is fast.

The important critical point is $S = 1, I = 0$. The population is well, but everyone is susceptible. Flu is coming. Is that critical point stable if a few people get sick?

$$\begin{bmatrix} \partial f/\partial S & \partial f/\partial I \\ \partial g/\partial S & \partial g/\partial I \end{bmatrix} = \begin{bmatrix} -\beta I & -\beta S \\ \beta I & \beta S - \alpha \end{bmatrix} = \begin{bmatrix} \mathbf{0} & -\beta \\ 0 & \boldsymbol{\beta - \alpha} \end{bmatrix} \quad \text{at } S = 1, I = 0$$

The eigenvalues of that matrix are 0 and $\beta - \alpha$. We certainly need $\beta < \alpha$ for stability. "*Sick must get well faster than well get sick.*" The other eigenvalue $\lambda = 0$ needs a closer analysis, and the model itself requires improvement.

A neutral eigenvalue like $\lambda = 0$ can be pushed either way by nonlinear terms. One way to establish nonlinear stability is to solve the equations—*after removing t*:

$$\frac{dI}{dS} = \frac{dI/dt}{dS/dt} = \frac{(\beta S - 1)I}{-\beta SI} = -1 + \frac{1}{\beta S} \quad \text{gives} \quad \boldsymbol{I = -S + \frac{\ln S}{\beta} + C}.$$

The moving point travels along the curve $I + S - (\ln S)/\beta = I(0) + S(0) - (\ln S(0))/\beta$.

An important fact about epidemics is the serious difficulty of estimating $\alpha$ and $\beta$. Their ratio $R_0 = \beta/\alpha$ controls the spread of disease: The epidemic dies out if $R_0 < 1$. One comment about estimating $\beta$: When the epidemic is over, you could compare $I + S - (\ln S)/\beta$ at $t = 0$ and $t = \infty$. Much more is in the books by Brauer and Castillo-Chavez, especially *Mathematical Models in Population Biology and Epidemiology*.

## The Law of Mass Action

When two chemical species react, the law of mass action decides the rate:

| | | |
|---|---|---|
| $\boldsymbol{S + E \to SE}$ | $\boldsymbol{\dfrac{dy}{dt} = kse}$ | $\boldsymbol{s}$ **= concentration of** $\boldsymbol{S}$<br>$\boldsymbol{e}$ **= concentration of** $\boldsymbol{E}$ |

This is like predator-prey and epidemics (multiply one population times the other, $s$ times $e$). Then $y$ is the concentration of $SE$. When $E$ is an enzyme, there is also a reverse reaction $SE \to S + E$ and a forward reaction $SE \to P + E$. For a chemist, the desired product is $P$. For us, there are three mass action laws with rates $k_1, k_{-1}, k_2$:

$$\frac{dy}{dt} = k_1 se - k_{-1}y - k_2 y \qquad \frac{ds}{dt} = -k_1 se + k_{-1}y \qquad \frac{de}{dt} = -k_1 se + k_{-1}y + k_2 y = -\frac{dy}{dt}$$

Life depends on enzymes: Very low concentrations $e(0) << s(0)$ and very fast reactions. Without $E$, blood would take years to clot. Steaks would take decades to digest. This math course might take a century to learn. The enzyme is the **catalyst** (like platinum in a catalytic converter).

After the fast reaction that uses $E$, the slower reactions bring the enzyme back. Beautifully, separating the two time scales leads to a separable equation for $y$:

$$\textbf{Michaelis-Menten equation} \qquad \frac{dy}{dt} = -\frac{cy}{y + K} \tag{14}$$

Maini and Baker have shown how matching fast time to slow time leads to (14).

This is just one example of the *nonlinear* differential equations of biology. Mathematics can reveal the main features of the solution. For a detailed picture we turn to accurate numerical methods—and those come in the next section.

## Continuous Chaos and Discrete Chaos

This section about stability will now end with extreme instability: **Chaos**. For this we need three differential equations (or two difference equations). Chaotic problems are a recent discovery, but now we know they are everywhere: Chaos is more common than stable equations and even more common than ordinary instability.

This is a deep subject, but you can see its remarkable features from simple experiments. Here are suggestions for one equation, then two, then the big one (Lorenz):

1. **Newton's method** on page 6 finds square roots by solving $f(x) = x^2 - c = 0$. Compute $x_1$, then $x_2$, then $x_3, \ldots$ **Then $x_n$ approaches $\pm\sqrt{c}$.**

$$x_{n+1} = x_n - \frac{f(x_n)}{f'(x_n)} = x_n - \frac{x_n^2 - c}{2x_n} = \frac{1}{2}\left(x_n + \frac{c}{x_n}\right).$$

   But if $c = -1$, these real $x$'s cannot approach the imaginary square roots $x = \pm i$. The $x_n$ will move around wildly when $x_{n+1} = \frac{1}{2}\left(x_n - x_n^{-1}\right)$. Try 100 steps from $x_0 = \sqrt{3}$ and $x_0 = 2$.

2. The **Hénon map** approaches a "strange attractor" in the $xy$ plane:

   **Stretching and folding** $\quad x_{n+1} = 1 + y_n - 1.4x_n^2$ and $y_n = 0.3x_n$

   Try four steps, starting from many different $x_0$, $y_0$ between $-1$ and 1.

3. The **Lorenz equations** arise in trying to predict atmospheric convection and weather:

$$x' = a(y - x) \qquad y' = x(b - z) - y \qquad z' = xy - cz$$

   Lorenz himself chose $a = 10$, $b = 28$, $c = 8/3$. The system becomes chaotic. The solutions are extremely sensitive to changes in the starting values. Harvey Mudd College has an ODE Architect Library that includes Lorenz and suggests great experiments. Try it!

## ■ REVIEW OF THE KEY IDEAS ■

1. The critical points of $y' = f(y,z), z' = g(y,z)$ solve $f(Y,Z) = g(Y,Z) = 0$. *Steady state* $y(t) = Y, z(t) = Z$.

2. Near that steady state, $f(y,z) \approx (\partial f/\partial y)(y - Y) + (\partial f/\partial z)(z - Z)$. Similarly $g(y,z)$ is "linearized" at $Y, Z$. These derivatives of $f$ and $g$ go in a $2 \times 2$ matrix $A$.

3. The equations $(y,z)' = (f,g)$ are stable at $Y$, $Z$ when the linearized equations $(y - Y, z - Z)' = A(y - Y, z - Z)$ are stable. Then $\lambda_1$ and $\lambda_2$ have real parts $< 0$.

4. Stability at $Y, Z$ requires $\dfrac{\partial f}{\partial y} + \dfrac{\partial g}{\partial z} < 0$ and $\dfrac{\partial f}{\partial y}\dfrac{\partial g}{\partial z} > \dfrac{\partial f}{\partial z}\dfrac{\partial g}{\partial y}$. This means that the eigenvalues have $\lambda_1 + \lambda_2 = a + d < 0$ and $\lambda_1\lambda_2 = ad - bc > 0$.

5. Boxes and books tumble unstably around their middle axes. Footballs are neutral.

6. Epidemics and kinetics are nonlinear when species 1 multiplies species 2 : $y' = kyz$.

## Problem Set 3.3

**1** If $y' = 2y + 3z + 4y^2 + 5z^2$ and $z' = 6z + 7yz$, how do you know that $Y = 0$, $Z = 0$ is a critical point ? What is the 2 by 2 matrix $A$ for linearization around $(0, 0)$ ? This steady state is certainly unstable because _____ .

**2** In Problem 1, change $2y$ and $6z$ to $-2y$ and $-6z$. What is now the matrix $A$ for linearization around $(0, 0)$ ? How do you know this steady state is stable ?

**3** The system $y' = f(y, z) = 1 - y^2 - z$, $z' = g(y, z) = -5z$ has a critical point at $Y = 1$, $Z = 0$. Find the matrix $A$ of partial derivatives of $f$ and $g$ at that point: stable or unstable ?

**4** This linearization is wrong but the zero derivatives are correct. *What is missing* ? $Y = 0$, $Z = 0$ is not a critical point of $y' = \cos(ay + bz)$, $z' = \cos(cy + dz)$.

$$\begin{bmatrix} y' \\ z' \end{bmatrix} = \begin{bmatrix} -a \sin 0 & -b \sin 0 \\ -c \sin 0 & -d \sin 0 \end{bmatrix} \begin{bmatrix} y \\ z \end{bmatrix} = \begin{bmatrix} 0 & 0 \\ 0 & 0 \end{bmatrix} \begin{bmatrix} y \\ z \end{bmatrix}.$$

**5** Find the linearized matrix $A$ at every critical point. Is that point stable ?

(a) $\begin{aligned} y' &= 1 - yz \\ z' &= y - z^3 \end{aligned}$ (b) $\begin{aligned} y' &= -y^3 - z \\ z' &= y + z^3 \end{aligned}$

**6** Can you create two equations $y' = f(y, z)$ and $z' = g(y, z)$ with four critical points: $(1, 1)$ and $(1, -1)$ and $(-1, 1)$ and $(-1, -1)$ ?

I don't think all four points could be stable ? This would be like a surface with four minimum points and no maximum.

**7** The second order nonlinear equation for a damped pendulum is $y'' + y' + \sin y = 0$. Write $z$ for the damping term $y'$, so the equation is $z' + z + \sin y = 0$.

Show that $Y = 0$, $Z = 0$ is a stable critical point at the bottom of the pendulum.
Show that $Y = \pi$, $Z = 0$ is an unstable critical point at the top of the pendulum.

**8** Those pendulum equations $y' = z$ and $z' = -\sin y - z$ have infinitely many critical points! What are two more and are they stable ?

**9** The Liénard equation $y'' + p(y)y' + q(y) = 0$ gives the first order system $y' = z$ and $z' =$ _____ . What are the equations for a critical point ? When is it stable ?

**10** Are these matrices stable or neutrally stable or unstable (source or saddle) ?

$$\begin{bmatrix} 2 & 1 \\ 0 & -3 \end{bmatrix} \quad \begin{bmatrix} 0 & 9 \\ -1 & 0 \end{bmatrix} \quad \begin{bmatrix} -1 & 2 \\ -1 & -1 \end{bmatrix} \quad \begin{bmatrix} -1 & -2 \\ -1 & -1 \end{bmatrix} \quad \begin{bmatrix} 0 & 9 \\ -1 & -1 \end{bmatrix}$$

**11** Suppose a predator $x$ eats a prey $y$ that eats a smaller prey $z$ :

$$\begin{aligned} dx/dt &= -x + xy \\ dy/dt &= -xy + y + yz \\ dz/dt &= -yz + 2z \end{aligned}$$

Find all critical points $X, Y, Z$
Find $A$ at each critical point
(9 partial derivatives)

**12** The damping in $y'' + (y')^3 + y = 0$ depends on the velocity $y' = z$. Then $z' + z^3 + y = 0$ completes the system. Damping makes this nonlinear system stable—is the linearized system stable ?

**13** Determine the stability of the critical points $(0, 0)$ and $(2, 1)$ :

(a) $\begin{aligned} y' &= -y + 4z + yz \\ z' &= -y - 2z + 2yz \end{aligned}$ (b) $\begin{aligned} y' &= -y^2 + 4z \\ z' &= y - 2x^4 \end{aligned}$

**Problems 14–17 are about Euler's equations for a tumbling box.**

**14** The correct coefficients involve the moments of inertia $I_1, I_2, I_3$ around the axes. The unknowns $x, y, z$ give the angular momentum around the three principal axes :

$$\begin{aligned} dx/dt &= ayz \quad &\text{with} \quad a &= (1/I_3 - 1/I_2) \\ dy/dt &= bxz \quad &\text{with} \quad b &= (1/I_1 - 1/I_3) \\ dz/dt &= cxy \quad &\text{with} \quad c &= (1/I_2 - 1/I_1). \end{aligned}$$

Multiply those equations by $x, y, z$ and add. This proves that $x^2 + y^2 + z^2$ is _____ .

**15** Find the 3 by 3 first derivative matrix from those three right hand sides $f$, $g$, $h$. What is the matrix $A$ in the 6 linearizations at the same 6 critical points ?

**16** You almost always catch an unstable tumbling book at a moment when it is flat. That tells us : The point $x(t), y(t), z(t)$ spends most of its time (near) (far from) the critical point $(0, 1, 0)$. This brings the travel time $t$ into the picture.

**17** In reality what happens when you

(a) throw a baseball with no spin (a knuckleball) ?

(b) hit a tennis ball with overspin ?

(c) hit a golf ball left of center ?

(d) shoot a basketball with underspin (a free throw) ?

## 3.4 The Basic Euler Methods

**For most differential equations, solutions are numerical**. We solve model equations to understand what to expect in more complicated problems. Then the numbers we need—close to exact but never perfect—come from finite time steps $\Delta t$.

This section will show you the key ideas. The approximations will be simple and clear, but not highly accurate. The next section comes closer to the reality of modern codes. The Runge-Kutta method is still frequently used, with refinements that those two creators certainly did not anticipate. The cycle of **predicting at $t + \Delta t$**, **correcting at $t + \Delta t$**, and **adjusting the stepsize $\Delta t$** for the next step is now highly developed.

Local accuracy comes from small steps, but speed comes from larger steps. The right balance depends on the particular equation and the user's need for accuracy. Always there is a requirement of *stability*—because small errors are unavoidable. But after the numerical errors enter the calculation, they must not grow faster than the solution itself.

### Euler's First Step $y_1 = y_0 + \Delta t\ f_0$

The equation to solve is $\boldsymbol{dy/dt = f(t,y)}$. The initial value $y(0)$ is given—this will be our starting $y_0$. A *difference equation* will go forward to $y_1$. That is our approximation to the exact solution at $t_1 = \Delta t$ (the end of the first time step and the start of the next step). By going forward in steps of size $\Delta t_1, \Delta t_2, \ldots$ we compute values $y_1, y_2, \ldots$ that are close to the exact solution.

We know two facts at $t = 0$. The value of $y$ is $y_0$ and the slope $dy/dt$ at that point is given by $f$ in the equation. *That slope is called* $f_0$. It is the right side $f(t,y)$ when $y = y_0$ and $t = 0$. With value $y_0$ and slope $f_0$, we know the tangent line $y = y_0 + tf_0$ to the curve $y(t)$. So we can take a step $\Delta t$ along that tangent line—not too large a step or we will wander too far from the exact curve $y(t)$.

| **Step $\Delta t$ along tangent line** | $y_1 = y_0 + \Delta t\ f_0$ | (1) |
|---|---|---|

Figure 3.10 shows $y_1$ for the model equation $y' = 2y$. At $y_0 = 1$ the slope is $f_0 = 2$ (since $f(y) = 2y$). **We follow that tangent line as far as $y_1 = 1 + 2\Delta t$.**

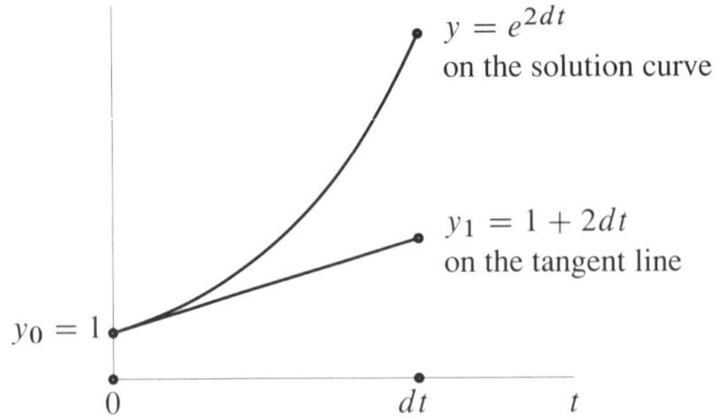


Figure 3.10: The tangent line $y = y_0 + tf_0$ starts at $y_0$. Euler stops at $\boldsymbol{y_1 = y_0 + \Delta t f_0}$.

## Euler's Method $y_{n+1} = y_n + \Delta t f_n$

On the graph, we are following pieces of tangent lines. This is the same as approximating the derivative $dy/dt$ (which changes during a time step) by the forward difference $\Delta y/\Delta t$ (which is held constant during a time step):

$$\frac{dy}{dt} = f(t, y) \qquad \text{becomes} \qquad \frac{\mathbf{y_1 - y_0}}{\mathbf{\Delta t}} = \mathbf{f_0}. \tag{2}$$

There is a new tangent line for the second time step. That step starts at $y_1$ (which we just computed). *The slope at that point in time is* $f_1 = f(\Delta t, y_1)$. We are using the differential equation $y' = f(t, y)$ to tell us the slopes $f_0, f_1, f_2, \ldots$ at the start of every time step:

$$n^{\text{th}} \textbf{ time step} \qquad \frac{\Delta y}{\Delta t} = f(t_n, y_n) \textbf{ is Euler's method } \frac{y_{n+1} - y_n}{\Delta t} = f_n \tag{3}$$

The model equation $\boldsymbol{dy/dt = 2y}$ has the exact solution $y(t) = e^{2t}$. Euler's method $y_{n+1} = y_n + \Delta t f_n$ will multiply $y_n$ at every step by the number $1 + 2\Delta t$:

$$y_{n+1} = y_n + \Delta t(2y_n) = (1 + 2\Delta t)y_n \qquad \text{leads to} \qquad \boldsymbol{y_n = (1 + 2\Delta t)^n y_0}. \tag{4}$$

We have seen powers of $(1 + \frac{1}{n})$ and $(1 + \frac{a}{n})$ in Section 1.3 from *compound interest*. The current balance was $y_n$ and the interest at rate $a$ was $a\Delta t y_n$. Then the new balance was $y_{n+1} = (1 + a\Delta t)y_n$. This is exactly Euler's method to solve $dy/dt = ay$, and our example has $a = 2$.

$$\textbf{Approximating } e^{2t} \qquad \boldsymbol{y_n = (1 + 2\Delta t)^n \approx (e^{2\Delta t})^n = e^{2n\Delta t}}. \tag{5}$$

The errors $y_n - y$ grow as $n$ increases. But the errors at each step also shrink as $\Delta t \to 0$. If we hold $n\,\Delta t$ fixed at some value $T$, then we are taking $n$ steps to reach that time $T$. As $n$ increases and $\Delta t$ decreases, the steps are smaller—the tangent lines stay closer. Then Euler's $y_n$ approaches the exact $y(T) = e^{2T}$.

## Euler's Error

The error $E_n$ is $y(n\,\Delta t) - y_n$. This is the exact solution minus the computed solution $y_n$ at time $n\,\Delta t$. It comes from accumulating small errors at every time step—the tangent lines move away from the true graph of $y(t)$.

First, estimate those small errors at the $n$ separate time steps. *How far is a tangent line from a curve, after a step* $\Delta t$? The answer comes from calculus.

$$\begin{array}{l}\textbf{Local error}\\ \textbf{Taylor series}\end{array} \qquad y(t + \Delta t) = y(t) + \Delta t\, y'(t) + \frac{\mathbf{1}}{\mathbf{2}}\boldsymbol{(\Delta t)^2 y''(t)} + \cdots \tag{6}$$

When we keep two terms and omit the third term, the error is $\le \frac{1}{2}(\Delta t)^2\, |y''|_{\max}$.

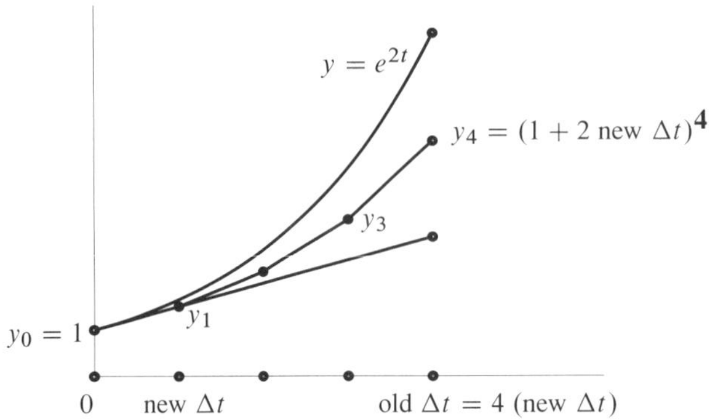


Figure 3.11: Euler's method converges to $y(T)$ as $n \to \infty$, with $n$ steps of size $\Delta t = T/n$.

The Mean Value Theorem would establish that bound of order $(\Delta t)^2$. This is the error in one step—a tangent line moving away from the curve. We will take $n$ steps to reach the time $n\ \Delta t = T$. If all goes well, the 1-step error $C(\Delta t)^2$ grows in $n$ steps to $CT\Delta t$.

$$\textbf{The error at time } T \textbf{ after } n \textbf{ steps is } |y(T) - y_n| \le Cn(\Delta t)^2 = CT\Delta t. \qquad (7)$$

Conclusion: Euler's method is *first-order accurate*. The error is proportional to $\Delta t$. If we take $2n$ steps of size $\Delta t/2$, and do twice as much work, that will divide the error by 2 (approximately). This is really minimum accuracy.

The Runge-Kutta method has error proportional to $(\Delta t)^4$. Then reducing $\Delta t$ to $\Delta t/2$ improves the error by a factor near 16. We will be matching many more terms in the Taylor series, where Euler only matched the first derivative. In the example $y' = 2y$, we know that $y(T) = e^{2T}$ :

$$\textbf{First-order accuracy} \qquad (1 + 2\ \Delta t)^n = \left(1 + \frac{2T}{n}\right)^n \approx e^{2T} \text{ with error } \frac{C}{n}. \qquad (8)$$

This table shows the slow improvement as $n$ increases, compared to the superfast improvement from keeping more terms in the Taylor series:

| $n$ | $\left(1 + \frac{1}{n}\right)^n$ **from Euler** | **Taylor series for $e$** |
|---|---|---|
| 1 | 2.0000000 | 2.0000000 |
| 2 | 2.2500000 | 2.5000000 |
| 3 | 2.3703704 | 2.6666667 |
| 4 | 2.4414062 | 2.7083333 |
| 5 | 2.4883200 | 2.7166667 |
| 6 | 2.5216264 | 2.7180556 |
| 7 | 2.5464997 | 2.7182540 |
| 8 | 2.5657845 | 2.7182788 |
| 9 | 2.5811748 | 2.7182815 |
| 10 | 2.5937425 | **2.7182818** |

## Stability

We jumped over an important point when we converted $n$ local errors of size $(\Delta t)^2$ to one global error of size $\Delta t$. The local errors occur in each step. The global error at $T$ is the composite of $n$ local errors. We assumed that local errors at early times would not grow much before the final time $T$.

Think of the local error as a small bank deposit every day. The global error at the end of a year ($T = 365\ \Delta t$) includes 365 small errors. Those small deposits should grow during the year (they earn interest too). The constant $C$ in equation (8) allows for this growth.

**What if the equation is $dy/dt = -100y$ ?** This shows decay, not growth. The solution starting at $y(0) = 1$ is $y(T) = \boldsymbol{e^{-100T}}$, very small. But does Euler's method show the same fast decay in the approximate solution, when the equation has $f_n = -100y_n$ ?

$$y_{n+1} = y_n + \Delta t f_n = (1 - 100\ \Delta t)y_n \qquad \boldsymbol{y_n = (1 - 100\ \Delta t)^n y_0} \tag{9}$$

If $100\Delta t$ is small, then $1 - 100\Delta t$ is less than 1 and its powers decay as they should. But we will have $\boldsymbol{100\Delta t = 3}$ when $\Delta t = 0.03$. That step seems small but *it is not*. The number $1 - 100\Delta t$ will be $-2$. Equation (9) shows that every step multiplies by $-2$. The powers of $-2$ grow exponentially!

$$y_n = 1, -2, 4, -8, \dots \qquad y_n = (1 - 100\Delta t)^n y_0 = (-2)^n y_0 \ \text{ is } \textit{exponentially unstable.}$$

*Conclusion*: Stability for $y' = -100y$ requires $|1 - 100\Delta t| \le 1$. **We need $\boldsymbol{\Delta t \le 2/100}$.**

In a way this limit on $\Delta t$ is acceptable. Euler is missing the $\frac{1}{2}(100\Delta t)^2$ term in the Taylor series for $e^{-100t}$. We would want $100\Delta t < 1$ just for reasonable accuracy. The stability requirement $100\Delta t < 2$ is not a heavy burden. But read further.

## Stiff Equations

Imagine an equation with solutions $e^{-t}$ and $e^{-100t}$. Then $e^{-t}$ will dominate, because it has much slower decay than $e^{-100t}$. We have decay rates $s = -1$ and $s = -100$:

$$y'' + 101y' + 100y = 0 \qquad \text{with} \qquad s^2 + 101s + 100 = (s+1)(s+100). \tag{10}$$

This is certainly *overdamped*. The roots $s = -1$ and $s = -100$ are real. Euler's method needs to follow $e^{-t}$ accurately, because that is the important solution. **But stability still requires $\boldsymbol{\Delta t \le 2/100}$.**

The unimportant solution $e^{-100t}$ is getting in the way. It reduces $\Delta t$ and therefore adds more work (many steps), beyond the ordinary demand of first order accuracy. A problem like equation (10) is called **stiff**: stability can be too expensive for ordinary Euler.

We can see this second order problem as two first order equations. Introduce $y'$ as a second unknown. As in Section 3.1, a "companion matrix" multiplies the vector $(y, y')$:

$$y'' + 101y' + 100y = 0 \text{ is the same as } \frac{d}{dt}\begin{bmatrix} y \\ y' \end{bmatrix} = \begin{bmatrix} 0 & 1 \\ -100 & -101 \end{bmatrix}\begin{bmatrix} y \\ y' \end{bmatrix}. \tag{11}$$

The eigenvalues of the matrix are the same roots $-1$ and $-100$. That is a ***stiff problem***: ***slow decay together with fast decay***.

Euler's method for this matrix equation is just like Euler for $y' = Ay$:

$$\frac{\boldsymbol{y}_{n+1} - \boldsymbol{y}_n}{\Delta t} = A\boldsymbol{y}_n \quad \text{or} \quad \boldsymbol{y}_{n+1} = (\boldsymbol{I} + \boldsymbol{A}\boldsymbol{\Delta t})\boldsymbol{y}_n. \tag{12}$$

Every step multiplies by $I + A\Delta t$. That matrix has eigenvalues $1 - \Delta t$ and $1 - 100\Delta t$. Normally $1 - \Delta t$ is more important and larger. But if $100\Delta t$ is greater than 2, then the second number $1 - 100\Delta t$ is below $-1$. Its powers will show extreme instability.

The cure for stiff systems is to switch to an **implicit method**.

## Backward Euler = Implicit Euler

**The idea of implicit methods is to use backward differences**. Go back from $y_{n+1}$ and $t_{n+1}$ and $f_{n+1}$, instead of going forward from $y_n$ and $t_n$ and $f_n$.

**Backward Euler**

$$\frac{\boldsymbol{y}^B_{n+1} - \boldsymbol{y}_n}{\boldsymbol{\Delta t}} = \boldsymbol{f}_{n+1} = \boldsymbol{f}(\boldsymbol{t}_{n+1}, \boldsymbol{y}^B_{n+1}). \tag{13}$$

The example $y' = -100y$ will divide by $1 + 100\Delta t$ instead of multiplying by $1 - 100\Delta t$:

$$\frac{y^B_{n+1} - y_n}{\Delta t} = -100\, y^B_{n+1} \quad \text{is} \quad \boldsymbol{(1 + 100\Delta t)y^B_{n+1} = y_n}.$$

That division happens at every time step. After $n$ steps this method remains very stable:

**"Implicit Euler"** $\quad y^B_n = \left(\dfrac{1}{1 + 100\Delta t}\right)^n y_0$ is decreasing correctly.

For this linear equation, division is no more expensive than multiplication. Implicit is the way to go. But we pay a much higher price for implicit when the problem is nonlinear. Instead of substituting the known $y_n$ to find $f_n = f(n\,\Delta t, y_n)$ in ordinary "explicit" Euler, we now have to solve a nonlinear equation to find the unknown $y^B_{n+1}$:

**Each step must solve for $\boldsymbol{y^B_{n+1}}$** $\quad y^B_{n+1} - \Delta t f(t_{n+1}, y^B_{n+1}) = y_n. \tag{14}$

If the forcing function $f$ is complicated, even an approximate solution for $y^B_{n+1}$ will be expensive. You see the struggle that is constantly presented: **Implicit methods are more stable but much slower**. For $\boldsymbol{y}' = A\boldsymbol{y}$, the matrix to invert is in $(I - \Delta t\, A)\boldsymbol{y}^B_{n+1} = \boldsymbol{y}_n$.

## Difference Equations vs Differential Equations

Compare $a^n$ with $e^{at}$: powers and exponentials. The powers come from a difference equation $Y_{n+1} = aY_n$. The exponentials come from a differential equation $y' = ay$. Stability means that those solutions *approach zero*. For ordinary numbers (this includes complex numbers) the test on $a$ is easy.

$$\boldsymbol{a^n \to 0 \text{ when } |a| < 1} \qquad \boldsymbol{e^{at} \to 0 \text{ when } \mathrm{Re}\, a < 0}.$$

When we have a matrix $A$, the same tests are applied to the eigenvalues:

$$\boldsymbol{A^n \to 0 \text{ when all } |\lambda| < 1} \qquad \boldsymbol{e^{At} \to 0 \text{ when all } \mathrm{Re}\, \lambda_i < 0}.$$

## ■ REVIEW OF THE KEY IDEAS ■

1. Euler's method is $(y_{n+1} - y_n)/\Delta t = f_n$ or $\mathbf{y_{n+1} = y_n + \Delta t\, f(n\,\Delta t, y_n)}$.

2. That step to $y_{n+1}$ follows the tangent line at $y_n$, not the curve $y(t)$. Error $\approx (\Delta t)^2$.

3. After $n$ steps to time $T = n\,\Delta t$, the error is proportional to $\Delta t$ : *First order accuracy.*

4. Stability requires $y_n$ to grow no faster than the exact $y(t)$ : Often a *size limit on* $\Delta t$.

5. **Backward Euler** is $y^B_{n+1} - y_n = \Delta t f(y^B_{n+1})$. Harder to find $y^B_{n+1}$ but more stable.

## Problem Set 3.4

**1** Apply Euler's method $y_{n+1} = y_n + \Delta t f_n$ to find $y_1$ and $y_2$ with $\Delta t = \frac{1}{2}$ :

(a) $y' = y$ (b) $y' = y^2$ (c) $y' = 2ty$ (all with $y(0) = y_0 = 1$)

**2** For the equations in Problem 1, find $y_1$ and $y_2$ with the step size reduced to $\Delta t = \frac{1}{4}$. Now the value $y_2$ is an approximation to the exact $y(t)$ at what time $t$ ? Then $y_2$ in this question corresponds to which $y_n$ in Problem 1 ?

**3** (a) For $dy/dt = y$ starting from $y_0 = 1$, what is Euler's $y_n$ when $\Delta t = 1$ ?

(b) Is it larger or smaller than the true solution $y = e^t$ at time $t = n$ ?

(c) What is Euler's $y_{2n}$ when $\Delta t = \frac{1}{2}$ ? This is closer to the true $y(n) = e^n$.

**4** For $dy/dt = -y$ starting from $y_0 = 1$, what is Euler's approximation $y_n$ after $n$ steps of size $\Delta t$ ? Find all the $y_n$'s when $\Delta t = 1$. Find all the $y_n$'s when $\Delta t = 2$. Those time steps are *too large* for this equation.

**5** The true solution to $y' = y^2$ starting from $y(0) = 1$ is $y(t) = 1/(1-t)$. This explodes at $t = 1$. Take 3 steps of Euler's method with $\Delta t = \frac{1}{3}$ and take 4 steps with $\Delta t = \frac{1}{4}$. Are you seeing any sign of explosion ?

**6** The true solution to $dy/dt = -2ty$ with $y(0) = 1$ is the bell-shaped curve $y = e^{-t^2}$. It decays quickly to zero. Show that step $n + 1$ of Euler's method gives $y_{n+1} = (1 - 2n\Delta t^2)y_n$. Do the $y_n$'s decay toward zero ? Do they stay there ?

**7** The equations $y' = -y$ and $z' = -10z$ are uncoupled. If we use Euler's method for both equations with the same $\Delta t$ between $\frac{2}{10}$ and 2, show that $y_n \to 0$ but $|z_n| \to \infty$. The method is failing on the solution $z = e^{-10t}$ that should decay fastest.

**8** What values $y_1$ and $y_2$ come from *backward Euler* for $dy/dt = -y$ starting from $y_0 = 1$ ? Show that $y^B_1 < 1$ and $y^B_2 < 1$ even if $\Delta t$ is very large. We have *absolute stability* : no limit on the size of $\Delta t$.

**9** The logistic equation $y' = y - y^2$ has an $S$-curve solution in Section 1.7 that approaches $y(\infty) = 1$. This is a steady state because $y' = 0$ when $y = 1$.

Write Euler's approximation $y_{n+1} =$ ______ to this logistic equation, with stepsize $\Delta t$. Show that this has the same steady state: $y_{n+1}$ equals $y_n$ if $y_n = 1$.

**10** The important question in Problem 9 is whether the steady state $y_n = 1$ is stable or unstable. Subtract 1 from both sides of Euler's $y_{n+1} = y_n + \Delta t(y_n - y_n^2)$:

$$y_{n+1} - 1 = y_n + \Delta t(y_n - y_n^2) - 1 = (y_n - 1)(1 - \Delta t y_n).$$

Each step multiplies the distance from 1 by $(1 - \Delta t y_n)$. Near the steady $y_\infty = 1$, $1 - \Delta t\, y_n$ has size $|1 - \Delta t|$. For which $\Delta t$ is this smaller than 1 to give stability?

**11** Apply backward Euler $y_{n+1}^B = y_n + \Delta t f_{n+1}^B = y_n + \Delta t \left[ y_{n+1}^B - \left(y_{n+1}^B\right)^2 \right]$ to the logistic equation $y' = f(y) = y - y^2$. What is $y_1^B$ if $y_0 = \frac{1}{2}$ and $\Delta t = \frac{1}{4}$? You have to solve a quadratic equation to find $y_1^B$. I am finding two answers for $y_1^B$. A computer code might choose the answer closer to $y_0$.

**12** For the bell-shaped curve equation $y' = -2ty$, show that backward Euler divides $y_n$ by $1 + 2n(\Delta t)^2$ to find $y_{n+1}^B$. As $n \to \infty$, what is the main difference from forward Euler in Problem 6?

**13** The equation $y' = \sqrt{|y|}$ has *many solutions* starting from $y(0) = 0$. One solution stays at $y(t) = 0$, another solution is $y = t^2/4$. (Then $y' = t/2$ agrees with $\sqrt{y}$.) Other solutions can stay at $y = 0$ up to $t = T$, and then switch to the parabola $y = (t - T)^2/4$. As soon as $y$ leaves the bad point $y = 0$, where $f(y) = y^{1/2}$ has infinite slope, the equation has only one solution.

Backward Euler $y_1 - \Delta t \sqrt{|y_1|} = y_0 = 0$ gives two correct values $y_1^B = 0$ and $y_1^B = (\Delta t)^2$. What are the three possible values of $y_2^B$?

**14** Every finite difference person will think of averaging forward and backward Euler:

**Centered Euler / Trapezoidal** $\quad \boldsymbol{y_{n+1}^C - y_n = \Delta t \left( \frac{1}{2} f_n + \frac{1}{2} f_{n+1}^C \right)}.$

For $y' = -y$ the key questions are **accuracy** and **stability**. Start with $y(0) = 1$.

$$y_1^C - y_0 = \Delta t \left( -\frac{1}{2} y_0 - \frac{1}{2} y_1^C \right) \quad \text{gives} \quad \boldsymbol{y_1^C = \frac{1 - \Delta t/2}{1 + \Delta t/2} y_0}.$$

**Stability** Show that $|1 - \Delta t/2| < |1 + \Delta t/2|$ for all $\Delta t$. *No stability limit on* $\Delta t$.

**Accuracy** For $y_0 = 1$ compare the exact $y_1 = e^{-\Delta t} = 1 - \Delta t + \frac{1}{2}\Delta t^2 - \cdots$ with $y_1^C = (1 - \frac{1}{2}\Delta t)/(1 - \frac{1}{2}\Delta t) = (1 - \frac{1}{2}\Delta t)(1 - \frac{1}{2}\Delta t + \frac{1}{4}\Delta t^2 - \cdots)$.

An extra power of $\Delta t$ is correct: *Second order accuracy*. A good method.

**The website has codes for Euler and Backward Euler and Centered Euler**. Those methods are slow and steady with first order and second order accuracy. The test problems give comparisons with faster methods like Runge-Kutta.

## 3.5 Higher Accuracy with Runge-Kutta

The section on basic Euler methods contained two messages. First, those methods are simple to understand (they follow a tangent line). Second, those methods are too simple to give good or even adequate accuracy. This section brings major improvements. The fourth order Runge-Kutta method is the basis for **ode45**, the workhorse among all of MATLAB's codes for solving $\boldsymbol{y}' = \boldsymbol{f}(t, \boldsymbol{y})$.

Notice that this equation—linear or more likely nonlinear—involves first derivatives $\boldsymbol{y}'$ and no higher derivatives. In case the original equation is $y'' = F(t, y, y')$, introduce $y' = y_2$ as a new equation together with the original $y_2' = F(t, y, y_2)$. The unknowns $y_1 = y$ and $y_2 = y'$ go into a vector $\boldsymbol{y}$. The right hand sides $y_2$ and $F$ go into a vector $\boldsymbol{f}$.

| | | |
|---|---|---|
| **$n$ equations for** <br> **$n$ unknown $y$'s** | $y_1' = y_2$ <br> $y_2' = F(t, y_1, y_2)$ | $y_1' = f_1(t, y_1, \ldots, y_n)$ <br> $\cdots\cdots\cdots$ <br> $y_n' = f_n(t, y_1, \ldots, y_n)$ |

In the middle is a system of two equations coming from $y'' = F$. On the right is a system of $n$ equations for the vector $\boldsymbol{y}$ of $n$ unknowns. The $n$ equations $\boldsymbol{y}' = \boldsymbol{f}(t, \boldsymbol{y})$ start from $n$ initial conditions $y_1(0), \ldots, y_n(0)$, and $\boldsymbol{f}$ is a vector of $n$ right hand sides.

We are ready for more accurate approximations to $y' = f(t, y)$ and $\boldsymbol{y}' = \boldsymbol{f}(t, \boldsymbol{y})$.

### Improved Euler = Simplified Runge-Kutta

Euler's first order method is $y_{n+1}^E = y_n + \Delta t f_n$. Let me describe an improvement to *second order accuracy*, which means an error of size $(\Delta t)^2$. This uses the Runge-Kutta idea: **Substitute Euler's $y_{n+1}^E$ once more into $f$. Use that output to get a better $y_{n+1}^S$:**

$$\textbf{Improved Euler} \quad \textbf{Simplified R-K} \qquad \frac{y_{n+1}^S - y_n}{\Delta t} = \frac{1}{2} f(t_n, y_n) + \frac{1}{2} f\left(t_{n+1}, y_{n+1}^E\right). \tag{1}$$

Let me show you the improvement for $y' = ay$. In this case $f(t, y)$ is $ay$. You can see $y^E$ as a **prediction** of the next value $y_{n+1}$ and $y^S$ as a **correction**:

$$\boldsymbol{y^E = y_n + a\,\Delta t\, y_n} \quad \text{goes into} \quad \boldsymbol{y^S} = y_n + \frac{1}{2} a\,\Delta t\, y_n + \frac{1}{2} a\,\Delta t(\boldsymbol{y_n + a\,\Delta t\, y_n}). \tag{2}$$

When that last term is multiplied out, we see the correct $(\Delta t)^2$ term included in $y_{n+1}^S$:

$$\textbf{Linear case } \boldsymbol{y' = ay} \qquad y_{n+1}^S = y_n + a\,\Delta t\, y_n + \frac{1}{2} a^2 (\Delta t)^2 y_n. \tag{3}$$

We are following the *tangent parabola* starting at $y_n$. The parabola stays much closer to the true $y(t)$ curve than the tangent line. This improvement means a $(\Delta t)^3$ error at each step. With stability, those errors produce a $(\Delta t)^2$ overall error after $n = T/\Delta t$ steps.

The exact $y(t + \Delta t)$ is $e^{a\Delta t} y(t)$. Equation (3) has *three* correct terms of $e^{a\Delta t}$. Euler uses the slope $y' = f(t, y)$ only at the *start* of the time step, but *the improvement $y^S$ in equation* (1) *averages the slope at the start and the end of the step.*

## Simplified Adams Method

Here is another way to achieve second order accuracy. **Save and reuse the computed value $y_{n-1}$ at the previous time $t - \Delta t$.** With the right coefficients 3/2 and −1/2, and essentially no extra work, we can again capture the term $\frac{1}{2}(\Delta t)^2 y''$ that Euler missed.

**Adams-Bashforth Multistep method**

$$y^{A}_{n+1} = y_n + \frac{3}{2}\Delta t f(t_n, y_n) - \frac{1}{2}\Delta t f(t_{n-1}, y_{n-1}). \qquad (4)$$

All we do is to save each computed value of $f_n$ for one more step. That number becomes the $f_{n-1}$ term in (4). The right hand side of (4) gives the correct $y'$ and $y''$ terms:

$$y_n + \frac{3}{2}\Delta t y'_n - \frac{1}{2}\Delta t y'_{n-1} \approx y_n + \frac{3}{2}\Delta t y'_n - \frac{1}{2}\Delta t(y'_n - \Delta t y''_n) = \boldsymbol{y_n + \Delta t y'_n + \frac{1}{2}(\Delta t)^2 y''_n}$$

*Each extra step back to $y_{n-2}, y_{n-3}, \ldots$ can increase the accuracy by* 1. Those multistep methods compete with Runge-Kutta and eventually they win. But fourth order is still mostly on the R-K side. One reason is that Adams needs a special effort to compute $y_{-1}$ before the first step can begin. Runge-Kutta starts cold.

Runge-Kutta easily changes $\Delta t$ from one step to the next. On the other hand, its four evaluations of $f(t, y)$ could be expensive. Stiff systems need backward differences.

## Fourth Order Runge-Kutta

The famous version of Runge-Kutta uses *four* evaluations of the right side. It starts at time $t_n$ with solution $y^{RK}_n$. It reaches time $t_{n+1} = t_n + \Delta t$ with approximate solution $y^{RK}_{n+1}$. On the way, Runge-Kutta stops twice for $\boldsymbol{k_2}$ and $\boldsymbol{k_3}$ at $t_{n+1/2} = t_n + \frac{1}{2}\Delta t$.

**At each step from $t_n$ to $t_{n+1}$ compute $k_1, k_2, k_3, k_4$**

$$\begin{aligned} \boldsymbol{k_1} &= f(t_n, y_n)/2 \\ \boldsymbol{k_2} &= f(t_{n+1/2}, y_n + \Delta t\, \boldsymbol{k_1})/2 \\ \boldsymbol{k_3} &= f(t_{n+1/2}, y_n + \Delta t\, \boldsymbol{k_2})/2 \\ \boldsymbol{k_4} &= f(t_{n+1}, y_n + 2\Delta t\, \boldsymbol{k_3})/2 \end{aligned}$$

A combination of those four $\boldsymbol{k}$'s gives fourth-order accuracy for $y^{RK}_{n+1}$:

**Runge-Kutta step**

$$\frac{y^{RK}_{n+1} - y_n}{\Delta t} = \frac{1}{3}(k_1 + 2k_2 + 2k_3 + k_4) \qquad (5)$$

That short line is one of the most important formulas in this book. Among highly accurate methods, Runge-Kutta is especially easy to code and run—probably the easiest there is. Before each step, we decide on $\Delta t$. For the model problem $y' = y$ the R-K combination produces five correct terms in the series for $e^{\Delta t}$. You can see evaluations of $f$ inside evaluations of $f$, starting with $\boldsymbol{k_1} = f_n/2 = y/2$:

$$\boldsymbol{k_2} = \frac{1}{2}\left(y + \frac{\Delta t}{2} y\right) \quad \boldsymbol{k_3} = \frac{1}{2}\left(y + \frac{\Delta t}{2}\left(y + \frac{\Delta t}{2} y\right)\right) \quad \boldsymbol{k_4} = \frac{1}{2}\left(y + \Delta t\left(y + \frac{\Delta t}{2}\left(y + \frac{\Delta t}{2} y\right)\right)\right)$$

Problem 1 will simplify $k_1 + 2k_2 + 2k_3 + k_4$. The new $y_{n+1}$ at the end of the step is $\boldsymbol{y_{n+1} = (1 + \Delta t + \cdots + \frac{1}{4!}(\Delta t)^4) y_n}$. All terms correct for $e^{\Delta t}$ and 4[th] order accuracy.

## The Stability of Runge-Kutta

To determine the limit of stability, apply the method to $y' = -y$. The true solution $y = e^{-t}y(0)$ will decrease. But if $\Delta t$ is too large, the approximations $y_n$ will *increase* in size. The first example of possible instability was Euler's method:

$$\textbf{Euler instability for } \boldsymbol{\Delta t > 2} \quad y^E_{n+1} = (1 - \Delta t)y_n \ \text{ has } \ |1 - \Delta t| > 1$$

When we apply the same test to Runge-Kutta, instability enters for $\boldsymbol{\Delta t > 2.78}$:

$$\textbf{RK instability for } \boldsymbol{\Delta t \geq 3} \qquad 1 - 3 + \frac{1}{2}9 - \frac{1}{6}27 + \frac{1}{24}81 = \frac{\mathbf{11}}{\mathbf{8}} > 1.$$

The full infinite series would give the small number $e^{-3}$. But these five terms give a multiplier $11/8$ that is larger than 1. If we take this over-large step $n$ times, the Runge-Kutta approximation $y_n = (11/8)^n$ will be enormous and completely wrong. The more exact stability limit is $a\,\Delta t < 2.78$ for $y' = ay$.

**Example 1** Apply all three methods to $\boldsymbol{dy/dt = y}$. The true solution $\boldsymbol{y = e^t}$ reaches $y = e = 2.71828\ldots$ at time $t = 1$. Try $\Delta t = 0.2$ and $0.1$.

| $\boldsymbol{\Delta t = 0.2}$ | $y^E$ | $y^S$ | $y^{RK}$ | $\boldsymbol{\Delta t = 0.1}$ | $y^E$ | $y^S$ | $y^{RK}$ |
|---|---|---|---|---|---|---|---|
| $t = 0$ | 1 | 1 | 1 | $t = 0$ | 1 | 1 | 1 |
| | | | | .1 | 1.10 | 1.1050 | 1.1051708 |
| $t = .2$ | 1.20 | 1.220 | 1.221400 | .2 | 1.21 | 1.2210 | 1.2214026 |
| | | | | .3 | 1.33 | 1.3492 | 1.3498585 |
| $t = .4$ | 1.44 | 1.488 | 1.491818 | .4 | 1.46 | 1.4909 | 1.4918242 |
| | | | | .5 | 1.61 | 1.6474 | 1.6487206 |
| $t = .6$ | 1.73 | 1.816 | 1.822106 | .6 | 1.77 | 1.8204 | 1.8221180 |
| | | | | .7 | 1.95 | 2.0116 | 2.0137516 |
| $t = .8$ | 2.07 | 2.215 | 2.225521 | .8 | 2.14 | 2.2228 | 2.2255396 |
| | | | | .9 | 2.36 | 2.4562 | 2.4596014 |
| $t = 1$ | 2.49 | 2.703 | 2.718251 | 1.0 | 2.59 | 2.7141 | 2.7182797 |

The error in $y^S$ is divided by 4 (from .015 to .004 at $t = 1$) when $\Delta t$ is cut in half. This indicates second order accuracy for simplified Runge-Kutta, as the theory predicted. The work is only doubled.

## ode 45 and ODEPACK and More

Runge-Kutta is accurate and easy to code. The final value $y_{n+1}$ can be made even better. With *six* evalutions of $f$ (not four) we can also compute a value $Y^5_{n+1}$ that has *fifth* order accuracy. By comparing with $y^{RK}_{n+1}$ we get an estimate of the error, which indicates whether a larger $\Delta t$ is possible or a smaller $\Delta t$ is necessary. This is the heart of Matlab's **ode 45** code. A good solver for stiff systems is **ode 15s**.

ODEPACK and SUNDIALS are open collections of Fortran 77 codes from Livermore Laboratory. Those emphasize Adams methods (backward differences for stiff problems).

*Mathematica* has DSolve for solution formulas and NDSolve for numerical solutions. Wolfram Alpha is remakable for the very wide range of problems it solves. SciPy and SymPy and Scilab are also free and high quality. **See the web** !

## ■ REVIEW OF THE KEY IDEAS ■

1. Higher order equations like $y'' + y' + y = F(t, y, y')$ reduce to $\boldsymbol{y}' = \boldsymbol{f}(t, \boldsymbol{y})$. Most finite difference methods prefer this first order system with $\boldsymbol{y} = (y, y')$.
2. $y^E_{n+1} = y_n + \Delta t f_n$ improves to second order accuracy by also using $f(t_{n+1}, y^E_{n+1})$.
3. Fourth order Runge-Kutta uses that substitution into $f(t, y)$ four times in each step.
4. The Runge-Kutta error is divided by almost $2^4 = 16$ when $\Delta t$ is divided by 2.
5. Stability for $y' = ay$ requires $a\Delta t^E > \mathbf{-2}$ and $a\Delta t^S > \mathbf{-2}$ and $a\Delta t^{RK} > \mathbf{-2.78}$. Otherwise disaster for $a < 0$: the approximations $Y_n$ will start to grow.

# Problem Set 3.5

**Runge-Kutta can only be appreciated by using it. A simple code is on math.mit.edu/dela. Professional codes are ode 45 (in MATLAB) and ODEPACK and many more.**

**1** For $y' = y$ with $y(0) = 1$, show that simplified Runge-Kutta and full Runge-Kutta give these approximations $y_1$ to the exact $y(\Delta t) = e^{\Delta t}$ :

$$y_1^S = 1 + \Delta t + \frac{1}{2}(\Delta t)^2 \qquad y_1^{RK} = 1 + \Delta t + \frac{1}{2}(\Delta t)^2 + \frac{1}{6}(\Delta t)^3 + \frac{1}{24}(\Delta t)^4$$

**2** With $\Delta t = 0.1$ compute those numbers $y_1^S$ and $y_1^{RK}$ and subtract from the exact $y = e^{\Delta t}$. The errors should be close to $(\Delta t)^3/6$ and $(\Delta t)^5/120$.

**3** Those values $y_1^S$ and $y_1^{RK}$ have errors of order $(\Delta t)^3$ and $(\Delta t)^5$. Errors of this size at every time step will produce total errors of size _____ and _____ at time $T$, from $N$ steps of size $\Delta t = T/n$.

Those estimates of total error are correct provided errors don't grow (*stability*).

**4** $dy/dt = f(t)$ with $y(0) = 0$ is solved by integration when $f$ does not involve $y$. From time $t = 0$ to $\Delta t$, simplified Runge-Kutta approximates the integral of $f(t)$ :

$$y_1^S = \Delta t\left(\frac{1}{2}f(0) + \frac{1}{2}f(\Delta t)\right) \text{ is close to } y(\Delta t) = \int_0^{\Delta t} f(t)\,dt$$

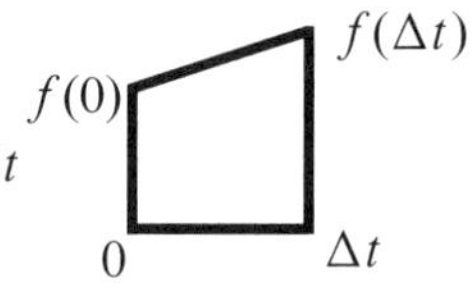

Suppose the graph of $f(t)$ is a straight line as shown. Then the region is a *trapezoid*. Check that its area is exactly $y_1^S$. Second order means exact for linear $f$.

**5** Suppose again that $f$ does not involve $y$, so $dy/dt = f(t)$ with $y(0) = 0$. Then full Runge-Kutta from $t = 0$ to $\Delta t$ approximates the integral of $f(t)$ by $y_1^{RK}$:

$$y_1^{RK} = \Delta t\,(c_1 f(0) + c_2 f(\Delta t/2) + c_3 f(\Delta t))\,. \qquad \textbf{Find } \boldsymbol{c_1, c_2, c_3}.$$

This approximation to $\int_0^{\Delta t} f(t)\,dt$ is called Simpson's Rule. It has 4th order accuracy.

**6** Reduce these second order equations to first order systems $\boldsymbol{y}' = \boldsymbol{f}(t, \boldsymbol{y})$ for the vector $\boldsymbol{y} = (y, y')$. Write the two components of $\boldsymbol{y}_1^E$ and $\boldsymbol{y}_1^S$.

(a) $y'' + yy' + y^4 = 1$ (b) $my'' + by' + ky = \cos t$

**7** When $my'' + by' + ky = \cos t$ in Problem 6 is reduced to a vector equation $\boldsymbol{y}' = A\boldsymbol{y} + \boldsymbol{f}$ find $\boldsymbol{y}_1^E$ and $\boldsymbol{y}_1^S$ from the initial vector $\boldsymbol{y}_0$.

**8** For $y' = -y$ and $y_0 = 1$ the exact solution $y = e^{-t}$ is approximated at time $\Delta t$ by 2 or 3 or 5 terms:

$$y_1^E = 1-\Delta t \quad y_1^S = 1-\Delta t+\frac{1}{2}(\Delta t)^2 \quad y_1^{RK} = 1-\Delta t+\frac{1}{2}(\Delta t)^2-\frac{1}{6}(\Delta t)^3+\frac{1}{24}(\Delta t)^4$$

(a) With $\Delta t = 1$ compare those three numbers to the exact $e^{-1}$. What error $E$?

(b) With $\Delta t = 1/2$ compare those three numbers to $e^{-1/2}$. Is the error near $E/16$?

**9** For $y' = ay$, simplified Runge-Kutta gives $y_{n+1}^S = (1 + a\Delta t + \frac{1}{2}(a\Delta t)^2)y_n$. This multiplier of $y_n$ reaches $1 - 2 + 2 = \mathbf{1}$ when $a\Delta t = -2$: *the stability limit.*

(**Computer experiment**) For $N = 1, 2, \ldots, 10$ discover the stability limit $L = L_N$ when the series for $e^{-L}$ is cut off after $N + 1$ terms:

$$\left|1 - L + \frac{1}{2}L^2 - \frac{1}{6}L^3 + \cdots \pm \frac{1}{N!}L^N\right| = 1.$$

We know $L = 2$ for $N = 1$ and $N = 2$. Runge-Kutta has $L = 2.78$ for $N = 4$.

## ■ CHAPTER 3 NOTES ■

**Proof that $y' = f(t, y)$ has a solution** **Functions $y_0, y_1, y_2, \ldots$ approach $y(t)$**

Section 3.1 stated a fact: $dy/dt = f(t, y)$ has one solution starting from $y(0)$, when $f$ is a good function: Assume $f$ and $df/dy$ are continuous at all points. Since we have no formula for $y$ (and we don't expect one), how can we know that a solution exists?

One good answer constructs $y_1$ from $y_0 = y(0)$, then $y_2$ from $y_1$, then $y_3$ from $y_2, \ldots$

**Equation** $\dfrac{d\boldsymbol{y_{n+1}}}{dt} = f(t, \boldsymbol{y_n}(t))$ **Solution** $\boldsymbol{y_{n+1}} = y_0 + \displaystyle\int_0^t f(s, \boldsymbol{y_n}(s))ds$ (6)

Let me practice with $y' = y$ and $y(0) = 1$. The solution is $e^t$. Take three steps to $y_3$:

$$\begin{array}{llll} y_0' = 0 & y_1' = y_0 & y_2' = y_1 & y_3' = y_2 \\ \boldsymbol{y(0) = 1} & y_1 = \boldsymbol{1 + t} & y_2 = \boldsymbol{1 + t + \dfrac{t^2}{2}} & y_3 = \boldsymbol{1 + t + \dfrac{t^2}{2} + \dfrac{t^3}{6}} \end{array}$$

The same construction of $e^t$ was in Section 1.3. Now we go much further, to solve nonlinear equations $y' = f(t, y)$. The key idea is to compare $y_{n+1} - y_n$ with the previous $y_n - y_{n-1}$. Subtract equation (6) for $y_n$ from equation (6) for $y_{n+1}$:

$$y_{n+1}(t) - y_n(t) = \int_0^t [f(s, y_n(s)) - f(s, y_{n-1}(s))]\, ds. \qquad (7)$$

When $|\partial f/\partial y| \le L$, the difference $|f(y_n) - f(y_{n-1})|$ is not larger than $L|y_n - y_{n-1}|$.

$$\begin{aligned} |y_2 - y_1| &\le \int_0^t L|y_1 - y_0|\, ds \le Lt|y_1 - y_0|_{\max} \\ |y_3 - y_2| &\le \int_0^t L|y_2 - y_1|\, ds \le \int_0^t L^2 t|y_1 - y_0|_{\max} = \frac{L^2t^2}{2}|y_1 - y_0|_{\max} \end{aligned}$$

We are seeing $Lt$ and $L^2t^2/2$ and next will be $L^3t^3/6$. Those numbers $L^nt^n/n!$ approach zero quickly because of $n!$ If $n$ is large and $N$ is larger, then

$$|y_N - y_n| \le |y_N - y_{N-1}| + |y_{N-1} - y_{N-2}| + \cdots + |y_{n+1} - y_n| \le C\frac{L^nt^n}{n!}$$

This is what we need to know: the differences $y_N(t) - y_n(t)$ approach zero. Cauchy showed that the numbers $y_n(t)$ *must approach a limit* $y(t)$. (Of course $y_{n+1}$ will approach the same limit.) That limiting function $y(t)$ will be our desired solution:

$$\boldsymbol{y_{n+1}}(t) = y_0 + \int_0^t f(s, \boldsymbol{y_n}(s))\, ds \;\rightarrow\; \boldsymbol{y}(t) = y_0 + \int_0^t f(s, \boldsymbol{y(s)})\, ds. \text{ Then } y' = f(t, y).$$

# Chapter 4

# Linear Equations and Inverse Matrices

## 4.1 Two Pictures of Linear Equations

The central problem of linear algebra is to solve a system of equations. Those equations are linear, which means that the unknowns are only multiplied by numbers—we never see $x^2$ or $x$ times $y$. Our first linear system is deceptively small, only "2 by 2." But you will see how far it leads:

$$\begin{array}{ll} \textbf{Two equations} & x - 2y = 1 \\ \textbf{Two unknowns} & 2x + y = 7 \end{array} \qquad (1)$$

We begin *a row at a time*. The first equation $\boldsymbol{x - 2y = 1}$ produces a straight line in the $xy$ plane. The point $x = 1$, $y = 0$ is on the line because it solves that equation. The point $x = 3$, $y = 1$ is also on the line because $3 - 2 = 1$. For $x = 101$ we find $y = 50$.

The slope of this line in Figure 4.1 is $\frac{1}{2}$, because $y$ increases by 1 when $x$ changes by 2. But slopes are important in calculus and this is linear algebra!

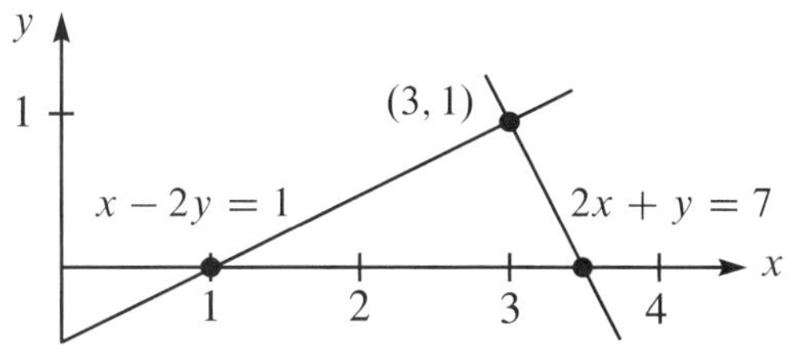


Figure 4.1: *Row picture*: The point (3, 1) where the two lines meet is the solution.

The second line in this "row picture" comes from the second equation $2x + y = 7$. You can't miss the intersection point where the two lines meet. *The point $x = 3$, $y = 1$ lies on both lines.* It solves both equations at once. This is the solution to our two equations.

**ROWS** ***The row picture shows two lines meeting at a single point (the solution).***

Turn now to the column picture. I want to recognize the same linear system as a "vector equation." Instead of numbers we need to see *vectors*. If you separate the original system into its columns instead of its rows, you get a vector equation:

**Combination equals $b$** $$x\begin{bmatrix}1\\2\end{bmatrix} + y\begin{bmatrix}-2\\1\end{bmatrix} = \begin{bmatrix}1\\7\end{bmatrix} = \boldsymbol{b}. \qquad (2)$$

This has two column vectors on the left side. The problem is *to find the combination of those vectors that equals the vector on the right*. We are multiplying the first column by $x$ and the second column by $y$, and adding vectors. With the right choices $x = 3$ and $y = 1$ (the same numbers as before), this produces $3(\textbf{\textit{column 1}}) + 1(\textbf{\textit{column 2}}) = \boldsymbol{b}$.

**COLUMNS** ***The column picture combines the column vectors on the left side of the equations to produce the vector b on the right side.***

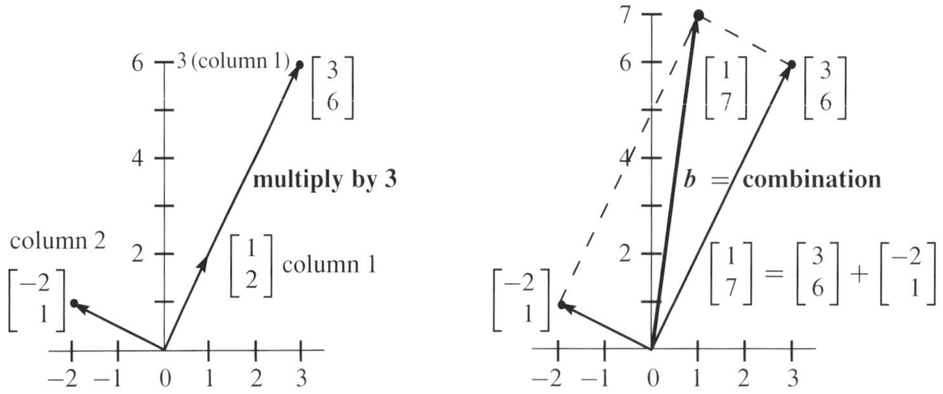


Figure 4.2: *Column picture*: A combination **3** (column 1) + **1** (column 2) gives the vector ***b***.

Figure 4.2 is the "column picture" of two equations in two unknowns. The left side shows the two separate columns, and column 1 is multiplied by 3. This multiplication by a *scalar* (a number) is one of the two basic operations in linear algebra:

**Scalar multiplication** $$3\begin{bmatrix}1\\2\end{bmatrix} = \begin{bmatrix}3\\6\end{bmatrix}.$$

If the components of a vector $\boldsymbol{v}$ are $v_1$ and $v_2$, then $c\boldsymbol{v}$ has components $cv_1$ and $cv_2$.

The other basic operation is *vector addition*. We add the first components and the second components separately. $3-2$ and $6+1$ give the vector sum $(1,7)$ as desired:

$$\textbf{Vector addition} \qquad \begin{bmatrix} 3 \\ 6 \end{bmatrix} + \begin{bmatrix} -2 \\ 1 \end{bmatrix} = \begin{bmatrix} 1 \\ 7 \end{bmatrix}.$$

The right side of Figure 4.2 shows this addition. The sum along the diagonal is the vector $\boldsymbol{b} = (1,7)$ on the right side of the linear equations.

To repeat: The left side of the vector equation is a ***linear combination*** of the columns. The problem is to find the right coefficients $x = 3$ and $y = 1$. We are combining scalar multiplication and vector addition into one step. That combination step is crucially important, because it contains both of the basic operations on vectors: *multiply and add.*

$$\begin{matrix}\textbf{Linear combination} \\ \textbf{of the 2 columns}\end{matrix} \qquad 3\begin{bmatrix} 1 \\ 2 \end{bmatrix} + \begin{bmatrix} -2 \\ 1 \end{bmatrix} = \begin{bmatrix} 1 \\ 7 \end{bmatrix}.$$

Of course the solution $x = 3, y = 1$ is the same as in the row picture. I don't know which picture you prefer! Two intersecting lines are more familiar at first. You may like the row picture better, but only for a day. My own preference is to combine column vectors. It is a lot easier to see a combination of four vectors in four-dimensional space, than to visualize how four "planes" might possibly meet at a point. (*Even one three-dimensional plane in four-dimensional space is hard enough...*)

The ***coefficient matrix*** on the left side of equation (1) is the 2 by 2 matrix $A$:

$$\textbf{Coefficient matrix} \qquad A = \begin{bmatrix} 1 & -2 \\ 2 & 1 \end{bmatrix}.$$

This is very typical of linear algebra, to look at a matrix by rows and also by columns. Its rows give the row picture and its columns give the column picture. Same numbers, different pictures, same equations. We write those equations as a matrix problem $A\boldsymbol{v} = \boldsymbol{b}$:

$$\textbf{Matrix multiplies vector} \qquad \begin{bmatrix} 1 & -2 \\ 2 & 1 \end{bmatrix}\begin{bmatrix} x \\ y \end{bmatrix} = \begin{bmatrix} 1 \\ 7 \end{bmatrix}.$$

The row picture deals with the two rows of $A$. The column picture combines the columns. The numbers $x = 3$ and $y = 1$ go into the solution vector $\boldsymbol{v}$. Here is matrix-vector multiplication, matrix $A$ times vector $\boldsymbol{v}$. Please look at this multiplication $A\boldsymbol{v}$!

$$\begin{matrix}\textbf{Dot products with rows} \\ \textbf{Combination of columns}\end{matrix} \qquad A\boldsymbol{v} = \boldsymbol{b} \quad \text{is} \quad \begin{bmatrix} 1 & -2 \\ 2 & 1 \end{bmatrix}\begin{bmatrix} 3 \\ 1 \end{bmatrix} = \begin{bmatrix} 1 \\ 7 \end{bmatrix}. \qquad (3)$$

## Linear Combinations of Vectors

Before I go to three dimensions, let me show you the most important operation on vectors. We can see a vector like $\boldsymbol{v} = (3, 1)$ as a pair of numbers, or as a point in the plane, or as an arrow that starts from $(0, 0)$. The arrow ends at the point $(3, 1)$ in Figure 4.3.

$$\boldsymbol{v} = \begin{bmatrix} 3 \\ 1 \end{bmatrix}$$

column vector

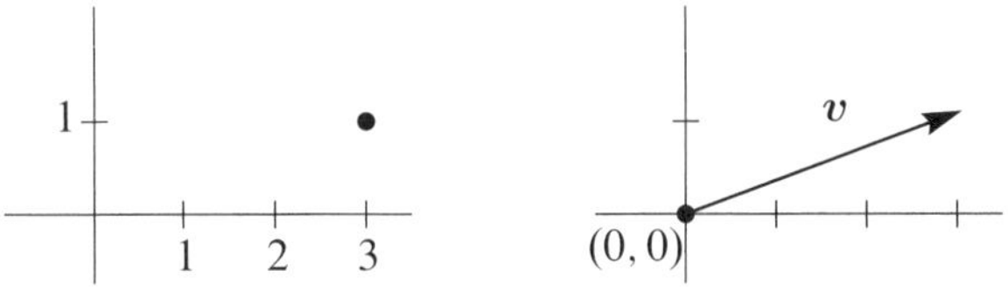


point (3, 1) arrow to (3, 1)

Figure 4.3: The vector $\boldsymbol{v}$ is given by two numbers or a point or an arrow from $(0, 0)$.

A first step is to multiply that vector by any number $c$. If $c = 2$ then the vector is doubled to $2\boldsymbol{v}$. If $c = -1$ then it changes direction to $-\boldsymbol{v}$. Always the "scalar" $c$ multiplies each separate component (here 3 and 1) of the vector $\boldsymbol{v}$. The arrow doubles the length to show $2\boldsymbol{v}$ and it reverses direction to show $-\boldsymbol{v}$ :

$$2\boldsymbol{v} = \begin{bmatrix} 6 \\ 2 \end{bmatrix} \qquad -\boldsymbol{v} = \begin{bmatrix} -3 \\ -1 \end{bmatrix}$$

column vectors

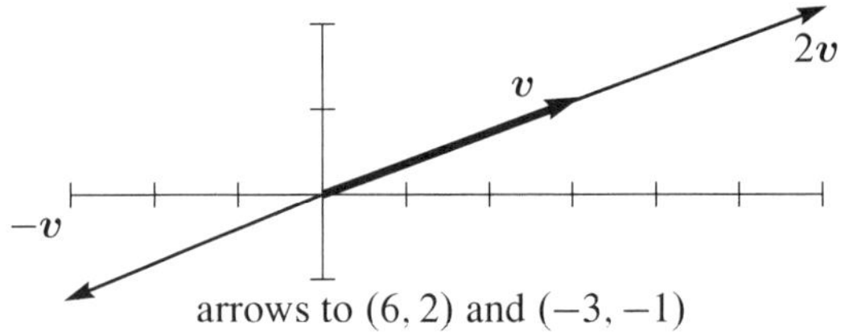


arrows to $(6, 2)$ and $(-3, -1)$

Figure 4.4: Multiply the vector $\boldsymbol{v} = (3, 1)$ by scalars $c = 2$ and $-1$ to get $c\boldsymbol{v} = (3c, c)$.

If we have another vector $\boldsymbol{w} = (-1, 1)$, we can add it to $\boldsymbol{v}$. Vector addition $\boldsymbol{v} + \boldsymbol{w}$ can use numbers (the normal way) or it can use the arrows (to visualize $\boldsymbol{v} + \boldsymbol{w}$). The arrows in Figure 4.5 go head to tail : **At the end of $\boldsymbol{v}$, place the start of $\boldsymbol{w}$.**

$$\boldsymbol{v} + \boldsymbol{w} = \begin{bmatrix} 3 \\ 1 \end{bmatrix} + \begin{bmatrix} -1 \\ 1 \end{bmatrix} = \begin{bmatrix} 2 \\ 2 \end{bmatrix}$$

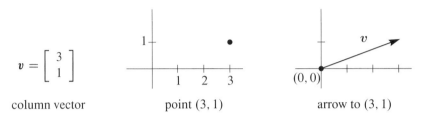

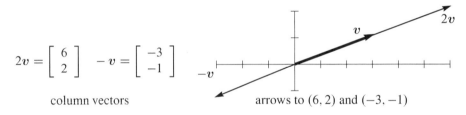

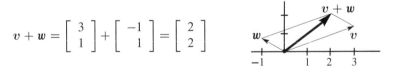

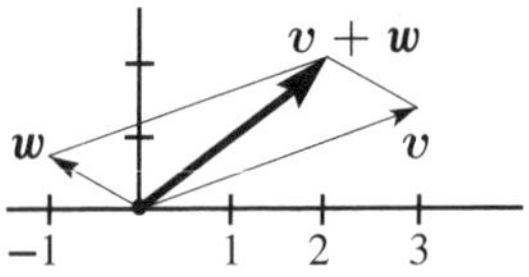


Figure 4.5: The sum of $\boldsymbol{v} = (3, 1)$ and $\boldsymbol{w} = (-1, 1)$ is $\boldsymbol{v} + \boldsymbol{w} = (2, 2)$. This is also $\boldsymbol{w} + \boldsymbol{v}$.

Allow me to say, adding $\boldsymbol{v} + \boldsymbol{w}$ and multiplying $c\boldsymbol{v}$ will soon be second nature. In themselves they are not impressive. What really counts is when you do both at once.

*Multiply $c\boldsymbol{v}$ and also $d\boldsymbol{w}$, then add to get the* **linear combination** $c\boldsymbol{v} + d\boldsymbol{w}$.

**Linear combination $2\boldsymbol{v} + 3\boldsymbol{w}$** $$2\begin{bmatrix} 3 \\ 1 \end{bmatrix} + 3\begin{bmatrix} -1 \\ 1 \end{bmatrix} = \begin{bmatrix} 3 \\ 5 \end{bmatrix}.$$

This is the basic operation of linear algebra! If you have two 5-dimensional vectors like $\boldsymbol{v} = (1, 1, 1, 1, 2)$ and $\boldsymbol{w} = (3, 0, 0, 1, 0)$, you can multiply $\boldsymbol{v}$ by 2 and $\boldsymbol{w}$ by 1. You can combine to get $2\boldsymbol{v} + \boldsymbol{w} = (5, 2, 2, 3, 4)$. Every combination $c\boldsymbol{v} + d\boldsymbol{w}$ is a vector in the big 5-dimensional space $\mathbf{R}^5$.

I admit that there is no picture to show these vectors in $\mathbf{R}^5$. Somehow I imagine arrows going to $\boldsymbol{v}$ and $\boldsymbol{w}$. If you think of all the vectors $c\boldsymbol{v}$, *they form a line in* $\mathbf{R}^5$. The line goes in both directions from $(0, 0, 0, 0, 0)$ because $c$ can be positive or negative or zero.

Similarly there is a line of all vectors $d\boldsymbol{w}$. The hard but all-important part is to imagine all the combinations $c\boldsymbol{v} + d\boldsymbol{w}$. Add all vectors on one line to all vectors on the other line, and what do you get? It is a "2-dimensional plane" inside the big 5-dimensional space. I don't lose sleep trying to visualize that plane. (There is no problem in working with the five numbers.) For linear combinations in high dimensions, algebra wins.

## Dot Product of $\boldsymbol{v}$ and $\boldsymbol{w}$

The other important operation on vectors is a kind of multiplication. This is not ordinary multiplication and we don't write $\boldsymbol{vw}$. The output from $\boldsymbol{v}$ and $\boldsymbol{w}$ will be one number and it is called the **dot product** $\boldsymbol{v} \cdot \boldsymbol{w}$.

**DEFINITION** The **dot product** of $\boldsymbol{v} = (v_1, v_2)$ and $\boldsymbol{w} = (w_1, w_2)$ is the number $\boldsymbol{v} \cdot \boldsymbol{w}$:

$$\boldsymbol{v} \cdot \boldsymbol{w} = v_1 w_1 + v_2 w_2. \tag{4}$$

The dot product of $\boldsymbol{v} = (3, 1)$ and $\boldsymbol{w} = (-1, 1)$ is $\boldsymbol{v} \cdot \boldsymbol{w} = (3)(-1) + (1)(1) = -2$.

**Example 1** The column vectors $(1, 2)$ and $(-2, 1)$ have a *zero* dot product:

**Dot product is zero**
**Perpendicular vectors** $$\begin{bmatrix} 1 \\ 2 \end{bmatrix} \cdot \begin{bmatrix} -2 \\ 1 \end{bmatrix} = -2 + 2 = 0.$$

In mathematics, zero is always a special number. For dot products, it means that *these two vectors are perpendicular*. The angle between them is $90°$.

The clearest example of two perpendicular vectors is $\boldsymbol{i} = (1, 0)$ along the $x$ axis and $\boldsymbol{j} = (0, 1)$ up the $y$ axis. Again the dot product is $\boldsymbol{i} \cdot \boldsymbol{j} = 0 + 0 = 0$. Those vectors $\boldsymbol{i}$ and $\boldsymbol{j}$ form a right angle. They are the columns of the 2 by 2 **identity matrix** $I$.

The dot product of $\boldsymbol{v} = (3, 1)$ and $\boldsymbol{w} = (1, 2)$ is 5. Soon $\boldsymbol{v} \cdot \boldsymbol{w}$ will reveal the angle between $\boldsymbol{v}$ and $\boldsymbol{w}$ (not $90°$). Please check that $\boldsymbol{w} \cdot \boldsymbol{v}$ is also 5.

## Multiplying a Matrix $A$ and a Vector $v$

Linear equations have the form $Av = b$. The right side $b$ is a column vector. On the left side, the coefficient matrix $A$ multiplies the unknown column vector $v$ (we don't use a "dot" for $Av$). The all-important fact is that $Av$ is computed by *dot products in the row picture*, and $Av$ is a **combination of the columns in the column picture**.

I put those words "combination of the columns" in boldface, because this is an essential idea that is sometimes missed. One definition is usually enough in linear algebra, but $Av$ has two definitions—the rows and the columns produce the same output vector $Av$.

The rules stay the same if $A$ has $n$ columns $a_1, \ldots, a_n$. Then $v$ has $n$ components. The vector $Av$ is still a combination of the columns, $Av = v_1 a_1 + v_2 a_2 + \cdots + v_n a_n$. **The numbers in $v$ multiply the columns in $A$.** Let me start with $n = 2$.

**By rows** $Av = \begin{bmatrix} (\text{row } 1) \cdot v \\ (\text{row } 2) \cdot v \end{bmatrix}$ **By columns** $Av = v_1(\text{column } 1) + v_2(\text{column } 2)$.

**Example 2** In equation (3) I wrote "dot products with rows" and "combination of columns." Now you know what those mean. They are the two ways to look at $Av$:

**Dot products with rows**
**Combination of columns**

$$\begin{bmatrix} a\,v_1 + b\,v_2 \\ c\,v_1 + d\,v_2 \end{bmatrix} = v_1 \begin{bmatrix} a \\ c \end{bmatrix} + v_2 \begin{bmatrix} b \\ d \end{bmatrix}. \qquad (5)$$

You might naturally ask, *which way to find $Av$*? My own answer is this: I compute by rows and I visualize (and understand) by columns. Combinations of columns are truly fundamental. But to calculate the answer $Av$, I have to find one component at a time. Those components of $Av$ are the dot products with the rows of $A$.

$$\begin{bmatrix} 2 & 3 \\ 4 & 5 \end{bmatrix} \begin{bmatrix} v_1 \\ v_2 \end{bmatrix} = \begin{bmatrix} 2v_1 + 3v_2 \\ 4v_1 + 5v_2 \end{bmatrix} = v_1 \begin{bmatrix} 2 \\ 4 \end{bmatrix} + v_2 \begin{bmatrix} 3 \\ 5 \end{bmatrix}.$$

## Singular Matrices and Parallel Lines

The row picture and column picture can fail—and they will fail together. For a 2 by 2 matrix, the row picture fails when the lines from row 1 and row 2 are parallel. The lines don't meet and $Av = b$ has no solution:

$$A = \begin{bmatrix} 2 & 3 \\ 4 & 6 \end{bmatrix} \qquad \begin{aligned} 2v_1 - 3v_2 &= 6 \\ 4v_1 - 6v_2 &= 0 \end{aligned}$$

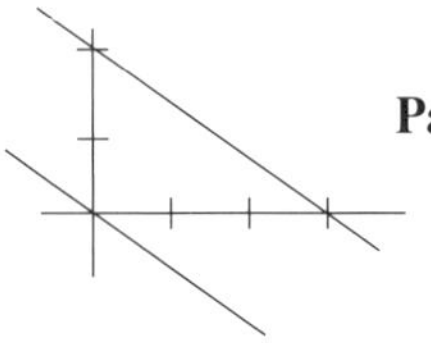

**Parallel lines**
**no solution**

The row picture shows the problem and so does the algebra: 2 times equation 1 produces $4v_1 - 6v_2 = \mathbf{12}$. But equation 2 requires $4v_1 - 6v_2 = \mathbf{0}$. Notice that this line goes through the center point $(0, 0)$ because the right side is zero.

How does the column picture fail? *Columns 1 and 2 point in the same direction.* When the rows are "dependent", the columns are also dependent. All combinations of the columns $(2,4)$ and $(3,6)$ lie in the same direction. Since the right side $\boldsymbol{b} = (6,0)$ is not on that line, $\boldsymbol{b}$ is *not* a combination of those two column vectors of $A$. Figure 4.6 (a) shows that there is *no solution* to the equation.

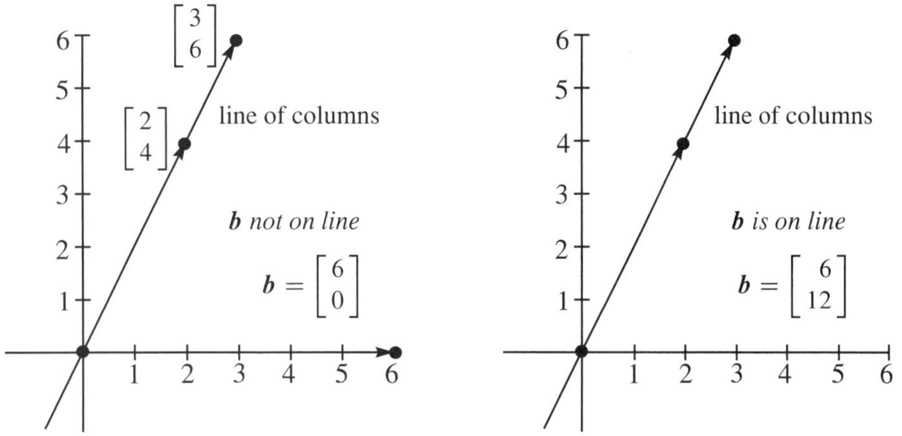


Figure 4.6: Column pictures (a) **No solution** (b) **Infinity of solutions**

**Example 3** Same matrix $A$, now $\boldsymbol{b} = (6, 12)$, infinitely many solutions to $A\boldsymbol{v} = \boldsymbol{b}$

$$A = \begin{bmatrix} 2 & 3 \\ 4 & 6 \end{bmatrix} \qquad \begin{aligned} 2v_1 - 3v_2 &= \mathbf{6} \\ 4v_1 - 6v_2 &= \mathbf{12} \end{aligned}$$

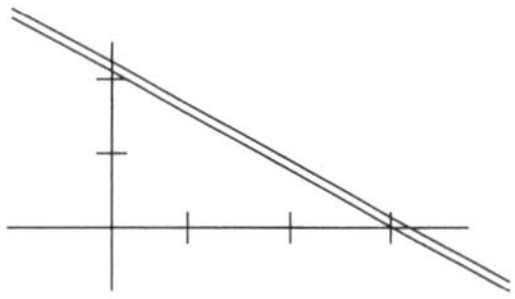

In the row picture, the two lines are the same. *All points* on that line solve both equations. Two times equation 1 gives equation 2. Those close lines are one line.

In the column picture above, the right side $\boldsymbol{b} = (6, 12)$ falls right onto the line of the columns. Later we will say: ***b** is in the column space of A*. There are infinitely many ways to produce $(6, 12)$ as a combination of the columns. They come from infinitely many ways to produce $\boldsymbol{b} = (0, 0)$ (**choose any $\boldsymbol{c}$**). Add one way to produce $\boldsymbol{b} = (6, 12) = 3(2, 4)$.

$$\begin{bmatrix} 0 \\ 0 \end{bmatrix} = \mathbf{3c}\begin{bmatrix} 2 \\ 4 \end{bmatrix} + \mathbf{2c}\begin{bmatrix} -3 \\ -6 \end{bmatrix} \qquad \begin{bmatrix} 6 \\ 12 \end{bmatrix} = \mathbf{3}\begin{bmatrix} 2 \\ 4 \end{bmatrix} + \mathbf{0}\begin{bmatrix} -3 \\ -6 \end{bmatrix}. \tag{6}$$

The vector $\boldsymbol{v}_n = (\mathbf{3c}, \mathbf{2c})$ is a **null solution** and $\boldsymbol{v}_p = (\mathbf{3}, \mathbf{0})$ is a **particular solution**. $\boldsymbol{A v_n}$ **equals zero and** $\boldsymbol{A v_p}$ **equals** $\boldsymbol{b}$. Then $A(\boldsymbol{v}_p + \boldsymbol{v}_n) = \boldsymbol{b}$. Together, $\boldsymbol{v}_p$ and $\boldsymbol{v}_n$ give the **complete solution**, all the ways to produce $\boldsymbol{b} = (6, 12)$ from the columns of $A$:

$$\textbf{Complete solution to } A\boldsymbol{v} = \boldsymbol{b} \qquad \boldsymbol{v}_{\text{complete}} = \boldsymbol{v}_p + \boldsymbol{v}_n = \begin{bmatrix} 3 \\ 0 \end{bmatrix} + \begin{bmatrix} 3c \\ 2c \end{bmatrix}. \tag{7}$$

## Equations and Pictures in Three Dimensions

In three dimensions, a linear equation like $x + y + 2z = 6$ produces a *plane*. The plane would go through $(0,0,0)$ if the right side were 0. In this case the "6" moves us to a parallel plane that misses the center point $(0,0,0)$.

A second linear equation will produce another plane. Normally the two planes meet in a *line*. Then a third plane (from a third equation) normally cuts through that line at a *point*. That point will lie on all three planes, so it solves all three equations.

This is the *row picture*, three planes in three–dimensional space. They meet at the solution. One big problem is that this row picture is hard to draw. Three planes are too many to see clearly how they meet (maybe Picasso could do it).

The *column picture* of $A\boldsymbol{v} = \boldsymbol{b}$ is easier. It starts with three column vectors in three-dimensional space. We want to combine those columns of $A$ to produce the vector $v_1(\text{column 1}) + v_2(\text{column 2}) + v_3(\text{column 3}) = \boldsymbol{b}$. Normally there is one way to do it. That gives the solution $(v_1, v_2, v_3)$ — which is also the meeting point in the row picture.

I want to give an example of success (one solution) and an example of failure (no solution). Both examples are simple, but they really go deeply into linear algebra.

**Example 4** Invertible matrix $A$, one solution $\boldsymbol{v}$ for any right side $\boldsymbol{b}$.

$$A\boldsymbol{v} = \boldsymbol{b} \quad \text{is} \quad \begin{bmatrix} 1 & 0 & 0 \\ -1 & 1 & 0 \\ 0 & -1 & 1 \end{bmatrix} \begin{bmatrix} v_1 \\ v_2 \\ v_3 \end{bmatrix} = \begin{bmatrix} 1 \\ 3 \\ 5 \end{bmatrix}. \tag{8}$$

This matrix is **lower triangular**. It has zeros above the main diagonal. Lower triangular systems are quickly solved by forward substitution, top to bottom. The top equation gives $v_1$, then move down. First $v_1 = \mathbf{1}$. Then $-v_1 + v_2 = 3$ gives $v_2 = \mathbf{4}$. Then $-v_2 + v_3 = 5$ gives $v_3 = \mathbf{9}$.

Figure 4.7 shows the three columns $\boldsymbol{a}_1, \boldsymbol{a}_2, \boldsymbol{a}_3$. When you combine them with $1, 4, 9$ you produce $\boldsymbol{b} = (1,3,5)$. In reverse, $\boldsymbol{v} = (1,4,9)$ must be the solution to $A\boldsymbol{v} = \boldsymbol{b}$.

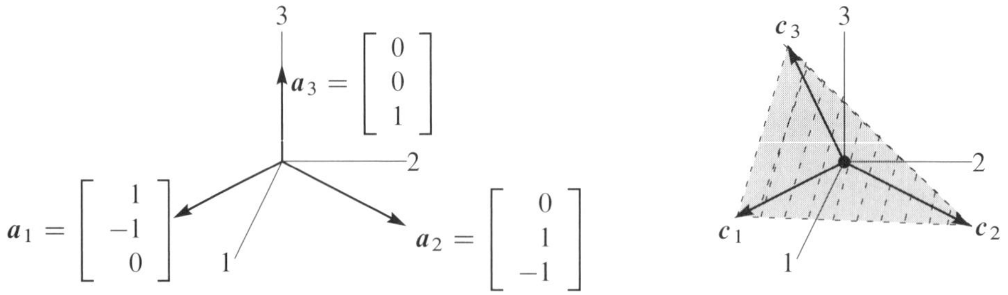


Figure 4.7: **Independent columns $\boldsymbol{a}_1, \boldsymbol{a}_2, \boldsymbol{a}_3$ not in a plane.** Dependent columns $\boldsymbol{c}_1, \boldsymbol{c}_2, \boldsymbol{c}_3$ are three vectors all in the same plane.

**Example 5** **Singular matrix**: no solution to $C\boldsymbol{v} = \boldsymbol{b}$ or infinitely many solutions (depending on $\boldsymbol{b}$).

$$\begin{aligned} \boldsymbol{w_1 - w_3 = b_1} \\ \boldsymbol{-w_1 + w_2 = b_2} \\ \boldsymbol{-w_2 + w_3 = b_3} \end{aligned} \quad \begin{bmatrix} 1 & 0 & -1 \\ -1 & 1 & 0 \\ 0 & -1 & 1 \end{bmatrix} \begin{bmatrix} w_1 \\ w_2 \\ w_3 \end{bmatrix} = \begin{bmatrix} 1 \\ 3 \\ 5 \end{bmatrix} \text{ or } \begin{bmatrix} 0 \\ 0 \\ 0 \end{bmatrix} \text{ or } \begin{bmatrix} 1 \\ 2 \\ -3 \end{bmatrix}. \quad (9)$$

This matrix $C$ is a "circulant." The diagonals are constants, all 1's or all 0's or all $-1$'s. The diagonals circle around so each diagonal has three equal entries. Circulant matrices will be perfect for the Fast Fourier Transform (**FFT**) in Chapter 8.

To see if $C\boldsymbol{w} = \boldsymbol{b}$ has a solution, add those three equations to get $0 = b_1 + b_2 + b_3$.

$$\textbf{Left side} \qquad (w_1 - w_3) + (-w_3 + w_2) + (-w_2 + w_3) = 0. \qquad (10)$$

$C\boldsymbol{w} = \boldsymbol{b}$ cannot have a solution unless $0 = b_1 + b_2 + b_3$. The components of $\boldsymbol{b} = (1, 3, 5)$ do not add to zero, so $C\boldsymbol{w} = (1, 3, 5)$ has no solution.

Figure 4.7 shows the problem. **The three columns of $C$ lie in a plane**. **All combinations $C\boldsymbol{w}$ of those columns will lie in that same plane**. If the right side vector $\boldsymbol{b}$ is not in the plane, then $C\boldsymbol{w} = \boldsymbol{b}$ cannot be solved. The vector $\boldsymbol{b} = (1, 3, 5)$ is off the plane, because the equation of the plane requires $b_1 + b_2 + b_3 = 0$.

Of course $C\boldsymbol{w} = (0, 0, 0)$ always has the zero solution $\boldsymbol{w} = (0, 0, 0)$. But when the columns of $C$ are in a plane (as here), there are additional nonzero solutions to $C\boldsymbol{w} = \boldsymbol{0}$. Those three equations are $w_1 = w_3$ and $w_1 = w_2$ and $w_2 = w_3$. The **null solutions** are $\boldsymbol{w}_n = (\boldsymbol{c}, \boldsymbol{c}, \boldsymbol{c})$. When all three components are equal, we have $C\boldsymbol{w}_n = \boldsymbol{0}$.

The vector $\boldsymbol{b} = (1, 2, -3)$ is also in the plane of the columns, because it does have $b_1 + b_2 + b_3 = 0$. In this good case there must be a **particular solution** to $C\boldsymbol{w}_p = \boldsymbol{b}$. There are many particular solutions $\boldsymbol{w}_p$, since any solution can be a particular solution. I will choose the particular $\boldsymbol{w}_p = (1, 3, 0)$ that ends in $w_3 = 0$:

$$C\boldsymbol{w}_p = \begin{bmatrix} 1 & 0 & -1 \\ -1 & 1 & 0 \\ 0 & -1 & 1 \end{bmatrix} \begin{bmatrix} 1 \\ 3 \\ 0 \end{bmatrix} = \begin{bmatrix} 1 \\ 2 \\ -3 \end{bmatrix}$$

> The **complete solution** is
> $\boldsymbol{w}_{\text{complete}} = \boldsymbol{w}_p + \text{ any } \boldsymbol{w}_n$

**Summary** These two matrices $A$ and $C$, with third columns $\boldsymbol{a}_3$ and $\boldsymbol{c}_3$, allow me to mention two key words of linear algebra: *independence and dependence*. This book will develop those ideas much further. I am happy if you see them early in the two examples:

| | | |
|---|---|---|
| **$\boldsymbol{a}_1, \boldsymbol{a}_2, \boldsymbol{a}_3$ are independent** | **$A$ is invertible** | **$A\boldsymbol{v} = \boldsymbol{b}$ has one solution $\boldsymbol{v}$** |
| **$\boldsymbol{c}_1, \boldsymbol{c}_2, \boldsymbol{c}_3$ are dependent** | **$C$ is singular** | **$C\boldsymbol{w} = \boldsymbol{0}$ has many solutions $\boldsymbol{w}_n$** |

Eventually we will have $n$ column vectors in $n$-dimensional space. The matrix will be $n$ by $n$. The key question is whether $A\boldsymbol{v} = \boldsymbol{0}$ has only the zero solution. Then the columns don't lie in any "hyperplane." When columns are independent, the matrix is invertible.

## Problem Set 4.1

**Problems 1–8 are about the row and column pictures of $A\boldsymbol{v} = \boldsymbol{b}$.**

**1** With $A = I$ (the identity matrix) draw the planes in the row picture. Three sides of a box meet at the solution $\boldsymbol{v} = (x, y, z) = (2, 3, 4)$:

$$\begin{aligned} 1x + 0y + 0z &= 2 \\ 0x + 1y + 0z &= 3 \\ 0x + 0y + 1z &= 4 \end{aligned} \qquad \text{or} \qquad \begin{bmatrix} 1 & 0 & 0 \\ 0 & 1 & 0 \\ 0 & 0 & 1 \end{bmatrix} \begin{bmatrix} x \\ y \\ z \end{bmatrix} = \begin{bmatrix} 2 \\ 3 \\ 4 \end{bmatrix}.$$

Draw the four vectors in the column picture. Two times column 1 plus three times column 2 plus four times column 3 equals the right side $\boldsymbol{b}$.

**2** If the equations in Problem 1 are multiplied by 2, 3, 4 they become $DV = \boldsymbol{B}$:

$$\begin{aligned} 2x + 0y + 0z &= 4 \\ 0x + 3y + 0z &= 9 \\ 0x + 0y + 4z &= 16 \end{aligned} \qquad \text{or} \qquad DV = \begin{bmatrix} 2 & 0 & 0 \\ 0 & 3 & 0 \\ 0 & 0 & 4 \end{bmatrix} \begin{bmatrix} x \\ y \\ z \end{bmatrix} = \begin{bmatrix} 4 \\ 9 \\ 16 \end{bmatrix} = \boldsymbol{B}$$

Why is the row picture the same? Is the solution $\boldsymbol{V}$ the same as $\boldsymbol{v}$? What is changed in the column picture—the columns or the right combination to give $\boldsymbol{B}$?

**3** If equation 1 is added to equation 2, which of these are changed: the planes in the row picture, the vectors in the column picture, the coefficient matrix, the solution? The new equations in Problem 1 would be $x = 2$, $x + y = 5$, $z = 4$.

**4** Find a point with $z = 2$ on the intersection line of the planes $x + y + 3z = 6$ and $x - y + z = 4$. Find the point with $z = 0$. Find a third point halfway between.

**5** The first of these equations plus the second equals the third:

$$\begin{aligned} x + \phantom{2}y + \phantom{2}z &= 2 \\ x + 2y + \phantom{2}z &= 3 \\ 2x + 3y + 2z &= 5. \end{aligned}$$

The first two planes meet along a line. The third plane contains that line, because if $x, y, z$ satisfy the first two equations then they also _____. The equations have infinitely many solutions (the whole line **L**). Find three solutions on **L**.

**6** Move the third plane in Problem 5 to a parallel plane $2x + 3y + 2z = 9$. Now the three equations have no solution—*why not*? The first two planes meet along the line **L**, but the third plane doesn't _____ that line.

**7** In Problem 5 the columns are $(1, 1, 2)$ and $(1, 2, 3)$ and $(1, 1, 2)$. This is a "singular case" because the third column is _____. Find two combinations of the columns that give $\boldsymbol{b} = (2, 3, 5)$. This is only possible for $\boldsymbol{b} = (4, 6, c)$ if $c =$ _____.

**8** Normally 4 "planes" in 4-dimensional space meet at a _____. Normally 4 vectors in 4-dimensional space can combine to produce $\boldsymbol{b}$. What combination of $(1,0,0,0)$, $(1,1,0,0)$, $(1,1,1,0)$, $(1,1,1,1)$ produces $\boldsymbol{b}=(3,3,3,2)$?

**Problems 9–14 are about multiplying matrices and vectors.**

**9** Compute each $A\boldsymbol{x}$ by dot products of the rows with the column vector:

$$\text{(a)} \begin{bmatrix} 1 & 2 & 4 \\ -2 & 3 & 1 \\ -4 & 1 & 2 \end{bmatrix}\begin{bmatrix} 2 \\ 2 \\ 3 \end{bmatrix} \quad \text{to} \quad \begin{bmatrix} 2 & 1 & 0 & 0 \\ 1 & 2 & 1 & 0 \\ 0 & 1 & 2 & 1 \\ 0 & 0 & 1 & 2 \end{bmatrix}\begin{bmatrix} 1 \\ 1 \\ 1 \\ 2 \end{bmatrix}$$

**10** Compute each $A\boldsymbol{x}$ in Problem 9 as a combination of the columns:

$$\text{9(a) becomes} \quad A\boldsymbol{x} = 2\begin{bmatrix} 1 \\ -2 \\ -4 \end{bmatrix} + 2\begin{bmatrix} 2 \\ 3 \\ 1 \end{bmatrix} + 3\begin{bmatrix} 4 \\ 1 \\ 2 \end{bmatrix} = \begin{bmatrix} \\ \\ \\ \end{bmatrix}.$$

How many separate multiplications for $A\boldsymbol{x}$, when the matrix is "3 by 3"?

**11** Find the two components of $A\boldsymbol{x}$ by rows or by columns:

$$\begin{bmatrix} 2 & 3 \\ 5 & 1 \end{bmatrix}\begin{bmatrix} 4 \\ 2 \end{bmatrix} \quad \text{and} \quad \begin{bmatrix} 3 & 6 \\ 6 & 12 \end{bmatrix}\begin{bmatrix} 2 \\ -1 \end{bmatrix} \quad \text{and} \quad \begin{bmatrix} 1 & 2 & 4 \\ 2 & 0 & 1 \end{bmatrix}\begin{bmatrix} 3 \\ 1 \\ 1 \end{bmatrix}.$$

**12** Multiply $A$ times $\boldsymbol{x}$ to find three components of $A\boldsymbol{x}$:

$$\begin{bmatrix} 0 & 0 & 1 \\ 0 & 1 & 0 \\ 1 & 0 & 0 \end{bmatrix}\begin{bmatrix} x \\ y \\ z \end{bmatrix} \quad \text{and} \quad \begin{bmatrix} 2 & 1 & 3 \\ 1 & 2 & 3 \\ 3 & 3 & 6 \end{bmatrix}\begin{bmatrix} 1 \\ 1 \\ -1 \end{bmatrix} \quad \text{and} \quad \begin{bmatrix} 2 & 1 \\ 1 & 2 \\ 3 & 3 \end{bmatrix}\begin{bmatrix} 1 \\ 1 \end{bmatrix}.$$

**13** (a) A matrix with $m$ rows and $n$ columns multiplies a vector with _____ components to produce a vector with _____ components.

(b) The planes from the $m$ equations $A\boldsymbol{x}=\boldsymbol{b}$ are in _____-dimensional space. The combination of the columns of $A$ is in _____-dimensional space.

**14** Write $2x+3y+z+5t=8$ as a matrix $A$ (how many rows?) multiplying the column vector $\boldsymbol{x}=(x,y,z,t)$ to produce $\boldsymbol{b}$. The solutions $\boldsymbol{x}$ fill a plane or "hyperplane" in 4-dimensional space. *The plane is 3-dimensional with no 4D volume.*

**Problems 15–22 ask for matrices that act in special ways on vectors.**

**15** (a) What is the 2 by 2 identity matrix? $I$ times $\begin{bmatrix} x \\ y \end{bmatrix}$ equals $\begin{bmatrix} x \\ y \end{bmatrix}$.

(b) What is the 2 by 2 exchange matrix? $P$ times $\begin{bmatrix} x \\ y \end{bmatrix}$ equals $\begin{bmatrix} y \\ x \end{bmatrix}$.

**16** (a) What 2 by 2 matrix $R$ rotates every vector by 90° ? $R$ times $\begin{bmatrix} x \\ y \end{bmatrix}$ is $\begin{bmatrix} y \\ -x \end{bmatrix}$.
(b) What 2 by 2 matrix $R^2$ rotates every vector by 180° ?

**17** Find the matrix $P$ that multiplies $(x, y, z)$ to give $(y, z, x)$. Find the matrix $Q$ that multiplies $(y, z, x)$ to bring back $(x, y, z)$.

**18** What 2 by 2 matrix $E$ subtracts the first component from the second component ? What 3 by 3 matrix does the same ?

$$E \begin{bmatrix} 3 \\ 5 \end{bmatrix} = \begin{bmatrix} 3 \\ 2 \end{bmatrix} \quad \text{and} \quad E \begin{bmatrix} 3 \\ 5 \\ 7 \end{bmatrix} = \begin{bmatrix} 3 \\ 2 \\ 7 \end{bmatrix}.$$

**19** What 3 by 3 matrix $E$ multiplies $(x, y, z)$ to give $(x, y, z + x)$ ? What matrix $E^{-1}$ multiplies $(x, y, z)$ to give $(x, y, z - x)$ ? If you multiply $(3, 4, 5)$ by $E$ and then multiply by $E^{-1}$, the two results are ( _____ ) and ( _____ ).

**20** What 2 by 2 matrix $P_1$ projects the vector $(x, y)$ onto the $x$ axis to produce $(x, 0)$ ? What matrix $P_2$ projects onto the $y$ axis to produce $(0, y)$ ? If you multiply $(5, 7)$ by $P_1$ and then multiply by $P_2$, you get ( _____ ) and ( _____ ).

**21** What 2 by 2 matrix $R$ rotates every vector through 45° ? The vector $(1, 0)$ goes to $(\sqrt{2}/2, \sqrt{2}/2)$. The vector $(0, 1)$ goes to $(-\sqrt{2}/2, \sqrt{2}/2)$. Those determine the matrix. Draw these particular vectors in the $xy$ plane and find $R$.

**22** Write the dot product of $(1, 4, 5)$ and $(x, y, z)$ as a matrix multiplication $A\boldsymbol{v}$. The matrix $A$ has one row. The solutions to $A\boldsymbol{v} = \mathbf{0}$ lie on a _____ perpendicular to the vector _____. The columns of $A$ are only in _____ -dimensional space.

**23** In MATLAB notation, write the commands that define this matrix $A$ and the column vectors $\boldsymbol{v}$ and $\boldsymbol{b}$. What command would test whether or not $A\boldsymbol{v} = \boldsymbol{b}$ ?

$$A = \begin{bmatrix} 1 & 2 \\ 3 & 4 \end{bmatrix} \qquad \boldsymbol{v} = \begin{bmatrix} 5 \\ -2 \end{bmatrix} \qquad \boldsymbol{b} = \begin{bmatrix} 1 \\ 7 \end{bmatrix}$$

**24** If you multiply the 4 by 4 all-ones matrix `A = ones(4)` and the column `v = ones(4,1)`, what is `A*v` ? (Computer not needed.) If you multiply `B = eye(4) + ones(4)` times `w = zeros(4,1) + 2*ones(4,1)`, what is `B*w` ?

**Questions 25–27 review the row and column pictures in 2, 3, and 4 dimensions.**

**25** Draw the row and column pictures for the equations $x - 2y = 0$, $x + y = 6$.

**26** For two linear equations in three unknowns $x, y, z$, the row picture will show (2 or 3) (lines or planes) in (2 or 3)-dimensional space. The column picture is in (2 or 3)-dimensional space. The solutions normally lie on a _____.

**27** For four linear equations in two unknowns $x$ and $y$, the row picture shows four _____. The column picture is in _____ -dimensional space. The equations have no solution unless the vector on the right side is a combination of _____.

## Challenge Problems

**28** Invent a 3 by 3 **magic matrix** $M_3$ with entries $1, 2, \ldots, 9$. All rows and columns and diagonals add to 15. The first row could be $8, 3, 4$. What is $M_3$ times $(1, 1, 1)$ ? What is $M_4$ times $(1, 1, 1, 1)$ if a 4 by 4 magic matrix has entries $1, \ldots, 16$ ?

**29** Suppose $\boldsymbol{u}$ and $\boldsymbol{v}$ are the first two columns of a 3 by 3 matrix $A$. Which third columns $\boldsymbol{w}$ would make this matrix singular ? Describe a typical column picture of $A\boldsymbol{v} = \boldsymbol{b}$ in that singular case, and a typical row picture (for a random $\boldsymbol{b}$).

**30** **Multiplying by $A$ is a "linear transformation".** Those important words mean:

If $\boldsymbol{w}$ is a combination of $\boldsymbol{u}$ and $\boldsymbol{v}$, then $A\boldsymbol{w}$ is the same combination of $A\boldsymbol{u}$ and $A\boldsymbol{v}$.

It is this "***linearity***" $A\boldsymbol{w} = cA\boldsymbol{u} + dA\boldsymbol{v}$ that gives us the name *linear algebra*.

If $\boldsymbol{u} = \begin{bmatrix} 1 \\ 0 \end{bmatrix}$ and $\boldsymbol{v} = \begin{bmatrix} 0 \\ 1 \end{bmatrix}$ then $A\boldsymbol{u}$ and $A\boldsymbol{v}$ are the columns of $A$.

Combine $\boldsymbol{w} = c\boldsymbol{u} + d\boldsymbol{v}$. **If** $\boldsymbol{w} = \begin{bmatrix} \mathbf{5} \\ \mathbf{7} \end{bmatrix}$ **how is $A\boldsymbol{w}$ connected to $A\boldsymbol{u}$ and $A\boldsymbol{v}$** ?

**31** A 9 by 9 ***Sudoku matrix*** $S$ has the numbers $1, \ldots, 9$ in every row and column, and in every 3 by 3 block. For the all-ones vector $\boldsymbol{v} = (1, \ldots, 1)$, what is $S\boldsymbol{v}$ ?

A better question is: **Which row exchanges will produce another Sudoku matrix** ? Also, which exchanges of block rows give another Sudoku matrix ?

Section 4.5 will look at all possible permutations (reorderings) of the rows. I see 6 orders for the first 3 rows, all giving Sudoku matrices. Also 6 permutations of the next 3 rows, and of the last 3 rows. And 6 block permutations of the block rows ?

**32** Suppose the second row of $A$ is some number $c$ times the first row :

$$A = \begin{bmatrix} a & b \\ ca & cb \end{bmatrix}.$$

Then if $a \neq 0$, the second column of $A$ is what number $d$ times the first column? **A square matrix with dependent rows will also have dependent columns**. This is a crucial fact coming soon.

## 4.2 Solving Linear Equations by Elimination

This section explains a systematic way to solve linear equations—the best way we know. The method is called ***"elimination"***, and you can see it in this 2 by 2 example. Before elimination, $x$ and $y$ appear in both equations. After elimination, the first unknown $x$ has disappeared from the second equation $5y = 5$.

$$\begin{aligned} x - 2y &= 1 \quad (\textit{multiply equation } 1 \textit{ by } 2) \\ 2x + y &= 7 \quad (\textit{subtract to eliminate } 2x) \end{aligned}$$

**After elimination**

$$\begin{aligned} \mathbf{x - 2y} &= \mathbf{1} \\ \mathbf{5y} &= \mathbf{5} \end{aligned}$$

The new equation $5y = 5$ instantly gives $y = 1$. Substituting $y = 1$ back into the first equation leaves $x - 2 = 1$. Therefore $x = 3$ and the solution $(x, y) = (3, 1)$ is complete.

Elimination produces an ***upper triangular system***—this is the goal. The nonzero coefficients $1, -2, 5$ form a triangle. That system is solved from the bottom upwards, first $y = 1$ and then $x = 3$. This quick process is called ***back substitution***. It is used for upper triangular systems of any size, after elimination produces a triangle.

Important point: The original equations have the same solution $x = 3$ and $y = 1$. Before and after elimination, the lines meet at the same point $(3, 1)$. Every step worked with both sides of correct equations.

The step that eliminated $x$ from equation 2 is the fundamental operation in this chapter. We use it so often that we look at it closely:

***To eliminate $\mathbf{2x}$ : Subtract a multiple of equation 1 from equation 2.***

Two times $x - 2y = 1$ gives $2x - 4y = 2$. When this is subtracted from $2x + y = 7$, the right side becomes $7 - 2 = 5$. The main point is that $2x$ cancels $2x$. **The system becomes triangular.**

Ask yourself how that multiplier $\ell = 2$ was found. The first equation contains $\mathbf{1}x$. ***So the first pivot was* 1** (the coefficient of $x$). The second equation contains $\mathbf{2}x$, **so the multiplier was 2**. Then subtraction $2x - 2x$ produced the zero and the triangle.

You will see the multiplier rule if I change the first equation to $3x - 6y = 3$. (Same straight line but the first pivot becomes 3.) The correct multiplier is now $\ell = \frac{2}{3}$. *To find that multiplier, divide the coefficient "**2**" to be eliminated by the pivot "**3**"* :

$$\begin{array}{lll} \mathbf{3}x - 6y = 3 & \textbf{Multiply equation 1 by } \tfrac{2}{3} & 3x - 6y = 3 \\ \mathbf{2}x + y \ = 7 & \textbf{Subtract from equation 2} & \qquad 5y = 5. \end{array}$$

The final system is triangular and the last equation still gives $y = 1$. Back substitution produces $3x - 6 = 3$ and $3x = 9$ and $x = 3$. We changed the numbers but not the lines or the solution. ***Divide by the pivot to find that multiplier $\boldsymbol{\ell} = \frac{2}{3}$*** :

| | | |
|---|---|---|
| ***Pivot*** | = | ***first nonzero in the row that does the elimination*** |
| ***Multiplier*** | = | ***(entry to eliminate) divided by (pivot)*** |

The new second equation starts with the second pivot, which is 5. We would use it to eliminate $y$ from the third equation if there were one. *To solve $n$ equations we want $n$ pivots.* ***The pivots are on the diagonal of the triangle after elimination.***

You could have solved those equations for $x$ and $y$ without reading this book. It is an extremely humble problem, but we stay with it a little longer. Even for a 2 by 2 system, elimination might break down. By understanding the possible breakdown (when we can't find a full set of pivots), you will understand the whole process of elimination.

## Breakdown of Elimination

Normally, elimination produces the pivots that take us to the solution. But failure is possible. At some point, the method might ask us to *divide by zero*. We can't do it. The process has to stop. There might be a way to adjust and continue—or failure may be unavoidable.

Example 1 fails with ***no solution*** to $0y = 5$. Example 2 fails with ***too many solutions*** to $0y = 0$. Example 3 succeeds by exchanging the equations.

**Example 1** ***Permanent failure with no solution.*** Elimination makes this clear:

$$\begin{array}{rlr} x - 2y = 1 & \text{Subtract 2 times} & x - 2y = 1 \\ 2x - 4y = 7 & \text{eqn. 1 from eqn. 2} & \mathbf{0}y = 5. \end{array}$$

There is *no* solution to $0y = 5$. *This system has no second pivot.* ***(Zero is never allowed as a pivot!)*** If there is no solution, elimination discovers that fact by reaching an impossible equation like $0y = 5$.

The row picture of failure shows parallel lines—which never meet. The column picture shows the two columns $(1, 2)$ and $(-2, -4)$ in the same direction. *All combinations of the columns lie along a line.* But the column from the right side is in a different direction $(1, 7)$. No combination of the columns can produce this right side—therefore no solution.

When we change the right side from $(1, 7)$ to $(1, 2)$, failure shows as a whole line of solution points. Instead of no solution, Example 2 changes to **infinitely many solutions**.

**Example 2** ***Failure with infinitely many solutions. Change $b = (1, 7)$ to $(1, 2)$.***

$$\begin{array}{rlrl} x - 2y = 1 & \text{Subtract 2 times} & x - 2y = 1 & \textbf{Too few pivots} \\ 2x - 4y = 2 & \text{eqn. 1 from eqn. 2} & \mathbf{0}y = \mathbf{0} & \textbf{Too many solutions} \end{array}$$

*Every $y$* satisfies $0y = 0$. There is really only one equation $x - 2y = 1$. The unknown $y$ is ***"free"***. After $y$ is freely chosen, $x$ is determined as $x = 1 + 2y$. I prefer to see a *particular solution* $\boldsymbol{v}_p = (1, 0)$ and a line of *null solutions* $\boldsymbol{v}_n = c\,(2, 1)$ in $\boldsymbol{v} = \boldsymbol{v}_p + \boldsymbol{v}_n$.

$$\textbf{Complete solution} \quad \begin{bmatrix} x \\ y \end{bmatrix} = \begin{bmatrix} 1 \\ 0 \end{bmatrix} + c \begin{bmatrix} 2 \\ 1 \end{bmatrix} = \textbf{particular } \boldsymbol{v}_p + \textbf{null } \boldsymbol{v}_n. \qquad (1)$$

In the row picture, the parallel lines have become the same line. Every point $(x, y)$ on that line satisfies both equations.

In the column picture, $\boldsymbol{b} = (1, 2)$ is now the same as column 1. So we can choose $x = 1$ and $y = 0$. We can also choose $x = 0$ and $y = -\frac{1}{2}$; column 2 times $-\frac{1}{2}$ equals $\boldsymbol{b}$. Every $(x, y)$ that solves the row problem also solves the column problem.

**Failure** For $n$ equations we do not get $n$ pivots. The rows combine into a zero row.

**Success** We do get $n$ pivots. **But we may have to exchange the *n* equations.**

Elimination can go wrong in a third way—but this time it can be fixed. *Suppose the first pivot position contains zero.* We refuse to allow zero as a pivot. When the first equation has no term involving $x$, we can *exchange* it with an equation below:

**Example 3** ***Temporary failure (zero in pivot). A row exchange produces two pivots*:**

$$\begin{aligned} 0x + 2y &= 4 \\ 3x - 2y &= 5 \end{aligned} \qquad \text{Exchange the two equations} \qquad \begin{aligned} 3x - 2y &= 5 \\ 2y &= 4. \end{aligned}$$

The new system is already triangular. This small example is ready for back substitution. The last equation gives $y = 2$, and then the first equation gives $x = 3$. The row picture is normal (two intersecting lines). The column picture is also normal (column vectors not in the same direction). The pivots 3 and 2 are normal—but a ***row exchange*** was required.

Examples 1 and 2 are ***singular***—there is no second pivot. Example 3 is ***nonsingular***—there is a full set of pivots and exactly one solution. Singular equations have no solution or infinitely many solutions. Pivots must be nonzero because we have to divide by them.

## Three Equations in Three Unknowns

To understand Gaussian elimination, you have to go beyond 2 by 2 systems. Three by three is enough to see the pattern. For now the matrices are square—an equal number of rows and columns. Here is a 3 by 3 system, specially constructed so that all steps lead to whole numbers and not fractions:

$$\begin{aligned} \mathbf{2}x + 4y - 2z &= 2 \\ 4x + 9y - 3z &= 8 \\ -2x - 3y + 7z &= 10 \end{aligned} \tag{2}$$

What are the steps? The first pivot is the boldface **2** (upper left). Below that pivot we want to eliminate the 4. *The first multiplier is the ratio* $4/2 = 2$. Multiply the pivot equation by $\ell_{21} = 2$ and subtract. Subtraction removes the $4x$ from the second equation:

**Step 1** Subtract 2 times equation 1 from equation 2. This leaves $y + z = 4$.

We also eliminate $-2x$ from equation 3, still using the first pivot. The quick way is to add equation 1 to equation 3. Then $2x$ cancels $-2x$. We do exactly that, but the rule in this book is to *subtract rather than add*. The systematic pattern has multiplier $\ell_{31} = -2/2 = -1$. Subtracting $-1$ times an equation is the same as adding:

**Step 2** Subtract $-1$ times equation 1 from equation 3. This leaves $y + 5z = 12$.

The two new equations involve only $y$ and $z$. The second pivot (in boldface) is 1 :

$$\textbf{x is eliminated} \qquad \begin{aligned} \mathbf{1}y + 1z &= 4 \\ 1y + 5z &= 12 \end{aligned}$$

*We have reached a* 2 *by* 2 *system.* The final step eliminates $y$ to make it 1 by 1 :

**Step 3** Subtract equation $2_{\text{new}}$ from $3_{\text{new}}$. The multiplier is $1/1 = 1$. Then $4z = 8$.

The original $A\boldsymbol{v} = \boldsymbol{b}$ has been converted into an upper triangular $U\boldsymbol{v} = \boldsymbol{c}$ :

$$\begin{aligned} 2x + 4y - 2z &= 2 \\ 4x + 9y - 3z &= 8 \\ -2x - 3y + 7z &= 10 \end{aligned} \qquad \begin{matrix} A\boldsymbol{v} = \boldsymbol{b} \\ \text{has become} \\ U\boldsymbol{v} = \boldsymbol{c} \end{matrix} \qquad \begin{aligned} \mathbf{2}x + 4y - 2z &= 2 \\ \mathbf{1}y + 1z &= 4 \\ \mathbf{4}z &= 8. \end{aligned} \tag{3}$$

The goal is achieved—forward elimination is complete from $A$ to $U$. ***The pivots are*** $\mathbf{2, 1, 4}$ ***on the diagonal of*** $U$. The pivots 1 and 4 were hidden in the original system. Elimination brought them out. $U\boldsymbol{v} = \boldsymbol{c}$ is ready for ***back substitution***, which is quick :

$$(4z = 8 \text{ gives } z = \mathbf{2}) \quad (y + z = 4 \text{ gives } y = \mathbf{2}) \quad (\text{equation 1 gives } x = \mathbf{-1})$$

*The solution is* $(x, y, z) = (-1, 2, 2)$. The row picture has three planes from the three equations. All the planes go through this solution. This picture is not easy to draw (it is totally impossible for larger systems).

The column picture shows a combination $A\boldsymbol{v}$ of column vectors producing the right side $\boldsymbol{b}$. The coefficients in that combination are $-1, 2, 2$ (the solution) :

$$A\boldsymbol{v} = (\mathbf{-1})\begin{bmatrix} 2 \\ 4 \\ -2 \end{bmatrix} + \mathbf{2}\begin{bmatrix} 4 \\ 9 \\ -3 \end{bmatrix} + \mathbf{2}\begin{bmatrix} -2 \\ -3 \\ 7 \end{bmatrix} \text{ equals } \begin{bmatrix} 2 \\ 8 \\ 10 \end{bmatrix} = \boldsymbol{b}. \tag{4}$$

The numbers $x, y, z$ multiply columns 1, 2, 3 in $A\boldsymbol{v} = \boldsymbol{b}$ and also in the triangular $U\boldsymbol{v} = \boldsymbol{c}$.

For a 4 by 4 problem, or an $n$ by $n$ problem, elimination proceeds the same way. Here is the whole idea, column by column from $A$ to $U$, when elimination succeeds.

**Column 1.** ***Use the first equation to create zeros below the first pivot.***

**Column 2.** ***Use the new equation 2 to create zeros below the second pivot.***

**Columns 3 to *n*.** ***Keep going to find all*** $n$ ***pivots and the triangular*** $U$.

$$\text{After column 2 we have} \begin{bmatrix} \boldsymbol{x} & x & x & x \\ 0 & \boldsymbol{x} & x & x \\ 0 & 0 & x & x \\ 0 & 0 & x & x \end{bmatrix}. \quad \text{We want } U = \begin{bmatrix} \boldsymbol{x} & x & x & x \\ & \boldsymbol{x} & x & x \\ & & \boldsymbol{x} & x \\ & & & \boldsymbol{x} \end{bmatrix}. \tag{5}$$

The result of forward elimination is an upper triangular system. The matrix will be nonsingular (= *invertible*) if and only if there is a full set of $n$ pivots (never zero!).

Here is a final example to show the original $A\boldsymbol{v} = \boldsymbol{b}$, the triangular system $U\boldsymbol{v} = \boldsymbol{c}$, and the solution $\boldsymbol{v} = (x, y, z)$ from back substitution:

$$\begin{array}{ll} x + \phantom{2}y + \phantom{2}z = 6 & \\ x + 2y + 2z = 9 & \textbf{Forward} \\ x + 2y + 3z = 10 & \textbf{Forward} \end{array} \qquad \begin{array}{r} x + y + z = 6 \\ y + z = 3 \\ z = 1 \end{array} \quad \begin{bmatrix} x \\ y \\ z \end{bmatrix} = \begin{bmatrix} 3 \\ 2 \\ 1 \end{bmatrix} \quad \begin{array}{l} \textbf{Back} \\ \textbf{Back} \\ \phantom{x} \end{array}$$

All multipliers are 1. All pivots are 1. All planes meet at the solution $\boldsymbol{v} = (3, 2, 1)$. The columns of $A$ combine with coefficients 3, 2, 1 to give $\boldsymbol{b} = (6, 9, 10)$:

$$A\boldsymbol{v} = \begin{bmatrix} 1 & 1 & 1 \\ 1 & 2 & 2 \\ 1 & 2 & 3 \end{bmatrix} \begin{bmatrix} 3 \\ 2 \\ 1 \end{bmatrix} = 3 \begin{bmatrix} 1 \\ 1 \\ 1 \end{bmatrix} + 2 \begin{bmatrix} 1 \\ 2 \\ 2 \end{bmatrix} + 1 \begin{bmatrix} 1 \\ 2 \\ 3 \end{bmatrix} = \begin{bmatrix} 6 \\ 9 \\ 10 \end{bmatrix}.$$

The numbers 6, 9, 10 are *dot products*. The first number 6 is the dot product of the first row (1, 1, 1) with $\boldsymbol{v} = (3, 2, 1)$.

*Question* What coefficient of $z$ in equation 3 would make the system singular?
*Answer* The third pivot would drop from 1 to 0 if the original $3z$ dropped to $2z$. Then the planes in the row picture have no point in common.

There is no solution to the new $\boldsymbol{A\,v = b}$. The three columns in the column picture would lie in the same plane, and $\boldsymbol{b} = (6, 9, 10)$ is not in that plane. So $\boldsymbol{b}$ will not be a combination of the columns, if the third column becomes (1, 2, 2). In this example column 3 becomes the same as column 2—useless, we need "independent" columns!

*Question* What coefficient of $y$ in equation 2 would become 0 in the first elimination step? Would the system become singular or not?

*Answer* Change equation 2 to $x + y + 2z = 7$ (for example). The coefficient of $y$ is now 1. Subtracting equation 1 leaves $\mathbf{0}y + z = 3$. **Now we can exchange equations 2 and 3.** This system is nonsingular. No problem except equations in the wrong order.

## ■ REVIEW OF THE KEY IDEAS ■

1. A linear system $A\boldsymbol{v} = \boldsymbol{b}$ becomes upper triangular ($U\boldsymbol{v} = \boldsymbol{c}$) by elimination.
2. We subtract $\ell_{ij}$ times equation $j$ from equation $i$, to make the $(i, j)$ entry zero.
3. The multiplier is $\ell_{ij} = \dfrac{\text{entry to eliminate in row } i}{\text{pivot in row } j}$. Pivots can not be zero!
4. A zero in the pivot position can be exchanged if there is a nonzero below it.
5. Back substitution solves the upper triangular system (bottom to top).
6. When breakdown is permanent, the system has no solution or infinitely many.

# Problem Set 4.2

**Problems 1–10 are about elimination on 2 by 2 systems.**

**1** What multiple $\ell_{21}$ of equation 1 should be subtracted from equation 2 ?

$$\begin{aligned} 2x + 3y &= 1 \\ 10x + 9y &= 11. \end{aligned}$$

After this step, solve the triangular system by back substitution, $y$ before $x$. Verify that $x$ times $(2, 10)$ plus $y$ times $(3, 9)$ equals $(1, 11)$. If the right side changes to $(4, 44)$, what is the new solution ?

**2** If you find solutions $\boldsymbol{v}$ and $\boldsymbol{w}$ to $A\boldsymbol{v} = \boldsymbol{b}$ and $A\boldsymbol{w} = \boldsymbol{c}$, what is the solution $\boldsymbol{u}$ to $A\boldsymbol{u} = \boldsymbol{b} + \boldsymbol{c}$ ? What is the solution $\boldsymbol{U}$ to $A\boldsymbol{U} = 3\boldsymbol{b} + 4\boldsymbol{c}$ ? (We saw superposition for linear differential equations, it works in the same way for all linear equations.)

**3** What multiple of equation 1 should be *subtracted* from equation 2 ?

$$\begin{aligned} 2x - 4y &= 6 \\ -x + 5y &= 0. \end{aligned}$$

After this elimination step, solve the triangular system. If the right side changes to $(-6, 0)$, what is the new solution ?

**4** What multiple $\ell$ of equation 1 should be subtracted from equation 2 to remove $cx$ ?

$$\begin{aligned} ax + by &= f \\ cx + dy &= g. \end{aligned}$$

The first pivot is $a$ (assumed nonzero). Elimination produces what formula for the second pivot ? The second pivot is missing when $ad = bc$: that is the *singular case*.

**5** Choose a right side which gives no solution and another right side which gives infinitely many solutions. What are two of those solutions ?

**Singular system**
$$\begin{aligned} 3x + 2y &= 10 \\ 6x + 4y &= \phantom{10} \end{aligned}$$

**6** Choose a coefficient $b$ that makes this system singular. Then choose a right side $g$ that makes it solvable. Find two solutions in that singular case.

$$\begin{aligned} 2x + by &= 16 \\ 4x + 8y &= g. \end{aligned}$$

**7** For which $a$ does elimination break down (1) permanently or (2) temporarily?

$$\begin{aligned} ax + 3y &= -3 \\ 4x + 6y &= \phantom{-}6. \end{aligned}$$

Solve for $x$ and $y$ after fixing the temporary breakdown by a row exchange.

**8** For which three numbers $k$ does elimination break down? Which is fixed by a row exchange? In these three cases, is the number of solutions 0 or 1 or $\infty$?

$$\begin{aligned} kx + 3y &= \phantom{-}6 \\ 3x + ky &= -6. \end{aligned}$$

**9** What test on $b_1$ and $b_2$ decides whether these two equations allow a solution? How many solutions will they have? Draw the column picture for $\boldsymbol{b} = (1, 2)$ and $(1, 0)$.

$$\begin{aligned} 3x - 2y &= b_1 \\ 6x - 4y &= b_2. \end{aligned}$$

**10** In the $xy$ plane, draw the lines $x + y = 5$ and $x + 2y = 6$ and the equation $y =$ ______ that comes from elimination. The line $5x - 4y = c$ will go through the solution of these equations if $c =$ ______.

**11** (Recommended) A system of linear equations can't have exactly two solutions. If $(x, y)$ and $(X, Y)$ are two solutions to $A\boldsymbol{v} = \boldsymbol{b}$, what is another solution?

**Problems 12–20 study elimination on 3 by 3 systems (and possible failure).**

**12** Reduce this system to upper triangular form by two row operations:

$$\begin{aligned} & & 2x+3y + z &= 8 \\ \text{Eliminate } x & \rightarrow & 4x+7y + 5z &= 20 \\ \text{Eliminate } y & \rightarrow & -2y + 2z &= 0. \end{aligned}$$

Circle the pivots. Solve by back substitution for $z, y, x$.

**13** Apply elimination (circle the pivots) and back substitution to solve

$$\begin{aligned} 2x - 3y \phantom{{}+ z} &= 3 \\ 4x - 5y + \phantom{3}z &= 7 \\ 2x - \phantom{5}y - 3z &= 5. \end{aligned}$$

List the three row operations: Subtract ______ times row ______ from row ______.

**14** Which number $d$ forces a row exchange? What is the triangular system (not singular) for that $d$? Which $d$ makes this system singular (no third pivot)?

$$\begin{aligned} 2x + 5y + z &= 0 \\ 4x + dy + z &= 2 \\ y - z &= 3. \end{aligned}$$

**15** Which number $b$ leads later to a row exchange? Which $b$ leads to a singular problem that row exchanges cannot fix? In that singular case find a nonzero solution $x, y, z$.

$$\begin{aligned} x + by \quad\quad &= 0 \\ x - 2y - z &= 0 \\ y + z &= 0. \end{aligned}$$

**16** (a) Construct a 3 by 3 system that needs two row exchanges to reach a triangular form.

(b) Construct a 3 by 3 system that needs a row exchange for pivot 2, but breaks down for pivot 3.

**17** If rows 1 and 2 are the same, how far can you get with elimination (allowing row exchange)? If columns 1 and 2 are the same, which pivot is missing?

$$\begin{array}{ll} \textbf{Equal} & 2x - y + z = 0 \\ \textbf{rows} & 2x - y + z = 0 \\ & 4x + y + z = 2 \end{array} \qquad \begin{array}{ll} 2x + 2y + z = 0 & \textbf{Equal} \\ 4x + 4y + z = 0 & \textbf{columns} \\ 6x + 6y + z = 2. & \end{array}$$

**18** Construct a 3 by 3 example that has 9 different coefficients on the left side, but rows 2 and 3 become zero in elimination. How many solutions to your system with $\boldsymbol{b} = (1, 10, 100)$ and how many with $\boldsymbol{b} = (0, 0, 0)$?

**19** Which number $q$ makes this system singular and which right side $t$ gives it infinitely many solutions? Find the solution that has $z = 1$.

$$\begin{aligned} x + 4y - 2z &= 1 \\ x + 7y - 6z &= 6 \\ 3y + qz &= t. \end{aligned}$$

**20** Three planes can fail to have an intersection point, *even if no planes are parallel.* The system is singular if row 3 is a combination of the first two rows. Find a third equation that can't be solved together with $x + y + z = 0$ and $x - 2y - z = 1$.

**21** Find the pivots and the solution for both systems ($A\boldsymbol{v} = \boldsymbol{b}$ and $S\boldsymbol{w} = \boldsymbol{b}$):

$$\begin{aligned} 2x + y \qquad\qquad &= 0 \\ x + 2y + z \qquad &= 0 \\ y + 2z + t &= 0 \\ z + 2t &= 5 \end{aligned} \qquad \begin{aligned} 2x - y \qquad\qquad &= 0 \\ -x + 2y - z \qquad &= 0 \\ -y + 2z - t &= 0 \\ -z + 2t &= 5. \end{aligned}$$

**22** If you extend Problem 21 following the $1, 2, 1$ pattern or the $-1, 2, -1$ pattern, what is the fifth pivot? What is the $n$th pivot? $S$ is my favorite matrix.

**23** If elimination leads to $x + y = 1$ and $2y = 3$, find three possible original problems.

**24** For which two numbers $a$ will elimination fail on $A = \begin{bmatrix} a & 2 \\ a & a \end{bmatrix}$?

**25** For which three numbers $a$ will elimination fail to give three pivots?

$$A = \begin{bmatrix} a & 2 & 3 \\ a & a & 4 \\ a & a & a \end{bmatrix} \text{ is singular for three values of } a.$$

**26** Look for a matrix that has row sums 4 and 8, and column sums 2 and $s$:

$$\text{Matrix } = \begin{bmatrix} a & b \\ c & d \end{bmatrix} \qquad \begin{matrix} a+b=4 & a+c=2 \\ c+d=8 & b+d=s \end{matrix}$$

The four equations are solvable only if $s =$ _____. Then find two different matrices that have the correct row and column sums. *Extra credit*: Write down the 4 by 4 system $A\boldsymbol{v} = (4, 8, 2, s)$ with $\boldsymbol{v} = (a, b, c, d)$ and make $A$ triangular by elimination.

**27** Elimination in the usual order gives what matrix $U$ and what solution $(x, y, z)$ to this "lower triangular" system? We are really solving by *forward substitution*:

$$\begin{aligned} 3x \qquad\quad &= 3 \\ 6x + 2y \quad\; &= 8 \\ 9x - 2y + z &= 9. \end{aligned}$$

**28** Create a MATLAB command `A(2, : ) = ...` for the new row 2, to subtract 3 times row 1 from the existing row 2 if the matrix $A$ is already known.

**29** If the last corner entry of $A$ is $A(5,5) = 11$ and the last pivot of $A$ is $U(5,5) = 4$, what different entry $A(5,5)$ would have made $A$ singular?

## Challenge Problems

**30** Suppose elimination takes $A$ to $U$ without row exchanges. Then row $i$ of $U$ is a combination of which rows of $A$? If $A\boldsymbol{v} = \mathbf{0}$, is $U\boldsymbol{v} = \mathbf{0}$? If $A\boldsymbol{v} = \boldsymbol{b}$, is $U\boldsymbol{v} = \boldsymbol{b}$?

**31** Start with 100 equations $A\boldsymbol{v} = \mathbf{0}$ for 100 unknowns $\boldsymbol{v} = (v_1, \ldots, v_{100})$. Suppose elimination reduces the 100th equation to $0 = 0$, so the system is "singular".

(a) Elimination takes linear combinations of the rows. So this singular system has the singular property: Some linear combination of the 100 **rows** is _____.

(b) Singular systems $A\boldsymbol{v} = \mathbf{0}$ have infinitely many solutions. This means that some linear combination of the 100 ***columns*** is _____.

(c) Invent a 100 by 100 singular matrix with no zero entries.

(d) For your matrix, describe in words the row picture and the column picture of $A\boldsymbol{v} = \mathbf{0}$. Not necessary to draw 100-dimensional space.

## 4.3 Matrix Multiplication

We know how to multiply $A$ times a column vector $\boldsymbol{v}$. Now we want to multiply $A$ times a matrix $B$ (matrix-matrix multiplication ). The rule is exactly what we would hope for :

> **Multiply $A$ times each column of $B$ to get a column of $AB$**
>
> The entry in row $i$, column $j$ of $AB$ is ( **row $i$ of $A$** ) $\cdot$ ( **column $j$ of $B$** )

If $B$ has only one column (call it $\boldsymbol{v}$), this is the same matrix-vector multiplication as before. When $B$ has $n$ columns, so has $AB$. The rule for matrix sizes makes dot products possible.

**Rule The number of columns in $A$ must match the number of rows in $B$.**

Figure 4.8 shows a typical (row $i$) $\cdot$ (column $j$) in the matrix multiplication $AB$.

$$\begin{bmatrix} * & & & \\ a_{i1} & a_{i2} & \cdots & a_{i5} \\ * & & & \\ * & & & \end{bmatrix} \begin{bmatrix} * & * & b_{1j} & * & * & * \\ & & b_{2j} & & & \\ & & \vdots & & & \\ & & b_{5j} & & & \end{bmatrix} = \begin{bmatrix} & & * & & & \\ * & * & (AB)_{ij} & * & * & * \\ & & * & & & \\ & & * & & & \end{bmatrix}$$

$A$ is **4** by **5** $\qquad$ $B$ is **5** by **6** $\qquad$ $AB$ is **4** by **6**

Figure 4.8: Here $i = 2$ and $j = 3$. Then $(AB)_{23}$ is (row 2 of $A$) $\cdot$ (column 3 of $B$).

Let me say right away that normally $AB$ is entirely different from $BA$. We can multiply in both orders only if $A$ and $B$ are square and the same size. But even the top left corner of $BA$ has nothing to do with the top left corner of $AB$ (and then $\boldsymbol{BA \neq AB}$).

**Top left** ( row 1 of $B$ ) $\cdot$ ( column 1 of $A$ ) $\neq$ ( row 1 of $A$ ) $\cdot$ ( column 1 of $B$ ).

**Example 1** Here $A$ has two columns and $B$ has two rows. We can multiply $AB$.

$$A_{\mathbf{2 \times 2}}\, B_{\mathbf{2 \times 3}} = (AB)_{\mathbf{2 \times 3}} \qquad \begin{bmatrix} a & b \\ c & d \end{bmatrix} \begin{bmatrix} 1 & 0 & \mathbf{1} \\ 0 & 1 & \mathbf{1} \end{bmatrix} = \begin{bmatrix} a & b & \boldsymbol{a+b} \\ c & d & \boldsymbol{c+d} \end{bmatrix}.$$

Column 3 of $B$ is $(1, 1)$. Then column 3 of $AB$ is $A$ times $(1, 1)$.

**Example 2** Here $B$ is the 3 by 3 **identity matrix** (very special, always written $B = I$).

$$\begin{array}{l} \boldsymbol{B} = \textbf{Identity matrix } \boldsymbol{I} \\ \boldsymbol{AI} = \boldsymbol{A} \textbf{ when sizes are right} \end{array} \qquad \begin{bmatrix} 1 & 1 & 1 \\ 1 & 2 & 2 \\ 1 & 2 & 3 \end{bmatrix} \begin{bmatrix} 1 & 0 & 0 \\ 0 & 1 & 0 \\ 0 & 0 & 1 \end{bmatrix} = \begin{bmatrix} 1 & 1 & 1 \\ 1 & 2 & 2 \\ 1 & 2 & 3 \end{bmatrix}$$

The first column of that answer is $A$ times the first column $(1, 0, 0)$ of $B = I$. This just reproduces the first column of $A$. Each column of $A$ is unchanged in $AI$.

Now put the identity matrix first, as in $IB$. Multiplication gives $IB = B$ for every $B$ (including $B = A$). We have here an unusual case, when the order $AI$ gives the same answer as $IA$. If $A$ is any square matrix and $I$ has the same size, then $\boldsymbol{AI} = \boldsymbol{IA} = \boldsymbol{A}$.

**Example 3** Another special matrix is the **inverse** of $A$. That matrix $B$ is written $A^{-1}$:

$$\textbf{A times } \boldsymbol{A^{-1}} \textbf{ is } \boldsymbol{I} \qquad \begin{bmatrix} 1 & 1 & 1 \\ 1 & 2 & 2 \\ 1 & 2 & 3 \end{bmatrix} \begin{bmatrix} 2 & -1 & 0 \\ -1 & 2 & -1 \\ 0 & -1 & 1 \end{bmatrix} = \begin{bmatrix} 1 & 0 & 0 \\ 0 & 1 & 0 \\ 0 & 0 & 1 \end{bmatrix}$$

The dot product of a row of $A$ with a column of $A^{-1}$ is 1 or 0. $A^{-1}$ times $A$ is also $I$.

To find that matrix $A^{-1}$, I had to look ahead to Section 4.4—this is a long calculation. We avoid computing $A^{-1}$ wherever possible, and so does any good linear algebra code.

$$\boxed{\textbf{The key fact about matrix multiplication is that} \quad (AB)C = A(BC).} \qquad (1)$$

To multiply three matrices $A, B, C$ you must keep them in order. But you can choose to multiply $AB$ first or $BC$ first. *Parentheses can be moved, and parentheses can be removed.*

**Example 4** Suppose $A$ and $C$ are 3 by 1 matrices ( those are column vectors ). Suppose $B$ is 1 by 3 ( a row vector ). Compute and compare $(AB)C$ and $A(BC)$.

**Solution** $BC$ is $(1 \times 3)$ times $(3 \times 1) = 1 \times 1$. One number $d$ from one dot product:

$$\textbf{A times BC} \qquad \begin{bmatrix} a_1 \\ a_2 \\ a_3 \end{bmatrix} \left( [\, b_1 \; b_2 \; b_3 \,] \begin{bmatrix} c_1 \\ c_2 \\ c_3 \end{bmatrix} \right) = \begin{bmatrix} a_1 d \\ a_2 d \\ a_3 d \end{bmatrix}. \qquad (2)$$

On the other hand, $AB$ is $(3 \times 1)$ times $(1 \times 3) = 3 \times 3$. This $AB$ is a full-size matrix !

$$\textbf{AB times C} \qquad \left( \begin{bmatrix} a_1 \\ a_2 \\ a_3 \end{bmatrix} [\, b_1 \; b_2 \; b_3 \,] \right) \begin{bmatrix} c_1 \\ c_2 \\ c_3 \end{bmatrix} = \begin{bmatrix} a_1 b_1 & a_1 b_2 & a_1 b_3 \\ a_2 b_1 & a_2 b_2 & a_2 b_3 \\ a_3 b_1 & a_3 b_2 & a_3 b_3 \end{bmatrix} \begin{bmatrix} c_1 \\ c_2 \\ c_3 \end{bmatrix}. \qquad (3)$$

If you multiply that first row of $AB$ times $C$, you will see $a_1 d$. Multiplying the other rows by $C$ gives $a_2 d$ and $a_3 d$. **$(AB)C$ in equation (3) equals $A(BC)$ in equation (2).**

## The Laws for Matrix Operations

May I put on record six laws that matrices do obey, while emphasizing an equation they don't obey ? The matrices can be square or rectangular, and the laws involving $A + B$ are all simple and all obeyed. Here are three addition laws :

$$\begin{array}{lll} A + B & = B + A & \text{(commutative law)} \\ c\,(A + B) & = c\,A + c\,B & \text{(distributive law)} \\ A + (B + C) & = (A + B) + C & \text{(associative law).} \end{array}$$

Three more laws hold for multiplication, but $AB = BA$ is not one of them :

| | |
|---|---|
| $AB \neq BA$ | (the commutative "law" is *usually broken*) |
| $A(B + C) = AB + AC$ | (distributive law from the left) |
| $(A + B)C = AC + BC$ | (distributive law from the right) |
| $A(BC) = (AB)C$ | (associative law for $ABC$) (***parentheses not needed***). |

When $A$ and $B$ are not square, $AB$ is a different size from $BA$. These matrices can't be equal—even if both multiplications are allowed. For square matrices, almost any example shows that $AB$ is different from $BA$:

$$AB = \begin{bmatrix} 0 & 0 \\ 1 & 0 \end{bmatrix}\begin{bmatrix} 0 & 1 \\ 0 & 0 \end{bmatrix} = \begin{bmatrix} 0 & 0 \\ 0 & \mathbf{1} \end{bmatrix} \quad \text{but} \quad BA = \begin{bmatrix} 0 & 1 \\ 0 & 0 \end{bmatrix}\begin{bmatrix} 0 & 0 \\ 1 & 0 \end{bmatrix} = \begin{bmatrix} \mathbf{1} & 0 \\ 0 & 0 \end{bmatrix}.$$

It is true that $AI = IA$. All square matrices commute with $I$ and also with $cI$. Only these matrices $cI$ commute with all other matrices.

The law $A(B + C) = AB + AC$ is proved a column at a time. Start with $A(\boldsymbol{b} + \boldsymbol{c}) = A\boldsymbol{b} + A\boldsymbol{c}$ for the first column. That is the key to everything—***linearity***. Say no more.

## Powers of Matrices

Look at the special case when $A = B = C$ = square matrix. Then ($A$ *times* $A^2$) is equal to ($A^2$ *times* $A$). The product in either order is $A^3$. The matrix powers $A^p$ follow the same rules as numbers:

$$A^p = AAA\cdots A \text{ ($p$ factors)} \qquad (A^p)(A^q) = A^{p+q} \qquad (A^p)^q = A^{pq}.$$

Those are the ordinary laws for exponents. $A^3$ times $A^4$ is $A^7$ (seven factors). $A^3$ to the fourth power is $A^{12}$ (twelve $A$'s). When $p$ and $q$ are zero or negative these rules still hold, provided $A$ has a "$-1$ power"—which is the *inverse matrix* $A^{-1}$. Then $A^0 = I$ is the identity matrix (no factors).

For a number, $a^{-1}$ is $1/a$. For a matrix, the inverse is written $A^{-1}$. (It is *never* $I/A$. But backslash $A\backslash I$ is allowed in MATLAB.) Every number has an inverse except $a = 0$. To decide when $A$ has an inverse is a central problem in linear algebra. This section is like a Bill of Rights for matrices, to say when $A$ and $B$ can be multiplied and how.

## Elimination Matrices

We now combine two ideas—elimination and matrices. The goal is to express all the steps of elimination in the clearest possible way. You will see how to subtract a multiple $\ell_{ij}$ times row $j$ from row $i$—using a matrix $E$.

The column vector $\boldsymbol{b}$ is multiplied by the elimination matrix $E$:

$$\textbf{Subtract } 2b_1 \textbf{ from } b_2 \qquad E\boldsymbol{b} = \begin{bmatrix} 1 & 0 & 0 \\ \mathbf{-2} & \mathbf{1} & 0 \\ 0 & 0 & 1 \end{bmatrix}\begin{bmatrix} b_1 \\ b_2 \\ b_3 \end{bmatrix} = \begin{bmatrix} b_1 \\ \boldsymbol{b_2 - 2b_1} \\ b_3 \end{bmatrix}. \tag{4}$$

Whatever we do to one side of $A\boldsymbol{v} = \boldsymbol{b}$, we do to the other side. *Elimination is multiplying both sides by $E$.* On the left side, we see row operations.

$$EA = \begin{bmatrix} 1 & 0 & 0 \\ \mathbf{-2} & \mathbf{1} & 0 \\ 0 & 0 & 1 \end{bmatrix}\begin{bmatrix} \text{row } 1 \\ \text{row } 2 \\ \text{row } 3 \end{bmatrix} = \begin{bmatrix} \text{row } 1 \\ \textbf{row 2} - \textbf{2 row 1} \\ \text{row } 3 \end{bmatrix}. \tag{5}$$

$EA$ will be our matrix after the first elimination step. The multiplier 2 was chosen to produce 0 in the 2, 1 position (row 2, column 1). This matrix $E$ should be named $E_{21}$ because it eliminates the original entry $a_{21}$ to leave zero.

The next step of elimination comes from a matrix $E_{31}$ (producing zero in place of $a_{31}$). Then $E_{32}$ produces zero in row 3, column 2, using a multiplier $\ell_{32}$. Altogether, the three steps from $A$ to the upper triangular $U$ come from three elimination matrices:

**Elimination by matrices** $\quad A$ becomes $\quad E_{32}E_{31}E_{21}A = U$ (upper triangular).

We do the same operations on the right side. $E_{32}E_{31}E_{21}\boldsymbol{b}$ becomes the new right side vector $\boldsymbol{c}$. Then back substitution solves $U\boldsymbol{v} = \boldsymbol{c}$.

**Example 5** Choose the multiplier $\ell_{21} = c/a$ to produce zero in $U_{21}$, using $E = E_{21}$:

$$EA = \begin{bmatrix} 1 & 0 \\ \boldsymbol{-c/a} & 1 \end{bmatrix}\begin{bmatrix} a & b \\ c & d \end{bmatrix} = \begin{bmatrix} a & b \\ 0 & d-(c/a)b \end{bmatrix} = U. \tag{6}$$

Undo this elimination by **adding** $c/a$ times row 1 of $U$ to row 2 of $U$:

$$E^{-1}U = \begin{bmatrix} 1 & 0 \\ \boldsymbol{c/a} & 1 \end{bmatrix}\begin{bmatrix} a & b \\ 0 & d-(c/a)b \end{bmatrix} = \begin{bmatrix} a & b \\ c & d \end{bmatrix} = A.$$

Thus $U = EA$ and $A = E^{-1}U$. Often we write this as $A = LU$.

## Four Ways to Multiply $AB$

I will end this section by writing down four different ways to compute $AB$. All four ways give the same answer. In the end we are doing the same calculations, but we are seeing those steps in different orders.

1. (Rows of $A$) times (columns of $B$) (*dot products*)
2. $A$ times (columns of $B$) (matrix-vector multiplications)
3. (Rows of $A$) times $B$ (vector-matrix multiplications )
4. (Columns of $A$) times (rows of $B$) (add up $n$ column-times-row matrices)

Let me look at the 1, 1 entry in the top corner of $AB$. The usual way is a dot product:

$$(\text{row 1 of } A) \cdot (\text{column 1 of } B) = (AB)_{11} = a_{11}b_{11} + a_{12}b_{21} + \cdots + a_{1n}b_{n1} \tag{7}$$

Orders **2** and **3** give that same dot product in $AB$. Here is order **4**, *columns times rows*:

$$(\text{column 1 of } A)(\text{row 1 of } B) = \begin{bmatrix} a_{11} \\ a_{21} \\ \cdot \end{bmatrix}\begin{bmatrix} b_{11} & b_{12} & \cdot \end{bmatrix} = \begin{bmatrix} a_{11}b_{11} & \cdot & \cdot \\ \cdot & \cdot & \cdot \\ \cdot & \cdot & \cdot \end{bmatrix} \tag{8}$$

The next column-times-row matrix is (column 2 of $A$)(row 2 of $B$). That starts with $a_{12}b_{21}$ in the top left corner. We get $a_{1j}b_{j1}$ when column $j$ of $A$ multiplies row $j$ of $B$.

Adding these simple matrices will produce the correct dot product (the sum of $a_{1j}b_{j1}$) in the top left corner—and in every entry of $AB$.

When $A$ and $B$ are $n$ by $n$ matrices, so is $AB$. It contains $n^2$ dot products. So it needs $n^3$ separate multiplications. For matrices of order $n = 100$ this is a million multiplications. No problem, that may only take one second (on the computer).

When $A$ is an $\boldsymbol{m}$ **by** $\boldsymbol{n}$ matrix and $B$ is $\boldsymbol{n}$ **by** $\boldsymbol{p}$, the product $AB$ is $\boldsymbol{m}$ **by** $\boldsymbol{p}$. It contains $mp$ dot products. So it needs $mnp$ separate multiplications.

Matrices of order $n = 10{,}000$ need a trillion ($10^{12}$) multiplications. Codes avoid multiplying full matrices whenever possible. And they watch especially for *sparse matrices*, when many of the entries (almost all) are zero. The codes don't waste time multiplying by zero.

## Problem Set 4.3

**Problems 1–16 are about the laws of matrix multiplication .**

**1** $A$ is 3 by 5, $B$ is 5 by 3, $C$ is 5 by 1, and $D$ is 3 by 1. *All entries are* 1. Which of these matrix operations are allowed, and what are the results?

$$BA \qquad AB \qquad ABD \qquad DBA \qquad A(B+C).$$

**2** What rows or columns or matrices do you multiply to find

(a) the third column of $AB$ ?

(b) the first row of $AB$ ?

(c) the entry in row 3, column 4 of $AB$ ?

(d) the entry in row 1, column 1 of $CDE$ ?

**3** Add $AB$ to $AC$ and compare with $A(B+C)$:

$$A = \begin{bmatrix} 1 & 5 \\ 2 & 3 \end{bmatrix} \quad \text{and} \quad B = \begin{bmatrix} 0 & 2 \\ 0 & 1 \end{bmatrix} \quad \text{and} \quad C = \begin{bmatrix} 3 & 1 \\ 0 & 0 \end{bmatrix}.$$

**4** In Problem 3, multiply $A$ times $BC$. Then multiply $AB$ times $C$.

**5** Compute $A^2$ and $A^3$. Make a prediction for $A^5$ and $A^n$ :

$$A = \begin{bmatrix} 1 & b \\ 0 & 1 \end{bmatrix} \quad \text{and} \quad A = \begin{bmatrix} 2 & 2 \\ 0 & 0 \end{bmatrix}.$$

**6** Show that $(A+B)^2$ is different from $A^2 + 2AB + B^2$, when

$$A = \begin{bmatrix} 1 & 2 \\ 0 & 0 \end{bmatrix} \quad \text{and} \quad B = \begin{bmatrix} 1 & 0 \\ 3 & 0 \end{bmatrix}.$$

Write down the correct rule for $(A+B)(A+B) = A^2 + \underline{\quad\quad} + B^2$.

**7** True or false. Give a specific example when false:

(a) If columns 1 and 3 of $B$ are the same, so are columns 1 and 3 of $AB$.

(b) If rows 1 and 3 of $B$ are the same, so are rows 1 and 3 of $AB$.

(c) If rows 1 and 3 of $A$ are the same, so are rows 1 and 3 of $ABC$.

(d) $(AB)^2 = A^2B^2$.

**8** How is each row of $DA$ and $EA$ related to the rows of $A$, when

$$D = \begin{bmatrix} 3 & 0 \\ 0 & 5 \end{bmatrix} \quad \text{and} \quad E = \begin{bmatrix} 0 & 1 \\ 0 & 1 \end{bmatrix} \quad \text{and} \quad A = \begin{bmatrix} a & b \\ c & d \end{bmatrix}?$$

How is each column of $AD$ and $AE$ related to the columns of $A$?

**9** Row 1 of $A$ is added to row 2. This gives $EA$ below. Then column 1 of $EA$ is added to column 2 to produce $(EA)F$. Notice $E$ and $F$ in boldface.

$$EA = \begin{bmatrix} \mathbf{1} & 0 \\ \mathbf{1} & \mathbf{1} \end{bmatrix} \begin{bmatrix} a & b \\ c & d \end{bmatrix} = \begin{bmatrix} a & b \\ a+c & b+d \end{bmatrix}$$

$$(EA)F = (EA) \begin{bmatrix} \mathbf{1} & \mathbf{1} \\ 0 & \mathbf{1} \end{bmatrix} = \begin{bmatrix} a & a+b \\ a+c & a+c+b+d \end{bmatrix}.$$

Do those steps in the opposite order, first multiply $AF$ and then $E(AF)$. Compare with $(EA)F$. What law is obeyed by matrix multiplication?

**10** Row 1 of $A$ is added to row 2 to produce $EA$. Then $F$ adds row 2 of $EA$ to row 1. Now $F$ is on the left, for row operations. The result is $F(EA)$:

$$F(EA) = \begin{bmatrix} 1 & 1 \\ 0 & 1 \end{bmatrix} \begin{bmatrix} a & b \\ a+c & b+d \end{bmatrix} = \begin{bmatrix} 2a+c & 2b+d \\ a+c & b+d \end{bmatrix}.$$

Do those steps in the opposite order: first add row 2 to row 1 by $FA$, then add row 1 of $FA$ to row 2. What law is or is not obeyed by matrix multiplication?

**11** (3 by 3 matrices) Choose the only $B$ so that for every matrix $A$

(a) $BA = 4A$

(b) $BA = 4B$ (tricky)

(c) $BA$ has rows 1 and 3 of $A$ reversed and row 2 unchanged

(d) All rows of $BA$ are the same as row 1 of $A$.

**12** Suppose $AB = BA$ and $AC = CA$ for these two particular matrices $B$ and $C$:

$$A = \begin{bmatrix} a & b \\ c & d \end{bmatrix} \quad \text{commutes with} \quad B = \begin{bmatrix} 1 & 0 \\ 0 & 0 \end{bmatrix} \quad \text{and} \quad C = \begin{bmatrix} 0 & 1 \\ 0 & 0 \end{bmatrix}.$$

Prove that $a = d$ and $b = c = 0$. Then $A$ is a multiple of $I$. The only matrices that commute with $B$ and $C$ and all other 2 by 2 matrices are $A =$ multiple of $I$.

**13** Which of the following matrices are guaranteed to equal $(A - B)^2$: $A^2 - B^2$, $(B - A)^2$, $A^2 - 2AB + B^2$, $A(A - B) - B(A - B)$, $A^2 - AB - BA + B^2$?

**14** True or false:

(a) If $A^2$ is defined then $A$ is necessarily square.

(b) If $AB$ and $BA$ are defined then $A$ and $B$ are square.

(c) If $AB$ and $BA$ are defined then $AB$ and $BA$ are square.

(d) If $AB = B$ then $A = I$.

**15** If $A$ is $m$ by $n$, how many separate multiplications are involved when

(a) $A$ multiplies a vector $\boldsymbol{x}$ with $n$ components?

(b) $A$ multiplies an $n$ by $p$ matrix $B$?

(c) $A$ multiplies itself to produce $A^2$? Here $m = n$ and $A$ is square.

**16** For $A = \begin{bmatrix} 2 & -1 \\ 3 & -2 \end{bmatrix}$ and $B = \begin{bmatrix} 1 & 0 & 4 \\ 1 & 0 & 6 \end{bmatrix}$, compute these answers *and nothing more*:

(a) column 2 of $AB$ (b) row 2 of $AB$ (c) row 2 of $A^2$

(d) row 2 of $A^3$.

**Problems 17–19 use $a_{ij}$ for the entry in row $i$, column $j$ of $A$.**

**17** Write down the 3 by 3 matrix $A$ whose entries are

(a) $a_{ij} =$ minimum of $i$ and $j$ (b) $a_{ij} = (-1)^{i+j}$ (c) $a_{ij} = i/j$.

**18** What words would you use to describe each of these classes of matrices? Give a 3 by 3 example in each class. Which matrix belongs to all four classes?

(a) $a_{ij} = 0$ if $i \neq j$ (b) $a_{ij} = 0$ if $i < j$ (c) $a_{ij} = a_{ji}$

(d) $a_{ij} = a_{1j}$.

**19** The entries of $A$ are $a_{ij}$. Assuming that zeros don't appear, what is

(a) the first pivot?

(b) the multiplier $\ell_{31}$ of row 1 to be subtracted from row 3?

(c) the new entry that replaces $a_{32}$ after that subtraction?

(d) the second pivot?

**Problems 20–24 involve powers of $A$.**

**20** Compute $A^2$, $A^3$, $A^4$ and also $A\boldsymbol{v}$, $A^2\boldsymbol{v}$, $A^3\boldsymbol{v}$, $A^4\boldsymbol{v}$ for

$$A = \begin{bmatrix} 0 & 2 & 0 & 0 \\ 0 & 0 & 2 & 0 \\ 0 & 0 & 0 & 2 \\ 0 & 0 & 0 & 0 \end{bmatrix} \quad \text{and} \quad \boldsymbol{v} = \begin{bmatrix} x \\ y \\ z \\ t \end{bmatrix}.$$

**21** Find all the powers $A^2, A^3, \ldots$ and $AB, (AB)^2, \ldots$ for

$$A = \begin{bmatrix} .5 & .5 \\ .5 & .5 \end{bmatrix} \quad \text{and} \quad B = \begin{bmatrix} 1 & 0 \\ 0 & -1 \end{bmatrix}.$$

**22** By trial and error find real nonzero 2 by 2 matrices such that

$$A^2 = -I \qquad BC = 0 \qquad DE = -ED \text{ (not allowing } DE = 0).$$

**23** (a) Find a nonzero matrix $A$ for which $A^2 = 0$.

(b) Find a matrix that has $A^2 \neq 0$ but $A^3 = 0$.

**24** By experiment with $n = 2$ and $n = 3$ predict $A^n$ for these matrices:

$$A_1 = \begin{bmatrix} 2 & 1 \\ 0 & 1 \end{bmatrix} \quad \text{and} \quad A_2 = \begin{bmatrix} 1 & 1 \\ 1 & 1 \end{bmatrix} \quad \text{and} \quad A_3 = \begin{bmatrix} a & b \\ 0 & 0 \end{bmatrix}.$$

**Problems 25–31 use column-row multiplication and block multiplication.**

**25** Multiply $A$ times $I$ using columns of $A$ (3 by 3) times rows of $I$.

**26** Multiply $AB$ using columns times rows:

$$AB = \begin{bmatrix} 1 & 0 \\ 2 & 4 \\ 2 & 1 \end{bmatrix} \begin{bmatrix} 3 & 3 & 0 \\ 1 & 2 & 1 \end{bmatrix} = \begin{bmatrix} 1 \\ 2 \\ 2 \end{bmatrix} \begin{bmatrix} 3 & 3 & 0 \end{bmatrix} + \underline{\qquad} = \underline{\qquad}.$$

**27** Show that the product of two upper triangular matrices is always upper triangular:

$$AB = \begin{bmatrix} x & x & x \\ 0 & x & x \\ 0 & 0 & x \end{bmatrix} \begin{bmatrix} x & x & x \\ 0 & x & x \\ 0 & 0 & x \end{bmatrix} = \begin{bmatrix} x & & \\ 0 & & \\ 0 & 0 & x \end{bmatrix}.$$

*Proof using dot products* (*Row-times-column*) (Row 2 of $A$) $\cdot$ (column 1 of $B$)$= 0$. Which other dot products give zeros?

*Proof using full matrices* (*Column-times-row*) Draw $x$'s and 0's in (column 2 of $A$) times (row 2 of $B$). Also show (column 3 of $A$) times (row 3 of $B$).

**28** If $A$ is 2 by 3 with rows 1, 1, 1 and 2, 2, 2, and $B$ is 3 by 4 with columns 1, 1, 1 and 2, 2, 2 and 3, 3, 3 and 4, 4, 4, use each of the four multiplication rules to find $AB$:

(1) Rows of $A$ times columns of $B$. **Inner products** (each entry in $AB$)

(2) Matrix $A$ times columns of $B$. **Columns of $AB$**

(3) Rows of $A$ times the matrix $B$. **Rows of $AB$**

(4) Columns of $A$ times rows of $B$. **Outer products** (3 matrices add to $AB$)

**29** Which matrices $E_{21}$ and $E_{31}$ produce zeros in the $(2,1)$ and $(3,1)$ positions of $E_{21}A$ and $E_{31}A$?

$$A = \begin{bmatrix} 2 & 1 & 0 \\ -2 & 0 & 1 \\ 8 & 5 & 3 \end{bmatrix}.$$

Find the single matrix $E = E_{31}E_{21}$ that produces both zeros at once. Multiply $EA$.

**30** **Block multiplication** produces zeros below the pivot in one big step:

$$EA = \begin{bmatrix} 1 & \mathbf{0} \\ -\boldsymbol{c}/a & I \end{bmatrix} \begin{bmatrix} a & \boldsymbol{b} \\ \boldsymbol{c} & D \end{bmatrix} = \begin{bmatrix} a & \boldsymbol{b} \\ \mathbf{0} & D - \boldsymbol{cb}/a \end{bmatrix} \text{ with vectors } \mathbf{0}, \boldsymbol{b}, \boldsymbol{c}.$$

In Problem 29, what are $\boldsymbol{c}$ and $D$ and what is the block $D - \boldsymbol{cb}/a$?

**31** With $i^2 = -1$, the product of $(A + iB)$ and $(\boldsymbol{x} + i\,\boldsymbol{y})$ is $A\boldsymbol{x} + iB\boldsymbol{x} + iA\boldsymbol{y} - B\boldsymbol{y}$. Use blocks to separate the real part without $i$ from the imaginary part that multiplies $i$:

$$\begin{bmatrix} A & -B \\ ? & ? \end{bmatrix} \begin{bmatrix} \boldsymbol{x} \\ \boldsymbol{y} \end{bmatrix} = \begin{bmatrix} A\boldsymbol{x} - B\boldsymbol{y} \\ ? \end{bmatrix} \begin{matrix} \text{real part} \\ \text{imaginary part} \end{matrix}$$

**32** (*Very important*) Suppose you solve $A\boldsymbol{v} = \boldsymbol{b}$ for three special right sides $\boldsymbol{b}$:

$$A\boldsymbol{v}_1 = \begin{bmatrix} 1 \\ 0 \\ 0 \end{bmatrix} \quad \text{and} \quad A\boldsymbol{v}_2 = \begin{bmatrix} 0 \\ 1 \\ 0 \end{bmatrix} \quad \text{and} \quad A\boldsymbol{v}_3 = \begin{bmatrix} 0 \\ 0 \\ 1 \end{bmatrix}.$$

If the three solutions $\boldsymbol{v}_1, \boldsymbol{v}_2, \boldsymbol{v}_3$ are the columns of a matrix $X$, what is $A$ times $X$?

**33** If the three solutions in Question 32 are $\boldsymbol{v}_1 = (1,1,1)$ and $\boldsymbol{v}_2 = (0,1,1)$ and $\boldsymbol{v}_3 = (0,0,1)$, solve $A\boldsymbol{v} = \boldsymbol{b}$ when $\boldsymbol{b} = (3,5,8)$. Challenge problem: What is $A$?

**34** **Practical question** Suppose $A$ is $m$ by $n$, $B$ is $n$ by $p$, and $C$ is $p$ by $q$. Then the multiplication count for $(AB)C$ is $mnp + mpq$. The same answer comes from $A$ times $BC$, now with $mnq + npq$ separate multiplications. Notice $npq$ for $BC$.

(a) If $A$ is 2 by 4, $B$ is 4 by 7, and $C$ is 7 by 10, do you prefer $(AB)C$ or $A(BC)$?

(b) With $N$-component vectors, would you choose $(\boldsymbol{u}^{\mathrm{T}}\boldsymbol{v})\boldsymbol{w}^{\mathrm{T}}$ or $\boldsymbol{u}^{\mathrm{T}}(\boldsymbol{v}\boldsymbol{w}^{\mathrm{T}})$?

(c) Divide by $mnpq$ to show that $(AB)C$ is faster when $n^{-1} + q^{-1} < m^{-1} + p^{-1}$.

**35** **Unexpected fact** A friend in England looked at powers of a $2 \times 2$ matrix:

$$A = \begin{bmatrix} 1 & 2 \\ 3 & 4 \end{bmatrix} \quad A^2 = \begin{bmatrix} 7 & 10 \\ 15 & 22 \end{bmatrix} \quad A^3 = \begin{bmatrix} 37 & 54 \\ 81 & 118 \end{bmatrix} \quad A^4 = \begin{bmatrix} A & B \\ C & D \end{bmatrix}$$

He noticed that the ratios $2/3$ and $10/15$ and $54/81$ are all the same. This is true for all powers. It doesn't work for an $n \times n$ matrix, unless $A$ is tridiagonal. One neat proof is to look at the equal $(1,1)$ entries of $A^nA$ and $AA^n$. Can you use that idea to show that $B/C = 2/3$ in this example?

## 4.4 Inverse Matrices

Suppose $A$ is a square matrix. We look for an "***inverse matrix***" $A^{-1}$ of the same size, so that $A^{-1}$ *times* $A$ *equals* $I$. Whatever $A$ does, $A^{-1}$ undoes. Their product is the identity matrix—which leaves all vectors unchanged, so $A^{-1}A\boldsymbol{v} = \boldsymbol{v}$. *But* $A^{-1}$ *might not exist.*

What a matrix mostly does is to multiply a vector $\boldsymbol{v}$. Multiplying $A\boldsymbol{v} = \boldsymbol{b}$ by $A^{-1}$ gives $A^{-1}A\boldsymbol{v} = A^{-1}\boldsymbol{b}$. ***This is*** $\boldsymbol{v} = A^{-1}\boldsymbol{b}$. The product $A^{-1}A$ is like multiplying by a number and then dividing by that number. A number has an inverse if it is not zero—matrices are more complicated and more interesting. The matrix $A^{-1}$ is called "$A$ inverse."

**DEFINITION** The matrix $A$ is ***invertible*** if there exists a matrix $A^{-1}$ such that

$$A^{-1}A = I \quad \text{and} \quad AA^{-1} = I. \tag{1}$$

***Not all matrices have inverses.*** This is the first question we ask about a square matrix: Is $A$ invertible? We don't mean that we immediately calculate $A^{-1}$. In most problems we never compute it! Here are six "notes" about $A^{-1}$.

**Note 1** $\boldsymbol{A^{-1}}$ ***exists if and only if elimination produces*** $n$ ***pivots*** (row exchanges are allowed). Elimination solves $A\boldsymbol{v} = \boldsymbol{b}$ without explicitly using the matrix $A^{-1}$.

**Note 2** The matrix $A$ cannot have two different inverses. Suppose $BA = I$ and also $AC = I$. Then $B = C$, according to this "proof by parentheses":

$$B(AC) = (BA)C \quad \text{gives} \quad BI = IC \quad \text{or} \quad B = C. \tag{2}$$

This shows that a *left-inverse* $B$ (multiplying from the left) and a *right-inverse* $C$ (multiplying $A$ from the right to give $AC = I$) must be the *same matrix*.

**Note 3** If $A$ is invertible, the one and only solution to $A\boldsymbol{v} = \boldsymbol{b}$ is $\boldsymbol{v} = A^{-1}\boldsymbol{b}$:

***Multiply*** $A\boldsymbol{v} = \boldsymbol{b}$ ***by*** $A^{-1}$. ***Then*** $\boldsymbol{v} = A^{-1}A\boldsymbol{v} = A^{-1}\boldsymbol{b}$.

**Note 4** (Important) ***Suppose there is a nonzero vector*** $\boldsymbol{v}$ ***such that*** $A\boldsymbol{v} = \boldsymbol{0}$. ***Then A cannot have an inverse***. No matrix can bring $\boldsymbol{0}$ back to $\boldsymbol{v}$.

*If A is invertible, then* $A\boldsymbol{v} = \boldsymbol{0}$ *can only have the zero solution* $\boldsymbol{v} = A^{-1}\boldsymbol{0} = \boldsymbol{0}$.

**Note 5** A 2 by 2 matrix is invertible if and only if $ad - bc$ is not zero:

**2 by 2 Inverse**
**Divide by $ad - bc$**

$$\begin{bmatrix} a & b \\ c & d \end{bmatrix}^{-1} = \frac{1}{ad - bc}\begin{bmatrix} \boldsymbol{d} & \boldsymbol{-b} \\ \boldsymbol{-c} & \boldsymbol{a} \end{bmatrix}. \tag{3}$$

This number $ad - bc$ is the **determinant** of $A$. A matrix is invertible if its determinant is not zero. $A^{-1}$ always involves a division by the determinant of $A$.

**Note 6** A diagonal matrix has an inverse provided no diagonal entries are zero:

$$\text{If} \quad A = \begin{bmatrix} d_1 & & \\ & \ddots & \\ & & d_n \end{bmatrix} \quad \text{then} \quad A^{-1} = \begin{bmatrix} 1/d_1 & & \\ & \ddots & \\ & & 1/d_n \end{bmatrix}.$$

**Example 1** The 2 by 2 matrix $A = \begin{bmatrix} 1 & 2 \\ 1 & 2 \end{bmatrix}$ is not invertible. It fails the test in Note 5, because $ad - bc$ equals $2 - 2 = 0$. It fails the test in Note 3, because $Av = \mathbf{0}$ when $v = (2, -1)$. It fails to have two pivots as required by Note 1.

Elimination turns the second row of this matrix $A$ into a zero row.

## The Inverse of a Product $AB$

For two nonzero numbers $a$ and $b$, the sum $a + b$ might or might not be invertible. The numbers $a = 3$ and $b = -3$ have inverses $\frac{1}{3}$ and $-\frac{1}{3}$. Their sum $a + b = 0$ has no inverse. But the product $ab = -9$ does have an inverse, which is $\frac{1}{3}$ times $-\frac{1}{3}$.

For two matrices $A$ and $B$, the situation is similar. It is hard to say much about the invertibility of $A + B$. But the *product* $AB$ has an inverse, if and only if the two factors $A$ and $B$ are separately invertible (and the same size). The important point is that $A^{-1}$ and $B^{-1}$ come in ***reverse order***:

If $A$ and $B$ are invertible then so is $AB$. The inverse of a product $AB$ is

$$(AB)^{-1} = B^{-1}A^{-1}. \tag{4}$$

To see why the order is reversed, multiply $AB$ times $B^{-1}A^{-1}$. Inside that is $BB^{-1} = I$:

$$\textbf{Inverse of } AB \qquad (AB)(B^{-1}A^{-1}) = AIA^{-1} = AA^{-1} = I.$$

We moved parentheses to multiply $BB^{-1}$ first. Similarly $B^{-1}A^{-1}$ times $AB$ equals $I$. This illustrates a basic rule of mathematics: Inverses come in reverse order. It is also common sense: If you put on socks and then shoes, the first to be taken off are the _____. The same reverse order applies to three or more matrices:

$$\textbf{Reverse order} \qquad (ABC)^{-1} = C^{-1}B^{-1}A^{-1}. \tag{5}$$

**Example 2** ***Inverse of an elimination matrix.*** If $E$ subtracts 5 times row 1 from row 2, then $E^{-1}$ *adds* 5 times row 1 to row 2:

$$\begin{matrix} \textbf{E subtracts} \\ \mathbf{E^{-1}} \textbf{ adds} \end{matrix} \qquad E = \begin{bmatrix} 1 & 0 & 0 \\ \mathbf{-5} & 1 & 0 \\ 0 & 0 & 1 \end{bmatrix} \quad \text{and} \quad E^{-1} = \begin{bmatrix} 1 & 0 & 0 \\ \mathbf{5} & 1 & 0 \\ 0 & 0 & 1 \end{bmatrix}.$$

Multiply $EE^{-1}$ to get the identity matrix $I$. Also multiply $E^{-1}E$ to get $I$. We are adding and subtracting the same 5 times row 1. Whether we add and then subtract (this is $EE^{-1}$) or subtract and then add (this is $E^{-1}E$), we are back at the start.

***For square matrices, an inverse on one side is automatically an inverse on the other side.***

If $AB = I$ then automatically $BA = I$ for square matrices. In that case $B$ is $A^{-1}$. This is extremely useful to know but we are not ready to prove it.

**Example 3** Suppose $F$ subtracts 4 times row 2 from row 3, and $F^{-1}$ adds it back :

$$F = \begin{bmatrix} 1 & 0 & 0 \\ 0 & 1 & 0 \\ 0 & -4 & 1 \end{bmatrix} \quad \text{and} \quad F^{-1} = \begin{bmatrix} 1 & 0 & 0 \\ 0 & 1 & 0 \\ 0 & 4 & 1 \end{bmatrix}.$$

Now multiply $F$ by the matrix $E$ in Example 2 to find $FE$. Also multiply $E^{-1}$ times $F^{-1}$ to find $(FE)^{-1}$. Notice the required order $(FE)^{-1} = E^{-1}F^{-1}$ for the inverses.

**Right order**
**Good inverse**

$$FE = \begin{bmatrix} 1 & 0 & 0 \\ -5 & 1 & 0 \\ 20 & -4 & 1 \end{bmatrix} \quad \text{and} \quad E^{-1}F^{-1} = \begin{bmatrix} 1 & 0 & 0 \\ \mathbf{5} & 1 & 0 \\ 0 & \mathbf{4} & 1 \end{bmatrix}. \tag{6}$$

The result is beautiful and correct. The product $FE$ contains "20" but its inverse doesn't. $E$ subtracts 5 times row 1 from row 2. Then $F$ subtracts 4 times the *new* row 2 (changed by row 1) from row 3. ***In this order FE, row 3 feels an effect from row* 1.**

In the order $E^{-1}F^{-1}$, that effect does not happen. First $F^{-1}$ adds 4 times row 2 to row 3. After that, $E^{-1}$ adds 5 times row 1 to row 2. There is no 20, because row 3 doesn't change again. ***In this order $E^{-1}F^{-1}$, row 3 feels no effect from row* 1.**

***$E^{-1}F^{-1}$ is quick. The multipliers* 5, 4 *fall into place below the diagonal of* 1's.**

## Calculating $A^{-1}$ by Gauss-Jordan Elimination

I hinted that $A^{-1}$ might not be explicitly needed. The equation $A\boldsymbol{v} = \boldsymbol{b}$ is solved by $\boldsymbol{v} = A^{-1}\boldsymbol{b}$. But it is not necessary or efficient to compute $A^{-1}$ and multiply it times $\boldsymbol{b}$. *Elimination goes directly to* $\boldsymbol{v}$. Elimination is also the way to find $A^{-1}$, as we now show.

**The Gauss-Jordan idea is to solve $AA^{-1} = I$. Find each column of $A^{-1}$.**

$A$ multiplies the first column of $A^{-1}$ (call that $\boldsymbol{v}_1$) to give the first column of $I$ (call that $\boldsymbol{e}_1$). This is our equation $A\boldsymbol{v}_1 = \boldsymbol{e}_1 = (1, 0, 0)$. There will be two more equations. Each of the columns $\boldsymbol{v}_1, \boldsymbol{v}_2, \boldsymbol{v}_3$ of $A^{-1}$ is multiplied by $A$ to produce a column of $I$ :

**3 columns of $A^{-1}$**

$$AA^{-1} = A\begin{bmatrix} \boldsymbol{v}_1 & \boldsymbol{v}_2 & \boldsymbol{v}_3 \end{bmatrix} = \begin{bmatrix} \boldsymbol{e}_1 & \boldsymbol{e}_2 & \boldsymbol{e}_3 \end{bmatrix} = I. \tag{7}$$

To invert a 3 by 3 matrix $A$, we have to solve three systems of equations: $A\boldsymbol{v}_1 = \boldsymbol{e}_1$ and $A\boldsymbol{v}_2 = \boldsymbol{e}_2 = (0, 1, 0)$ and $A\boldsymbol{v}_3 = \boldsymbol{e}_3 = (0, 0, 1)$. Gauss-Jordan finds $A^{-1}$ this way.

The ***Gauss-Jordan method*** computes $A^{-1}$ by solving ***all n equations together.***

Usually the "augmented matrix" $[A\ \ \boldsymbol{b}]$ has one extra column $\boldsymbol{b}$. Now we have three right sides (the columns of $I$). So the augmented matrix is the block matrix $[A\ \ I]$.

$$\begin{bmatrix} A & \boldsymbol{e}_1 & \boldsymbol{e}_2 & \boldsymbol{e}_3 \end{bmatrix} = \begin{bmatrix} 2 & -1 & 0 & 1 & 0 & 0 \\ -1 & 2 & -1 & 0 & 1 & 0 \\ 0 & -1 & 2 & 0 & 0 & 1 \end{bmatrix} \quad \textbf{Start Gauss-Jordan on } \begin{bmatrix} \boldsymbol{A} & \boldsymbol{I} \end{bmatrix}$$

$$\rightarrow \begin{bmatrix} 2 & -1 & 0 & 1 & 0 & 0 \\ \mathbf{0} & \frac{3}{2} & \mathbf{-1} & \frac{1}{2} & \mathbf{1} & \mathbf{0} \\ 0 & -1 & 2 & 0 & 0 & 1 \end{bmatrix} \quad \left(\tfrac{1}{2}\ \textbf{\textit{row}}\ \mathbf{1} + \textbf{\textit{row}}\ \mathbf{2}\right)$$

$$\rightarrow \begin{bmatrix} 2 & -1 & 0 & 1 & 0 & 0 \\ 0 & \frac{3}{2} & -1 & \frac{1}{2} & 1 & 0 \\ \mathbf{0} & \mathbf{0} & \frac{4}{3} & \frac{1}{3} & \frac{2}{3} & \mathbf{1} \end{bmatrix} \quad \left(\tfrac{2}{3}\ \textbf{\textit{row}}\ \mathbf{2} + \textbf{\textit{row}}\ \mathbf{3}\right)$$

We are halfway to $A^{-1}$. The matrix in the first three columns is $U$ (upper triangular). The pivots $2, \frac{3}{2}, \frac{4}{3}$ are on its diagonal. Gauss would finish by back substitution. *Jordan's idea is to continue with elimination*! He goes all the way to the ***identity matrix***.

Rows are subtracted from rows *above*, to produce ***zeros above the pivots***:

$$\begin{pmatrix} \text{Zero above} \\ \text{third pivot} \end{pmatrix} \quad \rightarrow \quad \begin{bmatrix} 2 & -1 & 0 & 1 & 0 & 0 \\ \mathbf{0} & \frac{3}{2} & \mathbf{0} & \frac{3}{4} & \frac{3}{2} & \frac{3}{4} \\ 0 & 0 & \frac{4}{3} & \frac{1}{3} & \frac{2}{3} & 1 \end{bmatrix} \quad \left(\tfrac{3}{4}\ \textbf{\textit{row}}\ \mathbf{3} + \textbf{\textit{row}}\ \mathbf{2}\right)$$

$$\begin{pmatrix} \text{Zero above} \\ \text{second pivot} \end{pmatrix} \quad \rightarrow \quad \begin{bmatrix} \mathbf{2} & \mathbf{0} & \mathbf{0} & \frac{3}{2} & \mathbf{1} & \frac{1}{2} \\ 0 & \frac{3}{2} & 0 & \frac{3}{4} & \frac{3}{2} & \frac{3}{4} \\ 0 & 0 & \frac{4}{3} & \frac{1}{3} & \frac{2}{3} & 1 \end{bmatrix} \quad \left(\tfrac{2}{3}\ \textbf{\textit{row}}\ \mathbf{2} + \textbf{\textit{row}}\ \mathbf{1}\right)$$

The last Gauss-Jordan step is to divide each row by its pivot. The new pivots are 1. We have reached $I$ in the first half of the matrix, because $A$ is invertible.

***The three columns of $A^{-1}$ are in the second half of $[\,I\ \ A^{-1}\,]$:***

$$\begin{matrix} \text{(divide by 2)} \\ \text{(divide by } \frac{3}{2}) \\ \text{(divide by } \frac{4}{3}) \end{matrix} \quad \begin{bmatrix} \mathbf{1} & 0 & 0 & \frac{3}{4} & \frac{1}{2} & \frac{1}{4} \\ 0 & \mathbf{1} & 0 & \frac{1}{2} & 1 & \frac{1}{2} \\ 0 & 0 & \mathbf{1} & \frac{1}{4} & \frac{1}{2} & \frac{3}{4} \end{bmatrix} = \begin{bmatrix} I & \boldsymbol{x}_1 & \boldsymbol{x}_2 & \boldsymbol{x}_3 \end{bmatrix} = \begin{bmatrix} I & A^{-1} \end{bmatrix}.$$

Starting from the 3 by 6 matrix $[\,A\ \ I\,]$, we ended with $[\,I\ \ A^{-1}\,]$. Here is the whole Gauss-Jordan process on one line for any invertible matrix $A$:

**Gauss-Jordan** ***Multiply $[\,A\ \ I\,]$ by $A^{-1}$ to get $[\,I\ \ A^{-1}\,]$.***

The elimination steps create the inverse matrix while changing $A$ to $I$. For large matrices, we probably don't want $A^{-1}$ at all. But for small matrices, it can be very worthwhile to know the inverse. We add three observations about this particular $A^{-1}$ because it is an important example. We introduce the words *symmetric*, *tridiagonal*, and *determinant*:

1. $A$ is ***symmetric*** across its main diagonal. So is $A^{-1}$.

2. $A$ is ***tridiagonal*** (only three nonzero diagonals). But $A^{-1}$ is a full matrix with no zeros. That is another reason we don't often compute inverse matrices. The inverse of a sparse matrix is generally a full matrix.

3. The *product of pivots* is $2(\frac{3}{2})(\frac{4}{3}) = 4$. This number 4 is the ***determinant*** of $A$.

$$A^{-1} \text{ involves division by the determinant} \qquad A^{-1} = \frac{1}{4}\begin{bmatrix} 3 & 2 & 1 \\ 2 & 4 & 2 \\ 1 & 2 & 3 \end{bmatrix}. \qquad (8)$$

**This is why an invertible matrix cannot have a zero determinant.**

**Example 4** Find $A^{-1}$ by Gauss-Jordan elimination starting from $A = \begin{bmatrix} 2 & 3 \\ 4 & 7 \end{bmatrix}$. There are two row operations and then a division to put 1's in the pivots:

$$\begin{bmatrix} A & I \end{bmatrix} = \begin{bmatrix} \mathbf{2} & \mathbf{3} & 1 & 0 \\ \mathbf{4} & \mathbf{7} & 0 & 1 \end{bmatrix} \rightarrow \begin{bmatrix} 2 & 3 & 1 & 0 \\ \mathbf{0} & \mathbf{1} & \mathbf{-2} & \mathbf{1} \end{bmatrix} \quad (\text{this is } \begin{bmatrix} U & L^{-1} \end{bmatrix})$$

$$\rightarrow \begin{bmatrix} \mathbf{2} & \mathbf{0} & \mathbf{7} & \mathbf{-3} \\ 0 & 1 & -2 & 1 \end{bmatrix} \rightarrow \begin{bmatrix} 1 & 0 & \mathbf{\frac{7}{2}} & \mathbf{-\frac{3}{2}} \\ 0 & 1 & \mathbf{-2} & \mathbf{1} \end{bmatrix} \quad (\text{this is } \begin{bmatrix} I & \mathbf{A^{-1}} \end{bmatrix}).$$

That $A^{-1}$ involves division by the determinant $ad - bc = 2 \cdot 7 - 3 \cdot 4 = 2$. The matrix $A$ must be invertible, or elimination cannot reduce it to $I$ (in the left half of $\begin{bmatrix} I & A^{-1} \end{bmatrix}$).

Gauss-Jordan shows why $A^{-1}$ is expensive. We must solve $n$ equations for its $n$ columns.

**To solve $A\boldsymbol{v} = \boldsymbol{b}$ without $A^{-1}$, we deal with *one* column $\boldsymbol{b}$ to find one column $\boldsymbol{v}$.**

In defense of $A^{-1}$, we want to say that its cost is not $n$ times the cost of one system. Surprisingly, the cost for $n$ columns is only multiplied by 3. This saving is because the $n$ equations $A\boldsymbol{v}_i = \boldsymbol{e}_i$ all involve the same matrix $A$. Working with the right sides is relatively cheap, because elimination only has to be done once on $A$.

**The complete $A^{-1}$ needs $n^3$ elimination steps, where one equation needs $n^3/3$.**

## Singular versus Invertible

We come back to the central question. Which matrices have inverses? The start of this section proposed the pivot test: ***$A^{-1}$ exists exactly when $A$ has a full set of $n$ pivots.*** (Row exchanges are allowed.) Now we can prove that by Gauss-Jordan elimination:

1. With $n$ pivots, elimination solves all the equations $A\boldsymbol{v}_i = \boldsymbol{e}_i$. The columns $\boldsymbol{v}_i$ go into $A^{-1}$. Then $AA^{-1} = I$ and $A^{-1}$ is at least a ***right-inverse***.

2. Elimination is really a sequence of multiplications by $E$'s and $P$'s and $D^{-1}$:

$$\textbf{Left-inverse of } A \qquad (D^{-1} \cdots E \cdots P \cdots E)A = I. \qquad (9)$$

$D^{-1}$ divides by the pivots. The matrices $E$ produce zeros below and above the pivots. Permutations $P$ will exchange rows if needed. The product matrix in equation (9) is a ***left-inverse***. With $n$ pivots we have reached $A^{-1}A = I$.

*The right-inverse equals the left-inverse.* That was Note 2 at the start of in this section. So a square matrix with a full set of pivots will always have a two-sided inverse.

Reasoning in reverse will now show that $A$ ***must have n pivots if*** $AC = I$. (Then we deduce that $C$ is also a left-inverse and $CA = I$.) Here is one route to those conclusions:

1. If $A$ doesn't have $n$ pivots, elimination will lead to a *zero row*.
2. Those elimination steps are taken by an invertible $M$. *So a row of $MA$ is zero.*
3. If $AC = I$ had been possible, then $MAC = M$. The zero row of $MA$, times $C$, gives a zero row of $M$ itself.
4. An invertible matrix $M$ can't have a zero row! So $A$ *must* have $n$ pivots if $AC = I$.

That argument took four steps, but the outcome is short and important.

Elimination gives a complete test for invertibility of a square matrix. $\boldsymbol{A^{-1}}$ ***exists (and Gauss-Jordan finds it) exactly when A has n pivots.*** The argument above shows more:

$$\text{If} \quad AC = I \quad \text{then} \quad CA = I \quad \text{and} \quad C = A^{-1}$$

**Example 5** Here $L$ is lower triangular with 1's on the diagonal. *Then $L^{-1}$ is too.*

***A triangular matrix is invertible if and only if no diagonal entries are zero.***

Here $L$ has 1's so $L^{-1}$ also has 1's. Use the Gauss-Jordan method to construct $L^{-1}$. Start by subtracting multiples of pivot rows from rows *below*. Normally this gets us halfway to the inverse, but for $L$ it gets us all the way. $L^{-1}$ appears on the right when $I$ appears on the left. Notice how $L^{-1}$ contains 11, from 3 times 5 minus 4.

**Gauss-Jordan on triangular** $L$

$$\begin{bmatrix} \mathbf{1} & \mathbf{0} & \mathbf{0} & 1 & 0 & 0 \\ \mathbf{3} & \mathbf{1} & \mathbf{0} & 0 & 1 & 0 \\ \mathbf{4} & \mathbf{5} & \mathbf{1} & 0 & 0 & 1 \end{bmatrix} = \begin{bmatrix} L & I \end{bmatrix}$$

$$\begin{matrix} \\ \to \\ \to \end{matrix} \begin{bmatrix} 1 & 0 & 0 & 1 & 0 & 0 \\ 0 & 1 & 0 & -3 & 1 & 0 \\ 0 & 5 & 1 & -4 & 0 & 1 \end{bmatrix} \quad \begin{matrix} \text{(3 times row 1 from row 2)} \\ \text{(4 times row 1 from row 3)} \\ \text{(then 5 times row 2 from row 3)} \end{matrix}$$

$$\to \begin{bmatrix} 1 & 0 & 0 & \mathbf{1} & \mathbf{0} & \mathbf{0} \\ 0 & 1 & 0 & \mathbf{-3} & \mathbf{1} & \mathbf{0} \\ 0 & 0 & 1 & \mathbf{11} & \mathbf{-5} & \mathbf{1} \end{bmatrix} = \begin{bmatrix} I & \boldsymbol{L^{-1}} \end{bmatrix}.$$

$L$ goes to $I$ by a product of elimination matrices $E_{32}E_{31}E_{21}$. So that product is $L^{-1}$. The 11 in $L^{-1}$ does not come into $L$, to spoil 3, 4, 5 in the good order $E_{21}^{-1}E_{31}^{-1}E_{32}^{-1} = L$.

## ■ REVIEW OF THE KEY IDEAS ■

1. The inverse matrix gives $AA^{-1} = I$ and $A^{-1}A = I$.

2. $A$ is invertible if and only if it has $n$ pivots (row exchanges allowed).

3. If $Av = \mathbf{0}$ for a nonzero vector $v$, then $A$ has no inverse.

4. The inverse of $AB$ is the reverse product $B^{-1}A^{-1}$. And $(ABC)^{-1} = C^{-1}B^{-1}A^{-1}$.

5. The Gauss-Jordan method solves $AA^{-1} = I$ to find the $n$ columns of $A^{-1}$. The augmented matrix $\begin{bmatrix} A & I \end{bmatrix}$ is row-reduced to $\begin{bmatrix} I & A^{-1} \end{bmatrix}$.

## Problem Set 4.4

**1** Find the inverses of $A, B, C$ (directly or from the 2 by 2 formula):

$$A = \begin{bmatrix} 0 & 3 \\ 4 & 0 \end{bmatrix} \quad \text{and} \quad B = \begin{bmatrix} 2 & 0 \\ 4 & 2 \end{bmatrix} \quad \text{and} \quad C = \begin{bmatrix} 3 & 4 \\ 5 & 7 \end{bmatrix}.$$

**2** For these "permutation matrices" find $P^{-1}$ by trial and error (with 1's and 0's):

$$P = \begin{bmatrix} 0 & 0 & 1 \\ 0 & 1 & 0 \\ 1 & 0 & 0 \end{bmatrix} \quad \text{and} \quad P = \begin{bmatrix} 0 & 1 & 0 \\ 0 & 0 & 1 \\ 1 & 0 & 0 \end{bmatrix}.$$

**3** Solve for the first column $(x, y)$ and second column $(t, z)$ of $A^{-1}$:

$$\begin{bmatrix} 10 & 20 \\ 20 & 50 \end{bmatrix}\begin{bmatrix} x \\ y \end{bmatrix} = \begin{bmatrix} 1 \\ 0 \end{bmatrix} \quad \text{and} \quad \begin{bmatrix} 10 & 20 \\ 20 & 50 \end{bmatrix}\begin{bmatrix} t \\ z \end{bmatrix} = \begin{bmatrix} 0 \\ 1 \end{bmatrix}.$$

**4** Show that $\left[\begin{smallmatrix} 1 & 2 \\ 3 & 6 \end{smallmatrix}\right]$ is not invertible by trying to solve $AA^{-1} = I$ for column 1 of $A^{-1}$:

$$\begin{bmatrix} 1 & 2 \\ 3 & 6 \end{bmatrix}\begin{bmatrix} x \\ y \end{bmatrix} = \begin{bmatrix} 1 \\ 0 \end{bmatrix}$$

(*For a different* $A$, could column 1 of $A^{-1}$ be possible to find but not column 2?)

**5** Find an upper triangular $U$ (not diagonal) with $U^2 = I$ which gives $U = U^{-1}$.

**6** (a) If $A$ is invertible and $AB = AC$, prove quickly that $B = C$.

(b) If $A = \left[\begin{smallmatrix} 1 & 1 \\ 1 & 1 \end{smallmatrix}\right]$, find two different matrices such that $AB = AC$.

**7** (Important) If $A$ has row 1 + row 2 = row 3, show that $A$ is not invertible:

(a) Explain why $A\boldsymbol{v} = (1, 0, 0)$ cannot have a solution.

(b) Which right sides $(b_1, b_2, b_3)$ might allow a solution to $A\boldsymbol{v} = \boldsymbol{b}$?

(c) What happens to row 3 in elimination?

**8** If $A$ has column 1 + column 2 = column 3, show that $A$ is not invertible:

(a) Find a nonzero solution $\boldsymbol{x}$ to $A\boldsymbol{x} = \mathbf{0}$. The matrix is 3 by 3.

(b) Elimination keeps column 1 + column 2 = column 3. Why is no third pivot?

**9** Suppose $A$ is invertible and you exchange its first two rows to reach $B$. Is the new matrix $B$ invertible and how would you find $B^{-1}$ from $A^{-1}$?

**10** Find the inverses (in any legal way) of

$$A = \begin{bmatrix} 0 & 0 & 0 & 2 \\ 0 & 0 & 3 & 0 \\ 0 & 4 & 0 & 0 \\ 5 & 0 & 0 & 0 \end{bmatrix} \quad \text{and} \quad B = \begin{bmatrix} 3 & 2 & 0 & 0 \\ 4 & 3 & 0 & 0 \\ 0 & 0 & 6 & 5 \\ 0 & 0 & 7 & 6 \end{bmatrix}.$$

**11** (a) Find invertible matrices $A$ and $B$ such that $A + B$ is not invertible.

(b) Find singular matrices $A$ and $B$ such that $A + B$ is invertible.

**12** If the product $C = AB$ is invertible ($A$ and $B$ are square), then $A$ itself is invertible. Find a formula for $A^{-1}$ that involves $C^{-1}$ and $B$.

**13** If the product $M = ABC$ of three square matrices is invertible, then $B$ is invertible. (So are $A$ and $C$.) Find a formula for $B^{-1}$ that involves $M^{-1}$ and $A$ and $C$.

**14** If you add row 1 of $A$ to row 2 to get $B$, how do you find $B^{-1}$ from $A^{-1}$?

$$\text{Notice the order.} \quad \text{The inverse of} \quad B = \begin{bmatrix} 1 & 0 \\ 1 & 1 \end{bmatrix} A \quad \text{is} \quad \underline{\qquad}.$$

**15** Prove that a matrix with a column of zeros cannot have an inverse.

**16** Multiply $\begin{bmatrix} a & b \\ c & d \end{bmatrix}$ times $\begin{bmatrix} d & -b \\ -c & a \end{bmatrix}$. What is the inverse of each matrix if $ad \neq bc$?

**17** (a) What 3 by 3 matrix $E$ has the same effect as these three steps? Subtract row 1 from row 2, subtract row 1 from row 3, then subtract row 2 from row 3.

(b) What single matrix $L$ has the same effect as these three reverse steps? Add row 2 to row 3, add row 1 to row 3, then add row 1 to row 2.

**18** If $B$ is the inverse of $A^2$, show that $AB$ is the inverse of $A$.

**19** (Recommended) $A$ is a 4 by 4 matrix with 1's on the diagonal and $-a, -b, -c$ on the diagonal above. Find $A^{-1}$ for this bidiagonal matrix.

**20** Find the numbers $a$ and $b$ that give the inverse of 5 ∗ eye(4) − ones(4,4) :

$$[5I-\text{ones}]^{-1} = \begin{bmatrix} 4 & -1 & -1 & -1 \\ -1 & 4 & -1 & -1 \\ -1 & -1 & 4 & -1 \\ -1 & -1 & -1 & 4 \end{bmatrix}^{-1} = \begin{bmatrix} a & b & b & b \\ b & a & b & b \\ b & b & a & b \\ b & b & b & a \end{bmatrix}.$$

What are $a$ and $b$ in the inverse of 6 ∗ eye(5) − ones(5,5) ? In MATLAB, $I =$ eye.

**21** Sixteen 2 by 2 matrices contain only 1's and 0's. How many of them are invertible?

**Questions 22–28 are about the Gauss-Jordan method for calculating $A^{-1}$.**

**22** Change $I$ into $A^{-1}$ as you reduce $A$ to $I$ (by row operations):

$$\begin{bmatrix} A & I \end{bmatrix} = \begin{bmatrix} 1 & 3 & 1 & 0 \\ 2 & 7 & 0 & 1 \end{bmatrix} \quad \text{and} \quad \begin{bmatrix} A & I \end{bmatrix} = \begin{bmatrix} 1 & 4 & 1 & 0 \\ 3 & 9 & 0 & 1 \end{bmatrix}$$

**23** Follow the 3 by 3 text example of Gauss-Jordan but with all plus signs in $A$. Eliminate above and below the pivots to reduce $[\,A \;\; I\,]$ to $[\,I \;\; A^{-1}\,]$:

$$\begin{bmatrix} A & I \end{bmatrix} = \begin{bmatrix} 2 & 1 & 0 & 1 & 0 & 0 \\ 1 & 2 & 1 & 0 & 1 & 0 \\ 0 & 1 & 2 & 0 & 0 & 1 \end{bmatrix}.$$

**24** Use Gauss-Jordan elimination on $[\,U \;\; I\,]$ to find the upper triangular $U^{-1}$:

$$\boldsymbol{UU^{-1} = I} \qquad \begin{bmatrix} 1 & a & b \\ 0 & 1 & c \\ 0 & 0 & 1 \end{bmatrix} \begin{bmatrix} & & \\ \boldsymbol{x}_1 & \boldsymbol{x}_2 & \boldsymbol{x}_3 \\ & & \end{bmatrix} = \begin{bmatrix} 1 & 0 & 0 \\ 0 & 1 & 0 \\ 0 & 0 & 1 \end{bmatrix}.$$

**25** Find $A^{-1}$ and $B^{-1}$ (*if they exist*) by elimination on $[\,A \;\; I\,]$ and $[\,B \;\; I\,]$:

$$A = \begin{bmatrix} 2 & 1 & 1 \\ 1 & 2 & 1 \\ 1 & 1 & 2 \end{bmatrix} \quad \text{and} \quad B = \begin{bmatrix} 2 & -1 & -1 \\ -1 & 2 & -1 \\ -1 & -1 & 2 \end{bmatrix}.$$

**26** What three matrices $E_{21}$ and $E_{12}$ and $D^{-1}$ reduce $A = \begin{bmatrix} \mathbf{1} & \mathbf{2} \\ \mathbf{2} & \mathbf{6} \end{bmatrix}$ to the identity matrix? Multiply $D^{-1}E_{12}E_{21}$ to find $A^{-1}$.

**27** Invert these matrices $A$ by the Gauss-Jordan method starting with $[\,A \quad I\,]$:

$$A = \begin{bmatrix} 1 & 0 & 0 \\ 2 & 1 & 3 \\ 0 & 0 & 1 \end{bmatrix} \quad \text{and} \quad A = \begin{bmatrix} 1 & 1 & 1 \\ 1 & 2 & 2 \\ 1 & 2 & 3 \end{bmatrix}.$$

**28** Exchange rows and continue with Gauss-Jordan to find $A^{-1}$:

$$[\,A \quad I\,] = \begin{bmatrix} 0 & 2 & 1 & 0 \\ 2 & 2 & 0 & 1 \end{bmatrix}.$$

**29** True or false (with a counterexample if false and a reason if true):

(a) A 4 by 4 matrix with a row of zeros is not invertible.

(b) Every matrix with 1's down the main diagonal is invertible.

(c) If $A$ is invertible then $A^{-1}$ and $A^2$ are invertible.

**30** For which three numbers $c$ is this matrix not invertible, and why not?

$$A = \begin{bmatrix} 2 & c & c \\ c & c & c \\ 8 & 7 & c \end{bmatrix}.$$

**31** Prove that $A$ is invertible if $a \neq 0$ and $a \neq b$ (find the pivots or $A^{-1}$):

$$A = \begin{bmatrix} a & b & b \\ a & a & b \\ a & a & a \end{bmatrix}.$$

**32** This matrix has a remarkable inverse. Find $A^{-1}$ by elimination on $[\,A \quad I\,]$. Extend to a 5 by 5 "alternating matrix" and guess its inverse; then multiply to confirm.

$$\text{Invert } A = \begin{bmatrix} 1 & -1 & 1 & -1 \\ 0 & 1 & -1 & 1 \\ 0 & 0 & 1 & -1 \\ 0 & 0 & 0 & 1 \end{bmatrix} \text{ and solve } A\boldsymbol{v} = \begin{bmatrix} 1 \\ 1 \\ 1 \\ 1 \end{bmatrix}.$$

**33** **(Puzzle)** Could a 4 by 4 matrix $A$ be invertible if every row contains the numbers $0, 1, 2, 3$ in some order? What if every row of $B$ contains $0, 1, 2, -3$ in some order?

**34** Find and check the inverses (assuming they exist) of these block matrices:

$$\begin{bmatrix} I & 0 \\ C & I \end{bmatrix} \quad \begin{bmatrix} A & 0 \\ C & D \end{bmatrix} \quad \begin{bmatrix} 0 & I \\ I & D \end{bmatrix}.$$

## 4.5 Symmetric Matrices and Orthogonal Matrices

This section introduces the **transpose** of a matrix. Start with any $m$ by $n$ matrix $A$. Then the rows of $A$ become the columns of $A^{\mathrm{T}}$ (called "*A transpose*"). The columns of $A$ are the rows of $A^{\mathrm{T}}$. The $m$ by $n$ matrix is flipped across its main diagonal. **Then $A^{\mathrm{T}}$ is $n$ by $m$.**

$$\textbf{Transpose} \qquad \text{If} \quad A = \begin{bmatrix} \mathbf{1} & \mathbf{2} & \mathbf{6} \\ 0 & 0 & 5 \end{bmatrix} \quad \text{then} \quad A^{\mathrm{T}} = \begin{bmatrix} \mathbf{1} & 0 \\ \mathbf{2} & 0 \\ \mathbf{6} & 5 \end{bmatrix}.$$

The entry in row $i$, column $j$ of $A^{\mathrm{T}}$ comes from row $j$, column $i$ of $A$. So $\boldsymbol{(A^{\mathrm{T}})_{ij} = A_{ji}}$.

The transpose of a lower triangular matrix is upper triangular. Two key rules:

| | | |
|---|---|---|
| **Products $AB$** | The transpose of $AB$ is $\boldsymbol{(AB)^{\mathrm{T}} = B^{\mathrm{T}}A^{\mathrm{T}}}$ | (1) |
| **Inverses $A^{-1}$** | The transpose of $A^{-1}$ is $\boldsymbol{(A^{-1})^{\mathrm{T}} = (A^{\mathrm{T}})^{-1}}$. | (2) |

Notice especially how $B^{\mathrm{T}}A^{\mathrm{T}}$ comes in reverse order. For inverses, this reverse order is quick to check: $B^{-1}A^{-1}$ times $AB$ produces $B^{-1}(A^{-1}A)B = I$. For transposes, rules (1) and (2) are tested and explained in the problem set. We want to move to the essential matrices of this section because they are the most important matrices in mathematics:

| | |
|---|---|
| ***Symmetric matrices*** | **$A^{\mathrm{T}}$ equals $A$**. Then $A$ is square and $a_{ij} = a_{ji}$. |
| ***Orthogonal matrices*** | **$A^{\mathrm{T}}$ equals $A^{-1}$**. Then $A$ is square and $A^{\mathrm{T}}A = I$. |

Here is a symmetric example $S$ and also an orthogonal example $Q$:

$$\textbf{Symmetric } S = \begin{bmatrix} 1 & \mathbf{4} \\ \mathbf{4} & 6 \end{bmatrix} \qquad \textbf{Orthogonal } Q = \begin{bmatrix} \cos\theta & -\sin\theta \\ \sin\theta & \cos\theta \end{bmatrix}$$

Symmetry of $S$ is easy to see: $\mathbf{4} = \mathbf{4}$. For orthogonality I will check that $Q^{\mathrm{T}}Q = I$:

$$\begin{array}{l}\textbf{Columns are orthogonal} \\ \textbf{Columns are unit vectors}\end{array} \quad \begin{bmatrix} \cos\theta & \sin\theta \\ -\sin\theta & \cos\theta \end{bmatrix} \begin{bmatrix} \cos\theta & -\sin\theta \\ \sin\theta & \cos\theta \end{bmatrix} = \begin{bmatrix} 1 & 0 \\ 0 & 1 \end{bmatrix}. \tag{3}$$

Those words at the left tell you the key facts about the columns $\boldsymbol{q}_1$ and $\boldsymbol{q}_2$:

$$\boldsymbol{Q^{\mathrm{T}}Q = I} \qquad \begin{bmatrix} \boldsymbol{q}_1^{\mathrm{T}} \\ \boldsymbol{q}_2^{\mathrm{T}} \end{bmatrix} \begin{bmatrix} \boldsymbol{q}_1 & \boldsymbol{q}_2 \end{bmatrix} = \begin{bmatrix} \boldsymbol{q}_1^{\mathrm{T}}\boldsymbol{q}_1 & \boldsymbol{q}_1^{\mathrm{T}}\boldsymbol{q}_2 \\ \boldsymbol{q}_2^{\mathrm{T}}\boldsymbol{q}_1 & \boldsymbol{q}_2^{\mathrm{T}}\boldsymbol{q}_2 \end{bmatrix} = \begin{bmatrix} 1 & 0 \\ 0 & 1 \end{bmatrix}. \tag{4}$$

Off the diagonal you see $\boldsymbol{q}_1^{\mathrm{T}}\boldsymbol{q}_2 = 0$ and $\boldsymbol{q}_2^{\mathrm{T}}\boldsymbol{q}_1 = 0$. The columns are orthogonal vectors. On the diagonal $\boldsymbol{q}_1^{\mathrm{T}}\boldsymbol{q}_1 = 1$ and $\boldsymbol{q}_2^{\mathrm{T}}\boldsymbol{q}_2 = 1$. The $\boldsymbol{q}$'s are unit column vectors: **length 1**.

Symmetric matrices will have the special letter $S$ and orthogonal matrices will be $Q$.

## Symmetric Matrices $S = A^{\mathrm{T}}A$

The full glory of symmetric matrices comes with their eigenvalues $\lambda$ and eigenvectors $\boldsymbol{x}$. Those strange words, half German and half English, are at the heart of Chapter 6. You will see the key equation $A\boldsymbol{x} = \lambda\boldsymbol{x}$ (this puts $A\boldsymbol{x}$ in the same direction as $\boldsymbol{x}$). Let me write here only two facts that show why symmetric matrices are special:

$S\boldsymbol{x} = \lambda\boldsymbol{x}$ Symmetric matrices have **real eigenvalues $\lambda$ and orthogonal eigenvectors $\boldsymbol{x}$**.

Those facts will be crucial in solving symmetric systems $\boldsymbol{y}' = S\boldsymbol{y}$ and $\boldsymbol{y}'' + S\boldsymbol{y} = \boldsymbol{0}$.

*It is equally important to know where symmetric matrices come from.* One part of applied mathematics and engineering mathematics is solving equations. We have solved $A\boldsymbol{v} = \boldsymbol{b}$ and we will soon solve $d\boldsymbol{y}/dt = A\boldsymbol{y}$. Solving is one half of our subject, the other half is discovering the equations in the first place.

Start with a physical or biological or economic problem. Model it by equations. Solving $F = ma$ and $e = mc^2$ may take thought, but we give first place to Newton and Einstein for *discovering* those equations.

To repeat: Where do symmetric matrices come from? In my experience, you start with a matrix $A$. Often this matrix is rectangular ($m$ by $n$). Its transpose is also rectangular ($A^{\mathrm{T}}$ is $n$ by $m$). Sooner or later, you are almost sure to see the matrix $A^{\mathrm{T}}A$. At that moment you have a square symmetric $n$ by $n$ matrix:

**$S = A^{\mathrm{T}}A$ is always symmetric. Its transpose is $S^{\mathrm{T}} = (A^{\mathrm{T}}A)^{\mathrm{T}} = A^{\mathrm{T}}A^{\mathrm{TT}} = S$.** (5)

This matrix $A^{\mathrm{T}}A$ is automatically square, because ($n$ by $m$) times ($m$ by $n$) is ($n$ by $n$).

**Example 1**
$$A^{\mathrm{T}}A = \begin{bmatrix} 1 & 1 & 3 \\ 0 & 0 & 4 \end{bmatrix}\begin{bmatrix} 1 & 0 \\ 1 & 0 \\ 3 & 4 \end{bmatrix} = \begin{bmatrix} 11 & \mathbf{12} \\ \mathbf{12} & 16 \end{bmatrix}.$$

The number 12 comes *twice* in $A^{\mathrm{T}}A$. It is (row 1 of $A^{\mathrm{T}}$) $\cdot$ (column 2 of $A$) and also (row 2 of $A^{\mathrm{T}}$) $\cdot$ (column 1 of $A$). The numbers 11 and 16 on the diagonal are dot products of a column with itself. So they give the *length squared* of the columns. These diagonal entries of $A^{\mathrm{T}}A$ cannot be negative.

*Comment.* Since $A$ is 3 by 2, the system $A\boldsymbol{v} = \boldsymbol{b}$ has three equations but only two unknowns $v_1$ and $v_2$. Almost surely there will be no solution. But if those numbers $b_1, b_2, b_3$ came from careful and expensive measurements, we cannot say "no solution" and stop. We want to find the "best solution" or "closest solution" to $A\boldsymbol{v} = \boldsymbol{b}$.

In practice we usually choose the vector $\widehat{\boldsymbol{v}}$ that makes $A\widehat{\boldsymbol{v}}$ as close as possible to $\boldsymbol{b}$. The error vector $\boldsymbol{e} = \boldsymbol{b} - A\widehat{\boldsymbol{v}}$ is as short as possible. We are minimizing $||\boldsymbol{e}||^2 = \boldsymbol{e}^{\mathrm{T}}\boldsymbol{e}$, the squared length of the error. The best vector $\widehat{\boldsymbol{v}}$ is the **least squares solution**.

In Section 7.1, minimizing the error is a calculus problem and also a linear algebra problem. Both approaches lead to the equation $A^{\mathrm{T}}A\widehat{\boldsymbol{v}} = A^{\mathrm{T}}\boldsymbol{b}$. The best $\widehat{\boldsymbol{v}}$ involves $A^{\mathrm{T}}A$.

## Difference Matrices

I want to show you larger examples of $A^{\mathrm{T}}A$ that are truly important. Start with a *backward difference matrix* $A$. It can have $n+1$ rows and $n$ columns. Here $n=3$:

$$\textbf{Difference matrix} \quad \textbf{Differences of } \boldsymbol{v}\textbf{'s} \qquad A = \begin{bmatrix} 1 & & \\ -1 & 1 & \\ & -1 & 1 \\ & & -1 \end{bmatrix} \qquad A\boldsymbol{v} = \begin{bmatrix} v_1 \\ v_2 - v_1 \\ v_3 - v_2 \\ - v_3 \end{bmatrix} \tag{6}$$

That vector $A\boldsymbol{v}$ in linear algebra corresponds to the derivative $dv/dx$ in calculus. You see backward differences $\Delta v = \left[\, v(x) - v(x-\Delta x) \,\right]/\Delta x$ in calculus. This is before the stepsize $\Delta x$ approaches zero and $\Delta v/\Delta x$ approaches $dv/dx$.

More often you see forward differences $\left[\, v(x+\Delta x) - v(x) \,\right]/\Delta x$, where the small $\Delta x$ goes forward from $x$. Those appear in linear algebra when we transpose the matrix $A$. But first differences are "anti-symmetric" and $A^{\mathrm{T}}$ will be *minus* a forward difference. So the vector $A^{\mathrm{T}}\boldsymbol{w}$ corresponds to the derivative $-dw/dx$:

$$\textbf{3 by 4 matrix} \quad \textbf{Differences of } \boldsymbol{w}\textbf{'s} \qquad A^{\mathrm{T}} = \begin{bmatrix} 1 & -1 & & \\ & 1 & -1 & \\ & & 1 & -1 \end{bmatrix} \qquad A^{\mathrm{T}}\boldsymbol{w} = \begin{bmatrix} w_1 - w_2 \\ w_2 - w_3 \\ w_3 - w_4 \end{bmatrix} \tag{7}$$

Now comes the symmetric matrix $S = A^{\mathrm{T}}A$. It will be 3 by 3. Since $A$ and $A^{\mathrm{T}}$ are "first differences" with 1 and $-1$, $A^{\mathrm{T}}A$ will be a **second difference matrix** with $-1$, 2, $-1$:

$$\textbf{Second differences} \quad S = \begin{bmatrix} 2 & -1 & 0 \\ -1 & 2 & -1 \\ 0 & -1 & 2 \end{bmatrix} \qquad S\boldsymbol{v} = \begin{bmatrix} 2v_1 - v_2 \\ -v_1 + 2v_2 - v_3 \\ - v_2 + 2v_3 \end{bmatrix} \tag{8}$$

The main diagonal of $S$ has 2's, because each column of $A$ produces $1^2 + (-1)^2 = 2$. The subdiagonal and superdiagonal of $S$ have $-1$'s, because this is the dot product of a column of $A$ with the next column.

Let me admit quietly that $S$ is my favorite matrix. You are seeing the 3 by 3 version, what I really like is $n$ by $n$. Chapter 7 makes the link with calculus, where the first derivative of the first derivative is the *second derivative*:

$$\boldsymbol{S v} \textbf{ corresponds to } -\frac{d^2 v}{dx^2} \qquad \frac{v(x+\Delta x) - 2v(x) + v(x-\Delta x)}{(\Delta x)^2} \approx \frac{d^2 v}{dx^2}. \tag{9}$$

All of Chapter 2 was about second order equations involving $y''$. Newton's Law $F = ma$ puts second derivatives (the acceleration $a$) at the heart of physics. When springs oscillate, and when current goes through a network, this matrix $S = A^{\mathrm{T}}A$ will appear.

The truth is that we need to know everything about $S$—its pivots, its determinant, its inverse, its eigenvalues, its eigenvectors. We will.

The matrix $L = AA^{\mathrm{T}}$ is almost as important. Please recognize that $L$ is also symmetric, but $L$ is different from $S$. When $A$ has $n$ columns and $n+1$ rows, $S = A^{\mathrm{T}}A$ is $n$ by $n$. But $L = AA^{\mathrm{T}}$ is square of size $n+1$. We keep $n = 3$ and $n+1 = 4$:

**Second differences in $L$**
**New boundary conditions**

$$L = AA^{\mathrm{T}} = \begin{bmatrix} 1 & -1 & & \\ -1 & 2 & -1 & \\ & -1 & 2 & -1 \\ & & -1 & 1 \end{bmatrix} \qquad (10)$$

This matrix has no inverse! Can you see a vector $\boldsymbol{w}$ that has $L\boldsymbol{w} = \mathbf{0}$? It is the vector of all ones, $\boldsymbol{w} = (1,1,1,1)$. Each row of $L$ adds to zero and that will produce $L\boldsymbol{w} = \mathbf{0}$.

## Permutation Matrices

A quick way to produce orthogonal matrices is to use the columns of the identity matrix. In any order, the columns of $I$ are orthonormal. The new order is called a "permutation" of the original order. So the new matrix is called a **permutation matrix**.

Important: We could put the *rows* of $I$ into the new order. That also produces a permutation matrix. If this row exchange matrix is $P$, then the column exchange matrix is $P^{\mathrm{T}}$. You can see the transpose in this 3 by 3 example starting from $I$:

**Rows** in the order 2, 3, 1 $\quad P = \begin{bmatrix} 0 & \mathbf{1} & 0 \\ 0 & 0 & \mathbf{1} \\ \mathbf{1} & 0 & 0 \end{bmatrix}$ $\qquad$ **Columns** in order 2, 3, 1 $\quad P^{\mathrm{T}} = \begin{bmatrix} 0 & 0 & \mathbf{1} \\ \mathbf{1} & 0 & 0 \\ 0 & \mathbf{1} & 0 \end{bmatrix}$. $\qquad$ (11)

When $P$ multiplies a vector $\boldsymbol{v}$, it puts the components of $\boldsymbol{v}$ in the new order $y, z, x$. Then $P^{\mathrm{T}}$ puts them back in the original order $x, y, z$:

$$P\begin{bmatrix} x \\ y \\ z \end{bmatrix} = \begin{bmatrix} y \\ z \\ x \end{bmatrix} \quad \text{and} \quad P^{\mathrm{T}}\begin{bmatrix} y \\ z \\ x \end{bmatrix} = \begin{bmatrix} x \\ y \\ z \end{bmatrix}.$$

These are orthogonal matrices, so $P^{-1}$ is the same as $P^{\mathrm{T}}$. Then $P^{\mathrm{T}}P = PP^{\mathrm{T}} = I$.

We can complete the list of all 3 by 3 permutation matrices (including the identity matrix itself, which exchanges nothing: the identity permutation). The other permutations exchange two rows or two columns of $I$. There are $P$ and $P^{\mathrm{T}}$ in (11), and four more.

$$I = \begin{bmatrix} 1 & & \\ & 1 & \\ & & 1 \end{bmatrix}, \; P_{12} = \begin{bmatrix} 0 & 1 & 0 \\ 1 & 0 & 0 \\ 0 & 0 & 1 \end{bmatrix}, \; P_{13} = \begin{bmatrix} 0 & 0 & 1 \\ 0 & 1 & 0 \\ 1 & 0 & 0 \end{bmatrix}, \; P_{23} = \begin{bmatrix} 1 & 0 & 0 \\ 0 & 0 & 1 \\ 0 & 1 & 0 \end{bmatrix}.$$

Altogether 6 permutation matrices when $n = 3$. And $n!$ permutation matrices of size $n$.

The effect of $P_{12}$ is to exchange (*permute*) rows 1 and 2, when we multiply $P_{12}A$ or $P_{12}\boldsymbol{b}$.

$$P_{12}\begin{bmatrix}\text{row 1 of } A\\ \text{row 2 of } A\\ \text{row 3 of } A\end{bmatrix}=\begin{bmatrix}\textbf{row 2 of } \boldsymbol{A}\\ \textbf{row 1 of } \boldsymbol{A}\\ \text{row 3 of } A\end{bmatrix} \qquad P_{12}\begin{bmatrix}b_1\\ b_2\\ b_3\end{bmatrix}=\begin{bmatrix}\boldsymbol{b_2}\\ \boldsymbol{b_1}\\ b_3\end{bmatrix}.$$

This is exactly what we do in elimination, when a zero appears in the first pivot position. If $a_{11}=0$ and $a_{21}\neq 0$, $P_{12}$ exchanges rows to produce a nonzero pivot.

**Elimination by matrices** *Eliminate by $E_{ij}$, exchange rows by $P_{jk}$.*

The elimination matrix $E_{ij}$ subtracts a multiple $\ell_{ij}$ of row $j$ from a lower row $i>j$. Before that, a permutation matrix $P_{jk}$ may put row $k$ into row $j$, to produce a better number (a larger number) in the pivot position.

We *must* use $P_{jk}$ to get a nonzero pivot. We *may* use $P_{jk}$ to get a larger pivot. The LAPACK code (open source) chooses the largest available number as the pivot. The $j$th pivot (in column $j$) will be the largest number in row $j$ or below. LAPACK is the foundation for the linear algebra part of many important software systems, including MATLAB.

## Orthogonal Matrices

When $A$ has orthogonal columns, the symmetric matrix $A^{\mathrm{T}}A$ is *diagonal*. The off-diagonal entries are dot products of different columns of $A$, so they are all zero.

When the columns of $A$ are *unit vectors* (length 1), all diagonal entries of $A^{\mathrm{T}}A$ are 1. Those entries are (row $i$ of $A^{\mathrm{T}}$) $\cdot$ (column $i$ of $A$) $=$ length squared $=1$. Dot products of columns with themselves are on the main diagonal of $A^{\mathrm{T}}A$.

The best case is **orthonormal columns**. Those are orthogonal unit vectors, both properties at the same time. In this case we write $\boldsymbol{q}$ for the vectors and $Q$ for the matrix:

$$\begin{array}{ll}\textbf{Orthogonal} & \boldsymbol{q}_i^{\mathrm{T}}\boldsymbol{q}_j=\mathbf{0}\\ \textbf{Unit vectors} & \boldsymbol{q}_i^{\mathrm{T}}\boldsymbol{q}_i=\mathbf{1}\end{array}\qquad Q^{\mathrm{T}}Q=\begin{bmatrix}\boldsymbol{q}_1^{\mathrm{T}}\\ \cdots\\ \boldsymbol{q}_n^{\mathrm{T}}\end{bmatrix}\begin{bmatrix}\boldsymbol{q}_1\cdots\boldsymbol{q}_n\end{bmatrix}=\begin{bmatrix}1&0&0\\0&1&0\\0&0&1\end{bmatrix}. \tag{12}$$

When $Q$ is square, I call it an **orthogonal matrix**. (The name "orthonormal matrix" might have been better.) I still use the letter $Q$ when the matrix is rectangular, with $m>n$. But a rectangular $Q^{\mathrm{T}}$ is only a ***left–inverse*** of $Q$:

$$(\boldsymbol{m}=\boldsymbol{n})\quad Q^{\mathrm{T}}Q=QQ^{\mathrm{T}}=I \qquad (\boldsymbol{m}>\boldsymbol{n})\quad Q^{\mathrm{T}}Q=I \text{ but } \boldsymbol{QQ^{\mathrm{T}}\neq I}. \tag{13}$$

$Q^{\mathrm{T}}Q=I$ is a very powerful property. When we multiply any vector by $Q$, its length will not change:

$$\textbf{Same length} \qquad \boldsymbol{||Qv||=||v||} \textbf{ for every vector } v. \tag{14}$$

The proof comes directly from $||Q\boldsymbol{v}||^2=(Q\boldsymbol{v})^{\mathrm{T}}(Q\boldsymbol{v})=\boldsymbol{v}^{\mathrm{T}}Q^{\mathrm{T}}Q\boldsymbol{v}$. The matrix $Q^{\mathrm{T}}Q$ is the identity. So we are left with $\boldsymbol{v}^{\mathrm{T}}\boldsymbol{v}=||\boldsymbol{v}||^2$.

The fact that lengths don't change makes orthogonal matrices very safe to compute with. Nothing blows up, nothing becomes too small (no overflow and no underflow). The basic computation in linear algebra is the solution of a linear system, and for (square) orthogonal matrices this is incredibly easy:

$$\boldsymbol{Q^{-1} = Q^{\mathrm{T}}} \qquad \text{The solution of } Q\boldsymbol{v} = \boldsymbol{b} \text{ is } \boldsymbol{v} = Q^{\mathrm{T}}\boldsymbol{b}. \tag{15}$$

To solve the equations, we just transpose the matrix. The greatest example is the **Fourier matrix**, which breaks up a signal $\boldsymbol{b}$ into separate pure frequencies. The vector $\boldsymbol{b}$ in the time domain is transformed to $\boldsymbol{v}$ in the frequency domain. The "energy" can be measured in either domain, because $||\boldsymbol{b}||^2$ is equal to $||\boldsymbol{v}||^2$—as we saw above.

The Fourier matrix $F$ is exceptional because multiplications by $F$ and $F^{-1}$ are extremely fast. They break up into diagonal matrices and permutation matrices. This is the insight behind the Fast Fourier Transform. (*The FFT is in Section 8.2.*)

The equation $Q\boldsymbol{v} = \boldsymbol{b}$ has a clear geometrical meaning when $Q$ is 2 by 2. $Q\boldsymbol{v}$ is expressing that vector $\boldsymbol{b}$ as a combination of the columns of $Q$. *Those columns $\boldsymbol{q}_1, \boldsymbol{q}_2$ give the perpendicular axes in* Figure 4.9. We are finding the component of $\boldsymbol{b}$ in each direction.

Those two components are $\boldsymbol{v}_1 = \boldsymbol{q}_1 \cdot \boldsymbol{b}$ and $\boldsymbol{v}_2 = \boldsymbol{q}_2 \cdot \boldsymbol{b}$. Solving $Q\boldsymbol{v} = \boldsymbol{b}$ by $\boldsymbol{v} = Q^{\mathrm{T}}\boldsymbol{b}$ is just a change from $x$, $y$ axes to $q_1, q_2$ axes.

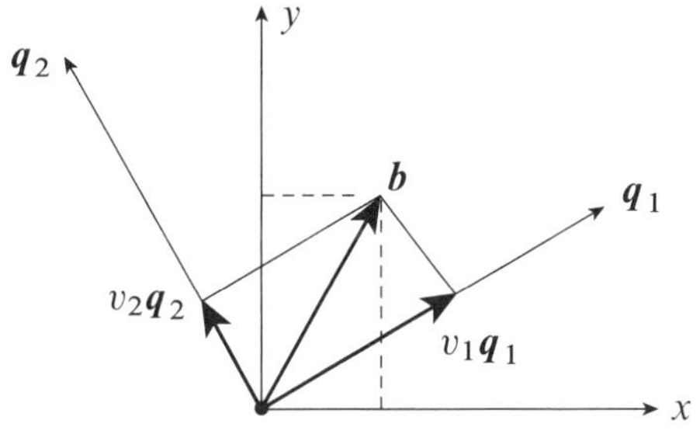


Figure 4.9: Every $\boldsymbol{b} = (x, y)$ splits into $\boldsymbol{b} = v_1\boldsymbol{q}_1 + v_2\boldsymbol{q}_2$. And $||\boldsymbol{b}||^2 = x^2 + y^2 = v_1^2 + v_2^2$.

## Both Symmetric and Orthogonal

Symmetric matrices are the best, they are everywhere in applied mathematics. Orthogonal matrices are a strong second, starting with rotation matrices and the Fourier matrix. Most symmetric matrices are not orthogonal and most orthogonal matrices are not symmetric. It is natural to wonder when and if we can have both properties at once.

Exchange and reflection and "Hadamard" matrices are symmetric and orthogonal:

$$\boldsymbol{P} = \begin{bmatrix} 0 & 1 \\ 1 & 0 \end{bmatrix} \quad \boldsymbol{R} = \begin{bmatrix} -\cos\theta & \sin\theta \\ \sin\theta & \cos\theta \end{bmatrix} \quad \boldsymbol{H} = \frac{1}{2}\begin{bmatrix} -1 & 1 & 1 & 1 \\ 1 & -1 & 1 & 1 \\ 1 & 1 & -1 & 1 \\ 1 & 1 & 1 & -1 \end{bmatrix}. \tag{16}$$

Notice that the columns of $H$ are unit vectors: $\frac{1}{4}((-1)^2 + 1^2 + 1^2 + 1^2) = 1$. Nobody knows which dimensions allow $n$ orthogonal vectors of 1's and $-1$'s (not odd dimensions!). Wikipedia describes this unsolved problem on its "Hadamard matrix" page.

To find more symmetric orthogonal matrices, and eventually all of them, we can use an important fact about orthogonal matrices:

**If $Q_1$ and $Q_2$ are orthogonal, so is their product $Q = Q_1 Q_2$.**

The test is always to check $Q^{\mathrm{T}}Q = I$. Here this is $(Q_1Q_2)^{\mathrm{T}}(Q_1Q_2) = Q_2^{\mathrm{T}}Q_1^{\mathrm{T}}Q_1Q_2$. In the middle is $Q_1^{\mathrm{T}}Q_1 = I$. Then the outside has $Q_2^{\mathrm{T}}Q_2 = I$.

*Conclusion*: We can multiply orthogonal matrices and stay orthogonal.

*Problem*: We can't always multiply symmetric matrices and stay symmetric.

Here is one approach that succeeds with both properties. Start with any diagonal matrix $D$ of 1's followed by $-1$'s:

$$\textbf{Symmetric and orthogonal} \qquad D = \operatorname{diag}(\mathbf{1}, \ldots, \mathbf{1}, \mathbf{-1}, \ldots, \mathbf{-1}). \tag{17}$$

Multiply $D$ on the left side by any orthogonal $Q$ and on the right side by $Q^{\mathrm{T}}$. That "symmetric multiplication" keeps the matrix $QDQ^{\mathrm{T}}$ symmetric:

$$\textbf{Symmetric and orthogonal} \qquad (QDQ^{\mathrm{T}})^{\mathrm{T}} = Q^{\mathrm{TT}}D^{\mathrm{T}}Q^{\mathrm{T}} = QDQ^{\mathrm{T}}. \tag{18}$$

This product of orthogonal matrices is also orthogonal. When you meet eigenvalues in Chapter 6, you will see that *all* symmetric orthogonal matrices have this form $QDQ^{\mathrm{T}}$. Possibly that small fact is appearing for the first time in a textbook.

## Factoring a Matrix

That was for fun, this is more important. "A symmetric matrix $S$ is like a real number $r$." "An orthogonal matrix $Q$ is like a complex number $e^{i\theta}$ with absolute value 1." Every complex number can be written in polar form $re^{i\theta}$, and what we hope for is true:

**Every square matrix $A$ can be written in polar form $A = SQ$.**

$A = SQ$ is equivalent to the Singular Value Decomposition (this is explained in Section 7.2). The SVD is the last and most remarkable step in the *Fundamental Theorem of Linear Algebra*. The polar form is in the Chapter 7 Notes.

## ■ REVIEW OF THE KEY IDEAS ■

1. The transpose has $A^{\mathrm{T}}_{ij} = A_{ji}$. Then $(AB)^{\mathrm{T}} = B^{\mathrm{T}}A^{\mathrm{T}}$ and $A\boldsymbol{v} \cdot \boldsymbol{w}$ equals $\boldsymbol{v} \cdot A^{\mathrm{T}}\boldsymbol{w}$.
2. Symmetric matrices have $S^{\mathrm{T}} = S$. Orthogonal matrices have $Q^{\mathrm{T}} = Q^{-1}$.
3. $A^{\mathrm{T}}A$ is always a symmetric matrix. Key examples are second difference matrices.
4. The columns of $Q$ are orthogonal vectors of length 1. Then $||Q\boldsymbol{x}|| = ||\boldsymbol{x}||$ for all $\boldsymbol{x}$.
5. The $n\,!$ permutation matrices $P$ reorder the rows of $I$ ($n$ by $n$), and $P^{\mathrm{T}} = P^{-1}$.

## Problem Set 4.5

**Questions 1–9 are about transposes $A^{\mathrm{T}}$ and symmetric matrices $S = S^{\mathrm{T}}$.**

1 Find $A^{\mathrm{T}}$ and $A^{-1}$ and $(A^{-1})^{\mathrm{T}}$ and $(A^{\mathrm{T}})^{-1}$ for

$$A = \begin{bmatrix} 1 & 0 \\ 9 & 3 \end{bmatrix} \quad \text{and also} \quad A = \begin{bmatrix} 1 & c \\ c & 0 \end{bmatrix}.$$

2 (a) Find 2 by 2 symmetric matrices $A$ and $B$ so that $AB$ is not symmetric.

(b) With $A^{\mathrm{T}} = A$ and $B^{\mathrm{T}} = B$, show that $AB = BA$ ensures that $AB$ will now be symmetric. The product is symmetric only when $A$ commutes with $B$.

3 (a) The matrix $((AB)^{-1})^{\mathrm{T}}$ comes from $(A^{-1})^{\mathrm{T}}$ and $(B^{-1})^{\mathrm{T}}$. *In what order?*

(b) If $U$ is upper triangular then $(U^{-1})^{\mathrm{T}}$ is _____ triangular.

4 Show that $A^2 = 0$ is possible but $A^{\mathrm{T}}A = 0$ is not possible (unless $A =$ zero matrix).

5 Every square matrix $A$ has a symmetric part and an antisymmetric part:

$$A = \text{symmetric} + \text{antisymmetric} = \left(\frac{A + A^{\mathrm{T}}}{2}\right) + \left(\frac{A - A^{\mathrm{T}}}{2}\right).$$

Transpose the antisymmetric part to get *minus* that part. Split these in two parts:

$$A = \begin{bmatrix} 3 & 5 \\ 7 & 9 \end{bmatrix} \qquad A = \begin{bmatrix} 1 & 4 & 8 \\ 0 & 2 & 6 \\ 0 & 0 & 3 \end{bmatrix}.$$

6 The transpose of a block matrix $M = \begin{bmatrix} \mathbf{A} & \mathbf{B} \\ \mathbf{C} & \mathbf{D} \end{bmatrix}$ is $M^{\mathrm{T}} =$ _____. Test an example to be sure. Under what conditions on $A$, $B$, $C$, $D$ is the block matrix symmetric?

7 True or false:

(a) The block matrix $\begin{bmatrix} \mathbf{0} & \mathbf{A} \\ \mathbf{A} & \mathbf{0} \end{bmatrix}$ is automatically symmetric.

(b) If $A$ and $B$ are symmetric then their product $AB$ is symmetric.

(c) If $A$ is not symmetric then $A^{-1}$ is not symmetric.

(d) When $A, B, C$ are symmetric, the transpose of $ABC$ is $CBA$.

8 (a) How many entries of $S$ can be chosen independently, if $S = S^{\mathrm{T}}$ is 5 by 5?

(b) How many entries can be chosen if $A$ is *skew-symmetric*? ($A^{\mathrm{T}} = -A$).

9 Transpose the equation $A^{-1}A = I$. The result shows that the inverse of $A^{\mathrm{T}}$ is _____. If $S$ is symmetric, **how does this show that $S^{-1}$ is also symmetric**?

**Questions 10–14 are about permutation matrices.**

**10** Why are there $n!$ permutation matrices of size $n$? They give $n!$ orders of $1, \ldots, n$.

**11** If $P_1$ and $P_2$ are permutation matrices, so is $P_1 P_2$. This still has the rows of $I$ in some order. Give examples with $P_1 P_2 \neq P_2 P_1$ and $P_3 P_4 = P_4 P_3$.

**12** There are 12 "*even*" permutations of $(1, 2, 3, 4)$, with an *even number of exchanges*. Two of them are $(1, 2, 3, 4)$ with no exchanges and $(4, 3, 2, 1)$ with two exchanges. List the other ten. Instead of writing each 4 by 4 matrix, just order the numbers.

**13** If $P$ has 1's on the antidiagonal from $(1, n)$ to $(n, 1)$, describe $PAP$. Is $P$ even?

**14** (a) Find a 3 by 3 permutation matrix with $P^3 = I$ (but not $P = I$).

(b) Find a 4 by 4 permutation with $P^4 \neq I$.

**Questions 15–18 are about first differences $A$ and second differences $A^{\mathrm{T}}A$ and $AA^{\mathrm{T}}$.**

**15** Write down the 5 by 4 backward difference matrix $A$.

(a) Compute the symmetric second difference matrices $S = A^{\mathrm{T}}A$ and $L = AA^{\mathrm{T}}$.

(b) Show that $S$ is invertible by finding $S^{-1}$. Show that $L$ is singular.

**16** In Problem 15, find the pivots of $S$ and $L$ (4 by 4 and 5 by 5). The pivots of $S$ in equation (8) are $2, 3/2, 4/3$. The pivots of $L$ in equation (10) are $1, 1, 1, 0$ (fail).

**17** (Computer problem) Create the 9 by 10 backward difference matrix $A$. Multiply to find $S = A^{\mathrm{T}}A$ and $L = AA^{\mathrm{T}}$. If you have linear algebra software, ask for the determinants $\det(S)$ and $\det(L)$.

*Challenge*: By experiment find $\det(S)$ when $S = A^{\mathrm{T}}A$ is $n$ by $n$.

**18** (Infinite computer problem) Imagine that the second difference matrix $S$ is infinitely large. The diagonals of 2's and $-1$'s go from minus infinity to plus infinity:

$$\textbf{Infinite tridiagonal matrix} \qquad S = \begin{bmatrix} \cdot & \cdot & & & \\ -1 & 2 & -1 & & \\ & -1 & 2 & -1 & \\ & & \cdot & \cdot & \end{bmatrix}$$

(a) Multiply $S$ times the infinite *all-ones* vector $\boldsymbol{v} = (\ldots, 1, 1, 1, 1, \ldots)$

(b) Multiply $S$ times the infinite *linear* vector $\boldsymbol{w} = (\ldots, 0, 1, 2, 3, \ldots)$

(c) Multiply $S$ times the infinite *squares* vector $\boldsymbol{u} = (\ldots, 0, 1, 4, 9, \ldots)$.

(d) Multiply $S$ times the infinite *cubes* vector $\boldsymbol{c} = (\ldots, 0, 1, 8, 27, \ldots)$.

The answers correspond to second derivatives (with minus sign) of 1 and $x^2$ and $x^3$.

**Questions 19–28 are about matrices with $Q^{\mathrm{T}}Q = I$. If $Q$ is square, then it is an orthogonal matrix and $Q^{\mathrm{T}} = Q^{-1}$ and $QQ^{\mathrm{T}} = I$.**

**19** Complete these matrices to be orthogonal matrices:

$$\text{(a)}\quad Q = \begin{bmatrix} 1/2 & \\ & 1/2 \end{bmatrix} \quad \text{(b)}\quad Q = \frac{1}{3}\begin{bmatrix} -1 & & \\ 2 & & \\ 2 & & \end{bmatrix} \quad \text{(c)}\quad Q = \frac{1}{2}\begin{bmatrix} 1 & 1 & & \\ 1 & 1 & & \\ 1 & -1 & & \\ 1 & -1 & & \end{bmatrix}.$$

**20** (a) Suppose $Q$ is an orthogonal matrix. Why is $Q^{-1} = Q^{\mathrm{T}}$ also an orthogonal matrix ?

(b) From $Q^{\mathrm{T}}Q = I$, the columns of $Q$ are orthogonal unit vectors (orthonormal vectors). Why are the rows of $Q$ (square matrix) also orthonormal vectors ?

**21** (a) Which vectors can be the first column of an orthogonal matrix ?

(b) If $Q_1^{\mathrm{T}}Q_1 = I$ and $Q_2^{\mathrm{T}}Q_2 = I$, is it true that $(Q_1Q_2)^{\mathrm{T}}(Q_1Q_2) = I$ ? Assume that the matrix shapes allow the multiplication $Q_1Q_2$.

**22** If $\boldsymbol{u}$ is a unit column vector (length 1, $\boldsymbol{u}^{\mathrm{T}}\boldsymbol{u} = 1$), show why $H = I - 2\boldsymbol{u}\boldsymbol{u}^{\mathrm{T}}$ is

(a) a symmetric matrix: $H = H^{\mathrm{T}}$ (b) an orthogonal matrix: $H^{\mathrm{T}}H = I$.

**23** If $\boldsymbol{u} = (\cos\theta, \sin\theta)$, what are the four entries in $H = I - 2\boldsymbol{u}\boldsymbol{u}^{\mathrm{T}}$? Show that $H\boldsymbol{u} = -\boldsymbol{u}$ and $H\boldsymbol{v} = \boldsymbol{v}$ for $\boldsymbol{v} = (-\sin\theta, \cos\theta)$. This $H$ is a **reflection matrix**: the $\boldsymbol{v}$-line is a mirror and the $\boldsymbol{u}$-line is reflected across that mirror.

**24** Suppose the matrix $Q$ is orthogonal and also upper triangular. What can $Q$ look like ? Must it be diagonal ?

**25** (a) To construct a 3 by 3 orthogonal matrix $Q$ whose first column is in the direction $\boldsymbol{w}$, what first column $\boldsymbol{q}_1 = c\boldsymbol{w}$ would you choose ?

(b) The next column $\boldsymbol{q}_2$ can be any unit vector perpendicular to $\boldsymbol{q}_1$. To find $\boldsymbol{q}_3$, choose a solution $\boldsymbol{v} = (v_1, v_2, v_3)$ to the two equations $\boldsymbol{q}_1^{\mathrm{T}}\boldsymbol{v} = 0$ and $\boldsymbol{q}_2^{\mathrm{T}}\boldsymbol{v} = 0$. *Why is there always a nonzero solution* $\boldsymbol{v}$ ?

**26** Why is every solution $\boldsymbol{v}$ to $A\boldsymbol{v} = \mathbf{0}$ orthogonal to every row of $A$ ?

**27** Suppose $Q^{\mathrm{T}}Q = I$ but $Q$ is not square. The matrix $P = QQ^{\mathrm{T}}$ is not $I$. But show that $P$ is symmetric and $P^2 = P$. This is a **projection matrix**.

**28** A 5 by 4 matrix $Q$ can have $Q^{\mathrm{T}}Q = I$ but *it cannot possibly have* $QQ^{\mathrm{T}} = I$. Explain in words why the four equations $Q^{\mathrm{T}}\boldsymbol{v} = \mathbf{0}$ must have a nonzero solution $\boldsymbol{v}$. Then $\boldsymbol{v}$ is not the same as $QQ^{\mathrm{T}}\boldsymbol{v}$ and $I$ is not the same as $QQ^{\mathrm{T}}$.

## Challenge Problems

**29** Can you find a rotation matrix $Q$ so that $QDQ^{\mathrm{T}}$ is a permutation ?

$$\begin{bmatrix}\cos\theta & -\sin\theta\\ \sin\theta & \cos\theta\end{bmatrix}\begin{bmatrix}1 & \\ & -1\end{bmatrix}\begin{bmatrix}\cos\theta & \sin\theta\\ -\sin\theta & \cos\theta\end{bmatrix} \quad \text{equals} \quad \begin{bmatrix}0 & 1\\ 1 & 0\end{bmatrix}.$$

**30** Split an orthogonal matrix ($Q^{\mathrm{T}}Q = QQ^{\mathrm{T}} = I$) into two rectangular submatrices:

$$Q = \begin{bmatrix} Q_1 \mid Q_2 \end{bmatrix} \quad \text{and} \quad Q^{\mathrm{T}}Q = \begin{bmatrix} Q_1^{\mathrm{T}}Q_1 & Q_1^{\mathrm{T}}Q_2\\ Q_2^{\mathrm{T}}Q_1 & Q_2^{\mathrm{T}}Q_2\end{bmatrix}$$

(a) What are those four blocks in $Q^{\mathrm{T}}Q = I$ ?

(b) $QQ^{\mathrm{T}} = Q_1Q_1^{\mathrm{T}} + Q_2Q_2^{\mathrm{T}} = I$ is column times row multiplication. Insert the diagonal matrix $D = \begin{bmatrix} I & 0\\ 0 & -I\end{bmatrix}$ and do the same multiplication for $QDQ^{\mathrm{T}}$.

*Note*: The description of all symmetric orthogonal matrices $S$ in (18) becomes $S = QDQ^{\mathrm{T}} = Q_1Q_1^{\mathrm{T}} - Q_2Q_2^{\mathrm{T}}$. This is exactly the reflection matrix $I - 2Q_2Q_2^{\mathrm{T}}$.

**31** The real reason that the transpose "flips $A$ across its main diagonal" is to obey this dot product law: $(A\boldsymbol{v}) \cdot \boldsymbol{w} = \boldsymbol{v} \cdot (A^{\mathrm{T}}\boldsymbol{w})$. That rule $(A\boldsymbol{v})^{\mathrm{T}}\boldsymbol{w} = \boldsymbol{v}^{\mathrm{T}}(A^{\mathrm{T}}\boldsymbol{w})$ **becomes integration by parts in calculus**, where $A = d/dx$ and $A^{\mathrm{T}} = -d/dx$.

(a) For 2 by 2 matrices, write out both sides (4 terms) and compare:

$$\left(\begin{bmatrix} a & \boldsymbol{b}\\ \boldsymbol{c} & d\end{bmatrix}\begin{bmatrix} v_1\\ v_2\end{bmatrix}\right)\cdot\begin{bmatrix} w_1\\ w_2\end{bmatrix} \quad \text{is equal to} \quad \begin{bmatrix} v_1\\ v_2\end{bmatrix}\cdot\left(\begin{bmatrix} a & \boldsymbol{c}\\ \boldsymbol{b} & d\end{bmatrix}\begin{bmatrix} w_1\\ w_2\end{bmatrix}\right).$$

(b) The rule $(AB)^{\mathrm{T}} = B^{\mathrm{T}}A^{\mathrm{T}}$ comes slowly but directly from part (a):

$$(AB)\,\boldsymbol{v} \cdot \boldsymbol{w} = A(B\boldsymbol{v}) \cdot \boldsymbol{w} = B\boldsymbol{v} \cdot A^{\mathrm{T}}\boldsymbol{w} = \boldsymbol{v} \cdot B^{\mathrm{T}}(A^{\mathrm{T}}\boldsymbol{w}) = \boldsymbol{v} \cdot (B^{\mathrm{T}}A^{\mathrm{T}})\boldsymbol{w}$$

Steps 1 and 4 are the ______ law. Steps 2 and 3 are the dot product law.

**32** How is a matrix $S = S^{\mathrm{T}}$ decided by its entries on and above the diagonal? How is $Q$ with orthonormal columns decided by its entries *below* the diagonal? Together this matches the number of entries in an $n$ by $n$ matrix. So it is reasonable that every matrix can be factored into $A = SQ$ (like $re^{i\theta}$).

## ■ CHAPTER 4 NOTES ■

*Important Question* Where do the rules for matrix-matrix multiplication $AB$ come from? *Answer* From matrix-vector multiplication $A\boldsymbol{v}$. The matrix $AB$ is defined so that

$$AB \text{ times } \boldsymbol{v} \text{ equals } A \text{ times } B\boldsymbol{v}. \qquad \text{Then } AB \text{ times } C \text{ equals } A \text{ times } BC.$$

Key idea: Choose the special vector $\boldsymbol{v} = (1, 0, \ldots, 0)$. Then $AB$ times this $\boldsymbol{v}$ is the first column of $AB$. And $B\boldsymbol{v}$ is the first column of $B$. *So column* 1 *of* $AB$ *equals* $A$ *times column* 1 *of* $B$. This was the $AB$ rule from the start. Every other column of $AB$ goes the same way, by moving the "1" in $\boldsymbol{v}$.

Thus $(AB)\boldsymbol{v} = A(B\boldsymbol{v})$. With several $v$'s in a matrix $C$, this becomes $(AB)C = A(BC)$.

**Elimination factors $A$ into $LU$ = (lower triangular) times (upper triangular).**

The MATLAB command $[L,\ U] = \boldsymbol{lu}(A)$ will output $L$ and $U$, unless there are row exchanges. $L$ and $U$ are a complete record of elimination on the left side of $A\boldsymbol{v} = \boldsymbol{b}$. The solution $\boldsymbol{v}$ comes from the right side $\boldsymbol{b}$ by solving the two triangular systems:

| | | | |
|---|---|---|---|
| **From $\boldsymbol{b}$ to $\boldsymbol{c}$**<br>**Forward substitution** | $L\boldsymbol{c} = \boldsymbol{b}$ | **From $\boldsymbol{c}$ to $\boldsymbol{v}$**<br>**Back substitution** | $U\boldsymbol{v} = \boldsymbol{c}$ |

Then $\boldsymbol{v}$ is the correct solution: $A\boldsymbol{v} = LU\boldsymbol{v} = L\boldsymbol{c} = \boldsymbol{b}$. The forward substitution is what happened to $\boldsymbol{b}$ as elimination went forward on $[\,A \quad \boldsymbol{b}\,]$.

Second difference matrices have beautiful inverses and $LU$ factors if the first diagonal entry is 1 instead of 2. Here is the 3 by 3 tridiagonal matrix $T$ and its inverse:

$$\boldsymbol{T_{11} = 1} \qquad T = \begin{bmatrix} 1 & -1 & 0 \\ -1 & 2 & -1 \\ 0 & -1 & 2 \end{bmatrix} \qquad T^{-1} = \begin{bmatrix} 3 & 2 & 1 \\ 2 & 2 & 1 \\ 1 & 1 & 1 \end{bmatrix}$$

One approach is Gauss-Jordan elimination on $[\,T \quad I\,]$. That seems too mechanical. I would rather write $T$ using first differences $L$ and $U$. The inverses are **sum matrices** $U^{-1}$ and $L^{-1}$:

$$T = \underset{\textbf{difference}}{\begin{bmatrix} 1 & & \\ -1 & 1 & \\ 0 & -1 & 1 \end{bmatrix}} \underset{\textbf{difference}}{\begin{bmatrix} 1 & -1 & 0 \\ & 1 & -1 \\ & & 1 \end{bmatrix}} \qquad T^{-1} = \underset{\textbf{sum}}{\begin{bmatrix} 1 & 1 & 1 \\ & 1 & 1 \\ & & 1 \end{bmatrix}} \underset{\textbf{sum}}{\begin{bmatrix} 1 & & \\ 1 & 1 & \\ 1 & 1 & 1 \end{bmatrix}}$$

**Question.** (**4 by 4**) What are the pivots of $T$? What is its 4 by 4 inverse?

# Chapter 5

# Vector Spaces and Subspaces

## 5.1 The Column Space of a Matrix

To a newcomer, matrix calculations involve a lot of numbers. To you, they involve vectors. The columns of $Av$ and $AB$ are linear combinations of $n$ vectors—the columns of $A$. This chapter moves from numbers and vectors to a third level of understanding (the highest level). Instead of individual columns, we look at "spaces" of vectors. Without seeing ***vector spaces*** and their ***subspaces***, you haven't understood everything about $A\boldsymbol{v} = \boldsymbol{b}$.

Since this chapter goes a little deeper, it may seem a little harder. That is natural. We are looking inside the calculations, to find the mathematics. The author's job is to make it clear. Section 5.5 will present the "*Fundamental Theorem of Linear Algebra*."

We begin with the most important vector spaces. They are denoted by $\mathbf{R}^1$, $\mathbf{R}^2$, $\mathbf{R}^3$, $\mathbf{R}^4$, . . .. Each space $\mathbf{R}^n$ consists of a whole collection of vectors. $\mathbf{R}^5$ contains all column vectors with five components. This is called "5-dimensional space."

**DEFINITION** ***The space $\mathbf{R}^n$ consists of all column vectors $v$ with $n$ components.***

The components of $v$ are real numbers, which is the reason for the letter $\mathbf{R}$. When the $n$ components are complex numbers, $\boldsymbol{v}$ lies in the space $\mathbf{C}^n$.

The vector space $\mathbf{R}^2$ is represented by the usual $xy$ plane. Each vector $v$ in $\mathbf{R}^2$ has two components. The word "*space*" asks us to think of all those vectors—the whole plane. Each vector gives the $x$ and $y$ coordinates of a point in the plane: $v = (x, y)$.

Similarly the vectors in $\mathbf{R}^3$ correspond to points $(x, y, z)$ in three-dimensional space. The one-dimensional space $\mathbf{R}^1$ is a line (like the $x$ axis). As before, we print vectors as a column between brackets, or along a line using commas and parentheses:

$$\begin{bmatrix} 4 \\ \pi \end{bmatrix} \text{ is in } \mathbf{R}^2, \quad (1, 1, 0, 1, 1) \text{ is in } \mathbf{R}^5, \quad \begin{bmatrix} 1+i \\ 1-i \end{bmatrix} \text{ is in } \mathbf{C}^2.$$

The great thing about linear algebra is that it deals easily with five-dimensional space. We don't draw the vectors, we just need the five numbers (or $n$ numbers).

To multiply $v$ by 7, multiply every component by 7. Here 7 is a "scalar." To add vectors in $\mathbf{R}^5$, add them a component at a time : five additions. The two essential vector operations go on *inside the vector space*, and they produce ***linear combinations***:

***We can add any vectors in*** $\mathbf{R}^n$***, and we can multiply any vector*** $v$ ***by any scalar*** $c$.

"Inside the vector space" means that ***the result stays in the space*** : This is crucial.

If $\boldsymbol{v}$ is in $\mathbf{R}^4$ with components $1, 0, 0, 1$, then $2\boldsymbol{v}$ is the vector in $\mathbf{R}^4$ with components $2, 0, 0, 2$. (In this case 2 is the scalar.) A whole series of properties can be verified in $\mathbf{R}^n$. The commutative law is $\boldsymbol{v} + \boldsymbol{w} = \boldsymbol{w} + \boldsymbol{v}$; the distributive law is $c(\boldsymbol{v} + \boldsymbol{w}) = c\boldsymbol{v} + c\boldsymbol{w}$. Every vector space has a unique "zero vector" satisfying $\mathbf{0} + \boldsymbol{v} = \boldsymbol{v}$. Those are three of the eight conditions listed in the Chapter 5 Notes.

These eight conditions are required of every vector space. There are vectors other than column vectors, and there are vector spaces other than $\mathbf{R}^n$. All vector spaces have to obey the eight reasonable rules.

*A real vector space is a set of* "vectors" *together with rules for vector addition and multiplication by real numbers.* The addition and the multiplication must produce vectors that are in the space. And the eight conditions must be satisfied (which is usually no problem). You need to see three vector spaces other than $\mathbf{R}^n$ :

| | |
|---|---|
| **M** | The vector space of ***all real*** **2 *by* 2 *matrices.*** |
| **Y** | The vector space of ***all solutions*** $y(t)$ to $Ay'' + By' + Cy = 0$. |
| **Z** | The vector space that consists only of a ***zero vector***. |

In **M** the "vectors" are really matrices. In **Y** the vectors are functions of $t$, like $y = e^{st}$. In **Z** the only addition is $\mathbf{0} + \mathbf{0} = \mathbf{0}$. In each space we can add: matrices to matrices, functions to functions, zero vector to zero vector. We can multiply a matrix by 4 or a function by 4 or the zero vector by 4. The result is still in **M** or **Y** or **Z**.

The space $\mathbf{R}^4$ is four-dimensional, and so is the space **M** of 2 by 2 matrices. Vectors in those spaces are determined by four numbers. The solution space **Y** is two-dimensional, because second order differential equations have two independent solutions. Section 5.4 will pin down those key words, *independence of vectors* and *dimension of a space*.

The space **Z** is zero-dimensional (by any reasonable definition of dimension). It is the smallest possible vector space. We hesitate to call it $\mathbf{R}^0$, which means no components—you might think there was no vector. *The vector space* **Z** *contains exactly one vector.* No space can do without that zero vector. Each space has its own zero vector—the zero matrix, the zero function, the vector $(0, 0, 0)$ in $\mathbf{R}^3$.

## Subspaces

At different times, we will ask you to think of matrices and functions as vectors. But at all times, the vectors that we need most are ordinary column vectors. They are vectors with $n$ components—but *maybe not all* of the vectors with $n$ components. There are important vector spaces ***inside*** $\mathbf{R}^n$. Those are ***subspaces*** of $\mathbf{R}^n$.

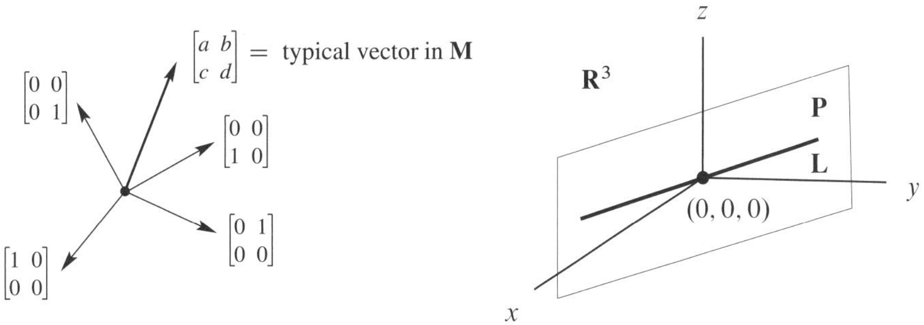


Figure 5.1: "4-dimensional" matrix space **M**. 3 subspaces of $\mathbf{R}^3$ : plane **P**, line **L**, point **Z**.

Start with the usual three-dimensional space $\mathbf{R}^3$. Choose a plane through the origin $(0, 0, 0)$. ***That plane is a vector space in its own right.*** If we add two vectors in the plane, their sum is in the plane. If we multiply an in-plane vector by 2 or $-5$, it is still in the plane. A plane in three-dimensional space is not $\mathbf{R}^2$ (even if it looks like $\mathbf{R}^2$). The vectors have three components and they belong to $\mathbf{R}^3$. The plane **P** is a vector space ***inside*** $\mathbf{R}^3$.

This illustrates one of the most fundamental ideas in linear algebra. The plane going through $(0, 0, 0)$ is a ***subspace*** of the full vector space $\mathbf{R}^3$.

**DEFINITION** A ***subspace*** of a vector space is a set of vectors (including **0**) that satisfies two requirements: ***If $v$ and $w$ are vectors in the subspace and $c$ is any scalar, then***

**(i)** $v + w$ is in the subspace and **(ii)** $cv$ is in the subspace.

In other words, the set of vectors is "closed" under addition $v + w$ and multiplication $cv$ (and $dw$). Those operations leave us in the subspace. We can also subtract, because $-w$ is in the subspace and its sum with $v$ is $v - w$. In short, ***all linear combinations $cv + dw$ stay in the subspace***.

First fact: ***Every subspace contains the zero vector***. The plane in $\mathbf{R}^3$ has to go through $(0, 0, 0)$. We mention this separately, for extra emphasis, but it follows directly from rule (**ii**). Choose $c = 0$, and the rule requires $0v$ to be in the subspace.

Planes that don't contain the origin fail those tests. When $v$ is on such a plane, $-v$ and $0v$ are *not* on the plane. A plane that misses the origin is not a subspace.

***Lines through the origin are also subspaces***. When we multiply by 5, or add two vectors on the line, we stay on the line. But the line must go through $(0, 0, 0)$.

Another subspace is all of $\mathbf{R}^3$. The whole space is a subspace (*of itself*). That is a fourth subspace in the figure. Here is a list of all the possible subspaces of $\mathbf{R}^3$ :

(**L**) Any line through $(0, 0, 0)$ — ($\mathbf{R}^3$) The whole space
(**P**) Any plane through $(0, 0, 0)$ — (**Z**) The single vector $(0, 0, 0)$

If we try to keep only *part* of a plane or line, the requirements for a subspace don't hold. Look at these examples in $\mathbf{R}^2$.

**Example 1** Keep only the vectors $(x, y)$ whose components are positive or zero (this is a quarter-plane). The vector $(2, 3)$ is included but $(-2, -3)$ is not. So rule (**ii**) is violated when we try to multiply by $c = -1$. ***The quarter-plane is not a subspace.***

**Example 2** Include also the vectors whose components are both negative. Now we have two quarter-planes. Requirement (**ii**) is satisfied; we can multiply by any $c$. But rule (**i**) now fails. The sum of $\boldsymbol{v} = (2, 3)$ and $\boldsymbol{w} = (-3, -2)$ is $(-1, 1)$, which is outside the quarter-planes. ***Two quarter-planes don't make a subspace.***

Rules (**i**) and (**ii**) involve vector addition $\boldsymbol{v} + \boldsymbol{w}$ and multiplication by scalars like $c$ and $d$. The rules can be combined into a single requirement—*the rule for subspaces*:

***A subspace containing $\boldsymbol{v}$ and $\boldsymbol{w}$ must contain all linear combinations $c\boldsymbol{v} + d\boldsymbol{w}$.***

**Example 3** Inside the vector space $\mathbf{M}$ of all 2 by 2 matrices, here are two subspaces:

(**U**) All upper triangular matrices $\begin{bmatrix} a & b \\ 0 & d \end{bmatrix}$ (**D**) All diagonal matrices $\begin{bmatrix} a & 0 \\ 0 & d \end{bmatrix}$.

Add any two matrices in **U**, and the sum is in **U**. Add diagonal matrices, and the sum is diagonal. In this case **D** is also a subspace of **U**! The zero matrix alone is also a subspace, when $a$, $b$, and $d$ all equal zero.

For a smaller subspace of diagonal matrices, we could require $a = d$. The matrices are multiples of the identity matrix $I$. These $aI$ form a "line of matrices" in **M** and **U** and **D**.

Is the matrix $I$ a subspace by itself? Certainly not. Only the zero matrix is. Your mind will invent more subspaces of 2 by 2 matrices—write them down for Problem 6.

## The Column Space of $A$

The most important subspaces are tied directly to a matrix $A$. We are trying to solve $A\boldsymbol{v} = \boldsymbol{b}$. If $A$ is not invertible, the system is solvable for some $\boldsymbol{b}$ and not solvable for other $\boldsymbol{b}$. We want to describe the good right sides $\boldsymbol{b}$—the vectors that *can* be written as $A$ times $\boldsymbol{v}$. Those $\boldsymbol{b}'s$ form the "*column space*" of $A$.

Remember that $A\boldsymbol{v}$ is a combination of the columns of $A$. To get every possible $\boldsymbol{b}$, we use every possible $\boldsymbol{v}$. Start with the columns of $A$, and ***take all their linear combinations. This produces the column space of A.*** It contains not just the $n$ columns of $A$!

**DEFINITION** ***The column space consists of all combinations of the columns.***

The combinations are all possible vectors $A\boldsymbol{v}$. They fill the column space $\boldsymbol{C}(A)$.

This column space is crucial to the whole book, and here is why. ***To solve $A\boldsymbol{v} = \boldsymbol{b}$ is to express b as a combination of the columns. The right side b has to be in the column space*** produced by $A$ on the left side. If $\boldsymbol{b}$ is not in $\boldsymbol{C}(A)$, $A\boldsymbol{v} = \boldsymbol{b}$ has no solution.

***The system $Av = b$ is solvable if and only if $b$ is in the column space of $A$.***

When $\boldsymbol{b}$ is in the column space, it is a combination of the columns. The coefficients in that combination give us a solution $\boldsymbol{v}$ to the system $A\boldsymbol{v} = \boldsymbol{b}$.

Suppose $A$ is an $m$ by $n$ matrix. Its columns have $m$ components (not $n$). So the columns belong to $\mathbf{R}^m$. ***The column space of $A$ is a subspace of $\mathbf{R}^m$ (not $\mathbf{R}^n$).*** The set of all column combinations $A\boldsymbol{x}$ satisfies rules (**i**) and (**ii**) for a subspace: When we add linear combinations or multiply by scalars, we still produce combinations of the columns. The word "subspace" is always justified *by taking all linear combinations.*

Here is a 3 by 2 matrix $A$, whose column space is a subspace of $\mathbf{R}^3$. The column space of $A$ is a plane in Figure 5.2.

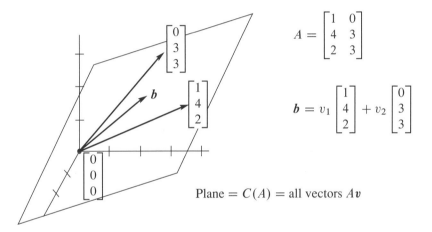

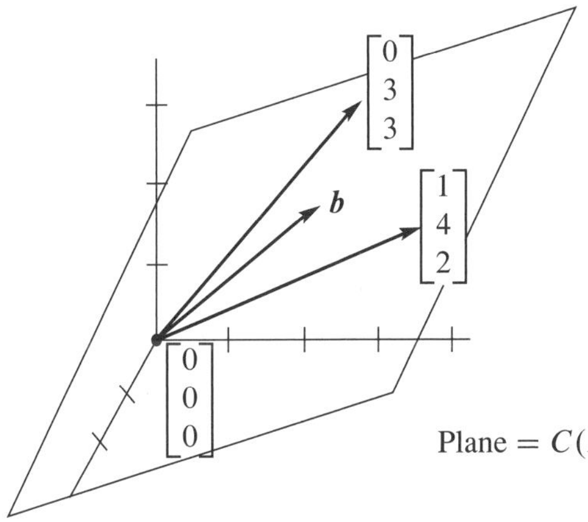


$$A = \begin{bmatrix} 1 & 0 \\ 4 & 3 \\ 2 & 3 \end{bmatrix}$$

$$\boldsymbol{b} = v_1 \begin{bmatrix} 1 \\ 4 \\ 2 \end{bmatrix} + v_2 \begin{bmatrix} 0 \\ 3 \\ 3 \end{bmatrix}$$

Plane $= C(A) =$ all vectors $A\boldsymbol{v}$

Figure 5.2: The column space $\boldsymbol{C}(A)$ is a plane containing the two columns of $A$. $A\boldsymbol{v} = \boldsymbol{b}$ is solvable when $\boldsymbol{b}$ is on that plane. Then $\boldsymbol{b}$ is a combination of the columns.

We drew one particular $\boldsymbol{b}$ (a combination of the columns). This $\boldsymbol{b} = A\boldsymbol{v}$ lies on the plane. The plane has zero thickness, so most right sides $\boldsymbol{b}$ in $\mathbf{R}^3$ are *not* in the column space. For most $\boldsymbol{b}$ there is no solution to our 3 equations in 2 unknowns.

Of course $(0, 0, 0)$ is in the column space. The plane passes through the origin. There is certainly a solution to $A\boldsymbol{v} = \mathbf{0}$. That solution, always available, is $\boldsymbol{v} =$ _____ .

To repeat, the attainable right sides $\boldsymbol{b}$ are exactly the vectors in the column space. One possibility is the first column itself—take $v_1 = 1$ and $v_2 = 0$. Another combination is the second column—take $v_1 = 0$ and $v_2 = 1$. The new level of understanding is to see *all* combinations—the whole subspace is generated by those two columns.

**Notation** The column space of $A$ is denoted by $\boldsymbol{C}(A)$. Start with the columns and take all their linear combinations. We might get the whole $\mathbf{R}^m$ or only a small subspace.

**Important** Instead of columns in $\mathbf{R}^m$, we could start with any set of vectors in a vector space $\mathbf{V}$. To get a subspace $\mathbf{SS}$ of $\mathbf{V}$, we take *all combinations* of the vectors in that set:

$$\begin{aligned} \mathbf{S} &= \text{set of vectors } \boldsymbol{s} \text{ in } \mathbf{V} \ (\mathbf{S} \text{ is probably } \textit{not} \text{ a subspace}) \\ \mathbf{SS} &= \text{all combinations of vectors in } \mathbf{S} \ (\mathbf{SS} \textit{ is} \text{ a subspace}) \end{aligned}$$

$$\mathbf{SS} = \boxed{\text{all } c_1\boldsymbol{s}_1 + \cdots + c_N\boldsymbol{s}_N} = \boxed{\textbf{the subspace of V “spanned” by S}}$$

When $\mathbf{S}$ is the set of columns, $\mathbf{SS}$ is the column space. When there is only one nonzero vector $\boldsymbol{v}$ in $\mathbf{S}$, the subspace $\mathbf{SS}$ is the line through $\boldsymbol{v}$. *Always* $\mathbf{SS}$ *is the smallest subspace containing* $\mathbf{S}$. This is a fundamental way to create subspaces and we will come back to it.

**The subspace SS is the “span” of S, containing all combinations of vectors in S.**

**Example 4** Describe the column spaces (they are subspaces of $\mathbf{R}^2$) for these matrices:

$$I = \begin{bmatrix} 1 & 0 \\ 0 & 1 \end{bmatrix} \quad \text{and} \quad A = \begin{bmatrix} 1 & 2 \\ 2 & 4 \end{bmatrix} \quad \text{and} \quad B = \begin{bmatrix} 1 & 2 & 3 \\ 0 & 0 & 4 \end{bmatrix}.$$

**Solution** The column space of $I$ is the *whole space* $\mathbf{R}^2$. Every vector is a combination of the columns of $I$. In vector space language, $\boldsymbol{C}(I)$ equals $\mathbf{R}^2$.

The column space of $A$ is only a line. The second column $(2,4)$ is a multiple of the first column $(1,2)$. Those vectors are different, but our eye is on vector ***spaces***. The column space contains $(1,2)$ and $(2,4)$ and all other vectors $(c,2c)$ along that line. The equation $A\boldsymbol{v} = \boldsymbol{b}$ is only solvable when $\boldsymbol{b}$ is on the line.

For the third matrix (with three columns) the column space $\boldsymbol{C}(B)$ is all of $\mathbf{R}^2$. Every $\boldsymbol{b}$ is attainable. The vector $\boldsymbol{b} = (5,4)$ is column 2 plus column 3, so $\boldsymbol{v}$ can be $(0,1,1)$. The same vector $(5,4)$ is also 2(column 1) + column 3, so another possible $\boldsymbol{v}$ is $(2,0,1)$. This matrix has the same column space as $I$—any $\boldsymbol{b}$ is allowed. But now $\boldsymbol{v}$ has extra components and $A\boldsymbol{v} = \boldsymbol{b}$ has more solutions—more combinations that give $\boldsymbol{b}$.

The next section creates the *nullspace* $\boldsymbol{N}(A)$, to describe all the solutions of $A\boldsymbol{v} = \mathbf{0}$. This section created the column space $\boldsymbol{C}(A)$, to describe all the attainable right sides $\boldsymbol{b}$.

## ■ REVIEW OF THE KEY IDEAS ■

1. $\mathbf{R}^n$ contains all column vectors with $n$ real components.
2. $\mathbf{M}$ (2 by 2 matrices) and $\mathbf{Y}$ (functions) and $\mathbf{Z}$ (zero vector alone) are vector spaces.
3. A subspace containing $\boldsymbol{v}$ and $\boldsymbol{w}$ must contain all their combinations $c\boldsymbol{v} + d\boldsymbol{w}$.
4. The combinations of the columns of $A$ form the ***column space*** $\boldsymbol{C}(A)$. Then the column space is “spanned” by the columns.

**5.** $Av = b$ has a solution exactly when $b$ is in the column space of $A$.

## ■ WORKED EXAMPLES ■

**5.1 A** We are given three different vectors $b_1, b_2, b_3$. Construct a matrix so that the equations $Av = b_1$ and $Av = b_2$ are solvable, but $Av = b_3$ is *not* solvable. How can you decide if this is possible ? How could you construct $A$ ?

**Solution** We want to have $b_1$ and $b_2$ in the column space of $A$. Then $Av = b_1$ and $Av = b_2$ will be solvable. *The quickest way is to make $b_1$ and $b_2$ the two columns of $A$.* Then the solutions are $v = (1, 0)$ and $v = (0, 1)$.

Also, we don't want $Av = b_3$ to be solvable. So don't make the column space any larger! Keeping only the columns $b_1$ and $b_2$, the question is: *Do we already have $b_3$ ?*

$$\text{Is } Av = \begin{bmatrix} b_1 & b_2 \end{bmatrix} \begin{bmatrix} v_1 \\ v_2 \end{bmatrix} = b_3 \text{ solvable?} \qquad \text{Is } b_3 \text{ a combination of } b_1 \text{ and } b_2\text{?}$$

If the answer is *no*, we have the desired matrix $A$. If $b_3$ *is* a combination of $b_1$ and $b_2$, then it is *not possible* to construct $A$. The column space $C(A)$ will have to contain $b_3$.

**5.1 B** Describe a subspace **S** of each vector space **V**, and then a subspace **SS** of **S**.

$$\begin{aligned} \mathbf{V}_3 &= \text{all combinations of } (1,1,0,0) \text{ and } (1,1,1,0) \text{ and } (1,1,1,1) \\ \mathbf{V}_2 &= \text{all vectors } v \text{ perpendicular to } u = (1,2,1), \text{ so } u \cdot v = 0 \\ \mathbf{V}_4 &= \text{all solutions } y(x) \text{ to the equation } d^4y/dx^4 = 0 \end{aligned}$$

Describe each **V** two ways: (1) *All combinations of ....* (2) *All solutions of ....*

**Solution** $\mathbf{V}_3$ starts with three vectors. A subspace **S** comes from all combinations of the first two vectors $(1, 1, 0, 0)$ and $(1, 1, 1, 0)$. A subspace **SS** of **S** comes from all multiples $(c, c, 0, 0)$ of the first vector. So many possibilities.

A subspace **S** of $\mathbf{V}_2$ is the line through $(1, -1, 1)$. This line is perpendicular to $u$. The zero vector $z = (0, 0, 0)$ is in **S**. The smallest subspace **SS** is **Z**.

$\mathbf{V}_4$ contains all cubic polynomials $y = a + bx + cx^2 + dx^3$, with $d^4y/dx^4 = 0$. The quadratic polynomials (without an $x^3$ term) give a subspace **S**. The linear polynomials are one choice of **SS**. The constants $y = a$ could be **SSS**.

In all three parts we could take **S** = **V** itself, and **SS** = the zero subspace **Z**.

Each **V** can be described as *all combinations of ....* and as *all solutions of ....*:

$$\begin{aligned} \mathbf{V}_3 &= \text{all combinations of the 3 vectors} & \mathbf{V}_3 &= \text{all solutions of } v_1 - v_2 = 0. \\ \mathbf{V}_2 &= \text{all combinations of } (1,0,-1) \text{ and } (1,-1,1) & \mathbf{V}_2 &= \text{all solutions of } u \cdot v = 0. \\ \mathbf{V}_4 &= \text{all combinations of } 1, x, x^2, x^3 & \mathbf{V}_4 &= \text{all solutions to } d^4y/dx^4 = 0. \end{aligned}$$

## Problem Set 5.1

**Questions 1–10 are about the "subspace requirements": $\boldsymbol{v} + \boldsymbol{w}$ and $c\boldsymbol{v}$ (and then all linear combinations $c\boldsymbol{v} + d\boldsymbol{w}$) stay in the subspace.**

**1** One requirement can be met while the other fails. Show this by finding

(a) A set of vectors in $\mathbf{R}^2$ for which $\boldsymbol{v} + \boldsymbol{w}$ stays in the set but $\frac{1}{2}\boldsymbol{v}$ may be outside.

(b) A set of vectors in $\mathbf{R}^2$ (other than two quarter-planes) for which every $c\boldsymbol{v}$ stays in the set but $\boldsymbol{v} + \boldsymbol{w}$ may be outside.

**2** Which of the following subsets of $\mathbf{R}^3$ are actually subspaces?

(a) The plane of vectors $(b_1, b_2, b_3)$ with $b_1 = b_2$.

(b) The plane of vectors with $b_1 = 1$.

(c) The vectors with $b_1 b_2 b_3 = 0$.

(d) All linear combinations of $\boldsymbol{v} = (1, 4, 0)$ and $\boldsymbol{w} = (2, 2, 2)$.

(e) All vectors that satisfy $b_1 + b_2 + b_3 = 0$.

(f) All vectors with $b_1 \leq b_2 \leq b_3$.

**3** Describe the smallest subspace of the matrix space $\mathbf{M}$ that contains

$$\text{(a)} \begin{bmatrix} 1 & 0 \\ 0 & 0 \end{bmatrix} \text{and} \begin{bmatrix} 0 & 1 \\ 0 & 0 \end{bmatrix} \quad \text{(b)} \begin{bmatrix} 1 & 1 \\ 0 & 0 \end{bmatrix} \quad \text{(c)} \begin{bmatrix} 1 & 0 \\ 0 & 0 \end{bmatrix} \text{and} \begin{bmatrix} 1 & 0 \\ 0 & 1 \end{bmatrix}.$$

**4** Let $\mathbf{P}$ be the plane in $\mathbf{R}^3$ with equation $x + y - 2z = 4$. The origin $(0, 0, 0)$ is not in $\mathbf{P}$! Find two vectors in $\mathbf{P}$ and check that their sum is not in $\mathbf{P}$.

**5** Let $\mathbf{P}_0$ be the plane through $(0, 0, 0)$ parallel to the previous plane $\mathbf{P}$. What is the equation for $\mathbf{P}_0$? Find two vectors in $\mathbf{P}_0$ and check that their sum is in $\mathbf{P}_0$.

**6** The subspaces of $\mathbf{R}^3$ are planes, lines, $\mathbf{R}^3$ itself, or $\mathbf{Z}$ containing only $(0, 0, 0)$.

(a) Describe the three types of subspaces of $\mathbf{R}^2$.

(b) Describe all subspaces of $\mathbf{D}$, the space of 2 by 2 diagonal matrices.

**7** (a) The intersection of two planes through $(0, 0, 0)$ is probably a _____ but it could be a _____. It can't be $\mathbf{Z}$!

(b) The intersection of a plane through $(0, 0, 0)$ with a line through $(0, 0, 0)$ is probably a _____ but it could be a _____.

(c) If $\mathbf{S}$ and $\mathbf{T}$ are subspaces of $\mathbf{R}^5$, prove that their intersection $\mathbf{S} \cap \mathbf{T}$ is a subspace of $\mathbf{R}^5$. Here $\mathbf{S} \cap \mathbf{T}$ consists of the vectors that lie in both subspaces. *Check the requirements on $\boldsymbol{v} + \boldsymbol{w}$ and $c\boldsymbol{v}$.*

**8** Suppose $\mathbf{P}$ is a plane through $(0, 0, 0)$ and $\mathbf{L}$ is a line through $(0, 0, 0)$. The smallest vector space $\mathbf{P} + \mathbf{L}$ containing both $\mathbf{P}$ and $\mathbf{L}$ is either _____ or _____.

**9** (a) Show that the set of *invertible* matrices in **M** is not a subspace.

(b) Show that the set of *singular* matrices in **M** is not a subspace.

**10** True or false (check addition in each case by an example):

(a) The symmetric matrices in **M** (with $A^{\mathrm{T}} = A$) form a subspace.

(b) The skew-symmetric matrices in **M** (with $A^{\mathrm{T}} = -A$) form a subspace.

(c) The unsymmetric matrices in **M** (with $A^{\mathrm{T}} \neq A$) form a subspace.

**Questions 11–19 are about column spaces $C(A)$ and the equation $A\boldsymbol{v} = \boldsymbol{b}$.**

**11** Describe the column spaces (lines or planes) of these particular matrices:

$$A = \begin{bmatrix} 1 & 2 \\ 0 & 0 \\ 0 & 0 \end{bmatrix} \quad B = \begin{bmatrix} 1 & 0 \\ 0 & 2 \\ 0 & 0 \end{bmatrix} \quad C = \begin{bmatrix} 1 & 0 \\ 2 & 0 \\ 0 & 0 \end{bmatrix}.$$

**12** For which right sides (find a condition on $b_1, b_2, b_3$) are these systems solvable?

$$\text{(a)} \begin{bmatrix} 1 & 4 & 2 \\ 2 & 8 & 4 \\ -1 & -4 & -2 \end{bmatrix} \begin{bmatrix} v_1 \\ v_2 \\ v_3 \end{bmatrix} = \begin{bmatrix} b_1 \\ b_2 \\ b_3 \end{bmatrix} \quad \text{(b)} \begin{bmatrix} 1 & 4 \\ 2 & 9 \\ -1 & -4 \end{bmatrix} \begin{bmatrix} v_1 \\ v_2 \end{bmatrix} = \begin{bmatrix} b_1 \\ b_2 \\ b_3 \end{bmatrix}$$

**13** Adding row 1 of $A$ to row 2 produces $B$. Adding column 1 to column 2 produces $C$. Which matrices have the same column space? Which have the same *row space*?

$$A = \begin{bmatrix} 1 & 3 \\ 2 & 6 \end{bmatrix} \quad \text{and} \quad B = \begin{bmatrix} 1 & 3 \\ 3 & 9 \end{bmatrix} \quad \text{and} \quad C = \begin{bmatrix} 1 & 4 \\ 2 & 8 \end{bmatrix}.$$

**14** For which vectors $(b_1, b_2, b_3)$ do these systems have a solution?

$$\begin{bmatrix} 1 & 1 & 1 \\ 0 & 1 & 1 \\ 0 & 0 & 1 \end{bmatrix} \begin{bmatrix} x_1 \\ x_2 \\ x_3 \end{bmatrix} = \begin{bmatrix} b_1 \\ b_2 \\ b_3 \end{bmatrix} \quad \text{and} \quad \begin{bmatrix} 1 & 1 & 1 \\ 0 & 1 & 1 \\ 0 & 0 & 0 \end{bmatrix} \begin{bmatrix} x_1 \\ x_2 \\ x_3 \end{bmatrix} = \begin{bmatrix} b_1 \\ b_2 \\ b_3 \end{bmatrix}$$

$$\text{and} \quad \begin{bmatrix} 1 & 1 & 1 \\ 0 & 0 & 1 \\ 0 & 0 & 1 \end{bmatrix} \begin{bmatrix} x_1 \\ x_2 \\ x_3 \end{bmatrix} = \begin{bmatrix} b_1 \\ b_2 \\ b_3 \end{bmatrix}.$$

**15** (Recommended) If we add an extra column $\boldsymbol{b}$ to a matrix $A$, then the column space gets larger unless ______. Give an example where the column space gets larger and an example where it doesn't. Why is $A\boldsymbol{v} = \boldsymbol{b}$ solvable exactly when the column space *doesn't* get larger? Then it is the same for $A$ and $\begin{bmatrix} A & \boldsymbol{b} \end{bmatrix}$.

**16** The columns of $AB$ are combinations of the columns of $A$. This means: *The column space of $AB$ is contained in* (possibly equal to) *the column space of $A$.* Give an example where the column spaces of $A$ and $AB$ are not equal.

**17** Suppose $Av = b$ and $Aw = b^*$ are both solvable. Then $Az = b + b^*$ is solvable. What is $z$? This translates into: If $b$ and $b^*$ are in the column space $C(A)$, then $b + b^*$ is also in $C(A)$.

**18** If $A$ is any 5 by 5 invertible matrix, then its column space is ____. Why?

**19** True or false (with a counterexample if false):

(a) The vectors $b$ that are not in the column space $C(A)$ form a subspace.

(b) If $C(A)$ contains only the zero vector, then $A$ is the zero matrix.

(c) The column space of $2A$ equals the column space of $A$.

(d) The column space of $A - I$ equals the column space of $A$ (test this).

**20** Construct a 3 by 3 matrix whose column space contains $(1, 1, 0)$ and $(1, 0, 1)$ but not $(1, 1, 1)$. Construct a 3 by 3 matrix whose column space is only a line.

**21** If the 9 by 12 system $Av = b$ is solvable for every $b$, then $C(A)$ must be ____.

## Challenge Problems

**22** Suppose $\mathbf{S}$ and $\mathbf{T}$ are two subspaces of a vector space $\mathbf{V}$. The **sum** $\mathbf{S} + \mathbf{T}$ contains all sums $s + t$ of a vector $s$ in $\mathbf{S}$ and a vector $t$ in $\mathbf{T}$. Then $\mathbf{S} + \mathbf{T}$ is a vector space.

If $\mathbf{S}$ and $\mathbf{T}$ are lines in $\mathbf{R}^m$, what is the difference between $\mathbf{S} + \mathbf{T}$ and $\mathbf{S} \cup \mathbf{T}$? That union contains all vectors from $\mathbf{S}$ and all vectors from $\mathbf{T}$. Explain this statement: *The span of* $\mathbf{S} \cup \mathbf{T}$ *is* $\mathbf{S} + \mathbf{T}$.

**23** If $\mathbf{S}$ is the column space of $A$ and $\mathbf{T}$ is $C(B)$, then $\mathbf{S} + \mathbf{T}$ is the column space of what matrix $M$? The columns of $A$ and $B$ and $M$ are all in $\mathbf{R}^m$. (I don't think $A + B$ is always a correct $M$.)

**24** Show that the matrices $A$ and $\begin{bmatrix} A & AB \end{bmatrix}$ (this has extra columns) have the same column space. But find a square matrix with $C(A^2)$ smaller than $C(A)$.

**25** An $n$ by $n$ matrix has $C(A) = \mathbf{R}^n$ exactly when $A$ is an ____ matrix.

## 5.2 The Nullspace of $A$ : Solving $A\boldsymbol{v} = \mathbf{0}$

This section is about the subspace containing all solutions to $A\boldsymbol{v} = \mathbf{0}$. The $m$ by $n$ matrix $A$ can be square or rectangular. *One immediate solution is* $\boldsymbol{v} = \mathbf{0}$. For invertible matrices this is the only solution. For other matrices, not invertible, there are nonzero solutions to $A\boldsymbol{v} = \mathbf{0}$. *Each solution* $\boldsymbol{v}$ *belongs to the nullspace of* $\boldsymbol{N}(A)$.

Elimination will find all solutions and identify this very important subspace.

***The nullspace of*** $A$ ***consists of all solutions to*** $A\boldsymbol{v} = \mathbf{0}$. **These vectors** $\boldsymbol{v}$ **are in** $\mathbf{R}^n$.

Check that the solution vectors form a subspace. Suppose $\boldsymbol{v}$ and $\boldsymbol{w}$ are in the nullspace, so that $A\boldsymbol{v} = \mathbf{0}$ and $A\boldsymbol{w} = \mathbf{0}$. The rules of matrix multiplication give $A(\boldsymbol{v} + \boldsymbol{w}) = \mathbf{0} + \mathbf{0}$. The rules also give $A(c\boldsymbol{v}) = c\mathbf{0}$. The right sides are still zero. Therefore $\boldsymbol{v} + \boldsymbol{w}$ and $c\boldsymbol{v}$ are also in the nullspace $\boldsymbol{N}(A)$. Since we can add and multiply without leaving the nullspace, it is a subspace.

The solution vectors $\boldsymbol{v}$ have $n$ components. They are vectors in $\mathbf{R}^n$, so *the nullspace* $\boldsymbol{N}(A)$ *is a subspace of* $\mathbf{R}^n$. The column space $\boldsymbol{C}(A)$ is a subspace of $\mathbf{R}^m$.

If the right side $\boldsymbol{b}$ is not zero, the solutions of $A\boldsymbol{v} = \boldsymbol{b}$ do *not* form a subspace. The vector $\boldsymbol{v} = \mathbf{0}$ is only a solution if $\boldsymbol{b} = \mathbf{0}$. When the set of solutions does not include $\boldsymbol{v} = \mathbf{0}$, it cannot be a subspace. Section 5.3 will show how the solutions to $A\boldsymbol{v} = \boldsymbol{b}$ (if there are any solutions) are shifted away from the origin by one particular solution $\boldsymbol{v}_p$.

**Example 1** $x + 2y + 3z = 0$ comes from the 1 by 3 matrix $A = [1 \;\; 2 \;\; 3]$. This equation $A\boldsymbol{v} = \mathbf{0}$ produces a plane through the origin $(0, 0, 0)$. The plane is a subspace of $\mathbf{R}^3$, and *it is the nullspace of* $A$.

The solutions to $x + 2y + 3z = 6$ also form a plane, but not a subspace.

**Example 2** Describe the nullspace of $A = \begin{bmatrix} 1 & 2 \\ 3 & 6 \end{bmatrix}$. This matrix is singular!

**Solution** Apply elimination to the linear equations $A\boldsymbol{v} = \mathbf{0}$:

$$\begin{aligned} v_1 + 2v_2 &= 0 \\ 3v_1 + 6v_2 &= 0 \end{aligned} \quad \rightarrow \quad \begin{aligned} v_1 + 2v_2 &= 0 \\ \mathbf{0} &= \mathbf{0} \end{aligned}$$

There is really only one equation. The second equation is the first equation multiplied by 3. In the row picture, the line $v_1 + 2v_2 = 0$ is the same as the line $3v_1 + 6v_2 = 0$. That line is the nullspace $\boldsymbol{N}(A)$. It contains all solutions $\boldsymbol{v} = (v_1, v_2)$.

To describe this line of solutions, here is an efficient way. Choose one point on the line (one "***special solution***"). Then all points on the line are multiples of this one. We choose the second component to be $v_2 = 1$ (a special choice). From the equation $v_1 + 2v_2 = 0$, the first component must be $v_1 = -2$. The special solution $\boldsymbol{s}$ is $(-2, 1)$:

**Special solution** The nullspace of $A = \begin{bmatrix} 1 & 2 \\ 3 & 6 \end{bmatrix}$ contains all multiples of $\boldsymbol{s} = \begin{bmatrix} -2 \\ 1 \end{bmatrix}$.

This is the best way to describe the nullspace, by computing special solutions to $A\boldsymbol{v} = \mathbf{0}$.

***The nullspace consists of all combinations of the special solutions.***

The plane $x + 2y + 3z = 0$ in Example 1 had *two* special solutions:

$$\begin{bmatrix} 1 & 2 & 3 \end{bmatrix}\begin{bmatrix} x \\ y \\ z \end{bmatrix} = 0 \text{ has the special solutions } \boldsymbol{s}_1 = \begin{bmatrix} -2 \\ 1 \\ 0 \end{bmatrix} \text{ and } \boldsymbol{s}_2 = \begin{bmatrix} -3 \\ 0 \\ 1 \end{bmatrix}.$$

Those vectors $\boldsymbol{s}_1$ and $\boldsymbol{s}_2$ lie on the plane $x + 2y + 3z = 0$, which is the nullspace of $A = \begin{bmatrix} 1 & 2 & 3 \end{bmatrix}$. All vectors on the plane are combinations of $\boldsymbol{s}_1$ and $\boldsymbol{s}_2$.

Notice what is special about $\boldsymbol{s}_1$ and $\boldsymbol{s}_2$. They have ones and zeros in the last two components. *Those components are "free" and we choose them specially as* 1 *and* 0. Then the first components $-2$ and $-3$ are determined by the equation $A\boldsymbol{v} = \mathbf{0}$.

The first column of $A = \begin{bmatrix} 1 & 2 & 3 \end{bmatrix}$ contains the *pivot*, so the first component $\boldsymbol{v}_1$ is *not free*. The free components correspond to columns without pivots. This description of special solutions will be completed after one more example.

The special choice (one or zero) is only for the free variables in the special solutions.

**Example 3** Describe the nullspaces $\boldsymbol{N}(A), \boldsymbol{N}(B), \boldsymbol{N}(C)$ of these three matrices:

$$A = \begin{bmatrix} 1 & 2 \\ 3 & 8 \end{bmatrix} \quad B = \begin{bmatrix} A \\ 2A \end{bmatrix} = \begin{bmatrix} 1 & 2 \\ 3 & 8 \\ 2 & 4 \\ 6 & 16 \end{bmatrix} \quad C = \begin{bmatrix} A & 2A \end{bmatrix} = \begin{bmatrix} 1 & 2 & 2 & 4 \\ 3 & 8 & 6 & 16 \end{bmatrix}.$$

**Solution** The equation $A\boldsymbol{v} = \mathbf{0}$ has only the zero solution $\boldsymbol{v} = \mathbf{0}$. *The nullspace is* $\mathbf{Z}$. It contains only the single point $\boldsymbol{v} = \mathbf{0}$ in $\mathbf{R}^2$. This comes from elimination:

$$\begin{bmatrix} 1 & 2 \\ 3 & 8 \end{bmatrix}\begin{bmatrix} v_1 \\ v_2 \end{bmatrix} = \begin{bmatrix} 0 \\ 0 \end{bmatrix} \text{ yields } \begin{bmatrix} 1 & 2 \\ 0 & 2 \end{bmatrix}\begin{bmatrix} v_1 \\ v_2 \end{bmatrix} = \begin{bmatrix} 0 \\ 0 \end{bmatrix} \text{ and } \begin{bmatrix} v_1 = 0 \\ v_2 = 0 \end{bmatrix}.$$

$A$ is invertible. There are no special solutions. All columns of this $A$ have pivots.

The rectangular matrix $B$ has the same nullspace $\mathbf{Z}$. The first two equations in $B\boldsymbol{v} = \mathbf{0}$ again require $\boldsymbol{v} = \mathbf{0}$. The last two equations would also force $\boldsymbol{v} = \mathbf{0}$. When we add extra equations, the nullspace certainly cannot become larger. The extra rows impose more conditions on the vectors $\boldsymbol{v}$ in the nullspace.

The rectangular matrix $C$ is different. It has extra columns instead of extra rows. The solution vector $\boldsymbol{v}$ has *four* components. Elimination will produce pivots in the first two columns of $C$, but the last two columns are "free". *They don't have pivots*:

**2 pivot columns**
**2 free columns**

$$C = \begin{bmatrix} 1 & 2 & 2 & 4 \\ 3 & 8 & 6 & 16 \end{bmatrix} \text{ becomes } U = \begin{bmatrix} 1 & 2 & 2 & 4 \\ 0 & 2 & 0 & 4 \end{bmatrix}$$

$$\uparrow \quad \uparrow \quad \uparrow \quad \uparrow$$

**pivot columns** **free columns**

For the free variables $v_3$ and $v_4$, we make special choices of ones and zeros. First $v_3 = 1$, $v_4 = 0$ and second $v_3 = 0$, $v_4 = 1$. Then the pivot variables $v_1$ and $v_2$ are determined.

Solve $Uv = \mathbf{0}$ to get two special solutions in the nullspace of $C$ (and $U$).

**Special solutions** $s_1$ **and** $s_2$

$$s_1 = \begin{bmatrix} -2 \\ 0 \\ 1 \\ 0 \end{bmatrix} \text{ and } s_2 = \begin{bmatrix} 0 \\ -2 \\ 0 \\ 1 \end{bmatrix} \begin{matrix} \leftarrow \text{pivot} \\ \leftarrow \quad \text{variables} \\ \leftarrow \text{free} \\ \leftarrow \quad \text{variables} \end{matrix}$$

One more comment to anticipate what is coming soon. Elimination will not stop at the upper triangular $U$ ! We can continue to make this matrix simpler, in two ways:

1. ***Produce zeros above the pivots.* Eliminate upward.**
2. ***Produce ones in the pivots.* Divide the whole row by its pivot.**

Those steps don't change the zero vector on the right side of the equation. The nullspace stays the same. This nullspace becomes easiest to see when we reach the ***reduced row echelon form*** $R$. It has $I$ in the pivot columns, when row 2 is divided by 2:

**Reduced form $R$**

$$U = \begin{bmatrix} 1 & 2 & 2 & 4 \\ 0 & 2 & 0 & 4 \end{bmatrix} \quad \text{becomes} \quad R = \begin{bmatrix} 1 & 0 & 2 & 0 \\ 0 & 1 & 0 & 2 \end{bmatrix}.$$

$\uparrow \ \uparrow$

**Now the pivot columns contain $I$**

I subtracted row 2 of $U$ from row 1, and then multiplied row 2 by $\frac{1}{2}$. The original two equations have simplified to $x_1 + 2x_3 = 0$ and $x_2 + 2x_4 = 0$.

The first special solution is still $s_1 = (-2, 0, 1, 0)$. All special solutions are unchanged. Special solutions are much easier to find from the reduced system $Rv = \mathbf{0}$.

Before moving to $m$ by $n$ matrices $A$ and their nullspaces $\boldsymbol{N}(A)$ and special solutions, allow me to repeat one comment. For many matrices, the only solution to $Av = \mathbf{0}$ is $v = \mathbf{0}$. Their nullspaces $\boldsymbol{N}(A) = \mathbf{Z}$ contain only that zero vector. The only combination of the columns that produces $\boldsymbol{b} = \mathbf{0}$ is then the "zero combination" or "trivial combination". The solution is trivial (just $v = \mathbf{0}$) but the idea is not trivial.

This case of a zero nullspace $\mathbf{Z}$ is of the greatest importance. It says that the columns of $A$ are **independent**. No combination of columns gives the zero vector (except the zero combination). All columns have pivots, and no columns are free. You will see this idea of independence again ...

## Solving $Av = \mathbf{0}$ by Elimination

This is important. ***$A$ is rectangular and we still use elimination***. We solve $m$ equations in $n$ unknowns. After $A$ is simplified to $U$ or to $R$, we read off the solution (or solutions). Remember the two stages (forward and back) in solving $Av = \mathbf{0}$:

1. **Elimination** takes $A$ to a triangular $U$ (or its reduced form $R$).

2. **Back substitution** in $U\boldsymbol{v} = \mathbf{0}$ or $R\boldsymbol{v} = \mathbf{0}$ produces $\boldsymbol{v}$.

You will notice a difference in back substitution, when $A$ and $U$ have fewer than $n$ pivots. *We are allowing all matrices in this chapter*, not just the nice ones (which are square matrices with inverses).

Pivots are still nonzero. The columns below the pivots are still zero. But it might happen that a column has no pivot. That free column doesn't stop the calculation. ***Go on to the next column***. The first example is a 3 by 4 matrix with two pivots:

$$\textbf{Elimination on} \qquad A = \begin{bmatrix} 1 & 1 & 2 & 3 \\ 2 & 2 & 8 & 10 \\ 3 & 3 & 10 & 13 \end{bmatrix}.$$

Certainly $a_{11} = 1$ is the first pivot. Clear out the 2 and 3 below that pivot:

$$A \to \begin{bmatrix} 1 & 1 & 2 & 3 \\ 0 & 0 & 4 & 4 \\ 0 & 0 & 4 & 4 \end{bmatrix} \qquad \begin{matrix} \text{(subtract } 2 \times \text{ row 1)} \\ \text{(subtract } 3 \times \text{ row 1)} \end{matrix}$$

The second column has a zero in the pivot position. We look below the zero for a nonzero entry, ready to do a row exchange. *The entry below that position is also zero.* Elimination can do nothing with the second column. This signals trouble, which we expect anyway for a rectangular matrix. There is no reason to quit, and we go on to the third column.

The second pivot is 4 (but it is in the third column). Subtracting row 2 from row 3 clears out that third column below the pivot. **The pivot columns are** 1 **and** 3:

$$\textbf{Triangular } \boldsymbol{U} \qquad U = \begin{bmatrix} \mathbf{1} & 1 & 2 & 3 \\ 0 & 0 & \mathbf{4} & 4 \\ 0 & 0 & 0 & 0 \end{bmatrix} \qquad \begin{matrix} \textit{Only two pivots} \\ \textit{The last equation} \\ \textit{became } 0 = 0 \end{matrix}$$

The fourth column also has a zero in the pivot position—but nothing can be done. There is no row below it to exchange, and forward elimination is complete. The matrix has three rows, four columns, and *only two pivots.* The third equation in $A\boldsymbol{v} = \mathbf{0}$ is the sum of the first two. It is automatically satisfied ($0 = 0$) when the first two equations are satisfied. Elimination reveals the inner truth about $A\boldsymbol{v} = \mathbf{0}$. Soon we push on from $U$ to $R$.

Now comes back substitution, to find all solutions to $U\boldsymbol{v} = \mathbf{0}$. With four unknowns and only two pivots, there are many solutions. The question is how to write them all down. A good method is to separate the ***pivot variables*** from the ***free variables***.

| | | |
|---|---|---|
| **P** | The ***pivot*** variables are $v_1$ and $v_3$. | **Columns 1 and 3 contain pivots.** |
| **F** | The ***free*** variables are $v_2$ and $v_4$. | **Columns 2 and 4 have no pivots.** |

The free variables $v_2$ and $v_4$ can be given any values whatsoever. Then back substitution finds the pivot variables $v_1$ and $v_3$. (In Chapter 2 no variables were free. When $A$ is invertible, all variables are pivot variables.) The simplest choices for the free variables are ones and zeros. Those choices give the *special solutions*.

***Special solutions*** to $v_1 + v_2 + 2v_3 + 3v_4 = 0$ and $4v_3 + 4v_4 = 0$

- Set $v_2 = 1$ and $v_4 = 0$. By back substitution $v_3 = 0$. Then $v_1 = -1$.
- Set $v_2 = 0$ and $v_4 = 1$. By back substitution $v_3 = -1$. Then $v_1 = -1$.

These special solutions solve $Uv = \mathbf{0}$ and therefore $Av = \mathbf{0}$. They are in the nullspace. The good thing is that *every solution is a combination of the special solutions*.

***Complete solution to*** $Av = \mathbf{0}$

$$v = v_2 \underset{\textbf{special}}{\begin{bmatrix} -1 \\ 1 \\ 0 \\ 0 \end{bmatrix}} + v_4 \underset{\textbf{special}}{\begin{bmatrix} -1 \\ 0 \\ -1 \\ 1 \end{bmatrix}} = \underset{\textbf{complete}}{\begin{bmatrix} -v_2 - v_4 \\ v_2 \\ -v_4 \\ v_4 \end{bmatrix}}. \qquad (1)$$

Please look again at that answer. It is the main goal of this section. The vector $s_1 = (-1, 1, 0, 0)$ is the special solution when $v_2 = 1$ and $v_4 = 0$. The second special solution has $v_2 = 0$ and $v_4 = 1$. ***All solutions are linear combinations of*** $s_1$ ***and*** $s_2$. The special solutions are in the nullspace $\boldsymbol{N}(A)$, and their combinations fill the whole nullspace.

There is a special solution for each free variable. If no variables are free—this means all $n$ columns have pivots—then the only solution to $Uv = \mathbf{0}$ and $Av = \mathbf{0}$ is the trivial solution $v = \mathbf{0}$. With no free variables, the nullspace is $\mathbf{Z}$.

**Example 4** Find the nullspace of $U = \begin{bmatrix} 1 & 5 & 7 \\ 0 & 0 & 9 \end{bmatrix}$.

The second column of $U$ has no pivot. So $v_2$ is free. The special solution has $v_2 = 1$. Back substitution into $9v_3 = 0$ gives $v_3 = 0$. Then $v_1 + 5v_2 = 0$ or $v_1 = -5$. The solutions to $Uv = \mathbf{0}$ are multiples of one special solution $s_1$ :

$$v = c \begin{bmatrix} -5 \\ 1 \\ 0 \end{bmatrix}$$

The nullspace of $U$ is a line in $\mathbf{R}^3$.
It contains multiples of the special solution $s_1 = (-5, 1, 0)$.
One variable is free.

**The matrix $R$ has zeros above and below the pivots, and ones in the pivots.** By continuing elimination on $U$, the 7 is removed and the pivot changes from 9 to 1. The final result will be the **reduced row echelon form $R$** :

$$U = \begin{bmatrix} 1 & 5 & 7 \\ 0 & 0 & 9 \end{bmatrix} \text{ reduces to } R = \begin{bmatrix} \mathbf{1} & 5 & 0 \\ 0 & 0 & \mathbf{1} \end{bmatrix} = \text{rref}(U).$$

## Echelon Matrices

Forward elimination goes from $A$ to $U$. It acts by row operations, including row exchanges. It goes on to the next column when no pivot is available in the current column. The $m$ by $n$ "staircase" $U$ is an ***echelon matrix***.

Here is a 4 by 7 echelon matrix with the three pivots $\boldsymbol{p}$ highlighted in boldface:

$$U = \begin{bmatrix} \boldsymbol{p} & \boldsymbol{x} & x & x & x & \boldsymbol{x} & x \\ 0 & \boldsymbol{p} & x & x & x & \boldsymbol{x} & x \\ 0 & 0 & 0 & 0 & 0 & \boldsymbol{p} & x \\ 0 & 0 & 0 & 0 & 0 & 0 & 0 \end{bmatrix}$$

**Three pivot variables** $\boldsymbol{v_1, v_2, v_6}$
**Four free variables** $\boldsymbol{v_3, v_4, v_5, v_7}$
**Four special solutions in $N(U)$**
**$R$ will have $p = 1$ and bold $x = 0$**

**Question** What are the column space and the nullspace for this matrix?

***Answer*** The columns have four components so they lie in $\mathbf{R}^4$. (Not in $\mathbf{R}^3$!) The fourth component of every column is zero. *The column space $\boldsymbol{C}(U)$ consists of all vectors of the form* $(b_1, b_2, b_3, 0)$. For those vectors we can solve $U\boldsymbol{v} = \boldsymbol{b}$ by back substitution. These vectors $\boldsymbol{b}$ are all possible combinations of the seven columns.

The nullspace $\boldsymbol{N}(U)$ is a subspace of $\mathbf{R}^7$. The solutions to $U\boldsymbol{v} = \mathbf{0}$ are all the combinations of the four special solutions—*one for each free variable*:

1. Columns 3, 4, 5, 7 have no pivots. The free variables are $v_3, v_4, v_5, v_7$.
2. Set one free variable to 1 and set the other free variables to zero.
3. Solve $U\boldsymbol{v} = \mathbf{0}$ for the pivot variables $v_1, v_2, v_6$ to get a special solution.

The nonzero rows of an echelon matrix go down in a staircase pattern. The pivots are the first nonzero entries in those rows. There is a column of zeros below every pivot.

## The Counting Theorem

Counting the pivots leads to an extremely important theorem. Suppose $A$ has more columns than rows. ***With $n > m$ there is at least one free variable.*** The system $A\boldsymbol{v} = \mathbf{0}$ has at least one special solution. This solution is *not zero*!

Suppose $A\boldsymbol{v} = \mathbf{0}$ has more unknowns than equations ($\boldsymbol{n > m}$, more columns than rows). Then there are **nonzero solutions** in $\boldsymbol{N}(A)$. There must be free columns, without pivots.

A short wide matrix ($n > m$) always has nonzero vectors in its nullspace. There must be at least $n - m$ free variables, since the number of pivots cannot exceed $m$. (The matrix only has $m$ rows, and a row never has two pivots.) Of course a row might have *no* pivot—which means an extra free variable. But here is the point: When there is a free variable, it can be set to 1. Then the equation $A\boldsymbol{v} = \mathbf{0}$ has a nonzero solution.

To repeat: There are at most $m$ pivots. With $n > m$, the system $Av = \mathbf{0}$ has a nonzero solution. Actually there are infinitely many solutions, since any multiple $cv$ is also a solution. The nullspace contains at least a line of solutions. With two free variables, there will be two special solutions and the nullspace will be even larger.

*The nullspace is a subspace. Its "dimension" is the number of special solutions.* This central idea—the ***dimension*** of a subspace—is defined and explained in this chapter.

| | | | | | | |
|---|---|---|---|---|---|---|
| Dimension of $\boldsymbol{C}(A)$ | = | **rank** | of matrix | = | number of pivot columns |
| Dimension of $\boldsymbol{N}(A)$ | = | **nullity** | of matrix | = | number of free columns. |

***Counting Theorem with n columns*** ***Rank r plus nullity n − r equals n.***

## The Reduced Row Echelon Matrix $R$

From an echelon matrix $U$ we go one more step. Continue with a 3 by 4 example:

$$U = \begin{bmatrix} 1 & 1 & 2 & 3 \\ 0 & 0 & 4 & 4 \\ 0 & 0 & 0 & 0 \end{bmatrix}.$$

*We can divide the second row by* 4. Then both pivots equal 1. *We can subtract* 2 *times this new row* $\begin{bmatrix} 0 & 0 & 1 & 1 \end{bmatrix}$ *from the row above.* ***The reduced row echelon matrix R has zeros above the pivots as well as below***:

| | | |
|---|---|---|
| **Reduced row echelon matrix** | $R = \text{rref}(A) = \begin{bmatrix} \mathbf{1} & 1 & \mathbf{0} & 1 \\ \mathbf{0} & 0 & \mathbf{1} & 1 \\ 0 & 0 & 0 & 0 \end{bmatrix}$ | **Pivot rows contain *I*** |

*R* ***has 1's as pivots. Zeros above pivots come from upward elimination.***

**Important** *If A is invertible, its reduced row echelon form is the identity matrix* $R = I$. This is the ultimate in row reduction. Of course the nullspace is then **Z**.

The zeros in $R$ make it easy to find the special solutions (the same as before):

**1.** Set $v_2 = 1$ and $v_4 = 0$. Solve $Rv = \mathbf{0}$. Then $v_1 = -1$ and $v_3 = 0$.

Those numbers $-1$ and $0$ are sitting in column 2 of $R$ (with plus signs).

**2.** Set $v_2 = 0$ and $v_4 = 1$. Solve $Rv = \mathbf{0}$. Then $v_1 = -1$ and $v_3 = -1$.

Those numbers $-1$ and $-1$ are sitting in column 4 (with plus signs).

*By reversing signs we can read off the special solutions directly from R. The nullspace* $\boldsymbol{N}(A) = \boldsymbol{N}(U) = \boldsymbol{N}(R)$ *contains all combinations of the special solutions:*

$$v = v_2 \begin{bmatrix} -1 \\ 1 \\ 0 \\ 0 \end{bmatrix} + v_4 \begin{bmatrix} -1 \\ 0 \\ -1 \\ 1 \end{bmatrix} = (\textbf{\textit{complete solution of }} Av = \mathbf{0}).$$

The next section of the book moves firmly from $U$ to the row reduced form $R$. The MATLAB command $[\,R, pivcol\,] = \text{rref}(A)$ produces $R$ and a list of the pivot columns.

## ■ REVIEW OF THE KEY IDEAS ■

1. The nullspace $\boldsymbol{N}(A)$ is a subspace of $\mathbf{R}^n$. It contains all solutions to $A\boldsymbol{v} = \mathbf{0}$.
2. Elimination produces an echelon matrix $U$, and then a row reduced $R$ (pivots $= 1$).
3. Every free column of $U$ or $R$ leads to a special solution. The free variable equals 1 and the other free variables equal 0. Back substitution solves $A\boldsymbol{v} = \mathbf{0}$.
4. The complete solution to $A\boldsymbol{v} = \mathbf{0}$ is a combination of the special solutions.
5. $A$ has at least one free column and one special solution if $n > m$ : $\boldsymbol{N}(A)$ is not $\boldsymbol{Z}$.
6. The count of pivot columns and free columns is $r + (n - r) = n$.

## ■ WORKED EXAMPLES ■

**3.2 A** Create a 3 by 4 matrix $R$ whose special solutions to $R\boldsymbol{v} = 0$ are $\boldsymbol{s}_1$ and $\boldsymbol{s}_2$ :

$$\boldsymbol{s}_1 = \begin{bmatrix} -3 \\ 1 \\ 0 \\ 0 \end{bmatrix} \quad \text{and} \quad \boldsymbol{s}_2 = \begin{bmatrix} -2 \\ 0 \\ -6 \\ 1 \end{bmatrix} \qquad \begin{array}{c} \text{pivot columns 1 and 3} \\ \text{free variables } v_2 \text{ and } v_4 \end{array}$$

Describe all matrices $A$ with this nullspace $\boldsymbol{N}(A) =$ combinations of $\boldsymbol{s}_1$ and $\boldsymbol{s}_2$.

**Solution** The reduced matrix $R$ has pivots $= 1$ in columns 1 and 3. There is no third pivot, so the third row of $R$ is all zeros. The free columns 2 and 4 will be combinations of the pivot columns:

$$R = \begin{bmatrix} 1 & 3 & 0 & 2 \\ 0 & 0 & 1 & 6 \\ 0 & 0 & 0 & 0 \end{bmatrix} \quad \text{has} \quad R\boldsymbol{s}_1 = \mathbf{0} \quad \text{and} \quad R\boldsymbol{s}_2 = \mathbf{0}.$$

The entries $3, 2, 6$ in $R$ are the negatives of $-3, -2, -6$ in the special solutions!

$R$ is only one matrix (one possible $A$) with the required nullspace. We could do any elementary operations on $R$—exchange rows, multiply a row by any $c \neq 0$, subtract any multiple of one row from another. ***R* can be multiplied** (*on the left*) **by any invertible matrix, without changing its nullspace.**

Every 3 by 4 matrix has at least one special solution. *These matrices have two.*

**3.2 B** Find the special solutions and the *complete solutions* to $A\boldsymbol{v} = \mathbf{0}$ and $A_2\boldsymbol{v} = \mathbf{0}$:

$$A = \begin{bmatrix} 3 & 6 \\ 1 & 2 \end{bmatrix} \qquad A_2 = \begin{bmatrix} A & A \end{bmatrix} = \begin{bmatrix} 3 & 6 & 3 & 6 \\ 1 & 2 & 1 & 2 \end{bmatrix}.$$

Which are the pivot columns? Which are the free variables? What is $R$ in each case?

**Solution** $A\boldsymbol{v} = \mathbf{0}$ has one special solution $\boldsymbol{s} = (-2, 1)$. The line of all $c\boldsymbol{s}$ is the complete solution. The first column of $A$ is its pivot column, and $v_2$ is the free variable:

$$A = \begin{bmatrix} 3 & 6 \\ 1 & 2 \end{bmatrix} \to R = \begin{bmatrix} 1 & 2 \\ 0 & 0 \end{bmatrix} \qquad \begin{bmatrix} A & A \end{bmatrix} \to R_2 = \begin{bmatrix} 1 & 2 & 1 & 2 \\ 0 & 0 & 0 & 0 \end{bmatrix}$$

Notice that $R_2$ has only one pivot column (the first column). All the variables $v_2, v_3, v_4$ are free. There are three special solutions to $A_2\,\boldsymbol{v} = \mathbf{0}$ (and also $R_2\,\boldsymbol{v} = \mathbf{0}$):

$\boldsymbol{s}_1 = (-2, 1, 0, 0)$ $\boldsymbol{s}_2 = (-1, 0, 1, 0)$ $\boldsymbol{s}_3 = (-2, 0, 0, 1)$ **Complete** $\boldsymbol{v} = c_1\boldsymbol{s}_1 + c_2\boldsymbol{s}_2 + c_3\boldsymbol{s}_3$.

**With $r$ pivots, $A$ has $n - r$ free variables and $A\boldsymbol{v} = \mathbf{0}$ has $n - r$ special solutions.**

## Problem Set 5.2

**Questions 1–4 and 5–8 are about the matrices in Problems 1 and 5.**

**1** Reduce these matrices to their ordinary echelon forms $U$:

$$A = \begin{bmatrix} 1 & 2 & 2 & 4 & 6 \\ 1 & 2 & 3 & 6 & 9 \\ 0 & 0 & 1 & 2 & 3 \end{bmatrix} \qquad B = \begin{bmatrix} 2 & 4 & 2 \\ 0 & 4 & 4 \\ 0 & 8 & 8 \end{bmatrix}.$$

Which are the free variables and which are the pivot variables?

**2** For the matrices in Problem 1, find a special solution for each free variable. (Set the free variable to 1. Set the other free variables to zero.)

**3** By combining the special solutions in Problem 2, describe every solution to $A\boldsymbol{v} = \mathbf{0}$ and $B\boldsymbol{v} = \mathbf{0}$. The nullspace contains only $\boldsymbol{v} = \mathbf{0}$ when there are no _____.

**4** By further row operations on each $U$ in Problem 1, find the reduced echelon form $R$. *True or false*: The nullspace of $R$ equals the nullspace of $U$.

**5** By row operations reduce this new $A$ and $B$ to triangular echelon form $U$. Write down a 2 by 2 lower triangular $L$ such that $B = LU$.

$$A = \begin{bmatrix} -1 & 3 & 5 \\ -2 & 6 & 10 \end{bmatrix} \qquad B = \begin{bmatrix} -1 & 3 & 5 \\ -2 & 6 & 7 \end{bmatrix}.$$

**6** For the same $A$ and $B$, find the special solutions to $Av = \mathbf{0}$ and $Bv = \mathbf{0}$. For an $m$ by $n$ matrix, the number of pivot variables plus the number of free variables is _____ .

**7** In Problem 5, describe the nullspaces of $A$ and $B$ in two ways. Give the equations for the plane or the line, and give all vectors $v$ that satisfy those equations as combinations of the special solutions.

**8** Reduce the echelon forms $U$ in Problem 5 to $R$. For each $R$ draw a box around the identity matrix that is in the pivot rows and pivot columns.

**Questions 9–17 are about free variables and pivot variables.**

**9** True or false (with reason if true or example to show it is false):

(a) A square matrix has no free variables.

(b) An invertible matrix has no free variables.

(c) An $m$ by $n$ matrix has no more than $n$ pivot variables.

(d) An $m$ by $n$ matrix has no more than $m$ pivot variables.

**10** Construct 3 by 3 matrices $A$ to satisfy these requirements (if possible):

(a) $A$ has no zero entries but $U = I$.

(b) $A$ has no zero entries but $R = I$.

(c) $A$ has no zero entries but $R = U$.

(d) $A = U = 2R$.

**11** Put as many 1's as possible in a 4 by 7 echelon matrix $U$ whose pivot columns are

(a) 2, 4, 5

(b) 1, 3, 6, 7

(c) 4 and 6.

**12** Put as many 1's as possible in a 4 by 8 *reduced* echelon matrix $R$ so that the free columns are

(a) 2, 4, 5, 6

(b) 1, 3, 6, 7, 8.

**13** Suppose column 4 of a 3 by 5 matrix is all zero. Then $v_4$ is certainly a _____ variable. The special solution for this variable is the vector $\boldsymbol{s}$ = _____ .

**14** Suppose the first and last columns of a 3 by 5 matrix are the same (not zero). Then _____ is a free variable. Find the special solution for this variable.

**15** Suppose an $m$ by $n$ matrix has $r$ pivots. The number of special solutions is _____ . The nullspace contains only $v = \mathbf{0}$ when $r$ = _____ . The column space is all of $\mathbf{R}^m$ when $r$ = _____ .

**16** The nullspace of a 5 by 5 matrix contains only $\boldsymbol{v} = \mathbf{0}$ when the matrix has ____ pivots. The column space is $\mathbf{R}^5$ when there are ____ pivots. Explain why.

**17** The equation $x - 3y - z = 0$ determines a plane in $\mathbf{R}^3$. What is the matrix $A$ in this equation? Which are the free variables? The special solutions are $(3, 1, 0)$ and ____.

**18** (Recommended) The plane $x - 3y - z = 12$ is parallel to the plane $x - 3y - z = 0$ in Problem 17. One particular point on this plane is $(12, 0, 0)$. All points on the plane have the form (fill in the first components)

$$\begin{bmatrix} x \\ y \\ z \end{bmatrix} = \begin{bmatrix} \\ 0 \\ 0 \end{bmatrix} + y \begin{bmatrix} \\ 1 \\ 0 \end{bmatrix} + z \begin{bmatrix} \\ 0 \\ 1 \end{bmatrix}.$$

**19** Prove that $U$ and $A = LU$ have the same nullspace when $L$ is invertible:

If $U\boldsymbol{v} = \mathbf{0}$ then $LU\boldsymbol{v} = \mathbf{0}$. If $LU\boldsymbol{v} = \mathbf{0}$, how do you know $U\boldsymbol{v} = \mathbf{0}$?

**20** Suppose column 1 + column 3 + column 5 = $\mathbf{0}$ in a 4 by 5 matrix with four pivots. Which column is sure to have no pivot (and which variable is free)? What is the special solution? What is the nullspace?

**Questions 21–28 ask for matrices (if possible) with specific properties.**

**21** Construct a matrix whose nullspace consists of all combinations of $(2, 2, 1, 0)$ and $(3, 1, 0, 1)$.

**22** Construct a matrix whose nullspace consists of all multiples of $(4, 3, 2, 1)$.

**23** Construct a matrix whose column space contains $(1, 1, 5)$ and $(0, 3, 1)$ and whose nullspace contains $(1, 1, 2)$.

**24** Construct a matrix whose column space contains $(1, 1, 0)$ and $(0, 1, 1)$ and whose nullspace contains $(1, 0, 1)$ and $(0, 0, 1)$.

**25** Construct a matrix whose column space contains $(1, 1, 1)$ and whose nullspace is the line of multiples of $(1, 1, 1, 1)$.

**26** Construct a 2 by 2 matrix whose nullspace equals its column space. This is possible.

**27** Why does no 3 by 3 matrix have a nullspace that equals its column space?

**28** (Important) If $AB = 0$ then the column space of $B$ is contained in the ____ of $A$. Give an example of $A$ and $B$.

**29** The reduced form $R$ of a 3 by 3 matrix with randomly chosen entries is almost sure to be ____. What reduced form $R$ is virtually certain if the random $A$ is 4 by 3?

**30** Show by example that these three statements are generally *false*:

(a) $A$ and $A^{\mathrm{T}}$ have the same nullspace.

(b) $A$ and $A^{\mathrm{T}}$ have the same free variables.

(c) If $R$ is the reduced form of $A$ then $R^{\mathrm{T}}$ is the reduced form of $A^{\mathrm{T}}$.

**31** If the nullspace of $A$ consists of all multiples of $\boldsymbol{v} = (2, 1, 0, 1)$, how many pivots appear in $U$? What is $R$?

**32** If the special solutions to $R\boldsymbol{v} = \mathbf{0}$ are in the columns of these $N$, go backward to find the nonzero rows of the reduced matrices $R$:

$$N = \begin{bmatrix} 2 & 3 \\ 1 & 0 \\ 0 & 1 \end{bmatrix} \quad \text{and} \quad N = \begin{bmatrix} 0 \\ 0 \\ 1 \end{bmatrix} \quad \text{and} \quad N = \begin{bmatrix} \; \end{bmatrix} \text{ (empty 3 by 1).}$$

**33** (a) What are the five 2 by 2 reduced echelon matrices $R$ whose entries are all 0's and 1's?

(b) What are the eight 1 by 3 matrices containing only 0's and 1's? Are all eight of them reduced echelon matrices $R$?

**34** Explain why $A$ and $-A$ always have the same reduced echelon form $R$.

## Challenge Problems

**35** If $A$ is 4 by 4 and invertible, describe all vectors in the nullspace of the 4 by 8 matrix $B = [A \;\; A]$.

**36** How is the nullspace $\boldsymbol{N}(C)$ related to the spaces $\boldsymbol{N}(A)$ and $\boldsymbol{N}(B)$, if $C = \begin{bmatrix} A \\ B \end{bmatrix}$?

**37** Kirchhoff's Law says that *current in* = *current out* at every node. This network has six currents $y_1, \ldots, y_6$ (the arrows show the positive direction, each $y_i$ could be positive or negative). Find the four equations $A\boldsymbol{y} = \mathbf{0}$ for Kirchhoff's Law at the four nodes. Reduce to $U\boldsymbol{y} = \mathbf{0}$. Find three special solutions in the nullspace of $A$.

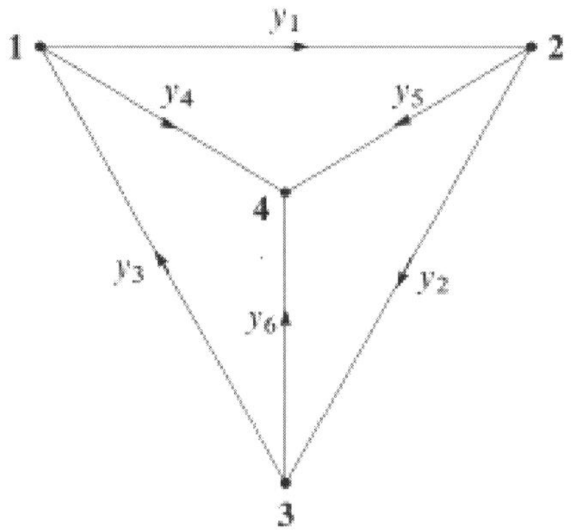

## 5.3 The Complete Solution to $A\boldsymbol{v} = \boldsymbol{b}$

To solve $A\boldsymbol{v} = \boldsymbol{b}$ by elimination, include $\boldsymbol{b}$ as a new column next to the $n$ columns of $A$. This "augmented matrix" is $\begin{bmatrix} A & \boldsymbol{b} \end{bmatrix}$. When the steps of elimination operate on $A$ (the left side of the equations), they also operate on the right side $\boldsymbol{b}$. So we always keep correct equations, and they become simple to solve.

There are still $r$ pivot columns and $n - r$ free columns in $A$. Each free column still gives a special solution to $A\boldsymbol{v} = \boldsymbol{0}$. The new question is to find a *particular solution* $\boldsymbol{v_p}$ with $A\boldsymbol{v}_p = \boldsymbol{b}$. That solution will exist unless elimination leads to an impossible equation (a zero row on the left side, a nonzero number on the right side). Then back substitution finds $\boldsymbol{v}_p$. **Every solution to $A\boldsymbol{v} = \boldsymbol{b}$ has the form $\boldsymbol{v_p} + \boldsymbol{v_n}$.**

In the process of elimination, we discover the **rank** of $A$. This is the number of pivots. The rank is also the number of nonzero rows after elimination. We start with $m$ equations $A\boldsymbol{v} = \boldsymbol{0}$, but *the true number of equations is the rank $r$*. We don't want to count repeated rows, or rows that are combinations of previous rows, or zero rows. You will soon see that *$r$ counts the number of independent rows.* And the great fact, still to prove and explain, is that the rank $r$ *also counts the number of independent columns*:

number of **pivots** = number of **independent rows** = number of **independent columns**.

This is part of the Fundamental Theorem of Linear Algebra in Section 5.5.

An example of $A\boldsymbol{v} = \boldsymbol{b}$ will make the possibilities clear.

$$\begin{bmatrix} 1 & 3 & 0 & 2 \\ 0 & 0 & 1 & 4 \\ 1 & 3 & 1 & 6 \end{bmatrix} \begin{bmatrix} v_1 \\ v_2 \\ v_3 \\ v_4 \end{bmatrix} = \begin{bmatrix} 1 \\ 6 \\ 7 \end{bmatrix} \quad \begin{array}{c} \text{has the} \\ \text{augmented} \\ \text{matrix} \end{array} \quad \begin{bmatrix} 1 & 3 & 0 & 2 & \mathbf{1} \\ 0 & 0 & 1 & 4 & \mathbf{6} \\ 1 & 3 & 1 & 6 & \mathbf{7} \end{bmatrix} = \begin{bmatrix} A & \boldsymbol{b} \end{bmatrix}.$$

***The augmented matrix is just*** $\begin{bmatrix} A & \boldsymbol{b} \end{bmatrix}$. When we apply the usual elimination steps to $A$ and $\boldsymbol{b}$, all the equations stay correct. Those steps produce $R$ and $\boldsymbol{d}$.

In this example we subtract row 1 from row 3 and then subtract row 2 from row 3. This produces a *row of zeros* in $R$, and it changes $\boldsymbol{b}$ to a new right side $\boldsymbol{d} = (1, 6, 0)$:

$$\begin{bmatrix} 1 & 3 & 0 & 2 \\ 0 & 0 & 1 & 4 \\ \mathbf{0} & \mathbf{0} & \mathbf{0} & \mathbf{0} \end{bmatrix} \begin{bmatrix} v_1 \\ v_2 \\ v_3 \\ v_4 \end{bmatrix} = \begin{bmatrix} 1 \\ 6 \\ \mathbf{0} \end{bmatrix} \quad \begin{array}{c} \text{has the} \\ \text{augmented} \\ \text{matrix} \end{array} \quad \begin{bmatrix} 1 & 3 & 0 & 2 & \mathbf{1} \\ 0 & 0 & 1 & 4 & \mathbf{6} \\ 0 & 0 & 0 & 0 & \mathbf{0} \end{bmatrix} = \begin{bmatrix} R & \boldsymbol{d} \end{bmatrix}.$$

That very last zero is crucial. The third equation has become $0 = 0$, and we are safe. *The equations can be solved.* In the original matrix $A$, the first row plus the second row equals the third row. If the equations are consistent, this must be true on the right side of the equations also ! The all-important property on the right side was $1 + 6 = 7$.

Here are the same augmented matrices for any vector $\boldsymbol{b} = (b_1, b_2, b_3)$ :

$$\begin{bmatrix} A & \boldsymbol{b} \end{bmatrix} = \begin{bmatrix} 1 & 3 & 0 & 2 & \boldsymbol{b}_1 \\ 0 & 0 & 1 & 4 & \boldsymbol{b}_2 \\ 1 & 3 & 1 & 6 & \boldsymbol{b}_3 \end{bmatrix} \longrightarrow \begin{bmatrix} 1 & 3 & 0 & 2 & \boldsymbol{b}_1 \\ 0 & 0 & 1 & 4 & \boldsymbol{b}_2 \\ 0 & 0 & 0 & 0 & \boldsymbol{b}_3 - \boldsymbol{b}_1 - \boldsymbol{b}_2 \end{bmatrix} = \begin{bmatrix} R & \boldsymbol{d} \end{bmatrix}$$

Now we get $0 = 0$ in the third equation provided $b_3 - b_1 - b_2 = 0$. This is $b_1 + b_2 = b_3$. The example satisfied this requirement with $1+6 = 7$. You see how elimination on $\begin{bmatrix} A & \boldsymbol{b} \end{bmatrix}$ brings out the test on $\boldsymbol{b}$ for $A\boldsymbol{v} = \boldsymbol{b}$ to be solvable.

## One Particular Solution

For an easy solution $\boldsymbol{v_p}$, *choose the free variables to be* $v_2 = v_4 = 0$. Then the two nonzero equations give the two pivot variables $v_1 = 1$ and $v_3 = 6$. Our particular solution to $A\boldsymbol{v} = \boldsymbol{b}$ (and also $R\boldsymbol{v} = \boldsymbol{d}$) is $\boldsymbol{v_p} = (1, 0, 6, 0)$. This particular solution is my favorite: *free variables are zero, pivot variables come from* $\boldsymbol{d}$. The method always works.

**For $R\boldsymbol{v} = \boldsymbol{d}$ to have a solution, zero rows in $R$ must also be zero in $\boldsymbol{d}$.**
**When $I$ is in the pivot rows and columns of $R$, the pivot variables are in $\boldsymbol{d}$:**

$$R\boldsymbol{v}_p = \boldsymbol{d} \qquad \begin{bmatrix} \mathbf{1} & 3 & \mathbf{0} & 2 \\ \mathbf{0} & 0 & \mathbf{1} & 4 \\ 0 & 0 & 0 & 0 \end{bmatrix} \begin{bmatrix} \mathbf{1} \\ 0 \\ \mathbf{6} \\ 0 \end{bmatrix} = \begin{bmatrix} \mathbf{1} \\ \mathbf{6} \\ 0 \end{bmatrix} \qquad \begin{matrix} \textbf{Pivot variables 1, 6} \\ \textbf{Free variables } 0, 0 \end{matrix}$$

Notice how we *choose* the free variables (as zero) and *solve* for the pivot variables. After the row reduction to $R$, those steps are quick. When the free variables are zero, the pivot variables for $\boldsymbol{v}_p$ are already seen in the right side vector $\boldsymbol{d}$.

| | | |
|---|---|---|
| $\boldsymbol{v}_{\text{particular}}$ | ***The particular solution $\boldsymbol{v_p}$ solves*** | $A\boldsymbol{v}_p = \boldsymbol{b}$ |
| $\boldsymbol{v}_{\text{nullspace}}$ | ***The $n - r$ special solutions solve*** | $A\boldsymbol{v}_n = \mathbf{0}$. |

That particular solution to $A\boldsymbol{v} = \boldsymbol{b}$ and $R\boldsymbol{v} = \boldsymbol{d}$ is $(1, 0, 6, 0)$. The two special (null) solutions to $R\boldsymbol{v} = \mathbf{0}$ come from the two free columns of $R$, by reversing signs of 3, 2, and 4. ***Please notice the form I use for the complete solution $\boldsymbol{v}_p + \boldsymbol{v}_n$ to $A\boldsymbol{v} = \boldsymbol{b}$:***

**Complete solution**
**one $\boldsymbol{v}_p$**
**many $\boldsymbol{v}_n$**

$$\boldsymbol{v} = \boldsymbol{v}_p + \boldsymbol{v}_n = \begin{bmatrix} 1 \\ 0 \\ 6 \\ 0 \end{bmatrix} + v_2 \begin{bmatrix} -3 \\ 1 \\ 0 \\ 0 \end{bmatrix} + v_4 \begin{bmatrix} -2 \\ 0 \\ -4 \\ 1 \end{bmatrix}.$$

***Question*** Suppose $A$ is a square invertible matrix, $m = n = r$. What are $\boldsymbol{v}_p$ and $\boldsymbol{v}_n$?
***Answer*** If $A^{-1}$ exists, the particular solution is the one and *only* solution $\boldsymbol{v} = A^{-1}\boldsymbol{b}$. There are no special solutions or free variables. $R = I$ has no zero rows. The only vector in the nullspace is $\boldsymbol{v}_n = \mathbf{0}$. The complete solution is $\boldsymbol{v} = \boldsymbol{v}_p + \boldsymbol{v}_n = A^{-1}\boldsymbol{b} + \mathbf{0}$.

This was the situation in Chapter 4. We didn't mention the nullspace in that chapter. $N(A)$ contained only the zero vector. Reduction goes from $\begin{bmatrix} A & \boldsymbol{b} \end{bmatrix}$ to $\begin{bmatrix} I & A^{-1}\boldsymbol{b} \end{bmatrix}$. The original $A\boldsymbol{v} = \boldsymbol{b}$ is reduced all the way to $\boldsymbol{v} = A^{-1}\boldsymbol{b}$ which is $\boldsymbol{d}$. This is a special case here, but square invertible matrices are the ones we see most often in practice. So they got their own chapter at the start of linear algebra.

For small examples we can reduce $\begin{bmatrix} A & \boldsymbol{b} \end{bmatrix}$ to $\begin{bmatrix} R & \boldsymbol{d} \end{bmatrix}$. For a large matrix, MATLAB does it better. One particular solution (not necessarily ours) is $A\backslash\boldsymbol{b}$ from the backslash command. Here is an example with *full column rank*. Both columns have pivots.

**Example 1** Find the condition on $(b_1, b_2, b_3)$ for $A\boldsymbol{v} = \boldsymbol{b}$ to be solvable, if

$$A = \begin{bmatrix} 1 & 1 \\ 1 & 2 \\ -2 & -3 \end{bmatrix} \quad \text{and} \quad \boldsymbol{b} = \begin{bmatrix} b_1 \\ b_2 \\ b_3 \end{bmatrix}.$$

This condition puts $\boldsymbol{b}$ in the column space of $A$. Find the complete $\boldsymbol{v} = \boldsymbol{v}_p + \boldsymbol{v}_n$.

*Solution* Use the augmented matrix, with its extra column $\boldsymbol{b}$. Subtract row 1 of $\begin{bmatrix} A & \boldsymbol{b} \end{bmatrix}$ from row 2, and add 2 times row 1 to row 3 to reach $\begin{bmatrix} R & \boldsymbol{d} \end{bmatrix}$:

$$\begin{bmatrix} 1 & 1 & b_1 \\ 1 & 2 & b_2 \\ -2 & -3 & b_3 \end{bmatrix} \rightarrow \begin{bmatrix} 1 & 1 & b_1 \\ 0 & 1 & b_2 - b_1 \\ 0 & -1 & b_3 + 2b_1 \end{bmatrix} \rightarrow \begin{bmatrix} 1 & 0 & 2b_1 - b_2 \\ 0 & 1 & b_2 - b_1 \\ 0 & 0 & b_3 + b_1 + b_2 \end{bmatrix}.$$

The last equation is $0 = 0$ provided $b_3 + b_1 + b_2 = 0$. This is the condition that puts $\boldsymbol{b}$ in the column space; then $A\boldsymbol{v} = \boldsymbol{b}$ will be solvable. The rows of $A$ add to the zero row. So for consistency (these are equations!) the entries of $\boldsymbol{b}$ must also add to zero. This example has no free variables since $n - r = 2 - 2$. Therefore no special solutions. The rank is $r = n$ so the only null solution is $\boldsymbol{v}_n = \boldsymbol{0}$. The unique particular solution to $A\boldsymbol{v} = \boldsymbol{b}$ and $R\boldsymbol{v} = \boldsymbol{d}$ is at the top of the augmented column $\boldsymbol{d}$ :

$$\textbf{Only one solution} \qquad \boldsymbol{v} = \boldsymbol{v}_p + \boldsymbol{v}_n = \begin{bmatrix} 2b_1 - b_2 \\ b_2 - b_1 \end{bmatrix} + \begin{bmatrix} 0 \\ 0 \end{bmatrix}.$$

If $b_3 + b_1 + b_2$ is not zero, there is *no* solution to $A\boldsymbol{v} = \boldsymbol{b}$ ($\boldsymbol{v}_p$ doesn't exist).

This example is typical of an extremely important case: $A$ has *full column rank*. Every column has a pivot. *The rank is $r = n$.* The matrix is tall and thin ($m \geq n$). Elimination puts $I$ at the top, when $A$ is reduced to $R$ with rank $n$ :

$$\textbf{Full column rank} \quad R = \begin{bmatrix} I \\ 0 \end{bmatrix} = \begin{bmatrix} n \text{ by } n \text{ identity matrix} \\ m - n \text{ rows of zeros} \end{bmatrix} \tag{1}$$

There are no free columns or free variables. The nullspace is $\boldsymbol{Z}$.

We will collect together the different ways of recognizing this type of matrix.

**Every matrix $A$ with full column rank ($r = n$) has all these properties:**

1. All columns of $A$ are pivot columns. They are independent.
2. There are no free variables or special solutions.
3. Only the zero vector $\boldsymbol{v} = \boldsymbol{0}$ solves $A\boldsymbol{v} = \boldsymbol{0}$ and is in the nullspace $\boldsymbol{N}(A)$.
4. If $A\boldsymbol{v} = \boldsymbol{b}$ has a solution (it might not) then it has only *one solution*.

In the essential language of the next section, $A$ has ***independent columns*** if $r = n$. $A\boldsymbol{v} = \mathbf{0}$ only happens when $\boldsymbol{v} = \mathbf{0}$. Eventually we will add one more fact to the list: *The square matrix $A^{\mathrm{T}}A$ is invertible when the columns are independent.*

In Example 1 the nullspace of $A$ (and $R$) has shrunk to the zero vector. The solution to $A\boldsymbol{v} = \boldsymbol{b}$ is *unique* (if it exists). There will be $m - n$ (here $3 - 2$) zero rows in $R$. There are $m - n$ conditions on $\boldsymbol{b}$ to have $0 = 0$ in those rows. Then $\boldsymbol{b}$ is in the column space.
With full column rank, $A\boldsymbol{v} = \boldsymbol{b}$ has *one* solution or *no* solution: $m > n$ is overdetermined.

## The Complete Solution

The other extreme case is full row rank. Now $A\boldsymbol{v} = \boldsymbol{b}$ has *one or infinitely many* solutions. In this case $A$ must be *short and wide* ($m \leq n$). A matrix has ***full row rank*** if $r = m$ ("*independent rows*"). Every row has a pivot, and here is an example.

**Example 2** There are $n = 3$ unknowns but only $m = 2$ equations:

$$\textbf{Full row rank} \qquad \begin{aligned} x + \;\; y + z &= 3 \\ x + 2y - z &= 4 \end{aligned} \qquad (\text{rank } r = m = 2)$$

These are two planes in $xyz$ space. The planes are not parallel so they intersect in a line. This line of solutions is exactly what elimination will find. ***The particular solution will be one point on the line. Adding the nullspace vectors $\boldsymbol{v}_n$ will move us along the line.*** Then $\boldsymbol{v} = \boldsymbol{v}_p + \boldsymbol{v}_n$ gives the whole line of solutions.

We find $\boldsymbol{v}_p$ and $\boldsymbol{v}_n$ by elimination on $\begin{bmatrix} A & \boldsymbol{b} \end{bmatrix}$. Subtract row 1 from row 2 and then subtract row 2 from row 1:

$$\begin{bmatrix} 1 & 1 & 1 & \mathbf{3} \\ 1 & 2 & -1 & \mathbf{4} \end{bmatrix} \to \begin{bmatrix} 1 & 1 & 1 & \mathbf{3} \\ 0 & 1 & -2 & \mathbf{1} \end{bmatrix} \to \begin{bmatrix} 1 & 0 & 3 & \mathbf{2} \\ 0 & 1 & -2 & \mathbf{1} \end{bmatrix} = \begin{bmatrix} R & \boldsymbol{d} \end{bmatrix}.$$

*The particular solution has free variable $\boldsymbol{v}_3 = 0$.* The special solution has $\boldsymbol{v}_3 = 1$:

$\boldsymbol{v}_{\textbf{particular}}$ comes directly from $\boldsymbol{d}$ on the right side: $\boldsymbol{v}_p = (2, 1, 0)$
$\boldsymbol{s}$ comes from the third column (free column) of $R$: $\boldsymbol{s} = (-3, 2, 1)$

It is wise to check that $\boldsymbol{v}_p$ and $\boldsymbol{s}$ satisfy the original equations $A\boldsymbol{v}_p = \boldsymbol{b}$ and $A\boldsymbol{s} = \mathbf{0}$:

$$\begin{aligned} 2+1 &= 3 \\ 2+2 &= 4 \end{aligned} \qquad\qquad \begin{aligned} -3+2+1 &= 0 \\ -3+4-1 &= 0 \end{aligned}$$

The nullspace solution $\boldsymbol{v}_n$ is any multiple of $\boldsymbol{s}$. It moves along the line of solutions, starting at $\boldsymbol{v}_{\textbf{particular}}$. ***Please notice again how to write the answer***:

$$\textbf{Complete solution} \qquad \boldsymbol{v} = \boldsymbol{v}_p + \boldsymbol{v}_n = \begin{bmatrix} 2 \\ 1 \\ 0 \end{bmatrix} + \boldsymbol{v}_3 \begin{bmatrix} -3 \\ 2 \\ 1 \end{bmatrix}.$$

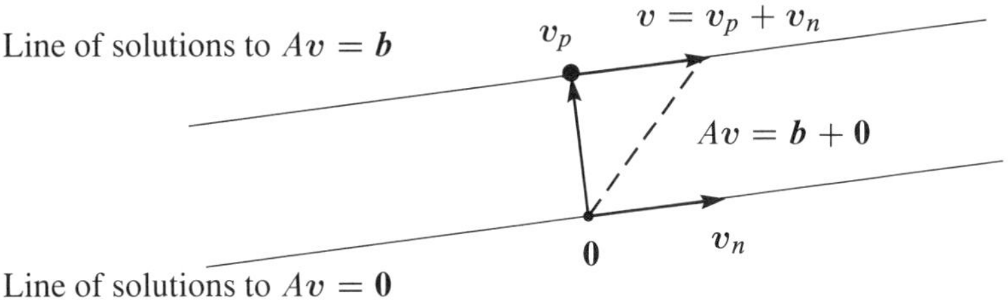


Figure 5.3: Complete solution = *one* particular solution + *all* nullspace solutions.

The line of solutions is drawn in Figure 5.3. Any point on the line *could* have been chosen as the particular solution; we chose the point with $v_3 = 0$.

The particular solution is *not* multiplied by an arbitrary constant! The special solution is, and you understand why.

Now we summarize this short wide case of *full row rank*. If $m < n$ the equations $Av = b$ are **underdetermined** (they have many solutions if they have one).

**Every matrix $A$ with *full row rank* ($r = m$) has all these properties:**

1. All $m$ rows have pivots, and $R$ has no zero rows.
2. $Av = b$ has a solution for every right side $b$.
3. The column space is the whole space $\mathbf{R}^m$.
4. There are $n - r = n - m$ special solutions in the nullspace of $A$.

In this case with $m$ pivots, the rows are "***linearly independent***." We are more than ready for the idea of linear independence, as soon as we summarize the four possibilities—which depend on the rank. Notice how $r, m, n$ are the critical numbers.

***The four possibilities for linear equations depend on the rank*** $r$.

| | | | | | |
|---|---|---|---|---|---|
| $\boldsymbol{r = m}$ | and | $\boldsymbol{r = n}$ | *Square and invertible* | $Av = b$ | has 1 solution |
| $\boldsymbol{r = m}$ | and | $r < n$ | *Short and wide* | $Av = b$ | has $\infty$ solutions |
| $r < m$ | and | $\boldsymbol{r = n}$ | *Tall and thin* | $Av = b$ | has 0 or 1 solution |
| $r < m$ | and | $r < n$ | *Not full rank* | $Av = b$ | has 0 or $\infty$ solutions |

The reduced $R$ will fall in the same category as the matrix $A$. They have the same rank.

In case the pivot columns happen to come first, we can display these four possibilities for $R$. For $Rv = d$ and $Av = b$ to be solvable, $d$ must end in $m - r$ zeros.

| | | | | |
|---|---|---|---|---|
| **Four types** | $R = \begin{bmatrix} I \end{bmatrix}$ | $\begin{bmatrix} I & F \end{bmatrix}$ | $\begin{bmatrix} I \\ 0 \end{bmatrix}$ | $\begin{bmatrix} I & F \\ 0 & 0 \end{bmatrix}$ |
| **Their ranks** | $r = m = n$ | $r = m < n$ | $r = n < m$ | $r < m, r < n$ |

Cases 1 and 2 have full row rank $r = m$. Cases 1 and 3 have full column rank $r = n$. Case 4 is the most general in theory and it is the least common in practice.

## ■ REVIEW OF THE KEY IDEAS ■

1. The rank $r$ is the number of pivots. The reduced matrix $R$ has $m - r$ zero rows.
2. $A\boldsymbol{v} = \boldsymbol{b}$ is solvable if and only if the last $m - r$ equations in $R\boldsymbol{v} = \boldsymbol{d}$ are $0 = 0$.
3. One particular solution $\boldsymbol{v}_p$ has all free variables equal to zero.
4. The $r$ pivot variables are determined after the $n - r$ free variables are chosen.
5. Full column rank $r = n$ means no free variables: one solution or no solution.
6. Full row rank $r = m$ means one solution if $m = n$ or infinitely many if $m < n$.

## ■ WORKED EXAMPLES ■

**5.3 A** This question connects elimination (**pivot columns and back substitution**) to **column space-nullspace-rank-solvability** (the full picture). $A$ is 3 by 4 with rank 2:

$$A\boldsymbol{v} = \boldsymbol{b} \quad \text{is} \quad \begin{aligned} v_1 + 2v_2 + 3v_3 + \ \ 5v_4 &= \mathbf{b_1} \\ 2v_1 + 4v_2 + 8v_3 + 12v_4 &= \mathbf{b_2} \\ 3v_1 + 6v_2 + 7v_3 + 13v_4 &= \mathbf{b_3} \end{aligned}$$

1. Reduce $[\, A \ \ \boldsymbol{b}\,]$ to $[\, U \ \ \boldsymbol{c}\,]$, so that $A\boldsymbol{v} = \boldsymbol{b}$ becomes a triangular system $U\boldsymbol{v} = \boldsymbol{c}$.
2. Find the condition on $b_1, b_2, b_3$ for $A\boldsymbol{v} = \boldsymbol{b}$ to have a solution.
3. Describe the column space of $A$. Which plane in $\mathbf{R}^3$ is the column space?
4. Describe the nullspace of $A$. What are the special solutions in $\mathbf{R}^4$?
5. Find a particular solution to $A\boldsymbol{v} = (0, 6, -6)$ and then the complete solution.

### Solution

1. The multipliers in elimination are 2 and 3 and $-1$. They take $[\, A \ \ \boldsymbol{b}\,]$ into $[\, U \ \ \boldsymbol{c}\,]$.

$$\begin{bmatrix} 1 & 2 & 3 & 5 & \mathbf{b_1} \\ 2 & 4 & 8 & 12 & \mathbf{b_2} \\ 3 & 6 & 7 & 13 & \mathbf{b_3} \end{bmatrix} \rightarrow \left[\begin{array}{cccc|l} 1 & 2 & 3 & 5 & \mathbf{b_1} \\ 0 & 0 & 2 & 2 & \mathbf{b_2 - 2b_1} \\ 0 & 0 & -2 & -2 & \mathbf{b_3 - 3b_1} \end{array}\right] \rightarrow \left[\begin{array}{cccc|l} 1 & 2 & 3 & 5 & \mathbf{b_1} \\ 0 & 0 & 2 & 2 & \mathbf{b_2 - 2b_1} \\ 0 & 0 & 0 & 0 & \mathbf{b_3 + b_2 - 5b_1} \end{array}\right]$$

2. The last equation shows the solvability condition $\mathbf{b_3 + b_2 - 5b_1} = 0$. Then $0 = 0$.
3. **First description**: The column space is the plane containing all combinations of the pivot columns $(1, 2, 3)$ and $(3, 8, 7)$. Those columns are in $A$, not in $U$ or $R$. **Second description**: The column space contains all vectors with $b_3 + b_2 - 5b_1 = 0$. That makes $A\boldsymbol{v} = \boldsymbol{b}$ solvable. *All columns of $A$ pass this test $b_3 + b_2 - 5b_1 = 0$. This is the equation for the plane in the first description of the column space.*
4. The special solutions have free variables $v_2 = 1, v_4 = 0$ and then $v_2 = 0, v_4 = 1$: $\boldsymbol{s}_1 = (-2, 1, 0, 0)$ and $\boldsymbol{s}_2 = (-2, 0, -1, 1)$. The nullspace contains all $c_1\boldsymbol{s}_1 + c_2\boldsymbol{s}_2$.

**5.** One particular solution $\boldsymbol{v}_p$ has free variables = zero. Back substitute in $U\boldsymbol{v} = \boldsymbol{c}$ :

**Particular solution to** $A\boldsymbol{v}_p = \boldsymbol{b} = (0, 6, -6)$
**This vector $\boldsymbol{b}$ satisfies** $b_3 + b_2 - 5b_1 = 0$
**The complete solution is** $\boldsymbol{v} = \boldsymbol{v}_p + \boldsymbol{v}_n$.

$$\boldsymbol{v}_p = \begin{bmatrix} -9 \\ 0 \\ 3 \\ 0 \end{bmatrix}$$

**5.3 B** Find the complete solution $\boldsymbol{v} = \boldsymbol{v}_p + \boldsymbol{v}_n$ by forward elimination on $[A \ \ \boldsymbol{b}]$ :

$$\begin{bmatrix} 1 & 2 & 1 & 0 \\ 2 & 4 & 4 & 8 \\ 4 & 8 & 6 & 8 \end{bmatrix} \begin{bmatrix} v_1 \\ v_2 \\ v_3 \\ v_4 \end{bmatrix} = \begin{bmatrix} 4 \\ 2 \\ 10 \end{bmatrix}.$$

Find numbers $y_1, y_2, y_3$ so that $y_1$ (row 1) + $y_2$ (row 2) + $y_3$ (row 3) = ***zero row.***

Check that $\boldsymbol{b} = (4, 2, 10)$ satisfies the condition $y_1 b_1 + y_2 b_2 + y_3 b_3 = 0$. Why is this the condition for the equations to be solvable and $\boldsymbol{b}$ to be in the column space?

**Solution** Forward elimination on $[A \ \ \boldsymbol{b}]$ produces a zero row in $[U \ \ \boldsymbol{c}]$. The third equation becomes $0 = 0$. The equations are consistent (and solvable because $0 = 0$) :

$$\begin{bmatrix} 1 & 2 & 1 & 0 & \mathbf{4} \\ 2 & 4 & 4 & 8 & \mathbf{2} \\ 4 & 8 & 6 & 8 & \mathbf{10} \end{bmatrix} \longrightarrow \begin{bmatrix} 1 & 2 & 1 & 0 & \mathbf{4} \\ 0 & 0 & 2 & 8 & \mathbf{-6} \\ 0 & 0 & 2 & 8 & \mathbf{-6} \end{bmatrix} \longrightarrow \begin{bmatrix} 1 & 2 & 1 & 0 & \mathbf{4} \\ 0 & 0 & 2 & 8 & \mathbf{-6} \\ 0 & 0 & 0 & 0 & \mathbf{0} \end{bmatrix}.$$

Columns 1 and 3 contain pivots. The variables $v_2$ and $v_4$ are free. If $v_2 = v_4 = 0$ we can solve (back substitution) for the particular solution $\boldsymbol{v}_p = (7, 0, -3, 0)$. The 7 and $-3$ appear again if elimination continues all the way to the row reduced $[R \ \ \boldsymbol{d}]$ :

$$\begin{bmatrix} 1 & 2 & 1 & 0 & \mathbf{4} \\ 0 & 0 & 2 & 8 & \mathbf{-6} \\ 0 & 0 & 0 & 0 & \mathbf{0} \end{bmatrix} \longrightarrow \begin{bmatrix} 1 & 2 & 1 & 0 & \mathbf{4} \\ 0 & 0 & 1 & 4 & \mathbf{-3} \\ 0 & 0 & 0 & 0 & \mathbf{0} \end{bmatrix} \longrightarrow \begin{bmatrix} 1 & 2 & 0 & -4 & \mathbf{7} \\ 0 & 0 & 1 & 4 & \mathbf{-3} \\ 0 & 0 & 0 & 0 & \mathbf{0} \end{bmatrix}.$$

For the nullspace part $\boldsymbol{v}_n$ with $\boldsymbol{b} = \mathbf{0}$, set the free variables $v_2, v_4$ to $1, 0$ and also $0, 1$ :

**Special solutions** $\quad \boldsymbol{s}_1 = (-2, 1, 0, 0)$ and $\boldsymbol{s}_2 = (4, 0, -4, 1)$

Then the complete solution to $A\boldsymbol{v} = \boldsymbol{b}$ (and $R\boldsymbol{v} = \boldsymbol{d}$) is $\boldsymbol{v}_{\text{complete}} = \boldsymbol{v}_p + c_1\boldsymbol{s}_1 + c_2\boldsymbol{s}_2$.

The rows of $A$ produced the zero row from $2(\text{row } 1) + (\text{row } 2) - (\text{row } 3) = (0, 0, 0, 0)$. Thus $\boldsymbol{y} = (2, 1, -1)$. The same combination for $\boldsymbol{b} = (4, 2, 10)$ gives $2(4) + (2) - (10) = 0$. Combinations that give $\boldsymbol{y}^{\mathrm{T}} A =$ zero must also give $\boldsymbol{y}^{\mathrm{T}} \boldsymbol{b} =$ zero. *Otherwise no solution.*

Later we will say this in different words: $\boldsymbol{y} = (2, 1, -1)$, **is in the nullspace of** $A^{\mathrm{T}}$. Then $\boldsymbol{y}$ *will be* perpendicular to every $\boldsymbol{b}$ in the column space of $A$. I am looking ahead...

## Problem Set 5.3

**1** (Recommended) Execute the six steps of Worked Example **3.4 A** to describe the column space and nullspace of $A$ and the complete solution to $A\boldsymbol{v} = \boldsymbol{b}$:

$$A = \begin{bmatrix} 2 & 4 & 6 & 4 \\ 2 & 5 & 7 & 6 \\ 2 & 3 & 5 & 2 \end{bmatrix} \qquad \boldsymbol{b} = \begin{bmatrix} b_1 \\ b_2 \\ b_3 \end{bmatrix} = \begin{bmatrix} 4 \\ 3 \\ 5 \end{bmatrix}$$

**2** Carry out the same six steps for this matrix $A$ with rank one. You will find *two* conditions on $b_1, b_2, b_3$ for $A\boldsymbol{v} = \boldsymbol{b}$ to be solvable. Together these two conditions put $\boldsymbol{b}$ into the _____ space.

$$A = \begin{bmatrix} 1 \\ 3 \\ 2 \end{bmatrix} \begin{bmatrix} 2 & 1 & 3 \end{bmatrix} = \begin{bmatrix} 2 & 1 & 3 \\ 6 & 3 & 9 \\ 4 & 2 & 6 \end{bmatrix} \qquad \boldsymbol{b} = \begin{bmatrix} b_1 \\ b_2 \\ b_3 \end{bmatrix} = \begin{bmatrix} 10 \\ 30 \\ 20 \end{bmatrix}$$

**Questions 3–15 are about the solution of $A\boldsymbol{v} = \boldsymbol{b}$. Follow the steps in the text to $\boldsymbol{v}_p$ and $\boldsymbol{v}_n$. Start from the augmented matrix $\begin{bmatrix} A & \boldsymbol{b} \end{bmatrix}$.**

**3** Write the complete solution as $\boldsymbol{v}_p$ plus any multiple of $\boldsymbol{s}$ in the nullspace:

$$\begin{aligned} x + 3y + 3z &= 1 \\ 2x + 6y + 9z &= 5 \\ -x - 3y + 3z &= 5. \end{aligned}$$

**4** Find the complete solution (also called the *general solution*) to

$$\begin{bmatrix} 1 & 3 & 1 & 2 \\ 2 & 6 & 4 & 8 \\ 0 & 0 & 2 & 4 \end{bmatrix} \begin{bmatrix} x \\ y \\ z \\ t \end{bmatrix} = \begin{bmatrix} 1 \\ 3 \\ 1 \end{bmatrix}.$$

**5** Under what condition on $b_1, b_2, b_3$ is this system solvable? Include $\boldsymbol{b}$ as a fourth column in elimination. Find all solutions when that condition holds:

$$\begin{aligned} x + 2y - 2z &= b_1 \\ 2x + 5y - 4z &= b_2 \\ 4x + 9y - 8z &= b_3. \end{aligned}$$

**6** What conditions on $b_1, b_2, b_3, b_4$ make each system solvable? Find $\boldsymbol{v}$ in that case:

$$\begin{bmatrix} 1 & 2 \\ 2 & 4 \\ 2 & 5 \\ 3 & 9 \end{bmatrix} \begin{bmatrix} v_1 \\ v_2 \end{bmatrix} = \begin{bmatrix} b_1 \\ b_2 \\ b_3 \\ b_4 \end{bmatrix} \qquad \begin{bmatrix} 1 & 2 & 3 \\ 2 & 4 & 6 \\ 2 & 5 & 7 \\ 3 & 9 & 12 \end{bmatrix} \begin{bmatrix} v_1 \\ v_2 \\ v_3 \end{bmatrix} = \begin{bmatrix} b_1 \\ b_2 \\ b_3 \\ b_4 \end{bmatrix}.$$

**7** Show by elimination that $(b_1, b_2, b_3)$ is in the column space if $b_3 - 2b_2 + 4b_1 = 0$.

$$A = \begin{bmatrix} 1 & 3 & 1 \\ 3 & 8 & 2 \\ 2 & 4 & 0 \end{bmatrix}.$$

What combination $y_1(\text{row } 1) + y_2(\text{row } 2) + y_3(\text{row } 3)$ gives the zero row?

**8** Which vectors $(b_1, b_2, b_3)$ are in the column space of $A$? Which combinations of the rows of $A$ give zero?

$$\text{(a)}\ A = \begin{bmatrix} 1 & 2 & 1 \\ 2 & 6 & 3 \\ 0 & 2 & 5 \end{bmatrix} \qquad \text{(b)}\quad A = \begin{bmatrix} 1 & 1 & 1 \\ 1 & 2 & 4 \\ 2 & 4 & 8 \end{bmatrix}.$$

**9** In Worked Example **5.3 A**, combine the pivot columns of $A$ with the numbers $-9$ and 3 in the particular solution $v_p$. What is that linear combination and why?

**10** Construct a 2 by 3 system $Av = b$ with particular solution $v_p = (2, 4, 0)$ and null (homogeneous) solution $v_n$ = any multiple of $(1, 1, 1)$.

**11** Why can't a 1 by 3 system have $v_p = (2, 4, 0)$ and $v_n$ = any multiple of $(1, 1, 1)$?

**12** (a) If $Av = b$ has two solutions $v_1$ and $v_2$, find two solutions to $Av = \mathbf{0}$.

(b) Then find another solution to $Av = b$.

**13** Explain why these are all false:

(a) The complete solution is any linear combination of $v_p$ and $v_n$.

(b) A system $Av = b$ has at most one particular solution.

(c) The solution $v_p$ with all free variables zero is the shortest solution (minimum length $\|v\|$). Find a 2 by 2 counterexample.

(d) If $A$ is invertible there is no solution $v_n$ in the nullspace.

**14** Suppose column 5 has no pivot. Then $v_5$ is a _____ variable. The zero vector (is) (is not) the only solution to $Av = \mathbf{0}$. If $Av = b$ has a solution, then it has _____ solutions.

**15** Suppose row 3 has no pivot. Then that row is _____. The equation $Rv = d$ is only solvable provided _____. The equation $Av = b$ (*is*) (*is not*) (*might not be*) solvable.

**Questions 16–21 are about matrices of "full rank" $r = m$ or $r = n$.**

**16** The largest possible rank of a 3 by 5 matrix is _____. Then there is a pivot in every _____ of $U$ and $R$. The solution to $Av = b$ (*always exists*) (*is unique*). The column space of $A$ is _____. An example is $A =$ _____.

**17** The largest possible rank of a 6 by 4 matrix is ____. Then there is a pivot in every ____ of $U$ and $R$. The solution to $A\boldsymbol{v} = \boldsymbol{b}$ (*always exists*) (*is unique*). The nullspace of $A$ is ____. An example is $A =$ ____.

**18** Find by elimination the rank of $A$ and also the rank of $A^{\mathrm{T}}$:

$$A = \begin{bmatrix} 1 & 4 & 0 \\ 2 & 11 & 5 \\ -1 & 2 & 10 \end{bmatrix} \quad \text{and} \quad A = \begin{bmatrix} 1 & 0 & 1 \\ 1 & 1 & 2 \\ 1 & 1 & q \end{bmatrix} \text{ (rank depends on } q\text{).}$$

**19** Find the rank of $A$ and also of $A^{\mathrm{T}}A$ and also of $AA^{\mathrm{T}}$:

$$A = \begin{bmatrix} 1 & 1 & 5 \\ 1 & 0 & 1 \end{bmatrix} \quad \text{and} \quad A = \begin{bmatrix} 2 & 0 \\ 1 & 1 \\ 1 & 2 \end{bmatrix}.$$

**20** Reduce $A$ to its echelon form $U$. Then find a triangular $L$ so that $A = LU$.

$$A = \begin{bmatrix} 3 & 4 & 1 & 0 \\ 6 & 5 & 2 & 1 \end{bmatrix} \quad \text{and} \quad A = \begin{bmatrix} 1 & 0 & 1 & 0 \\ 2 & 2 & 0 & 3 \\ 0 & 6 & 5 & 4 \end{bmatrix}.$$

**21** Find the complete solution in the form $\boldsymbol{v}_p + \boldsymbol{v}_n$ to these full rank systems:

(a) $x + y + z = 4$ (b) $\begin{aligned} x + y + z &= 4 \\ x - y + z &= 4. \end{aligned}$

**22** If $A\boldsymbol{v} = \boldsymbol{b}$ has infinitely many solutions, why is it impossible for $A\boldsymbol{v} = \boldsymbol{B}$ (new right side) to have only one solution? Could $A\boldsymbol{v} = \boldsymbol{B}$ have no solution?

**23** Choose the number $q$ so that (if possible) the ranks are (a) 1, (b) 2, (c) 3:

$$A = \begin{bmatrix} 6 & 4 & 2 \\ -3 & -2 & -1 \\ 9 & 6 & q \end{bmatrix} \quad \text{and} \quad B = \begin{bmatrix} 3 & 1 & 3 \\ q & 2 & q \end{bmatrix}.$$

**24** Give examples of matrices $A$ for which the number of solutions to $A\boldsymbol{v} = \boldsymbol{b}$ is

(a) 0 or 1, depending on $\boldsymbol{b}$

(b) $\infty$, regardless of $\boldsymbol{b}$

(c) 0 or $\infty$, depending on $\boldsymbol{b}$

(d) 1, regardless of $\boldsymbol{b}$.

**25** Write down all known relations between $r$ and $m$ and $n$ if $Av = b$ has

(a) no solution for some $\boldsymbol{b}$

(b) infinitely many solutions for every $\boldsymbol{b}$

(c) exactly one solution for some $\boldsymbol{b}$, no solution for other $\boldsymbol{b}$

(d) exactly one solution for every $\boldsymbol{b}$.

**Questions 26–33 are about Gauss-Jordan elimination (upwards as well as downwards) and the reduced echelon matrix $R$.**

**26** Continue elimination from $U$ to $R$. Divide rows by pivots so the new pivots are all 1. Then produce zeros *above* those pivots to reach $R$:

$$U = \begin{bmatrix} 2 & 4 & 4 \\ 0 & 3 & 6 \\ 0 & 0 & 0 \end{bmatrix} \quad \text{and} \quad U = \begin{bmatrix} 2 & 4 & 4 \\ 0 & 3 & 6 \\ 0 & 0 & 5 \end{bmatrix}.$$

**27** Suppose $U$ is square with $n$ pivots (an invertible matrix). *Explain why* $R = I$.

**28** Apply Gauss-Jordan elimination to $U\boldsymbol{v} = \boldsymbol{0}$ and $U\boldsymbol{v} = \boldsymbol{c}$. Reach $R\boldsymbol{v} = \boldsymbol{0}$ and $R\boldsymbol{v} = \boldsymbol{d}$:

$$\begin{bmatrix} U & \boldsymbol{0} \end{bmatrix} = \begin{bmatrix} 1 & 2 & 3 & \boldsymbol{0} \\ 0 & 0 & 4 & \boldsymbol{0} \end{bmatrix} \quad \text{and} \quad \begin{bmatrix} U & \boldsymbol{c} \end{bmatrix} = \begin{bmatrix} 1 & 2 & 3 & \boldsymbol{5} \\ 0 & 0 & 4 & \boldsymbol{8} \end{bmatrix}.$$

Solve $R\boldsymbol{v} = \boldsymbol{0}$ to find $\boldsymbol{v}_n$ (its free variable is $v_2 = 1$). Solve $R\boldsymbol{v} = \boldsymbol{d}$ to find $\boldsymbol{v}_p$ (its free variable is $v_2 = 0$).

**29** Apply Gauss-Jordan elimination to reduce to $R\boldsymbol{v} = \boldsymbol{0}$ and $R\boldsymbol{v} = \boldsymbol{d}$:

$$\begin{bmatrix} U & 0 \end{bmatrix} = \begin{bmatrix} 3 & 0 & 6 & \boldsymbol{0} \\ 0 & 0 & 2 & \boldsymbol{0} \\ 0 & 0 & 0 & \boldsymbol{0} \end{bmatrix} \quad \text{and} \quad \begin{bmatrix} U & \boldsymbol{c} \end{bmatrix} = \begin{bmatrix} 3 & 0 & 6 & \boldsymbol{9} \\ 0 & 0 & 2 & \boldsymbol{4} \\ 0 & 0 & 0 & \boldsymbol{5} \end{bmatrix}.$$

Solve $U\boldsymbol{v} = \boldsymbol{0}$ or $R\boldsymbol{v} = \boldsymbol{0}$ to find $\boldsymbol{v}_n$ (free variable $= 1$). What are the solutions to $R\boldsymbol{v} = \boldsymbol{d}$?

**30** Reduce to $U\boldsymbol{v} = \boldsymbol{c}$ (Gaussian elimination) and then $R\boldsymbol{v} = \boldsymbol{d}$ (Gauss-Jordan):

$$A\boldsymbol{v} = \begin{bmatrix} 1 & 0 & 2 & 3 \\ 1 & 3 & 2 & 0 \\ 2 & 0 & 4 & 9 \end{bmatrix} \begin{bmatrix} v_1 \\ v_2 \\ v_3 \\ v_4 \end{bmatrix} = \begin{bmatrix} 2 \\ 5 \\ 10 \end{bmatrix} = \boldsymbol{b}.$$

Find a particular solution $\boldsymbol{v}_p$ and all homogeneous (null) solutions $\boldsymbol{v}_n$.

**31** Find matrices $A$ and $B$ with the given property or explain why you can't:

(a) The only solution of $A\boldsymbol{v} = \begin{bmatrix} \mathbf{1} \\ \mathbf{2} \\ \mathbf{3} \end{bmatrix}$ is $\boldsymbol{v} = \begin{bmatrix} \mathbf{0} \\ \mathbf{1} \end{bmatrix}$.

(b) The only solution of $B\boldsymbol{v} = \begin{bmatrix} \mathbf{0} \\ \mathbf{1} \end{bmatrix}$ is $\boldsymbol{v} = \begin{bmatrix} \mathbf{1} \\ \mathbf{2} \\ \mathbf{3} \end{bmatrix}$.

**32** Reduce $\begin{bmatrix} A & \boldsymbol{b} \end{bmatrix}$ to $\begin{bmatrix} R & \boldsymbol{d} \end{bmatrix}$ and find the complete solution to $A\boldsymbol{v} = \boldsymbol{b}$:

$$A = \begin{bmatrix} 1 & 3 & 1 \\ 1 & 2 & 3 \\ 2 & 4 & 6 \\ 1 & 1 & 5 \end{bmatrix} \quad \text{and} \quad \boldsymbol{b} = \begin{bmatrix} 1 \\ 3 \\ 6 \\ 5 \end{bmatrix} \quad \text{and then} \quad \boldsymbol{b} = \begin{bmatrix} 1 \\ 0 \\ 0 \\ 0 \end{bmatrix}.$$

**33** The complete solution to $A\boldsymbol{v} = \begin{bmatrix} \mathbf{1} \\ \mathbf{3} \end{bmatrix}$ is $\boldsymbol{v} = \begin{bmatrix} \mathbf{1} \\ \mathbf{0} \end{bmatrix} + c \begin{bmatrix} \mathbf{0} \\ \mathbf{1} \end{bmatrix}$. Find $A$.

## Challenge Problems

**34** Suppose you know that the 3 by 4 matrix $A$ has the vector $\boldsymbol{s} = (2, 3, 1, 0)$ as the only special solution to $A\boldsymbol{v} = \mathbf{0}$.

**(a)** What is the *rank* of $A$ and the complete solution to $A\boldsymbol{v} = \mathbf{0}$?

**(b)** What is the exact row reduced echelon form $R$ of $A$? Good question.

**(c)** How do you know that $A\boldsymbol{v} = \boldsymbol{b}$ can be solved for all $\boldsymbol{b}$?

**35** If you have this information about the solutions to $A\boldsymbol{v} = \boldsymbol{b}$ for a specific $\boldsymbol{b}$, what does that tell you about the *shape* of $A$ ($m$ and $n$)? And possibly about $\boldsymbol{r}$ and $\boldsymbol{b}$.

1. There is exactly one solution.
2. All solutions to $A\boldsymbol{v} = \boldsymbol{b}$ have the form $\boldsymbol{v} = \begin{bmatrix} 2 \\ 1 \end{bmatrix} + c\begin{bmatrix} 1 \\ 1 \end{bmatrix}$.
3. There are no solutions.
4. All solutions to $A\boldsymbol{v} = \boldsymbol{b}$ have the form $\boldsymbol{v} = \begin{bmatrix} 1 \\ 1 \\ 0 \end{bmatrix} + c\begin{bmatrix} 1 \\ 0 \\ 1 \end{bmatrix}$
5. There are infinitely many solutions.

**36** Suppose $A\boldsymbol{v} = \boldsymbol{b}$ and $C\boldsymbol{v} = \boldsymbol{b}$ have the same (complete) solutions for every $\boldsymbol{b}$. Is it true that $A = C$?

## 5.4 Independence, Basis and Dimension

This important section is about the true size of a subspace. There are $n$ columns in an $m$ by $n$ matrix. But the true "dimension" of the column space is not necessarily $n$. The dimension is measured by counting *independent columns*—and we have to say what that means. We will see that **the true dimension of the column space is the rank $r$**.

The idea of independence applies to any vectors $\boldsymbol{u}_1, \ldots, \boldsymbol{u}_n$ in any vector space. Most of this section concentrates on the subspaces that we know and use—especially the column space and the nullspace of $A$. In the last part we also study "vectors" that are not column vectors. They can be matrices, or solutions to differential equations. They can be linearly independent (or dependent). First come the key examples using column vectors.

The goal is to understand a ***basis*** : **independent vectors that "span the space"**.

**Any basis Each vector in the space is a unique combination of the basis vectors.**

We are at the heart of our subject, and we cannot go on without a basis. The four essential ideas in this section (with first hints at their meaning) are :

| | |
|---|---|
| **1. Independent vectors** | *(no extra vectors)* |
| **2. Spanning a space** | *(their combinations produce the whole space)* |
| **3. Basis for a space** | *(independent and spanning : not too many or too few)* |
| **4. Dimension of a space** | *(the number of vectors in each and every basis)* |

### Bases for Important Spaces

Here are three examples to show you what a basis looks like (before the definition). A basis is a set of vectors that perfectly describes all vectors in the space. Take all combinations of the basis vectors to get every vector in the space.

1. *Basis for the column space of $A$*

   A natural choice is the $r$ pivot columns. Their combinations yield all columns.

2. *Basis for the nullspace of $A$*

   A natural choice is the set of $n - r$ special solutions to $A\boldsymbol{v} = \mathbf{0}$.

3. *Basis for the space of null solutions to $Ay'' + By' + Cy = 0$*

   A natural choice is the pair of solutions $y_1 = e^{s_1 t}$ and $y_2 = e^{s_2 t}$. These exponents $s_1$ and $s_2$ satisfy $As^2 + Bs + C = 0$, so $y_1$ and $y_2$ solve the differential equation.

   If $s$ is a double root of the quadratic, then $y_2 = te^{st}$ can be the second member of the basis. (Always two $y$'s for a linear second order equation.) All other solutions are combinations of $y_1$ and $y_2$. Then $y_1$ and $y_2$ **span** the solution space.

*The dimension of a space is easy.* Just count the number of basis vectors:

| **Column space** | **Nullspace** | **Solution space** |
|---|---|---|
| Dimension $r$ | Dimension $n - r$ | Dimension 2 |

Those bases were natural choices. They are not at all the only bases. A space has *many different bases.* The column space of this matrix $A$ is the whole space $\mathbf{R}^2$.

$$A = \begin{bmatrix} 1 & 3 & 7 \\ 2 & 5 & 9 \end{bmatrix} \qquad \textbf{Bases for } \boldsymbol{C(A)}$$

1. Pivot columns 1 and 2
2. Columns 1 and 3, or columns 2 and 3
3. Any independent $v$ and $w$ in $\mathbf{R}^2$

The vectors $(1, 0)$ and $(0, 1)$ are a perfectly good basis for the column space of this $A$.

## Linear Independence

Our first definition of independence is not so conventional, but you are ready for it.

**DEFINITION** The columns of $A$ are **linearly independent** when the only solution to $Av = \mathbf{0}$ is $v = \mathbf{0}$. ***No combination $Av$ of the columns is the zero vector, except $v = \mathbf{0}$.***

The columns are independent when the nullspace $N(A)$ contains only the zero vector. Let me illustrate linear independence (and dependence) with three vectors in $\mathbf{R}^3$:

**1.** If three vectors are *not* in the same plane, they are independent. No combination of $u_1, u_2, u_3$ in Figure 5.4 gives zero except the combination $0\,u_1 + 0\,u_2 + 0\,u_3$.

**2.** If three vectors $w_1, w_2, w_3$ are *in the same plane*, they are dependent.

This idea of independence applies to 7 vectors in 12-dimensional space. If they are the columns of $A$, and independent, the nullspace only contains $v = \mathbf{0}$. None of the vectors is a combination of the other six vectors.

Now we express the same idea in different words. The following definition of independence will apply to any sequence of vectors in any vector space. When the vectors are the columns of $A$, the two definitions say exactly the same thing.

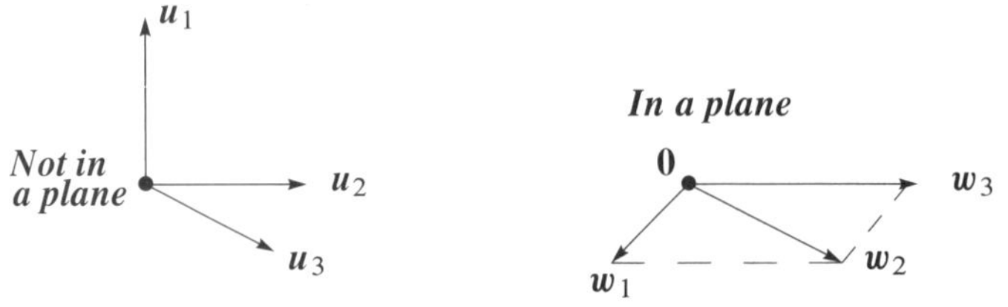


Figure 5.4: Independent vectors $u_1$, $u_2$, $u_3$. Only $0u_1 + 0u_2 + 0u_3$ gives the vector $\mathbf{0}$. Dependent vectors $w_1, w_2, w_3$. The combination $w_1 - w_2 + w_3$ is $(0, 0, 0)$.

**DEFINITION** The sequence of vectors $\boldsymbol{u}_1, \ldots, \boldsymbol{u}_n$ is ***linearly independent*** **if the only combination that gives the zero vector is** $0\boldsymbol{u}_1 + 0\boldsymbol{u}_2 + \cdots + 0\boldsymbol{u}_n$.

$$x_1\boldsymbol{u}_1 + x_2\boldsymbol{u}_2 + \cdots + x_n\boldsymbol{u}_n = \mathbf{0} \quad \text{only happens when all } x\text{'s are zero.} \tag{1}$$

If a combination gives $\mathbf{0}$, when the $x$'s are not all zero, the vectors are *dependent*.

*Correct language*: "The sequence of vectors is linearly independent." *Acceptable shortcut*: "The vectors are independent." *Not acceptable*: "The matrix is independent."

A sequence of vectors is either dependent or independent. They can be combined to give the zero vector (with nonzero $x$'s) or they can't. So the key question is: Which combinations of the vectors give zero? We begin with some small examples in $\mathbf{R}^2$:

(a) The vectors $(1, 0)$ and $(1, 0.00001)$ are independent.

(b) The vectors $(1, 1)$ and $(-1, -1)$ on the same line through $(0, 0)$ are *dependent*.

(c) The vectors $(1, 1)$ and $(0, 0)$ are *dependent* because of the zero vector.

(d) In $\mathbf{R}^2$, any three vectors $(a, b)$ and $(c, d)$ and $(e, f)$ are *dependent*.

The columns of $A$ are dependent exactly when *there is a nonzero vector in the nullspace*.

If one of the $\boldsymbol{u}$'s is the zero vector, independence has no chance. Why not?

Three vectors in $\mathbf{R}^2$ cannot be independent! The matrix $A$ with those three columns must have a free variable and then a special solution $As = \mathbf{0}$. The nullspace is larger than $\mathbf{Z}$. For three vectors in $\mathbf{R}^3$, we put them in a matrix and try to solve $A\boldsymbol{v} = \mathbf{0}$.

**Example 1** The columns of this $A$ are dependent. The nonzero vector $\boldsymbol{v}$ has $A\boldsymbol{v} = \mathbf{0}$.

$$A\boldsymbol{v} = \begin{bmatrix} 1 & 0 & 3 \\ 2 & 1 & 5 \\ 1 & 0 & 3 \end{bmatrix}\begin{bmatrix} -3 \\ 1 \\ 1 \end{bmatrix} \quad \text{is} \quad -3\begin{bmatrix} 1 \\ 2 \\ 1 \end{bmatrix} + 1\begin{bmatrix} 0 \\ 1 \\ 0 \end{bmatrix} + 1\begin{bmatrix} 3 \\ 5 \\ 3 \end{bmatrix} = \begin{bmatrix} 0 \\ 0 \\ 0 \end{bmatrix}.$$

The rank is only $r = 2$. *Independent columns produce full column rank $r = n$.*

In that matrix the rows are also dependent. Row 1 minus row 3 is the zero row. For a *square matrix*, we will show that dependent columns imply dependent rows.

**Question** How to find that solution to $A\boldsymbol{v} = \mathbf{0}$? The systematic way is elimination.

$$A = \begin{bmatrix} 1 & 0 & 3 \\ 2 & 1 & 5 \\ 1 & 0 & 3 \end{bmatrix} \text{ reduces to } R = \begin{bmatrix} 1 & 0 & 3 \\ 0 & 1 & -1 \\ 0 & 0 & 0 \end{bmatrix}.$$

The solution $\boldsymbol{v} = (-3, 1, 1)$ was exactly the special solution. It shows how the free column (column 3) is a combination of the pivot columns. That kills independence!

**Full column rank $n$.** The columns of $A$ are independent when the rank is $r = n$: $n$ pivots and no free variables. Only $\boldsymbol{v} = \mathbf{0}$ is in the nullspace.

**Dependent columns if $n > m$.** Suppose seven columns have five components each ($m = 5$ is less than $n = 7$). Then the columns *must be dependent.* Any seven vectors from $\mathbf{R}^5$ are dependent. The rank of $A$ cannot be larger than 5. There cannot be more than five pivots in five rows. $A\boldsymbol{v} = \mathbf{0}$ has at least $7 - 5 = 2$ free variables, so it has nonzero solutions—which means that the columns are dependent.

Any set of $n$ vectors in $\mathbf{R}^m$ must be linearly dependent if $n > m$.

This type of matrix has more columns than rows—it is short and wide. The columns are certainly dependent if $n > m$, because $A\boldsymbol{v} = \mathbf{0}$ has a nonzero solution. Elimination will reveal the $r$ pivot columns. *Those $r$ pivot columns are independent.*

*Note* Another way to describe linear dependence is this: "*One vector is a combination of the other vectors.*" That sounds clear. Why don't we say this? Our definition was longer: "*Some combination gives the zero vector, other than the trivial combination with every $v = 0$.*" Our definition doesn't pick out one particular vector as guilty.

All columns of $A$ are treated the same. We look at $A\boldsymbol{v} = \mathbf{0}$, and it has a nonzero solution or it hasn't. In the end that is better than asking if the last column (or the first, or a column in the middle) is a combination of the others.

## Spanning a Subspace

The first subspace in this book was the column space. Starting with columns $\boldsymbol{a}_1, \ldots, \boldsymbol{a}_n$, the subspace was filled out by including all their $\boldsymbol{v}$ combinations $v_1 a_1 + \cdots + v_n a_n$. *The column space consists of all combinations $A\boldsymbol{v}$ of the columns.* We now introduce the single word "span" to describe this: The column space is ***spanned*** by the columns.

**DEFINITION** A set of vectors ***spans*** a space if their linear combinations *fill* the space.

***The columns of a matrix span its column space. They might be dependent.***

**Example 2** $\boldsymbol{u}_1 = \begin{bmatrix} 1 \\ 0 \end{bmatrix}$ and $\boldsymbol{u}_2 = \begin{bmatrix} 0 \\ 1 \end{bmatrix}$ span the full two-dimensional space $\mathbf{R}^2$.

**Example 3** $\boldsymbol{u}_1 = \begin{bmatrix} 1 \\ 0 \end{bmatrix}, \boldsymbol{u}_2 = \begin{bmatrix} 0 \\ 1 \end{bmatrix}, \boldsymbol{u}_3 = \begin{bmatrix} 4 \\ 7 \end{bmatrix}$ also span the full space $\mathbf{R}^2$.

**Example 4** $\boldsymbol{w}_1 = \begin{bmatrix} 1 \\ 1 \end{bmatrix}$ and $\boldsymbol{w}_2 = \begin{bmatrix} -1 \\ -1 \end{bmatrix}$ only span a line in $\mathbf{R}^2$. So does $\boldsymbol{w}_1$ alone.

Think of two vectors coming out from $(0, 0, 0)$ in 3-dimensional space. Generally they span a plane. Your mind fills in that plane by taking linear combinations. Mathematically you know other possibilities: two vectors could span a line, three vectors could span all of $\mathbf{R}^3$, or they could span only a plane or a line or $\mathbf{Z}$.

It is possible that three vectors span only a line in $\mathbf{R}^5$, or ten vectors span only a plane. They are certainly not independent!

The columns span the column space. Here is a new subspace—*spanned by the rows.* ***The combinations of the rows produce the "row space".***

**DEFINITION** The ***row space*** of a matrix is the subspace of $\mathbf{R}^n$ spanned by the rows. ***The row space of*** $A$ ***is*** $\boldsymbol{C}(A^{\mathrm{T}})$. ***It is the column space of*** $A^{\mathrm{T}}$.

The rows of an $m$ by $n$ matrix have $n$ components. They are vectors in $\mathbf{R}^n$—or they would be if they were written as column vectors. There is a quick way to fix that: *Transpose the matrix.* Instead of the rows of $A$, look at the columns of $A^{\mathrm{T}}$. Same numbers, but now in the column space of $A^{\mathrm{T}}$. This row space $\boldsymbol{C}(A^{\mathrm{T}})$ is a subspace of $\mathbf{R}^n$.

**Example 5** The column space of $A$ is a plane. The row space is all of $\mathbf{R}^2$.

$$A = \begin{bmatrix} 1 & 4 \\ 2 & 7 \\ 3 & 5 \end{bmatrix} \text{ and } A^{\mathrm{T}} = \begin{bmatrix} 1 & 2 & 3 \\ 4 & 7 & 5 \end{bmatrix}. \text{ Here } m = 3 \text{ and } n = 2.$$

*The row space is spanned by the three rows of* $A$ (which are columns of $A^{\mathrm{T}}$). The columns are in $\mathbf{R}^m$ spanning the column space. Same numbers, different vectors, different spaces.

## A Basis for a Vector Space

Two vectors can't span all of $\mathbf{R}^3$, even if they are independent. Four vectors can't be independent, even if they span $\mathbf{R}^3$. We want ***enough independent vectors to span the space*** (and not more). A "***basis***" is just right.

**DEFINITION** A ***basis*** for a vector space is a sequence of vectors with two properties:

***The basis vectors are linearly independent and they span the space.***

This combination of properties is fundamental to linear algebra. Every vector $\boldsymbol{u}$ in the space is a combination of the basis vectors, because they span the space. More than that, the combination that produces $\boldsymbol{u}$ is *unique*, because the basis vectors $\boldsymbol{u}_1, \ldots, \boldsymbol{u}_n$ are independent:

**There is one and only one way to write $\boldsymbol{u}$ as a combination of the basis vectors.**

**Reason :** Suppose $\boldsymbol{u} = a_1\boldsymbol{u}_1 + \cdots + a_n\boldsymbol{u}_n$ and also $\boldsymbol{u} = b_1\boldsymbol{u}_1 + \cdots + b_n\boldsymbol{u}_n$. By subtraction $(a_1 - b_1)\boldsymbol{u}_1 + \cdots + (a_n - b_n)\boldsymbol{u}_n$ is the zero vector. From the independence of the $\boldsymbol{u}$'s, each $a_i - b_i = 0$. Hence $a_i = b_i$, and there are not two ways to produce $\boldsymbol{u}$.

**Example 6** The columns of the identity matrix $I$ are the "*standard basis*" for $\mathbf{R}^n$.

$$\text{The basis vectors} \quad \boldsymbol{i} = \begin{bmatrix} 1 \\ 0 \end{bmatrix} \quad \text{and} \quad \boldsymbol{j} = \begin{bmatrix} 0 \\ 1 \end{bmatrix} \quad \text{are independent. They span } \mathbf{R}^2.$$

Everybody thinks of this basis first. The vector $\boldsymbol{i}$ goes across and $\boldsymbol{j}$ goes straight up. The columns of the 3 by 3 identity matrix are the standard basis $\boldsymbol{i}, \boldsymbol{j}, \boldsymbol{k}$ for $\mathbf{R}^3$.

Now we find many other bases (infinitely many). The basis is not unique!

**Example 7** (Important) The columns of *every invertible n by n matrix* give a basis for $\mathbf{R}^n$:

$$\begin{matrix} \textbf{Invertible matrix} \\ \text{Independent columns} \\ \text{Column space is } \mathbf{R}^3 \end{matrix} \quad A = \begin{bmatrix} 1 & 0 & 0 \\ 1 & 1 & 0 \\ 1 & 1 & 1 \end{bmatrix} \quad \begin{matrix} \textbf{Singular matrix} \\ \text{Dependent columns} \\ \text{Column space} \neq \mathbf{R}^3 \end{matrix} \quad B = \begin{bmatrix} 1 & 0 & 1 \\ 1 & 1 & 2 \\ 1 & 1 & 2 \end{bmatrix}.$$

The only solution to $A\boldsymbol{v} = \mathbf{0}$ is $\boldsymbol{v} = A^{-1}\mathbf{0} = \mathbf{0}$. The columns are independent. They span the whole space $\mathbf{R}^n$—because every vector $\boldsymbol{b}$ is a combination of the columns. $A\boldsymbol{v} = \boldsymbol{b}$ can always be solved by $\boldsymbol{v} = A^{-1}\boldsymbol{b}$. Do you see how everything comes together for invertible matrices? Here it is in one sentence:

> The vectors $\boldsymbol{v}_1, \ldots, \boldsymbol{v}_n$ are a ***basis for*** $\mathbf{R}^n$ exactly when they are ***the columns of an n by n invertible matrix***. The vector space $\mathbf{R}^n$ has infinitely many different bases.

When the columns are dependent, we keep only the *pivot columns*—the first two columns of $B$ above, with its two pivots. They are independent and they span the column space.

> ***The pivot columns of A are a basis for its column space.*** The pivot rows are a basis for the row space. The pivot rows of the reduced $R$ are also a basis for the row space.

**Example 8** This matrix is not invertible. Its columns are not a basis for anything!

$$\begin{matrix} \textbf{One pivot column} \\ \textbf{One pivot row } (r = 1) \end{matrix} \qquad A = \begin{bmatrix} 2 & 4 \\ 3 & 6 \end{bmatrix} \text{ reduces to } R = \begin{bmatrix} 1 & 2 \\ 0 & 0 \end{bmatrix}.$$

Column 1 of $A$ is the pivot column. That column alone is a basis for its column space. Column 1 of $R$ is **not** a basis for the column space of $A$. That column $(1,0)$ in $R$ does not even belong to the column space of $A$. Elimination changes column spaces. (But the *dimension* remains the same: here dimension $= 1$.)

The row space of $A$ *is the same* as the row space of $R$. It contains $(2,4)$ and $(1,2)$ and all other multiples of those vectors. As always, there are infinitely many bases to choose from. One natural choice is to pick the nonzero rows of $R$ (rows with a pivot). So this matrix $A$ with rank one has only one vector in the basis:

$$\text{Basis for the column space: } \begin{bmatrix} 2 \\ 3 \end{bmatrix}. \quad \text{Basis for the row space: } \begin{bmatrix} 1 \\ 2 \end{bmatrix}.$$

**Example 9** Find bases for the column and row spaces of this rank two matrix:

$$R = \begin{bmatrix} 1 & 2 & 0 & 3 \\ 0 & 0 & 1 & 4 \\ 0 & 0 & 0 & 0 \end{bmatrix}.$$

Columns 1 and 3 are the pivot columns. They are a basis for the column space (of $R$!). The vectors in that column space all have the form $\boldsymbol{b} = (x, y, 0)$. This space is the "$xy$ plane" inside the full $xyz$ space. That plane is not $\mathbf{R}^2$, it is a subspace of $\mathbf{R}^3$. Columns 2 and 3 are also a basis for the same column space. Which pairs of columns of $R$ are *not* a basis for its column space?

The row space of $R$ is a subspace of $\mathbf{R}^4$. The simplest basis for that row space is the two nonzero rows of $R$. The third row (the zero vector) is in the row space too. But it is *not in a basis* for the row space. The basis vectors must be independent.

**Question** Given five vectors in $\mathbf{R}^7$, ***how do you find a basis for the space they span?***

*First answer* Make them the rows of $A$, and eliminate to find the nonzero rows of $R$.
*Second answer* Put the five vectors into the columns of $A$. Eliminate to find the pivot columns (of $A$ not $R$). Could another basis have more vectors, or fewer? This question has a good answer: *No*! ***All bases for a vector space contain the same*** *number* ***of vectors.***

## Dimension of a Vector Space

***The number of vectors, in any and every basis, is the "dimension" of the space.***

We have to prove what was stated above. There are many choices for the basis vectors, but the *number of basis vectors* doesn't change.

If $\boldsymbol{u}_1, \ldots, \boldsymbol{u}_m$ and $\boldsymbol{w}_1, \ldots, \boldsymbol{w}_n$ are both bases for the same vector space, then $m = n$.

***Proof*** Suppose that there are more $\boldsymbol{w}$'s than $\boldsymbol{u}$'s. From $n > m$ we want to reach a contradiction. The $\boldsymbol{u}$'s are a basis, so $\boldsymbol{w}_1$ must be a combination of the $\boldsymbol{u}$'s. If $\boldsymbol{w}_1$ equals $a_{11}\boldsymbol{u}_1 + \cdots + a_{m1}\boldsymbol{u}_m$, this is the first column of a matrix multiplication $UA$:

**Each $\boldsymbol{w}$ is a combination of the $\boldsymbol{u}$'s**

$$\begin{bmatrix} \boldsymbol{w}_1 & \boldsymbol{w}_2 & \ldots & \boldsymbol{w}_n \end{bmatrix} = \begin{bmatrix} \boldsymbol{u}_1 & \ldots & \boldsymbol{u}_m \end{bmatrix} \begin{bmatrix} a_{11} & & a_{1n} \\ \vdots & & \vdots \\ a_{m1} & & a_{mn} \end{bmatrix} = UA.$$

We don't know each number $a_{ij}$, but we know the shape of $A$ (it is $m$ by $n$). The second vector $\boldsymbol{w}_2$ is also a combination of the $\boldsymbol{u}$'s. The coefficients in that combination fill the second column of $A$. The key is that $A$ has a row for every $\boldsymbol{u}$ and a column for every $\boldsymbol{w}$. $A$ is a short wide matrix, since $n > m$. So $A\boldsymbol{v} = \mathbf{0}$ ***has a nonzero solution.***

$A\boldsymbol{v} = \mathbf{0}$ gives $UA\boldsymbol{v} = \mathbf{0}$ which is $W\boldsymbol{v} = \mathbf{0}$. *A combination of the $\boldsymbol{w}$'s gives zero*! Then the $\boldsymbol{w}$'s could not be a basis—our assumption $n > m$ is **not possible** for two bases.

If $m > n$ we exchange the $\boldsymbol{u}$'s and $\boldsymbol{w}$'s and repeat the same steps. The only way to avoid a contradiction is to have $m = n$. This completes the proof that $m = n$.

The number of basis vectors depends on the space—not on a particular basis. The number is the same for every basis, and it counts the "degrees of freedom" in the space. The dimension of the space $\mathbf{R}^n$ is $n$. We now introduce the important word ***dimension*** for other vector spaces too.

**DEFINITION** The ***dimension of a space*** is the ***number of vectors*** in every basis.

This matches our intuition. The line through $\boldsymbol{u} = (1, 5, 2)$ has dimension one. It is a subspace with this one vector $\boldsymbol{u}$ in its basis. Perpendicular to that line is the plane $x + 5y + 2z = 0$. This plane has dimension 2. To prove it, we find a basis $(-5, 1, 0)$ and $(-2, 0, 1)$. The dimension is 2 because the basis contains two vectors.

The plane is the nullspace of the matrix $A = \begin{bmatrix} 1 & 5 & 2 \end{bmatrix}$, which has two free variables. Our basis vectors $(-5, 1, 0)$ and $(-2, 0, 1)$ are the "special solutions" to $A\boldsymbol{v} = \mathbf{0}$. The $n-r$ special solutions give *a basis for the nullspace*, so the dimension of $\boldsymbol{N}(A)$ is $n - r$.

*Note about the language of linear algebra* We never say "the rank of a space" or "the dimension of a basis" or "the basis of a matrix". Those terms have no meaning. It is the ***dimension of the column space*** that equals the ***rank of the matrix.***

## Bases for Matrix Spaces and Function Spaces

The words "independence" and "basis" and "dimension" are not at all restricted to column vectors. We can ask whether three matrices $A_1, A_2, A_3$ are independent. When they are in the space of all 3 by 4 matrices, some combination might give the zero matrix. We can also ask the dimension of the full 3 by 4 matrix space. (It is 12.)

In differential equations, $d^2y/dx^2 = y$ has a space of solutions. One basis is $y = e^x$ and $y = e^{-x}$. Counting the basis functions gives the dimension 2 for the space of all solutions. (The dimension is 2 because of the second derivative.)

Matrix spaces and function spaces may look a little strange after $\mathbf{R}^n$. But in some way, you haven't got the ideas of basis and dimension straight until you can apply them to "vectors" other than column vectors.

**Example 10** Find a basis for the space of 3 by 3 symmetric matrices.

The basis vectors will be matrices! We need enough to span the space (then every $A = A^{\mathrm{T}}$ is a combination). The matrices must be independent (combinations don't give the zero matrix). Here is one basis for the symmetric matrices (many other bases).

$$\begin{bmatrix} 1 & 0 & 0 \\ 0 & 0 & 0 \\ 0 & 0 & 0 \end{bmatrix} \begin{bmatrix} 0 & 0 & 0 \\ 0 & 1 & 0 \\ 0 & 0 & 0 \end{bmatrix} \begin{bmatrix} 0 & 0 & 0 \\ 0 & 0 & 0 \\ 0 & 0 & 1 \end{bmatrix} \begin{bmatrix} 0 & 1 & 0 \\ 1 & 0 & 0 \\ 0 & 0 & 0 \end{bmatrix} \begin{bmatrix} 0 & 0 & 1 \\ 0 & 0 & 0 \\ 1 & 0 & 0 \end{bmatrix} \begin{bmatrix} 0 & 0 & 0 \\ 0 & 0 & 1 \\ 0 & 1 & 0 \end{bmatrix}$$

You could write every $A = A^{\mathrm{T}}$ as a combination of those six matrices. What coefficients would produce 1, 4, 5 and 4, 2, 8 and 5, 8, 9 in the rows? There is only one way to do this. The six matrices are independent. The *dimension* of symmetric matrix space (3 by 3 matrices) is **6**.

To push this further, think about the space of all $n$ by $n$ matrices. One possible basis uses matrices that have only a single nonzero entry (that entry is 1). There are $n^2$ positions for that 1, so there are $n^2$ basis matrices:

**The dimension of the whole $n$ by $n$ matrix space is $n^2$.**

**The dimension of the subspace of *upper triangular* matrices is $\frac{1}{2}n^2 + \frac{1}{2}n$.**

**The dimension of the subspace of *diagonal* matrices is $n$.**

**The dimension of the subspace of *symmetric* matrices is $\frac{1}{2}n^2 + \frac{1}{2}n$ (why?).**

**Function spaces** The equations $d^2y/dt^2 = 0$ and $d^2y/dt^2 = -y$ and $d^2y/dt^2 = y$ involve the second derivative. In calculus we solve to find the functions $y(t)$:

$$\begin{array}{ll} y'' = 0 & \text{is solved by any linear function } y = ct + d \\ y'' = -y & \text{is solved by any combination } y = c \sin t + d \cos t \\ y'' = y & \text{is solved by any combination } y = ce^t + de^{-t}. \end{array}$$

That solution space for $y'' = -y$ has two basis functions: $\sin t$ and $\cos t$. The space for $y'' = 0$ has $t$ and 1. It is the "nullspace" of the second derivative! The dimension is 2 in each case (these are second-order equations). We are finding the null solutions $y_n$.

The solutions of $y'' = 2$ don't form a subspace—the right side $b = 2$ is not zero. A particular solution is $y = t^2$. The complete solution is $y = y_p + y_n = t^2 + ct + d$.

That complete solution is one particular solution plus any function in the nullspace. A linear differential equation is like a linear matrix equation $A\boldsymbol{v} = \boldsymbol{b}$. But we solve it by calculus instead of linear algebra.

We end here with the space $\mathbf{Z}$ that contains only the zero vector. The dimension of this space is *zero*. ***The empty set*** (containing no vectors) ***is a basis for*** $\mathbf{Z}$. We can never allow the zero vector into a basis, because then linear independence is lost.

## ■ REVIEW OF THE KEY IDEAS ■

1. The columns of $A$ are ***independent*** if $\boldsymbol{v} = \mathbf{0}$ is the only solution to $A\boldsymbol{v} = \mathbf{0}$.

2. The vectors $\boldsymbol{u}_1, \ldots, \boldsymbol{u}_r$ ***span*** a space if their combinations fill that space. Spanning vectors can be dependent or independent.

3. ***A basis consists of linearly independent vectors that span the space.*** Every vector in the space is a *unique* combination of the basis vectors.

4. All bases for a space have the same number of vectors. This number of vectors in a basis is the ***dimension*** of the space.

5. The ***pivot columns*** are one basis for the column space. The dimension is the rank $r$.

6. The $n - r$ special solutions will be seen as a basis for the nullspace.

## ■ WORKED EXAMPLES ■

**5.4 A** Start with the vectors $\boldsymbol{u}_1 = (1,2,0)$ and $\boldsymbol{u}_2 = (2,3,0)$. **(a)** Are they linearly independent? **(b)** Are they a basis for any space? **(c)** What space $\mathbf{V}$ do they span? **(d)** What is the dimension of $\mathbf{V}$? **(e)** Which matrices $A$ have $\mathbf{V}$ as their column space? **(f)** Which matrices have $\mathbf{V}$ as their nullspace?

**Solution**

**(a)** $\boldsymbol{u}_1$ and $\boldsymbol{u}_2$ are independent—the only combination to give $\mathbf{0}$ is $0\boldsymbol{u}_1 + 0\boldsymbol{u}_2$.

**(b)** Yes, they are a basis for the space they span.

**(c)** That space $\mathbf{V}$ contains all vectors $(x, y, 0)$. It is the $xy$ plane in $\mathbf{R}^3$.

**(d)** The dimension of $\mathbf{V}$ is 2 since the basis contains two vectors.

**(e)** This $\mathbf{V}$ is the column space of any 3 by $n$ matrix $A$ of rank 2, if row 3 is all zero. In particular $A$ could just have columns $\boldsymbol{u}_1$ and $\boldsymbol{u}_2$.

**(f)** This $\mathbf{V}$ is the nullspace of any $m$ by 3 matrix $B$ of rank 1, if every row has the form $(0,0,c)$. In particular take $B = [0\ \ 0\ \ 1]$. Then $B\boldsymbol{u}_1 = \mathbf{0}$ and $B\boldsymbol{u}_2 = \mathbf{0}$.

**5.4 B** (***Important example***) Suppose $\boldsymbol{u}_1, \ldots, \boldsymbol{u}_n$ is a basis for $\mathbf{R}^n$ and the $n$ by $n$ matrix $A$ is invertible. Show that $A\boldsymbol{u}_1, \ldots, A\boldsymbol{u}_n$ is also a basis for $\mathbf{R}^n$.

**Solution** In *matrix language*: Put the basis vectors $\boldsymbol{u}_1, \ldots, \boldsymbol{u}_n$ in the columns of an invertible(!) matrix $U$. Then $A\boldsymbol{u}_1, \ldots, A\boldsymbol{u}_n$ are the columns of $AU$. Since $A$ and $U$ are invertible, so is $AU$ and its columns give a basis.

In *vector language*: Suppose $c_1 A\boldsymbol{u}_1 + \cdots + c_n A\boldsymbol{u}_n = \mathbf{0}$. This is $A\boldsymbol{v} = \mathbf{0}$ with $\boldsymbol{v} = c_1\boldsymbol{u}_1 + \cdots + c_n\boldsymbol{u}_n$. Multiply by $A^{-1}$ to reach $\boldsymbol{v} = \mathbf{0}$. Linear independence of the $\boldsymbol{u}$'s forces all $c_i = 0$. This shows that the $A\boldsymbol{u}$'s are independent.

To show that the $A\boldsymbol{u}$'s span $\mathbf{R}^n$, solve $c_1 A\boldsymbol{u}_1 + \cdots + c_n A\boldsymbol{u}_n = \boldsymbol{b}$. This is the same as $c_1\boldsymbol{u}_1 + \cdots + c_n\boldsymbol{u}_n = A^{-1}\boldsymbol{b}$. Since the $\boldsymbol{u}$'s are a basis, this must be solvable for all $\boldsymbol{b}$.

# Problem Set 5.4

**Questions 1–10 are about linear independence and linear dependence.**

**1** Show that $\boldsymbol{u}_1, \boldsymbol{u}_2, \boldsymbol{u}_3$ are independent but $\boldsymbol{u}_1, \boldsymbol{u}_2, \boldsymbol{u}_3, \boldsymbol{u}_4$ are dependent:

$$u_1 = \begin{bmatrix} 1 \\ 0 \\ 0 \end{bmatrix} \quad u_2 = \begin{bmatrix} 1 \\ 1 \\ 0 \end{bmatrix} \quad u_3 = \begin{bmatrix} 1 \\ 1 \\ 1 \end{bmatrix} \quad u_4 = \begin{bmatrix} 2 \\ 3 \\ 4 \end{bmatrix}.$$

Solve $c_1\boldsymbol{u}_1 + c_2\boldsymbol{u}_2 + c_3\boldsymbol{u}_3 + c_4\boldsymbol{u}_4 = \boldsymbol{0}$ or $A\boldsymbol{c} = \boldsymbol{0}$. The $\boldsymbol{u}$'s go in the columns of $A$.

**2** (Recommended) Find the largest possible number of independent vectors among

$$u_1 = \begin{bmatrix} 1 \\ -1 \\ 0 \\ 0 \end{bmatrix} \; u_2 = \begin{bmatrix} 1 \\ 0 \\ -1 \\ 0 \end{bmatrix} \; u_3 = \begin{bmatrix} 1 \\ 0 \\ 0 \\ -1 \end{bmatrix} \; u_4 = \begin{bmatrix} 0 \\ 1 \\ -1 \\ 0 \end{bmatrix} \; u_5 = \begin{bmatrix} 0 \\ 1 \\ 0 \\ -1 \end{bmatrix} \; u_6 = \begin{bmatrix} 0 \\ 0 \\ 1 \\ -1 \end{bmatrix}$$

**3** Prove that if $a = 0$ or $d = 0$ or $f = 0$ (3 cases), the columns of $U$ are dependent:

$$U = \begin{bmatrix} a & b & c \\ 0 & d & e \\ 0 & 0 & f \end{bmatrix}.$$

**4** If $a, d, f$ in Question 3 are all nonzero, show that the only solution to $U\boldsymbol{v} = \boldsymbol{0}$ is $\boldsymbol{v} = \boldsymbol{0}$. Then the upper triangular $U$ has independent columns.

**5** Decide the dependence or independence of

(a) the vectors $(1, 3, 2)$ and $(2, 1, 3)$ and $(3, 2, 1)$

(b) the vectors $(1, -3, 2)$ and $(2, 1, -3)$ and $(-3, 2, 1)$.

**6** Choose three independent columns of $U$ and $A$. Then make two other choices.

$$U = \begin{bmatrix} 2 & 3 & 4 & 1 \\ 0 & 6 & 7 & 0 \\ 0 & 0 & 0 & 9 \\ 0 & 0 & 0 & 0 \end{bmatrix} \quad \text{and} \quad A = \begin{bmatrix} 2 & 3 & 4 & 1 \\ 0 & 6 & 7 & 0 \\ 0 & 0 & 0 & 9 \\ 4 & 6 & 8 & 2 \end{bmatrix}.$$

**7** If $\boldsymbol{w}_1, \boldsymbol{w}_2, \boldsymbol{w}_3$ are independent vectors, show that the differences $\boldsymbol{v}_1 = \boldsymbol{w}_2 - \boldsymbol{w}_3$ and $\boldsymbol{v}_2 = \boldsymbol{w}_1 - \boldsymbol{w}_3$ and $\boldsymbol{v}_3 = \boldsymbol{w}_1 - \boldsymbol{w}_2$ are *dependent*. Find a combination of the $\boldsymbol{v}$'s that gives zero. Which singular matrix gives $[\,\boldsymbol{v}_1 \;\; \boldsymbol{v}_2 \;\; \boldsymbol{v}_3\,] = [\,\boldsymbol{w}_1 \;\; \boldsymbol{w}_2 \;\; \boldsymbol{w}_3\,]\,A$?

**8** If $\boldsymbol{w}_1, \boldsymbol{w}_2, \boldsymbol{w}_3$ are independent vectors, show that the sums $\boldsymbol{v}_1 = \boldsymbol{w}_2 + \boldsymbol{w}_3$ and $\boldsymbol{v}_2 = \boldsymbol{w}_1 + \boldsymbol{w}_3$ and $\boldsymbol{v}_3 = \boldsymbol{w}_1 + \boldsymbol{w}_2$ are *independent*. (Write $c_1\boldsymbol{v}_1 + c_2\boldsymbol{v}_2 + c_3\boldsymbol{v}_3 = \boldsymbol{0}$ in terms of the $\boldsymbol{w}$'s. Find and solve equations for the $c$'s, to show they are zero.)

9 Suppose $\boldsymbol{u}_1, \boldsymbol{u}_2, \boldsymbol{u}_3, \boldsymbol{u}_4$ are vectors in $\mathbf{R}^3$.

(a) These four vectors are dependent because ______.

(b) The two vectors $\boldsymbol{u}_1$ and $\boldsymbol{u}_2$ will be dependent if ______.

(c) The vectors $\boldsymbol{u}_1$ and $(0, 0, 0)$ are dependent because ______.

10 Find two independent vectors on the plane $x + 2y - 3z - t = 0$ in $\mathbf{R}^4$. Then find three independent vectors. Why not four? This plane is the nullspace of what matrix?

**Questions 11–14 are about the space *spanned* by a set of vectors. Take all linear combinations of the vectors, to find the space they span.**

11 Describe the subspace of $\mathbf{R}^3$ (is it a line or plane or $\mathbf{R}^3$?) spanned by

(a) the two vectors $(1, 1, -1)$ and $(-1, -1, 1)$

(b) the three vectors $(0, 1, 1)$ and $(1, 1, 0)$ and $(0, 0, 0)$

(c) all vectors in $\mathbf{R}^3$ with whole number components

(d) all vectors with positive components.

12 The vector $\boldsymbol{b}$ is in the subspace spanned by the columns of $A$ when ______ has a solution. The vector $\boldsymbol{c}$ is in the row space of $A$ when ______ has a solution.

*True or false*: If the zero vector is in the row space, the rows are dependent.

13 Find the dimensions of these 4 spaces. Which two of the spaces are the same? (a) column space of $A$ (b) column space of $U$ (c) row space of $A$ (d) row space of $U$:

$$A = \begin{bmatrix} 1 & 1 & 0 \\ 1 & 3 & 1 \\ 3 & 1 & -1 \end{bmatrix} \quad \text{and} \quad U = \begin{bmatrix} 1 & 1 & 0 \\ 0 & 2 & 1 \\ 0 & 0 & 0 \end{bmatrix}.$$

14 $\boldsymbol{v} + \boldsymbol{w}$ and $\boldsymbol{v} - \boldsymbol{w}$ are combinations of $\boldsymbol{v}$ and $\boldsymbol{w}$. Write $\boldsymbol{v}$ and $\boldsymbol{w}$ as combinations of $\boldsymbol{v} + \boldsymbol{w}$ and $\boldsymbol{v} - \boldsymbol{w}$. The two pairs of vectors ______ the same space. When are they a basis for the same space?

**Questions 15–25 are about the requirements for a basis.**

15 If $\boldsymbol{v}_1, \ldots, \boldsymbol{v}_n$ are linearly independent, the space they span has dimension ______. These vectors are a ______ for that space. If the vectors are the columns of an $m$ by $n$ matrix, then $m$ is ______ than $n$. If $m = n$, that matrix is ______.

16 Suppose $\boldsymbol{v}_1, \boldsymbol{v}_2, \ldots, \boldsymbol{v}_6$ are six vectors in $\mathbf{R}^4$.

(a) Those vectors (do) (do not) (might not) span $\mathbf{R}^4$.

(b) Those vectors (are) (are not) (might be) linearly independent.

(c) Any four of those vectors (are) (are not) (might be) a basis for $\mathbf{R}^4$.

**17** Find three different bases for the column space of $U = \begin{bmatrix} 1 & 0 & 1 & 0 & 1 \\ 0 & 1 & 0 & 1 & 0 \end{bmatrix}$. Then find two different bases for the row space of $U$.

**18** Find a basis for each of these subspaces of $\mathbf{R}^4$:

(a) All vectors whose components are equal.

(b) All vectors whose components add to zero.

(c) All vectors that are perpendicular to $(1, 1, 0, 0)$ and $(1, 0, 1, 1)$.

(d) The column space and the nullspace of $I$ (4 by 4).

**19** The columns of $A$ are $n$ vectors from $\mathbf{R}^m$. If they are linearly independent, what is the rank of $A$? If they span $\mathbf{R}^m$, what is the rank? If they are a basis for $\mathbf{R}^m$, what then? *Looking ahead*: The rank $r$ counts the number of _____ columns.

**20** Find a basis for the plane $x - 2y + 3z = 0$ in $\mathbf{R}^3$. Find a basis for the intersection of that plane with the $xy$ plane. Then find a basis for all vectors perpendicular to the plane.

**21** Suppose the columns of a 5 by 5 matrix $A$ are a basis for $\mathbf{R}^5$.

(a) The equation $A\boldsymbol{v} = \mathbf{0}$ has only the solution $\boldsymbol{v} = \mathbf{0}$ because _____.

(b) If $\boldsymbol{b}$ is in $\mathbf{R}^5$ then $A\boldsymbol{v} = \boldsymbol{b}$ is solvable because the basis vectors _____ $\mathbf{R}^5$.

Conclusion: $A$ is invertible. Its rank is 5. Its rows are also a basis for $\mathbf{R}^5$.

**22** Suppose $\mathbf{S}$ is a 5-dimensional subspace of $\mathbf{R}^6$. True or false (example if false):

(a) Every basis for $\mathbf{S}$ can be extended to a basis for $\mathbf{R}^6$ by adding one more vector.

(b) Every basis for $\mathbf{R}^6$ can be reduced to a basis for $\mathbf{S}$ by removing one vector.

**23** $U$ comes from $A$ by subtracting row 1 from row 3:

$$A = \begin{bmatrix} 1 & 3 & 2 \\ 0 & 1 & 1 \\ 1 & 3 & 2 \end{bmatrix} \quad \text{and} \quad U = \begin{bmatrix} 1 & 3 & 2 \\ 0 & 1 & 1 \\ 0 & 0 & 0 \end{bmatrix}.$$

Find bases for the two column spaces. Find bases for the two row spaces. Find bases for the two nullspaces. Which spaces stay fixed in elimination?

**24** True or false (give a good reason):

(a) If the columns of a matrix are dependent, so are the rows.

(b) The column space of a 2 by 2 matrix is the same as its row space.

(c) The column space of a 2 by 2 matrix has the same dimension as its row space.

(d) The columns of a matrix are a basis for the column space.

**25** For which numbers $c$ and $d$ do these matrices have rank 2?

$$A = \begin{bmatrix} 1 & 2 & 5 & 0 & 5 \\ 0 & 0 & c & 2 & 2 \\ 0 & 0 & 0 & d & 2 \end{bmatrix} \quad \text{and} \quad B = \begin{bmatrix} c & d \\ d & c \end{bmatrix}.$$

**Questions 26–28 are about spaces where the "vectors" are matrices.**

**26** Find a basis (and the dimension) for these subspaces of 3 by 3 matrices:

(a) All diagonal matrices.

(b) All skew-symmetric matrices ($A^{\mathrm{T}} = -A$).

**27** Construct six linearly independent 3 by 3 echelon matrices $U_1, \ldots, U_6$. What space of 3 by 3 matrices do they span?

**28** Find a basis for the space of all 2 by 3 matrices whose columns add to zero. Find a basis for the subspace whose rows also add to zero.

**Questions 29–32 are about spaces where the "vectors" are functions.**

**29** (a) Find all functions that satisfy $\frac{dy}{dx} = 0$.

(b) Choose a particular function that satisfies $\frac{dy}{dx} = 3$.

(c) Find all functions that satisfy $\frac{dy}{dx} = 3$.

**30** The cosine space $\mathbf{F}_3$ contains all combinations $y(x) = A\cos x + B\cos 2x + C\cos 3x$. Find a basis for the subspace $\boldsymbol{S}$ with $y(0) = 0$. What is the dimension of $\boldsymbol{S}$ ?

**31** Find a basis for the space of functions that satisfy

(a) $\frac{dy}{dx} - 2y = 0$  (b) $\frac{dy}{dx} - \frac{y}{x} = 0$.

**32** Suppose $y_1, y_2, y_3$ are three different functions of $x$. The space they span could have dimension 1, 2, or 3. Give an example of $y_1, y_2, y_3$ to show each possibility.

**33** Find a basis for the space $\mathbf{S}$ of vectors $(a, b, c, d)$ with $a + c + d = 0$ and also for the space $\mathbf{T}$ with $a + b = 0$ and $c = 2d$. What is the dimension of the intersection $\mathbf{S} \cap \mathbf{T}$?

**34** Which of the following are bases for $\mathbf{R}^3$?

(a) $(1, 2, 0)$ and $(0, 1, -1)$

(b) $(1, 1, -1)$, $(2, 3, 4)$, $(4, 1, -1)$, $(0, 1, -1)$

(c) $(1, 2, 2)$, $(-1, 2, 1)$, $(0, 8, 0)$

(d) $(1, 2, 2)$, $(-1, 2, 1)$, $(0, 8, 6)$

**35** Suppose $A$ is 5 by 4 with rank 4. Show that $A\boldsymbol{v} = \boldsymbol{b}$ has no solution when the 5 by 5 matrix $[\,A \;\; \boldsymbol{b}\,]$ is invertible. Show that $A\boldsymbol{v} = \boldsymbol{b}$ is solvable when $[\,A \;\; \boldsymbol{b}\,]$ is singular.

**36** (a) Find a basis for all solutions to $d^4y/dx^4 = y(x)$.

(b) Find a particular solution to $d^4y/dx^4 = y(x) + 1$. Find the complete solution.

## Challenge Problems

**37** Write the 3 by 3 identity matrix as a combination of the other five permutation matrices! Then show that those five matrices are linearly independent. (Assume a combination gives $c_1 P_1 + \cdots + c_5 P_5 =$ zero matrix, and prove that each $c_i = 0$.)

**38** Intersections and sums have $\dim(\mathbf{V}) + \dim(\mathbf{W}) = \dim(\mathbf{V} \cap \mathbf{W}) + \dim(\mathbf{V} + \mathbf{W})$. Start with a basis $\boldsymbol{u}_1, \ldots, \boldsymbol{u}_r$ for the intersection $\mathbf{V} \cap \mathbf{W}$. Extend with $\boldsymbol{v}_1, \ldots, \boldsymbol{v}_s$ to a basis for $\mathbf{V}$, and separately with $\boldsymbol{w}_1, \ldots, \boldsymbol{w}_t$ to a basis for $\mathbf{W}$. Prove that the $\boldsymbol{u}$'s, $\boldsymbol{v}$'s and $\boldsymbol{w}$'s together are ***independent***. The dimensions have $(r + s) + (r + t) = (r) + (r + s + t)$ as desired.

**39** Inside $\mathbf{R}^n$, suppose dimension $(\mathbf{V})$ + dimension $(\mathbf{W}) > n$. Why is some nonzero vector in both $\mathbf{V}$ and $\mathbf{W}$? Start with bases $\boldsymbol{v}_1, \ldots, \boldsymbol{v}_p$ and $\boldsymbol{w}_1, \ldots, \boldsymbol{w}_q$, $p + q > n$.

**40** Suppose $A$ is 10 by 10 and $A^2 = 0$ (zero matrix): $A$ times each column of $A$ is $\mathbf{0}$. This means that the column space of $A$ is contained in the ______. If $A$ has rank $r$, those subspaces have dimension $r \leq 10 - r$. So the rank of $A$ is $r \leq 5$, if $A^2 = 0$.

## 5.5 The Four Fundamental Subspaces

The figure on this page is the *big picture of linear algebra.* The Four Fundamental Subspaces are in position: Two orthogonal subspaces in $\mathbf{R}^n$ and two in $\mathbf{R}^m$. For any $\boldsymbol{b}$ in the column space, the complete solution to $A\boldsymbol{v} = \boldsymbol{b}$ has one particular solution $\boldsymbol{v}_p$ in the row space, plus any $\boldsymbol{v}_n$ in the nullspace.

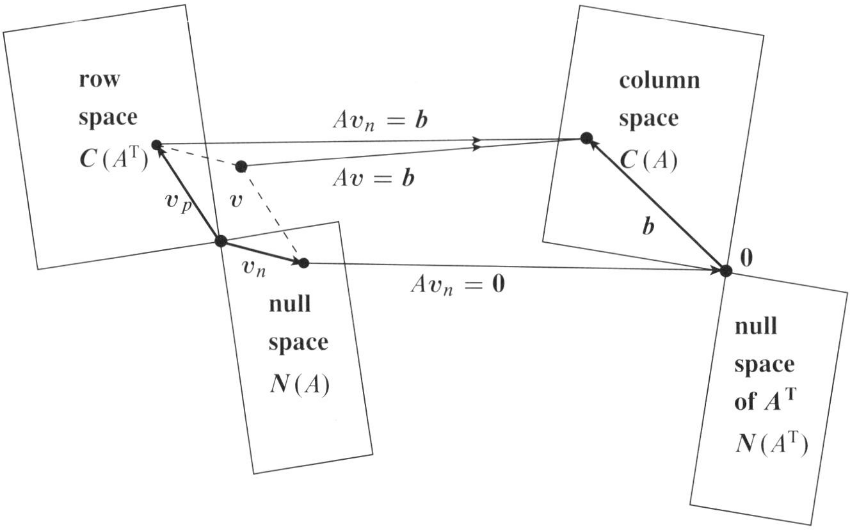


Figure 5.5: The Four Fundamental Subspaces. The complete solution $\boldsymbol{v}_p + \boldsymbol{v}_n$ to $A\boldsymbol{v} = \boldsymbol{b}$.

The main theorem in this chapter connects ***rank*** and ***dimension***. The ***rank*** of a matrix is the number of pivots. The ***dimension*** of a subspace is the number of vectors in a basis. We count pivots or we count basis vectors. *The rank of A reveals the dimensions of all four fundamental subspaces.* Here are the subspaces, including the new one.

Two subspaces come directly from $A$, and the other two come from $A^T$:

| ***Four Fundamental Subspaces*** | | **Dimensions** |
|---|---|---|
| **1.** The ***row space*** $C(A^T)$ | Subspace of $\mathbf{R}^n$. | $r$ |
| **2.** The ***column space*** $C(A)$ | Subspace of $\mathbf{R}^m$. | $r$ |
| **3.** The ***nullspace*** $N(A)$ | Subspace of $\mathbf{R}^n$. | $n - r$ |
| **4.** The ***left nullspace*** $N(A^T)$ | Subspace of $\mathbf{R}^m$. This is our new space. | $m - r$ |

In this book the column space and nullspace came first. We know $C(A)$ and $N(A)$ pretty well. Now the other two subspaces come forward. The row space contains all combinations of the rows. *This is the column space of* $A^T$.

For the left nullspace we solve $A^{\mathrm{T}}\boldsymbol{y} = \mathbf{0}$—that system is $n$ by $m$. *This is the nullspace* $\boldsymbol{N}(A^{\mathrm{T}})$. The vectors $\boldsymbol{y}$ go on the *left* side of $A$ when we transpose to get $\boldsymbol{y}^{\mathrm{T}}A = \mathbf{0}^{\mathrm{T}}$. The matrices $A$ and $A^{\mathrm{T}}$ are usually different. So are their column spaces and their nullspaces. But those spaces are connected in an absolutely beautiful way.

Part 1 of the Fundamental Theorem finds the dimensions of the four subspaces. One fact stands out: ***The row space and column space have the same dimension*** $r$. This is the rank of the matrix. The other important fact involves the two nullspaces:

***$\boldsymbol{N}(A)$ and $\boldsymbol{N}(A^{\mathrm{T}})$ have dimensions $n - r$ and $m - r$, to make up the full $n$ and $m$.***

Part 2 of the Fundamental Theorem will describe how the four subspaces fit together (two in $\mathbf{R}^n$ and two in $\mathbf{R}^m$). That completes the "right way" to understand every $A\boldsymbol{v} = \boldsymbol{b}$. Stay with it—you are doing real mathematics.

## The Four Subspaces for $R$

*Suppose $A$ is reduced to its row echelon form $R$.* For that special form, the four subspaces are easy to identify. We will find a basis for each subspace and check its dimension. Then we watch how the subspaces change (two of them don't change) as we look back at $A$. The main point will be that *the four dimensions are the same for $A$ and $R$.*

As a specific 3 by 5 example, look at the four subspaces for this echelon matrix $R$:

$$\begin{matrix} m = 3 \\ n = 5 \\ r = 2 \end{matrix} \qquad \begin{bmatrix} 1 & 3 & 5 & 0 & 7 \\ 0 & 0 & 0 & 1 & 2 \\ 0 & 0 & 0 & 0 & 0 \end{bmatrix} \qquad \begin{matrix} \textbf{pivot rows } 1 \textbf{ and } 2 \\ \\ \textbf{pivot columns } 1 \textbf{ and } 4 \end{matrix}$$

The rank of this matrix $R$ is $r = 2$ (*two pivots*). Take the four subspaces in order.

**1. The row space of $R$ has dimension 2, matching the rank.**

**Reason:** The first two rows are a basis. The row space contains combinations of all three rows, but the third row (the zero row) adds nothing new. So rows 1 and 2 span the row space. $\boldsymbol{C}(R^{\mathrm{T}})$.

The pivot rows 1 and 2 are independent. That is obvious for this example, and it is always true. If we look only at the pivot columns, we see the $r$ by $r$ identity matrix. There is no way to combine its rows to give the zero row (except by the combination with all coefficients zero). So the $r$ pivot rows are a basis for the row space.

***The dimension of the row space is the rank $r$. The nonzero rows of $R$ form a basis.***

**2. The column space of $R$ also has dimension $r = 2$, matching the rank.**

**Reason:** The pivot columns 1 and 4 form a basis for $\boldsymbol{C}(R)$. They are independent because they start with the $r$ by $r$ identity matrix. No combination of those pivot columns can give the zero column (except the combination with all coefficients zero). And they also span the column space. Every other (free) column is a combination of the pivot columns.

The combinations we need are revealed by the three special solutions:

Column 2 is 3 times column 1. The special solution is $(-3, 1, 0, 0, 0)$.

Column 3 is 5 times column 1. The special solution is $(-5, 0, 1, 0, 0, )$.

Column 5 is 7 (column 1) + 2 (column 4). That solution is $(-7, 0, 0, -2, 1)$.

The pivot columns are independent, and they span $\boldsymbol{C}(R)$, so they are a basis for $\boldsymbol{C}(R)$.

***The dimension of the column space is the rank $r$. The pivot columns form a basis.***

**3. The nullspace has dimension $n - r = 5 - 2$. There are $n - r = 3$ free variables.** $v_2, v_3, v_5$ are free (no pivots in those columns). They yield the three special solutions $\boldsymbol{s}_2$, $\boldsymbol{s}_3, \boldsymbol{s}_5$ to $R\boldsymbol{v} = \boldsymbol{0}$. Set a free variable to 1, and solve for the pivot variables $v_1$ and $v_4$.

$$\boldsymbol{s}_2 = \begin{bmatrix} -3 \\ 1 \\ 0 \\ 0 \\ 0 \end{bmatrix} \quad \boldsymbol{s}_3 = \begin{bmatrix} -5 \\ 0 \\ 1 \\ 0 \\ 0 \end{bmatrix} \quad \boldsymbol{s}_5 = \begin{bmatrix} -7 \\ 0 \\ 0 \\ -2 \\ 1 \end{bmatrix} \quad \begin{array}{l} R\boldsymbol{v} = \boldsymbol{0} \text{ has the} \\ \text{complete solution} \\ \boldsymbol{v} = v_2\boldsymbol{s}_2 + v_3\boldsymbol{s}_3 + v_5\boldsymbol{s}_5 \end{array}.$$

There is a special solution for each free variable. With $n$ variables and $r$ pivot variables, that leaves $n - r$ free variables and special solutions. $\boldsymbol{N}(R)$ has dimension $n - r$.

***The nullspace has dimension $n - r$. The special solutions form a basis.***

The special solutions are independent, because they contain the identity matrix in rows 2, 3, 5. All solutions are combinations of special solutions, $\boldsymbol{v} = v_2\boldsymbol{s}_2 + v_3\boldsymbol{s}_3 + v_5\boldsymbol{s}_5$, because this puts $v_2$, $v_3$ and $v_5$ in the correct positions. Then the pivot variables $v_1$ and $v_4$ are totally determined by the equations $R\boldsymbol{v} = \boldsymbol{0}$.

**4. The nullspace of $R^{\mathrm{T}}$ (*the left nullspace of $R$*) has dimension $m - r = 3 - 2$.**

**Reason:** The equation $R^{\mathrm{T}}\boldsymbol{y} = \boldsymbol{0}$ looks for combinations of the columns of $R^{\mathrm{T}}$ (*the rows of $R$*) that produce zero. You see why $y_1$ and $y_2$ must be zero, and $y_3$ *is free*.

$$\begin{array}{r} y_1\,[1,\ 3,\ 5,\ 0,\ 7] \\ +y_2\,[0,\ 0,\ 0,\ 1,\ 2] \\ +y_3\,[0,\ 0,\ 0,\ 0,\ 0] \\ \hline \textbf{Left nullspace} \qquad [0\ \ 0\ \ y_3]R = [0,\ 0,\ 0,\ 0,\ 0] \end{array} \tag{1}$$

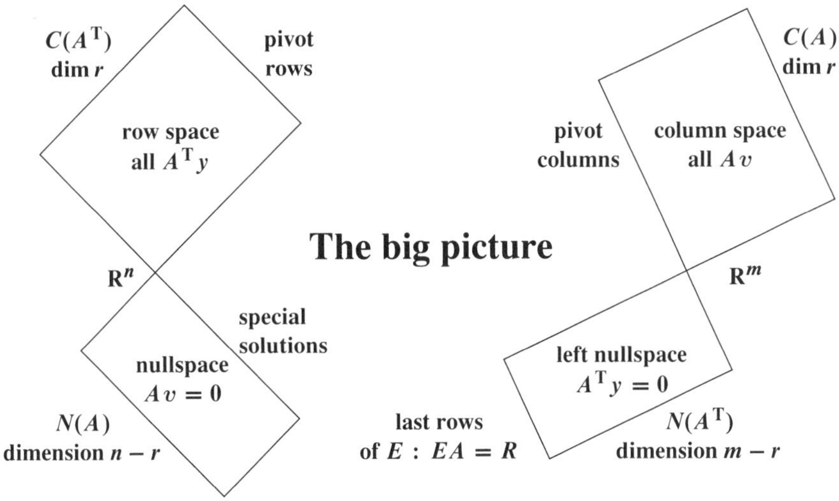


Figure 5.6: Bases and dimensions of the Four Fundamental Subspaces.

In all cases $R$ ends with $m - r$ zero rows. Every combination of these $m - r$ rows gives zero. These are the *only* combinations of the rows of $R$ that give zero, because the $r$ pivot rows are linearly independent. The left nullspace of $R$ contains all these solutions $y = (0, \ldots, 0, y_{r+1}, \ldots, y_m)$ to $R^{\mathrm{T}}y = \mathbf{0}$.

***If** $A$ **is** $m$ **by** $n$ **of rank** $r$**, its left nullspace has dimension** $m - r$.*

This subspace came fourth, and it completes the picture of linear algebra.

***In*** $\mathbf{R}^n$ ***the row space and nullspace have dimensions*** $r$ and $n - r$ (adding to $n$).
***In*** $\mathbf{R}^m$ ***the column space and left nullspace have dimensions*** $r$ and $m - r$ (total $m$).

So far this is proved for echelon matrices $R$. Figure 5.6 shows the same for $A$.

## The Four Subspaces for $A$

We have a job still to do. ***The subspace dimensions for A are the same as for R.*** The job is to explain why. $A$ is now any matrix that reduces to $R = \mathrm{rref}(A)$.

$$\textbf{This } A \textbf{ reduces to } R \qquad A = \begin{bmatrix} 1 & 3 & 5 & 0 & 7 \\ 0 & 0 & 0 & 1 & 2 \\ 1 & 3 & 5 & 1 & 9 \end{bmatrix} \quad \text{Notice } \boldsymbol{C}(A) \neq \boldsymbol{C}(R) \qquad (2)$$

An elimination matrix  takes $A$ to $R$. The big picture (Figure 5.6) applies to both. The invertible matrix $E$ is the product of the elementary matrices that reduce $A$ to $R$:

$$\boldsymbol{A} \textbf{ to } \boldsymbol{R} \textbf{ and back} \qquad EA = R \quad \text{and} \quad A = E^{-1}R \qquad (3)$$

1 ***A has the same row space as R*. Same dimension $r$ and same basis.**

*Reason*: Every row of $A$ is a combination of the rows of $R$. Also every row of $R$ is a combination of the rows of $A$. Elimination changes rows, but not row *spaces*.

Since $A$ has the same row space as $R$, we can choose the first $r$ rows of $R$ as a basis. *The first $r$ rows of $A$ could be dependent.* The good $r$ rows of $A$ end up as pivot rows.

2 ***The column space of A has dimension r*.** The $r$ pivot columns of $A$ are a basis.

**The number of independent columns equals the number of independent rows.**

*Wrong reason:* "$A$ and $R$ have the same column space." This is false. The columns of $R$ often end in zeros. The columns of $A$ don't often end in zeros. The column spaces can be different! But their *dimensions* are the same—both equal to $r$.

*Right reason:* The ***same combinations*** of the columns are zero (or nonzero) for $A$ and $R$. Say that another way: $A\boldsymbol{v} = \mathbf{0}$ *exactly when* $R\boldsymbol{v} = \mathbf{0}$. Pivot columns are independent.

We have just given one proof of the first great theorem of linear algebra: **Row rank equals column rank**. This was easy for $R$, and the ranks are the same for $A$. The Chapter 5 Notes propose three direct proofs not using $R$.

3 ***A has the same nullspace as R*. Same dimension $n - r$ and same basis.**

*Reason:* The elimination steps don't change the solutions. The special solutions are a basis for this nullspace (as we always knew). There are $n - r$ free variables, so the dimension of the nullspace is $n - r$. Notice that $r + (n - r)$ equals $n$:

$$\boxed{\textbf{(dimension of column space)} + \textbf{(dimension of nullspace)} = \textbf{dimension of } \mathbf{R}^n.}$$

That beautiful fact is the **Counting Theorem.** Now apply it also to $A^{\mathrm{T}}$.

4 ***The left nullspace of A*** (the nullspace of $A^{\mathrm{T}}$) ***has dimension m − r*.**

*Reason:* $A^{\mathrm{T}}$ is just as good a matrix as $A$. When we know the dimensions for every $A$, we also know them for $A^{\mathrm{T}}$. Its column space was proved to have dimension $r$. Since $A^{\mathrm{T}}$ is $n$ by $m$, the "whole space" is now $\mathbf{R}^m$. The counting rule for $A$ was $r + (n - r) = n$. The counting rule for $A^{\mathrm{T}}$ is $r + (m - r) = m$. We have all details of the main theorem:

***Fundamental Theorem of Linear Algebra*, Part 1**

***The column space and row space both have dimension r.***
***The nullspaces have dimensions n − r and m − r.***

By concentrating on *spaces* of vectors, not on individual numbers or vectors, we get these clean rules. You will soon take them for granted. But for an 11 by 17 matrix with 187 nonzero entries, I don't think most people would see why these facts are true:

**Two key facts**
$$\text{dimension of } \boldsymbol{C}(A) = \text{dimension of } \boldsymbol{C}(A^{\mathrm{T}}) = \text{rank of } A$$
$$\text{dimension of } \boldsymbol{C}(A) + \text{dimension of } \boldsymbol{N}(A) = 17.$$

**Example 1** $A = [1 \;\; 2 \;\; 3]$ has $m = 1$ and $n = 3$ and rank $r = 1$.

The row space is a line in $\mathbf{R}^3$. The nullspace is the plane $Av = x + 2y + 3z = 0$. This plane has dimension 2 (which is $3 - 1$). The dimensions add to $\mathbf{1} + \mathbf{2} = \mathbf{3}$.

The columns of this 1 by 3 matrix are in $\mathbf{R}^1$. The column space is all of $\mathbf{R}^1$. The left nullspace contains only the zero vector. The only solution to $A^{\mathrm{T}}\boldsymbol{y} = \mathbf{0}$ is $\boldsymbol{y} = \mathbf{0}$, no other multiple of $[1 \;\; 2 \;\; 3]$ gives the zero row. Thus $\boldsymbol{N}(A^{\mathrm{T}})$ is $\mathbf{Z}$, the zero space with dimension 0 (which is $m - r$). In $\mathbf{R}^m$ the dimensions add to $\mathbf{1} + \mathbf{0} = \mathbf{1}$.

**Example 2** $A = \begin{bmatrix} 1 & 2 & 3 \\ 2 & 4 & 6 \end{bmatrix}$ has $m = 2$ and $n = 3$ and rank $r = 1$.

The row space is the same line through $(1, 2, 3)$. The nullspace must be the same plane $x + 2y + 3z = 0$. The dimensions of those two spaces still add to $n : 1 + 2 = 3$.

All columns are multiples of the first column $(1, 2)$. Twice the first row minus the second row is the zero row. Therefore $A^{\mathrm{T}}\boldsymbol{y} = \mathbf{0}$ has the solution $\boldsymbol{y} = (2, -1)$. The column space and left nullspace are **perpendicular lines** in $\mathbf{R}^2$. Dimensions add to $m : 1 + 1 = 2$.

$$\text{Column space} = \text{line through } \begin{bmatrix} 1 \\ 2 \end{bmatrix} \qquad \text{Left nullspace} = \text{line through } \begin{bmatrix} 2 \\ -1 \end{bmatrix}.$$

If $A$ has three equal rows, its rank is ______. What are two of the $\boldsymbol{y}$'s in its left nullspace?

***The y's in the left nullspace combine with the rows to give the zero row.***

## Matrices of Rank One

Those examples had rank $r = 1$—and rank one matrices are special. We can describe them all. You will see again that dimension of row space = dimension of column space. When $r = 1$, every row is a multiple of the same row $\boldsymbol{r}^{\mathrm{T}}$:

$$\boldsymbol{A = c\,r^{\mathrm{T}}} \qquad A = \begin{bmatrix} 1 & 2 & 3 \\ 2 & 4 & 6 \\ -3 & -6 & -9 \\ 0 & 0 & 0 \end{bmatrix} \quad \text{is} \quad \boldsymbol{c} = \begin{bmatrix} 1 \\ 2 \\ -3 \\ 0 \end{bmatrix} \quad \text{times} \quad \begin{bmatrix} 1 & 2 & 3 \end{bmatrix} = \boldsymbol{r}^{\mathrm{T}}.$$

A column times a row (4 by 1 times 1 by 3) produces a matrix (4 by 3). All rows are multiples of the row $\boldsymbol{r}^{\mathrm{T}} = (1, 2, 3)$. All columns are multiples of the first column $\boldsymbol{c} = (1, 2, -3, 0)$. The row space is a line in $\mathbf{R}^n$, and the column space is a line in $\mathbf{R}^m$.

***Every rank one matrix has the special form $A = c\,r^{\mathrm{T}}$ = column times row.***

All columns are multiples of $\boldsymbol{c}$. All rows are multiples of $\boldsymbol{r}^{\mathrm{T}}$. *The nullspace is the plane perpendicular to* $\boldsymbol{r}$. ($A\boldsymbol{v} = \mathbf{0}$ means that $\boldsymbol{c}(\boldsymbol{r}^{\mathrm{T}}\boldsymbol{v}) = \mathbf{0}$ and then $\boldsymbol{r}^{\mathrm{T}}\boldsymbol{v} = 0$.) This **perpendicularity** of the subspaces will become Part 2 of the Fundamental Theorem.

A column vector $\boldsymbol{c}$ times a row vector $\boldsymbol{r}^{\mathrm{T}}$ is often called an *outer product*. The inner product $\boldsymbol{r}^{\mathrm{T}}\boldsymbol{c}$ is a number, the outer product $\boldsymbol{c}\boldsymbol{r}^{\mathrm{T}}$ is a matrix.

## Perpendicular Subspaces

Look at the equation $Av = \mathbf{0}$. This says that $v$ is in the nullspace of $A$. **It also says that $v$ is perpendicular to every row of $A$.** The first row multiplies $v$ to give the first zero in $Av = \mathbf{0}$:

$$Av = \begin{bmatrix} \textbf{row 1} \\ \cdots \\ \text{row } m \end{bmatrix} \begin{bmatrix} \\ v \\ \\ \end{bmatrix} = \begin{bmatrix} \mathbf{0} \\ \cdot \\ 0 \end{bmatrix} \qquad \begin{bmatrix} \mathbf{1} & \mathbf{1} & \mathbf{1} \\ 3 & 1 & 0 \\ 0 & 2 & 3 \end{bmatrix} \begin{bmatrix} \mathbf{1} \\ \mathbf{-3} \\ \mathbf{2} \end{bmatrix} = \begin{bmatrix} \mathbf{0} \\ 0 \\ 0 \end{bmatrix}$$

The vector $v = (1, -3, 2)$ in the nullspace is perpendicular to the first row $(1, 1, 1)$. Their dot product is $1 - 3 + 2 = 0$. That vector $v$ is also perpendicular to the rows $(3, 1, 0)$ and $(0, 2, 3)$—because of the zeros on the right hand side. The dot product of every row and every $v$ is zero.

**Every $v$ in the nullspace is perpendicular to the whole row space**. It is perpendicular to each row and it is perpendicular to all combinations of rows. We have found new words to describe the nullspace of $A$ :

**$N(A)$ contains all vectors $v$ that a perpendicular to the row space of $A$.**

These two fundamental subspaces $N(A)$ and $R(A^T)$ now have a *position in space*. They are "orthogonal subspaces" like the $xy$ plane and the $z$ axis in $R^3$. Tilt that picture and you still have orthogonal subspaces. Their dimensions 2 and 1 still add to 3 : the dimension of the whole space. For any matrix, the $r$-dimensional row space is perpendicular to the $(n-r)$-dimensional nullspace. If that matrix is $A^T$ instead of $A$, we have subspaces of $R^m$.

(In $R^n$) All solutions to $Av = \mathbf{0}$ are perpendicular to all *rows* of $A$.
(In $R^m$) All solutions to $A^T y = \mathbf{0}$ are perpendicular to all *columns* of $A$.

If $A$ is square and invertible, the two nullspaces are just $Z$ : only the zero vector. The row and column spaces are the whole space. These are the extreme in perpendicular subspaces : everything and nothing. No, *not nothing*, the zero vector is perpendicular to everything.

Let me draw the big picture using this new insight of perpendicular subspaces.

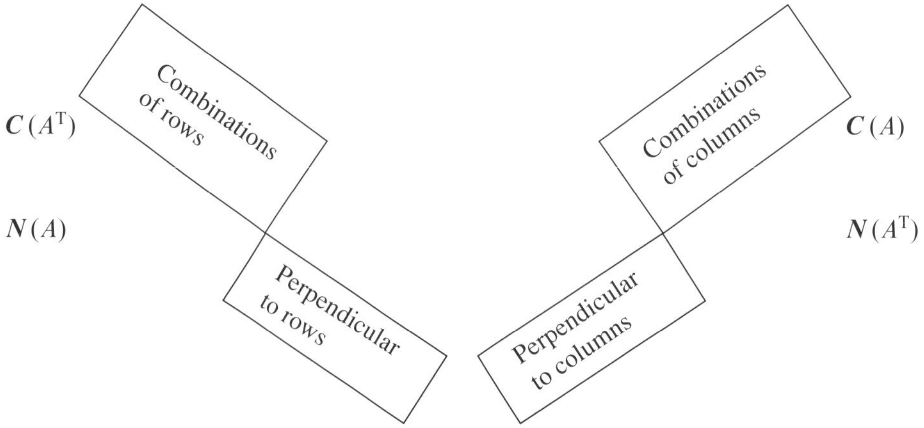

This perpendicularity is Part 2 of the Fundamental Theorem of Linear Algebra. We use a new symbol $S^{\perp}$ (*called* $S$ *perp*) for all vectors that are orthogonal to the subspace $S$.

$$\textbf{Fundamental Theorem, Part 2} : \boldsymbol{N}(A) = \boldsymbol{C}(A^{\mathrm{T}})^{\perp} \text{ and } \boldsymbol{N}(A^{\mathrm{T}}) = \boldsymbol{C}(A)^{\perp}.$$

We know we have *all* perpendicular vectors (not just some of them, like 2 lines in space). The dimensions $r$ and $n - r$ add to the full dimension $n$. For a line and plane in $\boldsymbol{R}^3$: (Line in space)$^{\perp}$ = (Plane in space) and $1 + 2 = 3$.

Here is Problem 37 in the problem set: Explain why $(S^{\perp})^{\perp} = S$.

## ■ REVIEW OF THE KEY IDEAS ■

1. The $r$ pivot rows of $R$ are a basis for the row spaces of $R$ and $A$ (same space).
2. The $r$ pivot columns of $A$ (not $R$) are a basis for its column space $\boldsymbol{C}(A)$.
3. The $n - r$ special solutions are a basis for the nullspaces of $A$ and $R$ (same space).
4. The last $m - r$ rows of $I$ are a basis for the left nullspace of $R$.
5. The last $m - r$ rows of $E$ are a basis for the left nullspace of $A$, if $EA = R$.
6. $\boldsymbol{R}(A^{\mathrm{T}})$ is perpendicular to $\boldsymbol{N}(A)$. And $\boldsymbol{C}(A)$ is perpendicular to $\boldsymbol{N}(A^{\mathrm{T}})$.

## ■ WORKED EXAMPLES ■

**5.5 A** Find bases and dimensions for all four fundamental subspaces if you know that

$$A = \begin{bmatrix} 1 & 0 & 0 \\ 2 & 1 & 0 \\ 5 & 0 & 1 \end{bmatrix} \begin{bmatrix} 1 & 3 & 0 & 5 \\ 0 & 0 & 1 & 6 \\ 0 & 0 & 0 & 0 \end{bmatrix} = E^{-1}R.$$

By changing only *one number* in $R$, change the dimensions of all four subspaces.

**Solution** This matrix has pivots in columns 1 and 3. Its rank is $r = 2$.

| | |
|---|---|
| **Row space** | Basis $(1, 3, 0, 5)$ and $(0, 0, 1, 6)$ from $R$. Dimension 2. |
| **Column space** | Basis $(1, 2, 5)$ and $(0, 1, 0)$ from $E^{-1}$ (and $A$). Dimension 2. |
| **Nullspace** | Basis $(-3, 1, 0, 0)$ and $(-5, 0, -6, 1)$ from $R$. Dimension 2. |
| **Nullspace of $A^{\mathrm{T}}$** | Basis $(-5, 0, 1)$ from row 3 of $E$. Dimension $3 - 2 = 1$. |

We need to comment on that left nullspace $\boldsymbol{N}(A^{\mathrm{T}})$. $EA = R$ says that the last row of $E$ combines the three rows of $A$ into the zero row of $R$. So that last row of $E$ is a basis vector for the left nullspace. If $R$ had *two* zero rows, then the last *two* rows of $E$ would be a basis. (Just like elimination, $\boldsymbol{y}^{\mathrm{T}}A = \boldsymbol{0}^{\mathrm{T}}$ combines rows of $A$ to give zero rows in $R$.)

To change all these dimensions we need to change the rank $r$. The way to do that is to change the zero row of $R$. **The best entry to change is $R_{34}$ in the corner.**

**5.5 B** How can you put four 1's into a 5 by 6 matrix of zeros, so that its *row space* has dimension 1? Describe all the ways to make its *column space* have dimension 1. Describe all the ways to make the dimension of its *nullspace* $\boldsymbol{N}(A)$ as small as possible. How would you make the *sum of the dimensions of all four subspaces small*?

**Solution** The rank is 1 if the four 1's go into the same row, or into the same column. They can also go into *two rows and two columns* (so $a_{ii} = a_{ij} = a_{ji} = a_{jj} = 1$). Since the column space and row space always have the same dimension, this answers the first two questions: The smallest dimension is 1.

The nullspace has its smallest possible dimension $6 - 4 = 2$ when the rank is $r = 4$. To achieve rank 4, the 1's must go into four different rows and columns.

You can't do anything about the sum $r + (n - r) + r + (m - r) = n + m$. It will be $6 + 5 = 11$ no matter how the 1's are placed. The sum is 11 even if there aren't any 1's...

If all the other entries of $A$ are 2's instead of 0's, how do these answers change?

## Problem Set 5.5

**1** (a) If a 7 by 9 matrix has rank 5, what are the dimensions of the four subspaces? What is the sum of all four dimensions?

(b) If a 3 by 4 matrix has rank 3, what are its column space and left nullspace?

**2** Find bases and dimensions for the four subspaces associated with $A$ and $B$:

$$A = \begin{bmatrix} 1 & 2 & 4 \\ 2 & 4 & 8 \end{bmatrix} \quad \text{and} \quad B = \begin{bmatrix} 1 & 2 & 4 \\ 2 & 5 & 8 \end{bmatrix}.$$

**3** Find a basis for each of the four subspaces associated with $A$:

$$A = \begin{bmatrix} 0 & 1 & 2 & 3 & 4 \\ 0 & 1 & 2 & 4 & 6 \\ 0 & 0 & 0 & 1 & 2 \end{bmatrix} = \begin{bmatrix} 1 & 0 & 0 \\ 1 & 1 & 0 \\ 0 & 1 & 1 \end{bmatrix} \begin{bmatrix} 0 & 1 & 2 & 3 & 4 \\ 0 & 0 & 0 & 1 & 2 \\ 0 & 0 & 0 & 0 & 0 \end{bmatrix}.$$

**4** Construct a matrix with the required property or explain why this is impossible:

(a) Column space contains $\begin{bmatrix} 1 \\ 1 \\ 0 \end{bmatrix}, \begin{bmatrix} 0 \\ 0 \\ 1 \end{bmatrix}$, row space contains $\begin{bmatrix} 1 \\ 2 \end{bmatrix}, \begin{bmatrix} 2 \\ 5 \end{bmatrix}$.

(b) Column space has basis $\begin{bmatrix} 1 \\ 1 \\ 3 \end{bmatrix}$, nullspace has basis $\begin{bmatrix} 3 \\ 1 \\ 1 \end{bmatrix}$.

(c) Dimension of nullspace $= 1 +$ dimension of left nullspace.

(d) Left nullspace contains $\begin{bmatrix} 1 \\ 3 \end{bmatrix}$, row space contains $\begin{bmatrix} 3 \\ 1 \end{bmatrix}$.

(e) Row space $=$ column space, nullspace $\neq$ left nullspace.

**5** If $V$ is the subspace spanned by $(1,1,1)$ and $(2,1,0)$, find a matrix $A$ that has $V$ as its row space. Find a matrix $B$ that has $V$ as its nullspace.

**6** Without elimination, find dimensions and bases for the four subspaces for

$$A = \begin{bmatrix} 0 & 3 & 3 & 3 \\ 0 & 0 & 0 & 0 \\ 0 & 1 & 0 & 1 \end{bmatrix} \quad \text{and} \quad B = \begin{bmatrix} 1 \\ 4 \\ 5 \end{bmatrix}.$$

**7** Suppose the 3 by 3 matrix $A$ is invertible. Write down bases for the four subspaces for $A$, and also for the 3 by 6 matrix $B = [\,A \quad A\,]$.

**8** What are the dimensions of the four subspaces for $A$, $B$, and $C$, if $I$ is the 3 by 3 identity matrix and 0 is the 3 by 2 zero matrix?

$$A = \begin{bmatrix} I & 0 \end{bmatrix} \quad \text{and} \quad B = \begin{bmatrix} I & I \\ 0^{\mathrm{T}} & 0^{\mathrm{T}} \end{bmatrix} \quad \text{and} \quad C = \begin{bmatrix} 0 \end{bmatrix}.$$

**9** Which subspaces are the same for these matrices of different sizes?

$$\text{(a)}\ [\,A\,] \text{ and } \begin{bmatrix} A \\ A \end{bmatrix} \qquad \text{(b)}\ \begin{bmatrix} A \\ A \end{bmatrix} \text{ and } \begin{bmatrix} A & A \\ A & A \end{bmatrix}.$$

Prove that all three of those matrices have the *same rank* $r$.

**10** If the entries of a 3 by 3 matrix are chosen randomly between 0 and 1, what are the most likely dimensions of the four subspaces? What if the matrix is 3 by 5?

**11** (Important) $A$ is an $m$ by $n$ matrix of rank $r$. Suppose there are right sides $\boldsymbol{b}$ for which $A\boldsymbol{v} = \boldsymbol{b}$ has *no solution*.

(a) What are all inequalities ($<$ or $\leq$) that must be true between $m, n$, and $r$?

(b) How do you know that $A^{\mathrm{T}}\boldsymbol{y} = \boldsymbol{0}$ has solutions other than $\boldsymbol{y} = \boldsymbol{0}$?

**12** Construct a matrix with $(1,0,1)$ and $(1,2,0)$ as a basis for its row space and its column space. Why can't this be a basis for the row space and nullspace?

**13** True or false (with a reason or a counterexample):

(a) If $m = n$ then the row space of $A$ equals the column space.

(b) The matrices $A$ and $-A$ share the same four subspaces.

(c) If $A$ and $B$ share the same four subspaces then $A$ is a multiple of $B$.

**14** Without computing $A$, find bases for its four fundamental subspaces:

$$A = \begin{bmatrix} 1 & 0 & 0 \\ 6 & 1 & 0 \\ 9 & 8 & 1 \end{bmatrix} \begin{bmatrix} 1 & 2 & 3 & 4 \\ 0 & 1 & 2 & 3 \\ 0 & 0 & 1 & 2 \end{bmatrix}.$$

**15** If you exchange the first two rows of $A$, which of the four subspaces stay the same? If $\boldsymbol{v} = (1, 2, 3, 4)$ is in the left nullspace of $A$, write down a vector in the left nullspace of the new matrix.

**16** Explain why $\boldsymbol{v} = (1, 0, -1)$ *cannot be a row of $A$ and also in the nullspace.*

**17** Describe the four subspaces of $\boldsymbol{R}^3$ associated with

$$A = \begin{bmatrix} 0 & 1 & 0 \\ 0 & 0 & 1 \\ 0 & 0 & 0 \end{bmatrix} \quad \text{and} \quad I + A = \begin{bmatrix} 1 & 1 & 0 \\ 0 & 1 & 1 \\ 0 & 0 & 1 \end{bmatrix}.$$

**18** (Left nullspace) Add the extra column $\boldsymbol{b}$ and reduce $A$ to echelon form:

$$\begin{bmatrix} A & \boldsymbol{b} \end{bmatrix} = \begin{bmatrix} 1 & 2 & 3 & b_1 \\ 4 & 5 & 6 & b_2 \\ 7 & 8 & 9 & b_3 \end{bmatrix} \rightarrow \begin{bmatrix} 1 & 2 & 3 & b_1 \\ 0 & -3 & -6 & b_2 - 4b_1 \\ 0 & 0 & 0 & b_3 - 2b_2 + b_1 \end{bmatrix}.$$

A combination of the rows of $A$ has produced the zero row. What combination is it? (Look at $b_3 - 2b_2 + b_1$ on the right side.) Which vectors are in the nullspace of $A^{\mathrm{T}}$ and which vectors are in the nullspace of $A$?

**19** Following the method of Problem 18, reduce $A$ to echelon form and look at the zero rows. The $\boldsymbol{b}$ column tells which combinations you have taken of the rows:

$$\text{(a)} \begin{bmatrix} 1 & 2 & b_1 \\ 3 & 4 & b_2 \\ 4 & 6 & b_3 \end{bmatrix} \qquad \text{(b)} \begin{bmatrix} 1 & 2 & b_1 \\ 2 & 3 & b_2 \\ 2 & 4 & b_3 \\ 2 & 5 & b_4 \end{bmatrix}$$

From the $\boldsymbol{b}$ column after elimination, read off $m-r$ basis vectors in the left nullspace. Those $\boldsymbol{y}$'s are combinations of rows that give zero rows.

**20** (a) Find the solutions to $A\boldsymbol{v} = \boldsymbol{0}$. Check that $\boldsymbol{v}$ is are perpendicular to the rows:

$$A = \begin{bmatrix} 1 & 0 & 0 \\ 2 & 1 & 0 \\ 3 & 4 & 1 \end{bmatrix} \begin{bmatrix} 4 & 2 & 0 & 1 \\ 0 & 0 & 1 & 3 \\ 0 & 0 & 0 & 0 \end{bmatrix} = ER.$$

(b) How many independent solutions to $A^{\mathrm{T}}\boldsymbol{y} = \boldsymbol{0}$? Why is $\boldsymbol{y}^{\mathrm{T}}$ the last row of $E^{-1}$?

**21** Suppose $A$ is the sum of two matrices of rank one: $A = \boldsymbol{u}\boldsymbol{v}^{\mathrm{T}} + \boldsymbol{w}\boldsymbol{z}^{\mathrm{T}}$.

(a) Which vectors span the column space of $A$?

(b) Which vectors span the row space of $A$?

(c) The rank is less than 2 if _____ or if _____.

(d) Compute $A$ and its rank if $\boldsymbol{u} = \boldsymbol{z} = (1, 0, 0)$ and $\boldsymbol{v} = \boldsymbol{w} = (0, 0, 1)$.

**22** Construct $A = \boldsymbol{u}\boldsymbol{v}^{\mathrm{T}} + \boldsymbol{w}\boldsymbol{z}^{\mathrm{T}}$ whose column space has basis $(1, 2, 4), (2, 2, 1)$ and whose row space has basis $(1, 0), (1, 1)$. Write $A$ as (3 by 2) times (2 by 2).

**23** Without multiplying matrices, find bases for the row and column spaces of $A$:

$$A = \begin{bmatrix} 1 & 2 \\ 4 & 5 \\ 2 & 7 \end{bmatrix} \begin{bmatrix} 3 & 0 & 3 \\ 1 & 1 & 2 \end{bmatrix}.$$

How do you know from these shapes that $A =$ (3 by 2) (2 by 3) cannot be invertible?

**24** (Important) $A^{\mathrm{T}}\boldsymbol{y} = \boldsymbol{d}$ is solvable when $\boldsymbol{d}$ is in which of the four subspaces? The solution $\boldsymbol{y}$ is unique when the _____ contains only the zero vector.

**25** True or false (with a reason or a counterexample):

(a) $A$ and $A^{\mathrm{T}}$ have the same number of pivots.

(b) $A$ and $A^{\mathrm{T}}$ have the same left nullspace.

(c) If the row space equals the column space then $A^{\mathrm{T}} = A$.

(d) If $A^{\mathrm{T}} = -A$ then the row space of $A$ equals the column space of $A$.

**26** (***Rank of $AB \leq$ ranks of $A$ and $B$***) If $AB = C$, the rows of $C$ are combinations of the rows of _____. So the rank of $C$ is not greater than the rank of _____. Since $B^{\mathrm{T}}A^{\mathrm{T}} = C^{\mathrm{T}}$, the rank of $C$ is also not greater than the rank of _____.

**27** If $a, b, c$ are given with $a \neq 0$, how would you choose $d$ so that $\begin{bmatrix} a & b \\ c & d \end{bmatrix}$ has rank 1? Find a basis for the row space and nullspace. Show they are perpendicular!

**28** Find the ranks of the 8 by 8 checkerboard matrix $B$ and the chess matrix $C$:

$$B = \begin{bmatrix} 1 & 0 & 1 & 0 & 1 & 0 & 1 & 0 \\ 0 & 1 & 0 & 1 & 0 & 1 & 0 & 1 \\ 1 & 0 & 1 & 0 & 1 & 0 & 1 & 0 \\ \cdot & \cdot & \cdot & \cdot & \cdot & \cdot & \cdot & \cdot \\ 0 & 1 & 0 & 1 & 0 & 1 & 0 & 1 \end{bmatrix} \quad \text{and} \quad C = \begin{bmatrix} r & n & b & q & k & b & n & r \\ p & p & p & p & p & p & p & p \\ & & & \text{four zero rows} & & & & \\ p & p & p & p & p & p & p & p \\ r & n & b & q & k & b & n & r \end{bmatrix}$$

The numbers $r, n, b, q, k, p$ are all different. Find bases for the row space and the left nullspace of $B$ and $C$. Challenge problem: Find a basis for the nullspace of $C$.

**29** Can tic-tac-toe be completed (5 ones and 4 zeros in $A$) so that rank $(A) = 2$ but neither side passed up a winning move?

**Problems 30-33 are about perpendicularity of the fundamental subspaces (two perpendicular pairs.)**

**30** The floor and a wall of your room are *not* perpendicular subspaces in $\boldsymbol{R}^3$. *Why not*? I am extending the floor and wall to be planes in $\boldsymbol{R}^3$.

**31** Explain why every $\boldsymbol{y}$ in $\boldsymbol{N}(A^{\mathrm{T}})$ is perpendicular to every column of $A$.

**32** Suppose $\boldsymbol{P}$ is the plane of vectors $\boldsymbol{R}^4$ satisfying $v_1 + v_2 + v_3 + v_4 = 0$. Find a basis for $\boldsymbol{P}^{\perp}$. Find a matrix $A$ with $\boldsymbol{N}(A) = \boldsymbol{P}$.

**33** Why can't $A$ have $(1, 4, 5)$ in its row space and $(4, 5, 1)$ in its nullspace?

## Challenge Problems

**34** If $A = \boldsymbol{u}\boldsymbol{v}^{\mathrm{T}}$ is a 2 by 2 matrix of rank 1, redraw Figure 5.6 to show clearly the Four Fundamental Subspaces in terms of $\boldsymbol{u}$ and $\boldsymbol{v}$. If another matrix $B$ produces those same four subspaces, what is the exact relation of $B$ to $A$?

**35** **M** is the 9-dimensional space of 3 by 3 matrices. Multiply every matrix $X$ by $A$:

$$A = \begin{bmatrix} 1 & 0 & -1 \\ -1 & 1 & 0 \\ 0 & -1 & 1 \end{bmatrix}. \quad \text{Notice: } A \begin{bmatrix} 1 \\ 1 \\ 1 \end{bmatrix} = \begin{bmatrix} 0 \\ 0 \\ 0 \end{bmatrix}.$$

(a) Which matrices $X$ lead to $AX =$ zero matrix?

(b) Which matrices have the form $AX$ for some matrix $X$?

(a) finds the "nullspace" of that operation $AX$ and (b) finds the "column space". What are the dimensions of those two subspaces of $\boldsymbol{M}$? Why do the dimensions add to $(n - r) + r = 9$?

**36** Suppose the $m$ by $n$ matrices $A$ and $B$ lead to *the same four subspaces*. If both matrices are already in row reduced echelon form, prove that $F$ must equal $G$:

$$A = \begin{bmatrix} I & F \\ 0 & 0 \end{bmatrix} \qquad B = \begin{bmatrix} I & G \\ 0 & 0 \end{bmatrix}.$$

**37** For any subspace $\boldsymbol{S}$ of $\boldsymbol{R}^n$, why is $(\boldsymbol{S}^{\perp})^{\perp} = \boldsymbol{S}$? "If $\boldsymbol{S}^{\perp}$ contains all vectors perpendicular to $\boldsymbol{S}$, then $\boldsymbol{S}$ contains all vectors perpendicular to $\boldsymbol{S}^{\perp}$." Dimensions add to $n$.

**38** If $A^{\mathrm{T}}A\boldsymbol{v} = \boldsymbol{0}$ then $A\boldsymbol{v} = \boldsymbol{0}$. Reason: This $A\boldsymbol{v}$ is in the nullspace of $A^{\mathrm{T}}$. Every $A\boldsymbol{v}$ is in the column space of $A$ (*why*?). Those spaces are perpendicular, and only $A\boldsymbol{v} = \boldsymbol{0}$ can be perpendicular to itself. *So $A^{\mathrm{T}}A$ has the same nullspace as $A$.*

## 5.6 Graphs and Networks

Over the years I have seen one model so often, and I found it so basic and useful, that I always put it first. The model consists of ***nodes connected by edges***. This is called a ***graph***.

Graphs of the usual kind display functions $f(x)$. Graphs of this node-edge kind lead to matrices. This section is about the ***incidence matrix*** of a graph—which tells how the $n$ nodes are connected by the $m$ edges. Normally $m > n$, there are more edges than nodes.

Every entry of an incidence matrix is 0 or 1 or $-1$. This continues to hold during elimination. All pivots and multipliers are $\pm 1$. Then the echelon matrix $R$ after elimination also contains $0, 1, -1$. So do the special solutions ! All four subspaces have basis vectors with these exceptionally simple components. The matrices are not concocted for a textbook, they come from a model that is absolutely essential in pure and applied mathematics.

For these incidence matrices, the four fundamental subspaces have meaning and importance. Up to now, I have created small matrix examples to show the column space and nullspace. I was claiming that all four subspaces need to be understood, but you wouldn't know their importance from such small examples. Now comes the chance to learn about the most valuable models in discrete mathematics—graphs and their matrices.

### Graphs and Incidence Matrices

Figure 5.7 displays a *graph* with $m = 6$ edges and $n = 4$ nodes. Its incidence matrix will be 6 by 4. This matrix $A$ tells which nodes are connected by which edges. The entries $-1$ and $+1$ also tell the direction of each arrow. *The first row* $-1, 1, 0, 0$ *of* $A$ (the incidence matrix) *shows that the first edge goes from node* 1 *to node* 2.

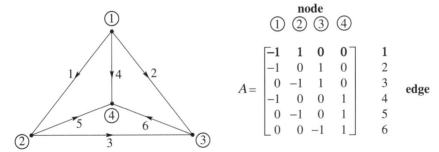

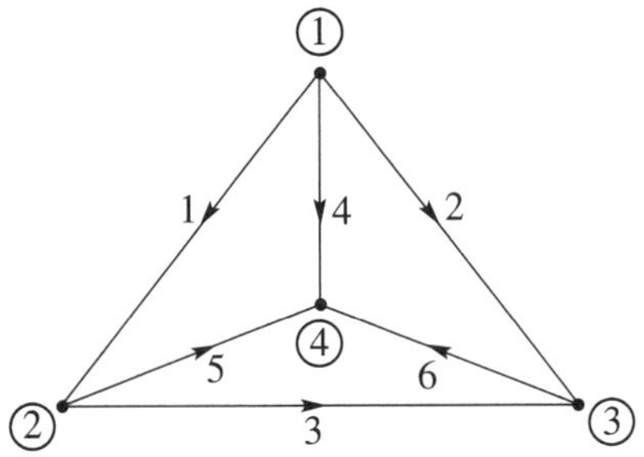


$$\begin{array}{c} \textbf{node} \\ \text{①}\ \ \text{②}\ \ \text{③}\ \ \text{④} \end{array}$$

$$A = \begin{bmatrix} \mathbf{-1} & \mathbf{1} & \mathbf{0} & \mathbf{0} \\ -1 & 0 & 1 & 0 \\ 0 & -1 & 1 & 0 \\ -1 & 0 & 0 & 1 \\ 0 & -1 & 0 & 1 \\ 0 & 0 & -1 & 1 \end{bmatrix} \begin{matrix} \mathbf{1} \\ 2 \\ 3 \\ 4 \\ 5 \\ 6 \end{matrix} \quad \textbf{edge}$$

Figure 5.7: Complete graph with $m = 6$ edges and $n = 4$ nodes. Edge 1 gives row 1.

Row numbers in $A$ are edge numbers on the graph. Column numbers are node numbers. This particular graph is *complete*—every pair of nodes is connected by an edge. You can write down $A$ immediately by looking at the graph. The graph and the matrix have the same information.

If edge 6 is removed from the graph, row 6 is removed from the matrix. The constant vector $(1, 1, 1, 1)$ is still in the nullspace of $A$. Our goal is to understand all four of the fundamental subspaces coming from $A$.

## The Nullspace and Row Space

For the nullspace we solve $Av = \mathbf{0}$. By writing down those $m$ equations we see that $A$ is a **difference matrix**:

$$Av = \begin{bmatrix} -1 & 1 & 0 & 0 \\ -1 & 0 & 1 & 0 \\ 0 & -1 & 1 & 0 \\ -1 & 0 & 0 & 1 \\ 0 & -1 & 0 & 1 \\ 0 & 0 & -1 & 1 \end{bmatrix} \begin{bmatrix} v_1 \\ v_2 \\ v_3 \\ v_4 \end{bmatrix} = \begin{bmatrix} v_2 - v_1 \\ v_3 - v_1 \\ v_3 - v_2 \\ v_4 - v_1 \\ v_4 - v_2 \\ v_4 - v_3 \end{bmatrix}. \tag{1}$$

The numbers $v_1, v_2, v_3, v_4$ can represent ***voltages*** at the nodes. Then $Av$ gives the ***voltage differences*** across the six edges. It is these differences that make currents flow.

The nullspace contains the solutions to $Av = \mathbf{0}$. All six voltage differences are zero. This means: All four voltages are *equal*. **Every $v$ in the nullspace is a constant vector** $v = (c, c, c, c)$. The nullspace of $A$ is a line in $\mathbf{R}^n$. Its dimension is $n - r = 1$, so $\boldsymbol{r} = \mathbf{3}$.

| **Counting Theorem** | $\boldsymbol{r + (n - r) = 3 + 1 = 4 =}$ **count of columns.** |
|---|---|

We can raise or lower all voltages by the same $c$, without changing the voltage *differences*. There is an "arbitrary constant" in $v$. For functions, we can raise or lower $f(x)$ by any constant amount $C$, without changing its derivative.

Calculus adds an arbitrary constant "$+C$" to indefinite integrals. Graph theory adds $(c, c, c, c)$ to the voltages. Linear algebra adds any vector $v_n$ in the nullspace to one particular solution of $Av = \boldsymbol{b}$.

The **row space** of $A$ is also a subspace of $\mathbf{R}^4$. Every row adds to zero, because $-1$ cancels $+1$ in each row. Then every combination of the rows also adds to zero. This is just saying that $v = (c, c, c, c)$ in the nullspace is orthogonal to every vector in the row space.

For any connected graph with $n$ nodes, the situation is the same. The vectors $v = (c, \ldots, c)$ fill the nullspace in $\mathbf{R}^n$. All rows are orthogonal to $v$; their components add to zero. **The row space $\boldsymbol{C}(A^{\mathrm{T}})$ has dimension** $n - 1$. This is the rank of $A$.

## The Column Space and Left Nullspace

The ***column space*** contains all combinations of the four columns. We expect three independent columns, since the rank is $r = n - 1 = 3$. The first three columns are independent (so are any three). But the four columns add to the zero vector, which says again that $(1, 1, 1, 1)$ is in the nullspace. ***How can we tell if a particular vector b is in the column space of an incidence matrix***?

**First answer** Apply elimination to $Av = \boldsymbol{b}$. On the left side, some combinations of rows will give zero rows. Then the same combination of $b$'s on the right side must be zero! Here is the first combination that elimination will discover:

Row 1 − Row 2 + Row 3 = *Zero row*. The right side $\boldsymbol{b}$ needs $b_1 - b_2 + b_3 = 0$. (2)

Since $A$ has $m = 6$ rows and its rank is $r = 3$, elimination leads to $6 - 3$ zero rows in the reduced matrix $R$. There will be *three tests* for the vector $\boldsymbol{b}$ to lie in the column space. Elimination will lead to *three conditions* on $\boldsymbol{b}$ for $A\boldsymbol{v} = \boldsymbol{b}$ to be solvable.

I want to find those conditions in a better way. The graph has three small loops.

**Second answer using loops** $A\boldsymbol{v}$ contains differences in $v$'s. If we add differences around a closed loop in the graph, the cancellation leaves zero. Around the big triangle formed by edges $1, 3, -2$ (the arrow goes backward on edge 2) the differences cancel out:

$$\textbf{Around a loop} \qquad (v_2 - v_1) + (v_3 - v_2) - (v_3 - v_1) = 0.$$

***The components of $A\boldsymbol{v}$ add to zero around every loop***. When $\boldsymbol{b}$ is in the column space of $A$, then $A\boldsymbol{v} = \boldsymbol{b}$. The vector $\boldsymbol{b}$ must obey the voltage law:

| **KVL** | ***Kirchhoff's Voltage Law*** (on a typical loop) | $b_1 + b_3 - b_2 = 0.$ |
|---|---|---|

By testing all the loops, we decide whether $\boldsymbol{b}$ is in the column space. $A\boldsymbol{v} = \boldsymbol{b}$ can be solved exactly when the components of $\boldsymbol{b}$ satisfy all the same dependencies as the rows of $A$. Then KVL is satisfied, elimination leads to $0 = 0$, and $A\boldsymbol{v} = \boldsymbol{b}$ is consistent.

*Question* I can see four loops in the graph, three small and one large. We are only expecting three tests, not four, for $\boldsymbol{b}$ to be in $\boldsymbol{C}(A)$. What is the explanation?

*Answer* Those four loops are not independent. If you combine the small loops in Figure 5.8, you get the large loop. So the tests from the small loops combine to give the test from the large loop. We only have to test KVL on the small loops.

We have described the column space of $A$ in two ways. First, $\boldsymbol{C}(A)$ contains all combinations of the columns (and $n - 1$ columns are enough , the $n$th column is dependent). Second, $\boldsymbol{C}(A)$ contains all vectors $\boldsymbol{b}$ that satisfy the Voltage Law. Around every loop the components of $\boldsymbol{b}$ add to zero. We will now see that this is requiring $\boldsymbol{b}$ to be orthogonal to every vector $\boldsymbol{y}$ in the nullspace of $A^{\mathrm{T}}$. *$\boldsymbol{C}(A)$ is orthogonal to the left nullspace $\boldsymbol{N}(A^{\mathrm{T}})$.*

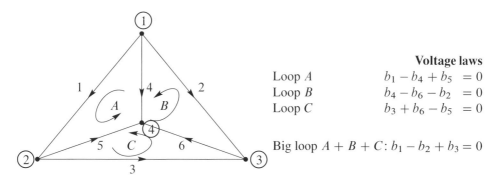

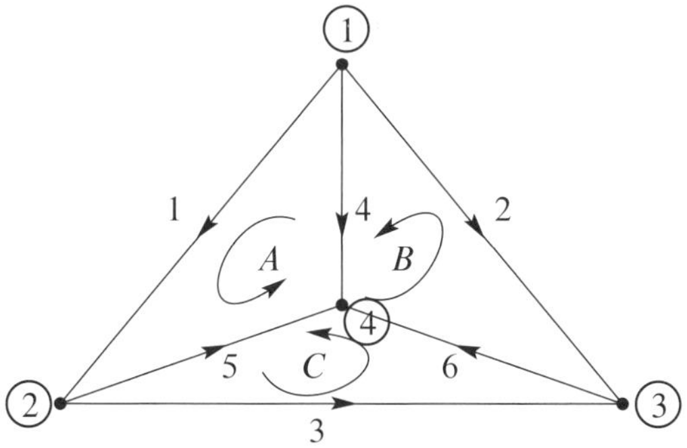


| | **Voltage laws** |
|---|---|
| Loop $A$ | $b_1 - b_4 + b_5 = 0$ |
| Loop $B$ | $b_4 - b_6 - b_2 = 0$ |
| Loop $C$ | $b_3 + b_6 - b_5 = 0$ |

Big loop $A + B + C$: $b_1 - b_2 + b_3 = 0$

Figure 5.8: Loops reveal the column space of $A$ and the nullspace of $A^{\mathrm{T}}$ and the tests on $\boldsymbol{b}$.

$N(A^{\mathrm{T}})$ contains all solutions to $A^{\mathrm{T}}\boldsymbol{y}=\mathbf{0}$. Its dimension is $m-r=6-3$: **three $\boldsymbol{y}$'s.**

$$A^{\mathrm{T}}y=\begin{bmatrix}-1&-1&0&-1&0&0\\1&0&-1&0&-1&0\\0&1&1&0&0&-1\\0&0&0&1&1&1\end{bmatrix}\begin{bmatrix}y_1\\y_2\\y_3\\y_4\\y_5\\y_6\end{bmatrix}=\begin{bmatrix}0\\0\\0\\0\end{bmatrix}. \qquad (3)$$

The true number of equations is $r=3$ and not $n=4$. Reason: The four equations add to $0=0$. The fourth equation follows automatically from the first three.

What do the equations mean? The first equation says that $-y_1-y_2-y_4=0$. ***The net flow into node* 1 *is zero***. The fourth equation says that $y_4+y_5+y_6=0$. ***Flow into the node minus flow out is zero***. These equations are famous and fundamental:

| ***Kirchhoff's Current Law*** | $A^{\mathrm{T}}\boldsymbol{y}=\mathbf{0}$ | ***Flow in equals flow out at each node.*** |
|---|---|---|

This law deserves first place among the equations of applied mathematics. It expresses "*conservation*" and "*continuity*" and "*balance*." Nothing is lost, nothing is gained. When currents or forces are balanced, the equation to solve is $A^{\mathrm{T}}\boldsymbol{y}=\mathbf{0}$. Notice the beautiful fact that the matrix in this balance equation is the transpose of the incidence matrix $A$.

What are the actual solutions to $A^{\mathrm{T}}\boldsymbol{y}=\mathbf{0}$? The currents must balance themselves. The easiest way is to **flow around a loop**. If a unit of current goes around the big triangle (forward on edge 1, forward on 3, backward on 2), the vector is $\boldsymbol{y}=(1,-1,1,0,0,0)$. This satisfies $A^{\mathrm{T}}\boldsymbol{y}=\mathbf{0}$. *Every loop current is a solution to Kirchhoff's Current Law.*

Around the loop, flow in equals flow out at every node. The smaller loop $A$ goes forward on edge 1, forward on 5, back on 4. Then $\boldsymbol{y}=(1,0,0,-1,1,0)$ will have $A^{\mathrm{T}}\boldsymbol{y}=\mathbf{0}$. **Each loop in the graph gives a vector $\boldsymbol{y}$ in $N(A^{\mathrm{T}})$.**

We expect three independent $\boldsymbol{y}$'s, since $6-3=3$. The three small loops in the graph are independent. The big triangle seems to give a fourth $\boldsymbol{y}$, but it is the sum of flows around the small loops. The small loops $A, B, C$ give a basis $\boldsymbol{y}_1, \boldsymbol{y}_2, \boldsymbol{y}_3$ for the nullspace of $A^{\mathrm{T}}$.

**Solutions to $A^{\mathrm{T}}y=\mathbf{0}$**
**Big loop**
**from three**
**small loops**

$$\boldsymbol{y}_1+\boldsymbol{y}_2+\boldsymbol{y}_3=\underset{A}{\begin{bmatrix}1\\0\\0\\-1\\1\\0\end{bmatrix}}+\underset{B}{\begin{bmatrix}0\\0\\1\\0\\-1\\1\end{bmatrix}}+\underset{C}{\begin{bmatrix}0\\-1\\0\\1\\0\\-1\end{bmatrix}}=\underset{A+B+C}{\begin{bmatrix}1\\-1\\1\\0\\0\\0\end{bmatrix}}$$

**Summary** The $m$ by $n$ incidence matrix $A$ comes from a connected graph with $n$ nodes and $m$ edges. The row space and column space have dimension $r = n - 1 =$ rank of $A$. The nullspaces of $A$ and $A^{\mathrm{T}}$ have dimension 1 and $m - r = m - n + 1$:

**1** The constant vectors $(c, c, \ldots, c)$ make up the nullspace $\boldsymbol{N}(A)$.

**2** There are $r = n - 1$ independent rows, from $n - 1$ edges with no loops (a tree).

**3** ***Voltage law gives*** $\boldsymbol{C}(A)$ : The components of $A\boldsymbol{v}$ add to zero around every loop.

**4** ***Current law*** $A^{\mathrm{T}}\boldsymbol{y} = \boldsymbol{0}$ : $\boldsymbol{N}(A^{\mathrm{T}})$ from currents on $m - r$ independent loops.

For every graph in a plane, linear algebra yields ***Euler's formula***:

$$\boxed{(\textit{number of nodes}) - (\textit{number of edges}) + (\textit{number of small loops}) = \mathbf{1}.}$$

This is $(\boldsymbol{n}) - (\boldsymbol{m}) + (\boldsymbol{m} - \boldsymbol{n} + \mathbf{1}) = \mathbf{1}$. The graph in our example has $4 - 6 + 3 = 1$.

A single triangle has (3 nodes) $-$ (3 edges) $+$ (1 loop). On a 10-node tree with 9 edges and no loops, Euler's count is $10 - 9 + 0 = 1$. All planar graphs lead to the answer 1.

## Trees

*A tree is a graph with no loops*. Figure 5.9 shows two trees with $n = 4$ nodes. These graphs (and all our graphs) are *connected*: Between every two nodes there is a path of edges, so the graph doesn't break into separate pieces. The tree must have $m = n - 1$ edges, to connect all $n$ nodes. The rank of the incidence matrix is also $r = n - 1$. Then the number of loops in a tree is confirmed as $m - r = 0$ (no loops).

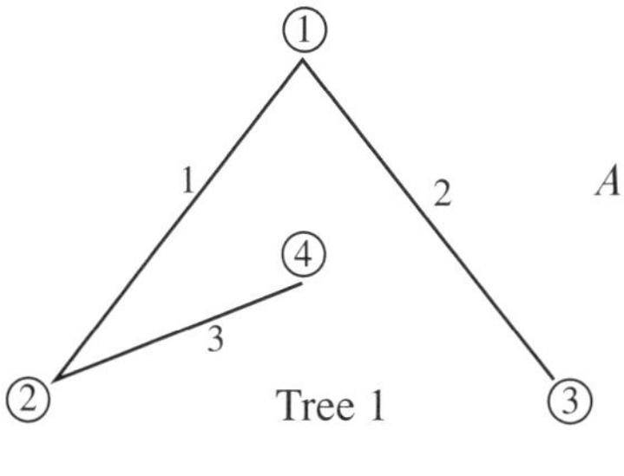


$$A_1 = \begin{bmatrix} -1 & 1 & 0 & 0 \\ -1 & 0 & 1 & 0 \\ 0 & -1 & 0 & 1 \end{bmatrix}$$

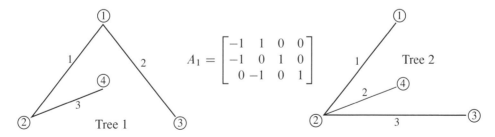

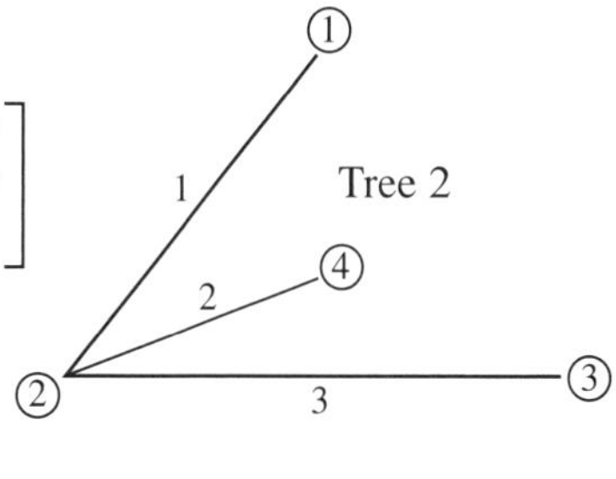


Figure 5.9: Two trees with $n = 4$ nodes and $m = 3$ edges. The rank of $A_1$ is $r = m$.

The incidence matrix $A$ of a tree has *independent rows*. In fact the three rows of $A_1$ are three independent rows $1, 2, 5$ of the previous 6 by 4 matrix (for the complete graph).

That original graph contains 16 different trees.

## The Adjacency Matrix and the Graph Laplacian

The adjacency matrix $W$ is square. With $n$ nodes in the graph, this matrix is $n$ by $n$. If there is an edge from node $i$ to node $j$, then $W_{ij} = 1$. If no edge, then $W_{ij} = 0$. Since our edges go both ways, $W$ is symmetric. The diagonal entries are zero.

All information about the graph is in the adjacency matrix $W$, except the numbering and arrow directions of the edges.

There are $m$ 1's above the diagonal of $W$, and also below. Section 7.5 will study the **graph Laplacian matrix $A^{\mathrm{T}}A$** ($A$ is the incidence matrix) and find this formula:

| **Graph Laplacian** $\quad A^{\mathrm{T}}A = D - W =$ **(degree matrix)−(adjacency matrix)**. |
|---|

The diagonal matrix $D$ tells the "degree" of every node. This is the number of edges that go in or out of that node. Here are $W$ and $A^{\mathrm{T}}A$ for the complete graph with six edges.

$$\textbf{Adjacency } W = \begin{bmatrix} 0 & 1 & 1 & 1 \\ 1 & 0 & 1 & 1 \\ 1 & 1 & 0 & 1 \\ 1 & 1 & 1 & 0 \end{bmatrix} \qquad \textbf{Graph Laplacian } A^{\mathrm{T}}A = \begin{bmatrix} 3 & -1 & -1 & -1 \\ -1 & 3 & -1 & -1 \\ -1 & -1 & 3 & -1 \\ -1 & -1 & -1 & 3 \end{bmatrix}$$

Every row of $A^{\mathrm{T}}A$ adds to zero. The degree 3 on the diagonal cancels the $-1$'s off the diagonal. The vector $(1,1,1,1)$ in the nullspace of $A$ is also in the nullspace of $A^{\mathrm{T}}A$.

**Challenge** Reconstruct a graph with arrows from $A$ and a graph without arrows from $W$.

$$A = \begin{bmatrix} 1 & 0 & 0 & -1 \\ 0 & -1 & 1 & 0 \\ 0 & 0 & -1 & 1 \\ 1 & -1 & 0 & 0 \end{bmatrix} \qquad W = \begin{bmatrix} 0 & 1 & 0 & 1 \\ 1 & 0 & 1 & 0 \\ 0 & 1 & 0 & 1 \\ 1 & 0 & 1 & 0 \end{bmatrix}$$

## ■ REVIEW OF THE KEY IDEAS ■

1. The $n$ nodes and $m$ edges of a graph give $n$ columns and $m$ rows in $A$.

2. Each row of the incidence matrix $A$ has $-1$ and $1$ (start and end of that edge).

3. **Voltage Law for $C(A)$**: The components of $A\boldsymbol{v}$ add to zero around any loop.

4. **Current Law for $N(A^{\mathrm{T}})$**: $A^{\mathrm{T}}\boldsymbol{y} =$ (flow in) minus (flow out) $=$ zero at every node.

5. Rank of $A = n-1$. Then $A^{\mathrm{T}}\boldsymbol{y} = \mathbf{0}$ for the currents $\boldsymbol{y}$ around $m-n+1$ small loops.

6. The adjacency matrix $W$ and the graph Laplacian $A^{\mathrm{T}}A$ are symmetric $n$ by $n$.

## Problem Set 5.6

**Problems 1–7 and 8–13 are about the incidence matrices for these two graphs.**

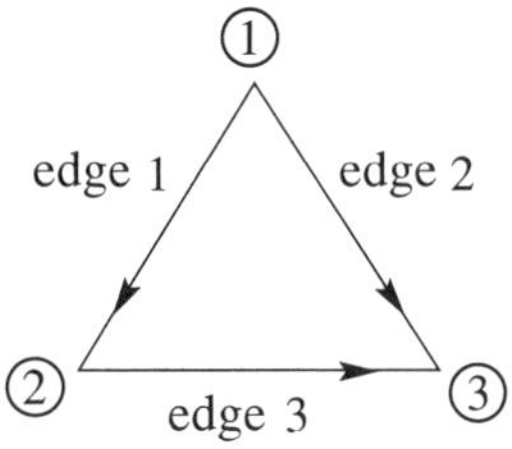


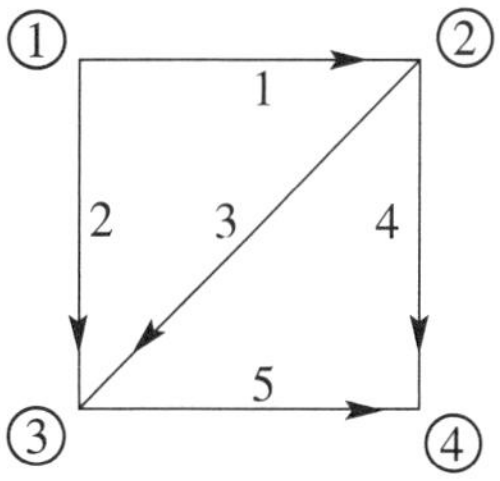


**1** Write down the 3 by 3 incidence matrix $A$ for the triangle graph. The first row has $-1$ in column 1 and $+1$ in column 2. What vectors $(v_1, v_2, v_3)$ are in its nullspace? How do you know that $(1, 0, 0)$ is not in its row space?

**2** Write down $A^{\mathrm{T}}$ for the triangle graph. Find a vector $\boldsymbol{y}$ in its nullspace. The components of $\boldsymbol{y}$ are currents on the edges—how much current is going around the triangle?

**3** By elimination on $A$ find the echelon matrix $R$. What tree corresponds to the two nonzero rows of $R$?

$$A\boldsymbol{v} = \boldsymbol{b} \qquad \begin{aligned} -v_1 + v_2 &= b_1 \\ -v_1 + v_3 &= b_2 \\ -v_2 + v_3 &= b_3. \end{aligned}$$

**4** Choose a vector $(b_1, b_2, b_3)$ for which $A\boldsymbol{v} = \boldsymbol{b}$ can be solved, and another vector $\boldsymbol{b}$ that allows no solution. What are the dot products $\boldsymbol{y}^{\mathrm{T}}\boldsymbol{b}$ for $\boldsymbol{y} = (1, -1, 1)$?

**5** Choose a vector $(f_1, f_2, f_3)$ for which $A^{\mathrm{T}}\boldsymbol{y} = \boldsymbol{f}$ can be solved, and a vector $\boldsymbol{f}$ that allows no solution. How are those $\boldsymbol{f}$'s related to $\boldsymbol{v} = (1, 1, 1)$? The equation $A^{\mathrm{T}}\boldsymbol{y} = \boldsymbol{f}$ is Kirchhoff's _____ law.

**6** Multiply matrices to find $A^{\mathrm{T}}A$. Choose a vector $\boldsymbol{f}$ for which $A^{\mathrm{T}}A\boldsymbol{v} = \boldsymbol{f}$ can be solved, and solve for $\boldsymbol{v}$. Put those voltages $\boldsymbol{v}$ and currents $\boldsymbol{y} = -A\boldsymbol{v}$ onto the triangle graph. The vector $f$ represents "current sources."

**7** Multiply $A^{\mathrm{T}}A$ (still for the first graph) and find its nullspace—it should be the same as $\boldsymbol{N}(A)$. Which vectors $\boldsymbol{f}$ are in its column space?

**8** Write down the 5 by 4 incidence matrix $A$ for the square graph with two loops. Find one solution to $A\boldsymbol{v} = \boldsymbol{0}$ and two solutions to $A^{\mathrm{T}}\boldsymbol{y} = \boldsymbol{0}$. The rank is _____.

**9** Find two requirements on the $b$'s for the five differences $v_2 - v_1, v_3 - v_1, v_3 - v_2, v_4 - v_2, v_4 - v_3$ to equal $b_1, b_2, b_3, b_4, b_5$. You have found Kirchhoff's _____ Law around the two _____ in the graph.

**10** By elimination, reduce $A$ to $U$. The three nonzero rows give the incidence matrix for what graph? You found one tree in the square graph—find the other seven trees.

**11** Multiply $A^{\mathrm{T}}A$ and explain how its entries come from columns of $A$ (and the graph).

(a) The diagonal of the Laplacian matrix $A^{\mathrm{T}}A$ counts edges into each node (the degree). Why is this the dot product of a column with itself?

(b) The off-diagonals $-1$ or $0$ tell which nodes $i$ and $j$ are connected. Why is $-1$ or $0$ the dot product of column $i$ with another column $j$?

**12** Find the rank and the nullspace of $A^{\mathrm{T}}A$. Why does $A^{\mathrm{T}}A\boldsymbol{v} = \boldsymbol{f}$ have a solution only if $f_1 + f_2 + f_3 + f_4 = 0$?

**13** Write down the 4 by 4 adjacency matrix $W$ for the square graph. Its entries 1 or 0 count paths of length 1 between nodes (those are just edges).

*Important.* Compute $W^2$ and check that its entries count the paths of length 2 between nodes. Why does $(A^2)_{ii}$ = degree of node $i$? Those paths go out and back.

**14** A connected graph with 7 nodes and 7 edges has how many loops?

**15** For the graph with 4 nodes, 6 edges, and 3 loops, add a new node. If you connect it to one old node, Euler's formula becomes ( ) − ( ) + ( ) = 1. If you connect it to two old nodes, Euler's formula becomes ( ) − ( ) + ( ) = 1.

**16** Suppose $A$ is a 12 by 9 incidence matrix from a connected (but unknown) graph.

(a) How many columns of $A$ are independent?

(b) What condition on $\boldsymbol{f}$ makes it possible to solve $A^{\mathrm{T}}\boldsymbol{y} = \boldsymbol{f}$?

(c) The diagonal entries of $A^{\mathrm{T}}A$ give the number of edges into each node. What is the sum of those diagonal entries?

**17** Why does a complete graph with $n = 6$ nodes have $m = 15$ edges? A tree that connects 6 nodes has only _____ edges and _____ loops.

**18** How do you know that *any* $n-1$ *columns* of the incidence matrix $A$ are independent? If they were dependent, the nullspace would contain a vector with a zero component. But the nullspace of $A$ actually contains _____.

**19** (a) Find the Laplacian $A^{\mathrm{T}}A$ for a complete graph with $n$ nodes.

(b) If the edge from node 1 to node 3 is removed, what is the change in $A^{\mathrm{T}}A$?

**20** Suppose batteries of strength $b_1, \ldots, b_m$ are inserted into the $m$ edges. Then the voltage differences across edges become $A\boldsymbol{v} - \boldsymbol{b}$. Unit resistances give currents $A\boldsymbol{v} - \boldsymbol{b}$ and Kirchhoff's Current Law is $A^{\mathrm{T}}(A\boldsymbol{v} - \boldsymbol{b}) = \boldsymbol{0}$. Solve this system for the square graph above when $\boldsymbol{b} = (1, 1, \ldots, 1)$.

# ■ CHAPTER 5 NOTES ■

**Vectors are not necessarily column vectors. In the definition of a *vector space*, addition $x + y$ and scalar multiplication $cx$ must obey the following eight rules:**

(1) $\boldsymbol{x} + \boldsymbol{y} = \boldsymbol{y} + \boldsymbol{x}$

(2) $\boldsymbol{x} + (\boldsymbol{y} + \boldsymbol{z}) = (\boldsymbol{x} + \boldsymbol{y}) + \boldsymbol{z}$

(3) There is a unique "zero vector" such that $\boldsymbol{x} + \mathbf{0} = \boldsymbol{x}$ for all $\boldsymbol{x}$

(4) For each $\boldsymbol{x}$ there is a unique vector $-\boldsymbol{x}$ such that $\boldsymbol{x} + (-\boldsymbol{x}) = \mathbf{0}$

(5) 1 times $\boldsymbol{x}$ equals $\boldsymbol{x}$

(6) $(c_1c_2)\boldsymbol{x} = c_1(c_2\boldsymbol{x})$

(7) $c(\boldsymbol{x} + \boldsymbol{y}) = c\boldsymbol{x} + c\boldsymbol{y}$

(8) $(c_1 + c_2)\boldsymbol{x} = c_1\boldsymbol{x} + c_2\boldsymbol{x}$.

Here are practice questions to bring out the meaning of those eight rules.

**1.** Suppose $(x_1, x_2) + (y_1, y_2)$ is defined to be $(x_1 + y_2, x_2 + y_1)$. With the usual multiplication $c\boldsymbol{x} = (cx_1, cx_2)$, which of the eight conditions are not satisfied?

**2.** Suppose the multiplication $c\boldsymbol{x}$ is defined to produce $(cx_1, 0)$ instead of $(cx_1, cx_2)$. With the usual addition in $\mathbf{R}^2$, are the eight conditions satisfied?

**3.** (a) Which rules are broken if we keep only the positive numbers $x > 0$ in $\mathbf{R}^1$? Every $c$ must be allowed. The half-line is not a subspace.

(b) The positive numbers with $\boldsymbol{x} + \boldsymbol{y}$ and $c\boldsymbol{x}$ redefined to equal the usual $xy$ and $x^c$ *do* satisfy the eight rules. Test rule 7 when $c = 3, x = 2, y = 1$. (Then $\boldsymbol{x} + \boldsymbol{y} = 2$ and $c\boldsymbol{x} = 8$.) Which number acts as the "zero vector"?

**4.** The matrix $A = \left[\begin{smallmatrix} 2 & -2 \\ 2 & -2 \end{smallmatrix}\right]$ is a "vector" in the space $\mathbf{M}$ of all 2 by 2 matrices. Write down the zero vector in this space, the vector $\frac{1}{2}A$, and the vector $-A$. What matrices are in the smallest subspace containing $A$?

**5.** The functions $\boldsymbol{f}(x) = x^2$ and $\boldsymbol{g}(x) = 5x$ are "vectors in function space." Which rule is broken if multiplying $\boldsymbol{f}(x)$ by $c$ gives $\boldsymbol{f}(cx)$ instead of $c\boldsymbol{f}(x)$? Keep the usual addition $\boldsymbol{f}(x) + \boldsymbol{g}(x)$.

**6.** If the sum of the "vectors" $\boldsymbol{f}(x)$ and $\boldsymbol{g}(x)$ is defined to be the function $f(g(x))$, then the "zero vector" is $\boldsymbol{g}(x) = x$. Keep the usual scalar multiplication $c\boldsymbol{f}(x)$ and find two rules that are broken.

## Row rank equals column rank : The first big theorem

The dimension of the row space $\boldsymbol{C}(A^{\mathrm{T}})$ equals the dimension of the column space $\boldsymbol{C}(A)$. Here I can outline four proofs (the fourth is neat). Proofs **2 , 3, 4** do not use elimination.

**Proof 1** Reduce $A$ to $R$ without changing the dimensions of the row and column spaces. The row space actually stays the same. The column space changes, going from $A$ to $R$, but its dimension stays the same. The theorem is clear for $R$ :

$$\begin{array}{lcl} r \text{ nonzero rows in } R & \leftrightarrow & r = \text{dimension of row space} \\ r \text{ pivot columns in } R & \leftrightarrow & r = \text{dimension of column space} \end{array}$$

**Proof 2** (G. Mackiw, *Mathematics Magazine* **68** 1996). Suppose $\boldsymbol{x}_1, \ldots, \boldsymbol{x}_r$ is a basis for the row space of $A$. The next paragraph will show that $A\boldsymbol{x}_1, \ldots, A\boldsymbol{x}_r$ are independent vectors in the column space. Then dim (row space) = $r \leq$ dim (column space). The same reasoning applies to $A^{\mathrm{T}}$, reversing that inequality. So the two dimensions must be equal.

$$\text{Suppose} \quad c_1 A\boldsymbol{x}_1 + \cdots + c_r A\boldsymbol{x}_r = A(c_1\boldsymbol{x}_1 + \cdots + c_r\boldsymbol{x}_r) = A\boldsymbol{v} = \boldsymbol{0}.$$

Then $\boldsymbol{v}$ is in the nullspace of $A$ and also in the row space (it is a combination of the $\boldsymbol{x}$'s). So $\boldsymbol{v}$ is orthogonal to itself and $\boldsymbol{v} = \boldsymbol{0}$. All the $c$'s must be zero since the $\boldsymbol{x}$'s are a basis.

This shows that $c_1 A\boldsymbol{x}_1 + \cdots + c_r A\boldsymbol{x}_r = 0$ requires that all $c_i = 0$. Therefore $A\boldsymbol{x}_1, \ldots, A\boldsymbol{x}_r$ are independent vectors in the column space : dimension of $\boldsymbol{C}(A) \geq r$.

**Proof 3** If $A$ has $r$ independent rows and $s$ independent columns, we can move those rows to the top of $A$ and those columns to the left. They meet in an $r$ by $s$ submatrix $B$ :

$$A = \begin{bmatrix} B & C \\ D & E \end{bmatrix} \; r \text{ rows} \qquad\qquad \begin{bmatrix} B & C \\ D & E \end{bmatrix} \begin{bmatrix} \boldsymbol{v} \\ \boldsymbol{0} \end{bmatrix} = \begin{bmatrix} \boldsymbol{0} \\ \boldsymbol{0} \end{bmatrix}.$$

Suppose $s > r$. Since $B\boldsymbol{v} = \boldsymbol{0}$ has $r$ equations in $s$ unknowns, it has a solution $\boldsymbol{v} \neq \boldsymbol{0}$. The upper part of the matrix has $B\boldsymbol{v} + C\boldsymbol{0} = \boldsymbol{0}$ as shown. The lower rows of $A$ are combinations of the upper rows, so they also have $D\boldsymbol{v} + E\boldsymbol{0} = \boldsymbol{0}$. But now a combination of the first $s$ independent columns $\begin{bmatrix} B \\ D \end{bmatrix}$ of $A$, with coefficients from $\boldsymbol{v}$, is producing zero. *Conclusion* : $s > r$ cannot happen. Thinking similarly for $A^{\mathrm{T}}$, $r > s$ cannot happen.

**Proof 4** Suppose $r$ column vectors $\boldsymbol{u}_1, \ldots, \boldsymbol{u}_r$ are a basis for the column space $\boldsymbol{C}(A)$. Then each column of $A$ is a combination of $\boldsymbol{u}$'s. Column 1 of $A$ is $w_{11}\boldsymbol{u}_1 + \cdots + w_{r1}\boldsymbol{u}_r$, with some coefficients $w$. The whole matrix $A$ equals $UW = (m \text{ by } r)(r \text{ by } n)$.

$$A = \begin{bmatrix} & & \\ \boldsymbol{u}_1 & \ldots & \boldsymbol{u}_r \\ & & \end{bmatrix} \begin{bmatrix} w_{11} & \ldots & w_{1n} \\ \vdots & & \vdots \\ w_{r1} & \ldots & w_{rn} \end{bmatrix} = UW.$$

*Now look differently at $A = UW$. Each row of $A$ is a combination of the $r$ rows of $W$!* Therefore the row space of $A$ has dimension $\leq r$.

This proves that (dimension of row space) $\leq$ (dimension of column space) for any $A$. Apply this reasoning to $A^{\mathrm{T}}$, and the two dimensions must be equal.

To my way of thinking, that is a really cool proof.

## The Transpose and Row Space of $d/dt$

This book is constantly emphasizing the parallels between linear differential equations and matrix equations. In both cases we have *null solutions* and *particular solutions*. The nullspace for a differential equation $Dy = 0$ contains the null solutions $y_n$ :

**Matrices** $A$ $\quad A\boldsymbol{v}_n = \mathbf{0} \qquad$ **Derivatives** $\boldsymbol{D} \quad Dy_n = y_n'' + By_n' + Cy_n = 0$

**The nullspace of this $D$ has dimension 2. This is the reason that $y$ needs two initial conditions.** We look for solutions $y_n = e^{st}$ and usually we find $e^{s_1 t}$ and $e^{s_2 t}$. These functions are a basis for the nullspace. In case $s_2 = s_1$, the second function is $te^{s_1 t}$. All is completely parallel to matrix equations, until we ask this question:

**What is the "row space" of $D$ when a differential operator has no rows ?**

I want to propose two answers to this question. They come from faithfully imitating the Fundamental Theorem of Linear Algebra. That theorem applies to $D$, because $D$ is linear.

**Answer 1** The row space of $D$ contains all functions $y_r(t)$ orthogonal to $e^{s_1 t}$ and $e^{s_2 t}$.

**Answer 2** The row space of $D$ contains all outputs $y_r(t) = D^{\mathrm{T}} q(t)$ from inputs $q(t)$.

This looks good, but when are functions "orthogonal" ? What is the "transpose" of $D$ ?

**Dot product of functions**
**Inner product of $y_n$ and $y_r$** $\quad \boldsymbol{(y_n(t), y_r(t)) = \int_{-\infty}^{\infty} y_n(t) y_r(t) dt}.$

Do you see this as reasonable ? For vectors, we add the products $v_j w_j$. For functions, we integrate $y_n y_r$. If the vectors or functions are complex, we add $\overline{v}_j w_j$ or integrate $\overline{y}_n y_r$. Then $(\boldsymbol{v}, \boldsymbol{v})$ and $(y_r, y_r)$ give the squared lengths $\|\boldsymbol{v}\|^2$ for vectors and $\|y_r\|^2$ for functions.

**The inner product tells us the correct meaning of the transpose.** For matrices, $A^{\mathrm{T}}$ is the matrix that obeys the inner product law $(Av, w) = (v, A^{\mathrm{T}} w)$. For differential equations,

$$(\boldsymbol{Df}, \boldsymbol{g}) = \int_{-\infty}^{\infty} (f'' + Bf' + Cf) g(t) dt = \int_{-\infty}^{\infty} f(t)(g'' - Bg' + Cg) dt = (\boldsymbol{f}, \boldsymbol{D}^{\mathrm{T}} \boldsymbol{g}).$$

Integration by parts gave $\int f'g = -\int fg'$. Two integrations gave $\int f''g = \int fg''$ with a plus sign (from two minus signs). Formally, that equation tells us $D^{\mathrm{T}}$ :

$$D = \frac{d^2}{dt^2} + B\frac{d}{dt} + C \quad \text{leads to} \quad D^{\mathrm{T}} = \frac{d^2}{dt^2} - B\frac{d}{dt} + C \quad \left(\frac{d}{dt} \textbf{ is antisymmetric}\right)$$

Now the row space of all $D^{\mathrm{T}} q(t)$ makes sense even when $D$ has no rows. Can we just verify that any row space function $D^{\mathrm{T}} q(t)$ is orthogonal to any nullspace function $y_n(t)$ ?

$$(y_n(t), D^{\mathrm{T}} q(t)) = (Dy_n(t), q(t)) = \int_{-\infty}^{\infty} (0)\, q(t)\, dt = 0.$$

Shakespeare said it best at the end of Hamlet: *The rest is silence.*

# Chapter 6

# Eigenvalues and Eigenvectors

## 6.1 Introduction to Eigenvalues

***Eigenvalues are the key to a system of n differential equations***: $dy/dt = ay$ becomes $d\boldsymbol{y}/dt = A\boldsymbol{y}$. Now $A$ is a matrix and $\boldsymbol{y}$ is a vector $(y_1(t), \ldots, y_n(t))$. The vector $\boldsymbol{y}$ changes with time. Here is a system of two equations with its 2 by 2 matrix $A$:

$$\begin{aligned} y_1{}' &= 4y_1 + y_2 \\ y_2{}' &= 3y_1 + 2y_2 \end{aligned} \qquad \text{is} \qquad \begin{bmatrix} y_1 \\ y_2 \end{bmatrix}' = \begin{bmatrix} 4 & 1 \\ 3 & 2 \end{bmatrix} \begin{bmatrix} y_1 \\ y_2 \end{bmatrix}. \tag{1}$$

How to solve this coupled system, $\boldsymbol{y}' = A\boldsymbol{y}$ with $y_1$ and $y_2$ in both equations? The good way is to find solutions that "uncouple" the problem. *We want $y_1$ and $y_2$ to grow or decay in exactly the same way* (with the same $e^{\lambda t}$):

$$\textbf{Look for} \quad \begin{aligned} y_1(t) &= e^{\lambda t} a \\ y_2(t) &= e^{\lambda t} b \end{aligned} \qquad \textbf{In vector notation this is} \quad \boldsymbol{y}(t) = e^{\lambda t}\boldsymbol{x} \tag{2}$$

That vector $\boldsymbol{x} = (a, b)$ is called an **eigenvector**. The growth rate $\lambda$ is an **eigenvalue**. This section will show how to find $\boldsymbol{x}$ and $\lambda$. Here I will jump to $\boldsymbol{x}$ and $\lambda$ for the matrix in (1).

$$\text{First eigenvector } \boldsymbol{x} = \begin{bmatrix} a \\ b \end{bmatrix} = \begin{bmatrix} 1 \\ 1 \end{bmatrix} \text{ and first eigenvalue } \boldsymbol{\lambda = 5} \text{ in } \boldsymbol{y} = e^{5t}\boldsymbol{x}$$

$$\begin{aligned} y_1 &= \boldsymbol{e^{5t}} \\ y_2 &= \boldsymbol{e^{5t}} \end{aligned} \qquad \text{has} \qquad \begin{aligned} y_1{}' &= 5\,e^{5t} = 4y_1 + \;\; y_2 \\ y_2{}' &= 5\,e^{5t} = 3y_1 + 2y_2 \end{aligned}$$

$$\text{Second eigenvector } \boldsymbol{x} = \begin{bmatrix} a \\ b \end{bmatrix} = \begin{bmatrix} 1 \\ -3 \end{bmatrix} \text{ and second eigenvalue } \boldsymbol{\lambda = 1} \text{ in } \boldsymbol{y} = e^{t}\boldsymbol{x}$$

**This $\boldsymbol{y = e^{\lambda t} x}$ is a second solution**

$$\begin{aligned} y_1 &= \boldsymbol{e^{t}} \\ y_2 &= \boldsymbol{-3\,e^{t}} \end{aligned} \qquad \text{has} \qquad \begin{aligned} y_1{}' &= \;\; e^{t} = 4y_1 + \;\; y_2 \\ y_2{}' &= -3\,e^{t} = 3y_1 + 2y_2 \end{aligned}$$

Those two $\boldsymbol{x}$'s and $\boldsymbol{\lambda}$'s combine with any $c_1, c_2$ to give the complete solution to $\boldsymbol{y}' = A\boldsymbol{y}$ :

$$\textbf{Complete solution} \quad \boldsymbol{y}(t) = c_1 \begin{bmatrix} e^{5t} \\ e^{5t} \end{bmatrix} + c_2 \begin{bmatrix} e^t \\ -3\,e^t \end{bmatrix} = c_1 e^{5t} \begin{bmatrix} 1 \\ 1 \end{bmatrix} + c_2 e^t \begin{bmatrix} 1 \\ -3 \end{bmatrix}. \quad (3)$$

This is exactly what we hope to achieve for other equations $\boldsymbol{y}' = A\boldsymbol{y}$ with constant $A$.

The solutions we want have the special form $\boldsymbol{y}(t) = e^{\lambda t}\boldsymbol{x}$. Substitute that solution into $\boldsymbol{y}' = A\boldsymbol{y}$, to see the equation $A\boldsymbol{x} = \lambda\boldsymbol{x}$ for an eigenvalue $\lambda$ and its eigenvector $\boldsymbol{x}$ :

$$\frac{d}{dt}(e^{\lambda t}\boldsymbol{x}) = A(e^{\lambda t}\boldsymbol{x}) \quad \text{is} \quad \lambda e^{\lambda t}\boldsymbol{x} = Ae^{\lambda t}\boldsymbol{x}. \qquad \text{Divide both sides by } e^{\lambda t}.$$

$$\textbf{Eigenvalue and eigenvector of } \boldsymbol{A} \qquad A\boldsymbol{x} = \lambda\boldsymbol{x} \qquad (4)$$

Those eigenvalues (5 and 1 for this $A$) are a new way to see into the heart of a matrix. This chapter enters a different part of linear algebra, based on $A\boldsymbol{x} = \lambda\boldsymbol{x}$. **The last page of Chapter 6 has eigenvalue-eigenvector information about many different matrices.**

## Finding Eigenvalues from det$(A - \boldsymbol{\lambda} I) = 0$

Almost all vectors change direction, when they are multiplied by $A$. ***Certain very exceptional vectors $x$ are in the same direction as $Ax$. Those are the "eigenvectors."*** The vector $A\boldsymbol{x}$ (in the same direction as $\boldsymbol{x}$) is a number $\lambda$ times the original $\boldsymbol{x}$.

The eigenvalue $\lambda$ tells whether the eigenvector $\boldsymbol{x}$ is stretched or shrunk or reversed or left unchanged—when it is multiplied by $A$. We may find $\lambda = 2$ or $\frac{1}{2}$ or $-1$ or 1. The eigenvalue $\lambda$ could be zero! $A\boldsymbol{x} = 0\boldsymbol{x}$ puts this eigenvector $\boldsymbol{x}$ in the nullspace of $A$.

If $A$ is the identity matrix, every vector has $A\boldsymbol{x} = \boldsymbol{x}$. All vectors are eigenvectors of $I$. Most 2 by 2 matrices have *two* eigenvector directions and *two* eigenvalues $\lambda_1$ and $\lambda_2$.

To find the eigenvalues, write the equation $A\boldsymbol{x} = \lambda\boldsymbol{x}$ in the good form $(A - \lambda I)\boldsymbol{x} = \boldsymbol{0}$.

If $(A - \lambda I)\boldsymbol{x} = \boldsymbol{0}$, then $A - \lambda I$ is a **singular matrix**. Its determinant must be **zero**.

$$\text{The determinant of } A - \lambda I = \begin{bmatrix} a - \lambda & b \\ c & d - \lambda \end{bmatrix} \quad \text{is} \quad \boldsymbol{(a - \lambda)(d - \lambda) - bc = 0}.$$

Our goal is to shift $A$ by the right amount $\lambda I$, so that $(A - \lambda I)\boldsymbol{x} = \boldsymbol{0}$ has a solution. Then $\boldsymbol{x}$ is the eigenvector, $\lambda$ is the eigenvalue, and $A - \lambda I$ is not invertible. So we look for numbers $\lambda$ that make $\mathbf{det}(\boldsymbol{A} - \boldsymbol{\lambda I}) = \mathbf{0}$. I will start with the matrix $A$ in equation (1).

**Example 1** For $A = \begin{bmatrix} 4 & 1 \\ 3 & 2 \end{bmatrix}$, subtract $\lambda$ from the diagonal and find the determinant:

$$\mathbf{det}(\boldsymbol{A} - \boldsymbol{\lambda I}) = \det \begin{bmatrix} 4 - \lambda & 1 \\ 3 & 2 - \lambda \end{bmatrix} = \lambda^2 - 6\lambda + 5 = \boldsymbol{(\lambda - 5)(\lambda - 1)}. \qquad (5)$$

I factored the quadratic, to see the two eigenvalues $\lambda_1 = 5$ and $\lambda_2 = 1$. The matrices $A - 5I$ and $A - I$ are *singular*. We have found the $\lambda$'s from $\det(A - \lambda I) = 0$.

For each of the eigenvalues 5 and 1, we now find an **eigenvector** $\boldsymbol{x}$ :

$$(A-5I)\,\boldsymbol{x}=\boldsymbol{0} \quad \text{is} \quad \begin{bmatrix} -1 & 1 \\ 3 & -3 \end{bmatrix}\begin{bmatrix} \boldsymbol{x} \end{bmatrix}=\begin{bmatrix} 0 \\ 0 \end{bmatrix} \quad \text{and} \quad \boldsymbol{x}=\begin{bmatrix} 1 \\ 1 \end{bmatrix}$$

$$(A-1I)\,\boldsymbol{x}=\boldsymbol{0} \quad \text{is} \quad \begin{bmatrix} 3 & 1 \\ 3 & 1 \end{bmatrix}\begin{bmatrix} \boldsymbol{x} \end{bmatrix}=\begin{bmatrix} 0 \\ 0 \end{bmatrix} \quad \text{and} \quad \boldsymbol{x}=\begin{bmatrix} 1 \\ -3 \end{bmatrix}$$

Those were the vectors $(a,b)$ in our special solutions $\boldsymbol{y}=e^{\lambda t}\boldsymbol{x}$. Both components of $\boldsymbol{y}$ have the growth rate $\lambda$, so the differential equation was easily solved: $\boldsymbol{y}=e^{\lambda t}\boldsymbol{x}$.

Two eigenvectors gave two solutions. Combinations $c_1\boldsymbol{y}_1+c_2\boldsymbol{y}_2$ give all solutions.

**Example 2** Find the eigenvalues and eigenvectors of the *Markov matrix* $A=\begin{bmatrix} .8 & .3 \\ .2 & .7 \end{bmatrix}$.

$$\det(A-\lambda I)=\det\begin{bmatrix} .8-\lambda & .3 \\ .2 & .7-\lambda \end{bmatrix}=\lambda^2-\frac{3}{2}\lambda+\frac{1}{2}=(\boldsymbol{\lambda}-\mathbf{1})\left(\boldsymbol{\lambda}-\tfrac{\mathbf{1}}{\mathbf{2}}\right).$$

I factored the quadratic into $\lambda-1$ times $\lambda-\frac{1}{2}$, to see the two eigenvalues $\boldsymbol{\lambda}=\mathbf{1}$ and $\frac{\mathbf{1}}{\mathbf{2}}$. The eigenvectors $\boldsymbol{x}_1$ and $\boldsymbol{x}_2$ are in the nullspaces of $A-I$ and $A-\frac{1}{2}I$.

$$\begin{array}{llll} (A-I)\,\boldsymbol{x}_1=\boldsymbol{0} & \text{is} \quad A\boldsymbol{x}_1=\boldsymbol{x}_1 & \text{The first eigenvector is} & \boldsymbol{x}_1=\mathbf{(.6,.4)} \\ (A-\frac{1}{2}I)\,\boldsymbol{x}_2=\boldsymbol{0} & \text{is} \quad A\boldsymbol{x}_2=\frac{1}{2}\boldsymbol{x}_2 & \text{The second eigenvector is} & \boldsymbol{x}_2=\mathbf{(1,-1)} \end{array}$$

$$\boldsymbol{x}_1=\begin{bmatrix} .6 \\ .4 \end{bmatrix} \quad \text{and} \quad A\boldsymbol{x}_1=\begin{bmatrix} .8 & .3 \\ .2 & .7 \end{bmatrix}\begin{bmatrix} .6 \\ .4 \end{bmatrix}=\boldsymbol{x}_1 \quad (A\boldsymbol{x}=\boldsymbol{x}\text{ means that }\lambda_1=1)$$

$$\boldsymbol{x}_2=\begin{bmatrix} 1 \\ -1 \end{bmatrix} \quad \text{and} \quad A\boldsymbol{x}_2=\begin{bmatrix} .8 & .3 \\ .2 & .7 \end{bmatrix}\begin{bmatrix} 1 \\ -1 \end{bmatrix}=\begin{bmatrix} .5 \\ -.5 \end{bmatrix} \quad (\text{this is }\tfrac{1}{2}\,\boldsymbol{x}_2\text{ so }\lambda_2=\tfrac{1}{2}).$$

If $\boldsymbol{x}_1$ is multiplied again by $A$, we still get $\boldsymbol{x}_1$. Every power of $A$ will give $A^n\boldsymbol{x}_1=\boldsymbol{x}_1$. Multiplying $\boldsymbol{x}_2$ by $A$ gave $\frac{1}{2}\boldsymbol{x}_2$, and if we multiply again we get $(\frac{1}{2})^2$ times $\boldsymbol{x}_2$.

***When A is squared, the eigenvectors x stay the same.*** $\boldsymbol{A^2x=A(\lambda x)=\lambda(Ax)=\lambda^2x}$.

Notice $\lambda^2$. This pattern keeps going, because the eigenvectors stay in their own directions. They never get mixed. The eigenvectors of $A^{100}$ are the same $\boldsymbol{x}_1$ and $\boldsymbol{x}_2$. The eigenvalues of $A^{100}$ are $1^{100}=1$ and $(\frac{1}{2})^{100}=$ very small number.

We mention that this particular $A$ is a ***Markov matrix***. Its entries are positive and every column adds to 1. Those facts guarantee that the largest eigenvalue must be $\lambda=1$.

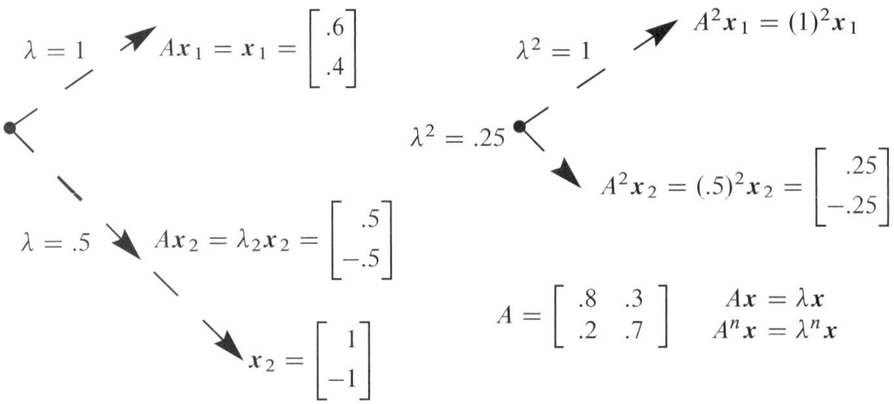


Figure 6.1: The eigenvectors keep their directions. $A^2$ has eigenvalues $1^2$ and $(.5)^2$.

The eigenvector $Ax_1 = x_1$ is the *steady state*—which all columns of $A^k$ will approach.

Giant Markov matrices are the key to Google's search algorithm. It ranks web pages. Linear algebra has made Google one of the most valuable companies in the world.

## Powers of a Matrix

When the eigenvalues of $A$ are known, we immediately know the eigenvalues of all powers $A^k$ and shifts $A + cI$ and all functions of $A$. Each eigenvector of $A$ is also an eigenvector of $A^k$ and $A^{-1}$ and $A + cI$ :

$$\text{If } Ax = \lambda x \text{ then } A^k x = \lambda^k x \text{ and } A^{-1}x = \frac{1}{\lambda}x \text{ and } (A + cI)x = (\lambda + c)x. \quad (6)$$

Start again with $A^2 x$, which is $A$ times $Ax = \lambda x$. Then $A\lambda x$ is the same as $\lambda A x$ for any number $\lambda$, and $\lambda A x$ is $\lambda^2 x$. We have proved that $A^2 x = \lambda^2 x$.

For higher powers $A^k x$, continue multiplying $Ax = \lambda x$ by $A$. Step by step you reach $A^k x = \lambda^k x$. For the eigenvalues of $A^{-1}$, first multiply by $A^{-1}$ and then divide by $\lambda$ :

$$\textbf{Eigenvalues of } A^{-1} \textbf{ are } \frac{1}{\lambda} \qquad Ax = \lambda x \qquad x = \lambda A^{-1} x \qquad A^{-1} x = \frac{1}{\lambda} x \quad (7)$$

We are assuming that $A^{-1}$ exists ! If $A$ is invertible then $\lambda$ will never be zero.

**Invertible matrices have all $\lambda \neq 0$. Singular matrices have the eigenvalue $\lambda = 0$.**

The shift from $A$ to $A + cI$ just adds $c$ to every eigenvalue (*don't change* $x$):

$$\textbf{Shift of } A \qquad \text{If } Ax = \lambda x \text{ then } (A + cI)x = Ax + cx = (\lambda + c)x. \quad (8)$$

As long as we keep the same eigenvector $x$, we can allow any function of $A$ :

$$\textbf{Functions of } A \qquad (A^2 + 2A + 5I)x = (\lambda^2 + 2\lambda + 5)x \qquad e^A x = e^\lambda x. \quad (9)$$

I slipped in $e^A = I + A + \frac{1}{2}A^2 + \cdots$ to show that infinite series produce matrices too.

Let me show you the powers of the Markov matrix $A$ in Example 2. That starting matrix is unrecognizable after a few steps.

$$\underset{A}{\begin{bmatrix} .8 & .3 \\ .2 & .7 \end{bmatrix}} \quad \underset{A^2}{\begin{bmatrix} .70 & .45 \\ .30 & .55 \end{bmatrix}} \quad \underset{A^3}{\begin{bmatrix} .650 & .525 \\ .350 & .475 \end{bmatrix}} \quad \cdots \quad \underset{A^{100}}{\begin{bmatrix} .6000 & .6000 \\ .4000 & .4000 \end{bmatrix}} \tag{10}$$

$A^{100}$ was found by using $\lambda = 1$ and its eigenvector $[.6, .4]$, not by multiplying 100 matrices. The eigenvalues of $A$ are 1 and $\frac{1}{2}$, so the eigenvalues of $A^{100}$ are 1 and $(\frac{1}{2})^{100}$. That last number is extremely small, and we can't see it in the first 30 digits of $A^{100}$.

How could you multiply $A^{99}$ times another vector like $\boldsymbol{v} = (.8, .2)$? This is not an eigenvector, but $\boldsymbol{v}$ is a *combination of eigenvectors*. This is a key idea, to express any vector $\boldsymbol{v}$ by using the eigenvectors.

$$\textbf{Separate into eigenvectors} \quad \boldsymbol{v} = \boldsymbol{x}_1 + (.2)\boldsymbol{x}_2 \qquad \boldsymbol{v} = \begin{bmatrix} .8 \\ .2 \end{bmatrix} = \begin{bmatrix} .6 \\ .4 \end{bmatrix} + \begin{bmatrix} .2 \\ -.2 \end{bmatrix}. \tag{11}$$

***Each eigenvector is multiplied by its eigenvalue, when we multiply the vector by A.*** After 99 steps, $\boldsymbol{x}_1$ is unchanged and $\boldsymbol{x}_2$ is multiplied by $(\frac{1}{2})^{99}$:

$$A^{99}\begin{bmatrix} .8 \\ .2 \end{bmatrix} \quad \text{is} \quad A^{99}(\boldsymbol{x}_1 + .2\boldsymbol{x}_2) = \boldsymbol{x}_1 + (.2)(\frac{1}{2})^{99}\boldsymbol{x}_2 = \begin{bmatrix} .6 \\ .4 \end{bmatrix} + \begin{bmatrix} \text{very} \\ \text{small} \\ \text{vector} \end{bmatrix}.$$

This is the first column of $A^{100}$, because $\boldsymbol{v} = (.8, .2)$ is the first column of $A$. The number we originally wrote as .6000 was not exact. We left out $(.2)(\frac{1}{2})^{99}$ which wouldn't show up for 30 decimal places.

The eigenvector $\boldsymbol{x}_1 = (.6, .4)$ is a "*steady state*" that doesn't change (because $\lambda_1 = 1$). The eigenvector $\boldsymbol{x}_2$ is a "*decaying mode*" that virtually disappears (because $\lambda_2 = 1/2$). The higher the power of $A$, the more closely its columns approach the steady state.

## Bad News About $AB$ and $A + B$

Normally the eigenvalues of $A$ and $B$ (separately) do not tell us the eigenvalues of $AB$. We also don't know about $A + B$. When $A$ and $B$ have different eigenvectors, our reasoning fails. The good results for $A^2$ are wrong for $AB$ and $A + B$, when $AB$ is different from $BA$. The eigenvalues won't come from $A$ and $B$ separately:

$$A = \begin{bmatrix} 0 & 1 \\ 0 & 0 \end{bmatrix} \quad B = \begin{bmatrix} 0 & 0 \\ 1 & 0 \end{bmatrix} \quad AB = \begin{bmatrix} 1 & 0 \\ 0 & 0 \end{bmatrix} \quad BA = \begin{bmatrix} 0 & 0 \\ 0 & 1 \end{bmatrix} \quad A + B = \begin{bmatrix} 0 & 1 \\ 1 & 0 \end{bmatrix}$$

All the eigenvalues of $A$ and $B$ are zero. But $AB$ has an eigenvalue $\lambda = 1$, and $A + B$ has eigenvalues 1 and $-1$. But one rule holds: **$AB$ and $BA$ have the same eigenvalues.**

## Determinants

The determinant is a single number with amazing properties. It is zero when the matrix has no inverse. That leads to the eigenvalue equation $\det(A - \lambda I) = 0$. When $A$ is invertible, the determinant of $A^{-1}$ is $1/(\det A)$. Every entry in $A^{-1}$ is a ratio of two determinants.

I want to summarize the algebra, leaving the details for my companion textbook *Introduction to Linear Algebra*. The difficulty with $\det(A - \lambda I) = 0$ is that an $n$ by $n$ determinant involves $n\,!$ terms. For $n = 5$ this is 120 terms—generally impossible to use.

For $n = 3$ there are six terms, three with plus signs and three with minus. Each of those six terms includes **one number from every row and every column** :

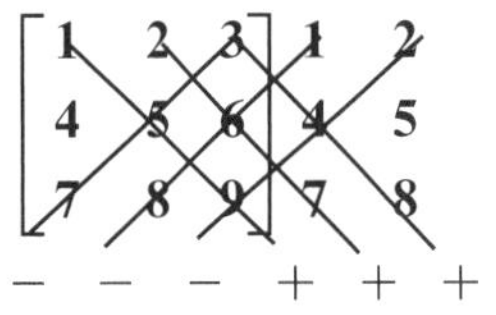


*Determinant from $n! = 6$ terms*

*Three plus signs, three minus signs*

$+(1)(5)(9) \quad +(2)(6)(7) \quad +(3)(4)(8)$

$-(3)(5)(7) \quad -(1)(6)(8) \quad -(2)(4)(9)$

That shows how to find the six terms. For this particular matrix the total must be $\det A = 0$, because the matrix happens to be singular : row 1 + row 3 equals 2(row 2).

Let me start with five useful properties of determinants, for all square matrices.

1. Subtracting a multiple of one row from another row leaves $\det A$ unchanged.
2. The determinant reverses sign when two rows are exchanged.
3. If $A$ is triangular then $\det A =$ product of diagonal entries.
4. The determinant of $AB$ equals ($\det A$) times ($\det B$).
5. The determinant of $A^{\mathrm{T}}$ equals the determinant of $A$.

By combining **1, 2, 3** you will see how the determinant comes from elimination :

**The determinant equals $\pm$ (product of the pivots).** (12)

Property **1** says that $A$ and $U$ have the same determinant, unless rows are exchanged.
Property **2** says that an odd number of exchanges would leave $\det A = -\det U$.
Property **3** says that $\det U$ is the product of the pivots on its main diagonal.

When elimination takes $A$ to $U$, we find $\det A = \pm$ (product of the pivots). This is how all numerical software (like MATLAB or Python or Julia ) would compute $\det A$.

Plus and minus signs play a big part in determinants. Half of the $n\,!$ terms have plus signs, and half come with minus signs. For $n = 3$, one row exchange puts $3 - 5 - 7$ or $1 - 6 - 8$ or $2 - 4 - 9$ on the main diagonal. A minus sign from one row exchange.

Two row exchanges (an even number) take you back to (2)(6)(7) and (3)(4)(8). This indicates how the 24 terms would go for $n = 4$, twelve terms with *plus* and twelve with *minus*.

Even permutation matrices have $\det P = 1$ and odd permutations have $\det P = -1$.

**Inverse of $A$** If $\det A \neq 0$, you can solve $A\boldsymbol{v} = \boldsymbol{b}$ and find $A^{-1}$ using determinants:

**Cramer's Rule** $$v_1 = \frac{\det B_1}{\det A} \quad v_2 = \frac{\det B_2}{\det A} \quad \cdots \quad v_n = \frac{\det B_n}{\det A} \tag{13}$$

The matrix $B_j$ replaces the $j^{\text{th}}$ column of $A$ by the vector $\boldsymbol{b}$. Cramer's Rule is expensive!

To find the columns of $A^{-1}$, we solve $AA^{-1} = I$. That is the Gauss-Jordan idea: For each column $\boldsymbol{b}$ in $I$, solve $A\boldsymbol{v} = \boldsymbol{b}$ to find a column $\boldsymbol{v}$ of $A^{-1}$.

In this special case, when $\boldsymbol{b}$ is a column of $I$, the numbers $\det B_j$ in Cramer's Rule are called **cofactors**. They reduce to determinants of size $n-1$, because $\boldsymbol{b}$ has so many zeros. Every entry of $A^{-1}$ is a cofactor of $A$ divided by the determinant of $A$.

I will close with three examples, to introduce the "trace" of a matrix and to show that real matrices can have imaginary (or complex) eigenvalues and eigenvectors.

**Example 3** Find the eigenvalues and eigenvectors of $S = \begin{bmatrix} 2 & 1 \\ 1 & 2 \end{bmatrix}$.

*Solution* You can see that $\boldsymbol{x} = (1, 1)$ will be in the same direction as $S\boldsymbol{x} = (3, 3)$. Then $\boldsymbol{x}$ is an eigenvector of $S$ with $\lambda = 3$. We want the matrix $S - \lambda I$ to be singular.

$$S = \begin{bmatrix} 2 & 1 \\ 1 & 2 \end{bmatrix} \qquad \det(S - \lambda I) = \det \begin{bmatrix} 2-\lambda & 1 \\ 1 & 2-\lambda \end{bmatrix} = \boldsymbol{\lambda^2 - 4\lambda + 3} = 0.$$

Notice that **3** is the determinant of $S$ (without $\lambda$). And **4** is the sum $2+2$ down the central diagonal of $S$. **The diagonal sum 4 is the "trace" of $A$. It equals $\lambda_1 + \lambda_2 = 3 + 1$.**

Now factor $\lambda^2 - 4\lambda + 3$ into $(\lambda - 3)(\lambda - 1)$. The matrix $S - \lambda I$ is singular (zero determinant) for $\lambda = 3$ and $\lambda = 1$. Each eigenvalue has an eigenvector:

$$\begin{aligned} \lambda_1 = 3 \quad (S - 3I)\boldsymbol{x}_1 &= \begin{bmatrix} -1 & 1 \\ 1 & -1 \end{bmatrix} \begin{bmatrix} 1 \\ 1 \end{bmatrix} = \begin{bmatrix} 0 \\ 0 \end{bmatrix} \\ \lambda_2 = 1 \quad (S - I)\boldsymbol{x}_2 &= \begin{bmatrix} 1 & 1 \\ 1 & 1 \end{bmatrix} \begin{bmatrix} 1 \\ -1 \end{bmatrix} = \begin{bmatrix} 0 \\ 0 \end{bmatrix} \end{aligned}$$

The eigenvalues 3 and 1 are *real*. The eigenvectors $(1, 1)$ and $(1, -1)$ are *orthogonal*. Those properties always come together for symmetric matrices (Section 6.5).

Here is an *antisymmetric* matrix with $A^{\mathrm{T}} = -A$. It rotates all real vectors by $\theta = 90°$. Real vectors can't be eigenvectors of a rotation matrix because it changes their direction.

**Example 4** This real matrix has imaginary eigenvalues $i$, $-i$ and complex eigenvectors:

$$A = \begin{bmatrix} 0 & -1 \\ 1 & 0 \end{bmatrix} = -A^{\mathrm{T}} \qquad \det(A - \lambda I) = \det \begin{bmatrix} -\lambda & -1 \\ 1 & -\lambda \end{bmatrix} = \boldsymbol{\lambda^2 + 1} = 0.$$

That determinant $\lambda^2 + 1$ is zero for $\lambda = i$ and $-i$. The eigenvectors are $(1, -i)$ and $(1, i)$:

$$\begin{bmatrix} 0 & -1 \\ 1 & 0 \end{bmatrix} \begin{bmatrix} 1 \\ -i \end{bmatrix} = \begin{bmatrix} i \\ 1 \end{bmatrix} = \boldsymbol{i} \begin{bmatrix} 1 \\ -i \end{bmatrix} \qquad \begin{bmatrix} 0 & -1 \\ 1 & 0 \end{bmatrix} \begin{bmatrix} 1 \\ i \end{bmatrix} = \begin{bmatrix} -i \\ 1 \end{bmatrix} = \boldsymbol{-i} \begin{bmatrix} 1 \\ i \end{bmatrix}$$

Somehow those complex vectors $\boldsymbol{x}_1$ and $\boldsymbol{x}_2$ don't get rotated (I don't really know how).

Multiplying the eigenvalues $(i)(-i)$ gives $\det A = 1$. Adding the eigenvalues gives $(i) + (-i) = 0$. This equals the sum $0 + 0$ down the diagonal of $A$.

**Product of eigenvalues = determinant** **Sum of eigenvalues = "trace"** (14)

Those are true statements for all square matrices. **The trace is the sum** $a_{11} + \cdots + a_{nn}$ **down the main diagonal of** $\boldsymbol{A}$. This sum and product are is especially valuable for 2 by 2 matrices, when the determinant $\lambda_1 \lambda_2 = \boldsymbol{ad - bc}$ and the trace $\lambda_1 + \lambda_2 = \boldsymbol{a + d}$ completely determine $\lambda_1$ and $\lambda_2$. Look now at rotation of a plane through any angle $\theta$.

**Example 5** Rotation comes from an orthogonal matrix $Q$. Then $\lambda_1 = e^{i\theta}$ and $\lambda_2 = e^{-i\theta}$:

$$Q = \begin{bmatrix} \cos\theta & -\sin\theta \\ \sin\theta & \cos\theta \end{bmatrix} \qquad \begin{aligned} \lambda_1 &= \cos\theta + i\sin\theta \\ \lambda_2 &= \cos\theta - i\sin\theta \end{aligned} \qquad \begin{aligned} \lambda_1 + \lambda_2 &= 2\cos\theta = \textbf{trace} \\ \lambda_1 \ \lambda_2 &= 1 = \textbf{determinant} \end{aligned}$$

I multiplied $(\lambda_1)(\lambda_2)$ to get $\cos^2\theta + \sin^2\theta = 1$. In polar form $e^{i\theta}$ times $e^{-i\theta}$ is 1. The eigenvectors of $Q$ are $(1, -i)$ and $(1, i)$ for all rotation angles $\theta$.

Before ending this section, I need to tell you the truth. It is not easy to find eigenvalues and eigenvectors of large matrices. The equation $\det(A - \lambda I) = 0$ is more or less limited to 2 by 2 and 3 by 3. For larger matrices, we can gradually make them triangular without changing the eigenvalues. *For triangular matrices the eigenvalues are on the diagonal.* A good code to compute $\lambda$ and $\boldsymbol{x}$ is free in LAPACK. The MATLAB command is eig $(A)$.

## ■ REVIEW OF THE KEY IDEAS ■

1. $A\boldsymbol{x} = \lambda\boldsymbol{x}$ says that eigenvectors $\boldsymbol{x}$ keep the same direction when multiplied by $A$.

2. $A\boldsymbol{x} = \lambda\boldsymbol{x}$ also says that $\det(A - \lambda I) = 0$. This equation determines $n$ eigenvalues.

3. The eigenvalues of $A^2$ and $A^{-1}$ are $\lambda^2$ and $\lambda^{-1}$, with the same eigenvectors as $A$.

4. Singular matrices have $\lambda = 0$. Triangular matrices have $\lambda$'s on their diagonal.

**5.** The sum down the main diagonal of $A$ (*the trace*) is the sum of the eigenvalues.

**6.** The determinant is the product of the $\lambda$'s. It is also $\pm$ (product of the pivots).

## Problem Set 6.1

**1** Example 2 has powers of this Markov matrix $A$:

$$A = \begin{bmatrix} .8 & .3 \\ .2 & .7 \end{bmatrix} \quad \text{and} \quad A^2 = \begin{bmatrix} .70 & .45 \\ .30 & .55 \end{bmatrix} \quad \text{and} \quad A^\infty = \begin{bmatrix} .6 & .6 \\ .4 & .4 \end{bmatrix}.$$

(a) $A$ has eigenvalues 1 and $\frac{1}{2}$. Find the eigenvalues of $A^2$ and $A^\infty$.

(b) What are the eigenvectors of $A^\infty$? One eigenvector is in the nullspace.

(c) Check the determinant of $A^2$ and $A^\infty$. Compare with $(\det A)^2$ and $(\det A)^\infty$.

**2** Find the eigenvalues and the eigenvectors of these two matrices:

$$A = \begin{bmatrix} 1 & 4 \\ 2 & 3 \end{bmatrix} \quad \text{and} \quad A + I = \begin{bmatrix} 2 & 4 \\ 2 & 4 \end{bmatrix}.$$

$A + I$ has the ______ eigenvectors as $A$. Its eigenvalues are ______ by 1.

**3** Compute the eigenvalues and eigenvectors of $A$ and also $A^{-1}$:

$$A = \begin{bmatrix} 0 & 2 \\ 1 & 1 \end{bmatrix} \quad \text{and} \quad A^{-1} = \begin{bmatrix} -1/2 & 1 \\ 1/2 & 0 \end{bmatrix}.$$

$A^{-1}$ has the ______ eigenvectors as $A$. When $A$ has eigenvalues $\lambda_1$ and $\lambda_2$, its inverse has eigenvalues ______. Check that $\lambda_1 + \lambda_2 =$ **trace of** $A = 0 + 1$.

**4** Compute the eigenvalues and eigenvectors of $A$ and $A^2$:

$$A = \begin{bmatrix} -1 & 3 \\ 2 & 0 \end{bmatrix} \quad \text{and} \quad A^2 = \begin{bmatrix} 7 & -3 \\ -2 & 6 \end{bmatrix}.$$

$A^2$ has the same ______ as $A$. When $A$ has eigenvalues $\lambda_1$ and $\lambda_2$, the eigenvalues of $A^2$ are ______. In this example, why is $\lambda_1^2 + \lambda_2^2 = 13$?

**5** Find the eigenvalues of $A$ and $B$ (easy for triangular matrices) and $A + B$:

$$A = \begin{bmatrix} 3 & 0 \\ 1 & 1 \end{bmatrix} \quad \text{and} \quad B = \begin{bmatrix} 1 & 1 \\ 0 & 3 \end{bmatrix} \quad \text{and} \quad A + B = \begin{bmatrix} 4 & 1 \\ 1 & 4 \end{bmatrix}.$$

Eigenvalues of $A + B$ (*are equal to*) (*might not be equal to*) eigenvalues of $A$ plus eigenvalues of $B$.

**6** Find the eigenvalues of $A$ and $B$ and $AB$ and $BA$:

$$A = \begin{bmatrix} 1 & 0 \\ 1 & 1 \end{bmatrix} \text{ and } B = \begin{bmatrix} 1 & 2 \\ 0 & 1 \end{bmatrix} \text{ and } AB = \begin{bmatrix} 1 & 2 \\ 1 & 3 \end{bmatrix} \text{ and } BA = \begin{bmatrix} 3 & 2 \\ 1 & 1 \end{bmatrix}.$$

(a) Are the eigenvalues of $AB$ equal to eigenvalues of $A$ times eigenvalues of $B$?

(b) Are the eigenvalues of $AB$ equal to the eigenvalues of $BA$? *Yes*!

**7** Elimination produces a triangular matrix $U$. The eigenvalues of $U$ are on its diagonal (*why*?). They are *not the eigenvalues of* $A$. Give a 2 by 2 example of $A$ and $U$.

**8** (a) If you know that $\boldsymbol{x}$ is an eigenvector, the way to find $\lambda$ is to _____.

(b) If you know that $\lambda$ is an eigenvalue, the way to find $\boldsymbol{x}$ is to _____.

**9** What do you do to the equation $A\boldsymbol{x} = \lambda\boldsymbol{x}$, in order to prove (a), (b), and (c)?

(a) $\lambda^2$ is an eigenvalue of $A^2$, as in Problem 4.

(b) $\lambda^{-1}$ is an eigenvalue of $A^{-1}$, as in Problem 3.

(c) $\lambda + 1$ is an eigenvalue of $A + I$, as in Problem 2.

**10** Find the eigenvalues and eigenvectors for both of these Markov matrices $A$ and $A^\infty$. Explain from those answers why $A^{100}$ is close to $A^\infty$:

$$A = \begin{bmatrix} .6 & .2 \\ .4 & .8 \end{bmatrix} \quad \text{and} \quad A^\infty = \begin{bmatrix} 1/3 & 1/3 \\ 2/3 & 2/3 \end{bmatrix}.$$

**11** A 3 by 3 matrix $B$ has eigenvalues 0, 1, 2. This information allows you to find:

(a) the rank of $B$ (b) the eigenvalues of $B^2$ (c) the eigenvalues of $(B^2 + I)^{-1}$.

**12** Find three eigenvectors for this matrix $P$. Projection matrices only have $\lambda = 1$ and 0. Eigenvectors are *in or orthogonal to* the subspace that $P$ projects onto.

**Projection matrix $P^2 = P = P^{\mathrm{T}}$**
$$P = \begin{bmatrix} .2 & .4 & 0 \\ .4 & .8 & 0 \\ 0 & 0 & 1 \end{bmatrix}.$$

If two eigenvectors $\boldsymbol{x}$ and $\boldsymbol{y}$ share the same repeated eigenvalue $\lambda$, so do all their combinations $c\boldsymbol{x} + d\boldsymbol{y}$. Find an eigenvector of $P$ with no zero components.

**13** From the unit vector $\boldsymbol{u} = \left(\frac{1}{6}, \frac{1}{6}, \frac{3}{6}, \frac{5}{6}\right)$ construct the rank one projection matrix $P = \boldsymbol{u}\boldsymbol{u}^{\mathrm{T}}$. This matrix has $P^2 = P$ because $\boldsymbol{u}^{\mathrm{T}}\boldsymbol{u} = 1$.

(a) Explain why $P\boldsymbol{u} = (\boldsymbol{u}\boldsymbol{u}^{\mathrm{T}})\boldsymbol{u}$ equals $\boldsymbol{u}$. Then $\boldsymbol{u}$ is an eigenvector with $\lambda = 1$.

(b) If $\boldsymbol{v}$ is perpendicular to $\boldsymbol{u}$ show that $P\boldsymbol{v} = \mathbf{0}$. Then $\lambda = 0$.

(c) Find three independent eigenvectors of $P$ all with eigenvalue $\lambda = 0$.

**14** Solve $\det(Q - \lambda I) = 0$ by the quadratic formula to reach $\lambda = \cos\theta \pm i\sin\theta$:

$$Q = \begin{bmatrix} \cos\theta & -\sin\theta \\ \sin\theta & \cos\theta \end{bmatrix} \quad \text{rotates the } xy \text{ plane by the angle } \theta. \text{ No real } \lambda\text{'s.}$$

Find the eigenvectors of $Q$ by solving $(Q - \lambda I)\boldsymbol{x} = \mathbf{0}$. Use $i^2 = -1$.

**15** Find three 2 by 2 matrices that have $\lambda_1 = \lambda_2 = 0$. The trace is zero and the determinant is zero. $A$ might not be the zero matrix but check that $A^2$ is all zeros.

**16** This matrix is singular with rank one. Find three $\lambda$'s and three eigenvectors:

$$\textbf{Rank one} \qquad A = \begin{bmatrix} 1 \\ 2 \\ 1 \end{bmatrix} \begin{bmatrix} 2 & 1 & 2 \end{bmatrix} = \begin{bmatrix} 2 & 1 & 2 \\ 4 & 2 & 4 \\ 2 & 1 & 2 \end{bmatrix}.$$

**17** When $a + b = c + d$ show that $(1, 1)$ is an eigenvector and find both eigenvalues:

$$\textit{Use the trace to find } \lambda_2 \qquad A = \begin{bmatrix} 5 & 1 \\ 2 & 4 \end{bmatrix} \qquad A = \begin{bmatrix} a & b \\ c & d \end{bmatrix}.$$

**18** If $A$ has $\lambda_1 = 4$ and $\lambda_2 = 5$ then $\det(A - \lambda I) = (\lambda - 4)(\lambda - 5) = \lambda^2 - 9\lambda + 20$. Find three matrices that have trace $a + d = 9$ and determinant 20, so $\lambda = 4$ and 5.

**19** Suppose $A\boldsymbol{u} = 0\boldsymbol{u}$ and $A\boldsymbol{v} = 3\boldsymbol{v}$ and $A\boldsymbol{w} = 5\boldsymbol{w}$. The eigenvalues are $0, 3, 5$.

(a) Give a basis for the nullspace of $A$ and a basis for the column space.

(b) Find a particular solution to $A\boldsymbol{x} = \boldsymbol{v} + \boldsymbol{w}$. Find all solutions.

(c) $A\boldsymbol{x} = \boldsymbol{u}$ has no solution. If it did then _____ would be in the column space.

**20** Choose the last row of $A$ to produce $(a)$ eigenvalues 4 and 7 $(b)$ any $\lambda_1$ and $\lambda_2$.

$$\textbf{Companion matrix} \qquad A = \begin{bmatrix} 0 & 1 \\ * & * \end{bmatrix}.$$

**21** ***The eigenvalues of A equal the eigenvalues of*** $A^{\mathrm{T}}$. This is because $\det(A - \lambda I)$ equals $\det(A^{\mathrm{T}} - \lambda I)$. That is true because _____. Show by an example that the eigenvectors of $A$ and $A^{\mathrm{T}}$ are *not* the same.

**22** Construct any 3 by 3 Markov matrix $M$: positive entries down each column add to 1. Show that $M^{\mathrm{T}}(1, 1, 1) = (1, 1, 1)$. By Problem 21, $\lambda = 1$ is also an eigenvalue of $M$. Challenge: A 3 by 3 singular Markov matrix with trace $\frac{1}{2}$ has what $\lambda$'s?

**23** Suppose $A$ and $B$ have the same eigenvalues $\lambda_1, \ldots, \lambda_n$ with the same independent eigenvectors $\boldsymbol{x}_1, \ldots, \boldsymbol{x}_n$. **Then $A = B$.** *Reason*: Any vector $\boldsymbol{v}$ is a combination $c_1\boldsymbol{x}_1 + \cdots + c_n\boldsymbol{x}_n$. What is $A\boldsymbol{v}$? What is $B\boldsymbol{v}$?

**24** The block $B$ has eigenvalues $1, 2$ and $C$ has eigenvalues $3, 4$ and $D$ has eigenvalues $5, 7$. Find the eigenvalues of the 4 by 4 matrix $A$:

$$A = \begin{bmatrix} B & C \\ 0 & D \end{bmatrix} = \begin{bmatrix} 0 & 1 & 3 & 0 \\ -2 & 3 & 0 & 4 \\ 0 & 0 & 6 & 1 \\ 0 & 0 & 1 & 6 \end{bmatrix}.$$

**25** Find the rank and the four eigenvalues of $A$ and $C$:

$$A = \begin{bmatrix} 1 & 1 & 1 & 1 \\ 1 & 1 & 1 & 1 \\ 1 & 1 & 1 & 1 \\ 1 & 1 & 1 & 1 \end{bmatrix} \quad \text{and} \quad C = \begin{bmatrix} 1 & 0 & 1 & 0 \\ 0 & 1 & 0 & 1 \\ 1 & 0 & 1 & 0 \\ 0 & 1 & 0 & 1 \end{bmatrix}.$$

**26** Subtract $I$ from the previous $A$. Find the eigenvalues of $B$ and $-B$:

$$B = A - I = \begin{bmatrix} 0 & 1 & 1 & 1 \\ 1 & 0 & 1 & 1 \\ 1 & 1 & 0 & 1 \\ 1 & 1 & 1 & 0 \end{bmatrix} \quad \text{and} \quad -B = \begin{bmatrix} 0 & -1 & -1 & -1 \\ -1 & 0 & -1 & -1 \\ -1 & -1 & 0 & -1 \\ -1 & -1 & -1 & 0 \end{bmatrix}.$$

**27** (Review) Find the eigenvalues of $A$, $B$, and $C$:

$$A = \begin{bmatrix} 1 & 2 & 3 \\ 0 & 4 & 5 \\ 0 & 0 & 6 \end{bmatrix} \quad \text{and} \quad B = \begin{bmatrix} 0 & 0 & 1 \\ 0 & 2 & 0 \\ 3 & 0 & 0 \end{bmatrix} \quad \text{and} \quad C = \begin{bmatrix} 2 & 2 & 2 \\ 2 & 2 & 2 \\ 2 & 2 & 2 \end{bmatrix}.$$

**28** Every permutation matrix leaves $\boldsymbol{x} = (1, 1, \ldots, 1)$ unchanged. Then $\lambda = 1$. Find two more $\lambda$'s (possibly complex) for these permutations, from $\det(P - \lambda I) = 0$:

$$P = \begin{bmatrix} 0 & 1 & 0 \\ 0 & 0 & 1 \\ 1 & 0 & 0 \end{bmatrix} \quad \text{and} \quad P = \begin{bmatrix} 0 & 0 & 1 \\ 0 & 1 & 0 \\ 1 & 0 & 0 \end{bmatrix}.$$

**29** **The determinant of $A$ equals the product $\lambda_1 \lambda_2 \cdots \lambda_n$.** Start with the polynomial $\det(A - \lambda I)$ separated into its $n$ factors (always possible). Then set $\lambda = 0$:

$$\det(A - \lambda I) = (\lambda_1 - \lambda)(\lambda_2 - \lambda) \cdots (\lambda_n - \lambda) \quad \text{so} \quad \det A = \underline{\qquad}.$$

**30** The sum of the diagonal entries (the *trace*) equals the sum of the eigenvalues:

$$A = \begin{bmatrix} a & b \\ c & d \end{bmatrix} \quad \text{has} \quad \det(A - \lambda I) = \lambda^2 - (a + d)\lambda + ad - bc = 0.$$

The quadratic formula gives the eigenvalues $\lambda = (a + d + \sqrt{\quad})/2$ and $\lambda =$ ______. Their sum is ______. **If $A$ has $\lambda_1 = 3$ and $\lambda_2 = 4$ then $\det(A - \lambda I) =$** ______.

## 6.2 Diagonalizing a Matrix

When $\boldsymbol{x}$ is an eigenvector, multiplication by $A$ is just multiplication by a number $\lambda$: $A\boldsymbol{x} = \lambda\boldsymbol{x}$. All the difficulties of matrices are swept away. Instead of an interconnected system, we can follow the eigenvectors separately. It is like having a *diagonal matrix*, with no off-diagonal interconnections. The 100th power of a diagonal matrix is easy.

The point of this section is very direct. ***The matrix A turns into a diagonal matrix* $\boldsymbol{\Lambda}$ *when we use the eigenvectors properly*.** This is the matrix form of our key idea. We start right off with that one essential computation.

**Diagonalization** Suppose the $n$ by $n$ matrix $A$ has $n$ linearly independent eigenvectors $\boldsymbol{x}_1, \ldots, \boldsymbol{x}_n$. Put them into the columns of an ***eigenvector matrix*** $V$. Then $V^{-1}AV$ is the ***eigenvalue matrix*** $\Lambda$, and $\Lambda$ is diagonal:

**Eigenvector matrix $V$**
**Eigenvalue matrix $\Lambda$**

$$V^{-1}AV = \Lambda = \begin{bmatrix} \lambda_1 & & \\ & \ddots & \\ & & \lambda_n \end{bmatrix}. \tag{1}$$

The matrix $A$ is "diagonalized." We use capital lambda for the eigenvalue matrix, because of the small $\lambda$'s (the eigenvalues) on its diagonal.

***Proof*** Multiply $A$ times its eigenvectors, which are the columns of $V$. The first column of $AV$ is $A\boldsymbol{x}_1$. That is $\lambda_1\boldsymbol{x}_1$. Each column of $V$ is multiplied by its eigenvalue $\lambda_i$:

**$A$ times $V$** $$AV = A\begin{bmatrix} & & \\ \boldsymbol{x}_1 & \cdots & \boldsymbol{x}_n \\ & & \end{bmatrix} = \begin{bmatrix} & & \\ \lambda_1\boldsymbol{x}_1 & \cdots & \lambda_n\boldsymbol{x}_n \\ & & \end{bmatrix}.$$

The trick is to split this matrix $AV$ into $V$ times $\Lambda$:

**$V$ times $\Lambda$** $$\begin{bmatrix} & & \\ \lambda_1\boldsymbol{x}_1 & \cdots & \lambda_n\boldsymbol{x}_n \\ & & \end{bmatrix} = \begin{bmatrix} & & \\ \boldsymbol{x}_1 & \cdots & \boldsymbol{x}_n \\ & & \end{bmatrix}\begin{bmatrix} \lambda_1 & & \\ & \ddots & \\ & & \lambda_n \end{bmatrix} = V\Lambda.$$

Keep those matrices in the right order! Then $\lambda_1$ multiplies the first column $\boldsymbol{x}_1$, as shown. The diagonalization is complete, and we can write $AV = V\Lambda$ in two good ways:

$$\boxed{AV = V\Lambda \quad \text{is} \quad V^{-1}AV = \Lambda \quad \text{or} \quad A = V\Lambda V^{-1}.} \tag{2}$$

The matrix $V$ has an inverse, because its columns (the eigenvectors of $A$) were assumed to be linearly independent. *Without n independent eigenvectors, we can't diagonalize.*

$A$ and $\Lambda$ have the same eigenvalues $\lambda_1, \ldots, \lambda_n$. The eigenvectors are different. The job of the original eigenvectors $\boldsymbol{x}_1, \ldots, \boldsymbol{x}_n$ was to diagonalize $A$. Those eigenvectors in $V$ produce $A = V\Lambda V^{-1}$. You will soon see the simplicity and importance and meaning of the $k$ th power $A^k = V\Lambda^k V^{-1}$.

**Sections 6.2 and 6.3 solve first order difference and differential equations.**

| | | |
|---|---|---|
| **6.2** | $u_{k+1} = Au_k$ | $u_k = A^k u_0 = c_1\lambda_1^k x_1 + \cdots + c_n\lambda_n^k x_n$ |
| **6.3** | $dy/dt = Ay$ | $y(t) = e^{At}y(0) = c_1 e^{\lambda_1 t}x_1 + \cdots + c_n e^{\lambda_n t}x_n.$ |

The idea is the same for both problems: ***n* independent eigenvectors give a basis.** We can write $\boldsymbol{u}_0$ and $\boldsymbol{y}(0)$ as combinations of eigenvectors. Then we follow each eigenvector as $k$ increases and $t$ increases: $\boldsymbol{A^k x}$ **is** $\boldsymbol{\lambda^k x}$ **and** $\boldsymbol{e^{At}x}$ **is** $\boldsymbol{e^{\lambda t}x}$.

Some matrices don't have $n$ independent eigenvectors (because of repeated $\lambda$'s). Then $A^k\boldsymbol{u}_0$ and $e^{At}\boldsymbol{y}(0)$ are still correct, but they lead to $k\lambda^k\,\boldsymbol{x}$ and $te^{\lambda t}\boldsymbol{x}$ : not so good.

**Example 1** Here $A$ is triangular so the $\lambda$'s are on its diagonal: $\lambda = 1$ and $\lambda = 6$.

$$\textbf{Eigenvectors in } V \qquad \underset{V^{-1}}{\begin{bmatrix} 1 & -1 \\ 0 & 1 \end{bmatrix}} \underset{A}{\begin{bmatrix} 1 & 5 \\ 0 & 6 \end{bmatrix}} \underset{V}{\begin{bmatrix} 1 & 1 \\ 0 & 1 \end{bmatrix}} = \underset{\Lambda}{\begin{bmatrix} \mathbf{1} & 0 \\ 0 & \mathbf{6} \end{bmatrix}}$$

In other words $A = V\Lambda V^{-1}$. Then watch $A^2 = V\Lambda V^{-1}V\Lambda V^{-1}$. When you remove $V^{-1}V = I$, this becomes $\boldsymbol{A^2 = V\Lambda^2 V^{-1}}$. ***The same eigenvectors for** $A$ **and** $A^2$ **are in** $V$**. The squared eigenvalues are in*** $\Lambda^2$.

The $k$ th power will be $A^k = V\Lambda^k V^{-1}$. And $\Lambda^k$ just contains $1^k$ and $6^k$ :

$$\textbf{Powers } A^k \qquad \begin{bmatrix} 1 & 5 \\ 0 & 6 \end{bmatrix}^k = \begin{bmatrix} 1 & 1 \\ 0 & 1 \end{bmatrix} \begin{bmatrix} 1 & \\ & 6^k \end{bmatrix} \begin{bmatrix} 1 & -1 \\ 0 & 1 \end{bmatrix} = \begin{bmatrix} 1 & 6^k - 1 \\ 0 & 6^k \end{bmatrix}.$$

*With* $k = 1$ *we get* $A$. *With* $k = 0$ *we get* $A^0 = I$ (eigenvalues $\lambda^0 = 1$). *With* $k = -1$ we get the inverse $A^{-1}$. You can see how $A^2 = [1\ \ 35;\ \ 0\ \ 36]$ fits the formula when $k = 2$.

Here are four remarks before we use $\Lambda$ again.

**Remark 1** When the eigenvalues $\lambda_1, \ldots, \lambda_n$ are all different, the eigenvectors $\boldsymbol{x}_1, \ldots, \boldsymbol{x}_n$ are independent. ***Any matrix that has no repeated eigenvalues can be diagonalized.***

**Remark 2** ***We can multiply eigenvectors by any nonzero constants.*** $A\boldsymbol{x} = \lambda\boldsymbol{x}$ will remain true. In Example 1, we can divide the eigenvector $(1, 1)$ by $\sqrt{2}$ to produce a unit vector.

**Remark 3** The eigenvectors in $V$ come in the same order as the eigenvalues in $\Lambda$. To reverse the order 1, 6 in $\Lambda$, put the eigenvector $(1, 1)$ before $(1, 0)$ in $V$ :

$$\begin{array}{l}\textbf{New order 6, 1}\\ \textbf{New order in } V\end{array} \qquad \begin{bmatrix} 0 & 1 \\ 1 & -1 \end{bmatrix} \begin{bmatrix} 1 & 5 \\ 0 & 6 \end{bmatrix} \begin{bmatrix} 1 & 1 \\ 1 & 0 \end{bmatrix} = \begin{bmatrix} 6 & 0 \\ 0 & 1 \end{bmatrix} = \Lambda_{\textbf{new}}$$

To diagonalize $A$ we *must* use an eigenvector matrix. From $V^{-1}AV = \Lambda$ we know that $AV = V\Lambda$. Suppose the first column of $V$ is $\boldsymbol{x}$. Then the first columns of $AV$ and $V\Lambda$ are $A\boldsymbol{x}$ and $\lambda_1\boldsymbol{x}$. For those to be equal, $\boldsymbol{x}$ must be an eigenvector.

**Remark 4** (**Warning for repeated eigenvalues**) Some matrices have too few eigenvectors (less than $n$). *Those matrices cannot be diagonalized.* Here are examples:

$$\textbf{Not diagonalizable} \quad \textbf{Only 1 eigenvector} \qquad A = \begin{bmatrix} 1 & -1 \\ 1 & -1 \end{bmatrix} \quad \text{and} \quad B = \begin{bmatrix} 0 & 1 \\ 0 & 0 \end{bmatrix}.$$

Their eigenvalues happen to be 0 and 0. The problem is the repetition of $\lambda$.

$$\textbf{Only one line of eigenvectors} \qquad Ax = 0x \quad \text{means} \quad \begin{bmatrix} 1 & -1 \\ 1 & -1 \end{bmatrix} \begin{bmatrix} \boldsymbol{x} \end{bmatrix} = \begin{bmatrix} 0 \\ 0 \end{bmatrix} \quad \text{and} \quad \boldsymbol{x} = c \begin{bmatrix} 1 \\ 1 \end{bmatrix}.$$

There is no second eigenvector, so the unusual matrix $A$ cannot be diagonalized.

Those matrices are the best examples to test any statement about eigenvectors. In many true-false questions, non-diagonalizable matrices lead to *false*.

Remember that there is no connection between invertibility and diagonalizability:

- ***Invertibility*** is concerned with the ***eigenvalues*** ($\lambda = 0$ or $\lambda \neq 0$).
- ***Diagonalizability needs n independent eigenvectors.***

Each eigenvalue has at least one eigenvector! $A - \lambda I$ is singular. If $(A - \lambda I)\boldsymbol{x} = \boldsymbol{0}$ leads you to $\boldsymbol{x} = \boldsymbol{0}$, $\lambda$ is *not* an eigenvalue. Look for a mistake in solving $\det(A - \lambda I) = 0$.

**Eigenvectors for $n$ different $\lambda$'s are independent. Then $V^{-1}AV = \Lambda$ will succeed.**
**Eigenvectors for repeated $\lambda$'s could be dependent. $V$ might not be invertible.**

**Example 2** **Powers of $A$** The Markov matrix $A$ in the last section had $\lambda_1 = 1$ and $\lambda_2 = .5$. Here is $A = V\Lambda V^{-1}$ with those eigenvalues in the matrix $\Lambda$:

$$\begin{bmatrix} .8 & .3 \\ .2 & .7 \end{bmatrix} = \begin{bmatrix} .6 & 1 \\ .4 & -1 \end{bmatrix} \begin{bmatrix} 1 & 0 \\ 0 & .5 \end{bmatrix} \begin{bmatrix} 1 & 1 \\ .4 & -.6 \end{bmatrix} = V\Lambda V^{-1}.$$

The eigenvectors $(.6, .4)$ and $(1, -1)$ are in the columns of $V$. They are also the eigenvectors of $A^2$. Watch how $A^2$ has the same $V$, and ***the eigenvalue matrix of $A^2$ is $\Lambda^2$***:

$$\textbf{Same } V \textbf{ for } A^2 \qquad \boxed{A^2} = V\Lambda V^{-1} V\Lambda V^{-1} = \boxed{V\Lambda^2 V^{-1}.} \tag{3}$$

Just keep going, and you see why the high powers $A^k$ approach a "steady state":

$$\textbf{Powers of } A \qquad A^k = \boldsymbol{V\Lambda^k V^{-1}} = \begin{bmatrix} .6 & 1 \\ .4 & -1 \end{bmatrix} \begin{bmatrix} 1^k & 0 \\ 0 & (.5)^k \end{bmatrix} \begin{bmatrix} 1 & 1 \\ .4 & -.6 \end{bmatrix}.$$

As $k$ gets larger, $(.5)^k$ gets smaller. In the limit it disappears completely. That limit is $A^\infty$:

$$\textbf{Limit } k \to \infty \qquad A^\infty = \begin{bmatrix} .6 & 1 \\ .4 & -1 \end{bmatrix} \begin{bmatrix} 1 & 0 \\ 0 & 0 \end{bmatrix} \begin{bmatrix} 1 & 1 \\ .4 & -.6 \end{bmatrix} = \begin{bmatrix} .6 & .6 \\ .4 & .4 \end{bmatrix}. \tag{4}$$

The limit has the steady state eigenvector $\boldsymbol{x}_1$ in both columns.

**Question** ***When does $A^k \to$ zero matrix?*** **Answer** ***All $|\lambda| < 1$.***

## Fibonacci Numbers

We present a famous example, where eigenvalues tell how fast the Fibonacci numbers grow. ***Every new Fibonacci number is the sum of the two previous F's*:**

***The sequence*** $0, 1, 1, 2, 3, 5, 8, 13, \ldots$ ***comes from*** $F_{k+2} = F_{k+1} + F_k$.

These numbers turn up in a fantastic variety of applications. Plants a grow in spirals, and a pear tree has 8 growths for every 3 turns. The champion is a sunflower that had 233 seeds in 144 loops. Those are the Fibonacci numbers $F_{13}$ and $F_{12}$. Our problem is more basic.

***Problem: Find the Fibonacci number $F_{100}$.*** The slow way is to apply the rule $F_{k+2} = F_{k+1} + F_k$ one step at a time. By adding $F_6 = 8$ to $F_7 = 13$ we reach $F_8 = 21$. Eventually we come to $F_{100}$. Linear algebra gives a better way.

The key is to begin with a matrix equation $\mathbf{u}_{k+1} = A\boldsymbol{u}_k$. That is a *one-step* rule for vectors, while Fibonacci gave a two-step rule for scalars. We match those rules by putting two Fibonacci numbers into a vector $\boldsymbol{u}_k$. Then you will see the matrix $A$.

$$\boldsymbol{u}_k = \begin{bmatrix} F_{k+1} \\ F_k \end{bmatrix}. \quad \text{The rule} \quad \begin{aligned} F_{k+2} &= F_{k+1} + F_k \\ F_{k+1} &= F_{k+1} \end{aligned} \quad \text{is} \quad \boldsymbol{u}_{k+1} = \begin{bmatrix} \mathbf{1} & \mathbf{1} \\ \mathbf{1} & \mathbf{0} \end{bmatrix} \boldsymbol{u}_k. \tag{5}$$

**Every step multiplies by** $A = \begin{bmatrix} 1 & 1 \\ 1 & 0 \end{bmatrix}$. After 100 steps we reach $\boldsymbol{u}_{100} = A^{100}\boldsymbol{u}_0$:

$$\boldsymbol{u}_0 = \begin{bmatrix} 1 \\ 0 \end{bmatrix}, \quad \boldsymbol{u}_1 = \begin{bmatrix} 1 \\ 1 \end{bmatrix}, \quad \boldsymbol{u}_2 = \begin{bmatrix} 2 \\ 1 \end{bmatrix}, \quad \boldsymbol{u}_3 = \begin{bmatrix} 3 \\ 2 \end{bmatrix}, \quad \ldots, \quad \boldsymbol{u}_{100} = \begin{bmatrix} F_{101} \\ F_{100} \end{bmatrix}.$$

This problem is just right for eigenvalues. To find them, subtract $\lambda I$ from $A$:

$$A - \lambda I = \begin{bmatrix} 1-\lambda & 1 \\ 1 & -\lambda \end{bmatrix} \quad \text{leads to} \quad \det(A - \lambda I) = \lambda^2 - \lambda - 1.$$

The equation $\lambda^2 - \lambda - 1 = 0$ is solved by the quadratic formula $\left(-b \pm \sqrt{b^2 - 4ac}\,\right)/2a$:

$$\textbf{Eigenvalues} \quad \lambda_1 = \frac{1+\sqrt{5}}{2} \approx 1.618 \quad \text{and} \quad \lambda_2 = \frac{1-\sqrt{5}}{2} \approx -.618.$$

These eigenvalues lead to eigenvectors $\boldsymbol{x}_1 = (\lambda_1, 1)$ and $\boldsymbol{x}_2 = (\lambda_2, 1)$. Step 2 finds the combination of those eigenvectors that gives $\boldsymbol{u}_0 = (1, 0)$:

$$\begin{bmatrix} 1 \\ 0 \end{bmatrix} = \frac{1}{\lambda_1 - \lambda_2}\left(\begin{bmatrix} \lambda_1 \\ 1 \end{bmatrix} - \begin{bmatrix} \lambda_2 \\ 1 \end{bmatrix}\right) \quad \text{or} \quad \boldsymbol{u}_0 = \frac{\boldsymbol{x}_1 - \boldsymbol{x}_2}{\lambda_1 - \lambda_2}. \tag{6}$$

Step 3 multiplies the eigenvectors $\boldsymbol{x}_1$ and $\boldsymbol{x}_2$ by $(\lambda_1)^{100}$ and $(\lambda_2)^{100}$:

$$\boldsymbol{A^{100}} \textbf{ times } \boldsymbol{u}_0 \qquad \boldsymbol{u}_{100} = \frac{(\lambda_1)^{100}\boldsymbol{x}_1 - (\lambda_2)^{100}\boldsymbol{x}_2}{\lambda_1 - \lambda_2}. \tag{7}$$

We want $F_{100}$ = second component of $\boldsymbol{u}_{100}$. The second components of $\boldsymbol{x}_1$ and $\boldsymbol{x}_2$ are 1. The difference between $(1+\sqrt{5})/2$ and $(1-\sqrt{5})/2$ is $\lambda_1 - \lambda_2 = \sqrt{5}$. We have $F_{100}$ :

$$F_{100} = \frac{1}{\sqrt{5}}\left[\left(\frac{1+\sqrt{5}}{2}\right)^{100} - \left(\frac{1-\sqrt{5}}{2}\right)^{100}\right] \approx 3.54 \cdot 10^{20}. \qquad (8)$$

Is this a whole number? *Yes.* The fractions and square roots must disappear, because Fibonacci's rule $F_{k+2} = F_{k+1} + F_k$ stays with integers. The second term in (8) is less than $\frac{1}{2}$, so it must move the first term to the nearest whole number:

$$k\text{th Fibonacci number} = \frac{\lambda_1^k - \lambda_2^k}{\lambda_1 - \lambda_2} = \text{nearest integer to } \frac{1}{\sqrt{5}}\left(\frac{1+\sqrt{5}}{2}\right)^k. \qquad (9)$$

The ratio of $F_6$ to $F_5$ is $8/5 = 1.6$. The ratio $F_{101}/F_{100}$ must be very close to the limiting ratio $\left(1+\sqrt{5}\right)/2$. The Greeks called this number the *"golden mean"*. For some reason a rectangle with sides 1.618 and 1 looks especially graceful.

## Matrix Powers $A^k$

Fibonacci's example is a typical difference equation $\boldsymbol{u}_{k+1} = A\mathbf{u}_k$. ***Each step multiplies by A.*** The solution is $\boldsymbol{u}_k = A^k \boldsymbol{u}_0$. We want to make clear how diagonalizing the matrix gives a quick way to compute $A^k$ and find $\boldsymbol{u}_k$ in three steps.

The eigenvector matrix $V$ produces $A = V\Lambda V^{-1}$. This is perfectly suited to computing powers, because ***every time $V^{-1}$ multiplies $V$ we get $I$*** :

$$\textbf{Powers of } A \qquad A^k \boldsymbol{u}_0 = (V\Lambda V^{-1})\cdots(V\Lambda V^{-1})\boldsymbol{u}_0 = V\Lambda^k V^{-1}\boldsymbol{u}_0$$

I will split $V\Lambda^k V^{-1}\boldsymbol{u}_0$ into three steps. Equation (10) puts those steps together in $\boldsymbol{u}_k$.

1. Write $\boldsymbol{u}_0$ as a combination $c_1\boldsymbol{x}_1 + \cdots + c_n\boldsymbol{x}_n$ of the eigenvectors. Then $\boldsymbol{c} = V^{-1}\boldsymbol{u}_0$.

2. Multiply each number $c_i$ by $(\lambda_i)^k$. Now we have $\Lambda^k V^{-1}\boldsymbol{u}_0$.

3. Add up the pieces $c_i(\lambda_i)^k \boldsymbol{x}_i$ to find the solution $\boldsymbol{u}_k = A^k\boldsymbol{u}_0$. This is $V\Lambda^k V^{-1}\boldsymbol{u}_0$.

$$\boldsymbol{u}_k = \boldsymbol{A}^k \boldsymbol{u}_0 = \boldsymbol{c}_1(\boldsymbol{\lambda}_1)^k \boldsymbol{x}_1 + \cdots + \boldsymbol{c}_n(\boldsymbol{\lambda}_n)^k \boldsymbol{x}_n. \qquad (10)$$

In matrix language $A^k\boldsymbol{u}_0$ equals $(V\Lambda V^{-1})^k\boldsymbol{u}_0$. The 3 steps are $V$ times $\Lambda^k$ times $V^{-1}\boldsymbol{u}_0$.

I am taking time with the three steps to compute $A^k\boldsymbol{u}_0$, because you will see exactly the same steps for differential equations and $e^{At}$. The equation will be $d\boldsymbol{y}/dt = A\boldsymbol{y}$. Please compare equation (10) for $A^k\boldsymbol{u}_0$ with this solution $e^{At}\boldsymbol{y}(0)$ from Section 6.3.

$$\textbf{Solve } d\boldsymbol{y}/dt = A\boldsymbol{y} \qquad \boldsymbol{y}(t) = e^{At}\boldsymbol{y}(0) = c_1 e^{\lambda_1 t}\boldsymbol{x}_1 + \cdots + c_n e^{\lambda_n t}\boldsymbol{x}_n. \qquad (11)$$

Those parallel equations (10) and (11) show the point of eigenvalues and eigenvectors. They split the solutions into $n$ simple pieces. By following each eigenvector separately—this is the result of diagonalizing the matrix—we have $n$ scalar equations.

**The growth factor $\lambda^k$ in (10) is like $e^{\lambda t}$ in (11).**

**Summary** I will display the matrices in those steps. Here is $\boldsymbol{u}_0 = Vc$ :

$$\textbf{Step 1} \qquad \boldsymbol{u}_0 = \begin{bmatrix} & & \\ \boldsymbol{x}_1 & \cdots & \boldsymbol{x}_n \\ & & \end{bmatrix} \begin{bmatrix} c_1 \\ \vdots \\ c_n \end{bmatrix}. \qquad \begin{array}{l} \text{This says that} \\ \boldsymbol{u}_0 = c_1\boldsymbol{x}_1 + \cdots + c_n\boldsymbol{x}_n \end{array} \tag{12}$$

The coefficients in Step 1 are $\boldsymbol{c} = V^{-1}\boldsymbol{u}_0$. Then Step 2 multiplies by $\Lambda^k$. Then Step 3 adds up all the $c_i(\lambda_i)^k\boldsymbol{x}_i$ to get the product of $V$ and $\Lambda^k$ and $V^{-1}\boldsymbol{u}_0$ :

$$A^k\boldsymbol{u}_0 = V\Lambda^k V^{-1}\boldsymbol{u}_0 = \begin{bmatrix} & & \\ \boldsymbol{x}_1 & \dots & \boldsymbol{x}_n \\ & & \end{bmatrix} \begin{bmatrix} (\lambda_1)^k & & \\ & \ddots & \\ & & (\lambda_n)^k \end{bmatrix} \begin{bmatrix} c_1 \\ \vdots \\ c_n \end{bmatrix}. \tag{13}$$

This result is exactly $\boldsymbol{u}_k = c_1(\lambda_1)^k\boldsymbol{x}_1 + \cdots + c_n(\lambda_n)^k\boldsymbol{x}_n$. It solves $\boldsymbol{u}_{k+1} = A\boldsymbol{u}_k$.

**Example 3** Start from $\boldsymbol{u}_0 = (1, 0)$. Compute $A^k\boldsymbol{u}_0$ when $V$ and $\Lambda$ contain these eigenvectors and eigenvalues:

$$A = \begin{bmatrix} 1 & 2 \\ 1 & 0 \end{bmatrix} \quad \text{has} \quad \lambda_1 = 2 \quad \text{and} \quad \boldsymbol{x}_1 = \begin{bmatrix} 2 \\ 1 \end{bmatrix}, \quad \lambda_2 = -1 \quad \text{and} \quad \boldsymbol{x}_2 = \begin{bmatrix} 1 \\ -1 \end{bmatrix}.$$

This matrix $A$ is like Fibonacci except the rule is changed to $F_{k+2} = F_{k+1} + \mathbf{2}F_k$. The new numbers $0, 1, 1, 3, \ldots$ grow faster because $\lambda = 2$ is larger than $(1 + \sqrt{5})/2$.

**Example 3 in three steps** Find $\boldsymbol{u}_0 = c_1\boldsymbol{x}_1 + c_2\boldsymbol{x}_2$ and $\boldsymbol{u}_k = c_1(\lambda_1)^k\boldsymbol{x}_1 + c_2(\lambda_2)^k\boldsymbol{x}_2$

**Step 1** $\qquad \boldsymbol{u}_0 = \begin{bmatrix} 1 \\ 0 \end{bmatrix} = \frac{1}{3}\begin{bmatrix} 2 \\ 1 \end{bmatrix} + \frac{1}{3}\begin{bmatrix} 1 \\ -1 \end{bmatrix}$ so $c_1 = c_2 = \frac{1}{3}$

**Step 2** $\qquad$ Multiply the two eigenvectors by $(\lambda_1)^k = 2^k$ and $(\lambda_2)^k = (-1)^k$

**Step 3** $\qquad$ Combine the pieces into $\boldsymbol{u}_k = \frac{1}{3}2^k\begin{bmatrix} 2 \\ 1 \end{bmatrix} + \frac{1}{3}(-1)^k\begin{bmatrix} 1 \\ -1 \end{bmatrix}$.

Behind these examples lies the fundamental idea: ***Follow each eigenvector***.

## Nondiagonalizable Matrices (Optional)

Suppose $\lambda$ is an eigenvalue of $A$. We discover that fact in two ways:

1. Eigenvectors (geometric) There are nonzero solutions to $A\boldsymbol{x} = \lambda\boldsymbol{x}$.
2. Eigenvalues (algebraic) The determinant of $A - \lambda I$ is zero.

The number $\lambda$ may be a simple eigenvalue or a multiple eigenvalue, and we want to know its ***multiplicity***. Most eigenvalues have multiplicity $M = 1$ (simple eigenvalues). Then there is a single line of eigenvectors, and $\det(A - \lambda I)$ does not have a double factor.

For exceptional matrices, an eigenvalue can be ***repeated***. Then there are two *different* ways to count its multiplicity. Always GM $\leq$ AM for each eigenvalue.

1. (Geometric Multiplicity = GM) Count the **independent eigenvectors** for $\lambda$. This is the dimension of the nullspace of $A - \lambda I$.

2. (Algebraic Multiplicity = AM) Count the **repetitions of the same $\lambda$** among the eigenvalues. Look at the $n$ roots of $\det(A - \lambda I) = 0$.

If $A$ has $\lambda = 4, 4, 4$, that eigenvalue has AM $= 3$ (triple root) and GM $=$ **1 or 2 or 3**.

The following matrix $A$ is the standard example of trouble. Its eigenvalue $\lambda = 0$ is repeated. It is a double eigenvalue (AM $= 2$) with only one eigenvector (GM $= 1$).

**AM = 2**
**GM = 1**
$$A = \begin{bmatrix} 0 & 1 \\ 0 & 0 \end{bmatrix} \quad \text{has} \quad \det(A - \lambda I) = \begin{vmatrix} -\lambda & 1 \\ 0 & -\lambda \end{vmatrix} = \lambda^2. \qquad \begin{matrix} \boldsymbol{\lambda = 0, 0} \textbf{ but} \\ \textbf{1 eigenvector} \end{matrix}$$

There "should" be two eigenvectors, because $\lambda^2 = 0$ has a double root. The double factor $\lambda^2$ makes AM $= 2$. But there is only one eigenvector $\boldsymbol{x} = (1, 0)$. ***This shortage of eigenvectors when*** GM ***is below*** AM ***means that*** $A$ ***is not diagonalizable***.

These three matrices have $\lambda = 5, 5$. Traces are 10, determinants are 25. They only have one eigenvector:

$$A = \begin{bmatrix} 5 & 1 \\ 0 & 5 \end{bmatrix} \quad \text{and} \quad A = \begin{bmatrix} 6 & -1 \\ 1 & 4 \end{bmatrix} \quad \text{and} \quad A = \begin{bmatrix} 7 & 2 \\ -2 & 3 \end{bmatrix}.$$

Those all have $\det(A - \lambda I) = (\lambda - 5)^2$. The algebraic multiplicity is AM $= 2$. But each $A - 5I$ has rank $r = 1$. The geometric multiplicity is GM $= 1$. There is only one line of eigenvectors for $\lambda = 5$, and these matrices are not diagonalizable.

## ■ REVIEW OF THE KEY IDEAS ■

1. If $A$ has $n$ independent eigenvectors $\boldsymbol{x}_1, \ldots, \boldsymbol{x}_n$, they go into the columns of $V$.

$$\boldsymbol{A} \textbf{ is diagonalized by } \boldsymbol{V} \qquad \boldsymbol{V^{-1}AV = \Lambda} \quad \textbf{and} \quad \boldsymbol{A = V\Lambda V^{-1}}.$$

2. The powers of $A$ are $A^k = V\Lambda^k V^{-1}$. The eigenvectors in $V$ are unchanged.

3. The eigenvalues of $A^k$ are $(\lambda_1)^k, \ldots, (\lambda_n)^k$ in the matrix $\Lambda^k$.

**4.** The solution to $\boldsymbol{u}_{k+1} = A\boldsymbol{u}_k$ starting from $\boldsymbol{u}_0$ is $\boldsymbol{u}_k = A^k\boldsymbol{u}_0 = V\Lambda^k V^{-1}\boldsymbol{u}_0$:

$$\boldsymbol{u}_k = c_1(\lambda_1)^k\boldsymbol{x}_1 + \cdots + c_n(\lambda_n)^k\boldsymbol{x}_n \quad \text{provided} \quad \boldsymbol{u}_0 = c_1\boldsymbol{x}_1 + \cdots + c_n\boldsymbol{x}_n.$$

That shows Steps $1, 2, 3$ ($c$'s from $V^{-1}\boldsymbol{u}_0$, powers $\lambda^k$ from $\Lambda^k$, and $\boldsymbol{x}$'s from $V$).

## ■ WORKED EXAMPLES ■

**6.2 A** Find the inverse and the eigenvalues and the determinant of $A$:

$$A = 5 * \textbf{eye}(4) - \textbf{ones}(4) = \begin{bmatrix} 4 & -1 & -1 & -1 \\ -1 & 4 & -1 & -1 \\ -1 & -1 & 4 & -1 \\ -1 & -1 & -1 & 4 \end{bmatrix}.$$

Describe an eigenvector matrix $V$ that gives $V^{-1}AV = \Lambda$.

**Solution** What are the eigenvalues of the all-ones matrix **ones**(4)? Its rank is certainly 1, so three eigenvalues are $\lambda = 0, 0, 0$. Its trace is 4, so the other eigenvalue is $\lambda = 4$. Subtract the all-ones matrix from $5I$ to get our matrix $A = 5I - \textbf{ones}(4)$:

**Subtract the eigenvalues 4, 0, 0, 0 from 5, 5, 5, 5. The eigenvalues of $A$ are 1, 5, 5, 5.**

**The $\lambda$'s add to 16. So does 4 + 4 + 4 + 4 from diag ($A$). Multiply $\lambda$'s: det $A$ = 125.**

The eigenvector for $\lambda = 1$ is $\boldsymbol{x} = (1, 1, 1, 1)$. The other eigenvectors are perpendicular to $\boldsymbol{x}$ (since $A$ is symmetric). The nicest eigenvector matrix $V$ is the symmetric orthogonal Hadamard matrix. Multiply by $1/2$ to have unit vectors in its columns.

$$\textbf{Orthonormal eigenvectors} \quad V = Q = \frac{1}{2}\begin{bmatrix} 1 & 1 & 1 & 1 \\ 1 & -1 & 1 & -1 \\ 1 & 1 & -1 & -1 \\ 1 & -1 & -1 & 1 \end{bmatrix} = Q^{\mathrm{T}} = Q^{-1}.$$

The eigenvalues of $A^{-1}$ are $1, \frac{1}{5}, \frac{1}{5}, \frac{1}{5}$. The eigenvectors are the same as for $A$. This inverse matrix $A^{-1} = Q\Lambda^{-1}Q^{-1}$ is surprisingly neat:

$$A^{-1} = \frac{1}{5} * (\textbf{eye}(4) + \textbf{ones}(4)) = \frac{1}{5}\begin{bmatrix} 2 & 1 & 1 & 1 \\ 1 & 2 & 1 & 1 \\ 1 & 1 & 2 & 1 \\ 1 & 1 & 1 & 2 \end{bmatrix}.$$

To check that $AA^{-1} = I$, use (**ones**) (**ones**) = 4 (**ones**). **Question: Can you find $A^3$?**

# Problem Set 6.2

**Questions 1–7 are about the eigenvalue and eigenvector matrices $\Lambda$ and $V$.**

1 (a) Factor these two matrices into $A = V\Lambda V^{-1}$:

$$A = \begin{bmatrix} 1 & 2 \\ 0 & 3 \end{bmatrix} \quad \text{and} \quad A = \begin{bmatrix} 1 & 1 \\ 3 & 3 \end{bmatrix}.$$

(b) If $A = V\Lambda V^{-1}$ then $A^3 = (\quad)(\quad)(\quad)$ and $A^{-1} = (\quad)(\quad)(\quad)$.

2 If $A$ has $\lambda_1 = 2$ with eigenvector $\boldsymbol{x}_1 = \begin{bmatrix} 1 \\ 0 \end{bmatrix}$ and $\lambda_2 = 5$ with $\boldsymbol{x}_2 = \begin{bmatrix} 1 \\ 1 \end{bmatrix}$, use $V\Lambda V^{-1}$ to find $A$. No other matrix has the same $\lambda$'s and $\boldsymbol{x}$'s.

3 Suppose $A = V\Lambda V^{-1}$. What is the eigenvalue matrix for $A + 2I$? What is the eigenvector matrix? Check that $A + 2I = (\quad)(\quad)(\quad)^{-1}$.

4 True or false: If the columns of $V$ (eigenvectors of $A$) are linearly independent, then

(a) $A$ is invertible (b) $A$ is diagonalizable

(c) $V$ is invertible (d) $V$ is diagonalizable.

5 If the eigenvectors of $A$ are the columns of $I$, then $A$ is a _____ matrix. If the eigenvector matrix $V$ is triangular, then $V^{-1}$ is triangular. Prove that $A$ is also triangular.

6 Describe all matrices $V$ that diagonalize this matrix $A$ (find all eigenvectors):

$$A = \begin{bmatrix} 4 & 0 \\ 1 & 2 \end{bmatrix}.$$

Then describe all matrices that diagonalize $A^{-1}$.

7 Write down the most general matrix that has eigenvectors $\begin{bmatrix} 1 \\ 1 \end{bmatrix}$ and $\begin{bmatrix} 1 \\ -1 \end{bmatrix}$.

**Questions 8–10 are about Fibonacci and Gibonacci numbers.**

8 Diagonalize the Fibonacci matrix by completing $V^{-1}$:

$$\begin{bmatrix} 1 & 1 \\ 1 & 0 \end{bmatrix} = \begin{bmatrix} \lambda_1 & \lambda_2 \\ 1 & 1 \end{bmatrix} \begin{bmatrix} \lambda_1 & 0 \\ 0 & \lambda_2 \end{bmatrix} \begin{bmatrix} & \\ & \end{bmatrix}.$$

Do the multiplication $V\Lambda^k V^{-1}\begin{bmatrix} 1 \\ 0 \end{bmatrix}$ to find its second component. This is the $k$th Fibonacci number $F_k = (\lambda_1^k - \lambda_2^k)/(\lambda_1 - \lambda_2)$.

9 Suppose $G_{k+2}$ is the *average* of the two previous numbers $G_{k+1}$ and $G_k$:

$$\begin{aligned} G_{k+2} &= \tfrac{1}{2}G_{k+1} + \tfrac{1}{2}G_k \\ G_{k+1} &= G_{k+1} \end{aligned} \quad \text{is} \quad \begin{bmatrix} G_{k+2} \\ G_{k+1} \end{bmatrix} = \begin{bmatrix} & A & \end{bmatrix} \begin{bmatrix} G_{k+1} \\ G_k \end{bmatrix}.$$

(a) Find $A$ and its eigenvalues and eigenvectors.

(b) Find the limit as $n \to \infty$ of the matrices $A^n = V\Lambda^n V^{-1}$.

(c) If $G_0 = 0$ and $G_1 = 1$ show that the Gibonacci numbers approach $\frac{2}{3}$.

**10** Prove that every third Fibonacci number in $0, 1, 1, 2, 3, \ldots$ is even.

**Questions 11–14 are about diagonalizability.**

**11** True or false: If the eigenvalues of $A$ are $2, 2, 5$ then the matrix is certainly

(a) invertible (b) diagonalizable (c) not diagonalizable.

**12** True or false: If the only eigenvectors of $A$ are multiples of $(1, 4)$ then $A$ has

(a) no inverse (b) a repeated eigenvalue (c) no diagonalization $V\Lambda V^{-1}$.

**13** Complete these matrices so that $\det A = 25$. Then check that $\lambda = 5$ is repeated—the trace is 10 so the determinant of $A - \lambda I$ is $(\lambda - 5)^2$. Find an eigenvector with $A\boldsymbol{x} = 5\boldsymbol{x}$. These matrices will not be diagonalizable because there is no second line of eigenvectors.

$$A = \begin{bmatrix} 8 & \\ & 2 \end{bmatrix} \quad \text{and} \quad A = \begin{bmatrix} 9 & 4 \\ & 1 \end{bmatrix} \quad \text{and} \quad A = \begin{bmatrix} 10 & 5 \\ -5 & \end{bmatrix}$$

**14** The matrix $A = \begin{bmatrix} 3 & 1 \\ 0 & 3 \end{bmatrix}$ is not diagonalizable because the rank of $A - 3I$ is ____. Change one entry to make $A$ diagonalizable. Which entries could you change?

**Questions 15–19 are about powers of matrices.**

**15** $A^k = V\Lambda^k V^{-1}$ approaches the zero matrix as $k \to \infty$ if and only if every $\lambda$ has absolute value less than ____. Which of these matrices has $A^k \to 0$?

$$A_1 = \begin{bmatrix} .6 & .9 \\ .4 & .1 \end{bmatrix} \quad \text{and} \quad A_2 = \begin{bmatrix} .6 & .9 \\ .1 & .6 \end{bmatrix}.$$

**16** (Recommended) Find $\Lambda$ and $V$ to diagonalize $A_1$ in Problem 15. What is the limit of $\Lambda^k$ as $k \to \infty$? What is the limit of $V\Lambda^k V^{-1}$? In the columns of this limiting matrix you see the ____.

**17** Find $\Lambda$ and $V$ to diagonalize $A_2$ in Problem 15. What is $(A_2)^{10}\boldsymbol{u}_0$ for these $\boldsymbol{u}_0$?

$$\boldsymbol{u}_0 = \begin{bmatrix} 3 \\ 1 \end{bmatrix} \quad \text{and} \quad \boldsymbol{u}_0 = \begin{bmatrix} 3 \\ -1 \end{bmatrix} \quad \text{and} \quad \boldsymbol{u}_0 = \begin{bmatrix} 6 \\ 0 \end{bmatrix}.$$

**18** Diagonalize $A$ and compute $V\Lambda^k V^{-1}$ to prove this formula for $A^k$:

$$A = \begin{bmatrix} 2 & -1 \\ -1 & 2 \end{bmatrix} \quad \text{has} \quad A^k = \frac{1}{2}\begin{bmatrix} 1+3^k & 1-3^k \\ 1-3^k & 1+3^k \end{bmatrix}.$$

**19** Diagonalize $B$ and compute $V\Lambda^k V^{-1}$ to prove this formula for $B^k$:

$$B = \begin{bmatrix} 5 & 1 \\ 0 & 4 \end{bmatrix} \quad \text{has} \quad B^k = \begin{bmatrix} 5^k & 5^k - 4^k \\ 0 & 4^k \end{bmatrix}.$$

**20** Suppose $A = V\Lambda V^{-1}$. Take determinants to prove $\det A = \det \Lambda = \lambda_1 \lambda_2 \cdots \lambda_n$. This quick proof only works when $A$ can be _____.

**21** Show that trace $VT$ = trace $TV$, by adding the diagonal entries of $VT$ and $TV$:

$$V = \begin{bmatrix} a & b \\ c & d \end{bmatrix} \quad \text{and} \quad T = \begin{bmatrix} q & r \\ s & t \end{bmatrix}.$$

Choose $T$ as $\Lambda V^{-1}$. Then $V\Lambda V^{-1}$ has the same trace as $\Lambda V^{-1} V = \Lambda$. The trace of $A$ equals the trace of $\Lambda$, which is certainly the sum of the eigenvalues.

**22** $AB - BA = I$ is impossible since the left side has trace = _____. But find an elimination matrix so that $A = E$ and $B = E^{\mathrm{T}}$ give

$$AB - BA = \begin{bmatrix} -1 & 0 \\ 0 & 1 \end{bmatrix} \quad \text{which has trace zero.}$$

**23** If $A = V\Lambda V^{-1}$, diagonalize the block matrix $B = \begin{bmatrix} A & \mathbf{0} \\ \mathbf{0} & 2A \end{bmatrix}$. Find its eigenvalue and eigenvector (block) matrices.

**24** Consider all 4 by 4 matrices $A$ that are diagonalized by the same fixed eigenvector matrix $V$. Show that the $A$'s form a subspace ($cA$ and $A_1 + A_2$ have this same $V$). What is this subspace when $V = I$? What is its dimension?

**25** Suppose $A^2 = A$. On the left side $A$ multiplies each column of $A$. Which of our four subspaces contains eigenvectors with $\lambda = 1$? Which subspace contains eigenvectors with $\lambda = 0$? From the dimensions of those subspaces, $A$ has a full set of independent eigenvectors. So every matrix with $A^2 = A$ can be diagonalized.

**26** (Recommended) Suppose $A\boldsymbol{x} = \lambda \boldsymbol{x}$. If $\lambda = 0$ then $\boldsymbol{x}$ is in the nullspace. If $\lambda \neq 0$ then $\boldsymbol{x}$ is in the column space. Those spaces have dimensions $(n - r) + r = n$. So why doesn't every square matrix have $n$ linearly independent eigenvectors?

**27** The eigenvalues of $A$ are 1 and 9, and the eigenvalues of $B$ are $-1$ and 9:

$$A = \begin{bmatrix} 5 & 4 \\ 4 & 5 \end{bmatrix} \quad \text{and} \quad B = \begin{bmatrix} 4 & 5 \\ 5 & 4 \end{bmatrix}.$$

Find a matrix square root of $A$ from $R = V\sqrt{\Lambda}\, V^{-1}$. Why is there no real matrix square root of $B$?

**28** The powers $A^k$ approach zero if all $|\lambda_i| < 1$ and they blow up if any $|\lambda_i| > 1$. Peter Lax gives these striking examples in his book *Linear Algebra*:

$$A = \begin{bmatrix} 3 & 2 \\ 1 & 4 \end{bmatrix} \quad B = \begin{bmatrix} 3 & 2 \\ -5 & -3 \end{bmatrix} \quad C = \begin{bmatrix} 5 & 7 \\ -3 & -4 \end{bmatrix} \quad D = \begin{bmatrix} 5 & 6.9 \\ -3 & -4 \end{bmatrix}$$

$$\|A^{1024}\| > 10^{700} \qquad B^{1024} = I \qquad C^{1024} = -C \qquad \|D^{1024}\| < 10^{-78}$$

Find the eigenvalues $\lambda = e^{i\theta}$ of $B$ and $C$ to show $B^4 = I$ and $C^3 = -I$.

**29** If $A$ and $B$ have the same $\lambda$'s with the same full set of independent eigenvectors, their factorizations into ______ are the same. So $A = B$.

**30** Suppose the same $V$ diagonalizes both $A$ and $B$. They have the same eigenvectors in $A = V\Lambda_1 V^{-1}$ and $B = V\Lambda_2 V^{-1}$. Prove that $AB = BA$.

**31** (a) If $A = \begin{bmatrix} a & b \\ 0 & d \end{bmatrix}$ then the determinant of $A - \lambda I$ is $(\lambda - a)(\lambda - d)$. Check the "Cayley-Hamilton Theorem" that $(A - aI)(A - dI) =$ *zero matrix*.

(b) Test the Cayley-Hamilton Theorem on Fibonacci's $A = \begin{bmatrix} 1 & 1 \\ 1 & 0 \end{bmatrix}$. The theorem predicts that $A^2 - A - I = 0$, since the polynomial $\det(A - \lambda I)$ is $\lambda^2 - \lambda - 1$.

**32** Substitute $A = V\Lambda V^{-1}$ into the product $(A - \lambda_1 I)(A - \lambda_2 I)\cdots(A - \lambda_n I)$ and explain why this produces the zero matrix. We are substituting the matrix $A$ for the number $\lambda$ in the polynomial $p(\lambda) = \det(A - \lambda I)$. The ***Cayley-Hamilton Theorem*** says that this product is always $p(A) =$ *zero matrix*, even if $A$ is not diagonalizable.

## Challenge Problems

**33** The $n$th power of rotation through $\theta$ is rotation through $n\theta$:

$$A^n = \begin{bmatrix} \cos\theta & -\sin\theta \\ \sin\theta & \cos\theta \end{bmatrix}^n = \begin{bmatrix} \cos n\theta & -\sin n\theta \\ \sin n\theta & \cos n\theta \end{bmatrix}.$$

Prove that neat formula by diagonalizing $A = V\Lambda V^{-1}$. The eigenvectors (columns of $V$) are $(1, i)$ and $(i, 1)$. You need to know Euler's formula $e^{i\theta} = \cos\theta + i\sin\theta$.

**34** The transpose of $A = V\Lambda V^{-1}$ is $A^{\mathrm{T}} = (V^{-1})^{\mathrm{T}}\Lambda V^{\mathrm{T}}$. The eigenvectors in $A^{\mathrm{T}}\boldsymbol{y} = \lambda\boldsymbol{y}$ are the columns of that matrix $(V^{-1})^{\mathrm{T}}$. They are often called ***left eigenvectors***.

How do you multiply three matrices $V\Lambda V^{-1}$ to find this formula for $A$?

$$\textbf{Sum of rank-1 matrices} \qquad A = V\Lambda V^{-1} = \lambda_1 \boldsymbol{x}_1 \boldsymbol{y}_1^{\mathrm{T}} + \cdots + \lambda_n \boldsymbol{x}_n \boldsymbol{y}_n^{\mathrm{T}}.$$

**35** The inverse of $A = \mathbf{eye}(n) + \mathbf{ones}(n)$ is $A^{-1} = \mathbf{eye}(n) + C * \mathbf{ones}(n)$. Multiply $AA^{-1}$ to find that number $C$ (depending on $n$).

## 6.3 Linear Systems $y' = Ay$

This section is about first order systems of linear differential equations. The key words are *systems* and *linear*. A system allows $n$ equations for $n$ unknown functions $y_1(t), \ldots, y_n(t)$. A linear system multiplies that unknown vector $\boldsymbol{y}(t)$ by a matrix $A$. Then a first order linear system can include a source term $\boldsymbol{q}(t)$, or not:

| **Without source** $\dfrac{d\boldsymbol{y}}{dt} = A\boldsymbol{y}(t)$ | **With source** $\dfrac{d\boldsymbol{y}}{dt} = A\boldsymbol{y}(t) + \boldsymbol{q}(t)$ |
|---|---|

Without a source term, the only input is $\boldsymbol{y}(0)$ at the start. With $\boldsymbol{q}(t)$ included, there is also a continuing input $\boldsymbol{q}(t)dt$ between times $t$ and $t + dt$. Forward from time $t$, this input grows or decays along with the $\boldsymbol{y}(t)$ that just arrived from the past. That is important.

The **transient solution** $\boldsymbol{y}_n(t)$ starts from $\boldsymbol{y}(0)$, when $\boldsymbol{q}(t) = \boldsymbol{0}$. The output coming from the source $\boldsymbol{q}(t)$ is one particular solution $\boldsymbol{y}_p(t)$. Linearity allows superposition! **The complete solution with source included is** $\boldsymbol{y}(t) = \boldsymbol{y}_n(t) + \boldsymbol{y}_p(t)$ **as always**.

The serious work of this section is to find $\boldsymbol{y}_n(t)$, the null solution to $\boldsymbol{y}_n{}' - A\boldsymbol{y}_n = \boldsymbol{0}$. Then Section 6.4 accounts for the source term $\boldsymbol{q}(t)$ and finds a particular solution.

We want to use the eigenvalues and eigenvectors of $A$. We don't want those to change with time. So we kept our equation linear time-invariant, with a constant matrix $A$. Fortunately, many important systems have $A$ = constant in the first place. The system is not changing, it is only the *state* of the system that changes: constant $A$, evolving state $\boldsymbol{y}(t)$.

We will express $\boldsymbol{y}(t)$ as a combination of eigenvectors of $A$. Section 6.4 uses $e^{At}$.

### Solution by Eigenvectors and Eigenvalues

Suppose the $n$ by $n$ matrix $A$ has $n$ independent eigenvectors. This is automatic if $A$ has $n$ different eigenvalues $\lambda$. Then the eigenvectors $\boldsymbol{x}_1, \ldots, \boldsymbol{x}_n$ are a basis in which we can express any starting vector $\boldsymbol{y}(0)$:

$$\textbf{Initial condition} \qquad \boldsymbol{y}(0) = c_1\boldsymbol{x}_1 + \cdots + c_n\boldsymbol{x}_n \quad \text{for some numbers } c_1, \ldots, c_n. \tag{1}$$

Computing the $c$'s is Step 1 in the solution, after finding the $\lambda$'s and $\boldsymbol{x}$'s.

Step 2 solves the equation $\boldsymbol{y}' = A\boldsymbol{y}$ using $\boldsymbol{y} = e^{\lambda t}\boldsymbol{x}$. *Start from any eigenvector*:

$$\text{If } A\boldsymbol{x} = \lambda\boldsymbol{x} \quad \text{then} \quad \boldsymbol{y(t) = e^{\lambda t}x} \quad \text{solves} \quad \frac{d\boldsymbol{y}}{dt} = A\boldsymbol{y}. \tag{2}$$

This solution $\boldsymbol{y} = e^{\lambda t}\boldsymbol{x}$ separates the time-dependent $e^{\lambda t}$ from the constant vector $\boldsymbol{x}$:

$$\frac{d\boldsymbol{y}}{dt} = A\boldsymbol{y} \quad \text{becomes} \quad \frac{d}{dt}(e^{\lambda t}\boldsymbol{x}) = \lambda e^{\lambda t}\boldsymbol{x} = A(e^{\lambda t}\boldsymbol{x}). \tag{3}$$

Step 3 is the final solution step. Add the $n$ separate solutions from the $n$ eigenvectors.

$$\textbf{Superposition} \qquad \boldsymbol{y}(t) = c_1 e^{\lambda_1 t}\boldsymbol{x}_1 + \cdots + c_n e^{\lambda_n t}\boldsymbol{x}_n. \tag{4}$$

At $t = 0$ this matches $\boldsymbol{y}(0)$ in equation (1). That was Step 1, where we chose the $c$'s.

**Example 1** Find all solutions to $\boldsymbol{y}' = \begin{bmatrix} -2 & 1 \\ 1 & -2 \end{bmatrix} \boldsymbol{y}$. Which solution has $\boldsymbol{y}(0) = \begin{bmatrix} 6 \\ 2 \end{bmatrix}$?

*Solution* First we find $\lambda = -1$ and $-3$. Their eigenvectors $\boldsymbol{x}_1$ and $\boldsymbol{x}_2$ go into $V$:

$$\det \begin{bmatrix} -2-\lambda & 1 \\ 1 & -2-\lambda \end{bmatrix} = \lambda^2 + 4\lambda + 3 \quad \text{factors into} \quad (\lambda+1)(\lambda+3)$$

$$\begin{matrix} A\boldsymbol{x}_1 = -1\,\boldsymbol{x}_1 \\ A\boldsymbol{x}_2 = -3\,\boldsymbol{x}_2 \end{matrix} \qquad \begin{bmatrix} -2 & 1 \\ 1 & -2 \end{bmatrix} \begin{bmatrix} \mathbf{1} \\ \mathbf{1} \end{bmatrix} = \begin{bmatrix} -1 \\ -1 \end{bmatrix} \qquad \begin{bmatrix} -2 & 1 \\ 1 & -2 \end{bmatrix} \begin{bmatrix} \mathbf{1} \\ \mathbf{-1} \end{bmatrix} = \begin{bmatrix} -3 \\ 3 \end{bmatrix}$$

**Step 1** Solve $\boldsymbol{y}(\mathbf{0}) = V\boldsymbol{c}$. Then $\boldsymbol{y}(0)$ is a mixture $\mathbf{4}\boldsymbol{x}_\mathbf{1} + \mathbf{2}\boldsymbol{x}_\mathbf{2}$ of the eigenvectors:

$$V\boldsymbol{c} = \begin{bmatrix} 1 & 1 \\ 1 & -1 \end{bmatrix} \begin{bmatrix} c_1 \\ c_2 \end{bmatrix} = \begin{bmatrix} 6 \\ 2 \end{bmatrix} \text{ gives } \begin{bmatrix} c_1 \\ c_2 \end{bmatrix} = \begin{bmatrix} 4 \\ 2 \end{bmatrix}. \text{ Then } \begin{bmatrix} 6 \\ 2 \end{bmatrix} = 4\begin{bmatrix} 1 \\ 1 \end{bmatrix} + 2\begin{bmatrix} 1 \\ -1 \end{bmatrix}.$$

**Step 2** finds the separate solutions $ce^{\lambda t}\boldsymbol{x}$ given by $4e^{-t}\boldsymbol{x}_1$ and $2e^{-3t}\boldsymbol{x}_2$. Now add:

**Step 3** $$\boldsymbol{y}(t) = 4e^{-t}\begin{bmatrix} 1 \\ 1 \end{bmatrix} + 2e^{-3t}\begin{bmatrix} 1 \\ -1 \end{bmatrix} = \begin{bmatrix} 4e^{-t} + 2e^{-3t} \\ 4e^{-t} - 2e^{-3t} \end{bmatrix}. \tag{5}$$

For a larger matrix the computations are harder. The idea doesn't change.

Now I want to show a matrix with complex eigenvalues and eigenvectors. This will lead us to complex numbers in $\boldsymbol{y}(t)$. But $A$ is real and $\boldsymbol{y}(0)$ is real, so $\boldsymbol{y}(t)$ must be real! Euler's formula $e^{it} = \cos t + i \sin t$ will get us back to real numbers.

**Example 2** Find all solutions to $\boldsymbol{y}' = \begin{bmatrix} -2 & 1 \\ -1 & -2 \end{bmatrix} \boldsymbol{y}$. Which solution has $\boldsymbol{y}(0) = \begin{bmatrix} 6 \\ 2 \end{bmatrix}$?

*Solution* Again we find the eigenvalues and eigenvectors, now complex:

$$\det(\boldsymbol{A} - \boldsymbol{\lambda I}) = \mathbf{0} \qquad \det \begin{bmatrix} -2-\lambda & 1 \\ -1 & -2-\lambda \end{bmatrix} = \boldsymbol{\lambda^2 + 4\lambda + 5} \text{ (no real factors)}$$

We use the quadratic formula to solve $\lambda^2 + 4\lambda + 5 = 0$. The eigenvectors are $x = (1, \pm i)$.

$$\begin{matrix} \boldsymbol{\lambda_1 = -2 + i} \\ \boldsymbol{\lambda_2 = -2 - i} \end{matrix} \qquad \lambda = \frac{-4 \pm \sqrt{4^2 - 4(5)}}{2} = \frac{-4 \pm 2i}{2} = -2 \pm i$$

$$\begin{bmatrix} -2 & 1 \\ -1 & -2 \end{bmatrix} \begin{bmatrix} \mathbf{1} \\ \boldsymbol{i} \end{bmatrix} = (-2+i)\begin{bmatrix} \mathbf{1} \\ \boldsymbol{i} \end{bmatrix} \qquad \begin{bmatrix} -2 & 1 \\ -1 & -2 \end{bmatrix} \begin{bmatrix} \mathbf{1} \\ \boldsymbol{-i} \end{bmatrix} = (-2-i)\begin{bmatrix} \mathbf{1} \\ \boldsymbol{-i} \end{bmatrix}.$$

To solve $y' = Ay$, Step 1 expresses $y(0) = (6, 2)$ as a combination of those eigenvectors:

$$y(0) = Vc = c_1 x_1 + c_2 x_2 \qquad \begin{bmatrix} 6 \\ 2 \end{bmatrix} = (3-i)\begin{bmatrix} 1 \\ i \end{bmatrix} + (3+i)\begin{bmatrix} 1 \\ -i \end{bmatrix}.$$

Step 2 finds the solutions $c_1 e^{\lambda_1 t} x_1$ and $c_2 e^{\lambda_2 t} x_2$. Step 3 combines them into $y(t)$:

**Solution** $$y(t) = c_1 e^{\lambda_1 t} x_1 + c_2 e^{\lambda_2 t} x_2 = (3-i)e^{(-2+i)t}\begin{bmatrix} 1 \\ i \end{bmatrix} + (3+i)e^{(-2-i)t}\begin{bmatrix} 1 \\ -i \end{bmatrix}.$$

As expected, this looks complex. As promised, it must be real. Factoring out $e^{-2t}$ leaves

$$(3-i)(\cos t + i \sin t)\begin{bmatrix} 1 \\ i \end{bmatrix} + (3+i)(\cos t - i \sin t)\begin{bmatrix} 1 \\ -i \end{bmatrix} = \begin{bmatrix} 6 \cos t + 2 \sin t \\ 2 \cos t - 6 \sin t \end{bmatrix}. \quad (6)$$

Put back the factor $e^{-2t}$ to find the (real) $y(t)$. It would be wise to check $y' = Ay$:

$$\boxed{y(0) = \begin{bmatrix} 6 \\ 2 \end{bmatrix} \quad \text{and} \quad y(t) = e^{-2t}\begin{bmatrix} 6 \cos t + 2 \sin t \\ 2 \cos t - 6 \sin t \end{bmatrix}} \quad (7)$$

The factor $e^{-2t}$ from the real part of $\lambda$ means decay. The $\cos t$ and $\sin t$ factors from the imaginary part mean oscillation. The oscillation frequency in $\cos t = \cos \omega t$ is $\omega = 1$.

*Note* The $-2$'s on the diagonal of $A$ (which is exactly $-2I$) are responsible for the real parts $-2$ of the $\lambda$'s. They give the decay factor $e^{-2t}$. Without the $-2$'s we would only have sines and cosines, which converts into **circular motion in the $y_1 - y_2$ plane.** That is a very important example to see by itself.

**Example 3** **Pure circular motion and pure imaginary eigenvalues**

$$y' = \begin{bmatrix} y_1' \\ y_2' \end{bmatrix} = \begin{bmatrix} 0 & 1 \\ -1 & 0 \end{bmatrix}\begin{bmatrix} y_1 \\ y_2 \end{bmatrix} = \begin{bmatrix} y_2 \\ -y_1 \end{bmatrix} \quad \text{sends } y \text{ around a circle.}$$

*Discussion* The equations are $y_1' = y_2$ and $y_2' = -y_1$. One solution is $y_1 = \sin t$ and $y_2 = \cos t$. A second solution is $y_1 = \cos t$ and $y_2 = -\sin t$. We need two solutions to match two required values $y_1(0)$ and $y_2(0)$. Those solutions would come in the usual way from the eigenvalues $\lambda = \pm i$ and the eigenvectors.

Figure 6.2 a shows the solution to Example 2 spiralling in to zero (because of $e^{-2t}$). Figure 6.2 b shows the solution to Example 3 staying on the circle (because of sine and cosine). These are good examples to see the "*phase plane*" with axes $y_1$ and $y_1{}' = y_2$.

Without the $-2$'s, the matrix $A = \begin{bmatrix} 0 & 1 \\ -1 & 0 \end{bmatrix}$ is a rotation by 90°. At every instant, $y'$ is at a 90° angle with $y$. That keeps $y$ moving in a circle. Its length is constant:

**Constant length**
**Circular orbit** $$\frac{d}{dt}(y_1^2 + y_2^2) = 2y_1 y_1' + 2y_2 y_2' = 2y_1 y_2 - 2y_2 y_1 = 0. \quad (8)$$

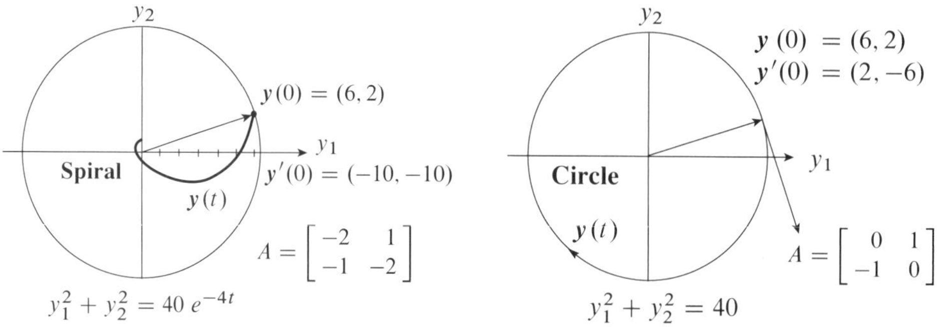


Figure 6.2: (a) The solution (7) including $e^{-2t}$. (b) The solution (6) without $e^{-2t}$.

## Conservative Motion

Travel around a circle is an example of conservative motion for $n = 2$. The length of $\boldsymbol{y}$ does not change. "Energy is conserved." For $n = 3$ this would become travel on a sphere. For $n > 3$ the vector $\boldsymbol{y}$ would move with constant length around a hypersphere.

Which linear differential equations produce this conservative motion ? We are asking for the squared length $||\boldsymbol{y}||^2 = \boldsymbol{y}^{\mathrm{T}}\boldsymbol{y}$ to *stay constant*. So its derivative is zero :

$$\frac{d}{dt}(\boldsymbol{y}^{\mathrm{T}}\boldsymbol{y}) = \left(\frac{d\boldsymbol{y}}{dt}\right)^{\mathrm{T}} \boldsymbol{y} + \boldsymbol{y}^{\mathrm{T}}\frac{d\boldsymbol{y}}{dt} = (A\boldsymbol{y})^{\mathrm{T}}\boldsymbol{y} + \boldsymbol{y}^{\mathrm{T}}(A\boldsymbol{y}) = \boldsymbol{y}^{\mathrm{T}}(\boldsymbol{A}^{\mathrm{T}} + \boldsymbol{A})\boldsymbol{y} = 0. \quad (9)$$

The first step was the product rule. Then $d\boldsymbol{y}/dt$ was replaced by $A\boldsymbol{y}$. **Conclusion** :

$$||\boldsymbol{y}||^2 \textbf{ is constant when } A \textbf{ is antisymmetric: } \boldsymbol{A}^{\mathrm{T}} + \boldsymbol{A} = \boldsymbol{0} \text{ and } \boldsymbol{A}^{\mathrm{T}} = -\boldsymbol{A}. \quad (10)$$

The simplest example is $A = \begin{bmatrix} 0 & 1 \\ -1 & 0 \end{bmatrix}$. Then $\boldsymbol{y}$ goes around the circle in Figure 6.2 b. The initial vector $\boldsymbol{y}(0)$ decides the size of the circle : $||\boldsymbol{y}(t)|| = ||\boldsymbol{y}(0)||$ for all time. When $A$ is antisymmetric, its eigenvalues are pure imaginary. This comes in Section 6.5.

## Stable Motion

Motion around a circle is only "neutral" stability. **For a truly stable linear system, the solution $\boldsymbol{y}(t)$ always goes to zero.** It is the spiral in Figure 6.2 a that shows stability :

$$A = \begin{bmatrix} -2 & 1 \\ -1 & -2 \end{bmatrix} \text{ has eigenvalues } \lambda = -2 \pm i. \text{ This } A \text{ is a } \textbf{stable matrix}.$$

The key is in the eigenvalues of $A$, which give the simple solutions $\boldsymbol{y} = e^{\lambda t}\boldsymbol{x}$. When $A$ is diagonalizable ($n$ independent eigenvectors), every solution is a combination of $e^{\lambda_1 t}\boldsymbol{x}_1, \ldots, e^{\lambda_n t}\boldsymbol{x}_n$. So we only have to ask when those simple solutions approach zero :

| **Stability** $\quad e^{\lambda t}x \to 0$ **when the real part of** $\lambda$ **is negative :** $\quad \mathbf{Re}\,\lambda < 0.$ |
|---|

The real parts $-2$ give the exponential decay factor $e^{-2t}$ in the solution $\boldsymbol{y}$. That factor produces the inward spiral in Figure 6.2 a and the stability of the equation $\boldsymbol{y}' = A\boldsymbol{y}$. The imaginary parts of $\lambda = -2 \pm i$ give oscillations : sines and cosines that stay bounded.

## Test for Stability When $n = 2$

For a 2 by 2 matrix, the trace and determinant tell us both eigenvalues. So the trace and determinant must decide stability. A real matrix $A$ has two possibilities **R** and **C** :

**R** Real eigenvalues $\lambda_1$ and $\lambda_2$
**C** Complex conjugate pair $\lambda_1 = s + i\omega$ and $\lambda_2 = s - i\omega$

Adding the eigenvalues gives the trace of $A$. Multiplying the eigenvalues gives the determinant of $A$. We check the two possibilities **R** and **C**, to see when Re $(\lambda) < 0$.

**R** If $\lambda_1 < 0$ and $\lambda_2 < 0$, then **trace** $= \lambda_1 + \lambda_2 < 0$ and **determinant** $= \lambda_1\lambda_2 > 0$
**C** If $s < 0$ in $\lambda = s \pm i\omega$, then **trace** $= 2s < 0$ and **determinant** $= s^2 + \omega^2 > 0$

Both cases give the same stability requirement : *Negative trace and positive determinant.*

$$A = \begin{bmatrix} a & b \\ c & d \end{bmatrix} \textbf{ is stable } \textbf{ exactly when } \quad \begin{array}{ll} \textbf{trace} = a + d & < 0 \\ \textbf{det} \ \ = ad - bc & > 0 \end{array} \qquad (11)$$

It was the quadratic formula that led us to the possibilities **R** and **C**, real or complex. Remember the equation $\det(A - \lambda I) = 0$ for the eigenvalues :

$$\det \begin{bmatrix} a - \lambda & b \\ c & d - \lambda \end{bmatrix} = \boldsymbol{\lambda}^2 - (a + d)\,\lambda + (ad - bc) = \boldsymbol{\lambda}^2 - (\textbf{trace})\,\boldsymbol{\lambda} + (\textbf{det}) = 0.$$

The quadratic formula for the two eigenvalues includes an all–important square root :

$$\textbf{Real or complex } \lambda \qquad \lambda = \frac{1}{2}\left[\text{trace} \ \pm \sqrt{(\text{trace})^2 - 4(\text{det})}\right]. \qquad (12)$$

The roots are real (case **R**) when $(\text{trace})^2 \geq 4\,(\text{det})$. The roots are complex (case **C**) when $(\text{trace})^2 < 4\,(\text{det})$. The line between **R** and **C** is the parabola in the stability picture :

$$\textbf{(Trace)}^{\mathbf{2}} \textbf{ = 4 (det)} \qquad \begin{bmatrix} -1 & 2 \\ 0 & -1 \end{bmatrix} \text{ is stable} \qquad \begin{bmatrix} 1 & 2 \\ 0 & 1 \end{bmatrix} \text{ is unstable}$$

Stable matrices only fill one quadrant of the trace-determinant plane : trace $< 0$, det $> 0$.

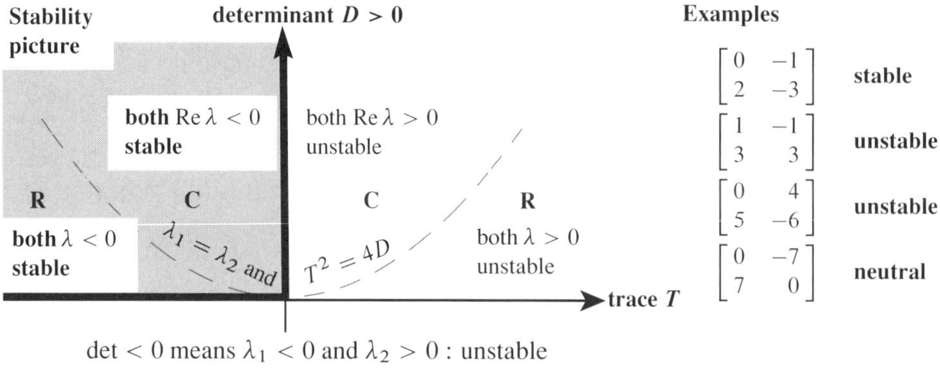


## Second Order Equation to First Order System

Chapter 2 of this book studied the second order equation $y'' + By' + Cy = 0$. Often this is oscillation with underdamping. The solutions $y = e^{(a+i\omega)t}$ and $e^{(a-i\omega)t}$ come from the quadratic equation $\boldsymbol{s^2 + Bs + C = 0}$, when we search for solutions $y = e^{st}$. If $B^2$ is larger than $4C$, then the roots are real and the solutions are $e^{s_1 t}$ and $e^{s_2 t}$. In that overdamped case, the oscillations are gone.

I want to show you exactly the same solutions in the language of $\boldsymbol{y}' = A\boldsymbol{y}$. Instead of one equation with $y''$ we will reach **two equations with** $\boldsymbol{y}' = (y_1', y_2')$. You have seen the key idea before: *The original $y$ and $y'$ become $y_1$ and $y_2$.* Then the matrix $A$ is a **companion matrix**.

$$\boldsymbol{y'' + By' + Cy = 0} \quad \begin{bmatrix} y_1 \\ y_2 \end{bmatrix}' = \begin{bmatrix} y' \\ y'' \end{bmatrix} = \begin{bmatrix} 0 & 1 \\ -C & -B \end{bmatrix} \begin{bmatrix} y \\ y' \end{bmatrix} = A\boldsymbol{y}. \qquad (13)$$

It is important to see why the roots $s_1$ and $s_2$ are also the eigenvalues $\lambda_1$ and $\lambda_2$. The reason is, these are still the roots of the same equation $s^2 + Bs + C = 0$. Only the letter $s$ is changed to $\lambda$.

$$\det(A - \lambda I) = \det \begin{bmatrix} -\lambda & 1 \\ -C & -B - \lambda \end{bmatrix} = \lambda^2 + B\lambda + C = 0. \qquad (14)$$

This was foreshadowed when we drew the six solution paths in Section 3.2: Sources, Sinks, Spirals, and Saddles. Those pictures were in the $y$, $y'$ plane (the phase plane). Now the same pictures are in the $y_1$, $y_2$ plane. I specially want to show you again the trace and determinant of $A$ and the whole new-old understanding of stability.

$$\begin{bmatrix} 0 & 1 \\ -C & -B \end{bmatrix} \text{ has trace} = -B \text{ and determinant} = C.$$

First the test for real roots of $s^2 + Bs + C = 0$ and for real eigenvalues of $A$:

| | | | |
|---|---|---|---|
| **R** | Real roots and real eigenvalues | $\boldsymbol{B^2 \geq 4C}$ | $\textbf{(trace)}^2 \geq \textbf{4(det)}$ |
| **C** | Complex roots and eigenvalues $\lambda = a \pm i\omega$ | $\boldsymbol{B^2 < 4C}$ | $\textbf{(trace)}^2 < \textbf{4(det)}$ |

In the picture, the dashed parabola $T^2 = 4D$ separates real from complex: **R** from **C**.

More than that, the highlighted quadrant displays the three possibilities for damping. These are all stable: $B > 0$ and $C > 0$.

| | | | |
|---|---|---|---|
| **Underdamping** | **Complex roots** | $B^2 < 4AC$ | **above the parabola** |
| **Critical damping** | **Equal roots** | $B^2 = 4AC$ | **on the parabola** |
| **Overdamping** | **Real roots** | $B^2 > 4AC$ | **below the parabola** |

The undamped case $B = 0$ is on the vertical axis: eigenvalues $\pm i\omega$ with $\omega^2 = C$. Everything comes together for 2 by 2 companion matrices. The eigenvectors are attractive too:

$$\boldsymbol{x}_1 = \begin{bmatrix} 1 \\ \lambda_1 \end{bmatrix} \quad \boldsymbol{x}_2 = \begin{bmatrix} 1 \\ \lambda_2 \end{bmatrix} \quad \text{agree with} \quad \begin{bmatrix} y \\ y' \end{bmatrix} = \begin{bmatrix} e^{\lambda t} \\ \lambda e^{\lambda t} \end{bmatrix} = \begin{bmatrix} 1 \\ \lambda \end{bmatrix} \quad \text{at } t = 0. \qquad (15)$$

The same method applies to systems with $n$ oscillators. $B$ and $C$ become matrices. The vectors $\boldsymbol{y}$ and $\boldsymbol{y}'$ have $n$ components and the joint vector $\boldsymbol{z} = (\boldsymbol{y}, \boldsymbol{y}')$ has $2n$ components. The network leads to $n$ second order equations for $\boldsymbol{y}$, or $2n$ first order equations for $\boldsymbol{z}$:

$$\boldsymbol{y}'' + B\boldsymbol{y}' + C\boldsymbol{y} = \boldsymbol{0} \qquad \boldsymbol{z}' = \begin{bmatrix} \boldsymbol{y}' \\ \boldsymbol{y}'' \end{bmatrix} = \begin{bmatrix} 0 & I \\ -C & -B \end{bmatrix} \begin{bmatrix} \boldsymbol{y} \\ \boldsymbol{y}' \end{bmatrix} = A\boldsymbol{z}. \qquad (16)$$

Eigenvectors give the null solutions $\boldsymbol{y}_n$. Real problems come with forcing terms $\boldsymbol{q} = \boldsymbol{F}e^{st}$.

Here I make just one point about repeated roots and repeated eigenvalues: **If $\lambda_1 = \lambda_2$ there is no second eigenvector of the companion matrix $A$.** That matrix can't be diagonalized and the eigenvector method fails. The next section will succeed with $e^{At}$, even without a full set of eigenvectors.

## Higher Order Equations Give First Order Systems

A third order (or higher order) equation reduces to first order in the same way. **Introduce derivatives of $y$ as new unknowns.** This is easy to see for a single third order equation with constant coefficients:

$$y''' + By'' + Cy' + Dy = 0 \qquad (17)$$

The idea is to create a vector unknown $\boldsymbol{z} = (y, y', y'')$. The first component $y$ satisfies a very simple equation: its derivative is the second component $y'$. Then the matrix below has $0, 1, 0$ in its first row. Similarly the derivative of $y'$ is $y''$. The second row of the companion matrix is $0, 0, 1$. The third row contains the original differential equation (17):

$$\boldsymbol{z}' = A\boldsymbol{z} \qquad \begin{bmatrix} y \\ y' \\ y'' \end{bmatrix}' = \begin{bmatrix} 0 & 1 & 0 \\ 0 & 0 & 1 \\ -D & -C & -B \end{bmatrix} \begin{bmatrix} y \\ y' \\ y'' \end{bmatrix}. \qquad (18)$$

**Companion matrices** have 1's on their superdiagonal. We want to know their eigenvalues.

## Eigenvalues of the Companion Matrix = Roots of the Polynomial

Start with the eigenvalues of the 2 by 2 companion matrix:

$$\det(A-\lambda I)=\det\begin{bmatrix}-\lambda & 1\\ -C & -B-\lambda\end{bmatrix}=\lambda^2+B\lambda+C=0. \tag{19}$$

Compare that with substituting $y=e^{\lambda t}$ in the single equation $y''+By'+Cy=0$:

$$\lambda^2e^{\lambda t}+B\lambda e^{\lambda t}+Ce^{\lambda t}\quad\text{gives}\quad \lambda^2+B\lambda+C=0. \tag{20}$$

**The equations are the same.** The $\lambda$'s in special solutions $y=e^{\lambda t}$ are the same as the eigenvalues in special solutions $\boldsymbol{z}=e^{\lambda t}\boldsymbol{x}$. This is our main point and it is true again for 3 by 3. The eigenvalue equation $\det(A-\lambda I)=0$ is exactly the polynomial equation from substituting $y=e^{\lambda t}$ in $y'''+By''+Cy'+Dy=0$:

$$\det\begin{bmatrix}-\lambda & 1 & 0\\ 0 & -\lambda & 1\\ -D & -C & -B-\lambda\end{bmatrix}=-(\lambda^3+B\lambda^2+C\lambda+D)=0. \tag{21}$$

**The eigenvectors of this companion matrix have the special form $\boldsymbol{x}=(1,\lambda,\lambda^2)$.** Fourth order equations become $\boldsymbol{z}'=A\boldsymbol{z}$ with $\boldsymbol{z}=(y,y',y'',y''')$. 4 by 4 companion matrix, eigenvalues from $\lambda^4+B\lambda^3+C\lambda^2+D\lambda+E=0$.

**Example 4** $(\lambda-2)^2=\lambda^2-4\lambda+4=0$ comes from $y''-4y'+4y=0$:

Companion matrix $A$
Repeated root $\lambda=2,2$

$$A=\begin{bmatrix}0 & 1\\ -4 & 4\end{bmatrix}\qquad \det(A-\lambda I)=\lambda^2-4\lambda+4.$$

$\lambda=2$ must have one eigenvector, and it is $\boldsymbol{x}=(1,2)$. *There is no second eigenvector.* The first order system $\boldsymbol{z}'=A\boldsymbol{z}$ and the second order equation $y''-4y'+4y=0$ are in (*the same*) trouble. **The only pure exponential solution is $y=e^{2t}$.**

The way out for $y$ is the solution $te^{2t}$. It needs that new form (including $t$). The way out for $\boldsymbol{z}$ is a "generalized eigenvector" but we are not going there.

## ■ REVIEW OF THE KEY IDEAS ■

1. The system $\boldsymbol{y}'=A\boldsymbol{y}$ is linear with constant coefficients, starting from $\boldsymbol{y}(0)$.
2. Its solution is usually a combination of exponentials $e^{\lambda t}$ times eigenvectors $\boldsymbol{x}$:

   ***n* independent eigenvectors** $\quad \boldsymbol{y}(t)=c_1e^{\lambda_1 t}\boldsymbol{x}_1+\cdots+c_ne^{\lambda_n t}\boldsymbol{x}_n.$

3. The constants $c_1,\ldots,c_n$ are determined by $\boldsymbol{y}(0)=c_1\boldsymbol{x}_1+\cdots+c_n\boldsymbol{x}_n$. *This is $V\boldsymbol{c}$!*
4. $\boldsymbol{y}(t)$ approaches zero (stability) if every $\lambda$ has negative real part: $\operatorname{Re}\lambda<0$.
5. 2 by 2 systems are stable if **trace $T=a+d<0$** and **det $D=ad-bc>0$**.
6. $y''+By'+Cy=0$ leads to a companion matrix with trace $=-B$ and det $=C$.

## Problem Set 6.3

1 Find all solutions $\boldsymbol{y} = c_1 e^{\lambda_1 t}\boldsymbol{x}_1 + c_2 e^{\lambda_2 t}\boldsymbol{x}_2$ to $\boldsymbol{y}' = \begin{bmatrix} 3 & 1 \\ 3 & 5 \end{bmatrix}\boldsymbol{y}$. Which solution starts from $\boldsymbol{y}(0) = c_1\boldsymbol{x}_1 + c_2\boldsymbol{x}_2 = (2, 2)$?

2 Find two solutions of the form $y = e^{\lambda t}\boldsymbol{x}$ to $\boldsymbol{y}' = \begin{bmatrix} 3 & 10 \\ 2 & 4 \end{bmatrix}\boldsymbol{y}$.

3 If $a \neq d$, find the eigenvalues and eigenvectors and the complete solution to $\boldsymbol{y}' = A\boldsymbol{y}$. This equation is stable when $a$ and $d$ are ______.

$$\boldsymbol{y}' = \begin{bmatrix} a & b \\ 0 & d \end{bmatrix}\boldsymbol{y}.$$

4 If $a \neq -b$, find the solutions $e^{\lambda_1 t}\boldsymbol{x}_1$ and $e^{\lambda_2 t}\boldsymbol{x}_2$ to $y' = A\boldsymbol{y}$:

$$A = \begin{bmatrix} a & b \\ a & b \end{bmatrix}. \quad \text{Why is } \boldsymbol{y}' = A\boldsymbol{y} \text{ not stable?}$$

5 Find the eigenvalues $\lambda_1, \lambda_2, \lambda_3$ and the eigenvectors $\boldsymbol{x}_1, \boldsymbol{x}_2, \boldsymbol{x}_3$ of $A$. Write $\boldsymbol{y}(0) = (0, 1, 0)$ as a combination $c_1\boldsymbol{x}_1 + c_2\boldsymbol{x}_2 + c_3\boldsymbol{x}_3 = V\boldsymbol{c}$ and solve $\boldsymbol{y}' = A\boldsymbol{y}$. What is the limit of $\boldsymbol{y}(t)$ as $t \to \infty$ (the steady state)? *Steady states come from* $\lambda = 0$.

$$A = \begin{bmatrix} -1 & 1 & 0 \\ 1 & -2 & 1 \\ 0 & 1 & -1 \end{bmatrix}.$$

6 The simplest 2 by 2 matrix without two independent eigenvectors has $\lambda = 0, 0$:

$$\begin{bmatrix} y_1 \\ y_2 \end{bmatrix}' = A\boldsymbol{y} = \begin{bmatrix} 0 & 1 \\ 0 & 0 \end{bmatrix}\begin{bmatrix} y_1 \\ y_2 \end{bmatrix} \text{ has a first solution } \begin{bmatrix} y_1 \\ y_2 \end{bmatrix} = e^{0t}\begin{bmatrix} 1 \\ 0 \end{bmatrix}.$$

Find a second solution to these equations $y_1{}' = y_2$ and $y_2{}' = 0$. That second solution starts with $t$ times the first solution to give $y_1 = t$. What is $y_2$?

**Note** A complete discussion of $\boldsymbol{y}' = A\boldsymbol{y}$ for all cases of repeated $\lambda$'s would involve the *Jordan form* of $A$: too technical. Section 6.4 shows that a triangular form is sufficient, as Problems 6 and 8 confirm. We can solve for $y_2$ and then $y_1$.

7 Find two $\lambda$'s and $\boldsymbol{x}$'s so that $\boldsymbol{y} = e^{\lambda t}\boldsymbol{x}$ solves

$$\frac{d\boldsymbol{y}}{dt} = \begin{bmatrix} 4 & 3 \\ 0 & 1 \end{bmatrix}\boldsymbol{y}.$$

What combination $\boldsymbol{y} = c_1 e^{\lambda_1 t}\boldsymbol{x}_1 + c_2 e^{\lambda_2 t}\boldsymbol{x}_2$ starts from $\boldsymbol{y}(0) = (5, -2)$?

**8** Solve Problem 7 for $\boldsymbol{y} = (y, z)$ by back substitution, $z$ before $y$:

$$\text{Solve } \frac{dz}{dt} = z \text{ from } z(0) = -2. \text{Then solve } \frac{dy}{dt} = 4y + 3z \text{ from } y(0) = 5.$$

The solution for $y$ will be a combination of $e^{4t}$ and $e^t$. The $\lambda$'s are 4 and 1.

**9** (a) If every column of $A$ adds to zero, why is $\lambda = 0$ an eigenvalue?

(b) With negative diagonal and positive off-diagonal adding to zero, $\boldsymbol{y}' = A\boldsymbol{y}$ will be a "continuous" Markov equation. Find the eigenvalues and eigenvectors, and the *steady state* as $t \to \infty$:

$$\text{Solve } \frac{d\boldsymbol{y}}{dt} = \begin{bmatrix} -2 & 3 \\ 2 & -3 \end{bmatrix} \boldsymbol{y} \quad \text{with} \quad \boldsymbol{y}(0) = \begin{bmatrix} 4 \\ 1 \end{bmatrix}. \quad \text{What is } \boldsymbol{y}(\infty)\,?$$

**10** A door is opened between rooms that hold $v(0) = 30$ people and $w(0) = 10$ people. The movement between rooms is proportional to the difference $v - w$:

$$\frac{dv}{dt} = w - v \quad \text{and} \quad \frac{dw}{dt} = v - w.$$

Show that the total $v + w$ is constant (40 people). Find the matrix in $d\boldsymbol{y}/dt = A\boldsymbol{y}$ and its eigenvalues and eigenvectors. What are $v$ and $w$ at $t = 1$ and $t = \infty$?

**11** Reverse the diffusion of people in Problem 10 to $d\boldsymbol{z}/dt = -A\boldsymbol{z}$:

$$\frac{dv}{dt} = v - w \quad \text{and} \quad \frac{dw}{dt} = w - v.$$

The total $v + w$ still remains constant. How are the $\lambda$'s changed now that $A$ is changed to $-A$? But show that $v(t)$ grows to infinity from $v(0) = 30$.

**12** $A$ has real eigenvalues but $B$ has complex eigenvalues:

$$A = \begin{bmatrix} a & 1 \\ 1 & a \end{bmatrix} \quad B = \begin{bmatrix} b & -1 \\ 1 & b \end{bmatrix} \quad (a \text{ and } b \text{ are real})$$

Find the stability conditions on $a$ and $b$ so that all solutions of $d\boldsymbol{y}/dt = A\boldsymbol{y}$ and $d\boldsymbol{z}/dt = B\boldsymbol{z}$ approach zero as $t \to \infty$.

**13** Suppose $P$ is the projection matrix onto the 45° line $y = x$ in $\mathbf{R}^2$. Its eigenvalues are 1 and 0 with eigenvectors $(1, 1)$ and $(1, -1)$. If $d\boldsymbol{y}/dt = -P\boldsymbol{y}$ (notice minus sign) can you find the limit of $\boldsymbol{y}(t)$ at $t = \infty$ starting from $\boldsymbol{y}(0) = (3, 1)$?

**14** The rabbit population shows fast growth (from $6r$) but loss to wolves (from $-2w$). The wolf population always grows in this model ($-w^2$ would control wolves):

$$\frac{dr}{dt} = 6r - 2w \quad \text{and} \quad \frac{dw}{dt} = 2r + w.$$

Find the eigenvalues and eigenvectors. If $r(0) = w(0) = 30$ what are the populations at time $t$? After a long time, what is the ratio of rabbits to wolves?

**15** (a) Write $(4, 0)$ as a combination $c_1\boldsymbol{x}_1 + c_2\boldsymbol{x}_2$ of these two eigenvectors of $A$:

$$\begin{bmatrix} 0 & 1 \\ -1 & 0 \end{bmatrix}\begin{bmatrix} 1 \\ i \end{bmatrix} = i\begin{bmatrix} 1 \\ i \end{bmatrix} \qquad \begin{bmatrix} 0 & 1 \\ -1 & 0 \end{bmatrix}\begin{bmatrix} 1 \\ -i \end{bmatrix} = -i\begin{bmatrix} 1 \\ -i \end{bmatrix}.$$

(b) The solution to $d\boldsymbol{y}/dt = A\boldsymbol{y}$ starting from $(4, 0)$ is $c_1 e^{it}\boldsymbol{x}_1 + c_2 e^{-it}\boldsymbol{x}_2$. Substitute $e^{it} = \cos t + i \sin t$ and $e^{-it} = \cos t - i \sin t$ to find $\boldsymbol{y}(t)$.

**Questions 16–19 reduce second-order equations to first-order systems for $(y, y')$.**

**16** Find $A$ to change the scalar equation $y'' = 5y' + 4y$ into a vector equation for $\boldsymbol{y} = (y, y')$:

$$\frac{d\boldsymbol{y}}{dt} = \begin{bmatrix} y' \\ y'' \end{bmatrix} = \begin{bmatrix} \quad & \quad \\ \quad & \quad \end{bmatrix}\begin{bmatrix} y \\ y' \end{bmatrix} = A\boldsymbol{y}.$$

What are the eigenvalues of $A$? Find them also by substituting $y = e^{\lambda t}$ into $y'' = 5y' + 4y$.

**17** Substitute $y = e^{\lambda t}$ into $y'' = 6y' - 9y$ to show that $\lambda = 3$ is a repeated root. This is trouble; we need a second solution after $e^{3t}$. The matrix equation is

$$\frac{d}{dt}\begin{bmatrix} y \\ y' \end{bmatrix} = \begin{bmatrix} 0 & 1 \\ -9 & 6 \end{bmatrix}\begin{bmatrix} y \\ y' \end{bmatrix}.$$

Show that this matrix has $\lambda = 3, 3$ and only one line of eigenvectors. *Trouble here too.* Show that the second solution to $y'' = 6y' - 9y$ is $y = te^{3t}$.

**18** (a) Write down two familiar functions that solve the equation $d^2y/dt^2 = -9y$. Which one starts with $y(0) = 3$ and $y'(0) = 0$?

(b) This second-order equation $y'' = -9y$ produces a vector equation $\boldsymbol{y}' = A\boldsymbol{y}$:

$$\boldsymbol{y} = \begin{bmatrix} y \\ y' \end{bmatrix} \qquad \frac{d\boldsymbol{y}}{dt} = \begin{bmatrix} y' \\ y'' \end{bmatrix} = \begin{bmatrix} 0 & 1 \\ -9 & 0 \end{bmatrix}\begin{bmatrix} y \\ y' \end{bmatrix} = A\boldsymbol{y}.$$

Find $\boldsymbol{y}(t)$ by using the eigenvalues and eigenvectors of $A$: $\boldsymbol{y}(0) = (3, 0)$.

**19** If $c$ is not an eigenvalue of $A$, substitute $\boldsymbol{y} = e^{ct}\boldsymbol{v}$ and find a particular solution to $d\boldsymbol{y}/dt = A\boldsymbol{y} - e^{ct}\boldsymbol{b}$. How does it break down when $c$ is an eigenvalue of $A$?

**20** A particular solution to $d\boldsymbol{y}/dt = A\boldsymbol{y} - \boldsymbol{b}$ is $\boldsymbol{y}_p = A^{-1}\boldsymbol{b}$, if $A$ is invertible. The usual solutions to $d\boldsymbol{y}/dt = A\boldsymbol{y}$ give $\boldsymbol{y}_n$. Find the complete solution $\boldsymbol{y} = \boldsymbol{y}_p + \boldsymbol{y}_n$:

(a) $\dfrac{dy}{dt} = y - 4$ (b) $\dfrac{d\boldsymbol{y}}{dt} = \begin{bmatrix} 1 & 0 \\ 1 & 1 \end{bmatrix}\boldsymbol{y} - \begin{bmatrix} 4 \\ 6 \end{bmatrix}.$

**21** Find a matrix $A$ to illustrate each of the unstable regions in the stability picture:

(a) $\lambda_1 < 0$ and $\lambda_2 > 0$ (b) $\lambda_1 > 0$ and $\lambda_2 > 0$ (c) $\lambda = a \pm ib$ with $a > 0$.

**22** Which of these matrices are stable ? Then Re $\lambda < 0$, trace $< 0$, and det $> 0$.

$$A_1 = \begin{bmatrix} -2 & -3 \\ -4 & -5 \end{bmatrix} \quad A_2 = \begin{bmatrix} -1 & -2 \\ -3 & -6 \end{bmatrix} \quad A_3 = \begin{bmatrix} -1 & 2 \\ -3 & -6 \end{bmatrix}.$$

**23** For an $n$ by $n$ matrix with trace$(A) = T$ and det$(A) = D$, find the trace and determinant of $-A$. Why is $\boldsymbol{z}' = -A\boldsymbol{z}$ unstable whenever $\boldsymbol{y}' = A\boldsymbol{y}$ is stable ?

**24** (a) For a real 3 by 3 matrix with stable eigenvalues (Re $\lambda < 0$), show that trace $< 0$ and det $< 0$. Either three real negative $\lambda$ or $\lambda_2 = \overline{\lambda}_1$ and $\lambda_3$ is real.

(b) The trace and determinant of a 3 by 3 matrix do not determine all three eigenvalues ! Show that $A$ is unstable even with trace $< 0$ and determinant $< 0$ :

$$A = \begin{bmatrix} 1 & 2 & 3 \\ 0 & 1 & 4 \\ 0 & 0 & -5 \end{bmatrix}.$$

**25** You might think that $\boldsymbol{y}' = -A^2\boldsymbol{y}$ would always be stable because you are squaring the eigenvalues of $A$. But why is that equation unstable for $A = \begin{bmatrix} 0 & 1 \\ -1 & 0 \end{bmatrix}$?

**26** Find the three eigenvalues of $A$ and the three roots of $s^3 - s^2 + s - 1 = 0$ (including $s = 1$). The equation $y''' - y'' + y' - y = 0$ becomes

$$\begin{bmatrix} y \\ y' \\ y'' \end{bmatrix}' = \begin{bmatrix} 0 & 1 & 0 \\ 0 & 0 & 1 \\ 1 & -1 & 1 \end{bmatrix} \begin{bmatrix} y \\ y' \\ y'' \end{bmatrix} \quad \text{or } \boldsymbol{z}' = A\boldsymbol{z}.$$

Each eigenvalue $\lambda$ has an eigenvector $\boldsymbol{x} = (1, \lambda, \lambda^2)$.

**27** Find the two eigenvalues of $A$ and the double root of $s^2 + 6s + 9 = 0$ :

$$y'' + 6y' + 9y = 0 \text{ becomes } \begin{bmatrix} y \\ y' \end{bmatrix}' = \begin{bmatrix} 0 & 1 \\ 9 & 6 \end{bmatrix} \begin{bmatrix} y \\ y' \end{bmatrix} \quad \text{or } \boldsymbol{z}' = A\boldsymbol{z}.$$

The repeated eigenvalue gives only one solution $\boldsymbol{z} = e^{\lambda t}\boldsymbol{x}$. Find a second solution $\boldsymbol{z}$ from the second solution $y = te^{\lambda t}$.

**28** Explain why a 3 by 3 companion matrix has eigenvectors $\boldsymbol{x} = (\boldsymbol{1}, \boldsymbol{\lambda}, \boldsymbol{\lambda}^2)$.

*First Way*: If the first component is $x_1 = 1$, the first row of $A\boldsymbol{x} = \lambda\boldsymbol{x}$ gives the second component $x_2 =$ _____. Then the second row of $A\boldsymbol{x} = \lambda\boldsymbol{x}$ gives the third component $x_3 = \lambda^2$.

*Second Way*: $\boldsymbol{y}' = A\boldsymbol{y}$ starts with $y_1' = y_2$ and $y_2' = y_3$. $\boldsymbol{y} = e^{\lambda t}\boldsymbol{x}$ solves those equations. At $t = 0$ the equations become $\lambda x_1 = x_2$ and _____.

**29** Find $A$ to change the scalar equation $y'' = 5y' - 4y$ into a vector equation for $z = (y, y')$:

$$\frac{dz}{dt} = \begin{bmatrix} y' \\ y'' \end{bmatrix} = \begin{bmatrix} \quad & \quad \\ \quad & \quad \end{bmatrix} \begin{bmatrix} y \\ y' \end{bmatrix} = Az.$$

What are the eigenvalues of the companion matrix $A$? Find them also by substituting $y = e^{\lambda t}$ into $y'' = 5y' - 4y$.

**30** (a) Write down two familiar functions that solve the equation $d^2y/dt^2 = -9y$. Which one starts with $y(0) = 3$ and $y'(0) = 0$?

(b) This second-order equation $y'' = -9y$ produces a vector equation $z' = Az$:

$$z = \begin{bmatrix} y \\ y' \end{bmatrix} \qquad \frac{dz}{dt} = \begin{bmatrix} y' \\ y'' \end{bmatrix} = \begin{bmatrix} 0 & 1 \\ -9 & 0 \end{bmatrix} \begin{bmatrix} y \\ y' \end{bmatrix} = Az.$$

Find $z(t)$ by using the eigenvalues and eigenvectors of $A$: $z(0) = (3, 0)$.

**31** (a) Change the third order equation $y''' - 2y'' - y' + 2y = 0$ to a first order system $z' = Az$ for the unknown $z = (y, y', y'')$. The companion matrix $A$ is 3 by 3.

(b) Substitute $y = e^{\lambda t}$ and also find det $(A - \lambda I)$. Those lead to the same $\lambda$'s.

(c) One root is $\lambda = 1$. Find the other roots and these complete solutions:

$$y = c_1 e^{\lambda_1 t} + c_2 e^{\lambda_2 t} + c_3 e^{\lambda_3 t} \qquad z = C_1 e^{\lambda_1 t} \boldsymbol{x}_1 + C_2 e^{\lambda_2 t} \boldsymbol{x}_2 + C_3 e^{\lambda_3 t} \boldsymbol{x}_3.$$

**32** These companion matrices have $\lambda = 2, 1$ and $\lambda = 4, 1$. Find their eigenvectors:

$$A = \begin{bmatrix} 0 & 1 \\ -2 & 3 \end{bmatrix} \text{ and } B = \begin{bmatrix} 0 & 1 \\ -4 & 5 \end{bmatrix} \qquad \text{Notice trace and determinant!}$$

## 6.4 The Exponential of a Matrix

This section expresses the solution to a system $dy/dt = Ay$ in a different way. Instead of combining eigenvector solutions $e^{\lambda t}x$, the new form uses the **matrix exponential** $e^{At}$ :

| | | |
|---|---|---|
| **Solution to** $y' = Ay$ | $y(t) = e^{At}y(0)$ | (1) |

This matrix $e^{At}$ matches $e^{at}$ when $n = 1$ : the scalar case. For matrices, we can still write the exponential as an infinite series. In one way this is better than depending on eigenvectors — but maybe not in practice :

*Advantage* We don't need $n$ independent eigenvectors for $e^{At}$.
*Disadvantage* An infinite series is usually not so practical.

The new way produces one short symbol $e^{At}$ for the "solution matrix." Still we often compute in the old way with eigenvectors. This is like a linear system $Av = b$, where $A^{-1}$ is the solution matrix but we compute $v$ by elimination.

For large matrices, $y' = Ay$ uses completely different ways — often finite differences.

### The Exponential Series

The most direct way to define the matrix $e^{At}$ is by an infinite series of powers of $A$ :

| | | |
|---|---|---|
| **Matrix exponential** | $e^{At} = I + At + \frac{1}{2}(At)^2 + \cdots = \sum_{n=0}^{\infty}(At)^n/n!$ | (2) |

This series always converges, like the scalar case $e^{at}$ in Chapter 1. $e^{At}$ is the great function of matrix calculus. The quickly growing factors $n!$ still assure convergence. The two key properties of $e^{at}$ continue to hold when $a$ becomes a matrix $A$ :

1. **The derivative of** $e^{At}$ **is** $Ae^{At}$ 2. $\left(e^{At}\right)\left(e^{AT}\right) = e^{A(t+T)}$

Property **1** says that $y(t) = e^{At}y(0)$ has derivative $y' = Ay$. And $y(t)$ starts correctly from $y(0)$ at $t = 0$, since $e^{A0} = I$ from equation (2). **So** $e^{At}y(0)$ **solves** $y' = Ay$.

Suppose we set $T = -t$ in Property **2**. Then $t + T = 0$ :

$$\textbf{The inverse of } e^{At} \textbf{ is } e^{-At} \qquad e^{At}e^{AT} = e^0 = I \text{ when } T \text{ is } -t. \tag{3}$$

$e^{At}$ has properties **1** and **2** even if $A$ cannot be diagonalized. When $A$ does have $n$ independent eigenvectors, the same eigenvector matrix $V$ diagonalizes $A$ and $e^{At}$. The next page shows that $e^{At} = Ve^{\Lambda t}V^{-1}$ : this is the good way to find $e^{At}$.

Assume $A$ has $n$ independent eigenvectors, so it is diagonalizable. Substitute $A = V\Lambda V^{-1}$ into the series for $e^{At}$. Whenever $V\Lambda V^{-1}V\Lambda V^{-1}$ appears, take out $V^{-1}V = I$.

$$\begin{array}{lll} \textbf{Use the series} & e^{At} & = I + V\Lambda V^{-1}t + \frac{1}{2}(V\Lambda V^{-1}t)(V\Lambda V^{-1}t) + \cdots \\ \textbf{Factor out } V \textbf{ and } V^{-1} & & = V\left[I + \Lambda t + \frac{1}{2}(\Lambda t)^2 + \cdots\right]V^{-1} \\ \textbf{Diagonalize } e^{At} & e^{At} & = \boldsymbol{V e^{\Lambda t} V^{-1}}. \end{array} \tag{4}$$

The numbers $e^{\lambda_i t}$ are on the diagonal of $e^{\Lambda t}$. Multiply $Ve^{\Lambda t}V^{-1}\boldsymbol{y}(0)$ to see $\boldsymbol{y}(t)$.

*Second Proof* $e^{At}$ has the same eigenvectors $\boldsymbol{x}$ as $A$. **The eigenvalues of $e^{At}$ are $e^{\lambda t}$** :

$$A^n\boldsymbol{x} = \lambda^n\boldsymbol{x} \quad \text{leads to} \quad e^{At}\boldsymbol{x} = \left(1 + \lambda t + \frac{1}{2}(\lambda t)^2 + \cdots\right)\boldsymbol{x} = e^{\lambda t}\boldsymbol{x}. \tag{5}$$

So the same eigenvector matrix $V$ diagonalizes both $A$ and $e^{At}$. The eigenvalue matrix for $e^{At}$ is diag $(e^{\lambda_1 t}, \ldots, e^{\lambda_n t})$. This is exactly $e^{\Lambda t}$. Again $\boldsymbol{e^{At} = Ve^{\Lambda t}V^{-1}}$.

The eigenvalues of the inverse matrix $e^{-At}$ are $e^{-\lambda t}$. This is $1/e^{\lambda t}$ as expected.

**Example 1** The rotation matrix $A = \begin{bmatrix} 0 & 1 \\ -1 & 0 \end{bmatrix}$ has eigenvalues $\lambda_1 = i$ and $\lambda_2 = -i$ :

$$e^{At} = Ve^{\Lambda t}V^{-1} = \begin{bmatrix} 1 & 1 \\ i & -i \end{bmatrix}\begin{bmatrix} e^{it} & 0 \\ 0 & e^{-it} \end{bmatrix}\frac{1}{2}\begin{bmatrix} 1 & -i \\ 1 & i \end{bmatrix} = \begin{bmatrix} \cos t & \sin t \\ -\sin t & \cos t \end{bmatrix}. \tag{6}$$

This produces $e^{At}$ without adding up an infinite series. We could also begin the series:

$$\begin{bmatrix} 1 & 0 \\ 0 & 1 \end{bmatrix} + \begin{bmatrix} 0 & t \\ -t & 0 \end{bmatrix} + \frac{1}{2}\begin{bmatrix} -t^2 & 0 \\ 0 & -t^2 \end{bmatrix} + \frac{1}{6}\begin{bmatrix} 0 & -t^3 \\ t^3 & 0 \end{bmatrix} = \begin{bmatrix} 1 - \frac{1}{2}t^2 & t - \frac{1}{6}t^3 \\ -t + \frac{1}{6}t^3 & 1 - \frac{1}{2}t^2 \end{bmatrix}.$$

The cosine series starts with $1 - \frac{1}{2}t^2$. The sine series starts with $t - \frac{1}{6}t^3$. The full series for $e^{At}$ gives the full series for $\cos t$ and $\sin t$ : very exceptional.

**Example 1 continued** What is the solution to $d\boldsymbol{y}/dt = A\boldsymbol{y}$ with $\boldsymbol{y}(0) = (1, 0)$?

*Answer* We know that $\boldsymbol{y}(t) = (y_1, y_2)$ is $e^{At}\boldsymbol{y}(0)$, and equation (6) gives $e^{At}$ :

$$\begin{array}{l} \boldsymbol{y_1' = y_2} \\ \boldsymbol{y_2' = -y_1} \end{array} \qquad \begin{bmatrix} y_1(t) \\ y_2(t) \end{bmatrix} = \begin{bmatrix} \cos t & \sin t \\ -\sin t & \cos t \end{bmatrix}\begin{bmatrix} 1 \\ 0 \end{bmatrix} = \begin{bmatrix} \cos t \\ -\sin t \end{bmatrix}. \tag{7}$$

Right! The derivative of $\cos t$ is $-\sin t$. The derivative of $y_2 = -\sin t$ is $-\cos t$. The equations $\boldsymbol{y}' = A\boldsymbol{y}$ are satisfied. When $t = 0$, we start correctly at $\boldsymbol{y}(0) = (1, 0)$.

This solution is important in physics and engineering. The point $\boldsymbol{y}(t)$ is on the unit circle $y_1^2 + y_2^2 = \cos^2 t + \sin^2 t = 1$. It goes around the circle with constant speed. The second derivative (acceleration) is $\boldsymbol{y}'' = (-\sin t, -\cos t)$ because $A^2 = -I$. This vector $\boldsymbol{y}''$ points in to the center $(0, 0)$. We have a planet going in a circle around the sun.

**Example 2** Suppose $A$ is triangular but we can't diagonalize it (only one eigenvector):

$$\boldsymbol{y}' = A\boldsymbol{y} = \begin{bmatrix} 1 & 1 \\ 0 & 1 \end{bmatrix} \begin{bmatrix} y_1 \\ y_2 \end{bmatrix} \qquad \begin{aligned} y_1' &= y_1 + y_2 \\ y_2' &= 0 + y_2 \end{aligned} \tag{8}$$

$A$ has no invertible eigenvector matrix $V$. How to find $\boldsymbol{y}(t)$ without two eigenvectors?

*Solution* Since $A$ is triangular, back substitution will solve $\boldsymbol{y}' = A\boldsymbol{y}$. Begin by solving the last equation $y_2{}' = y_2$. Then solve for $y_1$:

$$y_2(t) = e^t y_2(0) \qquad \text{Then } y_1{}' = y_1 + y_2 = y_1 + e^t y_2(0)$$

That equation for $y_1$ has a source term $q(t) = e^t y_2(0)$. Chapter 1 found the solution $y_1(t)$:

$$e^t y_1(0) + \int_0^t e^{t-s} q(s)\, ds = e^t y_1(0) + e^t y_2(0) \int_0^t ds = e^t y_1(0) + \boldsymbol{te^t}\, y_2(0). \tag{9}$$

**At last we have a reason for the extra factor $\boldsymbol{t}$.** The natural growth rate of $y_1$ is also the growth rate of $y_2$. This leads to "resonance" in $y_1{}' = y_1 + y_2$, and the growth of $te^t$ is extra fast. We saw resonance with $te^{st}$ in Chapter 2. Now we are seeing the $t$ in $e^{At}$.

$$\begin{aligned} y_1(t) &= \boldsymbol{e^t}\, y_1(0) + \boldsymbol{te^t}\, y_2(0) \\ y_2(t) &= \qquad\qquad\quad \boldsymbol{e^t}\, y_2(0) \end{aligned} \qquad \text{means that} \quad \boldsymbol{e^{At}} = \begin{bmatrix} \boldsymbol{e^t} & \boldsymbol{te^t} \\ 0 & \boldsymbol{e^t} \end{bmatrix}. \tag{10}$$

**Example 2 (using $\boldsymbol{e^{At}}$)** For this triangular matrix $A$, we can also add the series for $e^{At}$:

$$e^{At} = I + At + \frac{1}{2}(At)^2 + \frac{1}{6}(At)^3 + \cdots$$

$$= \begin{bmatrix} 1 & 0 \\ 0 & 1 \end{bmatrix} + \begin{bmatrix} t & \boldsymbol{t} \\ 0 & t \end{bmatrix} + \frac{1}{2}\begin{bmatrix} t^2 & \boldsymbol{2t^2} \\ 0 & t^2 \end{bmatrix} + \frac{1}{6}\begin{bmatrix} t^3 & \boldsymbol{3t^3} \\ 0 & t^3 \end{bmatrix} + \cdots \tag{11}$$

$$= \begin{bmatrix} e^t & \boldsymbol{te^t} \\ 0 & e^t \end{bmatrix} \qquad \text{because} \quad te^t = t + t^2 + \frac{1}{2}t^3 + \cdots$$

All the powers of a triangular matrix are triangular. So the diagonal entries of $A$ give the diagonal entries of $e^{At}$. Those are the eigenvalues of $e^{At}$ and here they are both $e^t$.

## Source Term in $\boldsymbol{y}' = A\boldsymbol{y} + \boldsymbol{q}$

We can solve $y' = ay + q$ for a single equation (1 by 1). Now allow a matrix $A$:

$$\textit{Old} \quad y(t) = e^{at} y(0) + \frac{e^{at} - 1}{a} q \qquad \textit{New} \quad \frac{d\boldsymbol{y}}{dt} = A\boldsymbol{y} + \boldsymbol{q} \tag{12}$$

**Change** $a$ **to** $A$! For constant $\boldsymbol{q}$, that is the only change in the formula for $\boldsymbol{y}$:

$$\boxed{\boldsymbol{y}' = A\boldsymbol{y} + \boldsymbol{q} \qquad \text{is solved by} \qquad \boldsymbol{y}(t) = e^{At}\boldsymbol{y}(0) + (e^{At} - I)A^{-1}\boldsymbol{q}.} \tag{13}$$

The derivative of $\boldsymbol{y}$ produces $A\boldsymbol{y}$, except for the constant $A^{-1}\boldsymbol{q}$ with derivative = zero. But this term $A^{-1}\boldsymbol{q}$ disappears safely in $A\boldsymbol{y}+\boldsymbol{q}$, because $-AA^{-1}\boldsymbol{q}+\boldsymbol{q}=\mathbf{0}$.

Chapter 1 was built on the growth factor $e^{at}$ in the integral for $y_p$. Now it is $e^{At}$ !

**Principle** *Each input $\boldsymbol{q}(s)$ has growth factor $e^{A(t-s)}$ from time $s$ to time $t$.* For constant $A$, the growth (or decay) over time $t-s$ is just multiplication by $e^{A(t-s)}$ :

$$\boxed{\boldsymbol{y}'=A\boldsymbol{y}+\boldsymbol{q}(t) \quad \text{is solved by} \quad \boldsymbol{y}(t)=e^{At}\boldsymbol{y}(0)+\int_0^t e^{A(t-s)}\boldsymbol{q}(s)\,ds.} \tag{14}$$

## Similar Matrices $A$ and $B$

To end this section, I will solve $\boldsymbol{y}'=A\boldsymbol{y}$ in one more way. Same result, new approach.

*Change of variables.* Write $\boldsymbol{y}=V\boldsymbol{z}$ to change from $\boldsymbol{y}(t)$ to the new variable $\boldsymbol{z}(t)$.

$$\frac{d\boldsymbol{y}}{dt}=A\boldsymbol{y} \quad \text{becomes} \quad V\frac{d\boldsymbol{z}}{dt}=AV\boldsymbol{z} \quad \text{which is} \quad \frac{d\boldsymbol{z}}{dt}=\boldsymbol{V^{-1}AV}\boldsymbol{z}. \tag{15}$$

The matrix $A$ has changed to $B=V^{-1}AV$. Then the solution for $\boldsymbol{z}$ involves $e^{Bt}$ :

$$\boldsymbol{B=V^{-1}AV} \qquad \boldsymbol{z}'=B\boldsymbol{z} \text{ produces } \boldsymbol{z}(t)=e^{Bt}\boldsymbol{z}(0) \tag{16}$$

Changing back to $\boldsymbol{y}=V\boldsymbol{z}$, that solution becomes $\boldsymbol{y}(t)=Ve^{Bt}\boldsymbol{z}(0)=\boldsymbol{Ve^{Bt}V^{-1}y(0)}$.

$$\textbf{The exponential of} \quad A=VBV^{-1} \quad \textbf{is} \quad e^{At}=Ve^{Bt}V^{-1}. \tag{17}$$

*Special case* : When $V$ is the eigenvector matrix, $B$ is the eigenvalue matrix $\Lambda$.

Here is my point. Equation (17) is true for any invertible matrix $V$. Choosing the eigenvector matrix of $A$ makes $B$ diagonal; in fact $B=V^{-1}AV=\Lambda$. This is the outstanding choice for $V$, to produce $B=\Lambda$ when $A$ has $n$ independent eigenvectors. But *any invertible $V$ is now allowed*, and we have a name for $B$ : **similar matrix**.

**Every matrix $B=V^{-1}AV$ is "similar" to $A$. They have the same eigenvalues.**

I can quickly prove that eigenvalues stay unchanged. **Eigenvectors change to $\boldsymbol{u}=V^{-1}\boldsymbol{x}$** :

$$\text{If } A\boldsymbol{x}=\lambda\boldsymbol{x} \quad \text{then} \quad V^{-1}A\boldsymbol{x}=\lambda V^{-1}\boldsymbol{x} \quad \text{which is} \quad V^{-1}AV\boldsymbol{u}=B\boldsymbol{u}=\lambda\boldsymbol{u}. \tag{18}$$

By allowing all invertible $V$, we have a whole family of matrices $B=V^{-1}AV$. All are similar to $A$, all have the same eigenvalues as $A$, only the eigenvectors change with $V$.

In case $A$ cannot be diagonalized, a good choice of $V$ makes $B$ upper triangular. $V$ is not easy to compute, but it greatly simplifies the problem. Example 2 showed how $\boldsymbol{z}(t)$ comes from back substitution in $\boldsymbol{z}'=\boldsymbol{B}\boldsymbol{z}$. Then $\boldsymbol{y}(t)=V\boldsymbol{z}(t)$ solves $y'=Ay$ without $n$ independent eigenvectors of $A$.

## Fundamental Matrices (Optional Topic)

A linear system $d\boldsymbol{y}/dt = A(t)\boldsymbol{y}$ is completely solved when you have $n$ independent solutions $\boldsymbol{y}_1(t)$ to $\boldsymbol{y}_n(t)$. Put those solutions into the columns of an $n$ by $n$ matrix $M(t)$:

**Fundamental matrix** $\quad M(t) = \left[ \boldsymbol{y}_1(t) \ldots \boldsymbol{y}_n(t) \right] \quad$ has $\quad \dfrac{d\boldsymbol{M}}{d\boldsymbol{t}} = \boldsymbol{AM(t)}. \qquad (19)$

Every column of $dM/dt$ has $d\boldsymbol{y}/dt = A\boldsymbol{y}$. All columns together give $dM/dt = AM$.

"Linear independence" means that $M$ is invertible. The determinant of $M$ is not zero. This determinant $W(t)$ is called the "*Wronskian*" of the $n$ solutions in the columns of $M$:

$$\boxed{W(t) = \text{Wronskian of } \boldsymbol{y}_1(t), \ldots, \boldsymbol{y}_n(t) = \text{Determinant of } M(t).} \qquad (20)$$

The beautiful fact is this: **If the Wronskian starts from $W \neq 0$ at time $t = 0$, then $W(t) \neq 0$ for all $t$.** Independence at the start means independence forever. A combination $\boldsymbol{y}(t) = c_1\boldsymbol{y}_1(t) + \cdots + c_n\boldsymbol{y}_n(t)$ can only be zero at time $t$ if it started from $\boldsymbol{y}(0) = 0$. Solutions to $\boldsymbol{y}' = A\boldsymbol{y}$ don't hit zero! So $W(t) = 0$ requires $W(0) = 0$, as in this neat formula discussed in the Chapter 6 Notes (exponentials are never zero).

$$\frac{dW}{dt} = (\text{trace}\, A(t))W \quad \text{and then} \quad W(t) = e^{\int \text{trace}\, A(t)\, dt}\, W(0). \qquad (21)$$

*What are $M(t)$ and $W(t)$ for a second order equation $y'' + B(t)y' + C(t)y = 0$?* We know how to convert this to a first order system $\boldsymbol{y}' = A(t)\boldsymbol{y}$. The vector unknown is $\boldsymbol{y} = (y, y')$ and $A(t)$ is a companion matrix containing $-B(t)$ and $-C(t)$. The two independent solutions in the columns of $M(t)$ are $(y_1, y_1{}')$ and $(y_2, y_2{}')$:

$$\textbf{Matrix } M(t) = \begin{bmatrix} y_1 & y_2 \\ y_1{}' & y_2{}' \end{bmatrix} \quad \textbf{Wronskian } W(t) = \det M = y_1 y_2{}' - y_2 y_1{}'. \qquad (22)$$

Again $W(t) \neq 0$ is the test for $y_1$ and $y_2$ to be independent. The test is passed for all $t$ if $W(0) \neq 0$. In the mysterious formula (21), the trace of $A(t)$ is $-B(t)$.

You will naturally ask: What is this fundamental matrix $M(t)$? Why are we only seeing it now? One answer is that you already know the *growth factor* $G$ from Chapter 1: $M = G(0,t) = \exp\left(\int a(t)dt\right)$. For systems, you also know $M = e^{At}$. That is the perfect answer when $A$ is constant. $e^{At}$ is the best possible $M(t)$ because it starts from $M(0) = I$.

It is often hard to find $M(t)$ when the matrix $A$ depends on $t$ (then nothing is easy). We know that $\boldsymbol{y}' = A(t)\boldsymbol{y}$ has $n$ independent solutions $\boldsymbol{y}(t)$. But in most cases we don't know what those solutions are. The point of fundamental matrices is that the solution $y(t)$ comes directly from $M(t)$, when and if we know $M$:

$$\boxed{\boldsymbol{y}(t) = M(t)M(0)^{-1}\boldsymbol{y}(0) \quad \text{for any} \quad M(t)} \qquad (23)$$

Let me say a little more about constant $A$ and varying $A(t)$, and then stop.

**Constant $A$ with $n$ independent eigenvectors in $V$** We know $n$ solutions $\boldsymbol{y} = e^{\lambda t}\boldsymbol{x}$ :

$$\text{Put those } \boldsymbol{y}\text{'s into} \quad M(t) = \begin{bmatrix} e^{\lambda_1 t}\boldsymbol{x}_1 & e^{\lambda_2 t}\boldsymbol{x}_2 & \ldots & e^{\lambda_n t}\boldsymbol{x}_n \end{bmatrix} = Ve^{\Lambda t}.$$

How does this differ from $e^{At}$? You can see everything at $t = 0$, when this $M(t)$ is $V$. If you want the fundamental matrix that equals $I$ at $t = 0$, just multiply by $M(0)^{-1} = V^{-1}$:

When $A = V\Lambda V^{-1}$, the best fundamental matrix is $M = Ve^{\Lambda t}V^{-1}$ which is $e^{At}$.

**Time-varying $A(t)$ with time-varying eigenvectors** The equation $\boldsymbol{y}' = A(t)\boldsymbol{y}$ is more difficult. The next page shows how the expected solution formula fails. The chain rule goes wrong. Finding even one solution $\boldsymbol{y}_1(t)$ is a big challenge. The optimistic point is that if we can find $\boldsymbol{y}_1(t)$, then "variation of parameters" will lead us to $\boldsymbol{y}_2 = C(t)\boldsymbol{y}_1$.

Let me focus on a famous equation that has been studied by great mathematicians:

$$\boxed{\textbf{Bessel's equation} \qquad x^2\frac{d^2y}{dx^2} + x\frac{dy}{dx} + (x^2 - p^2)y = 0.} \tag{24}$$

The solutions are *Bessel functions of order* $p$. When the order is $p = \frac{1}{2}$, these solutions $y_1$ and $y_2$ are quite special (the variable $t$ is usually changed to $x$).

$$y_1(x) = \sqrt{\frac{2}{\pi x}}\sin x \quad \text{and} \quad y_2(x) = \sqrt{\frac{2}{\pi x}}\cos x \quad \text{go into} \quad M = \begin{bmatrix} y_1 & y_2 \\ y_1{}' & y_2{}' \end{bmatrix}$$

Those are independent solutions and the Wronskian $W = y_1 y_2' - y_2 y_1'$ is never zero.

The most important Bessel functions have $p = 0, 1, 2, \ldots$ and whole books are written about these functions. They are not simple! The first and most famous Bessel function is $y = J_0(x)$, with order $p = 0$:

$$J_0(x) = 1 - \frac{x^2}{2^2} + \frac{x^4}{2^2 4^2} - \frac{x^6}{2^2 4^2 6^2} + \cdots \qquad \text{resembles a damped cosine.}$$

The second solution $Y_0$, independent of $J_0$, blows up at $x = 0$. When you divide Bessel's equation (24) by $x^2$, so as to start the equation with $y''$, you see that its coefficients are singular: $1/x$ and $1 - p^2/x^2$ also blow up at $x = 0$: A singular point.

## Failure of a Formula

A single equation $dy/dt = a(t)y$ has a neat solution $y = e^{P(t)}y(0)$. We choose $P(t)$ as the integral of $a(t)$. By the chain rule, $dy/dt$ has the desired factor $a(t) = dP/dt$. I am very sorry to say that $\boldsymbol{y} = e^{P(t)}\boldsymbol{y}(0)$ fails for matrices $A(t)$ and systems $\boldsymbol{y}' = A(t)\boldsymbol{y}$.

There is no doubt that the derivative of the integral of time-varying $A(t)$ is $A(t)$. Even for matrices, this part is true:

$$\textbf{Fundamental Theorem of Calculus} \qquad \frac{d}{dt}\int_0^t A(s)\,ds = \frac{dP}{dt} = A(t). \tag{25}$$

When $A$ is a constant matrix, that integral is $P = At$ and its derivative is $A$. Then the derivative of $e^{At}$ is $Ae^{At}$. This whole section is built on that true statement. We hope that the same chain rule will give the answer when $A(t)$ is varying and not constant:

$$\text{The derivative of } G = \exp\left(\int_0^t A(s)\,ds\right) \text{ "should be" } A(t)G. \textit{ Not always!} \qquad (26)$$

When the matrix $A(t)$ is changing with time, the chain rule in (26) can let us down. This leaves no simple formula for $\boldsymbol{y}(t)$. How can things go wrong?

*The difficulty is that* $e^A$ *times* $e^B$ *may not be the same as* $e^{A+B}$. Problem 7 gives an example of $A$ and $B$. Those matrices do not satisfy $AB = BA$ and this destroys the rule for exponents. It is true that $e^A e^B = e^{A+B}$ when $AB = BA$, but not here.

Let me use those matrices in Problem 7 to construct a two-part example:

$$\boldsymbol{y}' = B\,\boldsymbol{y} \quad \text{for } t \le 1 \qquad \text{and then} \qquad \boldsymbol{y}' = A\boldsymbol{y} \quad \text{for } t > 1 \qquad (27)$$

Our time-varying matrix $A(t)$ jumps from $B$ to $A$ at $t = 1$. The integral of $A(t)$ is $P(t)$:

$$P(t) = \int_0^t A(s)\,ds = B\,t \text{ (for } t \le 1) \quad \text{and} \quad A(t-1) + B \text{ (for } t > 1). \qquad (28)$$

But the exponential of $P(t)$ does not solve our differential equation (27) at $t = 2$:

$$P(2) = \int_0^2 A(s)\,ds = A + B \quad \text{is correct but} \quad \boldsymbol{y}(2) = e^{A+B}\boldsymbol{y}(0) \quad \text{is } \textbf{wrong}.$$

*The correct answer is* $\boldsymbol{y}(2) = e^A e^B \boldsymbol{y}(0)$. *First* $B$ *then* $A$. The solution is $e^{Bt}\boldsymbol{y}(0)$ up to time $t = 1$, when $B$ changes to $A$. After $t = 1$ the solution is $e^{A(t-1)}e^B y(0)$.

**The chain rule in (26) is wrong, because $e^A e^B$ is different from $e^{A+B}$.**

## ■ REVIEW OF THE KEY IDEAS ■

1. The exponential of $At$ is $e^{At} = I + At + \frac{1}{2}(At)^2 + \frac{1}{6}(At)^3 + \cdots$
2. The solution to $\boldsymbol{y}' = A\boldsymbol{y}$ is $\boldsymbol{y}(t) = e^{At}\boldsymbol{y}(0)$. This is $Ve^{\Lambda t}V^{-1}\boldsymbol{y}(0)$ if $V^{-1}$ exists.
3. That solution is the same as $c_1 e^{\lambda_1 t}\boldsymbol{x}_1 + \cdots + c_n e^{\lambda_n t}\boldsymbol{x}_n$ with $\boldsymbol{c} = V^{-1}\boldsymbol{y}(0)$.
4. The solution to $\boldsymbol{y}' = A\boldsymbol{y} + \boldsymbol{q}$ (constant source) is $\boldsymbol{y}(t) = e^{At}\boldsymbol{y}(0) + (e^{At} - I)A^{-1}\boldsymbol{q}$.
5. All similar matrices $B = VAV^{-1}$ (*with any* $V$) have the same eigenvalues as $A$.
6. If $A(t)$ is time-varying, easy formulas for the fundamental matrix $M(t)$ will fail.

## ■ WORKED EXAMPLE ■

**Show that $y(t) = e^{At}y(0)$ is exactly $c_1e^{\lambda_1 t}x_1 + \cdots + c_ne^{\lambda_n t}x_n$ if $y(0) = Vc$.**

Step **1** Write $\boldsymbol{y}(0) = c_1\boldsymbol{x}_1 + \cdots + c_n\boldsymbol{x}_n$. This is $\begin{bmatrix} \boldsymbol{x}_1 & \cdots & \boldsymbol{x}_n \end{bmatrix}\begin{bmatrix} c_1 \\ \vdots \\ c_n \end{bmatrix} = V\boldsymbol{c}$.

Step **2** Starting from an eigenvector $\boldsymbol{x}$, the solution is $\boldsymbol{y} = ce^{\lambda t}\boldsymbol{x}$.

Step **3** Add those $n$ solutions to get $Ve^{\Lambda t}\boldsymbol{c} = Ve^{\Lambda t}V^{-1}\boldsymbol{y}(0) = \boldsymbol{e^{At}y(0)}$.

Here are those steps for a triangular matrix $A$. Suppose $\boldsymbol{y}(0) = (5, 3)$. First $\Lambda$ and $V$:

$$A = \begin{bmatrix} 1 & 1 \\ 0 & 2 \end{bmatrix} \quad \text{has} \quad \lambda_1 = 1 \text{ and } \boldsymbol{x}_1 = \begin{bmatrix} 1 \\ 0 \end{bmatrix} \quad \lambda_2 = 2 \text{ and } \boldsymbol{x}_2 = \begin{bmatrix} 1 \\ 1 \end{bmatrix}$$

Step **1** $\boldsymbol{y}(0) = \begin{bmatrix} 5 \\ 3 \end{bmatrix} = \mathbf{2}\begin{bmatrix} 1 \\ 0 \end{bmatrix} + \mathbf{3}\begin{bmatrix} 1 \\ 1 \end{bmatrix} = \begin{bmatrix} 1 & 1 \\ 0 & 1 \end{bmatrix}\begin{bmatrix} \mathbf{2} \\ \mathbf{3} \end{bmatrix} = V\boldsymbol{c}$.

Step **2** The separate solutions $ce^{\lambda t}\boldsymbol{x}$ from eigenvectors are $\boldsymbol{2e^t x_1}$ and $\boldsymbol{3e^{2t}x_2}$.

Step **3** The final $\boldsymbol{y}(t) = e^{At}\boldsymbol{y}(0) = Ve^{\Lambda t}V^{-1}\boldsymbol{y}(0)$ is the sum $2e^t\boldsymbol{x}_1 + 3e^{2t}\boldsymbol{x}_2$.

**Challenge** **Find $e^{At}$ for the companion matrices $\begin{bmatrix} 0 & 1 \\ -C & 0 \end{bmatrix}$ and $\begin{bmatrix} 0 & 1 \\ -C & -B \end{bmatrix}$.**
Their eigenvectors in $Ve^{\Lambda t}V^{-1}$ are always $(1, \lambda)$.

# Problem Set 6.4

**1** If $A\boldsymbol{x} = \lambda\boldsymbol{x}$, find an eigenvalue and an eigenvector of $e^{At}$ and also of $-e^{-At}$.

**2** (a) From the infinite series $e^{At} = I + At + \cdots$ show that its derivative is $Ae^{At}$.

(b) The series for $e^{At}$ ends quickly if $A = \begin{bmatrix} 0 & 1 \\ 0 & 0 \end{bmatrix}$ because $A^2 = \begin{bmatrix} 0 & 0 \\ 0 & 0 \end{bmatrix}$.

Find $e^{At}$ and take its derivative (which should agree with $Ae^{At}$).

**3** For $A = \begin{bmatrix} 1 & 1 \\ 0 & 2 \end{bmatrix}$ with eigenvectors in $V = \begin{bmatrix} 1 & 1 \\ 0 & 1 \end{bmatrix}$, compute $e^{At} = Ve^{\Lambda t}V^{-1}$.

**4** Why is $e^{(A+3I)t}$ equal to $e^{At}$ multiplied by $e^{3t}$?

**5** Why is $e^{A^{-1}}$ *not* the inverse of $e^A$? What is the correct inverse of $e^A$?

**6** Compute $A^n = \begin{bmatrix} 1 & c \\ 0 & 0 \end{bmatrix}^n$. Add the series to find $e^{At} = \begin{bmatrix} e^t & c(e^t - 1) \\ 0 & 1 \end{bmatrix}$.

**7** Find $e^A$ and $e^B$ by using Problem 6 for $c = 4$ and $c = -4$. Multiply to show that the matrices $e^A e^B$ and $e^B e^A$ and $e^{A+B}$ are all different.

$$A = \begin{bmatrix} 1 & 4 \\ 0 & 0 \end{bmatrix} \qquad B = \begin{bmatrix} 1 & -4 \\ 0 & 0 \end{bmatrix} \qquad A + B = \begin{bmatrix} 2 & 0 \\ 0 & 0 \end{bmatrix}.$$

**8** Multiply the first terms $I + A + \frac{1}{2}A^2$ of $e^A$ by the first terms $I + B + \frac{1}{2}B^2$ of $e^B$. Do you get the correct first three terms of $e^{A+B}$ ? *Conclusion*: $e^{A+B}$ is not always equal to $(e^A)(e^B)$. The exponent rule only applies when $AB = BA$.

**9** Write $A = \left[\begin{smallmatrix} 1 & 4 \\ 0 & 0 \end{smallmatrix}\right]$ in the form $V\Lambda V^{-1}$. Find $e^{At}$ from $Ve^{\Lambda t}V^{-1}$.

**10** Starting from $\boldsymbol{y}(0)$ the solution at time $t$ is $e^{At}\boldsymbol{y}(0)$. Go an additional time $t$ to reach $e^{At}\, e^{At}\boldsymbol{y}(0)$. Conclusion: $e^{At}$ times $e^{At}$ equals _____ .

**11** Diagonalize $A$ by $V$ and confirm this formula for $e^{At}$ by using $Ve^{\Lambda t}V^{-1}$:

$$A = \begin{bmatrix} 2 & 4 \\ 0 & 3 \end{bmatrix} \qquad e^{At} = \begin{bmatrix} e^{2t} & 4(e^{3t} - e^{2t}) \\ 0 & e^{3t} \end{bmatrix} \quad \text{At } t = 0 \text{ this matrix is \_\_\_\_\_ .}$$

**12** (a) Find $A^2$ and $A^3$ and $A^n$ for $A = \begin{bmatrix} 1 & 1 \\ 0 & 1 \end{bmatrix}$ with repeated eigenvalues $\lambda = 1, 1$.

(b) Add the infinite series to find $e^{At}$. (The $Ve^{\Lambda t}V^{-1}$ method won't work.)

**13** (a) Solve $\boldsymbol{y}' = A\boldsymbol{y}$ as a combination of eigenvectors of this matrix $A$:

$$\boldsymbol{y}' = \begin{bmatrix} 0 & 1 \\ 1 & 0 \end{bmatrix} \boldsymbol{y} \qquad \text{with } \boldsymbol{y}(0) = \begin{bmatrix} 3 \\ 5 \end{bmatrix}$$

(b) Write the equations as $y_1' = y_2$ and $y_2' = y_1$. Find an equation for $y_1''$ with $y_2$ eliminated. Solve for $y_1(t)$ and compare with part (a).

**14** Similar matrices $A$ and $B = V^{-1}AV$ have the *same eigenvalues* if $V$ is invertible.

*Second proof* $\qquad \det(V^{-1}AV - \lambda I) = (\det V^{-1})\,(\det(A - \lambda I))\,(\det V).$

Why is this equation true ? Then both sides are zero when $\det(A - \lambda I) = 0$.

**15** If $B$ is *similar* to $A$, the growth rates for $\boldsymbol{z}' = B\boldsymbol{z}$ are the same as for $\boldsymbol{y}' = A\boldsymbol{y}$. That equation converts to the equation for $\boldsymbol{z}$ when $B = V^{-1}AV$ and $\boldsymbol{z} =$ _____ .

**16** If $A\boldsymbol{x} = \lambda\boldsymbol{x} \neq \boldsymbol{0}$, what is an eigenvalue and eigenvector of $(e^{At} - I)A^{-1}$ ?

**17** The matrix $B = \left[\begin{smallmatrix} 0 & -4 \\ 0 & 0 \end{smallmatrix}\right]$ has $B^2 = 0$. Find $e^{Bt}$ from a (short) infinite series. Check that the derivative of $e^{Bt}$ is $Be^{Bt}$.

**18** Starting from $\boldsymbol{y}(0) = \boldsymbol{0}$, solve $\boldsymbol{y}' = A\boldsymbol{y} + \boldsymbol{q}$ as a combination of the eigenvectors. Suppose the source is $\boldsymbol{q} = q_1\boldsymbol{x}_1 + \cdots + q_n\boldsymbol{x}_n$. Solve for one eigenvector at a time, using the solution $y(t) = (e^{at} - 1)q/a$ to the scalar equation $y' = ay + q$.

Then $\boldsymbol{y}(t) = (e^{At} - I)A^{-1}\boldsymbol{q}$ is a combination of eigenvectors when all $\lambda_i \neq 0$.

**19** Solve for $\boldsymbol{y}(t)$ as a combination of the eigenvectors $\boldsymbol{x}_1 = (1, 0)$ and $\boldsymbol{x}_2 = (1, 1)$:

$$\boldsymbol{y}' = A\boldsymbol{y} + \boldsymbol{q} \qquad \begin{bmatrix} y_1' \\ y_2' \end{bmatrix} = \begin{bmatrix} 1 & 1 \\ 0 & 2 \end{bmatrix}\begin{bmatrix} y_1 \\ y_2 \end{bmatrix} + \begin{bmatrix} 4 \\ 3 \end{bmatrix} \quad \text{with} \quad \begin{matrix} y_1(0) = 0 \\ y_2(0) = 0 \end{matrix}$$

**20** Solve $\boldsymbol{y}' = A\boldsymbol{y} = \begin{bmatrix} 2 & 3 \\ 2 & 1 \end{bmatrix}\boldsymbol{y}$ in three steps. First find the $\lambda$'s and $\boldsymbol{x}$'s.

(**1**) Write $\boldsymbol{y}(0) = (3, 1)$ as a combination $c_1\boldsymbol{x}_1 + c_2\boldsymbol{x}_2$

(**2**) Multiply $c_1$ and $c_2$ by $e^{\lambda_1 t}$ and $e^{\lambda_2 t}$.

(**3**) Add the solutions $c_1 e^{\lambda_1 t}\boldsymbol{x}_1 + c_2 e^{\lambda_2 t}\boldsymbol{x}_2$.

**21** Write five terms of the infinite series for $e^{At}$. Take the $t$ derivative of each term. Show that you have four terms of $Ae^{At}$. Conclusion: $e^{At}\boldsymbol{y}(0)$ solves $d\boldsymbol{y}/dt = A\boldsymbol{y}$.

**Problems 22-25 are about time-varying systems $\boldsymbol{y}' = A(t)\boldsymbol{y}$. Success then failure.**

**22** Suppose the constant matrix $C$ has $C\boldsymbol{x} = \lambda\boldsymbol{x}$, and $p(t)$ is the integral of $a(t)$. Substitute $\boldsymbol{y} = e^{\lambda p(t)}\boldsymbol{x}$ to show that $d\boldsymbol{y}/dt = a(t)C\boldsymbol{y}$. Eigenvectors still solve this special time-varying system: constant matrix $C$ multiplied by the scalar $a(t)$.

**23** Continuing Problem 22, show from the series for $M(t) = e^{p(t)C}$ that $dM/dt = a(t)CM$. Then $M$ is the fundamental matrix for the special system $\boldsymbol{y}' = a(t)C\boldsymbol{y}$. If $a(t) = 1$ then its integral is $p(t) = t$ and we recover $M = e^{Ct}$.

**24** The integral of $A = \begin{bmatrix} 1 & 2t \\ 0 & 0 \end{bmatrix}$ is $P = \begin{bmatrix} t & t^2 \\ 0 & 0 \end{bmatrix}$. The exponential of $P$ is $e^P = \begin{bmatrix} e^t & t(e^t - 1) \\ 0 & 1 \end{bmatrix}$. From the chain rule we might hope that the derivative of $e^{P(t)}$ is $P'e^{P(t)} = Ae^{P(t)}$. Compute the derivative of $e^{P(t)}$ and compare with the wrong answer $Ae^{P(t)}$. (One reason this feels wrong: Writing the chain rule as $(d/dt)e^P = e^P dP/dt$ would give $e^P A$ instead of $A\,e^P$. That is wrong too.)

**25** Find the solution to $\boldsymbol{y}' = A(t)\boldsymbol{y}$ in Problem 24 by solving for $y_2$ and then $y_1$:

$$\text{Solve} \quad \begin{bmatrix} dy_1/dt \\ dy_2/dt \end{bmatrix} = \begin{bmatrix} 1 & 2t \\ 0 & 0 \end{bmatrix}\begin{bmatrix} y_1 \\ y_2 \end{bmatrix} \quad \text{starting from} \quad \begin{bmatrix} y_1(0) \\ y_2(0) \end{bmatrix}.$$

Certainly $y_2(t)$ stays at $y_2(0)$. Find $y_1(t)$ by "undetermined coefficients" $A, B, C$: $y_1' = y_1 + 2ty_2(0)$ is solved by $y_1 = y_p + y_n = At + B + Ce^t$.

Choose $A, B, C$ to satisfy the equation and match the initial condition $y_1(0)$.

The wrong answer in Problem 24 included the incorrect factor $te^t$ in $e^{P(t)}$.

## 6.5 Second Order Systems and Symmetric Matrices

This section solves a differential equation that is crucial in engineering and physics:

**Oscillation equation** $$\frac{d^2 y}{dt^2} + S y = \mathbf{0}. \tag{1}$$

Since this is second order in time, we need two vectors as initial conditions at $t = 0$:

**Starting position and starting velocity** $\quad \boldsymbol{y}(0)$ and $\boldsymbol{v}(0) = \dfrac{d\boldsymbol{y}}{dt}(0)$ are given.

If $\boldsymbol{y}$ has $n$ components, we have $n$ second order equations and $2n$ initial conditions. This is the right number to find $\boldsymbol{y}(t)$. Allow me to say this early: The oscillation equation (1) is the most basic form of the **Fundamental Equation of Engineering**.

The more general equation includes a damping term $B\,d\boldsymbol{y}/dt$ and a forcing term $\boldsymbol{F} \cos \Omega t$. Those give *damped forced oscillations*, where equation (1) is about "free" oscillations. For one mass and one equation, Chapter 2 took that step to damping and forcing. Now we have $n$ masses and $n$ equations and three $n$ by $n$ matrices $M, B, K$.

**Fundamental Equation** $$M\frac{d^2\boldsymbol{y}}{dt^2} + B\frac{d\boldsymbol{y}}{dt} + K\boldsymbol{y} = \boldsymbol{F}\cos\Omega t. \tag{2}$$

The *mass matrix* is $M$, the *stiffness matrix* is $K$. Those are the pieces we always see and always need. When the damping matrix $B$ and the forcing vector $F$ are removed, that takes us to the heart of the fundamental equation: *free oscillations*.

**Mass and stiffness matrices** $$\boldsymbol{M}\boldsymbol{y}'' + \boldsymbol{K}\boldsymbol{y} = \mathbf{0}\,. \tag{3}$$

The matrix $S$ in equation (1) is $M^{-1}K$. Its symmetric form is $M^{-1/2}KM^{-1/2}$. In many applications the mass matrix $M$ is diagonal.

If we look for eigenvector solutions $\boldsymbol{y} = e^{i\omega t}\boldsymbol{x}$, the differential equation produces $K\boldsymbol{x} = \omega^2 M\boldsymbol{x}$. This "generalized" eigenvalue problem has an extra matrix $M$, but it is not more difficult than $S\boldsymbol{x} = \lambda\boldsymbol{x}$. The MATLAB command is $\mathsf{eig}(K, M)$. An essential point is that the eigenvalues are still real and positive, when both $M$ and $K$ are *positive definite*. Positive eigenvalues and positive energy are the key to Chapter 7.

When the forcing term is a constant $\boldsymbol{F}$, the damping brings us to a steady state $\boldsymbol{y}_\infty$. Then the time dependence is gone; those derivatives $d\boldsymbol{y}/dt$ and $d^2\boldsymbol{y}/dt^2$ are zero. The external force $\boldsymbol{F}$ is balanced by the internal force $K\boldsymbol{y}_\infty$. The system is in equilibrium:

**Steady state equation** $$K\boldsymbol{y}_\infty = \boldsymbol{F} = \text{constant}. \tag{4}$$

The central problem of computational mechanics is to create the stiffness matrix $K$ and force vector $\boldsymbol{F}$. Then the computer solves $M\boldsymbol{y}'' + K\boldsymbol{y} = 0$ and $K\boldsymbol{y}_\infty = \boldsymbol{F}$. For large problems, the *finite element method* is now the favorite way to take those steps. This is a sensational achievement by the collective efforts of thousands of engineers. [1]

## Solution by Eigenvalues

We want to solve $\boldsymbol{y}'' + S\boldsymbol{y} = \mathbf{0}$. This is a linear system with constant coefficients. Our solution method will be the same as for $\boldsymbol{y}' = A\boldsymbol{y}$. We use the eigenvectors and eigenvalues of $S$ to find special solutions, and we combine those to find the complete solution.

Each eigenvector of $S$ leads to two special solutions to $\boldsymbol{y}'' + S\boldsymbol{y} = \mathbf{0}$:

**Two solutions** If $S\boldsymbol{x} = \lambda\boldsymbol{x}$ then $\boldsymbol{y}(t) = (\cos\omega t)\boldsymbol{x}$ and $\boldsymbol{y}(t) = (\sin\omega t)\boldsymbol{x}$. (5)

The "frequency" $\omega$ is $\sqrt{\lambda}$. Substitute $\boldsymbol{y} = (\cos\omega t)\boldsymbol{x}$ into the differential equation:

$$\boldsymbol{\lambda = \omega^2 \textbf{ and } Sx = \omega^2 x} \qquad \boldsymbol{y}'' + S\boldsymbol{y} = -\omega^2(\cos\omega t)\boldsymbol{x} + S(\cos\omega t)\boldsymbol{x} = \mathbf{0}. \tag{6}$$

When $\cos\omega t$ is factored out, we see the requirement on $\boldsymbol{x}$. It must be an eigenvector of $S$. We expect $n$ eigenvectors (*normal modes of oscillation*). The eigenvectors don't interact. That is their beauty, each one goes its own way. And each eigenvector gives us two solutions from $(\cos\omega t)\boldsymbol{x}$ and $(\sin\omega t)\boldsymbol{x}$, so we have $2n$ special solutions.

A combination of those $2n$ solutions will match the $2n$ initial conditions ($n$ positions and $n$ velocities at $t = 0$). This determines the $2n$ constants $A_i$ and $B_i$ in the complete solution to $\boldsymbol{y}'' + S\boldsymbol{y} = 0$:

**Complete solution** $$\boldsymbol{y}(t) = \sum_{i=1}^{n} (A_i \cos\sqrt{\lambda_i}\,t + B_i \sin\sqrt{\lambda_i}\,t)\,\boldsymbol{x}_i. \tag{7}$$

Since $\sin 0 = 0$, it is the $A_i$ that match the vector $\boldsymbol{y}(0)$ of initial positions. It is the $B_i$ that match the vector $\boldsymbol{v}(0) = \boldsymbol{y}'(0)$ of initial velocities.

**Example 1** Two masses are connected by three identical springs in Figure 6.3. Find the stiffness matrix $S$ and its positive eigenvalues $\lambda_1 = \omega_1^2$ and $\lambda_2 = \omega_2^2$. If the system starts from rest, with the top spring unstretched ($y_1(0) = 0$) and the lower mass moved down ($y_2(0) = 2$), find the positions $\boldsymbol{y} = (y_1, y_2)$ at all later times:

$$m\,\frac{d^2\boldsymbol{y}}{dt^2} + S\boldsymbol{y} = \mathbf{0} \text{ with } \boldsymbol{y}(0) = \begin{bmatrix} 0 \\ 2 \end{bmatrix} \text{ and } \boldsymbol{y}'(0) = \begin{bmatrix} 0 \\ 0 \end{bmatrix}.$$

$\boldsymbol{y}(t)$ has eigenvectors $\boldsymbol{x}_1, \boldsymbol{x}_2$ times cosine and sine. Four conditions for $A_1, A_2, B_1, B_2$.

*Solution* Construct the matrix $S$ that expresses Newton's Law $m\boldsymbol{y}'' + S\boldsymbol{y} = \mathbf{0}$. The acceleration is $\boldsymbol{y}''$, and the force is $-S\boldsymbol{y}$.

[1] The finite element method is a key part of my textbook on *Computational Science and Engineering*. The foundations of the method and the reasons for its success are developed in *An Analysis of the Finite Element Method* (also published by Wellesley-Cambridge Press).

What force $F$ is acting on the upper mass ? The stretched top spring is pulling that mass up. The force is proportional to the stretch $y_1$. *This is Hooke's Law* $F = -ky_1$.

The middle spring is connected to both masses. It is stretched a distance $y_2 - y_1$. (No stretching if $y_2 = y_1$, the spring would just be shifted up or down.) **The difference $y_2 - y_1$ produces spring forces $k(y_2 - y_1)$**, pulling mass 1 down and mass 2 up.

The bottom spring with fixed end is stretched by $0 - y_2$, so the force is $-ky_2$.

$$\begin{array}{ll} \boldsymbol{F = ma} \textbf{ at the upper mass} & -ky_1 + k(y_2 - y_1) = my_1'' \\ \boldsymbol{F = ma} \textbf{ at the lower mass} & -k(y_2 - y_1) - ky_2 = my_2'' \end{array}$$

These equations $-S\boldsymbol{y} = m\boldsymbol{y}''$ or $m\boldsymbol{y}'' + S\boldsymbol{y} = \boldsymbol{0}$ have a symmetric matrix $S$. Take $k = m = 1$:

$$\boldsymbol{y}'' + S\boldsymbol{y} = \frac{d^2}{dt^2}\begin{bmatrix} y_1 \\ y_2 \end{bmatrix} + \begin{bmatrix} \mathbf{2} & \mathbf{-1} \\ \mathbf{-1} & \mathbf{2} \end{bmatrix}\begin{bmatrix} y_1 \\ y_2 \end{bmatrix} = \begin{bmatrix} 0 \\ 0 \end{bmatrix}. \tag{8}$$

The modeling part is complete, now for the solution part. The eigenvalues of that matrix are $\lambda_1 = 1$ and $\lambda_2 = 3$. The trace is $1 + 3 = 4$, the determinant is $(1)(3) = 3$. The first eigenvector $\boldsymbol{x}_1 = (1, 1)$ has the springs moving in the same direction in Figure 6.3. The second eigenvector $\boldsymbol{x}_2 = (1, -1)$ has the springs moving oppositely, with higher frequency because $\omega_2^2 = \lambda_2 = 3$.

Formula (7) for $\boldsymbol{y}(t)$ becomes a combination of eigenvectors times cosines:

$$\textbf{Solution} \quad \begin{bmatrix} y_1(t) \\ y_2(t) \end{bmatrix} = A_1 (\cos\sqrt{1}\,t)\begin{bmatrix} 1 \\ 1 \end{bmatrix} + A_2 (\cos\sqrt{3}\,t)\begin{bmatrix} 1 \\ -1 \end{bmatrix}. \tag{9}$$

I removed $B_1 \sin t$ and $B_2 \sin\sqrt{3}t$ because the example started from rest (zero velocity). **At time $t = 0$, cosines give position $\boldsymbol{y}(0)$ and sines give velocity $\boldsymbol{v}(0)$.**

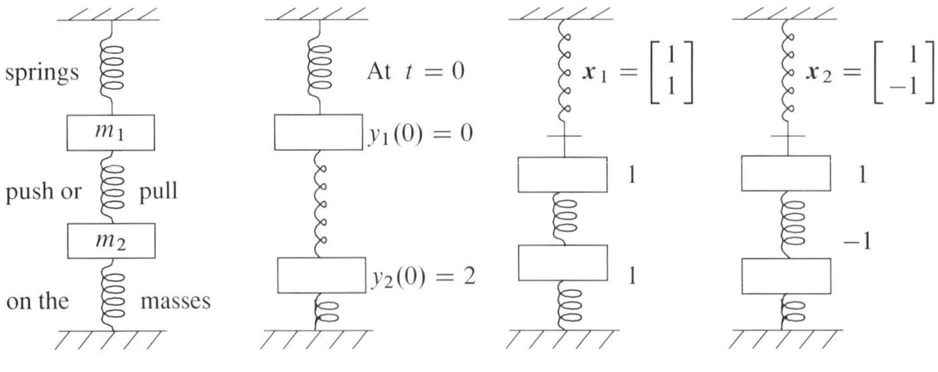


Figure 6.3: The masses oscillate up and down, $\boldsymbol{y}(t)$ combines $(\cos t)\,\boldsymbol{x}_1$ and $(\cos\sqrt{3}t)\,\boldsymbol{x}_2$.

The final step is to find $A_1$ and $A_2$ from the initial position $\boldsymbol{y}(0) = (0, 2)$ :

$$\textbf{Initial condition} \quad A_1 \begin{bmatrix} 1 \\ 1 \end{bmatrix} + A_2 \begin{bmatrix} 1 \\ -1 \end{bmatrix} = \begin{bmatrix} 0 \\ 2 \end{bmatrix} \quad \text{gives} \quad A_1 = 1 \text{ and } A_2 = -1.$$

**Final answer :** $\boldsymbol{y_1(t) = (\cos t - \cos \sqrt{3}t)}$ **and** $\boldsymbol{y_2(t) = (\cos t + \cos \sqrt{3}t)}$. The two masses oscillate forever. The solution part was easier than the modeling part. This is very typical.

## Symmetric Matrices

Example 1 led to a symmetric matrix $S$. *Many many examples* lead to symmetric matrices.

Perhaps this is an extension of Newton's third law, that every action produces an equal and opposite reaction. We really must focus on the special properties of symmetric matrices, because those properties are so useful and the matrices appear so often.

Eigenvalues and eigenvectors—this is the information we need from the matrix. For every class of matrices, we ask about $\lambda$ and $\boldsymbol{x}$. Are the eigenvalues *real*? Are they *positive*, so we can take square roots in $\lambda = \omega^2$ ? Are there *$n$ independent* eigenvectors ? Are the $\boldsymbol{x}$'s *orthogonal*? The example with $\lambda_1 = 1$ and $\lambda_2 = 3$ was perfect in all respects :

$$S = \begin{bmatrix} 2 & -1 \\ -1 & 2 \end{bmatrix} \text{ is } \textbf{symmetric positive definite} \qquad \begin{matrix} \textbf{Positive real } \boldsymbol{\lambda = 1} \textbf{ and } \boldsymbol{3} \\ \textbf{Orthogonal } \boldsymbol{x = (1,1), (1,-1)} \end{matrix}$$

**Real eigenvalues** **All the eigenvalues of a real symmetric matrix are real.**

*Proof* Suppose that $S\boldsymbol{x} = \lambda\boldsymbol{x}$. Until we know otherwise, $\lambda$ might be a complex number and $\boldsymbol{x}$ might be a complex vector. If that did happen, the rules for complex conjugates would give $\overline{S\boldsymbol{x}} = \overline{\lambda}\overline{\boldsymbol{x}}$. The key idea is to look at $\overline{\boldsymbol{x}}^{\mathrm{T}} S\boldsymbol{x}$ :

$$S \text{ is symmetric and real} \qquad \overline{\boldsymbol{x}}^{\mathrm{T}} S\boldsymbol{x} = \overline{\boldsymbol{x}}^{\mathrm{T}} S^{\mathrm{T}} \boldsymbol{x} = (\overline{S\boldsymbol{x}})^{\mathrm{T}} \boldsymbol{x}. \tag{10}$$

The left side is $\overline{\boldsymbol{x}}^{\mathrm{T}} \lambda\boldsymbol{x}$. The right side is $\overline{\boldsymbol{x}}^{\mathrm{T}}\overline{\lambda}\boldsymbol{x}$. One side has $\lambda$, the other side has $\overline{\lambda}$. They multiply $\overline{\boldsymbol{x}}^{\mathrm{T}}\boldsymbol{x}$ which is not zero—it is the squared length $|x_1|^2 + \cdots + |x_n|^2$. *Therefore* $\lambda = \overline{\lambda}$.

When $\lambda = a + ib$ equals $\overline{\lambda} = a - ib$, we know that $b = 0$ and $\lambda$ *is real.* Then the vector $\boldsymbol{x}$ in the nullspace of the real matrix $S - \lambda I$ can also be kept real.

**Orthogonal eigenvectors** If $S\boldsymbol{x} = \lambda_1\boldsymbol{x}$ and $S\boldsymbol{y} = \lambda_2\boldsymbol{y}$ and $\lambda_1 \neq \lambda_2$. **Then** $\boldsymbol{x}^{\mathrm{T}}\boldsymbol{y} = \boldsymbol{0}$.

*Proof* Take the dot product of the first equation with $\boldsymbol{y}$ and the second equation with $\boldsymbol{x}$ :

$$\text{Use } S^{\mathrm{T}} = S \qquad (S\boldsymbol{x})^{\mathrm{T}}\boldsymbol{y} = \boldsymbol{x}^{\mathrm{T}}S\boldsymbol{y} \text{ is } \lambda_1 \boldsymbol{x}^{\mathrm{T}}\boldsymbol{y} = \lambda_2 \boldsymbol{x}^{\mathrm{T}}\boldsymbol{y}. \tag{11}$$

Since $\lambda_1 \neq \lambda_2$, this proves that $\boldsymbol{x}^{\mathrm{T}}\boldsymbol{y} = 0$. The eigenvectors are perpendicular.

Remember: The main goal of eigenvectors is to *diagonalize a matrix*, $A = V\Lambda V^{-1}$. Here the matrix is $S$ and its eigenvectors are orthogonal. We can certainly make them unit vectors, so $\boldsymbol{x}^{\mathrm{T}}\boldsymbol{x} = 1$ and $\boldsymbol{x}^{\mathrm{T}}\boldsymbol{y} = 0$. The matrix $V$ with the eigenvectors in its columns has become an **orthogonal matrix**: $V^{\mathrm{T}}V = I$. The right letter for this orthogonal matrix $V$ is $Q$. **The eigenvector matrix $V$ in $V\Lambda V^{-1}$ can be orthogonal: $Q^{\mathrm{T}}Q = I$.**

**Spectral theorem / Principal axis theorem** $\quad \boldsymbol{S = Q\Lambda Q^{-1} = Q\Lambda Q^{\mathrm{T}}}$ (12)

In algebra, the eigenvectors are orthogonal. In geometry, the principal axes of an ellipse are orthogonal. If the ellipse equation is $2x^2 - 2xy + 2y^2 = 1$, this corresponds to the example matrix $S$. Its principal axes $(1,1)$ and $(1,-1)$ (eigenvectors) are at $+45°$ and $-45°$ from the $x$ axis. The ellipse is turned by $+45°$ from horizontal and vertical axes.

With repeated eigenvalues, $S = Q\Lambda Q^{\mathrm{T}}$ is still correct. Every symmetric $S$ has a full set of $n$ independent eigenvectors (Chapter 6 Notes) even if eigenvalues are repeated.

To summarize, $Q\Lambda Q^{\mathrm{T}}$ is a perfect description of symmetric matrices $S$. Every $S$ has those factors and every matrix of this form is sure to be symmetric: $(Q\Lambda Q^{\mathrm{T}})^{\mathrm{T}}$ equals $Q^{\mathrm{TT}}\Lambda^{\mathrm{T}}Q^{\mathrm{T}}$ which is $Q\Lambda Q^{\mathrm{T}}$. If we multiply columns of $Q$ times rows of $\Lambda Q^{\mathrm{T}}$, we see $S$ in a new way (a sum of rank one matrices):

**Matrices $\boldsymbol{\lambda x x^{\mathrm{T}}}$ with rank 1 add to $\boldsymbol{S}$**

$$S = \begin{bmatrix} & & \\ \boldsymbol{x}_1 & \cdots & \boldsymbol{x}_n \\ & & \end{bmatrix} \begin{bmatrix} \lambda_1 \boldsymbol{x}_1^{\mathrm{T}} \\ \vdots \\ \lambda_n \boldsymbol{x}_n^{\mathrm{T}} \end{bmatrix} = \lambda_1 \boldsymbol{x}_1 \boldsymbol{x}_1^{\mathrm{T}} + \cdots + \lambda_n \boldsymbol{x}_n \boldsymbol{x}_n^{\mathrm{T}}. \qquad (13)$$

This is the great factorization $S = Q\Lambda Q^{\mathrm{T}}$, in terms of eigenvalues and eigenvectors.

**Example 2** The eigenvectors $(1,1)$ and $(-1,1)$ with $\lambda = 16$ and $4$ give *unit* eigenvectors $\boldsymbol{x}_1 = (1,1)/\sqrt{2}$ and $\boldsymbol{x}_2 = (-1,1)/\sqrt{2}$:

$$S = \begin{bmatrix} 10 & -6 \\ -6 & 10 \end{bmatrix} \qquad Q\Lambda Q^{\mathrm{T}} = \frac{1}{\sqrt{2}}\begin{bmatrix} 1 & -1 \\ 1 & 1 \end{bmatrix}\begin{bmatrix} 16 & \\ & 4 \end{bmatrix}\frac{1}{\sqrt{2}}\begin{bmatrix} 1 & 1 \\ -1 & 1 \end{bmatrix}.$$

Those eigenvectors still point in the $45°$ direction and the $135°$ direction ($90°$ apart). They are the same as in Example 1, because this new $S$ is 6 times the original $S$, minus $2I$. Then the new eigenvalues 16 and 4 of $S$ must be 6 times the original 3 and 1, minus 2.

The eigenvectors in $Q$ are the principal axes of an ellipse $10x^2 - 12xy + 10y^2 = 1$.

If I change $-6$ and $-6$ off the diagonal to $6i$ and $-6i$, the determinant is still 64. The trace is still 20 and the eigenvalues are still 16 and 4 (**real**!). For complex matrices, we want a symmetric real part and an *antisymmetric* imaginary part. Let me explain why.

## Complex Matrices

*Important*: The squared length is $\overline{\boldsymbol{x}}^{\mathrm{T}}\boldsymbol{x}$ and not $\boldsymbol{x}^{\mathrm{T}}\boldsymbol{x}$ when $\boldsymbol{x}$ has complex components. We want $|x_1|^2 + \cdots + |x_n|^2$ because this is a positive number or zero. We don't want $x_1^2 + \cdots + x_n^2$ because that could be any complex number, and we are looking for $||\boldsymbol{x}||^2 =$ length squared $\geq 0$. When a component of $\boldsymbol{x}$ is $a + bi$, we want $a^2 + b^2$ and not $(a + bi)^2$. *The length squared of* $\boldsymbol{x} = (1, i)$ *is* $||\boldsymbol{x}||^2 = 1^2 + 1^2 = 2$ *and not* $1^2 + i^2 = 0$.

This changes all inner products (dot products) from $\boldsymbol{x}^{\mathrm{T}}\boldsymbol{y}$ to $\overline{\boldsymbol{x}}^{\mathrm{T}}\boldsymbol{y}$. Complex vectors $\boldsymbol{x}$ and $\boldsymbol{y}$ are perpendicular when $\overline{\boldsymbol{x}}^{\mathrm{T}}\boldsymbol{y} = 0$. This complex inner product forces us to replace the usual transpose by the **conjugate transpose** $(\overline{\boldsymbol{A}})^{\mathrm{T}} = \boldsymbol{A}^*$, when $A$ is complex:

$$\boldsymbol{A}^*_{ij} \text{ is } \overline{\boldsymbol{A}}_{ji} \qquad \text{Then } A\boldsymbol{x}\cdot\boldsymbol{y} = (\overline{A\boldsymbol{x}})^{\mathrm{T}}\boldsymbol{y} = \overline{\boldsymbol{x}}^{\mathrm{T}}\overline{A}^{\mathrm{T}}\boldsymbol{y} = \boldsymbol{x}\cdot A^*\boldsymbol{y}. \tag{14}$$

MATLAB automatically takes the conjugate transpose to give $A^*$, when you type $\boldsymbol{x}'$ or $A'$.

To keep the row space of $A$ perpendicular to the nullspace, we must use $\boldsymbol{C}(A^*)$ for the row space. This is the column space of $A^*$, not just the column space of $A^{\mathrm{T}}$. Replace every $i$ by $-i$. And an important name: the complex version of a *symmetric matrix* $A^{\mathrm{T}} = A$ is a "*Hermitian matrix*" $\boldsymbol{A}^* = \boldsymbol{A}$.

**Hermitian matrix** $\boldsymbol{A}_{ij} = \overline{\boldsymbol{A}}_{ji}$ Then $A\boldsymbol{x}\cdot\boldsymbol{y} = \boldsymbol{x}\cdot A^*\boldsymbol{y}$ becomes $A\boldsymbol{x}\cdot\boldsymbol{y} = \boldsymbol{x}\cdot A\boldsymbol{y}$.

**Example 3** This 2 by 2 complex matrix is Hermitian (notice $i$ and $-i$):

$$A = \begin{bmatrix} 3 & i \\ -i & 3 \end{bmatrix} = A^*$$

The determinant is 8 (real). The trace is 6 (the main diagonal of a Hermitian matrix is real). The eigenvalues of this matrix are 2 and 4 (*both real*!).

**Hermitian matrices $A = A^*$ have real eigenvalues and perpendicular eigenvectors.**

The eigenvectors of $A$ are $\boldsymbol{x}_1 = (1, i)$ and $\boldsymbol{x}_2 = (1, -i)$. They are perpendicular: $\boldsymbol{x}_1{}^*\boldsymbol{x}_2 = 1^2 + (-i)^2 = 0$. Divide by $\sqrt{2}$ to make them unit vectors. Then they are the columns of a complex orthogonal matrix $Q$. The right meaning of "complex orthogonal" is $Q^* = Q^{-1}$, and the right name when $Q$ is complex is *unitary*:

**Unitary matrix $\boldsymbol{Q}^*\boldsymbol{Q} = \boldsymbol{I}$** The columns of $Q$ are perpendicular unit vectors.

The great factorization $A = Q\Lambda Q^{\mathrm{T}}$ of real symmetric matrices becomes $\boldsymbol{A} = \boldsymbol{Q}\boldsymbol{\Lambda}\boldsymbol{Q}^*$.

## Orthogonal Matrices and Unitary Matrices

We have seen the big theorem: If $S$ is symmetric or Hermitian, its eigenvector matrix is orthogonal or unitary. The real case is $S = Q\Lambda Q^{\mathrm{T}} = S^{\mathrm{T}}$ and the complex case is $S = Q\Lambda Q^* = S^*$. The eigenvalues in $\Lambda$ are real.

What if our matrix is *anti*-symmetric or *anti*-Hermitian? Then $A^{\mathrm{T}} = -A$ or $\boldsymbol{A}^* = -\boldsymbol{A}$. The matrix $A$ could even be $i$ times $S$. (In that case $A^*$ will be $-i$ times $S^*$ which is exactly $-iS = -A$.) Multiplying by $i$ changes Hermitian to *anti*-Hermitian. The real eigenvalues $\lambda$ of $S$ change to the imaginary eigenvalues $i\lambda$ of $A$. The eigenvectors do *not* change: still orthogonal, still going into $Q$.

**Anti-Hermitian matrices have imaginary eigenvalues and orthogonal eigenvectors.**

Our standard examples are $A = \begin{bmatrix} 0 & 1 \\ -1 & 0 \end{bmatrix} = -A^{\mathrm{T}}$ and $A = \begin{bmatrix} 0 & i \\ i & 0 \end{bmatrix} = -A^*$. $\boldsymbol{\lambda = \pm i}$

Finally, what if our matrix is *orthogonal* or *unitary*? Then $Q^{\mathrm{T}}Q = I$ or $Q^*Q = I$. The eigenvalues of $Q$ are **complex numbers $\boldsymbol{\lambda = e^{i\theta}}$ on the unit circle**.

$$\textbf{If } \boldsymbol{Q^*Q = I} \textbf{ then all eigenvalues of } \boldsymbol{Q} \textbf{ have magnitude } \boldsymbol{|\lambda| = 1}.$$

The proof starts with $Q\boldsymbol{x} = \lambda\boldsymbol{x}$. The conjugate transpose is $\boldsymbol{x}^*Q^* = \overline{\lambda}\boldsymbol{x}^*$. Multiply the left hand sides using $Q^*Q = I$, and multiply the right hand sides using $\overline{\lambda}\lambda = |\lambda|^2$:

$$\boldsymbol{x}^*Q^*Q\boldsymbol{x} = \overline{\lambda}\boldsymbol{x}^*\lambda\boldsymbol{x} \quad \text{is the same as} \quad \boldsymbol{x}^*\boldsymbol{x} = |\lambda|^2\boldsymbol{x}^*\boldsymbol{x}. \quad \text{Then } |\lambda|^2 = 1 \text{ and } |\lambda| = 1.$$

The eigenvectors of $Q$, like the eigenvectors of $S$ and $A$, can be chosen orthogonal. *These are the essential facts about the best matrices.* The eigenvalues of $S$ and $A$ and $Q$ are on the *real axis*, the *imaginary axis*, and the *unit circle* in the complex plane.

In the eigenvalue-eigenvector world, a triangular matrix is not really one of the best. Its eigenvalues are easy (on the main diagonal). But its eigenvectors are not orthogonal. It may even fail to be diagonalizable. Matrices without $n$ eigenvectors are the worst.

## Symmetric and Orthogonal

At the end of Chapter 4, we looked at symmetric matrices that are also orthogonal: $\boldsymbol{A^{\mathrm{T}} = A}$ **and** $\boldsymbol{A^{\mathrm{T}} = A^{-1}}$. Every diagonal matrix $D$ of 1's and $-1$'s has both properties. Then every $A = QDQ^{\mathrm{T}}$ also has both properties. Symmetry is clear, and a product of orthogonal matrices $Q$ and $D$ and $Q^{\mathrm{T}}$ is sure to stay orthogonal.

The question we could not answer was: *Does $QDQ^{\mathrm{T}}$ give all possible examples*? The answer is yes, and now we can see why $A$ has this form—based on eigenvalues.

When $A$ is symmetric, its eigenvalues are real. When $A$ is orthogonal, its eigenvalues have $|\lambda| = 1$. The only possibilities for both are $\boldsymbol{\lambda = 1}$ **and** $\boldsymbol{\lambda = -1}$. The eigenvalue matrix $\Lambda = D$ is a diagonal matrix of 1's and $-1$'s. Then the great fact about symmetric matrices (the Spectral Theorem) guarantees that $A$ has the form $Q\Lambda Q^{\mathrm{T}}$ which is $QDQ^{\mathrm{T}}$.

## ■ REVIEW OF THE KEY IDEAS ■

1. A real symmetric matrix $S$ has *real eigenvalues* and *perpendicular eigenvectors*.

2. Diagonalization $S = V\Lambda V^{-1}$ becomes $S = Q\Lambda Q^{\mathrm{T}}$ with an orthogonal matrix $Q$.

3. A complex matrix is *Hermitian* if $\overline{S}^{\mathrm{T}} = S$ (often written $S^* = S$): *real* $\lambda$'s.

4. Every Hermitian matrix is $S = Q\Lambda\overline{Q}^{\mathrm{T}} = Q\Lambda Q^*$. Dot products are $\boldsymbol{x} \cdot \boldsymbol{y} = \boldsymbol{x}^*\boldsymbol{y}$.

5. All three matrices $S$ and $A = iS = -A^*$ and $Q$ have orthogonal eigenvectors.

6. Symmetric matrices in $\boldsymbol{y}'' + S\boldsymbol{y} = \boldsymbol{0}$ and $M\boldsymbol{y}'' + K\boldsymbol{y} = \boldsymbol{0}$ give oscillation.

# Problem Set 6.5

**Problems 1–14 are about eigenvalues. Then come differential equations.**

**1** Which of $A, B, C$ have two real $\lambda$'s ? Which have two independent eigenvectors ?

$$A = \begin{bmatrix} 7 & -11 \\ -11 & 7 \end{bmatrix} \quad B = \begin{bmatrix} 7 & -11 \\ 11 & 7 \end{bmatrix} \quad C = \begin{bmatrix} 7 & -11 \\ 0 & 7 \end{bmatrix}$$

**2** Show that $A$ has real eigenvalues if $b \geq 0$ and nonreal eigenvalues if $b < 0$:

$$A = \begin{bmatrix} 0 & b \\ 1 & 0 \end{bmatrix} \quad \text{and} \quad A = \begin{bmatrix} 1 & b \\ 1 & 1 \end{bmatrix}.$$

**3** Find the eigenvalues and the unit eigenvectors of the symmetric matrices

$$\text{(a) } S = \begin{bmatrix} 2 & 2 & 2 \\ 2 & 0 & 0 \\ 2 & 0 & 0 \end{bmatrix} \quad \text{and} \quad \text{(b)} \quad S = \begin{bmatrix} 1 & 0 & 2 \\ 0 & -1 & -2 \\ 2 & -2 & 0 \end{bmatrix}.$$

**4** Find an orthogonal matrix $Q$ that diagonalizes $S = \begin{bmatrix} \mathbf{-2} & \mathbf{6} \\ \mathbf{6} & \mathbf{7} \end{bmatrix}$. What is $\Lambda$?

**5** Show that this $A$ (**symmetric but complex**) has only one line of eigenvectors:

$$A = \begin{bmatrix} i & 1 \\ 1 & -i \end{bmatrix} \text{ is not even diagonalizable. Its eigenvalues are 0 and 0.}$$

$A^{\mathrm{T}} = A$ is not so special for complex matrices. *The good property is* $\overline{A}^{\mathrm{T}} = A$.

**6** Find *all* orthogonal matrices from all $\boldsymbol{x}_1, \boldsymbol{x}_2$ to diagonalize $S = \begin{bmatrix} 9 & 12 \\ 12 & 16 \end{bmatrix}$.

**7** (a) Find a symmetric matrix $S = \begin{bmatrix} 1 & b \\ b & 1 \end{bmatrix}$ that has a negative eigenvalue.

(b) How do you know that $S$ must have a negative pivot?

(c) How do you know that $S$ can't have two negative eigenvalues?

**8** If $A^2 = 0$ then the eigenvalues of $A$ must be ______. Give an example with $A \neq 0$. But if $A$ is symmetric, diagonalize it to prove that the matrix is $A = 0$.

**9** If $\lambda = a + ib$ is an eigenvalue of a real matrix $A$, then its conjugate $\overline{\lambda} = a - ib$ is also an eigenvalue. (If $A\boldsymbol{x} = \lambda\boldsymbol{x}$ then also $A\overline{\boldsymbol{x}} = \overline{\lambda}\overline{\boldsymbol{x}}$.) Prove that every real 3 by 3 matrix has at least one real eigenvalue.

**10** Here is a quick "proof" that the eigenvalues of *all* real matrices are real:

$$\textbf{False proof} \quad Ax = \lambda x \quad \text{gives} \quad x^{\mathrm{T}}Ax = \lambda x^{\mathrm{T}}x \quad \text{so} \quad \lambda = \frac{x^{\mathrm{T}}Ax}{x^{\mathrm{T}}x} \quad \text{is real.}$$

Find the flaw in this reasoning—a hidden assumption that is not justified. You could test those steps on the 90° rotation matrix $[0 \;\; -1;\;\; 1 \;\; 0]$ with $\lambda = i$ and $x = (i, 1)$.

**11** Write $A$ and $B$ in the form $\lambda_1 x_1 x_1^{\mathrm{T}} + \lambda_2 x_2 x_2^{\mathrm{T}}$ of the spectral theorem $Q\Lambda Q^T$:

$$A = \begin{bmatrix} 3 & 1 \\ 1 & 3 \end{bmatrix} \qquad B = \begin{bmatrix} 9 & 12 \\ 12 & 16 \end{bmatrix} \qquad (\text{keep } \|x_1\| = \|x_2\| = 1).$$

**12** What number $b$ in $\left[\begin{smallmatrix} 2 & b \\ 1 & 0 \end{smallmatrix}\right]$ makes $A = Q\Lambda Q^{\mathrm{T}}$ possible? What number makes $A = V\Lambda V^{-1}$ impossible? What number makes $A^{-1}$ impossible?

**13** This $A$ is nearly symmetric. But its eigenvectors are far from orthogonal:

$$A = \begin{bmatrix} 1 & 10^{-15} \\ 0 & 1 + 10^{-15} \end{bmatrix} \quad \text{has eigenvectors} \quad \begin{bmatrix} 1 \\ 0 \end{bmatrix} \quad \text{and} \quad [\;?\;]$$

What is the dot product of the two unit eigenvectors ? A small angle !

**14** (Recommended) This matrix $M$ is skew-symmetric and also orthogonal. Then all its eigenvalues are pure imaginary and they also have $|\lambda| = 1$. They can only be $i$ or $-i$. Find all four eigenvalues from the trace of $M$:

$$M = \frac{1}{\sqrt{3}} \begin{bmatrix} 0 & 1 & 1 & 1 \\ -1 & 0 & -1 & 1 \\ -1 & 1 & 0 & -1 \\ -1 & -1 & 1 & 0 \end{bmatrix} \quad \text{can only have eigenvalues } i \text{ or } -i.$$

**15** The complete solution to equation (8) for two oscillating springs (Figure 6.3) is

$$y(t) = (A_1 \cos t + B_1 \sin t)\begin{bmatrix} 1 \\ 1 \end{bmatrix} + (A_2 \cos \sqrt{3}t + B_2 \sin \sqrt{3}t)\begin{bmatrix} 1 \\ -1 \end{bmatrix}.$$

Find the numbers $A_1, A_2, B_1, B_2$ if $y(0) = (3, 5)$ and $y'(0) = (2, 0)$.

**16** If the springs in Figure 6.3 have different constants $k_1, k_2, k_3$ then $y'' + Sy = \mathbf{0}$ is

$$\begin{array}{ll} \text{Upper mass} & y_1'' + k_1 y_1 - k_2(y_2 - y_1) = 0 \\ \text{Lower mass} & y_2'' + k_2(y_2 - y_1) + k_3 y_2 = 0 \end{array} \qquad S = \begin{bmatrix} k_1 + k_2 & -k_2 \\ -k_2 & k_2 + k_3 \end{bmatrix}$$

For $k_1 = 1, k_2 = 4, k_3 = 1$ find the eigenvalues $\lambda = \omega^2$ of $S$ and the complete sine/cosine solution $y(t)$ in equation (7).

**17** Suppose the third spring is removed ($k_3 = 0$ and nothing is below mass 2). With $k_1 = 3, k_2 = 2$ in Problem 16, find $S$ and its real eigenvalues and orthogonal eigenvectors. What is the sine/cosine solution $\boldsymbol{y}(t)$ if $\boldsymbol{y}(0) = (1, 2)$ gives the cosines and $\boldsymbol{y}'(0) = (2, -1)$ gives the sines?

**18** Suppose the top spring is also removed ($k_1 = 0$ and also $k_3 = 0$). $S$ is singular! Find its eigenvalues and eigenvectors. If $\boldsymbol{y}(0) = (1, -1)$ and $\boldsymbol{y}' = (0, 0)$ find $\boldsymbol{y}(t)$. If $\boldsymbol{y}(0)$ changes from $(1, -1)$ to $(1, 1)$ what is $\boldsymbol{y}(t)$?

**19** The matrix in this question is skew-symmetric ($A^{\mathrm{T}} = -A$). Energy is conserved.

$$\frac{d\boldsymbol{y}}{dt} = \begin{bmatrix} 0 & c & -b \\ -c & 0 & a \\ b & -a & 0 \end{bmatrix} \boldsymbol{y} \quad \text{or} \quad \begin{aligned} y_1' &= cy_2 - by_3 \\ y_2' &= ay_3 - cy_1 \\ y_3' &= by_1 - ay_2. \end{aligned}$$

The derivative of $\|\boldsymbol{y}(t)\|^2 = y_1^2 + y_2^2 + y_3^2$ is $2y_1y_1' + 2y_2y_2' + 2y_3y_3'$. Substitute $y_1', y_2', y_3'$ to get *zero*. The energy $\|\boldsymbol{y}(t)\|^2$ stays equal to $\|\boldsymbol{y}(0)\|^2$.

**20** *When $A = -A^{\mathrm{T}}$ is skew-symmetric, $e^{At}$ is* ***orthogonal***. Prove $(e^{At})^{\mathrm{T}} = e^{-At}$ from the series $e^{At} = I + At + \frac{1}{2}A^2t^2 + \cdots$.

**21** The mass matrix $M$ can have masses $m_1 = 1$ and $m_2 = 2$. Show that the eigenvalues for $K\boldsymbol{x} = \lambda M\boldsymbol{x}$ are $\lambda = 2 \pm \sqrt{2}$, starting from $\det(K - \lambda M) = 0$:

$$M = \begin{bmatrix} 1 & 0 \\ 0 & 2 \end{bmatrix} \text{ and } K = \begin{bmatrix} 2 & -2 \\ -2 & 4 \end{bmatrix} \text{ are positive definite.}$$

Find the two eigenvectors $\boldsymbol{x}_1$ and $\boldsymbol{x}_2$. Show that $\boldsymbol{x}_1^{\mathrm{T}}\boldsymbol{x}_2 \neq 0$ but $\boldsymbol{x}_1^{\mathrm{T}}M\boldsymbol{x}_2 = 0$.

**22** What difference equation would you use to solve $\boldsymbol{y}'' = -S\boldsymbol{y}$?

**23** The second order equation $\boldsymbol{y}'' + S\boldsymbol{y} = \boldsymbol{0}$ reduces to a first order system $\boldsymbol{y}_1{}' = \boldsymbol{y}_2$ and $\boldsymbol{y}_2{}' = -S\boldsymbol{y}_1$. If $S\boldsymbol{x} = \omega^2\boldsymbol{x}$ show that the companion matrix $A = [0\ I\ ;\ -S\ 0]$ has eigenvalues $i\omega$ and $-i\omega$ with eigenvectors $(\boldsymbol{x}, i\omega\boldsymbol{x})$ and $(\boldsymbol{x}, -i\omega\boldsymbol{x})$.

**24** Find the eigenvalues $\lambda$ and eigenfunctions $y(x)$ for the differential equation $y'' = \lambda y$ with $y(0) = y(\pi) = 0$. There are infinitely many!

## Table of Eigenvalues and Eigenvectors

How are the properties of a matrix reflected in its eigenvalues and eigenvectors? This question is fundamental throughout Chapter 6. A table that organizes the key facts may be helpful. Here are the special properties of the eigenvalues $\lambda_i$ and the eigenvectors $\boldsymbol{x}_i$.

| | | |
|---|---|---|
| **Symmetric:** $S^{\mathrm{T}} = S$ | real $\lambda$'s | orthogonal $\boldsymbol{x}_i^{\mathrm{T}}\boldsymbol{x}_j = 0$ |
| **Orthogonal:** $Q^{\mathrm{T}} = Q^{-1}$ | all $\lvert\lambda\rvert = 1$ | orthogonal $\overline{\boldsymbol{x}}_i^{\mathrm{T}}\boldsymbol{x}_j = 0$ |
| **Skew-symmetric:** $A^{\mathrm{T}} = -A$ | imaginary $\lambda$'s | orthogonal $\overline{\boldsymbol{x}}_i^{\mathrm{T}}\boldsymbol{x}_j = 0$ |
| **Complex Hermitian:** $\overline{S}^{\mathrm{T}} = S$ | real $\lambda$'s | orthogonal $\overline{\boldsymbol{x}}_i^{\mathrm{T}}\boldsymbol{x}_j = 0$ |
| **Positive Definite:** $\boldsymbol{x}^{\mathrm{T}}S\boldsymbol{x} > 0$ | all $\lambda > 0$ | orthogonal since $S^{\mathrm{T}} = S$ |
| **Markov:** $m_{ij} > 0, \sum_{i=1}^{n} m_{ij} = 1$ | $\lambda_{\max} = 1$ | steady state $\boldsymbol{x} > 0$ |
| **Similar:** $B = V^{-1}AV$ | $\lambda(B) = \lambda(A)$ | $\boldsymbol{x}(B) = V^{-1}\boldsymbol{x}(A)$ |
| **Projection:** $P = P^2 = P^{\mathrm{T}}$ | $\lambda = 1;\ 0$ | column space; nullspace |
| **Plane Rotation**: $\cos\theta, \sin\theta$ | $e^{i\theta}$ and $e^{-i\theta}$ | $\boldsymbol{x} = (1, i)$ and $(1, -i)$ |
| **Reflection:** $I - 2\boldsymbol{u}\boldsymbol{u}^{\mathrm{T}}$ | $\lambda = -1;\ 1, .., 1$ | $\boldsymbol{u}$; whole plane $\boldsymbol{u}^{\perp}$ |
| **Rank One:** $\boldsymbol{u}\boldsymbol{v}^{\mathrm{T}}$ | $\lambda = \boldsymbol{v}^{\mathrm{T}}\boldsymbol{u};\ 0, .., 0$ | $\boldsymbol{u}$; whole plane $\boldsymbol{v}^{\perp}$ |
| **Inverse:** $A^{-1}$ | $1/\lambda(A)$ | keep eigenvectors of $A$ |
| **Shift:** $A + cI$ | $\lambda(A) + c$ | keep eigenvectors of $A$ |
| **Function:** any $f(A)$ | $f(\lambda_1), \ldots, f(\lambda_n)$ | keep eigenvectors of $A$ |
| **Stable Powers:** $A^n \to 0$ | all $\lvert\lambda\rvert < 1$ | any eigenvectors |
| **Stable Exponential:** $e^{At} \to 0$ | all Re $\lambda < 0$ | any eigenvectors |
| **Tridiagonal:** diagonals $-1, 2, -1$ | $\lambda_k = 2 - 2\cos\frac{k\pi}{n+1}$ | $\boldsymbol{x}_k = \left(\sin\frac{k\pi}{n+1}, \sin\frac{2k\pi}{n+1}, \ldots\right)$ |

## Factorizations Based on Eigenvalues (Singular Values in $\Sigma$)

| | | |
|---|---|---|
| **Diagonalizable:** $A = V\Lambda V^{-1}$ | diagonal of $\Lambda$ has $\lambda_i$ | eigenvectors in $V$ |
| **Symmetric:** $S = Q\Lambda Q^{\mathrm{T}}$ | diagonal of $\Lambda$ (real $\lambda_i$) | orthonormal eigenvectors in $Q$ |
| **Jordan form:** $J = V^{-1}AV$ | diagonal of $J$ is $\Lambda$ | each block gives $\boldsymbol{x} = (0, .., 1, .., 0)$ |
| **SVD for any $A$:** $A = U\Sigma V^{\mathrm{T}}$ | $\operatorname{rank}(A) = \operatorname{rank}(\Sigma)$ | eigenvectors of $A^{\mathrm{T}}A$, $AA^{\mathrm{T}}$ in $V, U$ |

## ■ CHAPTER 6 NOTES ■

**A symmetric matrix $S$ has perpendicular eigenvectors.** Suppose $S\boldsymbol{x} = \lambda_1\boldsymbol{x}$ and $S\boldsymbol{y} = \lambda_2\boldsymbol{y}$ and $\lambda_1 \neq \lambda_2$. Subtract $\lambda_1 I$ from both equations:

$$(S - \lambda_1 I)\boldsymbol{x} = \boldsymbol{0} \qquad \text{and} \qquad (S - \lambda_1 I)\boldsymbol{y} = (\lambda_2 - \lambda_1)\boldsymbol{y}.$$

This puts $\boldsymbol{x}$ in the nullspace and $\boldsymbol{y}$ in the column space of $S - \lambda_1 I$. That matrix is real symmetric, so its column space is also its row space. Then $\boldsymbol{x}$ in the nullspace is sure to be perpendicular to $\boldsymbol{y}$ in the row space. A new proof that $\boldsymbol{x}^{\mathrm{T}}\boldsymbol{y} = 0$.

Several proofs that $S$ has a full set of $n$ independent (and orthogonal) eigenvectors—even in the case of repeated eigenvalues—are on the course website for linear algebra: **web.mit.edu/18.06** (Proofs of the Spectral Theorem).

### Similar Matrices and the Jordan Form

For every $A$, we want to choose $V$ so that $V^{-1}AV$ is as ***nearly diagonal as possible***. When $A$ has a full set of $n$ eigenvectors, they go into the columns of $V$. Then the matrix $V^{-1}AV$ is diagonal, period. This matrix $\Lambda$ is the Jordan form of $A$—when $A$ can be diagonalized. But if eigenvectors are missing, $\Lambda$ can't be reached.

Suppose $A$ has $s$ independent eigenvectors. Then it is similar to a matrix with $s$ blocks. *Each block has the eigenvalue* $\lambda$ *on the diagonal with* 1's *just above it*. This block accounts for one eigenvector. When there are $n$ eigenvectors and $n$ blocks, $J$ is $\Lambda$.

**(Jordan form)** If $A$ has $s$ independent eigenvectors, it is similar to a matrix $J$ that has Jordan blocks $J_1$ to $J_s$ on its diagonal. Some matrix $V$ puts $A$ into its Jordan form $J$:

**Jordan form**
$$V^{-1}AV = \begin{bmatrix} J_1 & & \\ & \ddots & \\ & & J_s \end{bmatrix} = J.$$

Each block in $J$ has one eigenvalue $\lambda_i$, one eigenvector, and 1's above the diagonal:

**Jordan block**
$$J_i = \begin{bmatrix} \lambda_i & 1 & & \\ & \cdot & \cdot & \\ & & \cdot & 1 \\ & & & \lambda_i \end{bmatrix}.$$

***$A$ is similar to $B$ if they share the same Jordan form $J$—not otherwise.***

The Jordan form $J$ has an off-diagonal 1 for each missing eigenvector (and the 1's are next to the eigenvalues). This is the big theorem about matrix similarity. In every family of similar matrices, we are picking one outstanding member called $J$. It is nearly diagonal (or if possible completely diagonal). We can solve $dz/dt = Jz$ by back substitution. Then we have solved $d\boldsymbol{y}/dt = A\boldsymbol{y}$ with $\boldsymbol{y} = V\boldsymbol{z}$.

Jordan's Theorem is proved in my textbook *Linear Algebra and Its Applications*. The reasoning is rather intricate and the Jordan form is not at all popular in computations. A slight change in $A$ will separate the repeated eigenvalues and bring a diagonal $\Lambda$.

**Time-varying systems $y' = A(t)y$ : Wrong formula and correct formula for $y(t)$**

Section 6.4 recognized that linear systems are more difficult when the matrix depends on $t$. The formula $\boldsymbol{y}(t) = \exp(\int A(t)dt)\boldsymbol{y}(0)$ is not correct. The underlying reason is that $e^{A+B}$ (the wrong matrix) is generally different from $e^A e^B$ (the correct matrix at $t = 2$, when the system jumps from $\boldsymbol{y}' = B\boldsymbol{y}$ to $\boldsymbol{y}' = A\boldsymbol{y}$ at $t = 1$.) Go forward in time : $e^B$ *and then* $e^A$.

It is not usual for a basic textbook to attempt a correct formula. But this is a chance to emphasize that Euler's difference equation goes forward in the right order. It steps from $\boldsymbol{Y}_n$ at time $n\Delta t$ to $\boldsymbol{Y}_{n+1}$ at time $(n+1)\Delta t$, using the current matrix $A$ at time $n\Delta t$.

$$\textbf{Euler's method} \quad \Delta Y/\Delta t = AY \text{ or } \boldsymbol{Y_{n+1} = E_n Y_n} \quad \text{with } E_n = I + \Delta t A(n\Delta t).$$

When we reach $\boldsymbol{Y}_N$, we have multiplied $\boldsymbol{Y}_0$ by $N$ matrices $E_0$ to $E_{N-1}$ *in the right order* :

$$\boldsymbol{Y}_N = E_{N-1}E_{N-2}\ldots E_1 E_0 \boldsymbol{Y}_0.$$

Basic theory says that Euler's $\boldsymbol{Y}_N$ approaches the correct $\boldsymbol{y}(t)$, when $\Delta t = t/N$ and $N \to \infty$. That product of $E$'s approaches the correct replacement for $e^{At}$. When $A$ is a constant matrix, not changing with time, all $E$'s are the same and we reach $e^{At}$ from $E^N$ :

$$\textbf{Constant matrix } \boldsymbol{A} \qquad e^{\boldsymbol{A}t} = \text{limit of } (I + \Delta t A)^N = \text{limit of } \left(I + \frac{At}{N}\right)^N.$$

This came from compound interest in Section 1.3, when $A$ was a number (1 by 1 matrix).

The limit of $E_{N-1}E_{N-2}\ldots E_1 E_0$ is called a **product integral**. An ordinary "*sum integral*" $\int A(t)dt$ is the limit of a sum of $N$ terms $\Delta t A$ (each term going to zero). Now we are multiplying $N$ terms $I + \Delta t A$ (each term going to $I$). Term by term, $I + \Delta t A$ is close to $e^{\Delta t A}$. But matrices don't always commute, and $\exp \int A(t)dt$ is *wrong*. Matrix products $E_{N-1}\ldots E_1 E_0$ approach a *product integral* and the correct $\boldsymbol{y}(t)$.

**Product integral** $M(t) = \text{limit of } E_{N-1}E_{N-2}\ldots E_1 E_0$. Then $\boldsymbol{y}(t) = M(t)\boldsymbol{y}(0)$.

One final good note. The *determinant* $W(t)$ of the matrix $M(t)$ has a nice formula. This succeeds because numbers $\det A$ (but not matrices $A$) can be multiplied in any order. Here is the beautiful fact that gives the equation for the Wronskian determinant $W(t)$ :

$$\boxed{\text{If } \frac{dM}{dt} = AM \text{ then } \frac{dW}{dt} = (\text{trace}\,(A))W. \text{ Therefore } W(t) = e^{\int \text{trace}\,(A(t))dt}\,W(0).}$$

This is equation (21) in Section 6.4. We see again that the Wronskian $W(t)$ is never zero, because exponentials are never zero. For $y'' + B(t)y' + C(t)y = 0$, the companion matrix has trace $-B(t)$. The Wronskian is $W(t) = e^{-\int B(t)dt}\,W(0)$ as Abel discovered.

# Chapter 7

# Applied Mathematics and $A^{\mathrm{T}}A$

A chapter title that includes the symbols $A^{\mathrm{T}}A$ is not usual. Most textbooks deal with $A$ and its eigenvalues, and stop. When the original problem involves a rectangular matrix, as so many problems do, the steps to reach a square matrix are omitted. In reality, rectangular matrices are everywhere—they connect current and voltage, displacement and force, position and momentum, prices and income, *pairs of unknowns*.

It is true that the eventual equation contains a square matrix (very often symmetric). We start from $A$ and we reach $A^{\mathrm{T}}A$. Those two matrices have the same nullspace. We want $A^{\mathrm{T}}A$ to be invertible so we can solve the problem. Then $A$ must have independent columns (no nullspace except the zero vector) as we now assume: $A$ must be "tall and thin" with $m \geq n$ and full column rank $r = n$.

$S = A^{\mathrm{T}}A$ has positive eigenvalues. It is a **positive definite symmetric matrix**. Its eigenvectors lead us to the **Singular Value Decomposition** of $A$. The SVD in Section 7.2 is the best way to discover what is important, when a large matrix is filled with data. The singular vectors are like eigenvectors for a square matrix, with the extra guarantee of orthogonality.

The chapter starts with $m$ equations in $n$ unknowns—too many equations, too few unknowns, and *no solution to* $A\boldsymbol{v} = \boldsymbol{b}$. This is a major application of linear algebra (and geometry and calculus). A sensor or a scanner or a counter makes thousands of measurements. Often we are overwhelmed with data. If it lies close to a straight line, that line $v_1 + v_2 t$ or $C + Dt$ has only $n = 2$ parameters. Those are the two numbers we want, coming from $m = 1000$ or $1000000$ measurements.

Our first applications, are *least squares* and *weighted least squares*. The 2 by 2 matrix $A^{\mathrm{T}}A$ or $A^{\mathrm{T}}CA$ will appear ($C$ contains the weights). This is the symmetric matrix $S$ of Section 6.5 and Section 7.1, and the stiffness matrix $K$ of Section 7.4, and the conductance matrix of Section 7.5, and the second derivative $A^{\mathrm{T}}A = -d^2/dx^2$ in 7.3. (A minus sign is included, because if $A = d/dx$ is the first derivative then $-d/dx$ is its transpose.)

"Symmetric positive definite"—those are three important words in linear algebra. And they are key ideas in applied mathematics, to be presented in this chapter.

## 7.1 Least Squares and Projections

Start with $Av = \boldsymbol{b}$. The matrix $A$ has $n$ independent columns; its rank is $n$. But $A$ has $m$ rows, and $m$ is greater than $n$. We have $m$ measurements in $\boldsymbol{b}$, and we want to choose $n < m$ parameters $v$ that fit those measurements. An exact fit $Av = \boldsymbol{b}$ is generally impossible. We look for the closest fit to the data—the best solution $\widehat{v}$.

The *error vector* $\boldsymbol{e} = \boldsymbol{b} - A\widehat{v}$ tells how close we are to solving $Av = \boldsymbol{b}$. The errors in the $m$ equations are $e_1, \ldots, e_m$. Make the *sum of squares* as small as possible.

**Least squares solution** $\widehat{v}$ **Minimize** $||e||^2 = e_1^2 + \cdots + e_m^2 = ||b - Av||^2$.

This is our goal, to reduce $\boldsymbol{e}$. If $Av = \boldsymbol{b}$ has a solution (and possibly it could), then the best $\widehat{v}$ is certainly that solution vector $v$. In this case the error is $\boldsymbol{e} = \boldsymbol{0}$, certainly a minimum. But normally there is no exact solution to the $m$ equations $Av = \boldsymbol{b}$. The column space of $A$ is only an $n$-dimensional subspace of $\mathbf{R}^m$. Almost all vectors $\boldsymbol{b}$ are outside that subspace—they are not combinations of the columns of $A$. We reduce the error $E = ||\boldsymbol{e}||^2$ as far as possible, but we cannot reach zero error.

**Example 1** Find the straight line $b = C + Dt$ that goes through 4 points : $b = 1, 9, 9, 21$ at $t = 0, 1, 3, 4$. Those are four equations for $C$ and $D$, and they have *no solution*. The four crosses in Figure 7.1 are not on a straight line :

$$\begin{array}{l}\boldsymbol{Av = b} \textbf{ has}\\ \textbf{no solution}\end{array} \qquad \begin{aligned} C + 0D &= 1 \\ C + 1D &= 9 \\ C + 3D &= 9 \\ C + 4D &= 21 \end{aligned} \quad \text{is} \quad \begin{bmatrix} 1 & 0 \\ 1 & 1 \\ 1 & 3 \\ 1 & 4 \end{bmatrix} \begin{bmatrix} C \\ D \end{bmatrix} = \begin{bmatrix} 1 \\ 9 \\ 9 \\ 21 \end{bmatrix}. \qquad (1)$$

$C = 1$ solves the first equation, then $D = 8$ solves the second equation. Then the other equations fail by a lot. We want a better balance, where no equation is exact but the total squared error $E = e_1^2 + e_2^2 + e_3^2 + e_4^2$ from all four equations is as small as possible.

**The best $C$ and $D$ are 2 and 4. The best $v$ is $\widehat{v} = (2, 4)$. The best line is $2 + 4t$.** At the four measurement times $t = 0, 1, 3, 4$, this best line has heights $2, 6, 14, 18$. In other words, $A\widehat{v}$ is $\boldsymbol{p} = (2, 6, 14, 18)$ which is as close as possible to $\boldsymbol{b} = (1, 9, 9, 21)$.

For that vector $\boldsymbol{p} = (2, 6, 14, 18)$, the four bullets in Figure 7.1 fall on the line $2 + 4t$. How do we find that best solution $\widehat{v} = (C, D) = (2, 4)$ ? *It has the smallest error* $E$ :

$$E = e_1^2 + e_2^2 + e_3^2 + e_4^2 = (1 - C - 0D)^2 + (9 - C - 1D)^2 + (9 - C - 3D)^2 + (21 - C - 4D)^2.$$

We can use pure linear algebra to find $C = 2$ and $D = 4$, or pure calculus. To use calculus, set two partial derivatives to zero : $\partial E/\partial C = 0$ and $\partial E/\partial D = 0$. Solve for $C$ and $D$.

Linear algebra gives the right triangle in Figure 7.1. The vector $\boldsymbol{b}$ is split into $\boldsymbol{p} + \boldsymbol{e}$. The heights $\boldsymbol{p}$ lie on a line and the errors $\boldsymbol{e}$ are as small as possible. I will use calculus first, and then the linear algebra that I prefer—because it produces a right triangle $\boldsymbol{p} + \boldsymbol{e} = \boldsymbol{b}$.

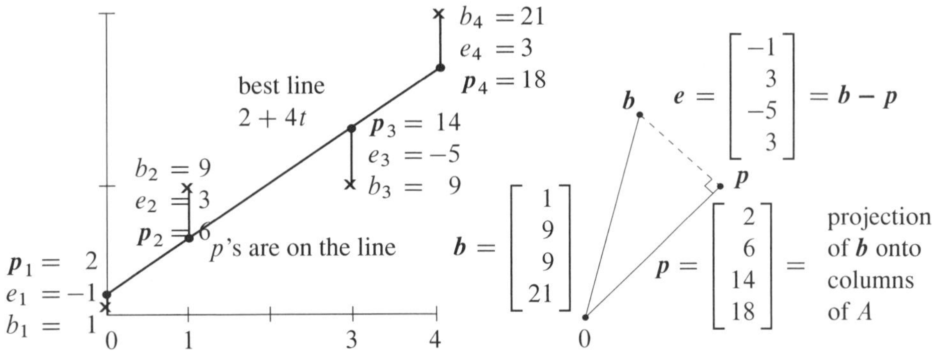


Figure 7.1: Two pictures! The best line has $\boldsymbol{e}^{\mathrm{T}}\boldsymbol{e} = 1 + 9 + 25 + 9 = 44 = ||\boldsymbol{b} - \boldsymbol{p}||^2$.

Let me give away the answer immediately (the equation for $C$ and $D$). Then you can compute the best solution $\widehat{\boldsymbol{v}}$ and the projection $\boldsymbol{p} = A\widehat{\boldsymbol{v}}$ and the error $\boldsymbol{e} = \boldsymbol{b} - A\widehat{\boldsymbol{v}}$. **The best least squares estimate $\widehat{\boldsymbol{v}} = (C, D)$ solves the "normal equations" using the square symmetric invertible matrix $A^{\mathrm{T}}A$ :**

**Normal equations to find $\widehat{\boldsymbol{v}}$** $\qquad$ $A^{\mathrm{T}}A\widehat{\boldsymbol{v}} = A^{\mathrm{T}}\boldsymbol{b}$. $\qquad$ (2)

In short, multiply the unsolvable equations $A\boldsymbol{v} = \boldsymbol{b}$ by $A^{\mathrm{T}}$ to get $A^{\mathrm{T}}A\widehat{\boldsymbol{v}} = A^{\mathrm{T}}\boldsymbol{b}$.

**Example 1** (completed) The normal equations $A^{\mathrm{T}}A\widehat{\boldsymbol{v}} = A^{\mathrm{T}}\boldsymbol{b}$ are

$$\begin{bmatrix} 1 & 1 & 1 & 1 \\ 0 & 1 & 3 & 4 \end{bmatrix} \begin{bmatrix} 1 & 0 \\ 1 & 1 \\ 1 & 3 \\ 1 & 4 \end{bmatrix} \begin{bmatrix} \widehat{C} \\ \widehat{D} \end{bmatrix} = \begin{bmatrix} 1 & 1 & 1 & 1 \\ 0 & 1 & 3 & 4 \end{bmatrix} \begin{bmatrix} 1 \\ 9 \\ 9 \\ 21 \end{bmatrix}. \qquad (3)$$

After multiplication this matrix $A^{\mathrm{T}}A$ is square and symmetric and positive definite:

$$\boldsymbol{A^{\mathrm{T}}A\widehat{v} = A^{\mathrm{T}}b} \qquad \begin{bmatrix} 4 & 8 \\ 8 & 26 \end{bmatrix} \begin{bmatrix} \widehat{C} \\ \widehat{D} \end{bmatrix} = \begin{bmatrix} 40 \\ 120 \end{bmatrix} \quad \text{gives} \quad \begin{bmatrix} \widehat{C} \\ \widehat{D} \end{bmatrix} = \begin{bmatrix} 2 \\ 4 \end{bmatrix}. \qquad (4)$$

At $t = 0, 1, 3, 4$ this best line $2 + 4t$ in Figure 7.1 has heights $\boldsymbol{p} = 2, 6, 14, 18$. The minimum error $\boldsymbol{b} - \boldsymbol{p}$ is $\boldsymbol{e} = (-1, 3, -5, 3)$. The picture on the right is the "linear algebra way" to see least squares. We project $\boldsymbol{b}$ to $\boldsymbol{p}$ in the column space of $A$ (you see how $\boldsymbol{p}$ is perpendicular to the error vector $\boldsymbol{e}$). Then $A\widehat{\boldsymbol{v}} = \boldsymbol{p}$ has the best possible right side $\boldsymbol{p}$.

The solution $\widehat{\boldsymbol{v}} = (\widehat{C}, \widehat{D}) = (2, 4)$ is the least squares choice of $C$ and $D$.

**Normal equations using calculus** The two equations are $\partial E/\partial C = 0$ and $\partial E/\partial D = 0$.

The first column shows the four terms $e_1^2 + e_2^2 + e_3^2 + e_4^2$ that add to $E$. Next to them are the derivatives that add to $\partial E/\partial C$ and $\partial E/\partial D$. Notice how the chain rule brings factors 0, 1, 3, 4 in the third column for $\partial E/\partial D$.

$$\textbf{Add each column} \quad E = \begin{matrix} (C+0D-1)^2 \\ (C+1D-9)^2 \\ (C+3D-9)^2 \\ (C+4D-21)^2 \end{matrix} \qquad \frac{\partial \boldsymbol{E}}{\partial \boldsymbol{C}} = \begin{matrix} 2(C+0D-1) \\ 2(C+1D-9) \\ 2(C+3D-9) \\ 2(C+4D-21) \end{matrix} \qquad \frac{\partial \boldsymbol{E}}{\partial \boldsymbol{D}} = \begin{matrix} 2(C+0D-1)(0) \\ 2(C+1D-9)(1) \\ 2(C+3D-9)(3) \\ 2(C+4D-21)(4) \end{matrix}$$

No problem to divide all derivatives by 2, when $\partial E/\partial C = 0$ and $\partial E/\partial D = 0$. The last two columns are added by matrix multiplication (notice the numbers 0, 1, 3, 4 in $\partial E/\partial D$).

$$\frac{1}{2}\begin{bmatrix} \partial E/\partial C \\ \partial E/\partial D \end{bmatrix} = \begin{bmatrix} 1 & 1 & 1 & 1 \\ 0 & 1 & 3 & 4 \end{bmatrix} \begin{bmatrix} C+0D & - & 1 \\ C+1D & - & 9 \\ C+3D & - & 9 \\ C+4D & - & 21 \end{bmatrix} = \begin{bmatrix} 0 \\ 0 \end{bmatrix}. \qquad (5)$$

The 2 by 4 matrix is $A^{\mathrm{T}}$. The 4 by 1 vector is $A\widehat{\boldsymbol{v}} - \boldsymbol{b}$. Calculus has found $A^{\mathrm{T}}A\boldsymbol{v} = A^{\mathrm{T}}\boldsymbol{b}$.

**Example 2** Suppose we have two equations for one unknown $\boldsymbol{v}$. Thus $n = 1$ but $m = 2$ (probably there is no solution). One unknown means only one column in $A$ :

$$A\boldsymbol{v} = \boldsymbol{b} \quad \text{is} \quad \begin{bmatrix} a_1 \\ a_2 \end{bmatrix} \boldsymbol{v} = \begin{bmatrix} b_1 \\ b_2 \end{bmatrix} \qquad \text{For example} \quad \begin{matrix} 2\boldsymbol{v} = 1 \\ 3\boldsymbol{v} = 8 \end{matrix}. \qquad (6)$$

The matrix $A$ is 2 by 1. The squared error is $E = e_1^2 + e_2^2 = (1 - 2\boldsymbol{v})^2 + (8 - 3\boldsymbol{v})^2$.

$$\textbf{Sum of squares} \qquad E(\boldsymbol{v}) = (b_1 - a_1\boldsymbol{v})^2 + (b_2 - a_2\boldsymbol{v})^2.$$

The graph of $E(\boldsymbol{v})$ is a parabola. Its bottom point is at the least squares solution $\widehat{\boldsymbol{v}}$. The minimum error occurs when $dE/d\boldsymbol{v} = 0$ :

$$\textbf{Equation for } \widehat{\boldsymbol{v}} \qquad \frac{dE}{d\boldsymbol{v}} = 2a_1\,(a_1\widehat{\boldsymbol{v}} - b_1) + 2a_2\,(a_2\widehat{\boldsymbol{v}} - b_2) = 0. \qquad (7)$$

Cancel the 2's, so $(a_1^2 + a_2^2)\widehat{\boldsymbol{v}} = (a_1b_1 + a_2b_2)$. The left side has $a_1^2 + a_2^2 = A^{\mathrm{T}}A$. The right side is $a_1b_1 + a_2b_2 = A^{\mathrm{T}}\boldsymbol{b}$. Calculus has again found $A^{\mathrm{T}}A\widehat{\boldsymbol{v}} = A^{\mathrm{T}}\boldsymbol{b}$ :

$$\begin{bmatrix} a_1 & a_2 \end{bmatrix} \begin{bmatrix} a_1 \\ a_2 \end{bmatrix} \widehat{\boldsymbol{v}} = \begin{bmatrix} a_1 & a_2 \end{bmatrix} \begin{bmatrix} b_1 \\ b_2 \end{bmatrix} \quad \text{produces} \quad \widehat{\boldsymbol{v}} = \frac{\boldsymbol{a}^{\mathrm{T}}\boldsymbol{b}}{\boldsymbol{a}^{\mathrm{T}}\boldsymbol{a}} = \frac{a_1b_1 + a_2b_2}{a_1^2 + a_2^2}. \qquad (8)$$

The numerical example has $\boldsymbol{a} = (2, 3)$ and $\boldsymbol{b} = (1, 8)$ and $\widehat{\boldsymbol{v}} = \boldsymbol{a}^{\mathrm{T}}\boldsymbol{b}/\boldsymbol{a}^{\mathrm{T}}\boldsymbol{a} = 26/13 = 2$.

**Example 3** The special case $a_1 = a_2 = 1$ has two measurements $v = b_1$ and $v = b_2$ of the same quantity (like pulse rate or blood pressure). The matrix has $A^{\mathrm{T}} = [1 \quad 1]$. To minimize $(v - b_1)^2 + (v - b_2)^2$, the best $\widehat{v}$ is just the average measurement:

$$\text{If} \quad a_1 = a_2 = 1 \quad \text{then} \quad A^{\mathrm{T}}A = 2 \quad \text{and} \quad A^{\mathrm{T}}\boldsymbol{b} = b_1 + b_2 \quad \text{and} \quad \widehat{v} = (b_1 + b_2)/2.$$

The linear algebra picture in Figure 7.2 shows the projection of $\boldsymbol{b}$ onto the line through $\boldsymbol{a}$. The projection is $\boldsymbol{p}$, the angle is 90°, and the other side of the right triangle is $\boldsymbol{e} = \boldsymbol{b} - \boldsymbol{p}$. **The normal equations are saying that $e$ is perpendicular to the line through $\boldsymbol{a}$.**

## Least Squares by Linear Algebra

Here is the linear algebra approach to $A^{\mathrm{T}}A\widehat{\boldsymbol{v}} = A^{\mathrm{T}}\boldsymbol{b}$. It takes one wonderful line :

**$\boldsymbol{e} = \boldsymbol{b} - A\widehat{\boldsymbol{v}}$ is perpendicular to the column space of $A$. So $\boldsymbol{e}$ is in the nullspace of $A^{\mathrm{T}}$.**

Then $A^{\mathrm{T}}\boldsymbol{b} = A^{\mathrm{T}}A\widehat{\boldsymbol{v}}$. That fourth subspace $\boldsymbol{N}(A^{\mathrm{T}})$ is exactly what least squares needs: $\boldsymbol{e}$ is perpendicular to the whole column space of $A$ and not just to $\boldsymbol{p} = A\widehat{\boldsymbol{v}} = A(A^{\mathrm{T}}A)^{-1}A^{\mathrm{T}}\boldsymbol{b}$.

Figure 7.2 shows the projection $\boldsymbol{p}$ as an $m$ by $m$ matrix $P$ multiplying $\boldsymbol{b}$. To project any vector onto the column space of $A$, multiply by the *projection matrix* $P$.

**Projection matrix gives $\boldsymbol{p} = P\boldsymbol{b}$** $$P = \frac{\boldsymbol{a}\boldsymbol{a}^{\mathrm{T}}}{\boldsymbol{a}^{\mathrm{T}}\boldsymbol{a}} \quad \text{or} \quad P = A(A^{\mathrm{T}}A)^{-1}A^{\mathrm{T}}. \tag{9}$$

The first form of $P$ gives the projection on the line through $\boldsymbol{a}$. Here $A$ has only one column and $A^{\mathrm{T}}A = \boldsymbol{a}^{\mathrm{T}}\boldsymbol{a}$. We can divide by that number, but for $n > 1$ the right notation is $(A^{\mathrm{T}}A)^{-1}$. The second form gives $P$ in all cases, provided only that $A^{\mathrm{T}}A$ is invertible:

**Two key properties of projection matrices** $$P^{\mathrm{T}} = P \quad \text{and} \quad P^2 = P. \tag{10}$$

The projection of $\boldsymbol{p}$ is $\boldsymbol{p}$ itself (because $\boldsymbol{p} = P\boldsymbol{b}$ is already in the column space). Then two projections give the same result as one projection : $P(P\boldsymbol{b}) = P\boldsymbol{b}$ and $P^2 = P$.

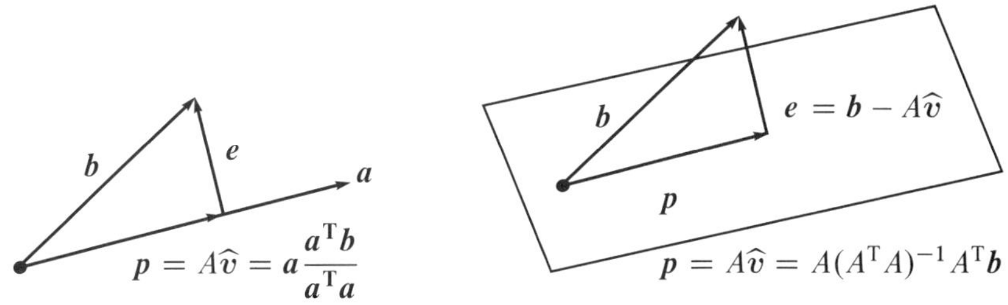


Figure 7.2: The projection $\boldsymbol{p}$ is the nearest point to $\boldsymbol{b}$ in the column space of $A$. Left ($n = 1$) : column space = line through $\boldsymbol{a}$. Right ($n = 2$) : Column space = plane.

Let me review the four essential equations of (unweighted) least squares :

| | | |
|---|---|---|
| **1.** | $Av = b$ | $m$ equations, $n$ unknowns, probably **no solution** |
| **2.** | $A^{\mathrm{T}}A\widehat{v} = A^{\mathrm{T}}b$ | **normal equations**, $\widehat{v} = (A^{\mathrm{T}}A)^{-1}A^{\mathrm{T}}b$ = best $v$ |
| **3.** | $p = A\widehat{v} = A(A^{\mathrm{T}}A)^{-1}A^{\mathrm{T}}b$ | **projection** $p$ of $b$ onto the column space of $A$ |
| **4.** | $P = A(A^{\mathrm{T}}A)^{-1}A^{\mathrm{T}}$ | **projection matrix** $P$ produces $p = Pb$ for any $b$ |

**Example 4** If $A = \begin{bmatrix} 1 & 0 \\ 1 & 1 \\ 1 & 2 \end{bmatrix}$ and $b = \begin{bmatrix} 6 \\ 0 \\ 0 \end{bmatrix}$ find $\widehat{v}$ and $p$ and the matrix $P$.

**Solution** Compute the square matrix $A^{\mathrm{T}}A$ and also the vector $A^{\mathrm{T}}b$ :

$$A^{\mathrm{T}}A = \begin{bmatrix} 1 & 1 & 1 \\ 0 & 1 & 2 \end{bmatrix}\begin{bmatrix} 1 & 0 \\ 1 & 1 \\ 1 & 2 \end{bmatrix} = \begin{bmatrix} 3 & 3 \\ 3 & 5 \end{bmatrix} \text{ and } \begin{bmatrix} 1 & 1 & 1 \\ 0 & 1 & 2 \end{bmatrix}\begin{bmatrix} 6 \\ 0 \\ 0 \end{bmatrix} = \begin{bmatrix} 6 \\ 0 \end{bmatrix}.$$

Now solve the normal equations $A^{\mathrm{T}}A\widehat{v} = A^{\mathrm{T}}b$ to find $\widehat{v}$ :

$$\begin{bmatrix} 3 & 3 \\ 3 & 5 \end{bmatrix}\begin{bmatrix} \widehat{v}_1 \\ \widehat{v}_2 \end{bmatrix} = \begin{bmatrix} 6 \\ 0 \end{bmatrix} \quad \text{gives} \quad \widehat{v} = \begin{bmatrix} \widehat{v}_1 \\ \widehat{v}_2 \end{bmatrix} = \begin{bmatrix} 5 \\ -3 \end{bmatrix}. \tag{11}$$

The combination $p = A\widehat{v}$ is the projection of $b$ onto the column space of $A$ :

$$p = 5\begin{bmatrix} 1 \\ 1 \\ 1 \end{bmatrix} - 3\begin{bmatrix} 0 \\ 1 \\ 2 \end{bmatrix} = \begin{bmatrix} 5 \\ 2 \\ -1 \end{bmatrix}. \text{ The error is } e = b - p = \begin{bmatrix} 1 \\ -2 \\ 1 \end{bmatrix}. \tag{12}$$

Two checks on the calculation. First, the error $e = (1, -2, 1)$ is perpendicular to both columns $(1, 1, 1)$ and $(0, 1, 2)$. Second, the projection matrix $P$ times $b = (6, 0, 0)$ correctly gives $p = (5, 2, -1)$. That solves the problem for one particular $b$.

To find $p = Pb$ for every $b$, compute $P = A(A^{\mathrm{T}}A)^{-1}A^{\mathrm{T}}$. The determinant of $A^{\mathrm{T}}A$ is $15 - 9 = 6$; then $(A^{\mathrm{T}}A)^{-1}$ is easy. Multiply $A$ times $(A^{\mathrm{T}}A)^{-1}$ times $A^{\mathrm{T}}$ to reach $P$ :

$$(A^{\mathrm{T}}A)^{-1} = \frac{1}{6}\begin{bmatrix} 5 & -3 \\ -3 & 3 \end{bmatrix} \text{ and } P = \frac{1}{6}\begin{bmatrix} 5 & 2 & -1 \\ 2 & 2 & 2 \\ -1 & 2 & 5 \end{bmatrix}. \tag{13}$$

We must have $P^2 = P$, because a second projection doesn't change the first projection.

**Warning** The matrix $P = A(A^{\mathrm{T}}A)^{-1}A^{\mathrm{T}}$ is deceptive. You might try to split $(A^{\mathrm{T}}A)^{-1}$ into $A^{-1}$ times $(A^{\mathrm{T}})^{-1}$. If you make that mistake, and substitute it into $P$, you will find $P = AA^{-1}(A^{\mathrm{T}})^{-1}A^{\mathrm{T}}$. Apparently everything cancels. This looks like $P = I$, the identity matrix. The next two lines explain why this is wrong.

***The matrix A is rectangular. It has no inverse matrix.*** We cannot split $(A^{\mathrm{T}}A)^{-1}$ into $A^{-1}$ times $(A^{\mathrm{T}})^{-1}$ because there is no $A^{-1}$ in the first place.

In our experience, a problem that involves a rectangular matrix almost always leads to $A^{\mathrm{T}}A$. When $A$ has independent columns, $A^{\mathrm{T}}A$ is invertible. This fact is so crucial that we state it clearly and give a proof.

**$A^{\mathrm{T}}A$ is invertible if and only if $A$ has linearly independent columns.**

**Proof** $A^{\mathrm{T}}A$ is a square matrix ($n$ by $n$). For every matrix $A$, we will now show that $A^{\mathrm{T}}A$ ***has the same nullspace as*** $A$. When $A$ has independent columns, its nullspace contains only the zero vector. Then $A^{\mathrm{T}}A$, with this same nullspace, is invertible.

Let $A$ be any matrix. If $\boldsymbol{x}$ is in its nullspace, then $A\boldsymbol{x} = \mathbf{0}$. Multiplying by $A^{\mathrm{T}}$ gives $A^{\mathrm{T}}A\boldsymbol{x} = \mathbf{0}$. So $\boldsymbol{x}$ is also in the nullspace of $A^{\mathrm{T}}A$.

Now start with the nullspace of $A^{\mathrm{T}}A$. **From** $A^{\mathrm{T}}A\boldsymbol{x} = \mathbf{0}$ **we must prove** $A\boldsymbol{x} = \mathbf{0}$. We can't multiply by $(A^{\mathrm{T}})^{-1}$, which generally doesn't exist. Just multiply by $\boldsymbol{x}^{\mathrm{T}}$:

$$(\boldsymbol{x}^{\mathrm{T}})A^{\mathrm{T}}A\boldsymbol{x} = 0 \quad \text{or} \quad (A\boldsymbol{x})^{\mathrm{T}}(A\boldsymbol{x}) = 0 \quad \text{or} \quad \|A\boldsymbol{x}\|^2 = 0.$$

This says : If $A^{\mathrm{T}}A\boldsymbol{x} = \mathbf{0}$ then $A\boldsymbol{x}$ has length zero. Therefore $A\boldsymbol{x} = \mathbf{0}$.

Every vector $\boldsymbol{x}$ in one nullspace is in the other nullspace. If $A^{\mathrm{T}}A$ has dependent columns, so has $A$. If $A^{\mathrm{T}}A$ has independent columns, so has $A$. This is the good case :

***When $A$ has independent columns, $A^{\mathrm{T}}A$ is square, symmetric, and invertible.***

To repeat for emphasis : $A^{\mathrm{T}}A$ is ($n$ by $m$) times ($m$ by $n$). Then $A^{\mathrm{T}}A$ is square ($n$ by $n$). It is symmetric, because its transpose is $(A^{\mathrm{T}}A)^{\mathrm{T}} = A^{\mathrm{T}}(A^{\mathrm{T}})^{\mathrm{T}}$ which equals $A^{\mathrm{T}}A$. We just proved that $A^{\mathrm{T}}A$ is invertible—provided $A$ has independent columns. Watch the difference between dependent columns and independent columns:

$$\underset{A^{\mathrm{T}}}{\begin{bmatrix} 1 & 1 & 0 \\ 2 & 2 & 0 \end{bmatrix}} \underset{\text{dependent}}{\overset{A}{\begin{bmatrix} 1 & 2 \\ 1 & 2 \\ 0 & 0 \end{bmatrix}}} = \underset{\text{singular}}{\overset{A^{\mathrm{T}}A}{\begin{bmatrix} 2 & 4 \\ 4 & 8 \end{bmatrix}}} \qquad \overset{A^{\mathrm{T}}}{\begin{bmatrix} 1 & 1 & 0 \\ 2 & 2 & 1 \end{bmatrix}} \underset{\text{indep.}}{\overset{A}{\begin{bmatrix} 1 & 2 \\ 1 & 2 \\ 0 & 1 \end{bmatrix}}} = \underset{\text{invertible}}{\overset{A^{\mathrm{T}}A}{\begin{bmatrix} 2 & 4 \\ 4 & 9 \end{bmatrix}}}$$

**Very brief summary** To find the projection $\boldsymbol{p} = \widehat{v}_1\boldsymbol{a}_1 + \cdots + \widehat{v}_n\boldsymbol{a}_n$, solve $A^{\mathrm{T}}A\widehat{\boldsymbol{v}} = A^{\mathrm{T}}\boldsymbol{b}$. This gives $\widehat{\boldsymbol{v}}$. The projection is $A\widehat{\boldsymbol{v}}$ and the error is $\boldsymbol{e} = \boldsymbol{b} - \boldsymbol{p} = \boldsymbol{b} - A\widehat{\boldsymbol{v}}$. The projection matrix $P = A(A^{\mathrm{T}}A)^{-1}A^{\mathrm{T}}$ multiplies $\boldsymbol{b}$ to give the projection $\boldsymbol{p} = P\boldsymbol{b}$.

**This matrix satisfies $P^2 = P$.** ***The distance from b to the subspace is $\|e\|$.***

## Weighted Least Squares

There is normally error in the measurements $\boldsymbol{b}$. That produces error in the output $\widehat{\boldsymbol{v}}$. Some measurements $b_i$ may be more reliable than others (from less accurate sensors). We should give heavier weight to those reliable $b_i$.

We assume that the expected error in each $b_i$ is zero. Then negative errors balance positive errors in the long run, and *the mean error is zero.* **The expected squared error in the measurement $b_i$ (the "mean squared error") is its variance ${\sigma_i}^2$:**

$$\textbf{Mean}\;\; m_i = E[e_i] = 0 \qquad \textbf{Variance}\;\; {\sigma_i}^2 = \text{expected squared error } \underline{E}[e_i^2] \qquad (14)$$

We should give equation $i$ more weight when $\sigma_i$ is small. Then $b_i$ is more reliable.

Statistically, the right weight is $w_i = 1/\sigma_i$. We multiply $A\boldsymbol{v} = \boldsymbol{b}$ by the diagonal matrix $W$ with those weights $w_1, \ldots, w_m$. Then solve $WA\boldsymbol{v} = W\boldsymbol{b}$ by ordinary least squares, using $WA$ and $W\boldsymbol{b}$ instead of $A$ and $\boldsymbol{b}$:

$$\textbf{Weighted least squares}\;\; (WA)^{\mathrm{T}}(WA)\widehat{\boldsymbol{v}} = (WA)^{\mathrm{T}}W\boldsymbol{b} \textbf{ is } A^{\mathrm{T}}CA\widehat{\boldsymbol{v}} = A^{\mathrm{T}}C\boldsymbol{b}. \qquad (15)$$

$C = W^{\mathrm{T}}W$ goes between $A^{\mathrm{T}}$ and $A$, to produce the weighted matrix $K = A^{\mathrm{T}}CA$.

**Example 5** Your pulse rate $v$ is measured twice. Using unweighted least squares ($w_1 = w_2 = 1$), the best estimate is $\widehat{v} = \frac{1}{2}(b_1 + b_2)$. Example 3 finds that least square solution $\widehat{v}$ to two equations $v = b_1$ and $v = b_2$. But if you were more nervous the first time, then $\sigma_1$ is larger than $\sigma_2$. The first measurement $b_1$ has a larger variance than $b_2$.

We should weight the two measurements by $w_1 = 1/\sigma_1$ and $w_2 = 1/\sigma_2$:

$$\textbf{With weights} \qquad \begin{aligned} w_1 v &= w_1 b_1 \\ w_2 v &= w_2 b_2 \end{aligned} \qquad \widehat{v} = \frac{w_1 b_1 + w_2 b_2}{w_1^2 + w_2^2} \qquad (16)$$

When $w_1 = w_2 = 1$, that answer $\widehat{v}$ reduces to the unweighted estimate $\frac{1}{2}(b_1 + b_2)$.

The weighted $K = A^{\mathrm{T}}CA$ has the same good properties as the unweighted $A^{\mathrm{T}}A$: square, symmetric, and invertible when $A$ has independent columns (as in the example). *Then all eigenvalues of $A^{\mathrm{T}}A$ and $A^{\mathrm{T}}CA$ have $\lambda > 0$*: positive definite matrices!

## ■ REVIEW OF THE KEY IDEAS ■

1. The least squares solution $\widehat{\boldsymbol{v}}$ minimizes $E = ||\boldsymbol{b} - A\boldsymbol{v}||^2$. Then $A^{\mathrm{T}}A\widehat{\boldsymbol{v}} = A^{\mathrm{T}}\boldsymbol{b}$.
2. To fit $m$ points by a line $C + Dt$, $A$ is $m$ by 2 and $\widehat{\boldsymbol{v}} = (\widehat{C}, \widehat{D})$ gives the best line.
3. The projection of $\boldsymbol{b}$ on the column space of $A$ is $\boldsymbol{p} = A\widehat{\boldsymbol{v}} = P\boldsymbol{b}$: closest point to $\boldsymbol{b}$.
4. The error is $e = \boldsymbol{b} - \boldsymbol{p}$. The projection matrix is $P = A(A^{\mathrm{T}}A)^{-1}A^{\mathrm{T}}$ with $P^2 = P$.
5. Weighted least squares has $A^{\mathrm{T}}CA\widehat{\boldsymbol{v}} = A^{\mathrm{T}}C\boldsymbol{b}$. Good weights $c_i$ are 1/variance of $b_i$.

## Problem Set 7.1

**1** Suppose your pulse is measured at $b_1 = 70$ beats per minute, then $b_2 = 120$, then $b_3 = 80$. The least squares solution to three equations $v = b_1, v = b_2, v = b_3$ with $A^{\mathrm{T}} = [1\ \ 1\ \ 1]$ is $\widehat{v} = (A^{\mathrm{T}}A)^{-1}A^{\mathrm{T}}\boldsymbol{b} =$ ______. Use calculus and projections:

(a) Minimize $E = (v-70)^2 + (v-120)^2 + (v-80)^2$ by solving $dE/dv = 0$.

(b) Project $\boldsymbol{b} = (70, 120, 80)$ onto $\boldsymbol{a} = (1, 1, 1)$ to find $\widehat{v} = \boldsymbol{a}^{\mathrm{T}}\boldsymbol{b}/\boldsymbol{a}^{\mathrm{T}}\boldsymbol{a}$.

**2** Suppose $Av = \boldsymbol{b}$ has $m$ equations $a_i v = b_i$ in *one unknown* $v$. For the sum of squares $E = (a_1 v - b_1)^2 + \cdots + (a_m v - b_m)^2$, find the minimizing $\widehat{v}$ by calculus. Then form $A^{\mathrm{T}}A\widehat{v} = A^{\mathrm{T}}\boldsymbol{b}$ with one column in $A$, and reach the same $\widehat{v}$.

**3** With $\boldsymbol{b} = (4, 1, 0, 1)$ at the points $x = (0, 1, 2, 3)$ set up and solve the normal equation for the coefficients $\widehat{\boldsymbol{v}} = (C, D)$ in the nearest line $C + Dx$. Start with the four equations $A\boldsymbol{v} = \boldsymbol{b}$ that would be solvable if the points fell on a line.

**4** In Problem 3, find the projection $\boldsymbol{p} = A\widehat{\boldsymbol{b}}$. Check that those four values on the line $C + Dx$. Compute the error $\boldsymbol{e} = \boldsymbol{b} - \boldsymbol{p}$ and verify that $A^{\mathrm{T}}\boldsymbol{e} = \boldsymbol{0}$.

**5** (Problem 3 by calculus) Write down $E = ||\boldsymbol{b} - A\boldsymbol{v}||^2$ as a sum of four squares: the last one is $(1 - C - 3D)^2$. Find the derivative equations $\partial E/\partial C = \partial E/\partial D = 0$. Divided by 2 to obtain $A^{\mathrm{T}}A\widehat{\boldsymbol{v}} = A^{\mathrm{T}}\boldsymbol{b}$.

**6** For the closest parabola $C + Dt + Et^2$ to the same four points, write down 4 unsolvable equations $A\boldsymbol{v} = \boldsymbol{b}$ for $\boldsymbol{v} = (C, D, E)$. Set up the normal equations for $\widehat{\boldsymbol{v}}$. If you fit the best cubic $C + Dt + Et^2 + Ft^3$ to those four points (thought experiment), what is the error vector $\boldsymbol{e}$ ?

**7** Write down three equations for the line $b = C + Dt$ to go through $b = 7$ at $t = -1, b = 7$ at $t = 1$, and $b = 21$ at $t = 2$. Find the least squares solution $\widehat{\boldsymbol{v}} = (C, D)$ and draw the closest line.

**8** Find the projection $\boldsymbol{p} = A\widehat{\boldsymbol{v}}$ in Problem 7. This gives the three heights of the closest line. Show that the error vector is $\boldsymbol{e} = (2, -6, 4)$.

**9** Suppose the measurements at $t = -1, 1, 2$ are the errors $2, -6, 4$ in Problem 8. Compute $\widehat{\boldsymbol{v}}$ and the closest line to these new measurements. Explain the answer: $\boldsymbol{b} = (2, -6, 4)$ is perpendicular to ______ so the projection is $\boldsymbol{p} = \boldsymbol{0}$.

**10** Suppose the measurements at $t = -1, 1, 2$ are $\boldsymbol{b} = (5, 13, 17)$. Compute $\widehat{\boldsymbol{v}}$ and the closest line $\boldsymbol{e}$. The error is $\boldsymbol{e} = \boldsymbol{0}$ because this $\boldsymbol{b}$ is ______.

**11** Find the best line $C + Dt$ to fit $\boldsymbol{b} = 4, 2, -1, 0, 0$ at times $t = -2, -1, 0, 1, 2$.

**12** Find the *plane* that gives the best fit to the 4 values $\boldsymbol{b} = (0, 1, 3, 4)$ at the corners $(1, 0)$ and $(0, 1)$ and $(-1, 0)$ and $(0, -1)$ of a square. At those 4 points, the equations $C + Dx + Ey = b$ are $A\boldsymbol{v} = \boldsymbol{b}$ with 3 unknowns $\boldsymbol{v} = (C, D, E)$.

**13** With $\boldsymbol{b} = 0, 8, 8, 20$ at $t = 0, 1, 3, 4$ set up and solve the normal equations $A^{\mathrm{T}}A\boldsymbol{v} = A^{\mathrm{T}}\boldsymbol{b}$. For the best straight line $C + Dt$, find its four heights $p_i$ and four errors $e_i$. What is the minimum value $E = e_1^2 + e_2^2 + e_3^2 + e_4^2$ ?

**14** (By calculus) Write down $E = ||\boldsymbol{b} - A\boldsymbol{v}||^2$ as a sum of four squares—the last one is $(C + 4D - 20)^2$. Find the derivative equations $\partial E/\partial C = 0$ and $\partial E/\partial D = 0$. Divide by 2 to obtain the normal equations $A^{\mathrm{T}}A\widehat{\boldsymbol{v}} = A^{\mathrm{T}}\boldsymbol{b}$.

**15** Which of the four subspaces contains the error vector $\boldsymbol{e}$ ? Which contains $p$ ? Which contains $\widehat{\boldsymbol{v}}$ ?

**16** Find the height $C$ of the best *horizontal line* to fit $\boldsymbol{b} = (0, 8, 8, 20)$. An exact fit would solve the four unsolvable equations $C = 0, C = 8, C = 8, C = 20$. Find the 4 by 1 matrix $A$ in these equations and solve $A^{\mathrm{T}}A\widehat{\boldsymbol{v}} = A^{\mathrm{T}}\boldsymbol{b}$.

**17** Write down three equations for the line $b = C + Dt$ to go through $b = 7$ at $t = -1, b = 7$ at $t = 1$, and $b = 21$ at $t = 2$. Find the least squares solution $\widehat{\boldsymbol{v}} = (C, D)$ and draw the closest line.

**18** Find the projection $\boldsymbol{p} = A\widehat{\boldsymbol{v}}$ in Problem 17. This gives the three heights of the closest line. Show that the error vector is $\boldsymbol{e} = (2, -6, 4)$. Why is $P\boldsymbol{e} = \boldsymbol{0}$ ?

**19** Suppose the measurements at $t = -1, 1, 2$ are the errors $2, -6, 4$ in Problem 18. Compute $\widehat{\boldsymbol{v}}$ and the closest line to these new measurements. Explain the answer: $\boldsymbol{b} = (2, -6, 4)$ is perpendicular to ______ so the projection is $\boldsymbol{p} = \boldsymbol{0}$.

**20** Suppose the measurements at $t = -1, 1, 2$ are $\boldsymbol{b} = (5, 13, 17)$. Compute $\widehat{\boldsymbol{v}}$ and the closest line and $\boldsymbol{e}$. The error is $\boldsymbol{e} = \boldsymbol{0}$ because this $\boldsymbol{b}$ is ______ ?

**Questions 21–26 ask for projections onto lines. Also errors $\boldsymbol{e} = \boldsymbol{b} - \boldsymbol{p}$ and matrices $\boldsymbol{p}$.**

**21** Project the vector $\boldsymbol{b}$ onto the line through $\boldsymbol{a}$. Check that $\boldsymbol{e}$ is perpendicular to $\boldsymbol{a}$ :

$$\text{(a)}\ \boldsymbol{b} = \begin{bmatrix} 1 \\ 2 \\ 3 \end{bmatrix} \text{ and } \boldsymbol{a} = \begin{bmatrix} 1 \\ 1 \\ 1 \end{bmatrix} \qquad \text{(b)}\ \boldsymbol{b} = \begin{bmatrix} 1 \\ 3 \\ 1 \end{bmatrix} \text{ and } \boldsymbol{a} = \begin{bmatrix} -1 \\ -3 \\ -1 \end{bmatrix}.$$

**22** Draw the projection of $\boldsymbol{b}$ onto $\boldsymbol{a}$ and also compute it from $\boldsymbol{p} = \widehat{v}\boldsymbol{a}$ :

$$\text{(a)}\ \boldsymbol{b} = \begin{bmatrix} \cos\theta \\ \sin\theta \end{bmatrix} \text{ and } \boldsymbol{a} = \begin{bmatrix} 1 \\ 0 \end{bmatrix} \qquad \text{(b)}\ \boldsymbol{b} = \begin{bmatrix} 1 \\ 1 \end{bmatrix} \text{ and } \boldsymbol{a} = \begin{bmatrix} 1 \\ -1 \end{bmatrix}.$$

**23** In Problem 22 find the projection matrix $P = \boldsymbol{a}\boldsymbol{a}^{\mathrm{T}}/\boldsymbol{a}^{\mathrm{T}}\boldsymbol{a}$ onto each vector $\boldsymbol{a}$. Verify in both cases that $P^2 = P$. Multiply $P\boldsymbol{b}$ in each case to find the projection $\boldsymbol{p}$.

**24** Construct the projection matrices $P_1$ and $P - 2$ onto the lines through the $\boldsymbol{a}$'s in Problem 22. Is it true that $(P_1 + P_2)^2 = P_1 + P_2$ ? This *would* be true if $P_1P_2 = 0$.

**25** Compute the projection matrices $\boldsymbol{a}\boldsymbol{a}^{\mathrm{T}}/\boldsymbol{a}^{\mathrm{T}}\boldsymbol{a}$ onto the lines through $\boldsymbol{a}_1 = (-1, 2, 2)$ and $\boldsymbol{a}_2 = (2, 2, -1)$. Multiply those two matrices $P_1P_2$ and explain the answer.

**26** Continuing Problem 25, find the projection matrix $P_3$ onto $\boldsymbol{a}_3 = (2, -1, 2)$. Verify that $P_1 + P_2 + P_3 = I$. The basis $\boldsymbol{a}_1, \boldsymbol{a}_2, \boldsymbol{a}_3$ is orthogonal!

**27** Project the vector $\boldsymbol{b} = (1, 1)$ onto the lines through $\boldsymbol{a}_1 = (1, 0)$ and $\boldsymbol{a}_2 = (1, 2)$. Draw the projections $\boldsymbol{p}_1$ and $\boldsymbol{p}_2$ and add $\boldsymbol{p}_1 + \boldsymbol{p}_2$. The projections do not add to $\boldsymbol{b}$ because the $\boldsymbol{a}$'s are not orthogonal.

**28** (Quick and recommended) Suppose $A$ is the 4 by 4 identity matrix with its last column removed. $A$ is 4 by 3. Project $\boldsymbol{b} = (1, 2, 3, 4)$ onto the column space of $A$. What shape is the projection matrix $P$ and what is $P$?

**29** If $A$ is doubled, then $P = 2A(4A^{\mathrm{T}}A)^{-1}2A^{\mathrm{T}}$. This is the same as $A(A^{\mathrm{T}}A)^{-1}A^{\mathrm{T}}$. The column space of $2A$ is the same as ______. Is $\widehat{\boldsymbol{v}}$ the same for $A$ and $2A$?

**30** What linear combination of $(1, 2, -1)$ and $(1, 0, 1)$ is closest to $\boldsymbol{b} = (2, 1, 1)$?

**31** (*Important*) If $P^2 = P$ show that $(I - P)^2 = I - P$. When $P$ projects onto the column space of $A$, $I - P$ projects onto which fundamental subspace ?

**32** If $P$ is the 3 by 3 projection matrix onto the line through $(1, 1, 1)$, then $I - P$ is the projection matrix onto ______.

**33** Multiply the matrix $P = A(A^{\mathrm{T}}A)^{-1}A^{\mathrm{T}}$ by itself. Cancel to prove that $P^2 = P$. Explain why $P(P\boldsymbol{b})$ always equals $P\boldsymbol{b}$: The vector $P\boldsymbol{b}$ is in the column space so its projection is ______.

**34** If $A$ is square and invertible, the warning against splitting $(A^{\mathrm{T}}A)^{-1}$ does not apply. Then $AA^{-1}(A^{\mathrm{T}})^{-1}A^{\mathrm{T}} = I$ is true. *When $A$ is invertible, why is $P = I$ and $\boldsymbol{e} = \boldsymbol{0}$?*

**35** An important fact about $A^{\mathrm{T}}A$ is this: ***If $A^{\mathrm{T}}A\boldsymbol{x} = \boldsymbol{0}$ then $A\boldsymbol{x} = \boldsymbol{0}$***. *New proof*: The vector $A\boldsymbol{x}$ is in the nullspace of ______. $A\boldsymbol{x}$ is always in the column space of ______. To be in both of those perpendicular spaces, $A\boldsymbol{x}$ must be zero.

### Notes on mean and variance and test grades

If all grades on a test are 90, the mean is $m = 90$ and the variance is $\sigma^2 = 0$. Suppose the expected grades are $g_1, \ldots, g_N$. Then $\sigma^2$ comes from *squaring distances to the mean*:

$$\textbf{Mean}\ \ m = \frac{g_1 + \cdots + g_N}{N} \qquad \textbf{Variance}\ \ \sigma^2 = \frac{(g_1 - m)^2 + \cdots + (g_N - m)^2}{N}$$

After every test my class wants to know $m$ and $\sigma$. My expectations are usually way off.

**36** Show that $\sigma^2$ also equals $\frac{1}{N}(g_1^2 + \cdots + g_N^2) - m^2$.

**37** If you flip a fair coin $N$ times (1 for heads, 0 for tails) what is the expected number $m$ of heads ? What is the variance $\sigma^2$ ?

## 7.2 Positive Definite Matrices and the SVD

This chapter about applications of $A^{\mathrm{T}}A$ depends on two important ideas in linear algebra. These ideas have big parts to play, we focus on them now.

**1. Positive definite symmetric matrices** (both $A^{\mathrm{T}}A$ and $A^{\mathrm{T}}CA$ are positive definite)

**2. Singular Value Decomposition** ($A = U\Sigma V^{\mathrm{T}}$ gives perfect bases for the 4 subspaces)

Those are orthogonal matrices $U$ and $V$ in the SVD. Their columns are orthonormal eigenvectors of $AA^{\mathrm{T}}$ and $A^{\mathrm{T}}A$. The entries in the diagonal matrix $\Sigma$ are the *square roots* of the eigenvalues. The matrices $AA^{\mathrm{T}}$ and $A^{\mathrm{T}}A$ have the same nonzero eigenvalues.

Section 6.5 showed that the eigenvectors of these symmetric matrices are orthogonal. I will show now that *the eigenvalues of $A^{\mathrm{T}}A$ are positive*, if $A$ has independent columns.

Start with $A^{\mathrm{T}}A\boldsymbol{x} = \lambda\boldsymbol{x}$. Then $\boldsymbol{x}^{\mathrm{T}}A^{\mathrm{T}}A\boldsymbol{x} = \lambda\boldsymbol{x}^{\mathrm{T}}\boldsymbol{x}$. Therefore $\lambda = ||A\boldsymbol{x}||^2/||\boldsymbol{x}||^2 > 0$

I separated $\boldsymbol{x}^{\mathrm{T}}A^{\mathrm{T}}A\boldsymbol{x}$ into $(A\boldsymbol{x})^{\mathrm{T}}(A\boldsymbol{x}) = ||A\boldsymbol{x}||^2$. We don't have $\lambda = 0$ because $A^{\mathrm{T}}A$ is invertible (since $A$ has independent columns). The eigenvalues must be positive.

Those are the key steps to understanding positive definite matrices. They give us three tests on $S$—three ways to recognize when a symmetric matrix $S$ is positive definite :

| | | |
|---|---|---|
| **Positive** | **1.** | All the eigenvalues of $S$ are positive. |
| **definite** | **2.** | The "energy" $\boldsymbol{x}^{\mathrm{T}}S\boldsymbol{x}$ is positive for all nonzero vectors $\boldsymbol{x}$. |
| **symmetric** | **3.** | $S$ has the form $S = A^{\mathrm{T}}A$ with independent columns in $A$. |

There is also a test on the pivots (all $> 0$) and a test on $n$ determinants (all $> 0$).

**Example 1** Are these matrices positive definite ? When their eigenvalues are positive, construct matrices $A$ with $S = A^{\mathrm{T}}A$ and find the positive energy $\boldsymbol{x}^{\mathrm{T}}S\boldsymbol{x}$.

$$\text{(a)}\quad S = \begin{bmatrix} 4 & 0 \\ 0 & 1 \end{bmatrix} \qquad \text{(b)}\quad S = \begin{bmatrix} 5 & 4 \\ 4 & 5 \end{bmatrix} \qquad \text{(c)}\quad S = \begin{bmatrix} 4 & 5 \\ 5 & 4 \end{bmatrix}$$

**Solution** The answers are *yes*, *yes*, and *no*. The eigenvalues of those matrices $S$ are

(a) 4 and 1 : *positive* (b) 9 and 1 : *positive* (c) 9 and $-1$ : *not positive*.

A quicker test than eigenvalues uses **two determinants** : the 1 by 1 determinant $S_{11}$ and the 2 by 2 determinant of $S$. Example (b) has $S_{11} = \mathbf{5}$ and det $S = 25 - 16 = \mathbf{9}$ (*pass*). Example (c) has $S_{11} = \mathbf{4}$ but det $S = 16 - 25 = \mathbf{-9}$ (*fail the test*).

**Positive energy is equivalent to positive eigenvalues**, when $S$ is symmetric. Let me test the energy $\boldsymbol{x}^{\mathrm{T}}S\boldsymbol{x}$ in all three examples. Two examples pass and the third fails :

$$[x_1 \quad x_2]\begin{bmatrix} 4 & 0 \\ 0 & 1 \end{bmatrix}\begin{bmatrix} x_1 \\ x_2 \end{bmatrix} = 4x_1^2 + x_2^2 > 0 \qquad \text{Positive energy when } \boldsymbol{x} \neq \boldsymbol{0}$$

$$[x_1 \quad x_2]\begin{bmatrix} 5 & 4 \\ 4 & 5 \end{bmatrix}\begin{bmatrix} x_1 \\ x_2 \end{bmatrix} = 5x_1^2 + 8x_1x_2 + 5x_2^2 \qquad \text{Positive energy when } \boldsymbol{x} \neq \boldsymbol{0}$$

$$[x_1 \quad x_2]\begin{bmatrix} 4 & 5 \\ 5 & 4 \end{bmatrix}\begin{bmatrix} x_1 \\ x_2 \end{bmatrix} = 4x_1^2 + 10x_1x_2 + 4x_2^2 \qquad \text{Energy } \mathbf{-2} \text{ when } \boldsymbol{x} = (1, -1)$$

Positive energy is a fundamental property. This is the best definition of *positive definiteness*.

When the eigenvalues are positive, there will be many matrices $A$ that give $A^{\mathrm{T}}A = S$. One choice of $A$ is symmetric and positive definite! Then $A^{\mathrm{T}}A$ is $A^2$, and this choice $A = \sqrt{S}$ is a true square root of $S$. The successful examples (a) and (b) have $\boldsymbol{S} = \boldsymbol{A}^2$:

$$\boxed{\begin{bmatrix} 4 & 0 \\ 0 & 1 \end{bmatrix} = \begin{bmatrix} 2 & 0 \\ 0 & 1 \end{bmatrix}\begin{bmatrix} 2 & 0 \\ 0 & 1 \end{bmatrix} \quad \text{and} \quad \begin{bmatrix} 5 & 4 \\ 4 & 5 \end{bmatrix} = \begin{bmatrix} 2 & 1 \\ 1 & 2 \end{bmatrix}\begin{bmatrix} 2 & 1 \\ 1 & 2 \end{bmatrix}}$$

We know that all symmetric matrices have the form $S = V\Lambda V^{\mathrm{T}}$ with orthonormal eigenvectors in $V$. The diagonal matrix $\Lambda$ has a square root $\sqrt{\Lambda}$, when all eigenvalues are positive. In this case $A = \sqrt{S} = V\sqrt{\Lambda}V^{\mathrm{T}}$ is the symmetric positive definite square root:

$$\boldsymbol{A}^{\mathrm{T}}\boldsymbol{A} = \sqrt{S}\sqrt{S} = (V\sqrt{\Lambda}V^{\mathrm{T}})(V\sqrt{\Lambda}V^{\mathrm{T}}) = V\sqrt{\Lambda}\sqrt{\Lambda}V^{\mathrm{T}} = \boldsymbol{S} \;\text{ because } V^{\mathrm{T}}V = I.$$

Starting from this unique square root $\sqrt{S}$, other choices of $A$ come easily. Multiply $\sqrt{S}$ by any matrix $Q$ that has orthonormal columns (so that $Q^{\mathrm{T}}Q = I$). Then $Q\sqrt{S}$ is another choice for $A$ (not a symmetric choice). In fact all choices come this way :

$$A^{\mathrm{T}}A = (Q\sqrt{S})^{\mathrm{T}}\,(Q\sqrt{S}) = \sqrt{S}Q^{\mathrm{T}}Q\sqrt{S} = S. \tag{1}$$

I will choose a particular $Q$ in Example 1, to get particular choices of $A$.

**Example 1** (*continued*) Choose $Q = \begin{bmatrix} 0 & -1 \\ 1 & 0 \end{bmatrix}$ to multiply $\sqrt{S}$. Then $A = Q\sqrt{S}$.

$$A = \begin{bmatrix} 0 & -1 \\ 1 & 0 \end{bmatrix}\begin{bmatrix} 2 & 0 \\ 0 & 1 \end{bmatrix} = \begin{bmatrix} 0 & -1 \\ 2 & 0 \end{bmatrix} \qquad \text{has} \quad S = A^{\mathrm{T}}A = \begin{bmatrix} 4 & 0 \\ 0 & 1 \end{bmatrix}$$

$$A = \begin{bmatrix} 0 & -1 \\ 1 & 0 \end{bmatrix}\begin{bmatrix} 2 & 1 \\ 1 & 2 \end{bmatrix} = \begin{bmatrix} -1 & -2 \\ 2 & 1 \end{bmatrix} \qquad \text{has} \quad S = A^{\mathrm{T}}A = \begin{bmatrix} 5 & 4 \\ 4 & 5 \end{bmatrix}.$$

## Positive Semidefinite Matrices

Positive *semidefinite* matrices include positive definite matrices, and more. Eigenvalues of $S$ can be zero. Columns of $A$ can be dependent. The energy $\boldsymbol{x}^{\mathrm{T}}S\boldsymbol{x}$ can be zero—*but not negative*. This gives new equivalent conditions on a (possibly singular) matrix $S = S^{\mathrm{T}}$.

**1′** All eigenvalues of $S$ satisfy $\lambda \geq 0$ (semidefinite allows zero eigenvalues).

**2′** The energy is nonnegative for every $\boldsymbol{x}$ : $\boldsymbol{x}^{\mathrm{T}}S\boldsymbol{x} \geq 0$ (zero energy is allowed).

**3′** $S$ has the form $A^{\mathrm{T}}A$ (every $A$ is allowed; its columns can be dependent).

**Example 2** The first two matrices are singular and positive semidefinite—but not the third :

$$\text{(d)} \quad S = \begin{bmatrix} 0 & 0 \\ 0 & 1 \end{bmatrix} \qquad \text{(e)} \quad S = \begin{bmatrix} 4 & 4 \\ 4 & 4 \end{bmatrix} \qquad \text{(f)} \quad S = \begin{bmatrix} -4 & 4 \\ 4 & -4 \end{bmatrix}.$$

The eigenvalues are $1, 0$ and $8, 0$ and $-8, 0$. The energies $\boldsymbol{x}^{\mathrm{T}}S\boldsymbol{x}$ are $x_2^2$ and $4(x_1 + x_2)^2$ and $-4(x_1 - x_2)^2$. So the third matrix is actually *negative* semidefinite.

## Singular Value Decomposition

Now we start with $A$, square or rectangular. Applications also start this way—the matrix comes from the model. The SVD splits any matrix into *orthogonal* $U$ times *diagonal* $\Sigma$ times *orthogonal* $V^{\mathrm{T}}$. Those orthogonal factors will give orthogonal bases for the four fundamental subspaces associated with $A$.

Let me describe the goal for any $m$ by $n$ matrix, and then how to achieve that goal.

**Find orthonormal bases** $\boldsymbol{v}_1, \ldots, \boldsymbol{v}_n$ **for** $\mathbf{R}^n$ **and** $\boldsymbol{u}_1, \ldots, \boldsymbol{u}_m$ **for** $\mathbf{R}^m$ **so that**

$$A\boldsymbol{v}_1 = \sigma_1\boldsymbol{u}_1 \quad \ldots \quad A\boldsymbol{v}_r = \sigma_r\boldsymbol{u}_r \qquad A\boldsymbol{v}_{r+1} = \mathbf{0} \quad \ldots \quad A\boldsymbol{v}_n = \mathbf{0} \qquad (2)$$

The rank of $A$ is $r$. Those requirements in (4) are expressed by a multiplication $\boldsymbol{AV} = \boldsymbol{U\Sigma}$. The $r$ nonzero singular values $\sigma_1 \geq \sigma_2 \geq \ldots \geq \sigma_r > 0$ are on the diagonal of $\Sigma$ :

$$\boldsymbol{AV} = \boldsymbol{U\Sigma} \qquad A\begin{bmatrix} \\ \boldsymbol{v}_1 \ \ldots \ \boldsymbol{v}_r \ \ldots \ \boldsymbol{v}_n \\ \\ \end{bmatrix} = \begin{bmatrix} \\ \boldsymbol{u}_1 \ \ldots \ \boldsymbol{u}_r \ \ldots \ \boldsymbol{u}_m \\ \\ \end{bmatrix}\begin{bmatrix} \sigma_1 & & & 0 \\ & \ddots & & \\ & & \sigma_r & \\ 0 & & & 0 \end{bmatrix} \qquad (3)$$

The last $n - r$ vectors in $V$ are a basis for the nullspace of $A$. The last $m - r$ vectors in $U$ are a basis for the nullspace of $A^{\mathrm{T}}$. The diagonal matrix $\Sigma$ is $m$ by $n$, with $r$ nonzeros.

Remember that $V^{-1} = V^{\mathrm{T}}$, because the columns $\boldsymbol{v}_1, \ldots, \boldsymbol{v}_n$ are orthonormal in $\mathbf{R}^n$ :

$$\textbf{Singular Value Decomposition} \qquad \boldsymbol{AV} = \boldsymbol{U\Sigma} \quad \textbf{becomes} \quad \boldsymbol{A} = \boldsymbol{U\Sigma V}^{\mathrm{T}}. \qquad (4)$$

The SVD has orthogonal matrices $U$ and $V$, containing eigenvectors of $AA^{\mathrm{T}}$ and $A^{\mathrm{T}}A$.

*Comment.* A square matrix is diagonalized by its eigenvectors : $A\boldsymbol{x}_i = \lambda_i \boldsymbol{x}_i$ is like $A\boldsymbol{v}_i = \sigma_i \boldsymbol{u}_i$. But even if $A$ has $n$ eigenvectors, they may not be orthogonal. We need *two bases*—an input basis of $\boldsymbol{v}$'s in $\mathbf{R}^n$ and an output basis of $\boldsymbol{u}$'s in $\mathbf{R}^m$. With two bases, any $m$ by $n$ matrix can be diagonalized. The beauty of those bases is that they can be chosen orthonormal. Then $U^{\mathrm{T}}U = I$ and $V^{\mathrm{T}}V = I$.

The $\boldsymbol{v}$'s are eigenvectors of the symmetric matrix $S = A^{\mathrm{T}}A$. We can guarantee their orthogonality, so that $\boldsymbol{v}_j^{\mathrm{T}}\boldsymbol{v}_i = 0$ for $j \neq i$. That matrix $S$ is positive semidefinite, so its eigenvalues are $\sigma_i^2 \geq 0$. **The key to the SVD is that $A\boldsymbol{v}_j$ is orthogonal to $A\boldsymbol{v}_i$ :**

$$\textbf{Orthogonal } \boldsymbol{u}\textbf{'s} \quad (A\boldsymbol{v}_j)^{\mathrm{T}}(A\boldsymbol{v}_i) = \boldsymbol{v}_j^{\mathrm{T}}(A^{\mathrm{T}}A\boldsymbol{v}_i) = \boldsymbol{v}_j^{\mathrm{T}}(\sigma_i^2\boldsymbol{v}_i) = \begin{cases} \sigma_i^2 & \text{if } j = i \\ \mathbf{0} & \text{if } j \neq i \end{cases} \tag{5}$$

This says that the vectors $\boldsymbol{u}_i = A\boldsymbol{v}_i/\sigma_i$ are orthonormal for $i = 1, \ldots, r$. They are a basis for the column space of $A$. And the $\boldsymbol{u}$'s are eigenvectors of the symmetric matrix $AA^{\mathrm{T}}$, which is usually different from $S = A^{\mathrm{T}}A$ (but the eigenvalues $\sigma_1^2, \ldots, \sigma_r^2$ are the same).

**Example 3** Find the input and output eigenvectors $\boldsymbol{v}$ and $\boldsymbol{u}$ for the rectangular matrix $A$ :

$$A = \begin{bmatrix} 2 & 2 & 0 \\ -1 & 1 & 0 \end{bmatrix} = U\Sigma V^{\mathrm{T}}.$$

**Solution** Compute $S = A^{\mathrm{T}}A$ and its unit eigenvectors $\boldsymbol{v}_1, \boldsymbol{v}_2, \boldsymbol{v}_3$. The eigenvalues $\sigma^2$ are $8, 2, 0$ so the positive singular values are $\sigma_1 = \sqrt{8}$ and $\sigma_2 = \sqrt{2}$:

$$A^{\mathrm{T}}A = \begin{bmatrix} 5 & 3 & 0 \\ 3 & 5 & 0 \\ 0 & 0 & 0 \end{bmatrix} \quad \text{has} \quad \boldsymbol{v}_1 = \frac{1}{2}\begin{bmatrix} \sqrt{2} \\ \sqrt{2} \\ 0 \end{bmatrix}, \quad \boldsymbol{v}_2 = \frac{1}{2}\begin{bmatrix} \sqrt{2} \\ -\sqrt{2} \\ 0 \end{bmatrix}, \quad \boldsymbol{v}_3 = \begin{bmatrix} 0 \\ 0 \\ 1 \end{bmatrix}.$$

The outputs $\boldsymbol{u}_1 = A\boldsymbol{v}_1/\sigma_1$ and $\boldsymbol{u}_2 = A\boldsymbol{v}_2/\sigma_2$ are also orthonormal, with $\sigma_1 = \sqrt{8}$ and $\sigma_2 = \sqrt{2}$. Those vectors $\boldsymbol{u}_1$ and $\boldsymbol{u}_2$ are in the column space of $A$ :

$$\boldsymbol{u}_1 = \begin{bmatrix} 2 & 2 & 0 \\ -1 & 1 & 0 \end{bmatrix}\frac{\boldsymbol{v}_1}{\sqrt{8}} = \begin{bmatrix} 1 \\ 0 \end{bmatrix} \quad \text{and} \quad \boldsymbol{u}_2 = \begin{bmatrix} 2 & 2 & 0 \\ -1 & 1 & 0 \end{bmatrix}\frac{\boldsymbol{v}_2}{\sqrt{2}} = \begin{bmatrix} 0 \\ 1 \end{bmatrix}.$$

Then $U = I$ and the Singular Value Decomposition for this 2 by 3 matrix is $U\Sigma V^{\mathrm{T}}$ :

$$A = \begin{bmatrix} 2 & 2 & 0 \\ -1 & 1 & 0 \end{bmatrix} = \begin{bmatrix} 1 & 0 \\ 0 & 1 \end{bmatrix}\begin{bmatrix} \boldsymbol{\sqrt{8}} & 0 & 0 \\ 0 & \boldsymbol{\sqrt{2}} & 0 \end{bmatrix}\frac{1}{2}\begin{bmatrix} \sqrt{2} & \sqrt{2} & 0 \\ \sqrt{2} & -\sqrt{2} & 0 \\ 0 & 0 & 2 \end{bmatrix}^{\mathrm{T}}.$$

## The Fundamental Theorem of Linear Algebra

I think of the SVD as the final step in the Fundamental Theorem. First come the *dimensions* of the four subspaces in Figure 7.3. Then come the *orthogonality* of those pairs of subspaces. Now come the *orthonormal bases of* $\boldsymbol{v}$*'s and* $\boldsymbol{u}$*'s that diagonalize* $A$:

**SVD**

$$\boxed{\begin{aligned} A\boldsymbol{v}_j &= \sigma_j \boldsymbol{u}_j \quad \text{for } j \le r \\ A\boldsymbol{v}_j &= \mathbf{0} \quad\quad \text{for } j > r \end{aligned}} \qquad \boxed{\begin{aligned} A^{\mathrm{T}}\boldsymbol{u}_j &= \sigma_j \boldsymbol{v}_j \quad \text{for } j \le r \\ A^{\mathrm{T}}\boldsymbol{u}_j &= \mathbf{0} \quad\quad \text{for } j > r \end{aligned}}$$

Multiplying $A\boldsymbol{v}_j = \sigma_j \boldsymbol{u}_j$ by $A^{\mathrm{T}}$ and dividing by $\sigma_j$ gives that equation $\boldsymbol{A^{\mathrm{T}}u_j = \sigma_j v_j}$.

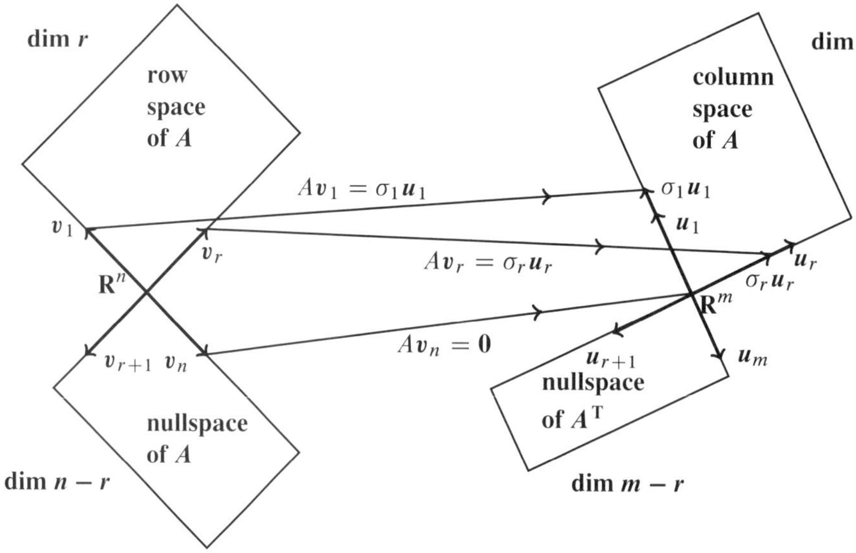


Figure 7.3: Orthonormal bases of $\boldsymbol{v}$'s and $\boldsymbol{u}$'s that diagonalize $A$: $m$ by $n$ with rank $r$.

The "**norm**" of $A$ is its largest singular value : $\boldsymbol{||A|| = \sigma_1}$. This measures the largest possible ratio of $||A\boldsymbol{v}||$ to $||\boldsymbol{v}||$. That ratio of lengths is a maximum when $\boldsymbol{v} = \boldsymbol{v}_1$ and $A\boldsymbol{v} = \sigma_1\boldsymbol{u}_1$. This singular value $\sigma_1$ is a much better measure for the size of a matrix than the largest eigenvalue. An extreme case can have zero eigenvalues and just one eigenvector $(1, 1)$ for $A$. But $A^{\mathrm{T}}A$ can still be large : if $\boldsymbol{v} = (1, -1)$ then $A\boldsymbol{v}$ is 200 times larger.

$$A = \begin{bmatrix} 100 & -100 \\ 100 & -100 \end{bmatrix} \text{ has } \lambda_{\max} = \mathbf{0}. \quad \text{But } \boldsymbol{\sigma_{\max}} = \textbf{norm of } \boldsymbol{A} = \mathbf{200}. \qquad (6)$$

## The Condition Number

A valuable property of $A = U\Sigma V^{\mathrm{T}}$ is that it puts the pieces of $A$ *in order of importance.* Multiplying a column $\boldsymbol{u}_i$ times a row $\sigma_i \boldsymbol{v}_i^{\mathrm{T}}$ produces one piece of the matrix. There will be $r$ nonzero pieces from $r$ nonzero $\sigma$'s, when $A$ has rank $r$. The pieces add up to $A$, when we multiply columns of $U$ times rows of $\Sigma V^{\mathrm{T}}$ :

**The pieces have rank 1**
$$A = \begin{bmatrix} & & \\ \boldsymbol{u}_1 & \ldots & \boldsymbol{u}_r \\ & & \end{bmatrix} \begin{bmatrix} \sigma_1 \boldsymbol{v}_1^{\mathrm{T}} \\ \ldots\ldots \\ \sigma_r \boldsymbol{v}_r^{\mathrm{T}} \end{bmatrix} = \boldsymbol{u}_1(\sigma_1 \boldsymbol{v}_1^{\mathrm{T}}) + \cdots + \boldsymbol{u}_r(\sigma_r \boldsymbol{v}_r^{\mathrm{T}}). \qquad (7)$$

The first piece gives the norm of $A$ which is $\sigma_1$. The last piece gives the norm of $A^{-1}$, which is $1/\sigma_n$ when $A$ is invertible. The **condition number** is $\sigma_1$ times $1/\sigma_n$ :

$$\boxed{\textbf{Condition number of } A \qquad c(A) = ||A||\ ||A^{-1}|| = \frac{\sigma_1}{\sigma_n}.} \qquad (8)$$

This number $c(A)$ is the key to numerical stability in solving $A\boldsymbol{v} = \boldsymbol{b}$. When $A$ is an orthogonal matrix, the symmetric $S = A^{\mathrm{T}}A$ is the identity matrix. So all singular values of an orthogonal matrix are $\sigma = 1$. At the other extreme, a singular matrix has $\sigma_n = 0$. In that case $c = \infty$. Orthogonal matrices have the best condition number $c = 1$.

## Data Matrices : Application of the SVD

"*Big data*" is the linear algebra problem of this century (and we won't solve it here). Sensors and scanners and imaging devices produce enormous volumes of information. Making decisive sense of that data is *the* problem for a world of analysts (mathematicians and statisticians of a new type). Most often the data comes in the form of a matrix.

The usual approach is by PCA—*Principal Component Analysis.* That is essentially the SVD. The first piece $\sigma_1 \boldsymbol{u}_1 \boldsymbol{v}_1^{\mathrm{T}}$ holds the most information (in statistics this piece has the greatest variance). It tells us the most. The Chapter 7 Notes include references.

## ■ REVIEW OF THE KEY IDEAS ■

1. Positive definite symmetric matrices have positive eigenvalues and pivots and energy.
2. $S = A^{\mathrm{T}}A$ is positive definite if and only if $A$ has independent columns.
3. $\boldsymbol{x}^{\mathrm{T}}A^{\mathrm{T}}A\boldsymbol{x} = (A\boldsymbol{x})^{\mathrm{T}}(A\boldsymbol{x})$ is zero when $A\boldsymbol{x} = \boldsymbol{0}$. $A^{\mathrm{T}}A$ can be positive *semidefinite*.
4. The SVD is a factorization $A = U\Sigma V^{\mathrm{T}} =$ (*orthogonal*) (*diagonal*) (*orthogonal*).
5. The columns of $V$ and $U$ are eigenvectors of $A^{\mathrm{T}}A$ and $AA^{\mathrm{T}}$ (singular vectors of $A$).
6. Those orthonormal bases achieve $A\boldsymbol{v}_i = \sigma_i \boldsymbol{u}_i$ and $A$ is diagonalized.
7. The largest piece of $A = \sigma_1 \boldsymbol{u}_1 \boldsymbol{v}_1^{\mathrm{T}} + \cdots + \sigma_r \boldsymbol{u}_r \boldsymbol{v}_r^{\mathrm{T}}$ gives the norm $||A|| = \sigma_1$.

# Problem Set 7.2

**1** For a 2 by 2 matrix, suppose the 1 by 1 and 2 by 2 determinants $a$ and $ac - b^2$ are positive. Then $c > b^2/a$ is also positive.

(i) $\lambda_1$ and $\lambda_2$ have the *same sign* because their product $\lambda_1\lambda_2$ equals ______.

(i) That sign is positive because $\lambda_1 + \lambda_2$ equals ______.

*Conclusion*: The tests $a > 0, ac - b^2 > 0$ guarantee positive eigenvalues $\lambda_1, \lambda_2$.

**2** Which of $S_1$, $S_2$, $S_3$, $S_4$ has two positive eigenvalues? Use $a$ and $ac - b^2$, don't compute the $\lambda$'s. Find an $\boldsymbol{x}$ with $\boldsymbol{x}^{\mathrm{T}}S_1\boldsymbol{x} < 0$, confirming that $A_1$ fails the test.

$$S_1 = \begin{bmatrix} 5 & 6 \\ 6 & 7 \end{bmatrix} \quad S_2 = \begin{bmatrix} -1 & -2 \\ -2 & -5 \end{bmatrix} \quad S_3 = \begin{bmatrix} 1 & 10 \\ 10 & 100 \end{bmatrix} \quad S_4 = \begin{bmatrix} 1 & 10 \\ 10 & 101 \end{bmatrix}.$$

**3** For which numbers $b$ and $c$ are these matrices positive definite?

$$S = \begin{bmatrix} 1 & b \\ b & 9 \end{bmatrix} \qquad S = \begin{bmatrix} 2 & 4 \\ 4 & c \end{bmatrix} \qquad S = \begin{bmatrix} c & b \\ b & c \end{bmatrix}.$$

**4** What is the energy $q = ax^2 + 2bxy + cy^2 = \boldsymbol{x}^{\mathrm{T}}S\boldsymbol{x}$ for each of these matrices? Complete the square to write $q$ as a sum of squares $d_1(\quad)^2 + d_2(\quad)^2$.

$$S = \begin{bmatrix} 1 & 2 \\ 2 & 9 \end{bmatrix} \quad \text{and} \quad S = \begin{bmatrix} 1 & 3 \\ 3 & 9 \end{bmatrix}.$$

**5** $\boldsymbol{x}^{\mathrm{T}}S\boldsymbol{x} = 2x_1x_2$ certainly has a saddle point and not a minimum at $(0, 0)$. What symmetric matrix $S$ produces this energy? What are its eigenvalues?

**6** Test to see if $A^{\mathrm{T}}A$ is positive definite in each case:

$$A = \begin{bmatrix} 1 & 2 \\ 0 & 3 \end{bmatrix} \quad \text{and} \quad A = \begin{bmatrix} 1 & 1 \\ 1 & 2 \\ 2 & 1 \end{bmatrix} \quad \text{and} \quad A = \begin{bmatrix} 1 & 1 & 2 \\ 1 & 2 & 1 \end{bmatrix}.$$

**7** Which 3 by 3 symmetric matrices $S$ and $T$ produce these quadratic energies?

$\boldsymbol{x}^{\mathrm{T}}S\boldsymbol{x} = 2\left(x_1^2 + x_2^2 + x_3^2 - x_1x_2 - x_2x_3\right)$. Why is $S$ positive definite?

$\boldsymbol{x}^{\mathrm{T}}T\boldsymbol{x} = 2\left(x_1^2 + x_2^2 + x_3^2 - x_1x_2 - x_1x_3 - x_2x_3\right)$. Why is $T$ semidefinite?

**8** Compute the three upper left determinants of $S$ to establish positive definiteness. (The first is 2.) Verify that their ratios give the second and third pivots.

**Pivots = ratios of determinants** $\qquad S = \begin{bmatrix} 2 & 2 & 0 \\ 2 & 5 & 3 \\ 0 & 3 & 8 \end{bmatrix}.$

**9** For what numbers $c$ and $d$ are $S$ and $T$ positive definite? Test the 3 determinants:

$$S = \begin{bmatrix} c & 1 & 1 \\ 1 & c & 1 \\ 1 & 1 & c \end{bmatrix} \quad \text{and} \quad T = \begin{bmatrix} 1 & 2 & 3 \\ 2 & d & 4 \\ 3 & 4 & 5 \end{bmatrix}.$$

**10** *If $S$ is positive definite then $S^{-1}$ is positive definite.* Best proof: The eigenvalues of $S^{-1}$ are positive because ____. *Second proof* (only for 2 by 2):

$$\text{The entries of } S^{-1} = \frac{1}{ac - b^2}\begin{bmatrix} c & -b \\ -b & a \end{bmatrix} \quad \text{pass the determinant tests} \quad \_\_\_\_.$$

**11** *If $S$ and $T$ are positive definite, their sum $S + T$ is positive definite.* Pivots and eigenvalues are not convenient for $S + T$. Better to prove $\boldsymbol{x}^{\mathrm{T}}(S + T)\boldsymbol{x} > 0$.

**12** A positive definite matrix *cannot have a zero* (or even worse, a negative number) *on its diagonal*. Show that this matrix fails to have $\boldsymbol{x}^{\mathrm{T}}S\boldsymbol{x} > 0$:

$$\begin{bmatrix} x_1 & x_2 & x_3 \end{bmatrix}\begin{bmatrix} 4 & 1 & 1 \\ 1 & \mathbf{0} & 2 \\ 1 & 2 & 5 \end{bmatrix}\begin{bmatrix} x_1 \\ x_2 \\ x_3 \end{bmatrix} \text{ is not positive when } (x_1, x_2, x_3) = (\quad,\quad,\quad).$$

**13** A diagonal entry $a_{jj}$ of a symmetric matrix cannot be smaller than all the $\lambda$'s. If it were, then $A - a_{jj}I$ would have ____ eigenvalues and would be positive definite. But $A - a_{jj}I$ has a ____ on the main diagonal.

**14** Show that *if all $\lambda > 0$ then $\boldsymbol{x}^{\mathrm{T}}S\boldsymbol{x} > 0$.* We must do this for *every* nonzero $\boldsymbol{x}$, not just the eigenvectors. So write $\boldsymbol{x}$ as a combination of the eigenvectors and *explain why all "cross terms" are $\boldsymbol{x}_i^{\mathrm{T}}\boldsymbol{x}_j = 0$.* Then $\boldsymbol{x}^{\mathrm{T}}S\boldsymbol{x}$ is

$$(c_1\boldsymbol{x}_1 + \cdots + c_n\boldsymbol{x}_n)^{\mathrm{T}}(c_1\lambda_1\boldsymbol{x}_1 + \cdots + c_n\lambda_n x_n) = c_1^2\lambda_1\boldsymbol{x}_1^{\mathrm{T}}\boldsymbol{x}_1 + \cdots + c_n^2\lambda_n\boldsymbol{x}_n^{\mathrm{T}}\boldsymbol{x}_n > 0.$$

**15** Give a quick reason why each of these statements is true:

(a) Every positive definite matrix is invertible.

(b) The only positive definite projection matrix is $P = I$.

(c) A diagonal matrix with positive diagonal entries is positive definite.

(d) A symmetric matrix with a positive determinant might not be positive definite !

**16** With positive pivots in $D$, the factorization $S = LDL^{\mathrm{T}}$ becomes $L\sqrt{D}\sqrt{D}L^{\mathrm{T}}$. (Square roots of the pivots give $D = \sqrt{D}\sqrt{D}$.) Then $A = \sqrt{D}L^{\mathrm{T}}$ yields the ***Cholesky factorization*** $S = A^{\mathrm{T}}A$ which is "symmetrized $LU$":

$$\text{From } A = \begin{bmatrix} 3 & 1 \\ 0 & 2 \end{bmatrix} \text{ find } S. \qquad \text{From } S = \begin{bmatrix} 4 & 8 \\ 8 & 25 \end{bmatrix} \text{ find } A = \mathbf{chol}(S).$$

**17** Without multiplying $S = \begin{bmatrix} \cos\theta & -\sin\theta \\ \sin\theta & \cos\theta \end{bmatrix} \begin{bmatrix} 2 & 0 \\ 0 & 5 \end{bmatrix} \begin{bmatrix} \cos\theta & \sin\theta \\ -\sin\theta & \cos\theta \end{bmatrix}$, find

(a) the determinant of $S$ (b) the eigenvalues of $S$
(c) the eigenvectors of $S$ (d) a reason why $S$ is symmetric positive definite.

**18** For $F_1(x,y) = \frac{1}{4}x^4 + x^2y + y^2$ and $F_2(x,y) = x^3 + xy - x$ find the second derivative matrices $H_1$ and $H_2$:

$$\textbf{Test for minimum} \quad H = \begin{bmatrix} \partial^2 F/\partial x^2 & \partial^2 F/\partial x \partial y \\ \partial^2 F/\partial y \partial x & \partial^2 F/\partial y^2 \end{bmatrix} \text{ is positive definite}$$

$H_1$ is positive definite so $F_1$ is concave up ($=$ convex). Find the minimum point of $F_1$ and the saddle point of $F_2$ (look only where first derivatives are zero).

**19** The graph of $z = x^2 + y^2$ is a bowl opening upward. The graph of $z = x^2 - y^2$ is *a saddle*. The graph of $z = -x^2 - y^2$ is a bowl opening downward. What is a test on $a, b, c$ for $z = ax^2 + 2bxy + cy^2$ to have a saddle point at $(0,0)$ ?

**20** Which values of $c$ give a bowl and which $c$ give a saddle point for the graph of $z = 4x^2 + 12xy + cy^2$ ? Describe this graph at the borderline value of $c$.

**21** When $S$ and $T$ are symmetric positive definite, $ST$ might not even be symmetric. But its eigenvalues are still positive. Start from $ST\boldsymbol{x} = \lambda\boldsymbol{x}$ and take dot products with $T\boldsymbol{x}$. Then prove $\lambda > 0$.

**22** Suppose $C$ is positive definite (so $\boldsymbol{y}^{\mathrm{T}}C\boldsymbol{y} > 0$ whenever $\boldsymbol{y} \neq \boldsymbol{0}$) and $A$ has independent columns (so $A\boldsymbol{x} \neq \boldsymbol{0}$ whenever $\boldsymbol{x} \neq \boldsymbol{0}$). Apply the energy test to $\boldsymbol{x}^{\mathrm{T}}A^{\mathrm{T}}CA\boldsymbol{x}$ to show that $A^{\mathrm{T}}CA$ is positive definite: *the crucial matrix in engineering*.

**23** Find the eigenvalues and unit eigenvectors $\boldsymbol{v}_1, \boldsymbol{v}_2$ of $A^{\mathrm{T}}A$. Then find $\boldsymbol{u}_1 = A\boldsymbol{v}_1/\sigma_1$:

$$A = \begin{bmatrix} 1 & 2 \\ 3 & 6 \end{bmatrix} \text{ and } A^{\mathrm{T}}A = \begin{bmatrix} 10 & 20 \\ 20 & 40 \end{bmatrix} \text{ and } AA^{\mathrm{T}} = \begin{bmatrix} 5 & 15 \\ 15 & 45 \end{bmatrix}.$$

Verify that $\boldsymbol{u}_1$ is a unit eigenvector of $AA^{\mathrm{T}}$. Complete the matrices $U, \Sigma, V$.

$$\textbf{SVD} \quad \begin{bmatrix} 1 & 2 \\ 3 & 6 \end{bmatrix} = \begin{bmatrix} \boldsymbol{u}_1 & \boldsymbol{u}_2 \end{bmatrix} \begin{bmatrix} \sigma_1 & \\ & 0 \end{bmatrix} \begin{bmatrix} \boldsymbol{v}_1 & \boldsymbol{v}_2 \end{bmatrix}^{\mathrm{T}}.$$

**24** Write down orthonormal bases for the four fundamental subspaces of this $A$.

**25** (a) Why is the trace of $A^{\mathrm{T}}A$ equal to the sum of all $a_{ij}^2$ ?

(b) For every rank-one matrix, why is $\sigma_1^2 =$ sum of all $a_{ij}^2$ ?

**26** Find the eigenvalues and unit eigenvectors of $A^{\mathrm{T}}A$ and $AA^{\mathrm{T}}$. Keep each $A\boldsymbol{v} = \sigma\boldsymbol{u}$:

$$\textbf{Fibonacci matrix} \qquad A = \begin{bmatrix} 1 & 1 \\ 1 & 0 \end{bmatrix}$$

Construct the singular value decomposition and verify that $A$ equals $U\Sigma V^{\mathrm{T}}$.

**27** Compute $A^{\mathrm{T}}A$ and $AA^{\mathrm{T}}$ and their eigenvalues and unit eigenvectors for $V$ and $U$.

$$\textbf{Rectangular matrix} \qquad A = \begin{bmatrix} 1 & 1 & 0 \\ 0 & 1 & 1 \end{bmatrix}.$$

Check $AV = U\Sigma$ (this will decide $\pm$ signs in $U$). $\Sigma$ *has the same shape as* $A$.

**28** Construct the matrix with rank one that has $A\boldsymbol{v} = 12\boldsymbol{u}$ for $\boldsymbol{v} = \frac{1}{2}(1,1,1,1)$ and $\boldsymbol{u} = \frac{1}{3}(2,2,1)$. Its only singular value is $\sigma_1 =$ _____.

**29** Suppose $A$ is invertible (with $\sigma_1 > \sigma_2 > 0$). Change $A$ by ***as small a matrix as possible*** to produce a singular matrix $A_0$. Hint: $U$ and $V$ do not change.

$$\text{From} \qquad A = \begin{bmatrix} \boldsymbol{u}_1 & \boldsymbol{u}_2 \end{bmatrix} \begin{bmatrix} \sigma_1 & \\ & \sigma_2 \end{bmatrix} \begin{bmatrix} \boldsymbol{v}_1 & \boldsymbol{v}_2 \end{bmatrix}^{\mathrm{T}} \qquad \text{find the nearest } A_0.$$

**30** The SVD for $A + I$ doesn't use $\Sigma + I$. *Why is* $\sigma(A+I)$ *not just* $\sigma(A) + I$ ?

**31** Multiply $A^{\mathrm{T}}A\boldsymbol{v} = \sigma^2\boldsymbol{v}$ by $A$. Put in parentheses to show that $A\boldsymbol{v}$ is an eigenvector of $AA^{\mathrm{T}}$. We divide by its length $||A\boldsymbol{v}|| = \sigma$ to get the unit eigenvector $\boldsymbol{u}$.

**32** My favorite example of the SVD is when $Av(x) = dv/dx$, with the endpoint conditions $v(0) = 0$ and $v(1) = 0$. We are looking for orthogonal functions $v(x)$ so that their derivatives $Av = dv/dx$ are also orthogonal. The perfect choice is $v_1 = \sin \pi x$ and $v_2 = \sin 2\pi x$ and $v_k = \sin k\pi x$. Then each $u_k$ is a cosine.

The derivative of $v_1$ is $Av_1 = \pi \cos \pi x = \pi u_1$. The singular values are $\sigma_1 = \pi$ and $\sigma_k = k\pi$. Orthogonality of the sines (and orthogonality of the cosines) is the foundation for Fourier series.

*You may object to* $AV = U\Sigma$. The derivative $A = d/dx$ is not a matrix! The orthogonal factor $V$ has functions $\sin k\pi x$ in its columns, not vectors. The matrix $U$ has cosine functions $\cos k\pi x$. Since when is this allowed? One answer is to refer you to the **chebfun** package on the web. This extends linear algebra to matrices whose columns are functions—not vectors.

Another answer is to replace $d/dx$ by a first difference matrix $A$. Its shape will be $N+1$ by $N$. $A$ has 1's down the diagonal and $-1$'s on the diagonal below. Then $AV = U\Sigma$ has discrete sines in $V$ and discrete cosines in $U$. For $N = 2$ those will be sines and cosines of 30° and 60° in $\boldsymbol{v}_1$ and $\boldsymbol{u}_1$.

****** Can you construct the difference matrix $A$ (3 by 2) and $A^{\mathrm{T}}A$ (2 by 2)? The discrete sines are $\boldsymbol{v}_1 = (\sqrt{3}/2, \sqrt{3}/2)$ and $\boldsymbol{v}_2 = (\sqrt{3}/2, -\sqrt{3}/2)$. Test that $A\boldsymbol{v}_1$ is orthogonal to $A\boldsymbol{v}_2$. What are the singular values $\sigma_1$ and $\sigma_2$ in $\Sigma$ ?

## 7.3 Boundary Conditions Replace Initial Conditions

This section is about steady-state problems, not initial-value problems. The time variable $t$ is replaced by the space variable $x$. Instead of two initial conditions at $t = 0$, we have one boundary condition at $x = 0$ and *another boundary condition at* $x = 1$.

Here is the simplest two-point boundary value problem for $y(x)$. Start with $f(x) = 1$.

| **Two boundary conditions** $\quad -\dfrac{d^2y}{dx^2} = f(x)$ with $y(0) = 0$ and $y(1) = 0$. | (1) |
|---|---|

One particular solution $y_p(x)$ will come from integrating $f(x)$ twice. If $f(x) = 1$ then two integrations give $x^2/2$, and the minus sign in (1) leads to $y_p = -x^2/2$.

The null solutions $y_n(x)$ solve the equation with zero force: $-y_n'' = 0$. The second derivative is zero for any linear function $y_n = Cx + D$. These are the null solutions.

We can use those two constants $C$ and $D$ to satisfy the two boundary conditions on the complete solution $\boldsymbol{y(x) = y_p + y_n = -x^2/2 + Cx + D}$.

$$\mathbf{y(0) = 0 \text{ and } y(1) = 0} \qquad \text{Set } x = 0 \text{ and } x = 1 \qquad D = 0 \text{ and } -\frac{1}{2} + C + D = 0$$

The boundary conditions give $D = 0$ and $C = \frac{1}{2}$. Then the solution is $y = y_p + y_n$:

**Solution to $-y'' = 1$** $\qquad y(x) = -\dfrac{x^2}{2} + \dfrac{x}{2} = \dfrac{x - x^2}{2}$

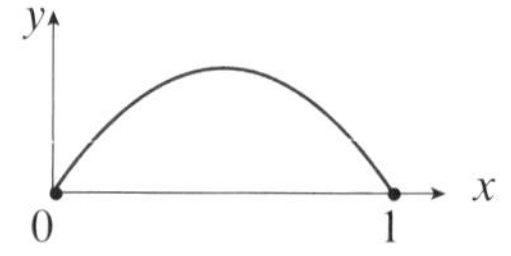


The graph of the parabola starts at $y = 0$ and returns (**fixed ends**). The slope $y' = \frac{1}{2} - x$ is decreasing. The second derivative is $y'' = -1$ and the parabola is bending down.

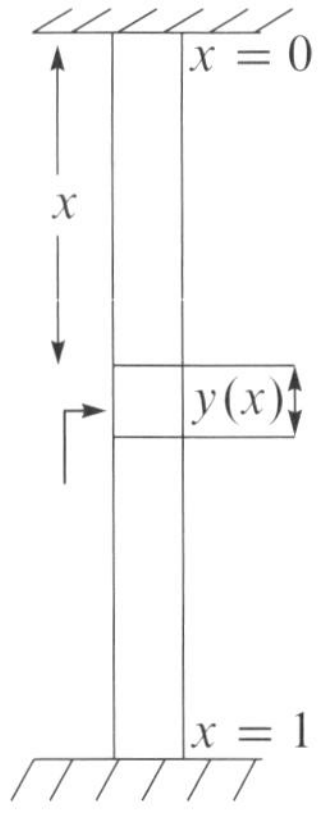


This boundary-value problem describes a bar that has its top and bottom both fixed. The weight of the bar stretches it downward. At point $x$ down the bar, the displacement is $y(x)$. So this fixed-fixed bar has $y(0) = 0$ and $y(1) = 0$. The force of gravity can be $f(x) = 1$. *The bar stretches in the top half where* $dy/dx > 0$. The bottom half is compressed because $dy/dx < 0$. Halfway down at $x = \frac{1}{2}$ is the largest displacement (top of the parabola). That halfway point has $y_{\max} = \frac{1}{2}(x - x^2) = \frac{1}{8}$.

I think of this elastic bar as one long spring. If we pulled it down in the middle, it would start to oscillate. *That is not our problem now.* Our bar is not moving—the oscillation is all damped out. The stretching comes from the bar's own weight.

## A Delta Function

This is my chance to introduce again the mysterious but extremely useful function $f(x) = \delta(x-a)$. This **delta function** is zero except at $x = a$. The bar is now so light that we can ignore its weight. All the force on the bar is at *one point* $x = a$. At that point a unit weight is stretching the bar above $x = a$ and compressing the bar below.

Here is an informal definition of the delta function (the symbol $\infty$ doesn't carry enough information by itself). The good definition is based on *integrating the function across the point* $x = a$. The integral is 1.

**Delta function** $\quad \delta(x-a) = \begin{cases} 0 & x \neq a \\ \infty & x = a \end{cases} \qquad \int \delta(x-a)\,dx = 1, \quad \int \delta(x-a)\,F(x)\,dx = F(a)$

The graph of $\delta(x-a)$ has an *infinite spike* at $x = a$. That spike is at $x = a = 0$ for the standard delta function $\delta(x)$. The function is zero away from the spike and infinite at that one point. **The area under this one-point spike is 1.**

This tells us that $\delta(x)$ cannot be a true function. It is somehow a limit of box functions $B_N(x)$ that have height $N$ over short intervals of width $1/N$. The area of each box is 1 :

$$\textbf{Box functions} \quad B_N(x) = \begin{cases} 0 & |x| > 1/2N \\ N & |x| < 1/2N \end{cases} \qquad \int B_N(x) = \text{ box area } = 1 \qquad \int B_N(x)F(x)\,dx \text{ approaches } F(0)$$

Mathematically, $\delta(x)$ and its shifts $\delta(x-a)$ are not functions. Physically, they represent action that is concentrated at a single point. In reality that action is probably over a very short interval, like the box functions, but the width of that interval is of no importance. What matters is the total impulse when a bat hits a ball, or the total force when a weight hangs on a bar.

The shifted delta function $\delta(x-a)$ is the derivative of the step function $H(x-a)$. The step function jumps from 0 to 1 at $x = a$. Then $\delta$ must integrate to 1.

## Response to a Delta Function is a Ramp Function

How to solve the differential equation $-y'' = \delta(x-a)$? One integration of the delta function gives a step function. *A second integration gives a ramp function or corner function.* The solution $y(x)$ must be linear (straight line graph) to the left of $x = a$, because $d^2y/dx^2 = 0$. And $y(x)$ is also linear to the right of $x = a$: constant slope.

**The slope of $y(x)$ drops by 1 at the point $x = a$.** To see why $-1$ is the jump in slope (there is no jump in $y$ !), integrate $y''$ across the point $x = a$ to get the change $-1$ in $y'$ :

$$\boldsymbol{y'' = -\delta\,(x-a)} \qquad \int y''\,dx = \left[\frac{dy}{dx}\right]_{\text{left of } a}^{\text{right of } a} = \int -\delta(x-a)\,dx = \mathbf{-1} \qquad (2)$$

The solution $y(x)$ starts with a fixed slope $s$. At $x = a$ it changes to slope $s - 1$ (the slope drops by 1). At the point $x = 1$, the bottom of the bar is fixed at $y(1) = 0$.

The constant upward slope $s$ over a distance $a$ and the downward slope $s - 1$ over the remaining distance $1 - a$ must bring the function $y(x)$ to zero:

$$sa + (s-1)(1-a) = 0 \text{ gives } sa + s - sa - 1 + a = 0. \text{ Then } \boldsymbol{s = 1 - a}. \tag{3}$$

The graph of $y = sx$ goes up to $sa = (1 - a)a$. Then $y(x)$ goes back down to zero.

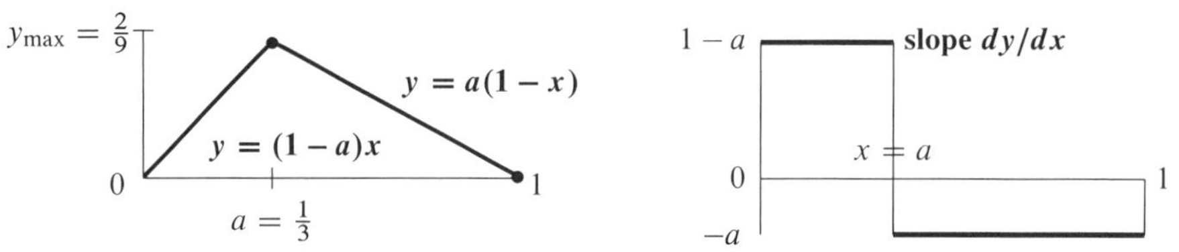


Figure 7.4: $-y'' = \delta(x - a)$ is solved by a **ramp function** that has a corner at $x = a$. At that corner point the slope $y'$ (which is a step function) drops by 1. Then $y'' = -\delta$.

How is the elastic bar stretched and compressed by this point load at $x = a = \frac{1}{3}$? The top third of the bar is stretched, the lower two thirds are compressed. The point $x = a$ shows the highest point on the graph of $y(x)$ and the greatest displacement. That downward displacement is $y(a) = a(1 - a) = \frac{2}{9}$.

Uniform stretching above the point load. Uniform compression below the point load.

## Eigenvalues and Eigenfunctions

For a square matrix, the eigenvector equation is $Ax = \lambda x$. For the second derivative (with a minus sign) and for a boundary condition at both endpoints, the eigenvector $\boldsymbol{x}$ becomes an **eigenfunction** $\boldsymbol{y(x)}$:

**Eigenvalues of** $\boldsymbol{-\dfrac{d^2}{dx^2}}$

$$-\frac{d^2y}{dx^2} = \lambda y \quad \text{with } y(0) = 0 \text{ and } y(1) = 0. \tag{4}$$

We can find these eigenfunctions $y(x)$. The solutions to the second order equation $y'' + \lambda y = 0$ are sines and cosines when $\lambda \geq 0$. The boundary conditions choose sines:

$$y(x) = A\cos(\sqrt{\lambda}\,x) + B\sin(\sqrt{\lambda}\,x) \text{ before applying the boundary conditions}$$
$$y(0) = 0 \text{ requires } \boldsymbol{A = 0} \qquad y = \sin\sqrt{\lambda} = 0 \text{ at } x = 1 \text{ requires } \boldsymbol{\sqrt{\lambda} = n\pi}$$

**The eigenfunction is $\boldsymbol{y(x) = \sin n\pi x}$. The eigenvalue is $\boldsymbol{\lambda = n^2\pi^2}$ for $\boldsymbol{n = 1, 2, 3, \ldots}$** Then $-y'' = \lambda y$. We have infinitely many $y$ and $\lambda$, not surprising since $S = -d^2/dx^2$ is not a matrix. It is an " operator" and it acts on functions $y(x)$.

## The Second Derivative $-d^2/dx^2$ is Symmetric Positive Definite

The derivatives $Ay = dy/dx$ and $Sy = -d^2y/dx^2$ are linear operators. The first derivative $A$ is **antisymmetric**. The second derivative $S$ is **symmetric**. $S$ is also **positive definite**, because of that minus sign. Its eigenvalues $\lambda = n^2\pi^2$ are all positive.

We will use the symbols $A^{\mathrm{T}}$ and $S^{\mathrm{T}}$, even though $A$ and $S$ are not matrices. To give meaning to $A^{\mathrm{T}} = -A$ and $S^{\mathrm{T}} = S$, we need *the inner product* $(f, g)$ *of two functions*:

**Inner product of $f$ and $g$** $$(f(x), g(x)) = \int_0^1 f(x)\, g(x)\, dx. \qquad (5)$$

This is the continuous form of the dot product $\boldsymbol{u} \cdot \boldsymbol{v} = \boldsymbol{u}^{\mathrm{T}}\boldsymbol{v}$ of two vectors. For $\boldsymbol{u} \cdot \boldsymbol{v}$ we multiply the components $u_i$ and $v_i$, and add. For functions we multiply the values of $f(x)$ and $g(x)$, and then integrate as in (5).

A matrix is symmetric if $S\boldsymbol{u} \cdot \boldsymbol{v}$ equals $\boldsymbol{u} \cdot S\boldsymbol{v}$ for all vectors. Then $(S\boldsymbol{u})^{\mathrm{T}}\boldsymbol{v} = \boldsymbol{u}^{\mathrm{T}}\left(S^{\mathrm{T}}\boldsymbol{v}\right)$ agrees with $\boldsymbol{u}^{\mathrm{T}}(S\boldsymbol{v})$. An operator is symmetric if $(Sf, g)$ equals $(f, Sg)$ for all functions that satisfy the boundary conditions. Use two integrations by parts to shift the second derivative operator $S$ from $f$ onto $g$:

**Integration by parts twice** $$\int_0^1 -\frac{d^2f}{dx^2}\, g(x)\, dx = \int_0^1 \frac{df}{dx}\frac{dg}{dx}\, dx = \int_0^1 f(x)\left(-\frac{d^2g}{dx^2}\right) dx. \qquad (6)$$

The integrated terms $[g\, df/dx]_0^1$ and $[f\, dg/dx]_0^1$ in the two integrations by parts are zero because $f = g = 0$ at both endpoints.

The left side and right side of (6) are the inner products $(Sf, g)$ and $(f, Sg)$. Moving $S$ from $f$ onto $g$ always produces $S^{\mathrm{T}}$. Here we have $\boldsymbol{S = S^{\mathrm{T}}}$ and symmetry is confirmed.

Thus the second derivative $S = -d^2/dx^2$ is symmetric positive definite (this is why we included the minus sign). Section 7.2 gave two other tests, in addition to positive eigenvalues. One test is *positive energy*, and that test is also passed. Choose $g = f$:

**Positive energy $f^{\mathrm{T}}Sf$** $$(Sf, f) = \int_0^1 -\frac{d^2f}{dx^2}\, f(x)\, dx = \int_0^1 \left(\frac{df}{dx}\right)^2 dx > \mathbf{0}. \qquad (7)$$

Zero energy requires $df/dx = 0$. Then the boundary conditions ensures $f(x) = 0$.

The third test for a positive definite $S$ looks for $A$ so that $S = A^{\mathrm{T}}A$. Here $A$ is the **first derivative** $(\boldsymbol{Af = df/dx})$. The boundary conditions are still $f(0) = 0$ and $f(1) = 0$. Problem 1 will show that $A^{\mathrm{T}}g$ is $-dg/dx$, with a minus sign from *one* integration by parts. Altogether $\boldsymbol{S = -d^2/dx^2 = (-d/dx)(d/dx) = A^{\mathrm{T}}A}$.

## Solving the Heat Equation

Differential equations in time give a chance to use all the eigenfunctions $\sin(n\pi x)$. An outstanding example is the **heat equation** $\boldsymbol{\partial u/\partial t = \partial^2 u/\partial x^2 = -Su}$. The eigenvalues of $-S$ are $-n^2\pi^2$, and the negative definite $-S$ leads to decay in time and not growth. Temperatures die out exponentially when there is no fire. Here are the two steps (developed much further in Section 8.3) to solve the heat equation $u_t = u_{xx}$ :

**1.** Write the initial function $u(0, x)$ as a combination of the eigenfunctions $\sin n\pi x$ :

$$\textbf{Fourier sine series} \quad u_{\text{start}} = b_1 \sin \pi x + b_2 \sin 2\pi x + \cdots + b_n \sin n\pi x + \cdots \tag{8}$$

**2.** With $\lambda = -n^2\pi^2$, every eigenfunction decays. Superposition gives $u$ at time $t$ :

$$u(t, x) = b_1 e^{-\pi^2 t} \sin \pi x + b_2 e^{-4\pi^2 t} \sin 2\pi x + \cdots = \sum_1^{\infty} \boldsymbol{b_n e^{-n^2\pi^2 t}} \sin \boldsymbol{n\pi x} \tag{9}$$

This is the famous **Fourier series solution** to the heat equation. Section 8.1 will show how to compute the Fourier coefficients $b_1, b_2, \ldots$ (a simple formula even when there are infinitely many $b$'s). You see how the solution is exactly analogous to $y(t) = \boldsymbol{c_1 e^{-\lambda_1 t} x_1 + c_2 e^{-\lambda_2 t} x_2}$. That solves an ODE, the heat equation is a PDE.

## Second Difference Matrix $K$

These pages will take a crucial first step in scientific computing. This is where differential equations meet matrix equations. The continuous problem (here continuous in $x$, previously in $t$) becomes discrete. Chapter 3 took that step for initial value problems, starting with Euler's forward difference $y(t + \Delta t) - y(t)$. Now we have problems $-y'' = f(x)$ with second derivatives. So we use *second differences* $y(x + \Delta x) - 2y(x) + y(x - \Delta x)$.

The second derivative is the derivative of $dy/dx$. The second difference is the difference of $\Delta y/\Delta x$. For first differences we have choices—forward or backward or centered differences. To approximate the second derivative $Sy = -y''$ there is *one* outstanding centered choice. This uses the **tridiagonal second difference matrix $K$** :

$$\boldsymbol{-\frac{d^2 y}{dx^2} \approx \frac{KY}{(\Delta x)^2}} \qquad \boldsymbol{-1 \quad 2 \quad -1} \text{ from } \boldsymbol{-Y_{i+1} + 2Y_i - Y_{i-1}} \qquad KY = \begin{bmatrix} 2 & -1 & & & \\ -1 & 2 & -1 & & \\ & -1 & \cdot & \cdot & \\ & & \cdot & \cdot & -1 \\ & & & -1 & 2 \end{bmatrix} \begin{bmatrix} Y_1 \\ Y_2 \\ \cdot \\ \cdot \\ Y_N \end{bmatrix} \tag{10}$$

The numbers $Y_1$ to $Y_N$ are approximations to the true values $y(\Delta x), \ldots, y(N\Delta x)$ in the continuous problem. The boundary conditions $y(0) = 0$ and $y(1) = 0$ become $Y_0 = 0$ and $Y_{N+1} = 0$. The step $\Delta x$ has length $1/(N + 1)$. The matrix $K$ correctly takes $Y_0$ and $Y_{N+1}$ to be zero, by working only with $Y_1$ to $Y_N$.

## The Matrix $K$ is Positive Definite

We know that the operator $S = -d^2/dx^2$ is positive definite. All of its eigenvectors $\sin n\pi x$ have positive eigenvalues $\lambda = n^2\pi^2$. So we hope that the matrix $K$ is also positive definite. That is true—and most unusually for a matrix of any large size $N$, we can find every eigenvector and eigenvalue of $K$.

The eigenvectors are the key. It doesn't happen often that *sampling the continuous eigenfunctions at N points produces the discrete eigenvectors.* This is the most important example in all of applied mathematics, of this unprecedented sampling for $y = \sin n\pi x$ :

$$\textbf{The } N \textbf{ eigenvectors of } K \textbf{ are } \boldsymbol{y}_n = (\sin n\pi\Delta x, \sin 2n\pi\Delta x, \ldots, \sin Nn\pi\Delta x). \quad (11)$$

$$\textbf{The } N \textbf{ eigenvalues of } K \textbf{ are the positive numbers } \lambda_n = 2 - 2\cos\frac{n\pi}{N+1}. \quad (12)$$

The 2 in every eigenvalue $\lambda$ comes from the 2's along the diagonal of $K$ (that diagonal is $2I$). The cosine in $\lambda$ and in the equation $K\boldsymbol{y}_n = \lambda_n \boldsymbol{y}_n$ are checked in Problem 12. All eigenvalues are positive because the cosines are below 1. Then $K$ is **positive definite**.

It is natural to try the other positive definite tests too (we don't have to do this, $\lambda > 0$ is enough). With a rectangular first difference matrix $A$, we have $K = A^{\mathrm{T}}A$ :

$$\boldsymbol{A}^{\mathrm{T}}\boldsymbol{A} = \boldsymbol{K} \qquad \begin{bmatrix} 1 & -1 & & \\ & 1 & -1 & \\ & & 1 & -1 \end{bmatrix} \begin{bmatrix} 1 & & \\ -1 & 1 & \\ & -1 & 1 \\ & & -1 \end{bmatrix} = \begin{bmatrix} 2 & -1 & \\ -1 & 2 & -1 \\ & -1 & 2 \end{bmatrix} \quad (13)$$

The three columns of that matrix $A$ are certainly independent. Therefore $A^{\mathrm{T}}A$ is a positive definite matrix, now proved twice.

Notice that $A^{\mathrm{T}}$ is *minus* the usual forward difference matrix. $A$ is *plus* a backward difference matrix. That sign change reflects the continuous case (for derivatives) where the "transpose" of $d/dx$ is $-d/dx$. For every vector $\boldsymbol{f}$, the energy $\boldsymbol{f}^{\mathrm{T}}K\boldsymbol{f}$ is the same as $\boldsymbol{f}^{\mathrm{T}}A^{\mathrm{T}}A\boldsymbol{f} = (A\boldsymbol{f})^{\mathrm{T}}(A\boldsymbol{f}) > 0$:

$$\textbf{The energy} \int_0^1 \left(\frac{d\boldsymbol{f}}{d\boldsymbol{x}}\right)^2 d\boldsymbol{x} \quad \textbf{becomes} \quad \boldsymbol{f}^{\mathrm{T}}\boldsymbol{K}\boldsymbol{f} = (\boldsymbol{A}\boldsymbol{f})^{\mathrm{T}}(\boldsymbol{A}\boldsymbol{f}) = \sum_{n=1}^{N+1}(f_n - f_{n-1})^2 > 0.$$

The test of positive energy $\boldsymbol{f}^{\mathrm{T}}K\boldsymbol{f}$ is passed, and $K$ is again proved to be positive definite.

## Boundary Conditions on the Slope

The fixed-fixed boundary conditions are $y(0) = 0$ and $y(1) = 0$. One or both of those conditions can change to a *slope condition* on $y' = dy/dx$. If the left condition changes to $y'(0) = 0$, the top of our elastic bar is *free* instead of fixed. This is like a tall building; $x = 0$ is up in the air (**free**) and $x = 1$ is down at the ground (**fixed**).

A fixed-free hanging bar combines $y(0) = 0$ at the top with $y'(1) = 0$ at the bottom. Its matrix is still positive definite. **But a free-free bar has no supports : semidefinite !**

$$\textbf{Free-free } Sy = f \qquad -\frac{d^2y}{dx^2} = f(x) \text{ with } \frac{dy}{dx}(0) = 0 \text{ and } \frac{dy}{dx}(1) = 0. \quad (14)$$

You will see that this problem generally has no solution. **One eigenvalue is now $\lambda = 0$.**

$$\textbf{Free-free } Sy = \lambda y \qquad -\frac{d^2y}{dx^2} = \lambda y(x) \text{ with } \frac{dy}{dx} = 0 \text{ at } x = 0 \text{ and } x = 1. \quad (15)$$

The fixed-fixed problem had eigenfunctions $y(x) = \sin n\pi x$ and eigenvalues $\lambda = n^2\pi^2$. This free-free problem will have $\boldsymbol{y(x) = \cos n\pi x}$ and again $\lambda = n^2\pi^2$. Those cosines start and end with zero slope. Also very important: The free-free problem has an extra eigenfunction $y = \cos 0x$ (which is the constant function $y = 1$). And then $\lambda = 0$:

$$\textbf{Constant } y \textbf{ and zero } \lambda \qquad y = 1 \text{ solves } -\frac{d^2y}{dx^2} = \lambda y \text{ with eigenvalue } \lambda = 0$$

Conclusion: The free-free problem (14) is only positive *semidefinite*. The eigenvalues include $\lambda = 0$. The problem is *singular* and for most loads $f(x)$ there is no solution.

**Example with $f(x) = x$** Show that $-y'' = x$ has no solution with $y'(0) = y'(1) = 0$.

*Solution* Integrate both sides of $-y'' = x$ from $x = 0$ to $x = 1$. The right side gives $\int x\,dx = \frac{1}{2}$. The left side gives $-\int y''\,dx = y'(0) - y'(1)$. But the boundary conditions make this zero and there can be no solution to $0 = \frac{1}{2}$. An operator with a zero eigenvalue is not invertible.

## Free-free Difference Matrix $B$

This problem $-y'' = f(x)$ with free-free conditions $y'(0) = y'(1) = 0$ leads to a **singular matrix** (not invertible). This is still a second difference matrix, to approximate the second derivative. *But row 1 and row N of the matrix are changed by the free-free boundary conditions*:

**Free-free matrix $B$**

**Change $K_{11} = 2$ to $B_{11} = 1$**

**Change $K_{NN} = 2$ to $B_{NN} = 1$**

$$B = \begin{bmatrix} \mathbf{1} & -1 & & \\ -1 & 2 & -1 & \\ & -1 & 2 & -1 \\ & & -1 & \mathbf{1} \end{bmatrix} \text{ is not invertible.}$$

The slope $dy/dx$ is approximated by a first difference in row 1 and row $N$. All other rows still contain the second difference $-1, 2, -1$. The usual $1, -2, 1$ has signs reversed because the differential equation has $-d^2y/dx^2$.

How to see that $B$ is not invertible ? MATLAB would find pivots $1, 1, \ldots, 1, 0$ from elimination. The zero in the last pivot position means failure. We can see this failure directly by solving $By = \mathbf{0}$. This is the fast way to show that a matrix is singular.

**To show that $B$ is not invertible, find the constant solution to $By$ = zero vector.**

**$y$ = constant vector**
**$B$ = singular matrix**

$$By = \begin{bmatrix} 1 & -1 & & \\ -1 & 2 & -1 & \\ & -1 & 2 & -1 \\ & & -1 & 1 \end{bmatrix} \begin{bmatrix} 1 \\ 1 \\ 1 \\ 1 \end{bmatrix} = \begin{bmatrix} 0 \\ 0 \\ 0 \\ 0 \end{bmatrix}. \qquad (16)$$

If $B^{-1}$ existed, we could multiply $By = \mathbf{0}$ by $B^{-1}$ to find $y = \mathbf{0}$. But this $y$ is not zero.

$B$ is positive semidefinite but it is not positive definite. We can still write the matrix $B$ as $A^{\mathrm{T}}A$, but in this free-free case the columns of $A$ will not be independent.

$$B = A^{\mathrm{T}}A \qquad \begin{bmatrix} 1 & -1 & & \\ -1 & 2 & -1 & \\ & -1 & 2 & -1 \\ & & -1 & 2 \end{bmatrix} = \begin{bmatrix} 1 & & \\ -1 & 1 & \\ & -1 & 1 \\ & & -1 \end{bmatrix} \begin{bmatrix} 1 & -1 & & \\ & 1 & -1 & \\ & & 1 & -1 \end{bmatrix}.$$

With only 3 rows, the 4 columns of $A$ must be dependent. They add up to a zero column.

## ■ REVIEW OF THE KEY IDEAS ■

1. Two initial conditions for $y(0)$ and $y'(0)$ can change to two **boundary conditions**.

2. The fixed-fixed problem $-y'' = \lambda y$ with $y(0) = 0$ and $y(1) = 0$ has $\lambda = n^2\pi^2$.

3. The second difference matrix $K$ has $\lambda_n = 2 - 2\cos\frac{n\pi}{N+1} > 0$. *Positive definite.*

4. Eigenfunctions and eigenvectors are sines, from fixed-fixed boundary conditions.

5. The free-free problem with $y'(0) = y'(1) = 0$ has $y$ = cosines. This allows $\lambda = 0$.

6. The free-free matrix $B$ has $\lambda = 0$ with the eigenvector $y = (1, \dots, 1)$. Semidefinite.

# Problem Set 7.3

**1** *Transpose the derivative with integration by parts*: $(dy/dx, g) = -(y, dg/dx)$.
$Ay$ is $dy/dx$ with boundary conditions $y(0) = 0$ and $y(1) = 0$. Why is $\int y'g\,dx$ equal to $-\int yg'\,dx$ ? Then $A^{\mathrm{T}}$ (which is normally written as $A^*$) is $A^{\mathrm{T}}g = -dg/dx$ with **no** boundary conditions on $g$. $A^{\mathrm{T}}Ay$ is $-y''$ with $y(0) = 0$ and $y(1) = 0$.

**Problems 2-6 have boundary conditions at $x = 0$ and $x = 1$: no initial conditions.**

2 Solve this boundary value problem in two steps. Find the complete solution $y_p + y_n$ with two constants in $y_n$, and find those constants from the boundary conditions:

Solve $-y'' = 12x^2$ with $y(0) = 0$ and $y(1) = 0$ and $y_p = -x^4$.

3 Solve the same equation $-y'' = 12x^2$ with $y(0) = 0$ and $y'(1) = 0$ (zero slope).

4 Solve the same equation $-y'' = 12x^2$ with $y'(0) = 0$ and $y(1) = 0$. Then try for both slopes $y'(0) = 0$ and $y'(1) = 0$: *this has no solution* $y = -x^4 + Ax + B$.

5 Solve $-y'' = 6x$ with $y(0) = 2$ and $y(1) = 4$. Boundary values need not be zero.

6 Solve $-y'' = e^x$ with $y(0) = 5$ and $y(1) = 0$, starting from $y = y_p + y_n$.

**Problems 7-11 are about the LU factors and the inverses of second difference matrices.**

7 The matrix $T$ with $T_{11} = 1$ factors perfectly into $LU = A^{\mathrm{T}}A$ (all its pivots are 1).

$$\boldsymbol{T} = \begin{bmatrix} 1 & -1 & & \\ -1 & 2 & -1 & \\ & -1 & 2 & -1 \\ & & -1 & 2 \end{bmatrix} = \begin{bmatrix} 1 & & & \\ -1 & 1 & & \\ & -1 & 1 & \\ & & -1 & 1 \end{bmatrix} \begin{bmatrix} 1 & -1 & & \\ & 1 & -1 & \\ & & 1 & -1 \\ & & & 1 \end{bmatrix} = LU.$$

Each elimination step adds the pivot row to the next row (and $L$ subtracts to recover $T$ from $U$). The inverses of those difference matrices $L$ and $U$ are **sum matrices**. Then the inverse of $T = LU$ is $U^{-1}L^{-1}$:

$$T^{-1} = \begin{bmatrix} 1 & 1 & 1 & 1 \\ & 1 & 1 & 1 \\ & & 1 & 1 \\ & & & 1 \end{bmatrix} \begin{bmatrix} 1 & & & \\ 1 & 1 & & \\ 1 & 1 & 1 & \\ 1 & 1 & 1 & 1 \end{bmatrix} = U^{-1}L^{-1}.$$

Compute $T^{-1}$ for $N = 4$ (as shown) and for any $N$.

8 The matrix equation $TY = (0, 1, 0, 0) =$ *delta vector* is like the differential equation $-y'' = \delta(x - a)$ with $a = 2\Delta x = \frac{2}{5}$. The boundary conditions are $y'(0) = 0$ and $y(1) = 0$. Solve for $y(x)$ and graph it from 0 to 1. Also graph $Y =$ second column of $T^{-1}$ at the points $x = \frac{1}{5}, \frac{2}{5}, \frac{3}{5}, \frac{4}{5}$. The two graphs are ramp functions.

9 The matrix $B$ has $B_{11} = 1$ (like $T_{11} = 1$) and also $B_{NN} = 1$ (where $T_{NN} = 2$). Why does $B$ have the same pivots 1, 1, ... as $T$, except for zero in the last pivot position? The early pivots don't know $B_{NN} = 1$.

Then $B$ is not invertible: $-y'' = \delta(x - a)$ has no solution with $y'(0) = y'(1) = 0$.

**10** When you compute $K^{-1}$, multiply by $\det K = N+1$ to get nice numbers:

Column 2 of $5K^{-1}$ solves the equation $Kv = 5\boldsymbol{\delta}$ when the delta vector is $\boldsymbol{\delta} =$ ____

**We know from $KK^{-1} = I$ that $K$ times each column of $K^{-1}$ is a delta vector.**

$$5K^{-1} = \begin{bmatrix} 4 & \mathbf{3} & 2 & 1 \\ 3 & \mathbf{6} & 4 & 2 \\ 2 & \mathbf{4} & 6 & 3 \\ 1 & \mathbf{2} & 3 & 4 \end{bmatrix}$$

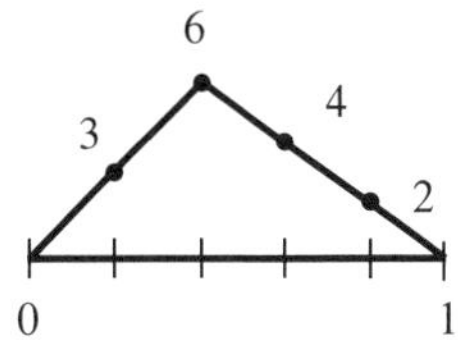


**graph of column 2**

**11** $K$ comes with two boundary conditions. $T$ only has $y(1) = 0$. $B$ has no boundary conditions on $y$. Verify that $K = A^{\mathrm{T}}A$. Then remove the first row of $A$ to get $T = A_1^{\mathrm{T}}A_1$. Then remove the last row to get dependent rows: $B = A_0^{\mathrm{T}}A_0$.

The backward first difference $A = \begin{bmatrix} 1 & & \\ -1 & 1 & \\ & -1 & 1 \\ & & -1 \end{bmatrix}$ gives $K = A^{\mathrm{T}}A$.

**12** Multiply $K_3$ by its eigenvector $\boldsymbol{y}_n = (\sin n\pi h, \sin 2n\pi h, \sin 3n\pi h)$ to verify that the eigenvalues $\lambda_1, \lambda_2, \lambda_3$ are $\lambda_n = 2 - 2\cos\frac{n\pi}{4}$ in $K\boldsymbol{y}_n = \lambda_n \boldsymbol{y}_n$. This uses the trigonometric identity $\sin(A+B) + \sin(A-B) = 2\sin A\cos B$.

**13** Those eigenvalues of $K_3$ are $2-\sqrt{2}$ and $2$ and $2+\sqrt{2}$. Those add to 6, which is the trace of $K_3$. Multiply those eigenvalues to get the determinant of $K_3$.

**14** The slope of a ramp function is a step function. The slope of a step function is a delta function. Suppose the ramp function is $r(x) = -x$ for $x \le 0$ and $r(x) = x$ for $x \ge 0$ (so $r(x) = |x|$). *Find $dr/dx$ and $d^2r/dx^2$.*

**15** Find the second differences $y_{n+1} - 2y_n + y_{n-1}$ of these infinitely long vectors $\boldsymbol{y}$:

| | |
|---|---|
| **Constant** | $(\ldots, 1, 1, 1, 1, 1, \ldots)$ |
| **Linear** | $(\ldots, -1, 0, 1, 2, 3, \ldots)$ |
| **Quadratic** | $(\ldots, 1, 0, 1, 4, 9, \ldots)$ |
| **Cubic** | $(\ldots, -1, 0, 1, 8, 27, \ldots)$ |
| **Ramp** | $(\ldots, 0, 0, 0, 1, 2, \ldots)$ |
| **Exponential** | $(\ldots, e^{-i\omega}, e^0, e^{i\omega}, e^{2i\omega}, \ldots)$. |

It is amazing how closely those second differences follow second derivatives for $y(x) = 1, x, x^2, x^3, \max(x,0)$, and $e^{i\omega x}$. From $e^{i\omega x}$ we also get $\cos\omega x$ and $\sin\omega x$.

## 7.4 Laplace's Equation and $A^{\mathrm{T}}A$

Section 7.3 solved the differential equation $-d^2y/dx^2 = \delta(x-a)$. Boundary values were given at $x = 0$ and $x = 1$ (our examples began with $y = 0$ at both endpoints). The solutions $y(x)$ went linearly up from zero and linearly back to zero. These boundary value problems correspond to a steady state—with no dependence on time.

Those are "1-dimensional Laplace equations"—certainly the simplest of their kind. This section is more ambitious, in three important ways:

**1** We will solve the 2-dimensional Laplace equation—our first PDE. The list of solutions is infinite, and they are particularly beautiful. Amazingly the imaginary number $i = \sqrt{-1}$ enters this real problem.

**Laplace's partial differential equation**
$$\frac{\partial^2 u}{\partial x^2} + \frac{\partial^2 u}{\partial y^2} = 0 \qquad (1)$$

**2** The discrete form of (1) is a matrix equation for a vector $U$. That vector has components $U_1, \ldots, U_n$ at the $n$ nodes of a graph. The graph could be a *line* in 1D or a *grid* in 2D, or any *network of nodes* connected by $m$ edges (Figure 7.5).

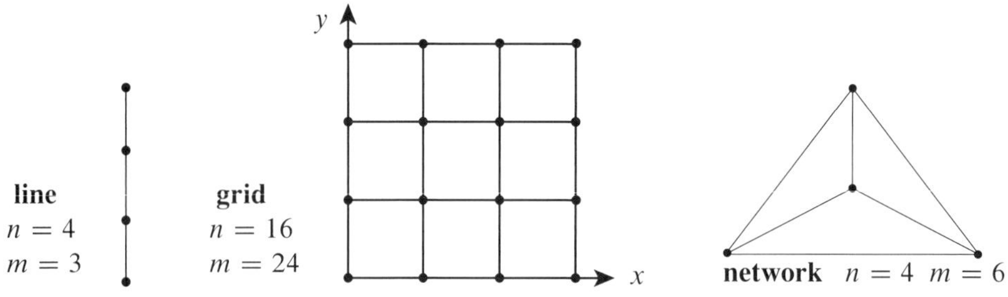


Figure 7.5: A 1D line graph, a 2D grid, and a complete graph: $n$ nodes and $m$ edges.

The natural discrete analog of Laplace's equation (1) is a "5-point scheme" on a grid:

$$\frac{\Delta_x^2 U}{(\Delta x)^2} + \frac{\Delta_y^2 U}{(\Delta y)^2} = \begin{array}{l}\text{2nd difference } \textit{across} \text{ grid} \\ +\text{2nd difference } \textit{down} \text{ grid}\end{array} = 0. \qquad (2)$$

For these equations we are given **boundary values of $u$ and $U$**. Instead of an interval like $0 \le x \le 1$, there is a region in the plane: $u$ is given along its boundary. $U$ is given at the 12 boundary points of the 4 by 4 grid. Equation (2) holds at each inside point.

**3** The continuous and discrete Laplace equations are good examples of $A^{\mathrm{T}}Au$. $A^{\mathrm{T}}A$ is symmetric with eigenvalues $\lambda \ge 0$. And one more matrix will produce $A^{\mathrm{T}}CA$ in Section 7.5. In engineering, $C$ contains the physical properties of the material: stiffness and conductivity and permeability. You will be seeing the structure of applied mathematics.

## Laplace's Equation is $A^{\mathrm{T}}Au = 0$

This is our first partial differential equation. It represents *equilibrium*, not change.

**Laplace's equation for $u(x, y)$** $$-\frac{\partial^2 u}{\partial x^2} - \frac{\partial^2 u}{\partial y^2} = 0 \qquad (3)$$

I have included minus signs to make the left side into $A^{\mathrm{T}}Au$. In one dimension, $A$ was $d/dx$ and $A^{\mathrm{T}}$ was $-d/dx$. Now we have two space variables $x$ and $y$, and two partial derivatives $\partial/\partial x$ and $\partial/\partial y$ will go into $A$. Then $-\partial/\partial x$ and $-\partial/\partial y$ go into $A^{\mathrm{T}}$.

The vector $Au$ has two components $\partial u/\partial x$ and $\partial u/\partial y$. This is the "*gradient vector.*" We are into the 2D world of multivariable calculus and partial derivatives:

**Gradient of $u$** $$Au = \text{grad}\, u(x, y) = \begin{bmatrix} \partial/\partial x \\ \partial/\partial y \end{bmatrix} u = \begin{bmatrix} \partial u/\partial x \\ \partial u/\partial y \end{bmatrix}. \qquad (4)$$

I will skip double integrals and the Divergence Theorem (which is the 2D form of the Fundamental Theorem of Calculus). Since $A$ is 2 by 1, you can guess that $A^{\mathrm{T}}$ is 1 by 2:

**Divergence** $$A^{\mathrm{T}}\boldsymbol{w} = -\,\text{div}\,\boldsymbol{w} = \begin{bmatrix} -\dfrac{\partial}{\partial x} & -\dfrac{\partial}{\partial y} \end{bmatrix} \begin{bmatrix} w_1(x, y) \\ w_2(x, y) \end{bmatrix} = -\frac{\partial w_1}{\partial x} - \frac{\partial w_2}{\partial y}. \qquad (5)$$

Then $A^{\mathrm{T}}Au$ is (minus) the divergence of the gradient of $u(x, y)$. This is the Laplacian:

**$A^{\mathrm{T}}Au = -$ div grad $u$** $$A^{\mathrm{T}}Au = \begin{bmatrix} -\dfrac{\partial}{\partial x} & -\dfrac{\partial}{\partial y} \end{bmatrix} \begin{bmatrix} \dfrac{\partial u}{\partial x} \\[2mm] \dfrac{\partial u}{\partial y} \end{bmatrix} = -\frac{\partial^2 u}{\partial x^2} - \frac{\partial^2 u}{\partial y^2}. \qquad (6)$$

You recognize $A^{\mathrm{T}}Au = 0$ as Laplace's equation. With zero on the right hand side, the minus sign can be included or not. We usually give Poisson's name when the equation has a nonzero source (or a sink) $f(x, y)$ on the right hand side:

$$u_{xx} + u_{yy} = f(x, y) \quad \textbf{is Poisson's equation}.$$

The subscripts in $u_{xx}$ and $u_{yy}$ indicate second partial derivatives: $u_{xx} = \partial^2 u/\partial x^2$ and $u_{yy} = \partial^2 u/\partial y^2$ . In this notation, $u_t$ indicates $\partial u/\partial t$. Previously that was $u'$, in the ordinary differential equations of earlier chapters. PDEs bring these new notations.

**Example 1** $u = xy$ solves Laplace's equation $u_{xx} + u_{yy} = \mathbf{0}$. And $u_p = x^2 + y^2$ solves Poisson's equation $u_{xx} + u_{yy} = \mathbf{4}$ with a constant source. The complete solution for Poisson is this particular solution $x^2 + y^2$ plus any null solution for Laplace.

## Solutions to Laplace's Equation

We want a complete set of solutions to $\boldsymbol{u_{xx} + u_{yy} = 0}$. The list will be infinitely long. Combinations of those solutions will also be solutions. Laplace's equation is linear, so superposition is allowed. Four solutions are easy to find: $u = \mathbf{1}, \boldsymbol{x}, \boldsymbol{y}, \boldsymbol{xy}$. For those four, $u_{xx}$ and $u_{yy}$ are both zero. To find further solutions, we need $u_{xx}$ to cancel $u_{yy}$.

Start with $u = x^2$, which has $u_{xx} = 2$. Then $u_{yy} = -2$ is achieved by $-y^2$. The combination $u = x^2 - y^2$ solves Laplace's equation. This solution has "*degree* 2" because if $x$ and $y$ are multiplied by $C$, then $u$ is multiplied by $C^2$. The same was true of $u = xy$, also degree 2 because $(Cx)(Cy)$ is $C^2$ times $xy$.

The real question starts with $x^3$. *Can this be completed to a solution of degree* 3? From $u = x^3$ we will have $u_{xx} = 6x$. To cancel $6x$, we need a piece that has $u_{yy} = -6x$. *That piece is* $-3xy^2$. The combination $u = x^3 - 3xy^2$ has degree 3 and goes into our list.

The hope is to find two solutions of every degree. Here is the list so far. I will write each pair of solutions in polar coordinates too, starting with $u = x = r\cos\theta$.

| | | | | | |
|---|---|---|---|---|---|
| degree | **1** | $\boldsymbol{x}$ | $\boldsymbol{y}$ | $r\cos\theta$ | $r\sin\theta$ |
| degree | **2** | $\boldsymbol{x^2 - y^2}$ | $\boldsymbol{2xy}$ | $r^2\cos 2\theta$ | $r^2\sin 2\theta$ |
| degree | **3** | $\boldsymbol{x^3 - 3xy^2}$ | ?? | $r^3\cos 3\theta$ | $r^3\sin 3\theta$ |

On the polar coordinate list, the pattern is clear. The pairs of solutions to Laplace's equation are $\boldsymbol{r^n \cos n\theta}$ and $\boldsymbol{r^n \sin n\theta}$. Those will be solutions also for $n = 4, 5, \ldots$

The first list (pairs of $x, y$ polynomials) also has a remarkable pattern. **Those are the real and imaginary parts of $\boldsymbol{(x+iy)^n}$.** Degree $n = 2$ shows the two parts clearly:

$$(x+iy)^2 \text{ is } \boldsymbol{x^2 - y^2} + i\,\boldsymbol{2xy} \quad \text{This is } \left(re^{i\theta}\right)^2 = r^2e^{2i\theta} = \boldsymbol{r^2\cos 2\theta} + i\,\boldsymbol{r^2\sin 2\theta}.$$

The polar pair $r^n\cos n\theta$ and $r^n\sin n\theta$ satisfy Laplace's equation for every $n$. The $x$-$y$ pair succeeds because $u_{yy}$ includes $i^2 = -1$, to cancel $u_{xx}$. We have two solutions for each $n$:

$$\textbf{Degree } n \quad \boldsymbol{u_n = \mathrm{Re}\,(x+iy)^n = r^n\cos n\theta \quad s_n = \mathrm{Im}(x+iy)^n = r^n\sin n\theta}. \tag{7}$$

All combinations of these solutions will also solve Laplace's equation. For ordinary differential equations (second order with $y''$), we had two solutions. All null solutions were combinations $c_1y_1 + c_2y_2$. By choosing $c_1$ and $c_2$ we matched the two initial conditions $y(0)$ and $y'(0)$. Now we have a partial differential equation with an infinite list of solutions, two of each degree.

By choosing the right coefficients $a_n$ and $b_n$ for every $n$, including the constant $a_0$, **we can match any function** $u = u_0(x, y)$ **around the boundary**:

**On the boundary** $\quad u_0(x, y) = a_0 + a_1 x + b_1 y + a_2(x^2 - y^2) + b_2(2xy) + \cdots$

**Circular boundary** $\quad u_0(1, \theta) = a_0 + a_1 \cos\theta + b_1 \sin\theta + a_2 \cos 2\theta + b_2 \sin 2\theta + \cdots$

**That last sum is a Fourier series**. It enters when we solve Laplace's equation inside a circle. The boundary condition $u = u_0$ is given on the circle $r = 1$. For 1D problems the boundary was the two endpoints $x = 0$ and $x = 1$. We only needed two solutions.

The right choice of all the Fourier coefficients $a_n$ and $b_n$ will come in Chapter 8, and it completes the solution to Laplace's equation inside a circle:

**Solution to** $\boldsymbol{u_{xx} + u_{yy} = 0}$ $\quad$ $$\boldsymbol{u = a_0 + \sum_{n=1}^{\infty} (a_n r^n \cos n\theta + b_n r^n \sin n\theta)}. \quad (8)$$

## Finite Differences and Finite Elements

Laplace's equation is often made discrete. The derivatives $u_{xx}$ and $u_{yy}$ are replaced by finite differences. That produces a large matrix $K2\mathrm{D}$, which is a two-dimensional analog of the tridiagonal $-1, 2, -1$ matrix $K$. For the square grid in Figure 7.5, there will be entries $-1, 2, -1$ in the $x$-direction and also in the $y$-direction. $K2\mathrm{D}$ has *five entries*: $2 + 2 = 4$ down its main diagonal and four entries of $-1$ on a typical inside row.

Suppose the region is not square but curved (like a circle). Then finite differences get complicated. The nodes of a square grid don't fall on circles. The favorite approach changes to the **finite element method**, which can divide the region into triangles of arbitrary shapes. (A triangle can even have a curved edge to fit a boundary.) These finite elements are described in my textbook *Computational Science and Engineering*, with codes that use linear functions $a + bx + cy$ inside each triangle of the mesh. The accuracy is studied in *An Analysis of the Finite Element Method*.

## Laplace's Difference Matrix $K2D$

The approach that fits with this book is finite differences. I want to construct the symmetric matrix $K2\mathrm{D}$ with rows like $-1, -1, 4, -1, -1$ and show that it is positive definite. $K2\mathrm{D}$ comes from second differences in the $x$ and $y$ directions. Each meshpoint needs two indices $i$ and $j$, to specify its row number and column number on the grid. Go across and up-down:

$$-\frac{\partial^2 u}{\partial x^2} \text{ becomes } \frac{-U_{i+1,j} + 2U_{i,j} - U_{i-1,j}}{(\Delta x)^2} \qquad -\frac{\partial^2 u}{\partial y^2} \text{ becomes } \frac{-U_{i,j+1} + 2U_{i,j} - U_{i,j-1}}{(\Delta y)^2}$$

The square grid has $\Delta x = \Delta y$. Combine $2U_{i,j}$ with $2U_{i,j}$. Then 4 goes on the diagonal of $K2\mathrm{D}$. The difference equation says that *each* $U_{ij}$ *is the average of its* 4 *neighbors*:

$$\Delta_x^2 U + \Delta_y^2 U = 0 \qquad \boldsymbol{4U_{i,j} - U_{i+1,j} - U_{i-1,j} - U_{i,j+1} - U_{i,j-1} = 0}. \quad (9)$$

If a neighbor of the $i, j$ node falls on the boundary of the square grid, that boundary value of $U$ will be known. Then that term moves to the right side of the difference equation. An entry of $-1$ disappears from $K$2D on boundary rows.

If we number the nodes a row at a time, the $u_{xx}$ term puts the 1D matrix $K$ in each block row. The $u_{yy}$ term connects three rows with $-I$ and $2I$ and $-I$.

$$\boldsymbol{K2D} = \begin{bmatrix} \boldsymbol{K} & & & \\ & \boldsymbol{K} & & \\ & & \cdot & \\ & & & \boldsymbol{K} \end{bmatrix} + \begin{bmatrix} \boldsymbol{2I} & \boldsymbol{-I} & & \\ \boldsymbol{-I} & \boldsymbol{2I} & \boldsymbol{-I} & \\ & \boldsymbol{-I} & \cdot & \cdot \\ & & \boldsymbol{-I} & \boldsymbol{2I} \end{bmatrix} = \mathsf{kron}\,(I, K) + \mathsf{kron}\,(K, I).$$

With $N$ interior points in each row, this block matrix $K$2D is $N^2$ by $N^2$. MATLAB's command $\mathsf{kron}(A, B)$ replaces each $A_{ij}$ by the block $A_{ij}B$, so the size grows to $N^2$.

Here is the matrix for a grid with $3 \times 3 = 9$ squares and $4 \times 4 = 16$ nodes. There are $2 \times 2 = 4$ interior nodes. The other $16 - 4 = 12$ nodes are around the square boundary, where $U$ is given by the boundary condition $u = u_0$. For a large grid, $N^2$ interior points will far outnumber $4N + 4$ boundary points.

**Laplace difference matrix**
**The interior mesh is 2 by 2**

$$K2\text{D} = \begin{bmatrix} 4 & -1 & 0 & -1 \\ -1 & 4 & -1 & 0 \\ 0 & -1 & 4 & -1 \\ -1 & 0 & -1 & 4 \end{bmatrix}.$$

Those rows lost two $-1$'s because each interior gridpoint is next to two boundary points. Normally we see four $-1$'s in almost every row of $K$2D.

Here is the solution to $K2\text{D}\,\boldsymbol{U} = \boldsymbol{0}$ in the square when boundary values are 0 and 4:

**Each bold value of $U$ is the average of 4 neighbors**

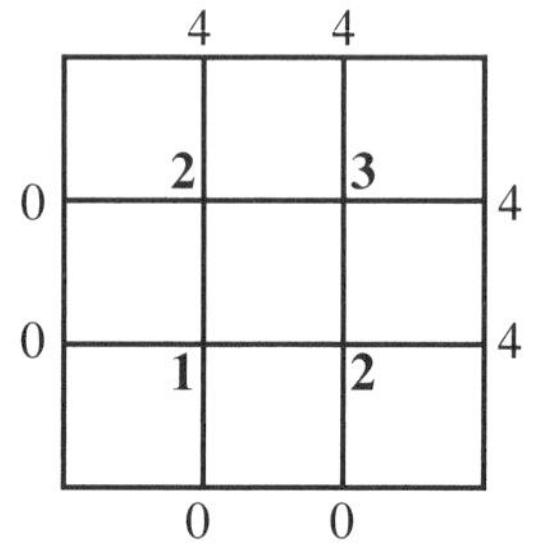


The eigenvalues of this matrix $K$2D are $\lambda = \mathbf{2, 4, 4, 6}$. They add to 16, which is the trace: the sum down the diagonal of $K2D$ above. The eigenvectors are orthogonal:

**Eigenvectors of $\boldsymbol{K}$2D** $(1, 1, 1, 1)$ and $(1, 1, -1, -1)$, $(1, -1, 1, -1)$ and $(1, -1, -1, 1)$.

Symmetry of $K$2D guaranteed orthogonal eigenvectors. Positive definiteness produced those positive eigenvalues 2, 4, 4, 6.

## Eigenvalues of the Laplacian : Continuous and Discrete

In one dimension, the eigenfunctions for $-u_{xx} = \lambda u$ are $\boldsymbol{u} = \sin \boldsymbol{n\pi x}$ with eigenvalue $\boldsymbol{\lambda = n^2\pi^2}$. These sine functions are zero at the endpoints $x = 0$ and $x = 1$. On a unit square in two dimensions, the eigenfunctions of the Laplacian are just products of sines: $\boldsymbol{u(x,y) = (\sin n\pi x)(\sin m\pi y)}$ with eigenvalue $\boldsymbol{\lambda = n^2\pi^2 + m^2\pi^2}$. Those functions are zero on the whole boundary of the square, where $x = 0$ or $x = 1$ or $y = 0$ or $y = 1$:

$$\boxed{-\left(\frac{\partial^2}{\partial x^2} + \frac{\partial^2}{\partial y^2}\right)(\sin n\pi x)(\sin m\pi y) = (\boldsymbol{n^2\pi^2 + m^2\pi^2})(\sin n\pi x)(\sin m\pi y).} \tag{10}$$

The problem on a square allows **separation of variables**. Each of the eigenvectors is a (function of $x$) times a (function of $y$). Two 1D problems, just what we hope for.

Equation (6) expressed $-u_{xx} - u_{yy}$ as $-\operatorname{div}(\operatorname{grad} u)$. **This is $\boldsymbol{A^{\mathrm{T}}A}$** ($A =$ gradient). The test $\lambda \geq 0$ is passed on non-square regions too, when the $x, y$ variables don't separate.

*Slope conditions* (a derivative of $u$ is zero instead of the function itself) allow the constant eigenfunction $u = 1$. Then $\lambda = 0$ and the Laplacian becomes *semidefinite*.

Turn now to the matrix Laplacian $K2\mathrm{D}$. In one dimension, the eigenvectors of $K$ are discrete sine vectors : Sample the continuous eigenfunction $\sin n\pi x$ at $N$ equally spaced points. The spacing is $\Delta x = 1/(N+1)$ inside the interval from 0 to 1. The eigenvalues of $K$ are $\lambda_n = 2 - 2\cos(n\pi\Delta x)$. We may hope and expect that the eigenvectors of $K2\mathrm{D}$ will contain products of sines, and the eigenvalues will be sums of 1D eigenvalues $\lambda(K)$.

*The $N^2$ eigenvalues of $K2D$ are positive. The $x$ and $y$ directions still separate.*

$$\lambda_{nm}(K2\mathrm{D}) = \lambda_n(K) + \lambda_m(K) = 4 - 2\cos\frac{n\pi}{N+1} - 2\cos\frac{m\pi}{N+1} > \mathbf{0}. \tag{11}$$

Thus $K2\mathrm{D}$ for a square is symmetric positive definite. This formula for the eigenvalues recovers $\lambda = 2, 4, 4, 6$ when $N = 2$, because the cosines of $\frac{\pi}{3}$ and $\frac{2\pi}{3}$ are $\frac{1}{2}$ and $-\frac{1}{2}$.

## ■ REVIEW OF THE KEY IDEAS ■

1. Laplace's equation is solved by the real and the imaginary part of every $(x+iy)^n$.

2. Those are $u = \boldsymbol{r^n \cos n\theta}$ and $s = \boldsymbol{r^n \sin n\theta}$. Their combinations are Fourier series.

3. The discrete equation is $\Delta_x^2 U + \Delta_y^2 U = 0$. The matrix $K2\mathrm{D}$ is positive definite.

4. Eigenvectors are (sines in $x$) (sines in $y$) : $-u_{xx} - u_{yy} = \lambda u$ and $(K2\mathrm{D})U = \lambda U$.

## Problem Set 7.4

**1** What solution to Laplace's equation completes "degree 3" in the table of pairs of solutions? We have one solution $u = x^3 - 3xy^2$, and we need another solution.

**2** What are the two solutions of degree 4, the real and imaginary parts of $(x + iy)^4$? Check $u_{xx} + u_{yy} = 0$ for both solutions.

**3** What is the second $x$-derivative of $(x + iy)^n$? What is the second $y$-derivative? Those cancel in $u_{xx} + u_{yy}$ because $i^2 = -1$.

**4** For the solved $2 \times 2$ example inside a $4 \times 4$ square grid, write the four equations (9) at the four interior nodes. Move the known boundary values 0 and 4 to the right hand sides of the equations. You should see $K2D$ on the left side multiplying the correct solution $\boldsymbol{U} = (U_{11}, U_{12}, U_{21}, U_{22}) = (1, 2, 2, 3)$.

**5** Suppose the boundary values on the $4 \times 4$ grid change to $U = 0$ on three sides and $U = 8$ on the fourth side. Find the four inside values so that each one is the average of its neighbors.

**6** (MATLAB) Find the inverse $(K2D)^{-1}$ of the 4 by 4 matrix $K2D$ displayed for the square grid.

**7** Solve this Poisson finite difference equation (right side $\neq 0$) for the inside values $U_{11}, U_{12}, U_{21}, U_{22}$. All boundary values like $U_{10}$ and $U_{13}$ are zero. The boundary has $i$ or $j$ equal to 0 or 3, the interior has $i$ *and* $j$ equal to 1 or 2:

$$4U_{ij} - U_{i-1,j} - U_{i+1,j} - U_{i,j-1} - U_{i,j+1} = \mathbf{1} \text{ at four inside points.}$$

**8** A $5 \times 5$ grid has a 3 by 3 interior grid: 9 unknown values $U_{11}$ to $U_{33}$. Create the $9 \times 9$ difference matrix $K2D$.

**9** Use eig($K2D$) to find the nine eigenvalues of $K2D$ in Problem 8. Those eigenvalues will be positive! The matrix $K2D$ is symmetric positive definite.

**10** If $u(x)$ solves $u_{xx} = 0$ and $v(y)$ solves $v_{yy} = 0$, verify that $u(x)v(y)$ solves Laplace's equation. Why is this only a 4-dimensional space of solutions? Separation of variables does not give all solutions—only the solutions with separable boundary conditions.

## 7.5 Networks and the Graph Laplacian

Start with a graph that has $n$ nodes and $m$ edges. Its $m$ by $n$ incidence matrix $\boldsymbol{A}$ was introduced in Section 5.6, with a row in the matrix for every edge in the graph. A single $-1$ and $1$ in the row indicates which two nodes are connected by that edge. Now we take the step to $\boldsymbol{L} = \boldsymbol{A}^{\mathrm{T}}\boldsymbol{A}$ and $\boldsymbol{K} = \boldsymbol{A}^{\mathrm{T}}\boldsymbol{C}\boldsymbol{A}$. These are symmetric positive semidefinite matrices that describe the whole network.

Those matrices $L$ and $K$ are the **graph Laplacians**. $L$ is unweighted (with $C = I$) and $K$ is weighted by $C$. These are the fundamental matrices for *flows in the networks.* They describe electrical networks and their applications go very much further. You see $A^{\mathrm{T}}A$ and $A^{\mathrm{T}}CA$ in descriptions of the brain and the Internet and our nervous system and the power grid.

Social networks and political networks and intellectual networks also use $L$ and $K$. Graphs have simply become **the most important model in discrete applied mathematics**. This is not a standard topic in teaching linear algebra. But it is today an essential topic in applying linear algebra. It belongs in this book.

### Examples of $A$ and $A^{\mathrm{T}}A$

We quickly review incidence matrices, by constructing $A$ for the planar graph and the line graph in Figure 7.6. You will see that every row of $A$ adds to $-1+1 = 0$. Then the all-ones vector $\boldsymbol{v} = (1, \ldots, 1)$ leads to $A\boldsymbol{v} = \mathbf{0}$. The columns of $A$ are *dependent*, because their sum is the zero column. $A\boldsymbol{v} = \mathbf{0}$ propagates to $A^{\mathrm{T}}A\boldsymbol{v} = \mathbf{0}$ and $A^{\mathrm{T}}CA\boldsymbol{v} = \mathbf{0}$, so $A^{\mathrm{T}}CA$ for this $A$ will be positive *semidefinite* (but not invertible and not positive definite).

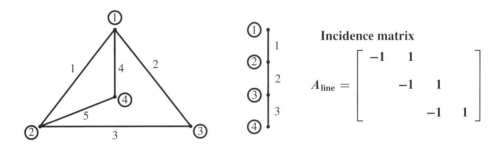

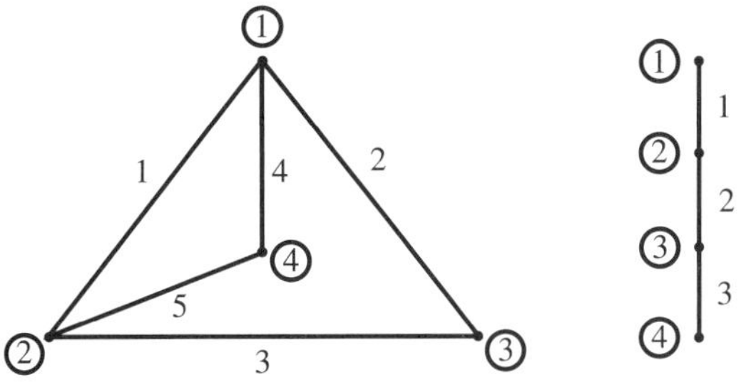


**Incidence matrix**

$$A_{\text{line}} = \begin{bmatrix} \mathbf{-1} & \mathbf{1} & & \\ & \mathbf{-1} & \mathbf{1} & \\ & & \mathbf{-1} & \mathbf{1} \end{bmatrix}$$

Figure 7.6: A planar graph and a line graph: $n = 4$ nodes and $m = 5$ or $3$ edges.

$\boldsymbol{A}_{\text{line}}$ is a 3 by 4 *difference matrix*. Then $A^{\mathrm{T}}A$ below contains second differences. Notice that the first and last entries of $A^{\mathrm{T}}A$ are 1 and not 2. **The diagonal 1, 2, 2, 1 counts the number of edges that meet at each node** (the "degrees" of the four nodes).

$$\begin{aligned} A\boldsymbol{v} &= \textbf{difference of } \boldsymbol{v}\textbf{'s} \\ A^{\mathrm{T}}A &= \textbf{line Laplacian} \end{aligned} \qquad A\boldsymbol{v} = \begin{bmatrix} v_2 - v_1 \\ v_3 - v_2 \\ v_4 - v_3 \end{bmatrix} \quad A^{\mathrm{T}}A = \begin{bmatrix} \mathbf{1} & -1 & 0 & 0 \\ -1 & \mathbf{2} & -1 & 0 \\ 0 & -1 & \mathbf{2} & -1 \\ 0 & 0 & -1 & \mathbf{1} \end{bmatrix} \tag{1}$$

For the planar graph, the incidence matrix $A$ again computes differences $\boldsymbol{v}_{\text{end}}-\boldsymbol{v}_{\text{start}}$ on every edge. The Laplacian matrix $L = A^{\mathrm{T}}A$ again has rows adding to zero. The diagonal of $L$ shows $3, 3, 2, 2$ edges into the four nodes. Everything in $A$ and $L$ can be copied directly from the graph! The missing pair of $-1$ entries in $L = A^{\mathrm{T}}A$ is because *no edge connects nodes* 3 *and* 4 on the 5-edge graph.

**Incidence matrix**
**Laplacian matrix**

$$A = \begin{bmatrix} -1 & 1 & 0 & 0 \\ -1 & 0 & 1 & 0 \\ 0 & -1 & 1 & 0 \\ -1 & 0 & 0 & 1 \\ 0 & -1 & 0 & 1 \end{bmatrix} \quad A^{\mathrm{T}}A = \begin{bmatrix} 3 & -1 & -1 & -1 \\ -1 & 3 & -1 & -1 \\ -1 & -1 & 2 & \mathbf{0} \\ -1 & -1 & \mathbf{0} & 2 \end{bmatrix} \qquad (2)$$

*Note* If any arrows change direction on the edges of the graph, this changes $A$. But $A^{\mathrm{T}}A$ *does not change*. The direction of arrows just multiplies $A$ by a $\pm$ diagonal sign matrix $S$. Then $(SA)^{\mathrm{T}}(SA)$ is the same as $A^{\mathrm{T}}A$ because $S^{\mathrm{T}}S = I$.

The eigenvalues of $L = A^{\mathrm{T}}A$ always include $\lambda = 0$, from the all-ones eigenvector. The energy $\boldsymbol{v}^{\mathrm{T}}(A^{\mathrm{T}}A)\boldsymbol{v}$ can also be written as $(A\boldsymbol{v})^{\mathrm{T}}(A\boldsymbol{v})$. This just adds up the squares of all the entries of $A\boldsymbol{v}$, which are differences across edges (*not the missing edge from* 3 *to* 4):

$$\textbf{Energy} = (v_2 - v_1)^2 + (v_3 - v_1)^2 + (v_3 - v_2)^2 + (v_4 - v_1)^2 + (v_4 - v_2)^2.$$

We see again that the all-ones vector $\boldsymbol{v} = (1, 1, 1, 1)$ has zero energy.

The Laplacian matrix $L = A^{\mathrm{T}}A$ is *not invertible*! A system of equations $A^{\mathrm{T}}A\boldsymbol{v} = \boldsymbol{f}$ has no solution (or infinitely many). To reach an invertible matrix, **we remove the last column and row of $A^{\mathrm{T}}A$**. This corresponds to "grounding a node" by setting the voltage at that node to be zero: $\boldsymbol{v_4 = 0}$. It is like fixing one temperature at zero, when the equations only tell us about differences of temperature.

When we know that $v_4 = 0$, column 4 is removed from $A$. That removes column 4 and also row 4 from $A^{\mathrm{T}}A$. *This reduced* 3 *by* 3 *matrix is positive definite:*

$$(A^{\mathrm{T}}A)_{\textbf{reduced}} = \begin{bmatrix} 3 & -1 & -1 \\ -1 & 3 & -1 \\ -1 & -1 & 2 \end{bmatrix} = (A_{\text{reduced}})^{\mathrm{T}}(A_{\text{reduced}}) = (3 \text{ by } 5)\,(5 \text{ by } 3). \qquad (3)$$

## The Weighted Laplacian $K = A^{\mathrm{T}}CA$

In many applications the edges come with positive weights $c_1, \ldots, c_m$. Those weights can be *conductances* (through $m$ resistors) or *stiffnesses* (of $m$ springs). In electrical engineering, Ohm's Law connects current $w$ to voltage difference $e$. In mechanical engineering, Hooke's Law connects spring force $w$ to the stretching $e$. **Those laws $\boldsymbol{w = ce}$ in every edge give a positive diagonal matrix** $C$ in $\boldsymbol{w} = C\boldsymbol{e} = CA\boldsymbol{v}$. The $m$ currents in $\boldsymbol{w}$ come from the $m$ voltage differences in $A\boldsymbol{v}$.

**Kirchhoff's Current Law is $\boldsymbol{A^{\mathrm{T}}w = 0}$.** That matrix $A^{\mathrm{T}}$ always enters the "balance of currents" and the "balance of forces" between springs. With current sources, or forces applied from outside, the balance equation is $A^{\mathrm{T}}\boldsymbol{w} = \boldsymbol{f}$.

When current sources enter the nodes, the Current Law $A^{\mathrm{T}}\boldsymbol{w} = \boldsymbol{f}$ is "*in equals out.*" Then $A^{\mathrm{T}}C\boldsymbol{e} = \boldsymbol{f}$ and $A^{\mathrm{T}}CA\boldsymbol{v} = \boldsymbol{f}$. Thus $K = A^{\mathrm{T}}CA$ is the **conductance matrix for the whole network**. Here is $A^{\mathrm{T}}CA$ for the line of resistors:

$$\begin{array}{lll} A^{\mathrm{T}}\boldsymbol{w} & = \boldsymbol{f} & \textbf{(Kirchhoff)} \\ A^{\mathrm{T}}C\boldsymbol{e} & = \boldsymbol{f} & \textbf{(Ohm)} \\ A^{\mathrm{T}}CA\boldsymbol{v} & = \boldsymbol{f} & \textbf{(System)} \end{array} \qquad (A^{\mathrm{T}}CA)_{\text{line}} = \begin{bmatrix} c_1 & -c_1 & 0 & 0 \\ -c_1 & c_1+c_2 & -c_2 & 0 \\ 0 & -c_2 & c_2+c_3 & -c_3 \\ 0 & 0 & -c_3 & c_3 \end{bmatrix}. \qquad (4)$$

The rows of $A^{\mathrm{T}}CA$ still add to zero. The matrix is still positive semidefinite. It becomes positive definite when row and column 4 are removed, which we must do to solve $A^{\mathrm{T}}CA\boldsymbol{v} = \boldsymbol{f}$. *This is a fundamental equation of discrete applied mathematics.*

A network can also have **voltage sources** (like batteries) on the edges. Those go into a vector $\boldsymbol{b}$ with $m$ components. From node to node the voltage drops are $-A\boldsymbol{v}$ (with a minus sign). But Ohm's Law applies to the voltage drops $\boldsymbol{e}$ *across the resistors*. By working with the matrix $C$ and including $\boldsymbol{b}$ in the vector $\boldsymbol{e} = \boldsymbol{b} - A\boldsymbol{v}$, Ohm's Law is simply $\boldsymbol{w} = C\boldsymbol{e}$. The inputs to the network are $\boldsymbol{f}$ and $\boldsymbol{b}$.

The three equations for $\boldsymbol{e}$, $\boldsymbol{w}$, $\boldsymbol{f}$ use the matrices $A, C, A^{\mathrm{T}}$. Those become *two* equations by eliminating $\boldsymbol{e} = C^{-1}\boldsymbol{w}$. We reach *one* equation by also eliminating $\boldsymbol{w}$.

| | 3 equations | 2 equations | 1 equation |
|---|---|---|---|
| **Drop** | $\boldsymbol{e} = \boldsymbol{b} - A\boldsymbol{v}$ | $\begin{bmatrix} C^{-1} & A \\ A^{\mathrm{T}} & 0 \end{bmatrix}\begin{bmatrix} \boldsymbol{w} \\ \boldsymbol{v} \end{bmatrix} = \begin{bmatrix} \boldsymbol{b} \\ \boldsymbol{f} \end{bmatrix}$ | $A^{\mathrm{T}}CA\boldsymbol{v} = A^{\mathrm{T}}C\boldsymbol{b} - \boldsymbol{f}$ |
| **Current** | $\boldsymbol{w} = C\boldsymbol{e}$ | | |
| **Balance** | $\boldsymbol{f} = A^{\mathrm{T}}\boldsymbol{w}$ | | |

I removed $\boldsymbol{e}$ by substituting $\boldsymbol{e} = C^{-1}\boldsymbol{w}$ into the first equation. The step from two equations to one equation substituted $\boldsymbol{w} = C(\boldsymbol{b} - A\boldsymbol{v})$ into $\boldsymbol{f} = A^{\mathrm{T}}\boldsymbol{w}$. Almost all entries of $A$ and $C$ will be zero. *The weighted graph Laplacian is $K = A^{\mathrm{T}}CA$.*

You see how the sources $\boldsymbol{b}$ and $\boldsymbol{f}$ produce the right side. They make the currents flow.

## A Framework for Applied Mathematics

The least squares equation $A^{\mathrm{T}}A\boldsymbol{v} = A^{\mathrm{T}}\boldsymbol{b}$ and the weighted least squares equation $A^{\mathrm{T}}CA\boldsymbol{v} = A^{\mathrm{T}}C\boldsymbol{b}$ are special cases with $\boldsymbol{f} = \boldsymbol{0}$. My experience is that all the symmetric steady state problems of applied mathematics fit into this $A^{\mathrm{T}}CA$ framework.

| | | |
|---|---|---|
| **Voltage Law** $\rightarrow$ $A$ | **Ohm's Law** $\rightarrow$ $C$ | **Current Law** $\rightarrow$ $A^{\mathrm{T}}$ |

I have learned to watch for $A^{\mathrm{T}}CA$ in every lecture about applied mathematics: it is there. Differential equations fit this framework too. Laplace's equation is $A^{\mathrm{T}}Au = 0$ when $Au$ is the gradient of $u(x, y)$. *A typical $A^{\mathrm{T}}CA$ equation is $-d/dx(c\, du/dx) = f(x)$.*

For matrices, those derivatives become differences. The graph analogy with Laplace's equation gave the name *graph Laplacian* to the matrix $A^{\mathrm{T}}A$.

Dynamic problems have time derivatives $du/dt$. This adds a new step to the $A^{\mathrm{T}}CA$ framework. The equation $du/dt = -A^{\mathrm{T}}Au$ is a matrix analog of the heat equation $\partial u/\partial t = \partial^2 u/\partial x^2$. The next chapter will solve the heat equation using the eigenvalues and eigenfunctions (sines and cosines) from $y'' = \lambda y$. **The solutions are Fourier series.**

**Example: A Network of Resistors**

I will add resistors to the five edges of our four-node graph. The conductances $1/R$ will be the numbers $c_1$ to $c_5$. The conductance matrix for the whole network is $A^{\mathrm{T}}CA$. The incidence matrix $A$ in equation (2) above is 5 by 4, and $A^{\mathrm{T}}CA$ is 4 by 4.

**Conductance matrix $K$ with five edges**

$$A^{\mathrm{T}}CA = \begin{bmatrix} c_1+c_2+c_4 & -c_1 & -c_2 & -c_4 \\ -c_1 & c_1+c_3+c_5 & -c_3 & -c_5 \\ -c_2 & -c_3 & c_2+c_3 & \mathbf{0} \\ -c_4 & -c_5 & \mathbf{0} & c_4+c_5 \end{bmatrix} \qquad (5)$$

Please compare this matrix to $A^{\mathrm{T}}A$ in equation (2), where all $c_i = 1$. The new matrix starts with $c_1 + c_2 + c_4$ because edges $1, 2, 4$ touch node 1. Along that row of $K$, the entries $-c_1, -c_2, -c_4$ produce *row sum = zero* as we expect. Then $A^{\mathrm{T}}CA$ is singular, not invertible. We must reduce the matrix to 3 by 3 by "grounding a node" and removing column 4 and row 4. The reduced matrix is symmetric positive definite.

Suppose the voltage $v_1 = V$ is fixed, as well as $v_4 = 0$ at the grounded node. Current will flow out of node 1 toward node 4 (with $\boldsymbol{b} = \boldsymbol{f} = \mathbf{0}$). *The terms $c_1V$ and $c_2V$ involving the known $v_1 = V$ move to the right hand side of $A^{\mathrm{T}}CA\boldsymbol{v} = \mathbf{0}$.* There are only two unknown voltages $v_2$ and $v_3$, and $V$ is like a boundary value:

**Reduced equations $\boldsymbol{v_1 = V}$ and $\boldsymbol{v_4 = 0}$**

$$\begin{bmatrix} c_1+c_3+c_5 & -c_3 \\ -c_3 & c_2+c_3 \end{bmatrix} \begin{bmatrix} v_2 \\ v_3 \end{bmatrix} = \begin{bmatrix} c_1V \\ c_2V \end{bmatrix}. \qquad (6)$$

When we solve for $v_2$ and $v_3$, we know all four voltages $\boldsymbol{v}$ and all five currents $\boldsymbol{w} = CA\boldsymbol{v}$.

**Summary**

The matrix $C$ changes an "ideal" $A^{\mathrm{T}}A$ problem into an "applied" $A^{\mathrm{T}}CA$ problem. You will see how this three-step framework appears all through applied mathematics. $Au$ is often a derivative of $u$, or a finite difference. Then $CAu$ comes from Ohm's Law or Hooke's Law. The material constants like conductance and stiffness go into $C$.

Finally $A^{\mathrm{T}}CA\boldsymbol{v} = \boldsymbol{f}$ is a continuity equation or a balance equation. It represents balance of forces, balance of inputs with outputs, balance of profits with losses. The combined matrix $K = A^{\mathrm{T}}CA$ is symmetric positive definite just like $A^{\mathrm{T}}A$.

**To find the forces or the flows inside the network, we solve for $\boldsymbol{v}$ and $\boldsymbol{e}$ and $\boldsymbol{w}$.**

## The Adjacency Matrix

The Laplacian matrices $L = A^{\mathrm{T}}A$ and $K = A^{\mathrm{T}}CA$ started with the incidence matrix $A$. The diagonal of $L$ has the degree of each node: the number of edges that touch the node. $A^{\mathrm{T}}A$ also comes directly from the **degree matrix $D$ minus the adjacency matrix $W$** :

$$A^{\mathrm{T}}A = \begin{bmatrix} 3 & -1 & -1 & -1 \\ -1 & 3 & -1 & -1 \\ -1 & -1 & 2 & \mathbf{0} \\ -1 & -1 & \mathbf{0} & 2 \end{bmatrix} = \begin{bmatrix} 3 & & & \\ & 3 & & \\ & & 2 & \\ & & & 2 \end{bmatrix} - \begin{bmatrix} 0 & 1 & 1 & 1 \\ 1 & 0 & 1 & 1 \\ 1 & 1 & 0 & \mathbf{0} \\ 1 & 1 & \mathbf{0} & 0 \end{bmatrix}. \qquad (7)$$

The degrees $3, 3, 2, 2$ in $D$ are the row sums in $W$. Then $D - W$ has zero row sums. When $L = A^{\mathrm{T}}A = D - W$ multiplies $(1, 1, 1, 1)$ the result will be $(0, 0, 0, 0)$.

**Question** The sum of the degrees is 10. How can this be predicted from the graph?

*Answer* The graph has five edges. Each edge produces two 1's in the adjacency matrix. There must be ten 1's in $W$. The degrees in $D$ must add to 10, to balance the 1's in $W$.

Since the *trace* of $L$ is $3 + 3 + 2 + 2$, the eigenvalues of $L$ must also add to 10.

**Question** What is the rule for $W$ and $D$ when there are weights $c_1, \ldots, c_m$ on the edges?

*Answer* Each entry $W_{ij} = 1$ comes from an edge between node $i$ and node $j$. When this edge $k$ has a weight $c_k$ (the conductance along the edge), the entry $W_{ij}$ changes from 1 to $c_k$. The weights produce $A^{\mathrm{T}}CA$ in equation (5) and also in equation (8).

**$A^{\mathrm{T}}CA = K$ with weights**

$$D - W = \begin{bmatrix} c_1 + c_2 + c_4 & & & \\ & \cdot & & \\ & & \cdot & \\ & & & c_4 + c_5 \end{bmatrix} - \begin{bmatrix} 0 & c_1 & c_2 & c_4 \\ c_1 & 0 & c_3 & c_5 \\ c_2 & c_3 & 0 & 0 \\ c_4 & c_5 & 0 & 0 \end{bmatrix}. \qquad (8)$$

Problems 1 – 5 will ask about a **complete graph**, when every pair of nodes is connected by an edge. All off-diagonal entries in the adjacency matrix $W$ are 1. All the degrees in the diagonal $D$ are $n - 1$. The Laplacians $L$ and $K$ have no zeros. Every question about $L = A^{\mathrm{T}}A = D - W$ has a good answer for this graph with all possible edges.

Here is a picture that summarizes this three-step vision of applied mathematics.

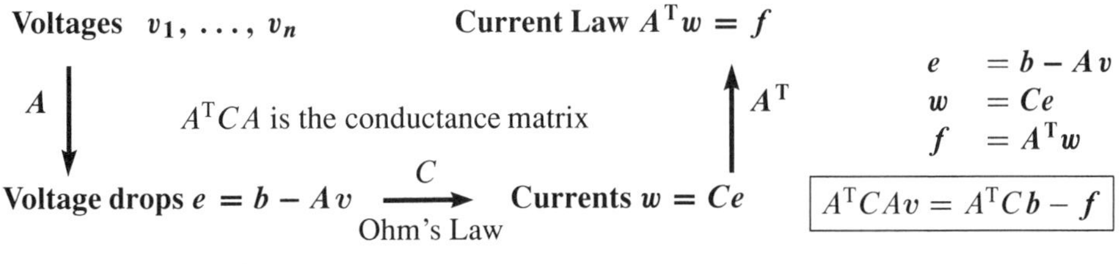


Figure 7.7: The $A^{\mathrm{T}}CA$ framework for steady state problems in science and engineering.

### Saddle-Point Matrix

The final matrix is $A^TCA$, after the edge currents $w_1, \ldots, w_m$ are eliminated. Before we took that step, the voltages $\boldsymbol{v}$ and the currents $\boldsymbol{w}$ were the two unknown vectors. With two equations we have a "saddle-point matrix" that contains $C^{-1}$ and $A$ and $A^T$:

$$\textbf{Saddle-point problem} \quad \textbf{Currents and voltages} \qquad \begin{bmatrix} C^{-1} & A \\ A^T & 0 \end{bmatrix} \begin{bmatrix} \boldsymbol{w} \\ \boldsymbol{v} \end{bmatrix} = \begin{bmatrix} \boldsymbol{b} \\ \boldsymbol{f} \end{bmatrix}. \tag{9}$$

Block matrices of this form appear when there is a constraint like Kirchhoff's Current Law $A^T\boldsymbol{w} = \boldsymbol{f}$. "Nature minimizes heat loss in the network subject to that constraint." The "KKT matrix" in (9) is symmetric but it is *not at all positive definite*.

A small example will show a positive and also a negative eigenvalue:

$$\begin{bmatrix} 3 & 2 \\ 2 & 0 \end{bmatrix} \text{ has eigenvalues 4 and } -1. \quad \text{The pivots are 3 and } -\frac{4}{3}.$$

Eigenvalues and pivots have the same signs! Multiply the eigenvalues or the pivots to reach the determinant $-4$. The zero on the diagonal rules out positive definiteness.

The saddle-point matrix has $m$ positive and $n$ negative eigenvalues. The energy in $(m+n)$-dimensional space goes upward in $m$ directions and downward in $n$ directions.

An important computational decision has voters on both sides. Is it better to eliminate $\boldsymbol{w}$ and work with one matrix $A^TCA$? Optimizers say no, finite element engineers say yes. Fluids calculations (with pressure dual to velocity) often look for the saddle point.

Computational science and engineering is a highly active subject, a mix of software and hardware and mathematics in solving $A^TCA$ equations with millions of unknowns.

## ■ REVIEW OF THE KEY IDEAS ■

1. Row $k$ of $A$ ($m$ by $n$) tells the start node and the end node of edge $k$ in the graph.
2. The Laplacian $L = A^TA$ has $L_{ij} = -1$ when an edge connects nodes $i$ and $j$.
3. The diagonal of $L = D - W$ shows the degrees of the nodes. Each row adds to zero.
4. With weights $c_k$ on the edges, $K = A^TCA$ is the weighted graph Laplacian.
5. Three steps $\boldsymbol{e} = \boldsymbol{b} - A\boldsymbol{v}$, $\boldsymbol{w} = C\boldsymbol{e}$, $\boldsymbol{f} = A^T\boldsymbol{w}$ combine into $A^TCA\boldsymbol{v} = A^TC\boldsymbol{b} - \boldsymbol{f}$.

# Problem Set 7.5

**Problems 1 – 5 are about complete graphs. Every pair of nodes has an edge.**

**1** With $n = 5$ nodes and all edges, find the diagonal entries of $A^TA$ (the degrees of the nodes). All the off-diagonal entries of $A^TA$ are $-1$. Show the reduced matrix $R$ without row 5 and column 5. Node 5 is "grounded" and $v_5 = 0$.

**2** Show that the *trace* of $A^{\mathrm{T}}A$ (sum down the diagonal = sum of eigenvalues) is $n^2 - n$. What is the trace of the reduced (and invertible) matrix $R$ of size $n-1$?

**3** For $n = 4$, write the 3 by 3 matrix $R = (A_{\text{reduced}})^{\mathrm{T}}(A_{\text{reduced}})$. Show that $RR^{-1} = I$ when $R^{-1}$ has all entries $\frac{1}{4}$ off the diagonal and $\frac{2}{4}$ on the diagonal.

**4** For every $n$, the reduced matrix $R$ of size $n-1$ is *invertible*. Show that $RR^{-1} = I$ when $R^{-1}$ has all entries $1/n$ off the diagonal and $2/n$ on the diagonal.

**5** Write the 6 by 3 matrix $M = A_{\text{reduced}}$ when $n = 4$. The equation $M\boldsymbol{v} = \boldsymbol{b}$ is to be solved by least squares. The vector $\boldsymbol{b}$ is like scores in 6 games between 4 teams (team 4 always scores zero; it is grounded). Knowing the inverse of $R = M^{\mathrm{T}}M$, what is the least squares ranking $\widehat{v}_1$ for team 1 from solving $M^{\mathrm{T}}M\widehat{\boldsymbol{v}} = M^{\mathrm{T}}\boldsymbol{b}$?

**6** For the tree graph with 4 nodes, $A^{\mathrm{T}}A$ is in equation (1). What is the 3 by 3 matrix $R = (A^{\mathrm{T}}A)_{\text{reduced}}$? How do we know it is positive definite?

**7** (a) If you are given the matrix $A$, how could you reconstruct the graph?

(b) If you are given $L = A^{\mathrm{T}}A$, how could you reconstruct the graph (no arrows)?

(c) If you are given $K = A^{\mathrm{T}}CA$, how could you reconstruct the weighted graph?

**8** Find $K = A^{\mathrm{T}}CA$ for a line of 3 resistors with conductances $c_1 = 1, c_2 = 4, c_3 = 9$. Write $K_{\text{reduced}}$ and show that this matrix is positive definite.

**9** A 3 by 3 square grid has $n = 9$ nodes and $m = 12$ edges. Number nodes by rows.

(a) How many nonzeros among the 81 entries of $L = A^{\mathrm{T}}A$?

(b) Write down the 9 diagonal entries in the degree matrix $D$ : they are not all 4.

(c) Why does the middle row of $L = D - W$ have four $-1$'s? Notice $L = K2D$!

**10** Suppose all conductances in equation (5) are equal to $c$. Solve equation (6) for the voltages $v_2$ and $v_3$ and find the current $I$ flowing out of node 1 (and into the ground at node 4). What is the "system conductance" $I/V$ from node 1 to node 4?

This overall conductance $I/V$ should be larger than the individual conductances $c$.

**11** The multiplication $A^{\mathrm{T}}A$ can be columns of $A^{\mathrm{T}}$ times rows of $A$. For the tree with $m = 3$ edges and $n = 4$ nodes, each (column times row) is $(4 \times 1)(1 \times 4) = 4 \times 4$. Write down those three column-times-row matrices and add to get $L = A^{\mathrm{T}}A$.

**12** A graph with two separate 3-node trees is *not connected*. Write its 6 by 4 incidence matrix $A$. Find *two* solutions to $A\boldsymbol{v} = \boldsymbol{0}$, not just one solution $\boldsymbol{v} = (1,1,1,1,1,1)$. To reduce $A^{\mathrm{T}}A$ we must ground *two* nodes and remove two rows and columns.

**13** "Element matrices" from column times row appear in the **finite element method**. Include the numbers $c_1, c_2, c_3$ in the element matrices $K_1, K_1, K_3$.

$$K_i = (\text{row } i \text{ of } A)^{\mathrm{T}} \ (\boldsymbol{c_i}) \ (\text{row } i \text{ of } A) \qquad K = A^{\mathrm{T}}CA = \boldsymbol{K_1 + K_2 + K_3}.$$

Write the element matrices that add to $A^{\mathrm{T}}A$ in (1) for the 4-node line graph.

$$A^{\mathrm{T}}A = \left[\begin{array}{cccc} \left[ K_1 \right. & & \left. \right] & \\ & \left[ K_2 \right. & & \left. \right] \\ & & \left[ K_3 \right. & \left. \right] \end{array}\right] = \begin{array}{l}\text{assembly of the nonzero} \\ \text{entries of } K_1 + K_2 + K_3 \\ \text{from edges } 1, 2, \text{ and } 3\end{array}$$

**14** An $n$ by $n$ grid has $n^2$ nodes. How many edges in this graph? How many interior nodes? How many nonzeros in $A$ and in $L = A^{\mathrm{T}}A$? *There are no zeros in* $L^{-1}$ !

**15** When only $\boldsymbol{e} = C^{-1}\boldsymbol{w}$ is eliminated from the 3-step framework, equation (9) shows

$$\begin{array}{l}\textbf{Saddle-point matrix} \\ \textbf{Not positive definite}\end{array} \qquad \begin{bmatrix} C^{-1} & A \\ A^{\mathrm{T}} & 0 \end{bmatrix} \begin{bmatrix} \boldsymbol{w} \\ \boldsymbol{v} \end{bmatrix} = \begin{bmatrix} \boldsymbol{b} \\ \boldsymbol{f} \end{bmatrix}.$$

Multiply the first block row by $A^{\mathrm{T}}C$ and subtract from the second block row:

$$\textbf{After block elimination} \qquad \begin{bmatrix} C^{-1} & A \\ 0 & -A^{\mathrm{T}}CA \end{bmatrix} \begin{bmatrix} \boldsymbol{w} \\ \boldsymbol{v} \end{bmatrix} = \begin{bmatrix} \boldsymbol{b} \\ \boldsymbol{f} - A^{\mathrm{T}}C\boldsymbol{b} \end{bmatrix}.$$

After $m$ positive pivots from $C^{-1}$, why does this matrix have negative pivots? The two-field problem for $\boldsymbol{w}$ and $\boldsymbol{v}$ is finding a saddle point, not a minimum.

**16** The least squares equation $A^{\mathrm{T}}A\boldsymbol{v} = A^{\mathrm{T}}\boldsymbol{b}$ comes from the projection equation $A^{\mathrm{T}}\boldsymbol{e} = \boldsymbol{0}$ for the error $\boldsymbol{e} = \boldsymbol{b} - A\boldsymbol{v}$. Write those two equations in the symmetric saddle point form of Problem 15 (with $\boldsymbol{f} = \boldsymbol{0}$).

In this case $\boldsymbol{w} = \boldsymbol{e}$ because the weighting matrix is $C = I$.

**17** Find the three eigenvalues and three pivots and the determinant of this saddle point matrix with $C = I$. One eigenvalue is negative because $A$ has one column:

$$m = 2, n = 1 \qquad \begin{bmatrix} C^{-1} & A \\ A^{\mathrm{T}} & 0 \end{bmatrix} = \begin{bmatrix} 1 & 0 & -1 \\ 0 & 1 & 1 \\ -1 & 1 & 0 \end{bmatrix}.$$

# ■ CHAPTER 7 NOTES ■

**Polar Form of an Invertible Matrix: $A = QS$ = (orthogonal) (positive definite).** This is like $re^{i\theta}$ for complex numbers (1 by 1 matrices). $|e^{i\theta}| = 1$ is the orthogonal $Q$ and $r > 0$ is the positive definite $S$. The matrix factors come directly from the Singular Value Decomposition of $A$:

$$A = U\Sigma V^{\mathrm{T}} = \left(UV^{\mathrm{T}}\right)\left(V\Sigma V^{\mathrm{T}}\right) = \text{(orthogonal)} \quad \text{times} \quad \text{(positive definite)}.$$

When $A$ is invertible, so is $\Sigma$. Then $\sigma_1$ to $\sigma_n$ are the (positive) eigenvalues of $V\Sigma V^{\mathrm{T}}$. In physical language, every motion combines a rotation/reflection $Q$ with a stretching $S$.

**Transpose of $A = d/dx$.** It is not enough to say that "the transpose is $-d/dx$." The boundary conditions on the functions $f$ and $g$ in $Af = df/dx$ and $A^{\mathrm{T}}g = -dg/dx$ are important parts of $A$ and $A^{\mathrm{T}}$. In Section 7.3 and especially Problem 1, $A$ comes with *two* conditions $f(0) = 0$ and $f(1) = 0$. Then $A^{\mathrm{T}} = -d/dx$ has no conditions on $g$. What we want is $(Af, g) = (f, A^{\mathrm{T}}g)$.

**Integration by parts is like transposing the operator $d/dx$.** The integrated term $fg$ is safely zero when $f(0) = f(1) = 0$. The *fixed-free* operator $d/dx$ with only one condition $f(0) = 0$ would transpose to the *free-fixed* operator $-d/dx$ with the other condition $g(1) = 0$. Then the integrated term is again $fg = 0$ at both ends. In each case, *boundary conditions on* $\boldsymbol{g}$ *make up for missing boundary conditions on* $f$.

**Principal Component Analysis (PCA) : Find the most significant (least random) data.**

Data often comes in rectangular matrices: A grade for each student in each course. Activity of each gene in each disease. Sales of each product in each store. Income in each age group in each city. An entry goes into each column and each row of the data matrix.

By subtracting off the means, we study the *variances*: measures of useful information as opposed to randomness. The SVD of the data matrix $A$ (showing the eigenvectors and eigenvalues of the correlation $A^{\mathrm{T}}A$) displays the **principal component**: the largest piece $\sigma_1\boldsymbol{u}_1\boldsymbol{v}_1^{\mathrm{T}}$ of the matrix. The orthogonal pieces $\sigma_i\boldsymbol{u}_i\boldsymbol{v}_i^{\mathrm{T}}$ are in order of importance. The largest $\sigma$ is the most significant. From a large matrix of partly random data, PCA and the SVD extract its most significant information.

*Wikipedia* lists many methods that are identical or closely related to PCA. The crucial singular vector $\boldsymbol{v}_1$ (which has $A^{\mathrm{T}}A\boldsymbol{v}_1 = \lambda_{\max}\boldsymbol{v}_1$) is also the vector that maximizes the Rayleigh quotient $(\boldsymbol{v}^{\mathrm{T}}A^{\mathrm{T}}A\boldsymbol{v})/\boldsymbol{v}^{\mathrm{T}}\boldsymbol{v}$. Computing the first few singular vectors does not require the whole SVD!

# Chapter 8

# Fourier and Laplace Transforms

This book began with linear differential equations. It will end that way. Those are the equations we can understand and solve—especially when the coefficients are constant. Even the heat equation and wave equation (*those are PDE's*) have good solutions.

These are extremely nice problems, no apologies for that. Almost every application starts with a linear response—current proportional to voltage, output proportional to input. For large voltages or large forces, the true law may become nonlinear. Even then, we often use a sequence of linear problems to deal with nonlinearity. The constant coefficient linear equation is the one we can solve.

This chapter introduces Fourier transforms and Laplace transforms. They express every input $f(x)$ and $f(t)$ and every output $y(x)$ and $y(t)$ as a **combination of exponentials**. For each exponential, the output multiplies the input by a constant that depends on the frequency: $y(t) = Y(s)e^{st}$ or $Y(\omega)e^{i\omega t}$. **That transfer function describes the system by its frequency response : the constants $Y$ that multiply exponentials.**

We have used the complex gain $1/(i\omega - a)$ to invert $y' - ay$, along with transfer functions in Chapters 1 and 2. Now we see them for every time-invariant and shift-invariant partial differential equation—with coefficients that are constant in time and space.

Naturally those ideas appear again for discrete problems with matrix equations. The matrices may be approximating derivatives (like the $-1, 2, -1$ second difference matrix). Or they come on their own from convolutions. Their eigenvectors will be discrete sines or cosines or complex exponentials. A combination of those eigenvectors is a *discrete Fourier series* (**DFT**). We find the coefficients in that combination by using the Fast Fourier Transform (**FFT**)— **the most important algorithm in modern applied mathematics**.

A note about sines and cosines versus complex exponentials. For real problems we may like sines and cosines. But they aren't perfect. We keep $\cos 0$ and we don't keep $\sin 0$. We want one of the highest frequency vectors $(1, -1, 1, -1, \ldots)$ and $(-1, 1, -1, 1, \ldots)$ but not both. In the end (and almost always for the FFT) *the complex exponentials win*. After all, they are eigenfunctions of the derivative $d/dx$. Transforms are based on combinations of those exponentials—and the derivative of $e^{i\omega x}$ is just $i\omega e^{i\omega x}$.

This page describes a specially nice function space. It is called "*Hilbert space.*" **The functions have dot products and lengths**. There are angles between functions, so two functions can be orthogonal (perpendicular). The functions in Hilbert space are just like vectors. In fact *they are vectors*—but Hilbert space is infinite-dimensional.

Here are parallels between real vectors $\boldsymbol{f} = (f_1, \ldots, f_N)$ and real functions $f(x)$. Physicists even separate $< f|$ (bra) from $|g >$ (ket). Not here!

| | | |
|---|---|---|
| **Inner product** | $\boldsymbol{f}^{\mathrm{T}}\boldsymbol{g} = f_1 g_1 + \cdots + f_N g_N$ | $< \boldsymbol{f}, \boldsymbol{g} > = \int_{-\pi}^{\pi} f(x)\, g(x) dx$ |
| **Length squared** | $\|\|\boldsymbol{f}\|\|^2 = \boldsymbol{f}^{\mathrm{T}}\boldsymbol{f} = \sum \|f_i\|^2$ | $\|\|\boldsymbol{f}\|\|^2 = < f, f > = \int_{-\pi}^{\pi} \|f(x)\|^2 dx$ |
| **Angle $\theta$** | $\cos\theta = \boldsymbol{f}^{\mathrm{T}}\boldsymbol{g}/\|\|\boldsymbol{f}\|\|\,\|\|\boldsymbol{g}\|\|$ | $\cos\theta = < f, g > /\|\|f\|\|\,\|\|g\|\|$ |
| **Orthogonality** | $\boldsymbol{f}^{\mathrm{T}}\boldsymbol{g} = \mathbf{0}$ | $< \boldsymbol{f}, \boldsymbol{g} > = \int_{-\pi}^{\pi} f(x)\, g(x) dx = \mathbf{0}$ |

A function is allowed into Hilbert space if it has a finite length: $\int |f(x)|^2 dx < \infty$. Thus $f(x) = 1/x$ and $f(x) = \delta(x)$ do *not* belong to Hilbert space. But a step function is good. And the function can even blow up at a point—just not too fast. For example $f(x) = 1/|x|^{1/4}$ belongs to Hilbert space and its length is $||f|| = 2\pi^{1/4}$:

$$f(0) \text{ is infinite but } ||f||^2 = \int_{-\pi}^{\pi} |x|^{-1/2} dx = 4\,|x|^{1/2}\Big]_0^{\pi} = 4\pi^{1/2}.$$

When $|f(x)| = |f(-x)|$, the integral from $-\pi$ to $\pi$ is twice the integral from 0 to $\pi$.

There is always an adjustment for *complex* vectors and functions:

**Inner product** $\quad \overline{\boldsymbol{f}}^{\mathrm{T}}\boldsymbol{g} = \overline{f}_1 g_1 + \cdots + \overline{f}_N g_N \qquad < \boldsymbol{f}, \boldsymbol{g} > = \displaystyle\int \overline{\boldsymbol{f}(\boldsymbol{x})}\boldsymbol{g}(\boldsymbol{x})\boldsymbol{dx}$

Orthogonality is still $< \boldsymbol{f}, \boldsymbol{g} > = \mathbf{0}$. The best examples are the complex exponentials:

**$e^{ikx}$ and $e^{inx}$ are orthogonal** $\qquad \displaystyle\int_{-\pi}^{\pi} e^{-ikx} e^{inx} dx = \left.\frac{e^{i(n-k)x}}{n-k}\right|_{-\pi}^{\pi} = \mathbf{0}.$

Those $e^{ikx}$ are an **orthogonal basis** for Hilbert space. Instead of $xyz$ axes, functions need infinitely many axes. Every $f(x)$ is a combination of the basis vectors $e^{ikx}$:

$$f(x) = \frac{e^{ix} - e^{-ix}}{1} + \frac{e^{3ix} - e^{-3ix}}{3} + \cdots \quad \text{has} \quad \int_{-\pi}^{\pi} |f(x)|^2 = 2\pi(1^2 + 1^2 + \frac{1}{3^2} + \frac{1}{3^2} + \cdots).$$

This particular $f(x)$ happens to be a step function. To Hilbert, step functions are vectors. Then Fourier "transformed" $f(x)$ into the numbers (like 1 and $\frac{1}{3}$) that multiply each $e^{ikx}$.

## 8.1 Fourier Series

This section explains three Fourier series: **sines, cosines, and exponentials $e^{ikx}$**. Square waves (1 or 0 or $-1$) are great examples, with delta functions in the derivative. We look at a spike, a step function, and a ramp—and smoother functions too.

Start with $\sin x$. It has period $2\pi$ since $\sin(x+2\pi) = \sin x$. It is an odd function since $\sin(-x) = -\sin x$, and it vanishes at $x = 0$ and $x = \pi$. Every function $\sin nx$ has those three properties, and Fourier looked at *infinite combinations of the sines*:

**Fourier sine series** $$S(x) = b_1 \sin x + b_2 \sin 2x + b_3 \sin 3x + \cdots = \sum_{n=1}^{\infty} \boldsymbol{b_n \sin nx} \quad (1)$$

If the numbers $b_1, b_2, b_3, \ldots$ drop off quickly enough (we are foreshadowing the importance of their decay rate) then the sum $S(x)$ will inherit all three properties:

**Periodic** $S(x+2\pi) = S(x)$ **Odd** $S(-x) = -S(x)$ $S(0) = S(\pi) = 0$

200 years ago, Fourier startled the mathematicians in France by suggesting that *any odd periodic function* $S(x)$ could be expressed as an infinite series of sines. This idea started an enormous development of Fourier series. Our first step is to **find the number $b_k$ that multiplies $\sin kx$. The function $S(x)$ is "transformed" to a sequence of $b$'s.**

*Suppose* $S(x) = \sum b_n \sin nx$. *Multiply both sides by* $\sin kx$. *Integrate from* $0$ *to* $\pi$ :

$$\int_0^{\pi} S(x) \sin kx \, dx = \int_0^{\pi} b_1 \sin x \sin kx \, dx + \cdots + \int_0^{\pi} \boldsymbol{b_k \sin kx \ \sin kx \, dx} + \cdots \quad (2)$$

On the right side, all integrals are zero except the highlighted one with $n = k$. This property of "**orthogonality**" will dominate the whole chapter. For sines, integral $= 0$ is a fact of calculus:

**Sines are orthogonal** $$\int_0^{\pi} \boldsymbol{\sin nx \ \sin kx \, dx = 0} \quad \text{if} \quad n \neq k. \quad (3)$$

Zero comes quickly if we integrate $\int \cos mx \, dx = \left[\frac{\sin mx}{m}\right]_0^{\pi} = 0 - 0$. So we use this:

**Product of sines** $$\sin nx \ \sin kx = \frac{1}{2}\cos(n-k)x - \frac{1}{2}\cos(n+k)x. \quad (4)$$

Integrating $\cos(n-k)x$ and $\cos(n+k)x$ gives zero, proving orthogonality of the sines.

*The exception is when* $n = k$. Then we are integrating $(\sin kx)^2 = \frac{1}{2} - \frac{1}{2}\cos 2kx$:

$$\int_0^{\pi} \sin kx \ \sin kx \, dx = \int_0^{\pi} \frac{1}{2} \, dx - \int_0^{\pi} \frac{1}{2} \cos 2kx \, dx = \frac{\pi}{2}. \quad (5)$$

The highlighted term in equation (2) is $\boldsymbol{(\pi/2) b_k}$. Multiply both sides by $2/\pi$ to find $b_k$.

**Sine coefficients** $S(-x) = -S(x)$ $\qquad b_k = \dfrac{2}{\pi}\displaystyle\int_0^{\pi} S(x)\sin kx\,dx = \dfrac{1}{\pi}\int_{-\pi}^{\pi} \boldsymbol{S(x)\sin kx\,dx}.$ (6)

Notice that $S(x)\sin kx$ is *even* (equal integrals from $-\pi$ to 0 and from 0 to $\pi$).

I will go immediately to the most important example of a Fourier sine series. $S(x)$ is an **odd square wave** with $SW(x) = 1$ for $0 < x < \pi$. It is drawn in Figure 8.1 as an odd function (with period $2\pi$) that vanishes at $x = 0$ and $x = \pi$.

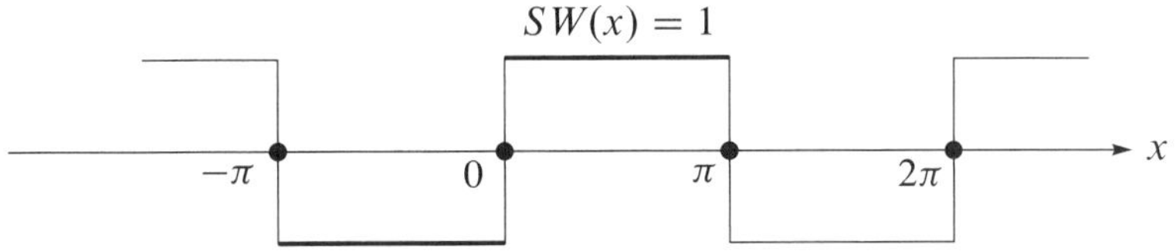


Figure 8.1: The odd square wave with $SW(x + 2\pi) = SW(x) = \{1 \text{ or } 0 \text{ or } -1\}$.

**Example 1** Find the Fourier sine coefficients $b_k$ of the odd square wave $SW(x)$.

**Solution** For $k = 1, 2, \ldots$ use formula (6) with $S(x) = 1$ between 0 and $\pi$:

$$b_k = \frac{2}{\pi}\int_0^{\pi} \sin kx\,dx = \frac{2}{\pi}\left[\frac{-\cos kx}{k}\right]_0^{\pi} = \frac{2}{\pi}\left\{\frac{\mathbf{2}}{\mathbf{1}}, \frac{0}{2}, \frac{\mathbf{2}}{\mathbf{3}}, \frac{0}{4}, \frac{\mathbf{2}}{\mathbf{5}}, \frac{0}{6}, \ldots\right\} \quad (7)$$

The even-numbered coefficients $b_{2k}$ are all zero because $\cos 2k\pi = \cos 0 = 1$. The odd-numbered coefficients $b_k = \boldsymbol{4/\pi k}$ decrease at the rate $1/k$. We will see that same $1/k$ decay rate for all functions formed from *smooth pieces and jumps.*

Put those coefficients $4/\pi k$ and zero into the Fourier sine series for $SW(x)$:

$$\textbf{Square wave} \qquad \boldsymbol{SW(x)} = \frac{4}{\pi}\left[\frac{\sin x}{\mathbf{1}} + \frac{\sin 3x}{\mathbf{3}} + \frac{\sin 5x}{\mathbf{5}} + \frac{\sin 7x}{\mathbf{7}} + \cdots\right] \quad (8)$$

Figure 8.2 graphs this sum after one term, then two terms, and then five terms. You can see the all-important **Gibbs phenomenon** appearing as these "partial sums" include more terms. Away from the jumps, we safely approach $SW(x) = 1$ or $-1$. At $x = \pi/2$, the series gives a beautiful alternating formula for the number $\pi$ :

$$1 = \frac{4}{\pi}\left[\frac{1}{1} - \frac{1}{3} + \frac{1}{5} - \frac{1}{7} + \cdots\right] \quad \text{so that} \quad \boldsymbol{\pi = 4\left(\frac{1}{1} - \frac{1}{3} + \frac{1}{5} - \frac{1}{7} + \cdots\right)}. \quad (9)$$

*The Gibbs phenomenon is the overshoot that moves closer and closer to the jumps.* Its height approaches $1.18\ldots$ and it does not decrease with more terms of the series. This overshoot is the one greatest obstacle to calculation of all discontinuous functions (like shock waves). We try hard to avoid Gibbs but sometimes we can't.

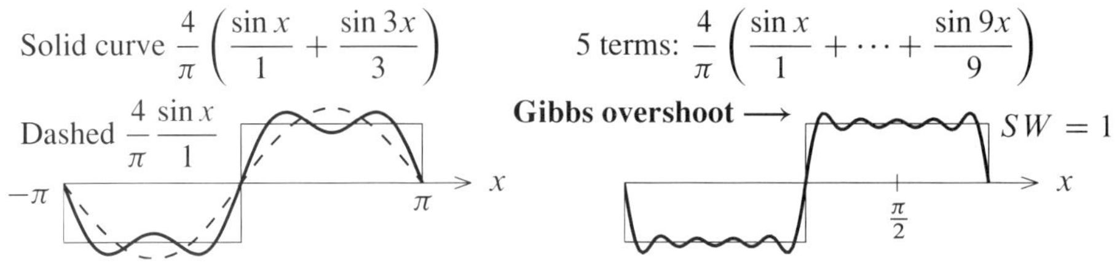


Figure 8.2: The sums $b_1 \sin x + \cdots + b_N \sin Nx$ overshoot the square wave near jumps.

## Fourier Cosine Series

The cosine series applies to ***even functions*** $C(x) = C(-x)$. They are symmetric across 0 :

$$\textbf{Cosine series} \quad C(x) = a_0 + a_1 \cos x + a_2 \cos 2x + \cdots = \boldsymbol{a_0} + \sum_{n=1}^{\infty} \boldsymbol{a_n \cos nx}. \qquad (10)$$

Every cosine has period $2\pi$. Figure 8.3 shows two even functions, the **repeating ramp** $RR(x)$ and the **up-down train** $UD(x)$ of delta functions. That sawtooth ramp $RR$ is the integral of the square wave. The delta functions in $UD$ give the derivative of the square wave. (For sines, the integral and derivative are cosines.) $RR$ and $UD$ will be valuable examples, one smoother than $SW$, one less smooth.

First we find formulas for the cosine coefficients $a_0$ and $a_k$. *The constant term $a_0$ is the average value of the function $C(x)$* :

$$\boldsymbol{a_0} = \textbf{average} \qquad \boldsymbol{a_0} = \frac{1}{\pi}\int_0^{\pi} C(x)\,dx = \frac{1}{2\pi}\int_{-\pi}^{\pi} C(x)\,dx. \qquad (11)$$

I just integrated every term in the cosine series (10) from 0 to $\pi$. On the right side, the integral of $a_0$ is $a_0\pi$ (divide both sides by $\pi$). All other integrals are zero :

$$\int_0^{\pi} \cos nx\,dx = \left[\frac{\sin nx}{n}\right]_0^{\pi} = 0 - 0 = 0. \qquad (12)$$

In words, the constant function 1 is orthogonal to $\cos nx$ over the interval $[0, \pi]$.

The other cosine coefficients $a_k$ come from the *orthogonality of cosines*. As with sines, we multiply both sides of (10) by $\cos kx$ and integrate from 0 to $\pi$:

$$\int_0^{\pi} C(x)\cos kx\,dx = \int_0^{\pi} a_0 \cos kx\,dx + \int_0^{\pi} a_1 \cos x \cos kx\,dx + \cdots + \int_0^{\pi} \boldsymbol{a_k (\cos kx)^2\,dx} + \cdots$$

You know what is coming. On the right side, only the highlighted term can be nonzero. For $k > 0$, that bold nonzero term is $\boldsymbol{a_k \pi/2}$. Multiply both sides by $2/\pi$ to find $a_k$ :

$$\begin{array}{l}\textbf{Cosine coefficients} \\ \boldsymbol{C(-x) = C(x)}\end{array} \qquad a_k = \frac{2}{\pi}\int_0^{\pi} C(x)\cos kx\,dx = \frac{1}{\pi}\int_{-\pi}^{\pi} \boldsymbol{C(x)\cos kx\,dx}. \qquad (13)$$

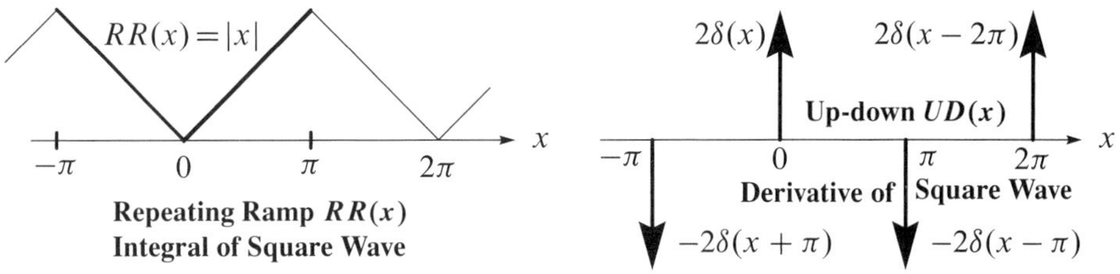


Figure 8.3: The repeating ramp $RR$ and the up-down $UD$ (periodic spikes) are even. The slope of $RR$ is $-1$ then $1$ : odd square wave $SW$. **The next derivative is $\boldsymbol{UD}$ :** $\pm 2\delta$.

**Example 2** Find the cosine coefficients of the ramp $RR(x)$ and the up-down $UD(x)$.

**Solution** The simplest way is to start with the sine series for the square wave:

$$SW(x) = \frac{4}{\pi}\left[\frac{\sin x}{1} + \frac{\sin 3x}{3} + \frac{\sin 5x}{5} + \frac{\sin 7x}{7} + \cdots\right] = \text{slope of } RR$$

Take the derivative of every term to produce cosines in the up-down delta function:

$$\textbf{Up-down spikes} \qquad \boldsymbol{UD(x)} = \frac{4}{\pi}\left[\cos x + \cos 3x + \cos 5x + \cos 7x + \cdots\right]. \tag{14}$$

Those coefficients don't decay at all. The terms in the series don't approach zero, so officially the series cannot converge. Nevertheless it is correct and important. At $x = 0$, the cosines are all 1 and their sum is $+\infty$. At $x = \pi$, the cosines are all $-1$. Then their sum is $-\infty$. (The downward spike is $-2\delta(x-\pi)$.) The true way to recognize $\delta(x)$ is by the integral test $\int \delta(x) f(x)\, dx = f(0)$ and Example 3 will do this.

For the repeating ramp, we integrate the square wave series for $SW(x)$ and add $a_0$. The average ramp height is $a_0 = \pi/2$, halfway from 0 to $\pi$ :

$$\textbf{Ramp series} \qquad \boldsymbol{RR(x)} = \frac{\pi}{2} - \frac{\pi}{4}\left[\frac{\cos x}{\mathbf{1^2}} + \frac{\cos 3x}{\mathbf{3^2}} + \frac{\cos 5x}{\mathbf{5^2}} + \frac{\cos 7x}{\mathbf{7^2}} + \cdots\right]. \tag{15}$$

The constant of integration is $a_0$. *Those coefficients $a_k$ drop off like $1/k^2$.* They could be computed directly from formula (13) using $\int x \cos kx\, dx$, and integration by parts (or an appeal to *Mathematica* or *Maple*). It was much easier to integrate every sine separately in $SW(x)$, which makes clear the crucial point: **Each "degree of smoothness" in the function brings a faster decay rate of its Fourier coefficients $\boldsymbol{a_k}$ and $\boldsymbol{b_k}$.** Every integration divides those numbers by $k$.

| | |
|---|---|
| **No decay** | **Delta** functions (with spikes) |
| **$\boldsymbol{1/k}$ decay** | **Step** functions (with jumps) |
| **$\boldsymbol{1/k^2}$ decay** | **Ramp** functions (with corners) |
| **$\boldsymbol{1/k^4}$ decay** | **Spline** functions (jumps in $f'''$) |
| **$\boldsymbol{r^k}$ decay with $\boldsymbol{r < 1}$** | **Analytic** functions like $1/(2-\cos x)$ |

## The Fourier Series for a Delta Function

**Example 3** Find the (cosine) coefficients of the ***delta function*** $\boldsymbol{\delta(x)}$, made $2\pi$-periodic.

**Solution** The spike in $\delta(x)$ occurs at $x = 0$. All the integrals are 1, because the cosine of 0 is 1. We divide by $2\pi$ for $a_0$ and by $\pi$ for the other cosine coefficients $a_k$.

$$\textbf{Average}\quad a_0 = \frac{1}{2\pi}\int_{-\pi}^{\pi} \delta(x)\,dx = \frac{\mathbf{1}}{\mathbf{2\pi}} \qquad \textbf{Cosines}\quad a_k = \frac{1}{\pi}\int_{-\pi}^{\pi} \delta(x)\cos kx\,dx = \frac{\mathbf{1}}{\boldsymbol{\pi}}$$

Then the series for the delta function has *all cosines in equal amounts*: **No decay**.

**Delta function**
$$\boldsymbol{\delta(x) = \frac{1}{2\pi} + \frac{1}{\pi}\left[\cos x + \cos 2x + \cos 3x + \cdots\right].} \tag{16}$$

This series cannot truly converge (its terms don't approach zero). But we can graph the sum after $\cos 5x$ and after $\cos 10x$. Figure 8.4 shows how these "partial sums" are doing their best to approach $\delta(x)$. They oscillate faster while going higher.

There is a neat formula for the sum $\delta_N$ that stops at $\cos Nx$. Start by writing each term $2\cos x$ as $e^{ix} + e^{-ix}$. We get a geometric progression from $e^{-iNx}$ up to $e^{iNx}$.

$$\delta_N = \frac{1}{2\pi}\left[1 + e^{ix} + e^{-ix} + \cdots + e^{iNx} + e^{-iNx}\right] = \frac{1}{2\pi}\,\frac{\sin(N+\frac{1}{2})x}{\sin\frac{1}{2}x}. \tag{17}$$

This is the function graphed in Figure 8.4.

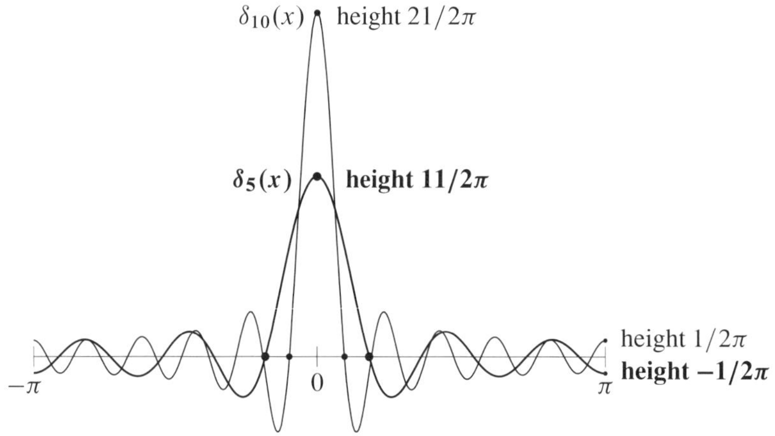


Figure 8.4: The sums $\delta_N(x) = (1 + 2\cos x + \cdots + 2\cos Nx)/2\pi$ try to approach $\delta(x)$.

## Complete Series: Sines and Cosines

Over the half-period $[0, \pi]$, the sines are not orthogonal to all the cosines. In fact the integral of $\sin x$ times 1 is not zero. So for functions $F(x)$ that are not odd or even, we must move to the *complete series* (*sines plus cosines*) on the full interval. Since our functions are periodic, that "full interval" can be $[-\pi, \pi]$ or $[0, 2\pi]$. We have both $a$'s and $b$'s.

**Complete Fourier series** $$F(x) = \boldsymbol{a_0} + \sum_{n=1}^{\infty} \boldsymbol{a_n \cos nx} + \sum_{n=1}^{\infty} \boldsymbol{b_n \sin nx}\,. \qquad (18)$$

On every "$2\pi$ interval" the sines and cosines are orthogonal. We find the Fourier coefficients $a_k$ and $b_k$ in the usual way: **Multiply** (18) **by** 1 **and** $\cos kx$ **and** $\sin kx$. **Then integrate both sides from** $-\pi$ **to** $\pi$ **to get** $a_0$ **and** $a_k$ **and** $b_k$.

$$\boldsymbol{a_0} = \frac{1}{2\pi}\int_{-\pi}^{\pi} F(x)\,dx \quad \boldsymbol{a_k} = \frac{1}{\pi}\int_{-\pi}^{\pi} F(x)\cos kx\,dx \quad \boldsymbol{b_k} = \frac{1}{\pi}\int_{-\pi}^{\pi} F(x)\sin kx\,dx$$

Orthogonality kills off infinitely many integrals and leaves only the one we want.

Another approach is to split $F(x) = C(x) + S(x)$ into an even part and an odd part. Then we can use the earlier cosine and sine formulas. The two parts are

$$C(x) = F_{\mathbf{even}}(x) = \frac{F(x) + F(-x)}{2} \qquad S(x) = F_{\mathbf{odd}}(x) = \frac{F(x) - F(-x)}{2}. \qquad (19)$$

The even part gives the $a$'s and the odd part gives the $b$'s. Test on a square pulse from $x = 0$ to $x = h$—this one-sided thin box function is not odd or even.

**Example 4** Find the $a$'s and $b$'s if $\boldsymbol{F(x)} =$ **tall box** $= \begin{cases} \mathbf{1/h} & \text{for } 0 < x < h \\ \mathbf{0} & \text{for } h < x < 2\pi \end{cases}$

**Solution** The integrals for $a_0$ and $a_k$ and $b_k$ stop at $x = h$ where $F(x)$ drops to zero. The coefficients decay like $1/k$ because of the jump at $x = 0$ and the drop at $x = h$:

**Coefficients of square pulse** $$\boldsymbol{a_0} = \frac{1}{2\pi}\int_0^h 1/h\,dx = \frac{\mathbf{1}}{\mathbf{2\pi}} = \textbf{average}$$

$$\boldsymbol{a_k} = \frac{1}{\pi h}\int_0^h \cos kx\,dx = \frac{\boldsymbol{\sin kh}}{\boldsymbol{\pi kh}} \qquad \boldsymbol{b_k} = \frac{1}{\pi h}\int_0^h \sin kx\,dx = \frac{\boldsymbol{1 - \cos kh}}{\boldsymbol{\pi kh}}.$$

**Important** As $h$ approaches zero, the box gets thinner and taller. Its width is $h$ and its height is $1/h$ and its area is 1. The box approaches a delta function! And its Fourier coefficients approach the coefficients of the delta function as $h \to 0$:

$$a_0 = \frac{\mathbf{1}}{\mathbf{2\pi}} \qquad a_k = \frac{\sin kh}{\pi kh} \text{ approaches } \frac{\mathbf{1}}{\boldsymbol{\pi}} \qquad b_k = \frac{1 - \cos kh}{\pi kh} \text{ approaches } \mathbf{0}. \quad (20)$$

## Energy in Function = Energy in Coefficients

There is an extremely important equation (***the energy identity***) that comes from integrating $(F(x))^2$. When we square the Fourier series of $F(x)$, and integrate from $-\pi$ to $\pi$, all the "cross terms" drop out. The only nonzero integrals come from $1^2$ and $\cos^2 kx$ and $\sin^2 kx$. Those integrals give $2\pi$ and $\pi$ and $\pi$, multiplied by $a_0^2$ and $a_k^2$ and $b_k^2$:

$$\textbf{Energy} \quad \int_{-\pi}^{\pi} (F(x))^2 dx = 2\pi a_0^2 + \pi(a_1^2 + b_1^2 + a_2^2 + b_2^2 + \cdots). \qquad (21)$$

The energy in $F(x)$ equals the energy in the coefficients. The left side is like the length squared of a vector, except *the vector is a function*. The right side comes from an infinitely long vector of $a$'s and $b$'s. The lengths are equal, which says that the Fourier transform from function to vector is like an orthogonal matrix. Normalized by $\sqrt{2\pi}$ and $\sqrt{\pi}$, **sines and cosines are an orthonormal basis in function space**.

## Complex Exponentials $c_k e^{ikx}$

This is a small step and we have to take it. In place of separate formulas for $a_0$ and $a_k$ and $b_k$, we will have *one formula* for all the complex coefficients $c_k$. And the function $F(x)$ might be complex (as in quantum mechanics). The Discrete Fourier Transform will be much simpler when we use $N$ complex exponentials for a vector.

We practice with the complex infinite series for a $2\pi$-periodic function:

$$\textbf{Complex Fourier series} \quad F(x) = c_0 + c_1 e^{ix} + c_{-1}e^{-ix} + \cdots = \sum_{n=-\infty}^{\infty} c_n e^{inx} \qquad (22)$$

If every $c_n = c_{-n}$, we can combine $e^{inx}$ with $e^{-inx}$ into $2\cos nx$. Then (22) is the cosine series for an even function. If every $c_n = -c_{-n}$, we use $e^{inx} - e^{-inx} = 2i \sin nx$. Then (22) is the sine series for an odd function and the $c$'s are pure imaginary.

**To find $c_k$, multiply (22) by $e^{-ikx}$** (not $e^{ikx}$) **and integrate from $-\pi$ to $\pi$:**

$$\int_{-\pi}^{\pi} F(x)e^{-ikx}dx = \int_{-\pi}^{\pi} c_0 e^{-ikx}dx + \int_{-\pi}^{\pi} c_1 e^{ix}e^{-ikx}dx + \cdots + \int_{-\pi}^{\pi} \boldsymbol{c_k e^{ikx} e^{-ikx} dx} + \cdots$$

The complex exponentials are orthogonal. **Every integral on the right side is zero,** except for the highlighted term (when $n = k$ and $e^{ikx}e^{-ikx} = 1$). The integral of 1 is $2\pi$. That surviving term gives the formula for $c_k$:

$$\textbf{Fourier coefficients} \quad \int_{-\pi}^{\pi} F(x)e^{-ikx}\,dx = 2\pi c_k \quad \text{for} \quad k = 0, \pm 1, \ldots l \qquad (23)$$

Notice that $c_0 = a_0$ is still the average of $F(x)$. The orthogonality of $e^{inx}$ and $e^{ikx}$ is checked by integrating $e^{inx}$ times $e^{-ikx}$. Remember to use that complex conjugate $e^{-ikx}$.

**Example 5** For a delta function, all integrals are 1 and every $c_k$ is $1/2\pi$. *Flat transform!*

**Example 6** Find $c_k$ for the $2\pi$-periodic shifted box $F(x) = \begin{cases} 1 & \text{for } s \le x \le s+h \\ 0 & \text{elsewhere in } [-\pi,\pi] \end{cases}$

**Solution** The integrals (23) have $F = 1$ from $s$ to $s+h$:

$$c_k = \frac{1}{2\pi}\int_s^{s+h} 1 \cdot e^{-ikx}\,dx = \frac{1}{2\pi}\left[\frac{e^{-ikx}}{-ik}\right]_s^{s+h} = \boldsymbol{e^{-iks}}\left(\frac{1-e^{-ikh}}{2\pi ik}\right). \qquad (24)$$

**Notice above all the simple effect of the shift by** $s$. *It "modulates" each* $c_k$ *by* $e^{-iks}$. The energy is unchanged, the integral of $|F|^2$ just shifts, and $|e^{-iks}| = 1$.

$$\boxed{\textbf{Shift } F(x) \textbf{ to } F(x-s) \longleftrightarrow \textbf{Multiply every } c_k \textbf{ by } e^{-iks}.} \qquad (25)$$

**Example 7** A centered box has shift $s = -h/2$. It becomes balanced around $x = 0$. This even function equals 1 on the interval from $-h/2$ to $h/2$:

**Centered by** $s = -\dfrac{h}{2}$ $\qquad c_k = e^{ikh/2}\,\dfrac{1-e^{-ikh}}{2\pi ik} = \dfrac{1}{2\pi}\,\dfrac{\sin(kh/2)}{k/2}.$

Divide by $h$ for a tall box. The ratio of $\sin(kh/2)$ to $kh/2$ is called the "**sinc**" of $kh/2$.

**Tall box** $\qquad \dfrac{F_{\text{centered}}}{h} = \dfrac{1}{2\pi}\displaystyle\sum_{-\infty}^{\infty} \text{sinc}\left(\frac{kh}{2}\right) e^{ikx} = \begin{cases} 1/h & \text{for } -h/2 \le x \le h/2 \\ 0 & \text{elsewhere in } [-\pi,\pi] \end{cases}$

That division by $h$ produces area $= 1$. Every coefficient approaches $\frac{1}{2\pi}$ as $h \to 0$. The Fourier series for the tall thin box again approaches the Fourier series for $\delta(x)$.

## The Rules for Derivatives and Integrals

*The derivative of* $e^{ikx}$ *is* $ike^{ikx}$. This great fact puts the Fourier functions $e^{ikx}$ in first place for applications. They are eigenfunctions for $d/dx$ (and the eigenvalues are $\lambda = ik$). Differential equations with constant coefficients are naturally solved by Fourier series.

**Multiply by $ik$** **The derivative of** $F(x) = \sum c_k e^{ikx}$ **is** $dF/dx = \sum ikc_k e^{ikx}$

The second derivative has coefficients $(ik)^2 c_k = -\boldsymbol{k^2} c_k$. High frequencies are growing stronger. And in the opposite direction (when we integrate), we divide by $ik$ and high frequencies get weaker. The solution becomes smoother. Please look at this example:

**Response $1/(k^2+1)$**
**to frequency $k$** $\qquad -\dfrac{d^2y}{dx^2} + y = e^{ikx}$ is solved by $y(x) = \dfrac{e^{ikx}}{\boldsymbol{k^2+1}}$

This was a typical problem in Chapter 2. The transfer function is $1/(k^2+1)$. There we learned: The forcing function $e^{ikx}$ is exponential so the solution is exponential.

**All we are doing now is superposition.** Allow all the exponentials at once!

$$-\frac{d^2y}{dx^2} + y = \sum c_k e^{ikx} \quad \text{is solved by} \quad y(x) = \sum \frac{c_k e^{ikx}}{\boldsymbol{k^2+1}}. \tag{26}$$

1. **Derivative rule $dF/dx$ has Fourier coefficients $ikc_k$** (energy moves to high $k$).
2. **Shift rule $F(x-s)$ has Fourier coefficients $e^{-iks}c_k$** (no change in energy).

## Application: Laplace's Equation in a Circle

Our first application is to Laplace's equation $u_{xx} + u_{yy} = 0$ (Section 7.4). The idea is to construct $u(x, y)$ as an infinite series, choosing its coefficients to match $u_0(x, y)$ along the boundary. The shape of the boundary is crucial, and we take a circle of radius 1.

Begin with the solutions $1,\ r\cos\theta,\ r\sin\theta,\ r^2\cos 2\theta,\ r^2\sin 2\theta,\ \ldots$ to Laplace's equation. Combinations of these special solutions give all solutions in the circle:

$$\boldsymbol{u(r,\theta) = a_0 + a_1 r\cos\theta + b_1 r\sin\theta + a_2 r^2\cos 2\theta + b_2 r^2\sin 2\theta + \cdots} \tag{27}$$

It remains to choose the constants $a_k$ and $b_k$ to make $u = u_0$ on the boundary. For a circle, $\theta$ and $\theta + 2\pi$ give the same point. This means that $u_0(\theta)$ is periodic:

**Set $r = 1$** $$u_0(\theta) = a_0 + a_1\cos\theta + b_1\sin\theta + a_2\cos 2\theta + b_2\sin 2\theta + \cdots \tag{28}$$

This is exactly the Fourier series for $u_0$. **The constants $a_k$ and $b_k$ must be the Fourier coefficients of $u_0(\theta)$.** Thus Laplace's boundary value problem is completely solved, if an infinite series (27) is acceptable as the solution.

**Example 8** **Point source** $u_0 = \delta(\theta)$. The boundary is held at $u_0 = 0$, except for the source at $x = 1$, $y = 0$ (where $\theta = 0$). Find the temperature $u(r, \theta)$ inside the circle.

**Delta function** $$u_0(\theta) = \frac{1}{2\pi} + \frac{1}{\pi}(\cos\theta + \cos 2\theta + \cos 3\theta + \cdots) = \frac{1}{2\pi}\sum_{-\infty}^{\infty} e^{in\theta}$$

Inside the circle, each $\cos n\theta$ is multiplied by $r^n$ to solve Laplace's equation:

**Inside the circle** $$u(r,\theta) = \frac{1}{2\pi} + \frac{1}{\pi}(r\cos\theta + r^2\cos 2\theta + r^3\cos 3\theta + \cdots) \tag{29}$$

Poisson managed to sum this infinite series! It involves a series of powers $(re^{i\theta})^n$. His sum gives the response at every $(r, \theta)$ to the point source at $r = 1, \theta = 0$:

**Temperature inside circle** $$u(r,\theta) = \frac{1}{2\pi}\frac{1 - r^2}{1 + r^2 - 2r\cos\theta} \tag{30}$$

At the center $r = 0$, this produces the average of $u_0 = \delta(\theta)$ which is $a_0 = 1/2\pi$. On the boundary $r = 1$, this gives $u = 0$ except $u = \infty$ at the point where $\cos 0 = 1$.

**Example 9** $\boldsymbol{u_0(\theta) = 1}$ **on the top half of the circle and** $\boldsymbol{u_0 = -1}$ **on the bottom half.**

**Solution** The boundary values $u_0$ are a square wave $SW$. We know its sine series:

$$\textbf{Square wave for } \boldsymbol{u_0(\theta)} \qquad SW(\theta) = \frac{4}{\pi}\left[\frac{\sin\theta}{1} + \frac{\sin 3\theta}{3} + \frac{\sin 5\theta}{5} + \cdots\right] \tag{31}$$

Inside the circle, multiplying by $r, r^3, r^5, \ldots$ gives fast decay of high frequencies:

$$\textbf{Rapid decay inside} \qquad u(r,\theta) = \frac{4}{\pi}\left[\frac{r\sin\theta}{1} + \frac{r^3\sin 3\theta}{3} + \frac{r^5\sin 5\theta}{5} + \cdots\right] \tag{32}$$

Laplace's equation has smooth solutions inside, even when $u_0(\theta)$ is not smooth.

## Problem Set 8.1

**1** (a) To prove that $\cos nx$ is orthogonal to $\cos kx$ when $k \neq n$, use the formula $(\cos nx)\,(\cos kx) = \frac{1}{2}\cos(n+k)x + \frac{1}{2}\cos(n-k)x$. Integrate from $x = 0$ to $x = \pi$. What is $\int \cos^2 kx\,dx$ ?

(b) From 0 to $\pi$, $\cos x$ is **not** orthogonal to $\sin x$. The period has to be $2\pi$ :

$$\text{Find } \int_0^{\pi} (\sin x)\,(\cos x)\,dx \quad \text{and} \quad \int_{-\pi}^{\pi} (\sin x)\,(\cos x)\,dx \quad \text{and} \quad \int_0^{2\pi} (\sin x)\,(\cos x)\,dx.$$

**2** Suppose $F(x) = x$ for $0 \le x \le \pi$. Draw graphs for $-2\pi \le x \le 2\pi$ to show three extensions of $F$: a $2\pi$-periodic even function and a $2\pi$-periodic odd function and a $\pi$-periodic function.

**3** Find the Fourier series on $-\pi \le x \le \pi$ for

(a) $f_1(x) = \sin^3 x$, an odd function (sine series, only two terms)

(b) $f_2(x) = |\sin x|$, an even function (cosine series)

(c) $f_3(x) = x$ for $-\pi \le x \le \pi$ (sine series with jump at $x = \pi$)

**4** Find the complex Fourier series $e^x = \sum c_k e^{ikx}$ on the interval $-\pi \le x \le \pi$. The even part of a function is $\frac{1}{2}(f(x) + f(-x))$, so that $f_{\text{even}}(x) = f_{\text{even}}(-x)$. Find the cosine series for $f_{\text{even}}$ and the sine series for $f_{\text{odd}}$. *Notice the jump at* $x = \pi$.

**5** From the energy formula (21), the square wave sine coefficients satisfy

$$\pi(b_1^2 + b_2^2 + \cdots) = \int_{-\pi}^{\pi} |SW(x)|^2\,dx = \int_{-\pi}^{\pi} 1\,dx = 2\pi.$$

Substitute the numbers $b_k$ from equation (8) to find that $\pi^2 = 8(1 + \frac{1}{9} + \frac{1}{25} + \cdots)$.

**6** If a square pulse is centered at $x = 0$ to give

$$f(x) = 1 \quad \text{for} \quad |x| < \frac{\pi}{2}, \qquad f(x) = 0 \quad \text{for} \quad \frac{\pi}{2} < |x| < \pi,$$

draw its graph and find its Fourier coefficients $a_k$ and $b_k$.

**7** Plot the first three partial sums and the function $x(\pi - x)$:

$$x(\pi - x) = \frac{8}{\pi}\left(\frac{\sin x}{1} + \frac{\sin 3x}{27} + \frac{\sin 5x}{125} + \cdots\right), 0 < x < \pi.$$

Why is $1/k^3$ the decay rate for this function? What is its second derivative?

**8** Sketch the $2\pi$-periodic half wave with $f(x) = \sin x$ for $0 < x < \pi$ and $f(x) = 0$ for $-\pi < x < 0$. Find its Fourier series.

**9** Suppose $G(x)$ has period $2L$ instead of $2\pi$. Then $G(x + 2L) = G(x)$. Integrals go from $-L$ to $L$ or from 0 to $2L$. The Fourier formulas change by a factor $\pi/L$:

The coefficients in $G(x) = \sum_{-\infty}^{\infty} C_k e^{ik\pi x/L}$ are $C_k = \frac{1}{2L}\int_{-L}^{L} G(x)e^{-ik\pi x/L}\,dx.$

Derive this formula for $C_k$: Multiply the first equation for $G(x)$ by ______ and integrate both sides. Why is the integral on the right side equal to $2LC_k$?

**10** For $G_{\text{even}}$, use Problem 9 to find the cosine coefficient $A_k$ from $(C_k + C_{-k})/2$:

$$G_{\text{even}}(x) = \sum_{0}^{\infty} A_k \cos\frac{k\pi x}{L} \quad \text{has} \quad A_k = \frac{1}{L}\int_{0}^{L} G_{\text{even}}(x)\cos\frac{k\pi x}{L}\,dx.$$

$G_{\text{even}}$ is $\frac{1}{2}(G(x) + G(-x))$. Exception for $A_0 = C_0$: Divide by $2L$ instead of $L$.

**11** Problem 10 tells us that $\boldsymbol{a_k = \frac{1}{2}(c_k + c_{-k})}$ on the usual interval from 0 to $\pi$. Find a similar formula for $b_k$ from $c_k$ and $c_{-k}$. In the reverse direction, find the complex coefficient $c_k$ in $F(x) = \sum c_k e^{ikx}$ from the real coefficients $a_k$ and $b_k$.

**12** Find the solution to Laplace's equation with $u_0 = \theta$ on the boundary. Why is this the imaginary part of $2(z - z^2/2 + z^3/3\cdots) = 2\log(1+z)$? Confirm that on the unit circle $z = e^{i\theta}$, the imaginary part of $2\log(1+z)$ agrees with $\theta$.

**13** If the boundary condition for Laplace's equation is $u_0 = 1$ for $0 < \theta < \pi$ and $u_0 = 0$ for $-\pi < \theta < 0$, find the Fourier series solution $u(r,\theta)$ inside the unit circle. What is $u$ at the origin $r = 0$?

**14** With boundary values $u_0(\theta) = 1 + \frac{1}{2}e^{i\theta} + \frac{1}{4}e^{2i\theta} + \cdots$, what is the Fourier series solution to Laplace's equation in the circle? Sum this geometric series.

**15** (a) Verify that the fraction in Poisson's formula (30) satisfies Laplace's equation.

(b) Find the response $u(r,\theta)$ to an impulse at $x = 0, y = 1$ (where $\theta = \frac{\pi}{2}$).

**16** With complex exponentials in $F(x) = \sum c_k e^{ikx}$, the energy identity (21) changes to $\int_{-\pi}^{\pi} |F(x)|^2\,dx = 2\pi\sum|c_k|^2$. Derive this by integrating $(\sum c_k e^{ikx})(\sum \overline{c}_k e^{-ikx})$.

**17** A centered square wave has $F(x) = 1$ for $|x| \le \pi/2$.

(a) Find its energy $\int |F(x)|^2\, dx$ by direct integration

(b) Compute its Fourier coefficients $c_k$ as specific numbers

(c) Find the sum in the energy identity (Problem 16).

**18** $F(x) = 1 + (\cos x)/2 + \cdots + (\cos nx)/2^n + \cdots$ is analytic : infinitely smooth.

(a) If you take 10 derivatives, what is the Fourier series of $d^{10}F/dx^{10}$?

(b) Does that series still converge quickly ? Compare $n^{10}$ with $2^n$ for $n = 2^{10}$.

**19** If $f(x) = 1$ for $|x| \le \pi/2$ and $f(x) = 0$ for $\pi/2 < |x| < \pi$, find its cosine coefficients. Can you graph and compute the Gibbs overshoot at the jumps ?

**20** Find all the coefficients $a_k$ and $b_k$ for $F$, $I$, and $D$ on the interval $-\pi \le x \le \pi$ :

$$F(x) = \delta\left(x - \frac{\pi}{2}\right) \qquad I(x) = \int_0^x \delta\left(x - \frac{\pi}{2}\right) dx \qquad D(x) = \frac{d}{dx}\delta\left(x - \frac{\pi}{2}\right).$$

**21** For the one-sided tall box function in Example 4, with $F = 1/h$ for $0 \le x \le h$, what is its odd part $\frac{1}{2}(F(x) - F(-x))$ ? I am surprised that the Fourier coefficients of this odd part disappear as $h$ approaches zero and $F(x)$ approaches $\delta(x)$.

**22** Find the series $F(x) = \sum c_k e^{ikx}$ for $F(x) = e^x$ on $-\pi \le x \le \pi$. That function $e^x$ looks smooth, but there must be a hidden jump to get coefficients $c_k$ proportional to $1/k$. Where is the jump ?

**23** (a) (Old particular solution) Solve $Ay'' + By' + Cy = e^{ikx}$.

(b) (New particular solution) Solve $Ay'' + By' + Cy = \sum c_k e^{ikx}$.

## 8.2 The Fast Fourier Transform

Fourier series apply to functions. But we compute with vectors. We need to replace the infinite sequence of coefficients $c_k$ (or $a_k$ and $b_k$) by a **finite sequence** $\boldsymbol{c_0, c_1, \ldots, c_{N-1}}$. We want to preserve and use orthogonality, so the computations will be fast. For the Discrete Fourier Transform, you will see how the FFT makes the computations extra fast.

This section describes two separate ideas. The DFT provides formulas for the $c$'s. *The FFT is an amazing algorithm to compute the $c$'s by rearranging those formulas.*

### Discrete Fourier Transform (DFT)

The DFT chooses $N$ orthogonal basis vectors $\boldsymbol{e}_0$ to $\boldsymbol{e}_{N-1}$ for $N$-dimensional space. The vector $\boldsymbol{e}_k$ comes from $e^{ikx}$, by sampling that function at $N$ points spaced by $2\pi/N$ :

**Basis vector $\boldsymbol{e_k}$**
**Discrete $\boldsymbol{e^{ikx}}$**
$$(e^{ik0}, e^{ik2\pi/N}, e^{ik4\pi/N}, \ldots) = (\mathbf{1}, \boldsymbol{w}^{\boldsymbol{k}}, \boldsymbol{w}^{\boldsymbol{2k}}, \ldots) \text{ with } \boldsymbol{w} = \boldsymbol{e}^{\boldsymbol{i2\pi/N}}.$$

The continuous Fourier series is $\sum c_k e^{ikx}$. The discrete Fourier series is $\sum c_k \boldsymbol{e}_k$. That sum is a multiplication $\boldsymbol{f} = F\boldsymbol{c}$ with the symmetric $N$ by $N$ **Fourier matrix** $F$. The basis vectors $\boldsymbol{e}_k$ go into the columns of $F$.

The matrix $F$ containing powers of $w$ is shown in detail in equation (4).

**Fourier matrix**
$\boldsymbol{f} = \boldsymbol{Fc}$
$$\boldsymbol{f} = c_0\boldsymbol{e}_0 + c_1\boldsymbol{e}_1 + \cdots = \begin{bmatrix} | & & | \\ \boldsymbol{e}_0 & \cdots & \boldsymbol{e}_{N-1} \\ | & & | \end{bmatrix} \begin{bmatrix} c_0 \\ \cdot \\ \cdot \\ c_{N-1} \end{bmatrix} \tag{1}$$

Inverting $\boldsymbol{f} = F\boldsymbol{c}$ gives $\boldsymbol{c} = F^{-1}\boldsymbol{f}$. The continuous case produced $e^{-ikx}$ in the Fourier coefficient formula $c_k = \int e^{-ikx} f(x)dx/2\pi$. The discrete case produces powers of $\overline{w} = e^{-i2\pi/N}$ in the inverse matrix. Those powers of $\overline{w}$ are displayed in equation (3).

**Inverse matrix**
$\boldsymbol{c} = F^{-1}\boldsymbol{f}$
$$\boldsymbol{c} = \frac{1}{N} \begin{bmatrix} — & \overline{\boldsymbol{e}}_0^{\mathrm{T}} & — \\ & \cdot & \\ & \cdot & \\ — & \overline{\boldsymbol{e}}_{N-1}^{\mathrm{T}} & — \end{bmatrix} \begin{bmatrix} f_0 \\ \cdot \\ \cdot \\ f_{N-1} \end{bmatrix} = \frac{1}{N} \overline{F}^{\mathrm{T}} \boldsymbol{f}. \tag{2}$$

The constant vector $\boldsymbol{e}_0 = (1, 1, \ldots, 1)$ has $||\boldsymbol{e}_0||^2 = 1 + 1 + \cdots + 1 = N$. Every basis vector has $||\boldsymbol{e}_k||^2 = N$ instead of $\int |e^{ikx}|^2 dx = 2\pi$.

Please notice that $F^{-1}$ produces the coefficients $c_k$ from the vector $\boldsymbol{f}$ : the *Fourier transform*. The Fourier matrix $F$ reconstructs $\boldsymbol{f}$ from the $c$'s (the *inverse transform*). The entries of $F^{-1}$ are like $e^{-ikx}$ and the entries of $F$ are like $e^{ikx}$. Thus $F^{-1} = \overline{F}/N$ contains powers of $\overline{w} = e^{-i2\pi/N}$, while $F$ contains powers of $w = e^{i2\pi/N}$.

**The** MATLAB **command $\boldsymbol{c}$ = fft($\boldsymbol{f}$) uses $\overline{\boldsymbol{w}}$ and the inverse Fourier matrix** $F^{-1}$. The opposite command $\boldsymbol{f}$ = **ifft**($\boldsymbol{c}$) adds up the $N$-term series $F\boldsymbol{c}$ to reconstruct $\boldsymbol{f}$ in (1).

**Example 1** The delta vector $\boldsymbol{f} = (1, 0, 0, \ldots)$ is like a delta function $\delta(x)$. The Fourier coefficients of a delta function are all equal to $c_k = 1/2\pi$. The discrete coefficients of a delta vector are all equal to $c_k = 1/N$. **The transform of $\boldsymbol{f}$ is a constant vector.**

**Fourier transform** $\boldsymbol{F^{-1} f = c}$

$$\frac{1}{N}\begin{bmatrix} 1 & 1 & \cdot & \cdot & 1 \\ 1 & \overline{w} & \cdot & \cdot & \overline{w}^{N-1} \\ 1 & \overline{w}^2 & \cdot & \cdot & \overline{w}^{2(N-1)} \\ \cdot & \cdot & \cdot & \cdot & \cdot \end{bmatrix}\begin{bmatrix} 1 \\ 0 \\ 0 \\ \cdot \end{bmatrix} = \frac{1}{N}\begin{bmatrix} \mathbf{1} \\ \mathbf{1} \\ \mathbf{1} \\ \cdot \end{bmatrix}. \quad (3)$$

**Example 2** The shifted vector $\boldsymbol{f} = (0, 1, 0, \ldots)$ is like a shifted delta function $\delta(x - \frac{2\pi}{N})$. The shifted vector $\boldsymbol{f}$ picks out the next column $(1, \overline{w}, \overline{w}^2, \ldots)$ of $F^{-1}$ in equation (3). The shifted delta function chooses the (same) values of $c_k = \boldsymbol{e}^{-ikx}$ at $x = 2\pi/N$.

The only difference between those discrete and continuous $c$'s is dividing by $N$ or $2\pi$.

**Example 3** **The constant vector $\boldsymbol{c} = (1, 1, \ldots)/N$ transforms back to the delta vector!**

**Fourier matrix** $\boldsymbol{Fc = f}$

$$\begin{bmatrix} 1 & 1 & \cdot & \cdot & 1 \\ 1 & w & \cdot & \cdot & w^{N-1} \\ 1 & w^2 & \cdot & \cdot & w^{2(N-1)} \\ \cdot & \cdot & \cdot & \cdot & \cdot \end{bmatrix}\frac{1}{N}\begin{bmatrix} 1 \\ 1 \\ 1 \\ \cdot \end{bmatrix} = \begin{bmatrix} \mathbf{1} \\ \mathbf{0} \\ \mathbf{0} \\ \cdot \end{bmatrix} \quad (4)$$

That equation says that $N - 1$ basis vectors starting with $(1, w, w^2, \ldots)$ are orthogonal to the first vector $(1, 1, \ldots, 1)$. ***The basis vectors $\boldsymbol{e}_k$ in the columns of $F$ are orthogonal.***

After a few words about the FFT, equation (7) will confirm this orthogonality.

**Fast Fourier Transform (FFT)**

The FFT is a brilliant rearrangement of those matrix-vector multiplications $\boldsymbol{f} = F\boldsymbol{c}$ and $\boldsymbol{c} = F^{-1}\boldsymbol{f}$. Normally, multiplying a vector by an $N$ by $N$ matrix takes $N^2$ separate multiplications. (Each entry in the square matrix is used once. There are $N^2$ entries.) The FFT computes $\boldsymbol{c}$ and $\boldsymbol{f}$ with only $\frac{1}{2}N \log_2 N$ separate multiplications.

For size $N = 1024 = 2^{10}$, the logarithm is 10. In this case $N^2$ (a million steps) are reduced to $5N$ (five thousand steps). The transform is speeded up by a factor near 200, which is truly astonishing.

In my opinion, the FFT is the most important algorithm in computational science. It has transformed whole industries. When your instruments measure the response to an input (like the pressure in an oil well), the DFT shows the response to each frequency. The FFT computes $N$ numbers from $N$ numbers, very fast.

## The Basis Vectors $\boldsymbol{e}_k$ in the Fourier Matrix $F$

A crucial point is that the basis vectors $\boldsymbol{e}_0, \ldots, \boldsymbol{e}_{N-1}$ are **orthogonal**. Those vectors are complex, just as the functions $e^{ikx}$ are complex. So their inner products $\overline{\boldsymbol{e}}_k^{\mathrm{T}}\boldsymbol{e}_n$ require the complex conjugate of one vector, just like $\int e^{inx}\, e^{-ikx}\, dx$.

Here is a typical basis vector $\boldsymbol{e}_k$, followed by the Fourier matrix that contains $\boldsymbol{e}_0, \boldsymbol{e}_1, \ldots, \boldsymbol{e}_{N-1}$ in its columns:

$$\boldsymbol{e}_k = \begin{bmatrix} 1 \\ e^{2\pi ik/N} \\ e^{4\pi ik/N} \\ \cdot \\ \cdot \end{bmatrix} = \begin{bmatrix} \mathbf{1} \\ \boldsymbol{w}^{\boldsymbol{k}} \\ \boldsymbol{w}^{\boldsymbol{2k}} \\ \cdot \\ \cdot \end{bmatrix} \qquad F = \begin{bmatrix} 1 & 1 & \cdot & \cdot & 1 \\ 1 & w & \cdot & \cdot & w^{N-1} \\ 1 & w^2 & \cdot & \cdot & w^{2(N-1)} \\ \cdot & \cdot & \cdot & \cdot & \cdot \\ 1 & w^{N-1} & \cdot & \cdot & w^{(N-1)^2} \end{bmatrix} \quad (5)$$

The number $w$ is $e^{2\pi i/N}$. We use the Greek letter $\omega$ for its conjugate $\overline{w} = e^{-2\pi i/N} = \omega$. It is the properties of $1, w, w^2, \ldots$ that make the basis vectors (columns of $F$) orthogonal. Our first step is to locate $w$ and $\overline{w}$ in the complex plane. In fact we can locate all the powers of $w$ up to $w^N = (e^{2\pi i/N})^N = e^{2\pi i} = 1$. For $N = 8$, the powers of $w$ produce 8 points evenly spaced around the unit circle . Notice that $w^8 = 1$.

For $N = 4$, the four powers will be $\boldsymbol{i}$, $i^2 = \mathbf{-1}$, $i^3 = \boldsymbol{-i}$, and $i^4 = \mathbf{1}$.

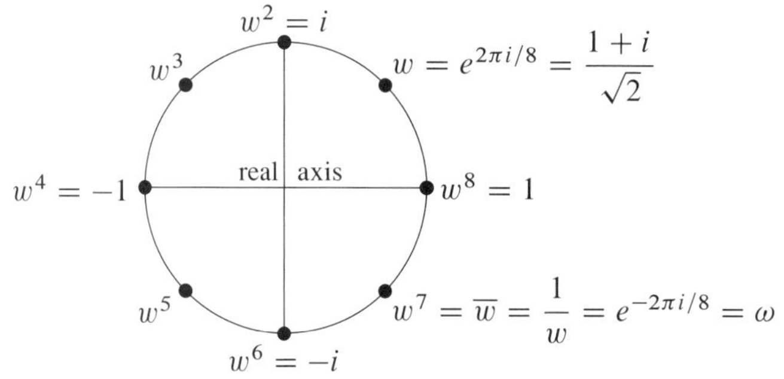


Figure 8.5: The eight powers of $w = \cos\dfrac{\pi}{4} + i\sin\dfrac{\pi}{4}$. The polar form $w = e^{2\pi i/8}$ is best.

### Orthogonality of the Discrete Fourier Basis

The key to good formulas for the Fourier coefficients $c_k$ is orthogonality. That property removes every term except term $k$, when we take a dot product with the basis vector $\boldsymbol{e}_k$ :

$$\boldsymbol{f} = c_0\boldsymbol{e}_0 + \cdots + c_{N-1}\boldsymbol{e}_{N-1} \quad \text{and} \quad \overline{\boldsymbol{e}}_{\boldsymbol{k}}^{\mathrm{T}}\boldsymbol{f} = \boldsymbol{c_k}\,\overline{\boldsymbol{e}}_{\boldsymbol{k}}^{\mathrm{T}}\,\boldsymbol{e_k} = \boldsymbol{Nc_k}. \qquad (6)$$

Since $\boldsymbol{e}_0 = (1, 1, 1, \ldots)$ and $\boldsymbol{e}_1 = (1, w, w^2, \ldots)$, the crucial step is their zero dot product: $1 + w + w^2 + \cdots = 0$. **The eight numbers around the circle in Figure 8.5 add to zero**.

Here is the statement and proof that every pair of $\boldsymbol{e}$'s is orthogonal:

**If $z^N = 1$ and $z \neq 1$, then the sum $S = 1 + z + z^2 + \cdots + z^{N-1}$ is zero.** (7)

*Proof.* Multiply $S$ times $z$. This gives $Sz = z + z^2 + z^3 + \cdots + z^N$. Since $z^N = 1$, $S$ times $z$ has all the same terms as the original sum $S$. Then $Sz = S$. Therefore $S = 0$.

Every dot product $\overline{e}_k^{\mathrm{T}} e_n$ is exactly our sum $S$. **The number $z$ is $\overline{w}^k w^n$.**

$$(1, \overline{w}^k, \overline{w}^{2k}, \ldots)^{\mathrm{T}}(1, w^n, w^{2n}, \ldots) = 1 + z + z^2 + \cdots = S \tag{8}$$

The $N$th power of $z = \overline{w}^k w^n$ is $z^N = \left(\overline{w}^N\right)^k \left(w^N\right)^n = (1)(1)$. Therefore $S = 0$.

**Conclusion** When we multiply $\overline{F}^{\mathrm{T}}$ times $F$, the diagonal entries are $\overline{e}_k^{\mathrm{T}} e_k = N$ (because this is a sum of $N$ ones). Off the diagonal we have $k \neq n$ and $\overline{e}_k^{\mathrm{T}} e_n = 0$. Therefore $\overline{F}^{\mathrm{T}} F = NI$. This confirms that **the inverse of the Fourier matrix is $F^{-1} = \frac{1}{N}\overline{F}^{\mathrm{T}}$.**

*Note* 1. Your eye sees right away that *the* 8 *numbers around the circle add to zero.* Each number cancels its opposite number: $1 + w^4$ is zero, $w + w^5$ is zero, $w^2 + w^6$ is zero, $w^3 + w^7$ is zero. But this proof won't work for $N = 7$ or 5 or 3. We can't pair off the points when $N$ is odd. They still add to zero by equation (8).

*Note 2.* A cool proof of orthogonality is to see the vectors $e_0, \ldots, e_{N-1}$ as *eigenvectors of a symmetric matrix.* Every symmetric matrix has orthogonal eigenvectors. Problem 14 will choose a suitable matrix (it is a **circulant matrix**) and pursue this idea.

Here are the components of $f = Fc$ and $c = F^{-1} f$ : **Discrete Fourier Transform**

$$f_j = e_j^{\mathrm{T}} c = \sum_{j=0}^{N-1} w^{jk} c_k \qquad c_k = \frac{1}{N}\overline{e}_k^{\mathrm{T}} f = \frac{1}{N}\sum_{k=0}^{N-1} \overline{w}^{jk} f_j \tag{9}$$

The symmetry of transform and inverse transform is beautiful. We didn't see this so clearly for Fourier series, where $c$ was a vector but $f$ was a periodic function. The elegant symmetry reappears when the transform is between *function $f(x)$ and function $c(k)$*:

**Fourier Integral Transform**

$$c(k) = \int_{-\infty}^{\infty} f(x)\, e^{-ikx}\, dx \qquad f(x) = \frac{1}{2\pi}\int_{-\infty}^{\infty} c(k)\, e^{ikx}\, dk. \tag{10}$$

Everybody notices $e^{-ikx}$ and $e^{ikx}$. Be sure to notice $dx$ and $dk$. The functions $f(x)$ and $c(k)$ are defined for $-\infty < x < \infty$ and $-\infty < k < \infty$. The transform connects $f(x)$ in the space domain to $c(k)$ in the frequency domain. $f(x) = \delta(x)$ transforms to $c(k) = 1$. Section 8.6 will solve $-y'' + y = f(x)$ (no boundaries!) using this integral transform.

Two more examples of the discrete transform are **cos** and **sin**.

**Example 4** Sample $\cos x$ and $\sin x$ at $0, \pi/2, \pi, 3\pi/2$ to get discrete vectors **cos** and **sin**. Transform those vectors by $F^{-1}$. Invert their transforms by $F$.

**Discrete cosine and sine** $\quad$ $\mathbf{cos} = (1, 0, -1, 0)$ and $\mathbf{sin} = (0, 1, 0, -1)$.

To transform $x$-space to $k$-space, we multiply $\boldsymbol{f}$ by $F^{-1}$. For $N = 4$, this matrix contains powers of $\overline{w} = -i$. We remember to divide by $N = 4$:

$$F^{-1}\ \textbf{cos} = \frac{1}{4}\begin{bmatrix} 1 & 1 & 1 & 1 \\ 1 & -i & -1 & i \\ 1 & -1 & 1 & -1 \\ 1 & i & -1 & -i \end{bmatrix}\begin{bmatrix} 1 \\ 0 \\ -1 \\ 0 \end{bmatrix} = \begin{bmatrix} 0 \\ \mathbf{1/2} \\ 0 \\ \mathbf{1/2} \end{bmatrix} \qquad F^{-1}\ \textbf{sin} = \begin{bmatrix} 0 \\ \boldsymbol{-i/2} \\ 0 \\ \boldsymbol{i/2} \end{bmatrix}$$

Multiplication by $F$ transforms back to **cos** and **sin**. This is exactly consistent with the famous formulas of Euler: $\textbf{cos}\,x = \frac{\mathbf{1}}{\mathbf{2}}(\boldsymbol{e^{ix} + e^{-ix}})$ and $\textbf{sin}\,x = \frac{\boldsymbol{-i}}{\mathbf{2}}(\boldsymbol{e^{ix} - e^{-ix}})$.

Let me also write **exp** for the samples $(1, w, w^2, w^3)$ of $e^{ix}$ at $x = 0, \pi/2, \pi, 3\pi/2$. Then we have Euler's great formulas for vectors:

$$\textbf{exp} = \textbf{cos} + i\,\textbf{sin} \qquad \textbf{cos} = \frac{1}{2}(\,\textbf{exp} + \overline{\textbf{exp}})$$

$$\overline{\textbf{exp}} = \textbf{cos} - i\,\textbf{sin} \qquad \textbf{sin} = \frac{-i}{2}(\,\textbf{exp} - \overline{\textbf{exp}})$$

## One Step of the Fast Fourier Transform

Multiplication by an $N$ by $N$ matrix takes $N^2$ multiplications and additions. Since the Fourier matrix has no zero entries, you might think it is impossible to do better. But the entries $w^{jk}$ are very special. **The FFT idea is to factor $\boldsymbol{F}$ into sparse matrices.**

If you prefer to think of the summation formulas $\sum w^{jk} c_k$ and $\sum \overline{w}^{jk} f_j$, each sum has $N$ terms and a vector needs $N$ sums. In summation language, the FFT idea is to rewrite and regroup the sums to have many fewer terms. I will try to use both languages.

***The key idea is to connect $F_N$ with the half-size Fourier matrix $F_{N/2}$.*** Assume that $N$ is a power of 2 (say $N = 1024$). We will connect $F_{1024}$ to ***two copies*** of $F_{512}$. When $N = 4$, we connect $F_4$ to two $F_2$'s:

$$F_4 = \begin{bmatrix} 1 & 1 & 1 & 1 \\ 1 & i & i^2 & i^3 \\ 1 & i^2 & i^4 & i^6 \\ 1 & i^3 & i^6 & i^9 \end{bmatrix} \quad \text{and} \quad \begin{bmatrix} F_2 & 0 \\ 0 & F_2 \end{bmatrix} = \begin{bmatrix} 1 & 1 & & \\ 1 & i^2 & & \\ & & 1 & 1 \\ & & 1 & i^2 \end{bmatrix}.$$

On the left is $F_4$, with no zeros. On the right is a matrix that is half zero. The work is cut in half. But wait, those matrices are not the same. The block matrix with $F_2$'s is only one piece of the factorization of $F_4$. The other pieces also have many zeros:

$$\textbf{Key idea} \quad F_4 = \begin{bmatrix} 1 & & 1 & \\ & 1 & & i \\ 1 & & -1 & \\ & 1 & & -i \end{bmatrix}\begin{bmatrix} 1 & 1 & & \\ 1 & i^2 & & \\ & & 1 & 1 \\ & & 1 & i^2 \end{bmatrix}\begin{bmatrix} 1 & & & \\ & & 1 & \\ & 1 & & \\ & & & 1 \end{bmatrix}. \tag{11}$$

The permutation matrix on the right puts $c_0$ and $c_2$ (evens) ahead of $c_1$ and $c_3$ (odds). The middle matrix performs separate half-size transforms on those evens and odds. The matrix at the left combines the two half-size outputs, and it produces the correct full-size output $\boldsymbol{f} = F_4\boldsymbol{c}$. You could multiply those three matrices to see $F_4$.

The same idea applies when $N = 1024$ and $M = \frac{1}{2}N = 512$. The number $w$ is $e^{2\pi i/1024}$. It is at the angle $\theta = 2\pi/1024$ on the unit circle. The Fourier matrix $F_{1024}$ is full of powers of $w$. The first stage of the FFT is the great factorization discovered by Cooley and Tukey (and foreshadowed in 1805 by Gauss):

**FFT (Step 1)**
$$F_{1024} = \begin{bmatrix} I_{512} & D_{512} \\ I_{512} & -D_{512} \end{bmatrix} \begin{bmatrix} F_{512} & \\ & F_{512} \end{bmatrix} \begin{bmatrix} \text{even-odd} \\ \text{permutation} \end{bmatrix} \quad (12)$$

$I_{512}$ is the identity matrix. $D_{512}$ is the diagonal matrix with entries $(1, w, \ldots, w^{511})$ using $w_{1024}$. The two copies of $F_{512}$ are what we expected. They use the 512th root of unity, which is nothing but $w_{512} = (w_{1024})^2$. The even-odd permutation matrix separates the incoming vector $\boldsymbol{c}$ into $\boldsymbol{c}' = (c_0, c_2, \ldots, c_{1022})$ and $\boldsymbol{c}'' = (c_1, c_3, \ldots, c_{1023})$.

Here are the algebra formulas which express this neat FFT factorization of $F_N$:

**(FFT)** Set $M = \frac{1}{2}N$. The components of $\boldsymbol{f} = F_N\boldsymbol{c}$ are combinations of the half-size transforms $\boldsymbol{f}' = F_M\boldsymbol{c}'$ and $\boldsymbol{f}'' = F_M\boldsymbol{c}''$. Equation (13) shows $I\boldsymbol{f}' + D\boldsymbol{f}''$ and $I\boldsymbol{f}' - D\boldsymbol{f}''$ with numbers $(w_N)^j$ on the main diagonal of $D$:

$$\begin{array}{llcll} \textbf{First half} & \boldsymbol{f}_j & = & \boldsymbol{f}_j' + (w_N)^j \boldsymbol{f}_j'', & j = 0, \ldots, M-1 \\ \textbf{Second half} & \boldsymbol{f}_{j+M} & = & \boldsymbol{f}_j' - (w_N)^j \boldsymbol{f}_j'', & j = 0, \ldots, M-1 \end{array} \quad (13)$$

Thus each FFT step has three parts: split $\boldsymbol{c}$ into $\boldsymbol{c}'$ and $\boldsymbol{c}''$, transform them separately by $F_M$ into $\boldsymbol{f}'$ and $\boldsymbol{f}''$, and reconstruct $\boldsymbol{f}$ from equation (13). $N$ must be even!

The algebra of (13) is a splitting into even numbers $2k$ and odd $2k+1$, with $w = w_N$:

$$\textbf{Even/Odd} \quad f_j = \sum_0^{N-1} w^{jk} c_k = \sum_0^{M-1} w^{2jk} c_{2k} + \sum_0^{M-1} w^{j(2k+1)} c_{2k+1} \text{ with } M = \tfrac{N}{2}. \quad (14)$$

The even $c$'s go into $\boldsymbol{c}' = (c_0, c_2, \ldots)$ and the odd $c$'s go into $\boldsymbol{c}'' = (c_1, c_3, \ldots)$. Then come the transforms $F_M\boldsymbol{c}'$ and $F_M\boldsymbol{c}''$. The key is $w_N^2 = w_M$. This gives $w_N^{2jk} = w_M^{jk}$.

$$\textbf{Rewrite} \quad f_j = \sum w_M^{jk} c_k' + (w_N)^j \sum w_M^{jk} c_k'' = f_j' + (w_N)^j f_j''. \quad (15)$$

For $j \geq M$, the minus sign in (13) comes from factoring out $(w_N)^M = -1$.

MATLAB easily separates even $c$'s from odd $c$'s. Then two half-size inverse transforms use ifft. The last step produces $f$ from the half-size $f'$ and $f''$.

Problem 2 shows that $F$ and $F^{-1}$ have the same rows, in different orders.

**FFT Step from $N$ to $N/2$ in MATLAB**

```
f' = ifft (c(0 : 2 : N − 2)) ∗ N/2; % evens
f'' = ifft (c(1 : 2 : N − 1)) ∗ N/2; % odds
D = w .^(0 : N/2 − 1)'; % diagonal of matrix D
f = [f' + D .∗ f''; f' − D .∗ f''];
```

The flow graph shows $\boldsymbol{c}'$ and $\boldsymbol{c}''$ going through the half-size $F_2$. Those steps are called "*butterflies*," from their shape. Then the outputs $\boldsymbol{f}'$ and $\boldsymbol{f}''$ are combined (multiplying $\boldsymbol{f}''$ by $1, i$ and also by $-1, -i$) to produce $\boldsymbol{f} = F_4\boldsymbol{c}$. The indices 0, 1, 2, 3 are in binary.

**Flow Graph $c$ to $f$ $N = 4$ $M = 2$**

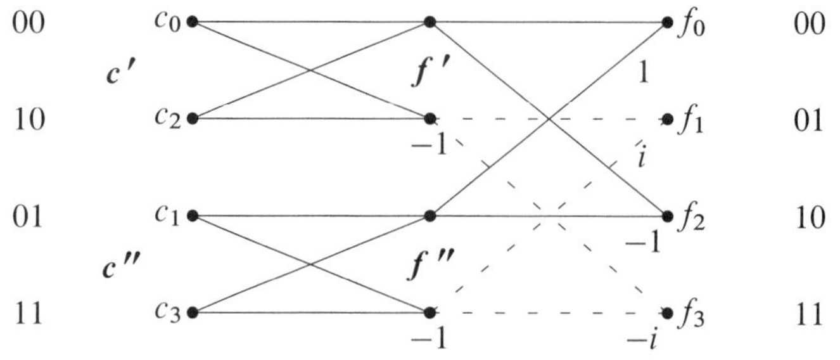


Figure 8.6: Flow graph from $\boldsymbol{c}$ to $\boldsymbol{f}$ for the Fast Fourier Transform with $N = 4$.

This reduction from $F_N$ to two $F_M$'s almost cuts the work in half—you see the zeros in the matrix factorization (12). That reduction is good but not great. The full idea of the FFT is much more powerful. It saves much more time than 50%.

## The Full FFT by Recursion

If you have read this far, you may have guessed what comes next. We reduced $F_N$ to $F_{N/2}$. **Keep going to $F_{N/4}$.** The two copies of $F_{512}$ lead to four copies of $F_{256}$. Then 256 leads to 128. *That is recursion.* It is a basic principle of many fast algorithms. Here is the second stage with $F = F_{256}$ and $D =$ diag $(1, w_{512}, \ldots, (w_{512})^{255})$:

$$\begin{bmatrix} F_{512} & 0 \\ 0 & F_{512} \end{bmatrix} = \begin{bmatrix} I & D & & \\ I & -D & & \\ & & I & D \\ & & I & -D \end{bmatrix} \begin{bmatrix} F & & & \\ & F & & \\ & & F & \\ & & & F \end{bmatrix} \begin{bmatrix} \text{pick } 0, 4, 8, \ldots \\ \text{pick } 2, 6, 10, \ldots \\ \text{pick } 1, 5, 9, \ldots \\ \text{pick } 3, 7, 11, \ldots \end{bmatrix}.$$

Before the FFT was invented, the operation count was $N^2 = (1024)^2$. This is about a million multiplications. I am not saying that they take a long time. The cost becomes large when we have many transforms to do—which is typical. Then the saving is also large:

$$\textbf{The final count for size } N = 2^L \textbf{ is reduced from } N^2 \textbf{ to } \frac{1}{2}NL.$$

Here is the reasoning behind $\frac{1}{2}NL$. There are $L$ levels, going from $N = 2^L$ down to $N = 1$. Each level has $\frac{1}{2}N$ multiplications from diagonal matrices $D$, to reassemble the half-size outputs. This yields the final count $\frac{1}{2}NL$, which is $\frac{1}{2}N \log_2 N$.

***Exactly the same idea gives a fast inverse transform.*** The matrix $F_N^{-1}$ contains powers of the conjugate $\overline{w}$. We just replace $w$ by $\overline{w}$ in the diagonal matrix $D$, and in formula (13). The fastest FFT will be adapted to the processor and cache capacity of each computer. For free software that automatically adjusts, we highly recommend the website fftw.org. This gives the "fastest Fourier transform in the west."

## ■ REVIEW OF THE KEY IDEAS ■

1. Multiplying coefficients $\boldsymbol{c}$ by the Fourier matrix $F$ adds the series $f_j = \sum w^{jk} c_k$.

2. The inverse matrix $F^{-1} = \overline{F}/N$ computes the coefficients $c_k = \sum \overline{w}^{jk} f_j / N$.

3. The **FFT** splits those sums in half: $\frac{N}{2}$ terms with powers of $w^2$. Then recombine.

4. By recursion the FFT has $\log_2 N$ steps with diagonal matrices: $N \log_2 N$ operations.

5. The columns $\boldsymbol{e}_k = (1, w^k, w^{2k}, \ldots)$ are orthogonal, when $w = e^{2\pi i/N}$ and $w^N = 1$.

## Problem Set 8.2

**1** Multiply the three matrices in equation (11) and compare with $F$. In which six entries do you need to know that $i^2 = -1$? This is $(w_4)^2 = w_2$. If $M = N/2$, why is $(w_N)^M = -1$?

**2** *Why is row $i$ of $\overline{F}$ the same as row $N - i$ of $F$* (numbered from 0 to $N - 1$)?

**3** From Problem 2, find the 4 by 4 permutation matrix $P$ so that $F = P\overline{F}$. Check that $P^2 = I$ so that $P = P^{-1}$. Then from $\overline{F}F = 4I$ show that $F^2 = 4P$.

It is amazing that $F^4 = 16P^2 = 16I$. *Four transforms of any* $\boldsymbol{c}$ *bring back* $\mathbf{16}\,\boldsymbol{c}$. For all $N$, $F^2/N$ is a permutation matrix $P$ and $\boldsymbol{F^4 = N^2 I}$.

**4** Invert the three factors in equation (11) to find a fast factorization of $F^{-1}$.

**5** $F$ is symmetric. Transpose equation (11) to find a new Fast Fourier Transform.

**6** All entries in the factorization of $F_6$ involve powers of $w$ = sixth root of 1:

$$F_6 = \begin{bmatrix} I & D \\ I & -D \end{bmatrix} \begin{bmatrix} F_3 & \\ & F_3 \end{bmatrix} \begin{bmatrix} & P & \end{bmatrix}.$$

Write down these factors with $1, w, w^2$ in $D$ and powers of $w^2$ in $F_3$. Multiply!

**7** Put the vector $\boldsymbol{c} = (1, 0, 1, 0)$ through the three steps of the FFT to find $\boldsymbol{y} = F\boldsymbol{c}$. Do the same for $\boldsymbol{c} = (0, 1, 0, 1)$.

**8** Compute $\boldsymbol{y} = F_8\boldsymbol{c}$ by the three FFT steps for $\boldsymbol{c} = (1, 0, 1, 0, 1, 0, 1, 0)$. Repeat the computation for $\boldsymbol{c} = (0, 1, 0, 1, 0, 1, 0, 1)$.

**9** If $w = e^{2\pi i/64}$ then $w^2$ and $\sqrt{w}$ are among the _____ and _____ roots of 1.

**10** $F$ is a symmetric matrix. Its eigenvalues aren't real. How is this possible?

**The three great symmetric tridiagonal matrices of applied mathematics are $K$, $B$, $C$.** The eigenvectors of $K$, $B$, and $C$ are discrete **sines**, **cosines**, and **exponentials**. The eigenvector matrices give the **DST**, **DCT**, and **DFT** — discrete transforms for signal processing. Notice that diagonals of the circulant matrix $C$ loop around to the far corners.

$$K = \begin{bmatrix} \mathbf{2} & -1 & & \\ -1 & 2 & -1 & \\ & \cdot & \cdot & \cdot \\ & & -1 & \mathbf{2} \end{bmatrix} \qquad B = \begin{bmatrix} \mathbf{1} & -1 & & \\ -1 & 2 & -1 & \\ & \cdot & \cdot & \cdot \\ & & -1 & \mathbf{1} \end{bmatrix}$$

$$C = \begin{bmatrix} 2 & -1 & \cdot & \mathbf{-1} \\ -1 & 2 & -1 & \\ & \cdot & \cdot & \cdot \\ \mathbf{-1} & \cdot & -1 & 2 \end{bmatrix} \qquad \begin{array}{l} K_{11} = K_{NN} = \mathbf{2} \\ B_{11} = B_{NN} = \mathbf{1} \\ C_{1N} = C_{N1} = \mathbf{-1} \end{array}$$

**11** The eigenvectors of $K_N$ and $B_N$ are the discrete sines $\boldsymbol{s}_1, \ldots, \boldsymbol{s}_N$ and the discrete cosines $\boldsymbol{c}_0, \ldots, \boldsymbol{c}_{N-1}$. Notice the eigenvector $\boldsymbol{c}_0 = (1, 1, \ldots, 1)$. Here are $\boldsymbol{s}_k$ and $\boldsymbol{c}_k$—these vectors are samples of $\sin kx$ and $\cos kx$ from 0 to $\pi$.

$$\left(\sin\frac{\pi k}{N+1}, \sin\frac{2\pi k}{N+1}, \ldots, \sin\frac{N\pi k}{N+1}\right) \text{ and } \left(\cos\frac{\pi k}{2N}, \cos\frac{3\pi k}{2N}, \ldots, \cos\frac{(2N-1)\pi k}{2N}\right)$$

For 2 by 2 matrices $K_2$ and $B_2$, verify that $\boldsymbol{s}_1, \boldsymbol{s}_2$ and $\boldsymbol{c}_0, \boldsymbol{c}_1$ are eigenvectors.

**12** Show that $C_3$ has eigenvalues $\lambda = 0, 3, 3$ with eigenvectors $\boldsymbol{e}_0 = (1, 1, 1)$, $\boldsymbol{e}_1 = (1, w, w^2)$, $\boldsymbol{e}_2 = (1, w^2, w^4)$. You may prefer the real eigenvectors $(1, 1, 1)$ and $(1, 0, -1)$ and $(1, -2, 1)$.

**13** Multiply to see the eigenvectors $\boldsymbol{e}_k$ and eigenvalues $\lambda_k$ of $C_N$. Simplify to $\lambda_k = 2 - 2\cos(2\pi k/N)$. Explain why $C_N$ is only semidefinite. It is not positive definite.

$$C\boldsymbol{e}_k = \begin{bmatrix} 2 & -1 & & -1 \\ -1 & 2 & -1 & \\ & -1 & 2 & -1 \\ -1 & & -1 & 2 \end{bmatrix} \begin{bmatrix} 1 \\ w^k \\ w^{2k} \\ w^{(N-1)k} \end{bmatrix} = (2 - w^k - w^{-k}) \begin{bmatrix} 1 \\ w^k \\ w^{2k} \\ w^{(N-1)k} \end{bmatrix}.$$

**14** The eigenvectors $\boldsymbol{e}_k$ of $C$ are automatically perpendicular because $C$ is a _____ matrix. (To tell the truth, $C$ has repeated eigenvalues as in Problem 12. There was a plane of eigenvectors for $\lambda = 3$ and we chose orthogonal $\boldsymbol{e}_1$ and $\boldsymbol{e}_2$ in that plane.)

**15** Write the 2 eigenvalues for $K_2$ and the 3 eigenvalues for $B_3$. Always $K_N$ and $B_{N+1}$ have the same $N$ eigenvalues, with the extra eigenvalue _____ for $B_{N+1}$. (This is because $K = A^{\mathrm{T}}A$ and $B = AA^{\mathrm{T}}$.)

## 8.3 The Heat Equation

The first partial differential equation in this book was $u_{xx} + u_{yy} = 0$ (Laplace's equation). This describes a steady state—time is not involved. There is no growth or oscillation or decay. The problem includes boundary conditions on $u(x, y)$, but not initial conditions. This is like a matrix equation $A\boldsymbol{u} = \boldsymbol{b}$ (where $\boldsymbol{b}$ comes from boundary conditions).

Now we move to the **heat equation** $\boldsymbol{u_t = u_{xx}}$. Time is very much involved. We think of $u$ as the temperature along a bar at time $t$. We are given the initial temperature $u(0, x)$ at time $t = 0$ and at each position $x$. Then heat begins to flow (from positions with higher temperature to neighbors at lower temperature). This is like a matrix equation $\boldsymbol{u}' = A\boldsymbol{u}$ with an initial condition $u(0)$. $Au$ is now the second derivative $u_{xx}$.

We have a PDE and not an ODE, a partial and not an ordinary differential equation, because the temperature $u$ is a function of both $x$ and $t$.

**Example 1** (*Infinite bar*) Suppose the bar goes from $x = -\infty$ to $x = \infty$. At time $t = 0$, the temperature is $u = -1$ on the left side $x < 0$ and $u = 1$ on the right side $x > 0$. *Heat will flow from the right side to the left side.* The temperature along the left half will go up from $u = -1$. The right half will go down from $u = 1$. *Solved in Example* 6.

**Example 2** (*Finite bar*) Suppose the bar goes from $x = 0$ to $x = 1$. The initial condition $u(0, x) = 1$ tells us the (constant) temperature along the bar at time $t = 0$. We also need boundary conditions like $u(t, 0) = 0$ and $u(t, 1) = 0$ at the ends of the bar. Then the ends stay at zero temperature for all time $t > 0$.

**Heat will flow out the ends.** Imagine a bar in a freezer, with the sides coated. Heat escapes only at $x = 0$ and $x = 1$. We solve the heat equation to find the temperature $u(t, x)$ at every position $0 < x < 1$ and every time $t > 0$.

$$\textbf{Heat equation} \quad \frac{\partial u}{\partial t} = \frac{\partial^2 u}{\partial x^2} \quad \text{with } u(0, x) = 1 \text{ and } u(t, 0) = u(t, 1) = 0. \tag{1}$$

A good form for the solution is a Fourier series. It is natural to choose a sine series, since every basis function $\sin k\pi x$ is zero at $x = 0$ and $x = 1$—exactly what the boundary conditions require: zero temperature at the ends of the bar.

The initial value $u(0, x)$ and the differential equation $u_t = u_{xx}$ will have to tell us the coefficients $b_1(t), b_2(t), \ldots$ in the Fourier sine series. Heat escapes and $b_k(t) \to 0$.

**Solution plan** The equation $u_t = u_{xx}$ looks different from $du/dt = Au$, but it's not. The solution still combines the eigenvectors. The pieces for the ODE were $ce^{\lambda t}\boldsymbol{x}$. The pieces for the PDE are $be^{\lambda t} \sin k\pi x$.

1. Eigenvectors of $A$ change to eigenfunctions of the second derivative: $(\sin k\pi x)'' = -k^2\pi^2 \sin k\pi x$.
2. $\boldsymbol{u}(0) = c_1\boldsymbol{x}_1 + c_2\boldsymbol{x}_2 + \cdots$ changes to $u(0, x) = b_1 \sin \pi x + b_2 \sin 2\pi x + \cdots$ (*with infinitely many* $b$'s)
3. The solution (7) adds up $b_k e^{\lambda_k t} \sin k\pi x$. It is an infinite Fourier series.

Infinity could make the problem difficult, but the $\sin k\pi x$ are orthogonal. Problem solved.

### Solution by Fourier Series

Everything comes from choosing the right form for the solution $u(t,x)$. Here it is :

$$\textbf{Sine series}\quad u(t,x)=b_1(t)\sin\pi x+b_2(t)\sin 2\pi x+\cdots=\sum_{k=1}^{\infty}\boldsymbol{b_k(t)\sin k\pi x}. \qquad (2)$$

This form shows **separation of variables**. Functions $\boldsymbol{b_k(t)}$ depending on $t$ multiply functions $\boldsymbol{\sin k\pi x}$ depending on $x$. When we substitute that product $b_k(t)\sin k\pi x$ into the heat equation, we get a differential equation for each of the coefficients $b_k$ :

$$\frac{\partial}{\partial t}(b_k\sin k\pi x)=\frac{\partial^2}{\partial x^2}(b_k\sin k\pi x)\quad\text{gives}\quad\frac{\partial b_k}{\partial t}\sin k\pi x=-k^2\pi^2 b_k\sin k\pi x. \qquad (3)$$

Then $\boldsymbol{b_k{}'=-k^2\pi^2 b_k}$. Solving this equation will produce every $b_k(t)$ from $b_k(0)$ :

$$\textbf{Decay comes from } e^{\lambda t}\qquad \boldsymbol{b_k(t)=e^{-k^2\pi^2 t}b_k(0)}. \qquad (4)$$

*Final step* : The starting values $b_k(0)$ are decided by the initial condition $u(0,x)=1$ :

$$\textbf{At } t=0\qquad u(0,x)=\sum_{k=1}^{\infty}b_k(0)\sin k\pi x=1\quad\text{for } 0<x<1. \qquad (5)$$

This is an ordinary Fourier series question: What are the coefficients of a square wave $\boldsymbol{SW}(x)$ ? Sines are odd functions, $\sin(-x)=-\sin x$. *The series in* (5) *must add to* $-1$ *for $x$ between* $-1$ *and* 0. So the square wave jumps from $-1$ to 1. It is negative on half of the interval and positive on the other half :

$$\boldsymbol{SW}(x)=\left\{\begin{array}{ll}-1 & \text{for}\quad -1<x<0\\ \phantom{-}1 & \text{for}\quad\phantom{-}0<x<1\end{array}\right\}=\frac{4}{\pi}\left(\frac{\sin\pi x}{1}+\frac{\sin 3\pi x}{3}+\cdots\right). \qquad (6)$$

The even coefficients $b_2, b_4, \ldots$ are all zero. The odd coefficients are $b_k=4/\pi k$. Those $b$'s were computed in Section 8.1, as the first example of a Fourier series. Now these numbers are giving the coefficients $b_k(0)$ at $t=0$. Then the equation $b_k{}'=-k^2\pi^2 b_k$ tells us the coefficients $\boldsymbol{e^{-k^2\pi^2 t}b_k(0)}$ at all future times $t>0$:

$$\textbf{Solution}\qquad u(t,x)=\sum_{k=1}^{\infty}e^{-k^2\pi^2 t}\,b_k(0)\sin k\pi x=\frac{4}{\pi}\left(\boldsymbol{e^{-\pi^2 t}}\sin\boldsymbol{\pi x}+\cdots\right) \qquad (7)$$

This completes the solution of the heat equation. The heat drops off quickly ! Those are powerful exponentials $e^{-\pi^2 t}$ and $e^{-9\pi^2 t}$. The bar will feel extremely cold when $t=1$.

*Note* The correct heat equation should be $\boldsymbol{u_t=cu_{xx}}$ with a **diffusion constant $c$**. Otherwise the equation is dimensionally wrong. The units of $c$ are (distance)$^2$/time, in order to balance $u_t$ with $u_{xx}$. Then $c$ is large for metals—heat flows easily—compared to its value for water or air. The factor $c$ enters the eigenvalues $-ck^2\pi^2$.

The heat equation is also the **diffusion equation**. A smokestack is almost a point source (a delta function). The smoke spreads out (diffuses into the air). This would involve two space dimensions $x$ and $y$, or even $x, y, z$. The PDE could become $\boldsymbol{u_t = c(u_{xx} + u_{yy})}$.

*Summary   We had a boundary value problem in $x$, and an initial value problem in $t$:*

1. The basis functions $S_k = \sin k\pi x$ **depend on $\boldsymbol{x}$**. They solve $u_{xx} = \lambda u$.
2. The coefficients $b_k$ **depend on $\boldsymbol{t}$**. They solve $b' = \lambda b$ with $b(0)$ coming from $u(0)$.

The basis functions $S_k(x)$ satisfy the *boundary conditions*.
Their coefficients $b_k(t)$ satisfy the *initial conditions*:

$$\textbf{Separation at } \boldsymbol{t = 0} \qquad u(0, x) = \sum b_k(0)\, S_k(x) \tag{8}$$

The PDE for $u(t, x)$ gives an ODE for each coefficient $b_k(t)$. Here are three more bars.

**Example 3** (*Insulated bar*) No heat escapes from the ends of the bar. The boundary conditions change to $\partial u/\partial x = 0$ at those ends. *The basis functions change to cosines.* The series (8) becomes a Fourier cosine series.

$$\begin{aligned} &\text{Initial condition} & u(0, x) &= \textstyle\sum a_k(0) \cos k\pi x \\ &\text{Equation for the } a_k & da_k/dt &= -k^2\pi^2 a_k \text{ for } k = 0, 1, 2, \ldots \end{aligned}$$

Notice that $k = 0$ is included. The first basis function is $\cos 0\pi x = 1$. Its coefficient is controlled by $da_0/dt = 0$. Thus $k = 0$ contributes a constant $a_0$ to the solution $u(t, x)$. The temperature approaches this constant everywhere along the bar, since $a_1, a_2, a_3, \ldots$ all die out exponentially fast.

**Example 4** (*Circular bar*) Now sines and cosines are both included. The basis functions can also be complex exponentials $e^{ikx}$. Again $u$ goes to a constant steady state $c_0$:

$$u(t, x) = \sum_{-\infty}^{\infty} c_k(t) e^{ik\pi x} \quad \text{and} \quad \frac{dc_k}{dt} = -k^2\pi^2 c_k. \tag{9}$$

When you have a separated form for the pieces of $u$, your problem is nearly solved.

**Example 5** (*Infinite bar*) This problem leads to something new and important. There are no boundaries. All exponentials $e^{ikx}$ (not just whole numbers $k$) are needed. By combining the solutions for $-\infty < k < \infty$ we can solve the heat equation starting from a delta function $\delta(x)$. This "**heat kernel**" is the key to chemical engineering. By a totally unexpected development it is also central to mathematical finance. The prices of stock options are modelled by the Black-Scholes partial differential equation.

To solve for each separate $e^{ikx}$, look for the right multiplier $e^{i\omega t}$:

$$\boldsymbol{u = e^{i\omega t} e^{ikx}} \ \textbf{ solves } \ \boldsymbol{u_t = u_{xx}} \ \textbf{ when } \ \boldsymbol{i\omega = (ik)^2}. \tag{10}$$

Then $i\omega t = (ik)^2 t = -k^2 t$. The solution $u(t, x)$ has a separated form, with these pieces:

$$\boldsymbol{u(t, x) = e^{-k^2 t} e^{ikx}} \ \textbf{ solves the heat equation.} \ \text{ It starts from } \ u(0, x) = e^{ikx}. \tag{11}$$

## The Heat Kernel $U(t, x)$

The delta function $\delta(x)$ contains all exponentials $e^{ikx}$ in equal amounts. By superposition, the solution $U$ to the heat equation starting from $\delta(x)$ will contain the solutions $e^{-k^2t}e^{ikx}$ in equal amounts. Integrate $e^{-k^2t}e^{ikx}$ over all $k$ to find the heat kernel $U$.

$$\text{The solution with } \boldsymbol{U(0,x)=\delta(x)} \quad \text{is} \quad U(t,x)=\frac{1}{2\pi}\int_{-\infty}^{\infty} e^{-k^2t}e^{ikx}\,dk. \tag{12}$$

Computing this integral is possible, but unexpected. No simple function of $k$ has the derivative $e^{-k^2t}$, or close. The neat way is to start with $\partial U/\partial x$. The derivative of $e^{ikx}$ brings the extra factor $ik$. Then integration by parts connects $dU/dx$ to $U$:

$$\frac{\boldsymbol{dU}}{\boldsymbol{dx}}=\frac{1}{2\pi}\int_{-\infty}^{\infty}(e^{-k^2t}\boldsymbol{k})(\boldsymbol{i}\,e^{ikx})\,dk=\frac{1}{4\pi t}\int_{-\infty}^{\infty}(e^{-k^2t})(xe^{ikx})\,dk=-\frac{\boldsymbol{xU}}{\boldsymbol{2t}}. \tag{13}$$

Now $dU/U$ equals $-x\,dx/2t$. Integration gives $-x^2/4t$ and then $U=ce^{-x^2/4t}$.

The total heat $\int u\,dx$ starts at $\int \delta(x)\,dx=1$. To stay at 1, we choose $c=1/\sqrt{4\pi t}$. Then we have the "fundamental solution" for a point source.

$$\textbf{Heat kernel} \quad \boldsymbol{U_t=U_{xx}} \ \textbf{with} \ \boldsymbol{U(0,x)=\delta(x)} \qquad \boldsymbol{U=\frac{1}{\sqrt{4\pi t}}e^{-x^2/4t}} \tag{14}$$

**Example 6** On an infinite bar, the heat kernel (14) solves $u_t=u_{xx}$ starting from $\delta(x)$ at $t=0$. Now solve Example 1, which started from $u=-1$ for negative $x$ and $u=1$ for positive $x$. Then solve for any initial function $u(0,x)$.

Here is the key idea for Example 1. The derivative of the jump from $-1$ to 1 at $x=0$ is $du/dx=\boldsymbol{2\delta(x)}$. The solution starting from $2\delta(x)$ has $du/dx=2U$, which cancels $\sqrt{4}$ in (14). *Then integrate* $2U$ *to undo the derivative and solve Example* 1 *for* $u$:

$$\begin{array}{l}\boldsymbol{u=}\textbf{ Error function}\\ \textbf{Integral of } \boldsymbol{2U}\end{array} \qquad u(t,x)=\frac{1}{\sqrt{\pi t}}\int_0^x e^{-X^2/4t}\,dX. \tag{15}$$

For $x>0$ this solution is positive. For $x<0$ it is negative (the integral in (15) goes backward). At $x=0$ the solution stays at zero, which we expect by symmetry. I wrote the words "error function" because this important integral has been computed and tabulated to high accuracy (no simple function has the derivative $e^{-x^2}$). We just change the variable of integration from $X$ to $Y=X/\sqrt{4t}$, to see the standard error function:

$$u=\frac{1}{\sqrt{\pi t}}\int_0^x e^{-X^2/4t}\,dX=\frac{2}{\sqrt{\pi}}\int_0^{x/\sqrt{4t}} e^{-Y^2}\,dY=\mathbf{erf}\left(\frac{x}{\sqrt{4t}}\right). \tag{16}$$

The integral is a cumulative probability for a normal distribution (this is the area under a bell-shaped curve). Statisticians need these integrals erf$(x)$ all the time. At $x=\infty$ we have the total probability $=$ total area under the curve $=1$.

Finally, we can solve $u_t = u_{xx}$ from *any starting function* $u(0,x)$. The key is to realize that every function of $x$ is **an integral of shifted delta functions $\delta(x-a)$**:

$$\text{Every function } u_0(x) \text{ has } \int_{-\infty}^{\infty} u_0(a)\,\delta(x-a)\,da = u_0(x). \tag{17}$$

By superposition, the solution to $u_t = u_{xx}$ must be an integral of shifted heat kernels.

$$\textbf{Temperature at time } t \qquad u(t,x) = \frac{1}{\sqrt{4\pi t}} \int_{-\infty}^{\infty} u_0(a) e^{-(x-a)^2/4t}\,da. \tag{18}$$

I have used the crucial fact that when the point source shifts by $a$ to become $\delta(x-a)$, *the solution also shifts by* $a$. So I just shifted the heat kernel $U$, by changing $x$ to $x-a$. The heat equation on the whole line $-\infty < x < \infty$ is **linear shift-invariant**.

The solution (18) is reduced to one infinite integral—still not simple. And for a more realistic finite bar, with boundary conditions at $x=0$ and $x=1$, we have to think again. There will also be changes when the diffusion coefficient $c$ in $u_t = (cu_x)_x$ is changing with $x$ or $t$ or $u$. This thinking probably leads us to finite differences.

## Separation of Variables

**The basis functions $\sin k\pi x$ are eigenfunctions**. The same is true for $\cos k\pi x$ and $e^{ik\pi x}$. Let me show this by substituting $u = B(t)\,A(x)$ into the equation $u_t = u_{xx}$. Right away **$u_t$ gives $B'$ and $u_{xx}$ gives $A''$**. The separated variables are connected by $u_t = u_{xx}$:

$$B'(t)\,A(x) = B(t)\,A''(x) \quad \textbf{leads to} \quad \frac{A''(x)}{A(x)} = \frac{B'(t)}{B(t)} = \textbf{constant} \tag{19}$$

Why a constant? Because $A''/A$ depends only on $x$ and $B'/B$ depends only on $t$. They are equal, so neither one can move. Call that constant $-\lambda$:

$$\frac{A''}{A} = -\lambda \text{ gives } A = \sin\sqrt{\lambda}\,x \text{ and } \cos\sqrt{\lambda}\,x \qquad \frac{B'}{B} = -\lambda \text{ gives } B = e^{-\lambda t} \tag{20}$$

The products $BA = e^{-\lambda t}\sin\sqrt{\lambda}\,x$ and $BA = e^{-\lambda t}\cos\sqrt{\lambda}\,x$ solve the heat equation for any number $\lambda$. But the boundary condition $u(t,0) = 0$ eliminates the cosines. Then $u=0$ at $x=1$ requires $\sin\sqrt{\lambda} = 0$ and $\lambda = k^2\pi^2$. Separation of variables has recovered the correct basis functions $\sin k\pi x$ as eigenfunctions for $A'' = -\lambda A$.

**Example 7** (Smokestack problem) We backed away from the heat equation in 2 + 1 dimensions. The solution to $\boldsymbol{u_t = u_{xx} + u_{yy}}$ involves three variables $t$, $x$, $y$. Put a smokestack at the center point $x = y = 0$, and suppose there is no wind. Then nothing depends on the direction angle $\theta$. Smoke will diffuse out from the center. The concentration depends only on the radial distance $r$, and we solve the radially symmetric heat equation. Our final solution is $u(t,r)$.

The heat equation is not quite $u_t = u_{rr}$ because $r$ = constant is *curved* (a circle). The correct radial equation is perfect for separation of variables $u = \boldsymbol{B(t)\,A(r)}$.

$$\frac{\partial u}{\partial t} = \frac{\partial^2 u}{\partial r^2} + \frac{1}{r}\frac{\partial u}{\partial r} \quad \text{leads to} \quad B'(t)\,A(r) = B(t)\,(A'' + \frac{1}{r}A'). \tag{21}$$

Again $B'/B = \text{constant} = -\lambda$ and $B = e^{-\lambda t}$ as before. But instead of $A''/A = -\lambda$, we have **Bessel's equation for the radial eigenfunction $A(r)$**:

$$\textbf{Basis functions } A(r) \qquad \frac{d^2A}{dr^2} + \frac{1}{r}\frac{dA}{dr} = -\lambda A \quad \text{has a variable coefficient } \frac{1}{r}. \tag{22}$$

The solutions are among the special functions that have been studied for centuries. They are not complex exponentials because the coefficient $1/r$ is not constant. *Bessel replaces Fourier*. This book can't go all the way to solve Bessel's equation, but see Section 6.5. A heat equation with symmetry led Bessel to new eigenfunctions.

## ■ REVIEW OF THE KEY IDEAS ■

1. The heat equation $u_t = u_{xx}$ is solved by $e^{-k^2\pi^2 t} \sin k\pi x$ for every $k = 1, 2, \ldots$
2. A combination of those solutions matches the initial $u(0, x)$ to its Fourier sine series.
3. With $u_x = 0$ at $x = 0$ and 1, use cosines. With an infinite bar, use all $e^{-k^2 t} e^{ikx}$.
4. The heat kernel $U = e^{-x^2/4t}/\sqrt{4\pi t}$ solves $U_t = U_{xx}$ starting from $U_0 = \delta(x)$.
5. Separation into $B(t)A(x)$ shows that $A(x)$ is an eigenfunction of the "$x$ part" $u_{xx}$.

# Problem Set 8.3

**1** Solve the heat equation $u_t = cu_{xx}$ on an infinite bar with coefficient $c$, starting from $u = e^{ikx}$ at $t = 0$. As in (10) the solution has the product form $u = e^{i\omega t} e^{ikx}$. With $c$ in the equation, *find $\omega$ for each $k$*.

**2** Solve the same equation $u_t = cu_{xx}$ starting from the point source $u = \delta(x) = \int e^{ikx}\, dk/2\pi$ at $t = 0$. By superposition, you integrate over all $k$ the solutions $u$ in Problem 1. The result is the heat kernel as in equation (14) but adjusted for $c$.

**3** To solve $u_t = cu_{xx}$ for a bar between $x = 0$ and $x = 1$, the basis functions are still $\sin k\pi x$ (with $u = 0$ at the ends). What are the eigenvalues $\lambda_k$ that go into the solution $\sum b_k(0)\, e^{-\lambda_k t} \sin k\pi x$ ?

**4** Following Problem 3, solve $u_t = cu_{xx}$ when the initial temperature is $u_0 = 1$ for $\frac{1}{4} \le x \le \frac{3}{4}$ (and $u_0 = 0$ on the first and last quarters of the bar). The problem is to find the coefficients $b_k(0)$ for that initial temperature.

**5** Solve the heat equation $u_t = u_{xx}$ from a point source $u(x,0) = \delta(x)$ with free boundary conditions $u'(\pi,t) = u'(-\pi,t) = 0$. Use the infinite cosine series $\delta(x) = (1 + 2\cos x + 2\cos 2x + \cdots)/2\pi$ multiplied by time decay factors $b_k(t)$.

**6** (Bar from $x = 0$ to $x = \infty$) Solve $u_t = u_{xx}$ on the positive half of an infinite bar, starting from the shifted delta function $u_0 = \delta(x-a)$ at a point $x = a > 0$. Here is a way to use the full-bar heat kernel $U$ in (14), and still keep $u = 0$ at $x = 0$.

Imagine a negative point source at $x = -a$. Solve the heat equation on the fully infinite bar, including both sources in $u_0 = \delta(x-a) - \delta(x+a)$ at $t = 0$. Your solution (a difference of heat kernels) will stay zero at the boundary $x = 0$ (*Why*?). Then it must be the correct solution on the half-bar, since it started correctly.

**7** Check that the basis functions $s_k = \sin\left(k + \frac{1}{2}\right)\pi x$ are orthogonal over $0 \le x \le 1$. Find a formula for the coefficient $B_4$ in the Fourier series $F(x) = \sum B_k s_k$. (Multiply by $s_4(x)$ and integrate, to isolate $B_4$.)

**8** The basis functions $\sin(k + \frac{1}{2})\pi x$ are for **fixed-free** boundaries ($u = 0$ at $x = 0$ and $u' = 0$ at $x = 1$). What are the basis functions for **free-fixed boundaries** ($u' = 0$ at $x = 0$ and $u = 0$ at $x = 1$)?

**9** Suppose $\boldsymbol{u_t = u_{xx} - u}$ with boundary condition $u = 0$ at $x = 0$ and $x = 1$. Find the new numbers $\lambda_k$ in the general solution $u = \sum b_k(0)\, e^{-\lambda_k t} \sin k\pi x$. (Previously $\lambda_k = -k^2\pi^2$, now there is a new term in $\lambda$ because of $-u$.)

**10** Explain each step in equation (13). Solve $dU/dx = -xU/2t$ to reach $U = e^{-x^2/4t}$. How do the known infinite integrals $\int e^{-x^2} dx = \sqrt{\pi}$ and $\int u\, dx = 1$ lead to the factor $1/\sqrt{4\pi t}$?

**11** (**Shift invariance**) What is the solution to $u_t = u_{xx}$ starting from $\boldsymbol{\delta(x-a)}$ at $t = 0$?

**12** What are basis functions $A(x,y)$ for heat flow in a square plate, when $u = 0$ along the four sides $x = 0, x = 1, y = 0, y = 1$ ? The heat equation is $u_t = u_{xx} + u_{yy}$. Find eigenfunctions for $A_{xx} + A_{yy} = \lambda A$ that satisfy the boundary conditions.

The first eigenfunction is $A_{11} = (\sin \pi x)(\sin \pi y)$. Find the eigenvalues $\lambda$.

**13** Substitute $U = e^{-x^2/4t}/\sqrt{4\pi t}$ to show that this heat kernel solves $U_t = U_{xx}$.

**Notes on a heat bath** (**This is the opposite problem to a hot bar in a freezer**.)
The bar is initially at $\boldsymbol{U = 0}$. It is placed into a heat bath at the fixed temperature $U_B = 1$. The boundary conditions are no longer zero and the bar will get hot.

The difference $V = U - U_B$ has zero boundary values, and its initial values are $V = -1$. Now the eigenfunction method (separation of variables) solves for $V$. The series in (7) is multiplied by $-1$ to account for $V(x, 0) = -1$. Adding back $U_B$ solves the heat bath problem: $U = U_B + V = 1 - u(x, t)$.

Here $U_B \equiv 1$ is the *steady state* solution at $t = \infty$, and $V$ is the *transient* solution. The transient starts at $V = -1$ and decays quickly to $V = 0$.

**Heat bath at one end** This problem is different in another way too. The fixed "Dirichlet" boundary condition is replaced by the free "Neumann" condition on the slope: $\boldsymbol{u'(1, t) = 0}$. Only the left end is in the heat bath. Heat flows down the metal bar and out at the far end, now located at $x = 1$. How does the solution change for fixed-free?

Again $U_B = 1$ is a steady state. The boundary conditions apply to $V = 1 - U_B$:

**Fixed-free eigenfunctions** $V(0) = 0$ and $V'(1) = 0$ lead to $A(x) = \sin\left(k + \frac{1}{2}\right)\pi x.$

Those new eigenfunctions (adjusted to $A'(1) = 0$) give a new product form $B_k(t)\, A_k(x)$:

**Fixed-free solution**
$$V(x, t) = \sum_{\text{odd } k} B_k(0)\, e^{-(k+\frac{1}{2})^2\pi^2 t} \sin\,(k + \tfrac{1}{2})\pi x.$$

All frequencies shift by $\frac{1}{2}$ and multiply by $\pi$, because $A'' = -\lambda A$ has a free end at $x = 1$. The crucial question is: **Does orthogonality still hold for these new eigenfunctions sin** $\boldsymbol{\left(k + \frac{1}{2}\right)\pi x}$? The answer to Problem 7 is *yes* because $A'' = -\lambda A$ is symmetric.

**Notes on stochastic equations and models for stock prices with Brownian motion.**
A "stochastic differential equation" has a random term on the right hand side. Instead of a smooth forcing term $q(t)$, or even a delta function $\delta(t)$, the models for stock prices include Brownian motion $dW$. The idea is subtle and important, and I will just write it down. *A random step has* $dW = Z\sqrt{dt}$. Here $Z$ has a normal Gaussian distribution with mean zero and variance $\sigma^2 = 1$. But a new $Z$ is chosen randomly *at every instant.*

The step size $\sqrt{\Delta t}$ produces a random walk $W(t)$ with wild oscillations. You could see a discrete random walk from $W(t + \Delta t) = W(t) + Z\sqrt{\Delta t}$, and then let $\Delta t$ approach zero. The true random walk is *nowhere continuous.*

A steady return $S(t)$ on an investment has $S' = aS$. The growth is $S(t) = e^{at}S(0)$ exactly as in Chapter 1. But stock prices also respond to a stochastic part $\sigma\, dW$, where the number $\sigma$ measures the **volatility of the market**. This mixes ups and downs from Brownian motion $\sigma\, dW$ with steady growth (drift) from $dS = aS\, dt$:

**"Diffusion" and "drift"**
$$\frac{dS}{S} = \sigma\, dW + a\, dt.$$

Then the basic model for the value of a call option leads to the Black-Scholes equation. The solution comes by a change of variables to reach the heat equation. When they are buying and selling options, traders would have that solution available at all times.

## 8.4 The Wave Equation

Heat travels with *infinite speed*. Waves travel with *finite speed*. Start both of them from a point source $u_0(x) = \delta(x)$. Compare the solutions at time $t$ :

| | | |
|---|---|---|
| **Heat equation** $u_t = u_{xx}$ | $u(t,x) = \frac{1}{\sqrt{4\pi t}}\, e^{-x^2/4t}$ | is a *smooth function* |
| **Wave equation** $u_{tt} = c^2 u_{xx}$ | $u(t,x) = \frac{1}{2}\delta(x-ct) + \frac{1}{2}\delta(x+ct)$ | *has spikes* |

We are starting from a big bang $u = \delta(x)$ at $x = 0$. At a later time $t$, the bang reaches the two points $x = ct$ and $x = -ct$. That represents travel to the right and to the left with velocities $dx/dt = c$ and $-c$. The speed of sound in air is $c = 342$ meters/second.

Notice another difference from the heat equation. After the bang passes point $x = c$ at time $t = 1$, *silence returns*: $\delta(x - ct) = 0$ when $ct > x$. For the heat equation, temperatures like $e^{-x^2/4t}$ never return to zero. A wavefront passes by and we hear it only once. There is no echo or our ears would be full of sound.

In reality the heat equation is often mixed in with the wave equation. The sound diffuses as it travels. Then we do hear noise forever, but not much: the intensity decays fast.

### The One-Way Wave Equation

We begin with a problem that will be particularly clear. It is first order in time ($t \geq 0$) and first order in space ($-\infty < x < \infty$). The velocity is still $c$ :

$$\textbf{One-way wave} \qquad \frac{\partial u}{\partial t} = c\,\frac{\partial u}{\partial x} \quad \text{with } u = u_0(x) \text{ at } t = 0. \tag{1}$$

One solution is $u = e^{x+ct}$. Its time derivative $\partial u/\partial t$ brings a factor $c$. The same will be true for $\sin(x + ct)$ and $\cos(x + ct)$ and *any function of* $x + ct$. The right function is $u_0(x + ct)$ because this gives the correct start $u_0(x)$ at time $t = 0$:

$$\boxed{\textbf{Solution to } u_t = cu_x \qquad u(t,x) = u_0(x + ct).} \tag{2}$$

Suppose $u_0(x)$ is a step function (a wall of water). We have $u_0(x) = 0$ for negative $x$ and $u_0(x) = 1$ for positive $x$. Then the dam breaks. A wall of water moves to the left with velocity $c$. At time $t$, the water reaches the point $x = -ct$ where $x + ct = 0$.

$$\textbf{Wall at } x = -ct \qquad \begin{aligned} u &= u_0(x+ct) = \mathbf{0} \quad \text{for} \quad x + ct < 0 \\ u &= u_0(x+ct) = \mathbf{1} \quad \text{for} \quad x + ct > 0 \end{aligned} \tag{3}$$

The line $x + ct = 0$ is called a "characteristic." The signal travels (with signal speed $c$) along that line in space-time, to tell about the jump from $u = 0$ to $u = 1$.

For any initial function $u_0(x)$, the solution $u = u_0(x + ct)$ is a shift of the graph. It is a one-way wave, no change in shape. The waves from $u_{tt} = c^2 u_{xx}$ go both ways.

## Waves in Space

Now we solve the wave equation $\boldsymbol{\partial^2 u/\partial t^2 = c^2\, \partial^2 u/\partial x^2}$. The three-dimensional form would be $u_{tt} = c^2(u_{xx} + u_{yy} + u_{zz})$. This is the equation satisfied by light as it travels in empty space: a vacuum. The speed of light $c$ is about 300 million meters per second $(186,000$ miles/second). This is the fastest possible speed in Einstein's relativity theory.

The atmosphere slows down light. Positioning by GPS uses the speed $c$ and the travel time to find the distance from satellite to receiver. (It includes many other extremely small effects.) In fact GPS is the only everyday technology I know that requires both special relativity and general relativity. Amazing that your cell phone can include GPS.

The wave equation is second order in time because of $\partial^2 u/\partial t^2$. We are given the initial velocity $v_0(x)$ as well as the initial position $u_0(x)$.

$$\textbf{At } t = 0 \textbf{ and all } x \qquad \boldsymbol{u = u_0(x)} \quad \textbf{and} \quad \boldsymbol{\partial u/\partial t = v_0(x)}. \tag{4}$$

Look for functions that have $u_{tt}$ equal to $c^2 u_{xx}$. Now $e^{x+ct}$ and $e^{x-ct}$ will both succeed. Two time derivatives produce a factor $c$ twice (or a factor $-c$ twice, both cases give $c^2$). *All functions $f(x+ct)$ and all functions $g(x-ct)$ satisfy the wave equation.* The wave equation is linear, so we can combine those solutions.

$$\textbf{Complete solution to } \boldsymbol{u_{tt} = c^2 u_{xx}} \qquad \boldsymbol{u(t,x) = f(x+ct) + g(x-ct)} \tag{5}$$

Two functions $f(x+ct)$ and $g(x-ct)$ are exactly what we need to match two conditions $u_0$ and $v_0$ at $t = 0$:

$$\begin{array}{lrl} \textbf{Position} & u_0(x) = & f(x) + g(x) \\ \textbf{Velocity} & v_0(x) = & cf'(x) - cg'(x) \end{array} \quad \text{and then} \quad \frac{1}{c}\int_0^x v_0\, dx = f(x) - g(x).$$

Add those equations to find $2f(x)$. Subtract those equations to find $2g(x)$. Divide by 2:

$$f(x) = \frac{1}{2}u_0(x) + \frac{1}{2c}\int_0^x v_0\, dx \qquad g(x) = \frac{1}{2}u_0(x) - \frac{1}{2c}\int_0^x v_0\, dx \tag{6}$$

Then d'Alembert's solution $u$ to the wave equation has a wave traveling to the left with shape $f$ and a wave traveling to the right with shape $g$:

$$\boxed{u = f(x+ct) + g(x-ct) = \frac{u_0(x+ct) + u_0(x-ct)}{2} + \frac{1}{2c}\int_{x-ct}^{x+ct} v_0(x)\, dx} \tag{7}$$

**Example 1** Start from rest (velocity $v_0 = 0$) with a sine wave $u_0(x) = \sin \omega x$. That wave splits into two waves:

$$u(t,x) = \frac{u_0(x+ct) + u_0(x-ct)}{2} = \frac{1}{2}\sin(\omega x + c\omega t) + \frac{1}{2}\sin(\omega x - c\omega t). \qquad (8)$$

The trigonometry formula $\sin A + \sin B = 2\sin\frac{A+B}{2}\cos\frac{A-B}{2}$ produces a short answer:

$\boldsymbol{u(t,x) = (\sin\omega x)(\cos c\omega t)}$ **Two traveling waves produce one standing wave**.

You sometimes see standing waves in the ocean. Not what a surfer wants to find.

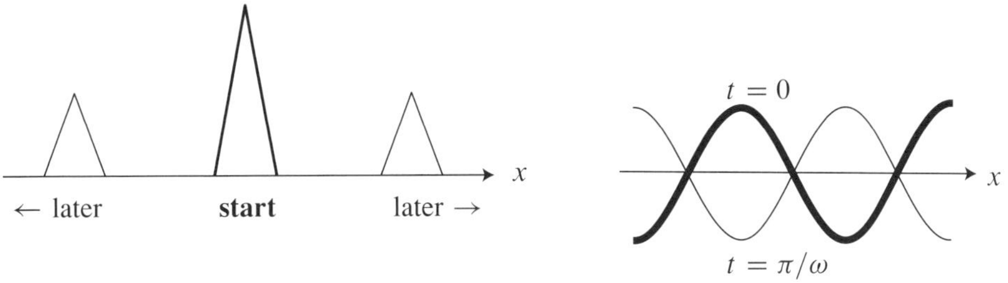


Figure 8.7: Always two traveling waves. Sometimes their sum is a standing wave.

## The Wave Equation from $x = 0$ to $x = 1$

Now we leave infinite space-time. The waves we know best are on a finite Earth. They may come from a violin string, fixed at both ends. They could also be water waves (even a tsunami). They may be electromagnetic waves: light or X-rays or TV signals. Or they may be sound waves that our ears convert into words. All these waves are bringing information to our brains, and they are essential to life as we know it.

Start with a violin string of length 1. The velocity $c$ depends on the tension in the string. The ends at $x = 0$ and 1 are assumed to remain fixed:

**Boundary conditions at the ends** $\quad u(t,0) = 0$ and $u(t,1) = 0$. $\qquad$ (9)

If we pluck the string with our finger at time $t = 0$, we give a vertical displacement $u_0$ and a vertical velocity $v_0$ (this might be zero):

**Initial conditions at the start** $\quad u(0,x) = u_0(x)$ and $\dfrac{\partial u}{\partial t}(0,x) = v_0(x)$. $\qquad$ (10)

If we remove our finger after time zero, waves move along the string. They are reflected back at the ends of the string. The sound is not a single beautiful note (it is a mixture of waves with many frequencies). Still a composer can include this plucking sound in a symphony and a guitarist uses it all the time.

The usual sound from violins comes from a *continuous source*—which is the bow. Now we are solving $\boldsymbol{u_{tt} = u_{xx} + f(t,x)}$. When the violinist puts a finger on the string, *that changes the length and it changes the frequencies*. Instead of waves of length 1 we will have waves of length $L$ and higher notes.

With several strings the violinist or cellist or guitarist is producing several waves of different frequencies to form chords. Let me stay with one string of length 1.

### Separation of Variables

We will use the most important method of solving partial differential equations by hand. The wave equation $u_{tt} = c^2 u_{xx}$ has two variables $t$ and $x$. **The simplest solutions are functions of $x$ multiplied by functions of $t$.**

$$\text{If } \boldsymbol{u = X(x)T(t)} \quad \text{then} \quad u_{tt} = c^2 u_{xx} \quad \text{is} \quad \boldsymbol{X(x)T''(t) = c^2 X''(x)T(t)}. \tag{11}$$

$T''$ and $X''$ are ordinary second derivatives. We can divide equation (11) by $c^2 XT$:

$$\textbf{Separation of variables} \qquad \frac{T''}{c^2 T} = \frac{X''}{X} = -\omega^2. \tag{12}$$

The function $T''/T$ depends only on $t$. The function $X''/X$ depends only on $x$. So both functions are constant and they are equal. By writing $-\omega^2$ for the constant, the two separated equations have the right form:

$$X'' = -\omega^2 X \qquad\qquad \boldsymbol{X = A\cos\omega x + B\sin\omega x} \tag{13}$$

$$T'' = -\omega^2 c^2 T \qquad\qquad \boldsymbol{T = C\cos\omega ct + D\sin\omega ct} \tag{14}$$

Key question: Which frequencies $\omega$ are allowed? The boundary values at $x = 0$ and $x = 1$ decide this perfectly. We want sines and not cosines, in order to have $X(0) = 0$. We want frequencies that are multiples of $\pi$ in order to have $X(1) = B\sin\omega = 0$. This gives very specific frequencies $\boldsymbol{\omega = \pi, 2\pi, 3\pi, \ldots}$ and no others.

The base frequency of the violin string is $\pi$ and the harmonics are multiples $\omega = n\pi$. If we touch the string and reduce its length to $L$, we want $\sin\omega L = 0$. Then the permitted frequencies increase to $\boldsymbol{\omega = n\pi/L}$. The notes go up the scale, separated by an octave.

Those frequencies $\omega$ also go into the time function $T(t)$. The initial condition is $T' = 0$ if the initial velocity is $v_0 = 0$. Only the cosine survives in the time direction:

$$X = B\sin n\pi x \qquad T = C\cos n\pi ct \qquad \boldsymbol{u = XT = b(\sin n\pi x)(\cos n\pi ct)}. \tag{15}$$

With length $L$, the *natural frequencies* in time are $\omega = n\pi c/L$. The *wavelengths* in space are $2L/n$. The displacement of the string is a combination of solutions $X(x)T(t)$:

$$\boldsymbol{u(t,x) = \sum_{n=1}^{\infty} b_n \left(\sin\frac{n\pi x}{L}\right)\left(\cos\frac{n\pi ct}{L}\right)}. \tag{16}$$

You see immediately that $u_{tt} = c^2 u_{xx}$ for every one of those terms, and any combination.

Final question: *What are the numbers $b_n$*? Those are decided by the remaining condition:

$$\textbf{Initial condition} \qquad \boldsymbol{u(0,x)} = u_0(x) = \sum_{n=1}^{\infty} \boldsymbol{b_n \sin \frac{n\pi x}{L}}. \tag{17}$$

This is a Fourier sine series! The formula for $b_k$ comes from multiplying both sides by $\sin k\pi x/L$ and integrating from 0 to $L$ along the string. Only one term $n = k$ survives:

$$\int_0^L u_0(x) \sin k\pi x \, dx = \int_0^L b_k (\sin k\pi x)^2 dx = \frac{L}{2} b_k. \tag{18}$$

Inserting each $b_k$ into (16) completes the solution of the wave equation on $0 \le x \le L$.

**Example 2** Suppose the length is $L = 3$ and the initial displacement is a *hat function*:

$$u_0(x) = x \text{ for } 0 \le x \le 1 \text{ and } u_0(x) = \frac{1}{2}(3 - x) \text{ for } 1 \le x \le 3.$$

The integrals in (18) lead in *Mathematica* to $b_k = 3/2k^2\pi^2$. The decay rate is $1/k^2$ for this function $u_0(x)$ with a corner. The slope drops from 1 to $-\frac{1}{2}$ at $x = 1$. The infinite series (16) will converge at every point in space-time to the correct solution $u(t,x)$.

Notice also that every piece of $u$ splits into $f + g$, by the formula for $\sin A \cos B$:

$$\sin \frac{n\pi x}{L} \cos \frac{n\pi ct}{L} = 2 \sin \frac{n\pi(x + ct)}{2L} + 2 \sin \frac{n\pi(x - ct)}{2L} = \boldsymbol{f(x + ct) + g(x - ct)}.$$

We get two wave functions as always, specially chosen to fit the string length $L$. If the initial velocity $v_0$ is not zero, then the solution $u(t,x)$ also contains sine functions of $t$.

*Our functions $X(x) = \sin n\pi x/L$ are actually eigenfunctions of the string*:

$\boldsymbol{Ax = \lambda x}$ becomes $\boldsymbol{X'' = -\omega^2 X}$ The matrix $A$ changes to a second derivative.

Again linear algebra and differential equations go hand in hand. For *linear* equations.

## ■ REVIEW OF THE KEY IDEAS ■

1. The one-way wave equation $u_t = cu_x$ is solved by $u(t,x) = u_0(x + ct)$.
2. The two-way equation $u_{tt} = c^2 u_{xx}$ allows two waves $f(x + ct)$ and $g(x - ct)$.
3. At $t = 0$, the d'Alembert solution (7) matches $u_0(x)$ and $v_0(x)$ on the whole line.
4. The Fourier solution (16) chooses $b_k$ so that $u(0,x) = u_0(x)$ for $0 \le x \le L$.
5. **Separation of variables** into $u = X(x)T(t)$ gives $\boldsymbol{X'' = -\omega^2 X}$ and $\boldsymbol{T'' = -\omega^2 c^2 T}$.
6. Zero boundary conditions give $\omega = n\pi/L$ and eigenfunctions $\boldsymbol{X(x) = \sin n\pi x/L}$.

## Problem Set 8.4

**Problems 1 – 4 are about the one-way wave equation $\partial u/\partial t = c\,\partial u/\partial x$.**

1 Suppose $u(0, x) = \sin 2x$. What is the solution to $u_t = cu_x$? At which times $t_1, t_2, \ldots$ will the solution return to the initial condition $\sin 2x$?

2 Suppose $u_0(x) = \delta(x)$, a big bang at the origin of the one-dimensional universe. At time $t$ the bang is heard at the point $x =$ ______. For $u_{tt} = c^2 u_{xx}$ the bang will reach the two points $x =$ ______ and $x =$ ______ at time $t$.

3 (a) Integrate both sides of $u_t = cu_x$ from $x = -\infty$ to $\infty$ to prove that the total mass $M = \int u\,dx$ is constant: $dM/dt = 0$.

(b) Multiply by $u$ and integrate both sides of $uu_t = cuu_x$ to prove that $E = \int u^2\,dx$ is constant.

4 Is the wave $u(t, x) = u_0(x + ct)$ traveling left or right if $c > 0$? To solve $u_t = cu_x$ on the halfline $0 \le x \le \infty$, why is a boundary condition $u(t, 0) = 0$ *not wanted*? With $c < 0$ and waves in the opposite direction, that condition is appropriate.

**Problems 5 – 9 are about the one-dimensional wave equation $\partial^2 u/\partial t^2 = c^2 \partial^2 u/\partial x^2$.**

5 A "box of water" has $u_0(x) = 1$ for $-1 \le x \le 1$. Starting with zero velocity $v_0(x)$, the wave equation $u_{tt} = c^2 u_{xx}$ is solved by $u(t, x) = \frac{1}{2}u_0(x + ct) + \frac{1}{2}u_0(x - ct)$. Graph this solution for small $t = \frac{1}{2}c$ and large $t = 3c$.

6 Under a flat ocean with $u_0(x) = 1$, an earthquake produces $v_0(x) = \delta(x)$. A one-dimensional tsunami starts moving with speed $c$. What is the solution (7) at time $t$?

7 Separation of variables gives $u(t, x) = (\sin nx)(\sin nct)$ and three other similar solutions to $u_{tt} = c^2 u_{xx}$. *What are those three*? Which complex functions $e^{ikx}e^{i\omega t}$ solve the wave equation?

8 The 3D wave equation $u_{tt} = u_{xx} + u_{yy} + u_{zz}$ becomes 1D when $u$ has spherical symmetry: $u$ depends only on $r$ and $t$.

$$r = \sqrt{x^2 + y^2 + z^2} \quad \text{and} \quad \frac{\partial^2 u}{\partial t^2} = \frac{\partial^2 u}{\partial r^2} + \frac{2}{r}\frac{\partial u}{\partial r}.$$

(a) *Multiply by $r$ to find $(ru)_{tt} = (ru)_{rr}$*! Then $ru$ is a function of $r + t$ and $r - t$.

(b) Describe the solution $ru = \delta(r - t - 1)$. This spherical sound wave has the radius $r =$ ______ at $t = 8$.

9 The wave equation along a bar with density $\rho$ and stiffness $k$ is $(\rho u_t)_t = (ku_x)_x$.

What is the velocity $c$ in $u_{tt} = c^2 u_{xx}$? What is $\omega$ in $u = \sin(\pi x/L)\cos\omega t$?

**10** The small vibrations of a beam satisfy the fourth order equation $u_{tt} = -c^2 u_{xxxx}$. Look for solutions $u = X(x)T(t)$ and find separate equations for the functions $X$ and $T$. Then find four solutions $X(x)$ when $T(t) = \cos \omega t$.

**11** If that beam is clamped ($u = 0$ and $\partial u/\partial x = 0$ at both ends $x = 0$ and $x = L$), show that the frequencies $\omega$ in Problem 10 must have $(\cos \omega L)(\cosh \omega L) = 1$.

**Problems 12 – 16 solve the wave equation with boundary conditions at $x = 0$ and $x = L$.**

**12** A string plucked halfway along has $u_0(x) = \delta(x - \frac{L}{2})$ and $v_0(x) = 0$. Find the Fourier coefficients $b_k$ from equation (18). Write the first three terms of the Fourier series solution in (16).

**13** Suppose the string starts with zero velocity $v_0(x)$ from a *hat function*: $u_0(x) = 2x/L$ for $x < L/2$ and $u_0(x) = 2(L - x)/L$ for $x > L/2$. Find the Fourier coefficients $b_k$ from (18) and the first two nonzero terms of $u(t, x)$ in (16).

**14** Suppose the string starts with zero velocity $v_0(x)$ from a *box function*: $u_0(x) = 1$ for $x < L/2$. Find all the $b_k$ in the solution $u = \sum b_k \sin(n\pi x/L) \cos(n\pi ct/L)$.

**15** The boundary condition at a *free end* $x = L$ is $\partial u/\partial x = 0$ instead of $u = 0$.

Solve $X'' + \omega^2 X = 0$ to find $X(x)$ and all allowable $\omega$'s with this new condition.

Then solve $T'' + \omega^2 c^2 T = 0$ to complete the solution $u = \sum a_n\, X(x)\, T(t)$.

**16** What is the solution $u(t, x)$ on a string of length $L = 2$ if $u(0, x) = \delta(x - 1)$? The end $x = 0$ is fixed by $u(t, 0) = 0$ and the end $x = 2$ is free: $\partial u/\partial x(t, 0) = 0$.

## 8.5 The Laplace Transform

When it succeeds, the Laplace transform can turn a linear differential equation into an algebra problem. Laplace transforms are applied to initial value problems ($t > 0$). Fourier transforms are for boundary value problems. Laplace has $e^{-st}$ instead of $e^{ikx}$.

When does this transform method succeed ? I see two desirable situations :

1. The linear equation should have constant coefficients, as in $Ay'' + By' + Cy = f(t)$.
2. The driving function $f(t)$ should have a "convenient" transform.

Our list of good functions includes $f(t) = \boldsymbol{e^{at}}$ and its transform $F(s) = \mathbf{1/(s-a)}$. Then the differential equation will tell us the transform $Y(s)$ of the solution. The final step is to discover which function $y(t)$ has this transform $Y(s)$. Using our list of transforms and especially the rules for finding new transforms, this becomes a problem in algebra: *Invert the transform* $\mathbf{Y(s)}$ *to find the solution* $\boldsymbol{y(t)}$. These pages complete Section 2.6.

Particular solutions are easy with $f(t) = e^{at}$. The method of undetermined coefficients taught us to look for $y_p(t) = Ye^{at}$. The Laplace transform is not strictly needed when $f(t) = e^{at}$ or $t^n$ or $\sin \omega t$ or $\cos \omega t$. But for driving functions that turn on and off, and functions that jump or explode (step functions and delta functions and worse), the algebra becomes more systematic and better organized by the Laplace transform. Examples **1**, **2**, **3** with real, imaginary, and complex poles show you the key ideas.

### The Transform $F(s)$

Start with a function $f(t)$ defined for $t \geq 0$. Multiply by $e^{-st}$ and integrate from $t = 0$ to $t = \infty$. The result is the Laplace transform $F(s)$ and it depends on the exponent $s$ :

**Laplace transform** $$\mathscr{L}[f(t)] = F(s) = \int_{t=0}^{\infty} f(t)\, e^{-st}\, dt. \qquad (1)$$

The number $s$ can be real or complex. The one key requirement on $s$ is that the infinite integral in (1) must give a finite answer. Here are examples needing $s > 0$ and $s > a$.

$$\boldsymbol{f(t) = 1} \qquad \boldsymbol{F(s)} = \int_0^{\infty} e^{-st}\, dt = \left[\frac{e^{-st}}{-s}\right]_{t=0}^{t=\infty} = \frac{\mathbf{1}}{\boldsymbol{s}}. \qquad (2)$$

$$\boldsymbol{f(t) = e^{at}} \qquad \boldsymbol{F(s)} = \int_0^{\infty} e^{at}\, e^{-st}\, dt = \left[\frac{e^{(a-s)t}}{a-s}\right]_0^{\infty} = \frac{\mathbf{1}}{\boldsymbol{s-a}}. \qquad (3)$$

The integral of $e^{-st}$ is finite when $s$ is positive. More than that, it is finite when *the real part of $s$ is positive.* A factor $e^{-i\omega t}$ from the imaginary part $i\omega$ has absolute value 1. Laplace transforms are defined when the real part of $s$ exceeds some value $s_0$. Here $s_0 = a$.

**Important All functions in this section have $f(t) = 0$ for $t < 0$. They start at $t = 0$.** So the constant function $f(t) = 1$ is actually the unit step function. It jumps from 0 to 1 at $t = 0$. Its derivative is the delta function $\delta(t)$; this includes the spike at $t = 0$. In this way, the initial value problem $y' + y = 1$ ignores all $t < 0$ and starts from $y(0)$.

You will see that the Laplace transform of that equation is $sY(s) - y(0) + Y(s) = 1/s$. Then algebra gives $Y(s)$ and the inverse Laplace transform gives $y(t)$.

The second example $f = e^{at}$ includes the first example $f = 1$, which has $a = 0$. Then $1/(s-a)$ becomes $1/s$. We need $\operatorname{Re} s > a$ to drive $e^{at}e^{-st}$ to zero at $t = \infty$. There are decreasing functions like $f(t) = e^{-t^2}$ that allow every complex number $s$. There are also rapidly increasing functions like $f(t) = e^{t^2}$ that allow no $s$ at all.

For a delta function located at $t = T \geq 0$, the integral picks out the transform $e^{-sT}$:

$$\boldsymbol{f(t) = \delta(t - T)} \qquad \boldsymbol{F(s)} = \int_0^\infty \delta(t-T)\,e^{-st}\,dt = \boldsymbol{e^{-sT}}. \tag{4}$$

To complete this group of examples (the all-star functions), a simple trick gives the transforms of $\cos\omega t$ and $\sin\omega t$. Write Euler's formula $e^{i\omega t} = \cos\omega t + i\sin\omega t$. Take the Laplace transform of every term:

$$\textbf{Linearity} \qquad \mathscr{L}[e^{i\omega t}] = \mathscr{L}[\cos\omega t] + i\,\mathscr{L}[\sin\omega t]$$

The left side is $1/(s - i\omega)$. Multiply by $(s + i\omega)/(s + i\omega)$ to see real and imaginary parts:

$$\frac{1}{s - i\omega}\,\frac{s + i\omega}{s + i\omega} = \frac{s + i\omega}{s^2 + \omega^2} \qquad \boldsymbol{\mathscr{L}[\cos\omega t] = \frac{s}{s^2 + \omega^2}} \text{ and } \boldsymbol{\mathscr{L}[\sin\omega t] = \frac{\omega}{s^2 + \omega^2}} \tag{5}$$

## Exponents in $f(t)$ are Poles in $F(s)$

Let me pause one minute, before using Laplace transforms to solve differential equations. We can already see the key connection between a function $f(t)$ and its transform $F(s)$. Look at this *Table of Transforms*:

| $f(t)$ | $1$ | $e^{at}$ | $\delta(t-T)$ | $\cos\omega t$ | $\sin\omega t$ | $t^n e^{ct}$ |
|---|---|---|---|---|---|---|
| $F(s)$ | $\frac{1}{s}$ | $\frac{1}{s-a}$ | $e^{-sT}$ | $\frac{s}{s^2+\omega^2}$ | $\frac{\omega}{s^2+\omega^2}$ | $\frac{n!}{(s-c)^{n+1}}$ |

Here is the important message. **If $f(t)$ includes $e^{at}$ then $F(s)$ has a "pole" at $s = a$.** A pole is an isolated point $a$, real or complex, where the function $F(s)$ blows up. Some integer power $(s-a)^m$ will cancel the pole and leave an "analytic" function $(s-a)^m F(s)$.

An example shows this matchup of exponents in $f(t)$ to poles in the transform $F(s)$:

$$f(t) = e^{0t} + e^{at} + e^{i\omega t} + e^{-i\omega t} + te^{ct} \quad \text{has exponents } \mathbf{0},\ \boldsymbol{a},\ \boldsymbol{i\omega},\ \boldsymbol{-i\omega},\ \boldsymbol{c}$$

$$\boldsymbol{F(s)} = \frac{1}{s} + \frac{1}{s-a} + \frac{2s}{(s-i\omega)(s+i\omega)} + \frac{1}{(s-c)^2} = \frac{\text{something}}{\boldsymbol{s(s-a)(s-i\omega)(s+i\omega)(s-c)^2}}.$$

The first term $1/s$ has exponent 0 in $f(t)$ and blowup at the pole $s = 0$. The last term $1/(s-c)^2$ has exponent $c$ and double blowup (*double pole*) at $s = c$. In the middle, $2\cos\omega t$ contains two exponents $i\omega$ and $-i\omega$, so the transform $F(s)$ has those two poles.

At the very end you see all the pieces of $F(s)$ tangled together in one big fraction. This is how $F(s)$ comes to us from a differential equation. Normally we must factor the denominator to see five separate poles at $s = 0, a, i\omega, -i\omega, c$. Then $F(s)$ splits into its simple pieces (called partial fractions). The inverse Laplace transform of each piece of $F(s)$ gives a piece of $f(t)$. **PF2** and **PF3** in Section 2.6 allowed two or three pieces.

An engineer moves poles by changing the design. Then the exponents move. The system becomes more stable if their real parts become more negative. A quick accurate picture of stability comes from the poles of $F(s)$. If all those poles are in the left half of the complex plane, where $\operatorname{Re} s < 0$, the function will decay to zero (asymptotic stability).

The new function in this example is $\boldsymbol{te^{ct}}$. We remember that the extra factor $t$ appears in the solution $y(t)$ when the exponent $c$ is repeated ($c$ is a *double root* of the polynomial $s^2 - 2cs + c^2$ that comes from $y'' - 2cy' + c^2y$). The double root becomes a *double pole* in the transform, when $(s-c)^2$ shows up in the denominator of $F(s)$. Here is the required step, to confirm that **the transform of $\boldsymbol{f(t) = te^{ct}}$ is $\boldsymbol{F(s) = 1/(s-c)^2}$.**

$$\textbf{The derivative of } \boldsymbol{F(s) = \int_0^\infty f(t)e^{-st}\,dt} \textbf{ is } \boldsymbol{\frac{dF}{ds} = \int_0^\infty -tf(t)e^{-st}\,dt}.$$

**Rule: If the function $\boldsymbol{f(t)}$ transforms to $\boldsymbol{F(s)}$, then $\boldsymbol{tf(t)}$ transforms to $\boldsymbol{-dF/ds}$.**

When this rule is applied to $f(t) = e^{ct}$ with $F(s) = 1/(s-c)$, we learn that $te^{ct}$ transforms to $dF/ds = 1/(s-c)^2$.

This rule extends directly to higher powers of $t$ in $t^n f(t)$. Each time you multiply by $t$, take the derivative of $F(s)$. Remember to multiply by $-1$:

$$\boldsymbol{t^2 f(t) \longrightarrow (-1)^2 \frac{d^2F}{ds^2}} \qquad t^2e^{ct} \longrightarrow \frac{d^2}{ds^2}\left(\frac{1}{s-c}\right) = \frac{d}{ds}\frac{-1}{(s-c)^2} = \boldsymbol{\frac{2}{(s-c)^3}}.$$

Continuing this way, the transform of $t^n e^{ct}$ is $n!/(s-c)^{n+1}$. This was the last entry in our Table of Transforms. In the special case $c = 0$, **the transform of $\boldsymbol{t^n}$ is $\boldsymbol{n!/s^{n+1}}$.**

Now we can work with any real poles $c$ or imaginary poles $i\omega$ in $F(s)$. Example 3 will allow complex poles $c + i\omega$. This solves all equations $Ay'' + By' + Cy = 0$.

### Transforms of Derivatives

Differential equations involve $dy/dt$. We must connect the transform $\mathscr{L}[dy/dt]$ to $\mathscr{L}[y]$. This step was especially easy for Fourier transforms—just multiply by $ik$. *For Laplace transforms we expect to multiply $Y(s)$ by $s$ to get $\mathscr{L}[dy/dt]$, but another term appears.*

The reason this happens is that Laplace completely ignores $t < 0$. The integral starts at $t = 0$ and the number $y(0)$ is important. A good thing that $y(0)$ enters the Laplace transform, because we certainly expect it to enter the solution to a differential equation.

It is integration by parts that connects $\mathscr{L}[dy/dt]$ to $\mathscr{L}[y]$. Two minus signs cancel:

$$\mathscr{L}\left[\frac{dy}{dt}\right] = \int_0^\infty \frac{dy}{dt}e^{-st}\,dt = \int_0^\infty y(t)(se^{-st})\,dt + \left[y(t)e^{-st}\right]_0^\infty = \boldsymbol{s\,\mathscr{L}[y] - y(0)}. \quad (6)$$

This is the key fact that turns a differential equation for $y(t)$ into an algebra problem for $Y(s)$. If we repeat this step (apply it now to $dy/dt$), you will see the transform of the second derivative. **Use equations (6) and (7) to transform differential equations.**

$$\mathscr{L}\left[\frac{d^2y}{dt^2}\right] = s\mathscr{L}\left[\frac{dy}{dt}\right] - \frac{dy}{dt}(0) = \boldsymbol{s^2\mathscr{L}[y] - sy(0) - \frac{dy}{dt}(0)}. \quad (7)$$

Let me use this rule right away to solve three differential equations. The first has **real poles.** The second has **imaginary poles**. The third has **complex poles** $s = -1 \pm i$.

**Example 1** Solve $\boldsymbol{y' - y = 2e^{-t}}$ starting from $y(0) = 1$.

**Solution** Take the Laplace transform of both sides. We know $\mathscr{L}[2e^{-t}] = 2/(s+1)$:

$$s\mathscr{L}[y] - y(0) - \mathscr{L}[y] = \mathscr{L}[2e^{-t}] \quad \text{is the same as} \quad \boldsymbol{(s-1)Y(s) = 1 + \tfrac{2}{s+1}}.$$

Then algebra gives $Y(s)$ and we split into "partial fractions" to recognize $y(t)$.

$$Y(s) = \frac{1}{s-1} + \frac{2}{(s-1)(s+1)} = \frac{1}{s-1} + \left(\frac{1}{s-1} - \frac{1}{s+1}\right) = \boldsymbol{\frac{2}{s-1} - \frac{1}{s+1}}$$

**The inverse transform of $\boldsymbol{Y(s)}$ is $\boldsymbol{y(t) = 2e^t - e^{-t}}$**

I always check that $y(0) = 2 - 1 = 1$ and $y'(t) = 2e^t + e^{-t}$ agrees with $y + 2e^{-t}$. And don't forget our usual method. A particular solution is $y_p = -e^{-t}$. It has the same form as the driving function $f(t) = e^{-t}$. The null solution is $y_n = Ce^t$.

**From Chapter 2** $\quad y = y_p + y_n = \boldsymbol{-e^{-t} + Ce^t} \quad y(0) = 1$ gives $C = 2$

Maybe the earlier method is simpler for this example? The next examples give practice with second order equations. The complex poles of $Y(s)$ give oscillations $e^{i\omega t}$ in $y(t)$.

**Example 2** Solve the equation $y'' + y = \frac{1}{2}\sin 2t$ starting from rest: $y(0) = y'(0) = 0$. The transform of $y''$ is $s^2 Y(s)$ from (7):

$$s^2 Y(s) + Y(s) = \frac{1}{s^2 + 2^2} \quad \text{and then} \quad Y(s) = \frac{1}{(s^2+1)(s^2+4)}$$

Partial fractions will rewrite that transform $Y(s)$ as

$$Y(s) = \frac{1}{(s^2+1)(s^2+4)} = \frac{1}{3}\frac{(s^2+4)-(s^2+1)}{(s^2+1)(s^2+4)} = \frac{\mathbf{1/3}}{\mathbf{s^2+1}} - \frac{\mathbf{1/3}}{\mathbf{s^2+4}}. \tag{8}$$

We recognize those fractions as transforms of sine functions with $\omega = 1$ and $\omega = 2$:

**Solution** $\quad \boldsymbol{y(t) = \frac{1}{3}\sin t - \frac{1}{6}\sin 2t}$ has initial values $y(0) = 0$ and $y'(0) = 0$.

The transform of $\sin 2t$ is $2/(s^2+4)$, which explains why $1/3$ becomes $1/6$.

In Chapter 2 we would have found $y_p(t)$ and $y_n(t)$ to reach the same $y(t)$:

$$\boldsymbol{y = y_p + y_n} = -\frac{1}{6}\sin 2t + c_1 \cos t + c_2 \sin t.$$

Then $c_1 = 0$ because $y(0) = 0$, and $c_2 = \frac{1}{3}$ because $y'(0) = 0$. Both ways are good.

**Example 3** $\quad y'' + 2y' + 2y = 0$ with $y(0) = y'(0) = 1$ has $Y(s) = \dfrac{\mathbf{s-1}}{\mathbf{s^2+2s+2}}$.

Then the roots of $s^2 + 2s + 2$ are the complex poles $\boldsymbol{s = -1 \pm i}$.

This $Y(s)$ is not yet in our table. But we know the complex solutions $e^{(-1+i)t}$ and $e^{(-1-i)t}$. Their real and imaginary parts are $e^{-t}\cos t$ and $e^{-t}\sin t$. The combination that has $y(0) = y'(0) = 1$ is $y = \boldsymbol{e^{-t}\cos t + 2e^{-t}\sin t}$. This must be the function $y(t)$ that transforms to $Y(s)$.

The real and imaginary parts of $e^{ct}e^{i\omega t}$ transform to the real and imaginary parts of $1/(s - c - i\omega)$. Those two new transforms solve Example 3 when $c = -1$ and $\omega = 1$. We can now solve every equation $Ay'' + By' + Cy = 0$.

| | |
|---|---|
| $\boldsymbol{e^{ct}\cos\omega t}$ transforms to $\dfrac{\boldsymbol{s-c}}{\boldsymbol{(s-c)^2+\omega^2}}$ | $\boldsymbol{e^{ct}\sin\omega t}$ transforms to $\dfrac{\boldsymbol{\omega}}{\boldsymbol{(s-c)^2+\omega^2}}$. |

## Shifts and Step Functions and Cutoffs

Suppose the driving function $f(t)$ in a differential equation turns on at time $T$. Or suppose it turns off. Or it jumps to a different function. All these jumps in $f(t)$ are realistic in practical problems, and they are automatically handled by the Laplace transform.

Essentially, *we need the transform of a step function.* The basic example is a unit step that jumps from $f = 0$ for $t < T$ to $f = 1$ for $t \geq T$. The transform is an easy integral:

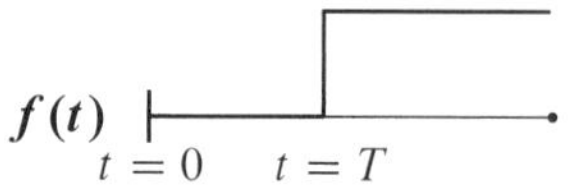


$$\boldsymbol{F(s)} = \int_T^\infty e^{-st}\,dt = \left[\frac{e^{-st}}{-s}\right]_T^\infty = \frac{\boldsymbol{e^{-sT}}}{\boldsymbol{s}}. \tag{9}$$

A step function at $T$ transforming to $e^{-sT}/s$ is an example of a new rule.

**The step at $T$ is a time shift of the step at $t = 0$. Multiply the transform by $e^{-sT}$.**

The original $f(t)$ has the transform $F(s)$. The shifted function is zero until $t = T$, and then it is $f(t-T)$. For the example of a unit step, the shifted step is zero for $t < T$.

Here is the proof of the transform rule for the shifted function: **multiply by $e^{-sT}$**.

**$f(t)$ shifts to $f(t-T)$**
**$F(s)$ becomes $e^{-sT}F(s)$**

$$\int_T^\infty f(t-T)\,e^{-st}\,dt = \int_0^\infty f(\tau)\,e^{-s(\tau+T)}\,d\tau = \boldsymbol{e^{-sT}F(s)}.$$

The first integral has $T \le t < \infty$. The second integral has $0 \le \tau < \infty$. The new variable $\tau = t - T$ shifts the lower limit on the integral back to $\tau = 0$, and it produces the all-important factor $e^{-sT}$. We end with two examples that need this shift rule.

**Example 4** **(Unit step function)** Solve $y' - ay = H(t-T) = \left\{\begin{matrix} \mathbf{0} & t<T \\ \mathbf{1} & t \ge T \end{matrix}\right\}$.

The transform of every term (with $y(0) = 1$) will give the transform $Y(s)$ of the solution:

$$s\,Y(s) - 1 - a\,Y(s) = \frac{e^{-sT}}{s} \qquad \boldsymbol{Y(s) = \frac{1}{s-a} + \frac{e^{-sT}}{(s-a)s}}. \tag{10}$$

The inverse transform of $1/(s-a)$ is $\boldsymbol{e^{at}}$. Split the other fraction into two parts:

$$\frac{1}{(s-a)s} = \frac{1}{a}\left(\frac{1}{s-a} - \frac{1}{s}\right) \quad \text{has inverse transform} \quad \boldsymbol{\frac{1}{a}\left(e^{at}-1\right)}. \tag{11}$$

The factor $e^{-sT}$ in (10) will shift that function in (11). The final solution is

**Jump in $y'$**
**Corner in $y$**

$$\boldsymbol{y(t)} = \begin{cases} \boldsymbol{e^{at}} & \text{for } t \le T \\ \boldsymbol{e^{at} + \frac{1}{a}\left(e^{a(t-T)} - 1\right)} & \text{for } t \ge T \end{cases} \tag{12}$$

The first part $y = e^{at}$ has $y' = ay$ as required. This meets the second part correctly at $t = T$ (*no jump in $y$*). Then the second part of $y(t)$ continues with $y' = ay + 1$:

$$\textit{Check} \quad y' = ae^{at} + e^{a(t-T)} = a\left[e^{at} + \frac{1}{a}e^{a(t-T)} - \frac{1}{a} + \frac{1}{a}\right] = ay + 1.$$

*Question* Could we have solved this problem without Laplace transforms? Certainly $y = e^{at}$ solves the first part starting from $y(0) = 1$. This is $y_n$ since $f = 0$, and it reaches $e^{aT}$ at time $T$. Starting from there, we need to add on a particular solution $y_p$. This $y_p$ will match the driving function $f = 1$ that begins to act at $t = T$:

$$y_p{}' - ay_p = 1 \quad \text{starting from} \quad y_p(T) = 0.$$

Eventually, and somehow, we would find the particular solution $y_p = \left(e^{a(t-T)} - 1\right)/a$. Combined with $y_n = e^{at}$, the complete solution $y_n + y_p$ agrees with equation (12).

**Example 5** **Suppose the driving function $f(t) = 1$ turns off instead of on at time $T$ :**

$$\text{Solve } \boldsymbol{y' - ay} = \begin{cases} \mathbf{1} & t \le T \\ \mathbf{0} & t > T \end{cases} \quad \text{with } y(0) = 1.$$

**Solution** Instead of the previous $H(t-T)$, this new driving function is $1 - H(t-T)$. The step function drops from 1 to 0. We still take the Laplace transform of every term in the differential equation:

$$s\,Y(s) - 1 - a\,Y(s) = \text{ transform of } [1 - H(t-T)] = \frac{1}{s} - \frac{e^{-sT}}{s}.$$

Solve this equation for $Y(s)$ and begin to recognize the inverse transform:

$$Y(s) = \frac{1}{s-a} + \frac{1}{(s-a)s} - \frac{e^{-sT}}{(s-a)s} \text{ has the new term } \frac{1}{(s-a)s} \text{ compared to (10).}$$

The inverse transform of this new term is $(e^{at}-1)/a$, according to (11). Since the last term in $Y(s)$ now has a minus sign, the final solution has two pieces meeting at $t = T$ :

$$y(t) = \begin{cases} e^{at} + \frac{1}{a}(e^{at} - 1) & \text{for } t \le T \\ e^{at} + \frac{1}{a}(e^{at} - 1) - \frac{1}{a}(e^{a(t-T)} - 1) & \text{for } t \ge T. \end{cases}$$

That first part for $t \le T$ would be our standard $y_n + y_p$, starting from $y(0) = 1$. The second part matches the first part at $t = T$ (*no jump in* $y$). That second part simplifies to

$$y(t) = e^{at} + \frac{e^{at} - e^{a(t-T)}}{a} \quad \text{and we verify that} \quad y' = ay.$$

## Rules for the Laplace Transform

Part of this section is about specific functions $f(t)$. We made a Table of Transforms $F(s)$. The other part of the section is about rules. (This is like calculus. You learn the derivatives of $t^n$ and $\sin t$ and $\cos t$ and $e^t$. Then you learn the product rule and quotient rule and chain rule.) We need a Table of Rules for the Laplace transform, when we know that $F(s)$ and $G(s)$ are the transforms of $f(t)$ and $g(t)$.

| | | |
|---|---|---|
| **Addition Rule** | The transform of $\boldsymbol{f(t) + g(t)}$ is | $\boldsymbol{F(s) + G(s)}$ |
| **Shifting Rule** | The transform of $\boldsymbol{f(t - T)}$ is | $\boldsymbol{e^{-sT}\,F(s)}$ |
| **Derivative of $f$** | The transform of $\boldsymbol{df/dt}$ is | $\boldsymbol{sF(s) - f(0)}$ |
| **Derivative of $F$** | The transform of $\boldsymbol{tf(t)}$ is | $\boldsymbol{-dF/ds}$ |
| **Convolution Rule** | Section 8.6 will transform $\boldsymbol{f(t)g(t)}$ and invert $\boldsymbol{F(s)G(s)}$ | |

## Problem Set 8.5

**1** When the driving function is $f(t) = \delta(t)$, the solution starting from rest is the **impulse response**. The impulse is $\delta(t)$, the response is $y(t)$. Transform this equation to find the **transfer function** $Y(s)$. Invert to find the impulse response $y(t)$.

$$y'' + y = \delta(t) \text{ with } y(0) = 0 \text{ and } y'(0) = 0$$

**2** (Important) Find the first derivative and second derivative of $f(t) = \sin t$ for $t \geq 0$. Watch for a jump at $t = 0$ which produces a spike (delta function) in the derivative.

**3** Find the Laplace transform of the unit box function $b(t) = \{1 \text{ for } 0 \leq t < 1\} = H(t) - H(t-1)$. The unit step function is $H(t)$ in honor of Oliver Heaviside.

**4** If the Fourier transform of $f(t)$ is defined by $\widehat{f}(k) = \int f(t)e^{-ikt}\,dt$ and $f(t) = 0$ for $t < 0$, what is the connection between $\widehat{f}(k)$ and the Laplace transform $F(s)$?

**5** What is the Laplace transform $R(s)$ of the standard **ramp function** $r(t) = t$? For $t < 0$ all functions are zero. The derivative of $r(t)$ is the unit step $H(t)$. Then multiplying $R(s)$ by $s$ gives ______.

**6** Find the Laplace transform $F(s)$ of each $f(t)$, and the poles of $F(s)$:

(a) $f = 1 + t$ (b) $f = t\cos\omega t$ (c) $f = \cos(\omega t - \theta)$
(d) $f = \cos^2 t$ (e) $f = e^{-2t}\cos t$ (f) $f = te^{-t}\sin\omega t$

**7** Find the Laplace transform $s$ of $f(t) =$ next integer above $t$ and $f(t) = t\,\delta(t)$.

**8** *Inverse Laplace Transform*: Find the function $f(t)$ from its transform $F(s)$:

(a) $\dfrac{1}{s - 2\pi i}$ (b) $\dfrac{s+1}{s^2+1}$ (c) $\dfrac{1}{(s-1)(s-2)}$
(d) $1/(s^2 + 2s + 10)$ (e) $e^{-s}/(s-a)$ (f) $2s$

**9** Solve $y'' + y = 0$ from $y(0)$ and $y'(0)$ by expressing $Y(s)$ as a combination of $s/(s^2+1)$ and $1/(s^2+1)$. Find the inverse transform $y(t)$ from the table.

**10** Solve $y'' + 3y' + 2y = \delta$ starting from $y(0) = 0$ and $y'(0) = 1$ by Laplace transform. Find the poles and partial fractions for $Y(s)$ and invert to find $y(t)$.

**11** Solve these initial-value problems by Laplace transform:

(a) $y' + y = e^{i\omega t}$, $y(0) = 8$ (b) $y'' - y = e^t$, $y(0) = 0$, $y'(0) = 0$
(c) $y' + y = e^{-t}$, $y(0) = 2$ (d) $y'' + y = 6t$, $y(0) = 0$, $y'(0) = 0$
(e) $y' - i\omega y = \delta(t)$, $y(0) = 0$ (f) $my'' + cy' + ky = 0$, $y(0) = 1$, $y'(0) = 0$

**12** The transform of $e^{At}$ is $(sI - A)^{-1}$. Compute that matrix (the transfer function) when $A = [1\ \ 1;\ 1\ \ 1]$. Compare the poles of the transform to the eigenvalues of $A$.

**13** If $dy/dt$ decays exponentially, show that $sY(s) \to y(0)$ as $s \to \infty$.

**14** Transform Bessel's time-varying equation $ty''+y'+ty = 0$ using $\mathscr{L}[ty] = -dY/ds$ to find a first-order equation for $Y$. By separating variables or by substituting $Y(s) = C/\sqrt{1+s^2}$, find the Laplace transform of the Bessel function $y = J_0$.

**15** Find the Laplace transform of a single arch of $f(t) = \sin \pi t$.

**16** Your acceleration $v' = c(v^* - v)$ depends on the velocity $v^*$ of the car ahead:

(a) Find the ratio of Laplace transforms $V^*(s)/V(s)$.

(b) If that car has $v^* = t$ find your velocity $v(t)$ starting from $v(0) = 0$.

**17** A line of cars has $v_n' = c[v_{n-1}(t-T) - v_n(t-T)]$ with $v_0(t) = \cos \omega t$ in front.

(a) Find the growth factor $A = 1/(1 + i\omega e^{i\omega T}/c)$ in oscillation $v_n = A^n e^{i\omega t}$.

(b) Show that $|A| < 1$ and the amplitudes are safely decreasing if $cT < \frac{1}{2}$.

(c) If $cT > \frac{1}{2}$ show that $|A| > 1$ (dangerous) for small $\omega$. (Use $\sin\theta < \theta$.) Human reaction time is $T \geq 1$ sec and aggressiveness is $c = 0.4/\text{sec}$. Danger is pretty close. Probably drivers adjust to be barely safe.

**18** For $f(t) = \delta(t)$, the transform $F(s) = 1$ is the limit of transforms of tall thin box functions $b(t)$. The boxes have width $\epsilon \to 0$ and height $1/\epsilon$ and area 1.

$$\text{Inside integrals, } b(t) = \begin{cases} 1/\epsilon & \text{for } 0 \le t < \epsilon \\ 0 & \text{otherwise} \end{cases} \text{ approaches } \delta(t).$$

Find the transform $B(s)$, depending on $\epsilon$. Compute the limit of $B(s)$ as $\epsilon \to 0$.

**19** The transform $1/s$ of the unit step function $H(t)$ comes from the limit of the transforms of short steep ramp functions $r_\epsilon(t)$. These ramps have slope $1/\epsilon$:

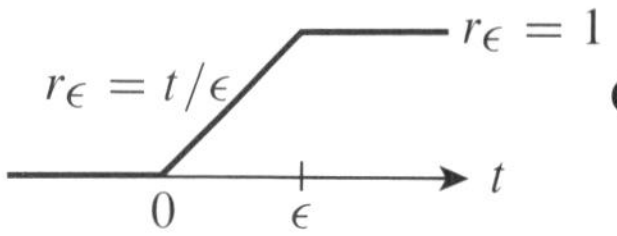


$$\text{Compute } R_\epsilon(s) = \int_0^\epsilon \frac{t}{\epsilon} e^{-st}\,dt + \int_\epsilon^\infty e^{-st}\,dt. \text{ Let } \epsilon \to 0.$$

**20** In Problems 18 and 19, show that the derivative of the ramp function $r_\epsilon(t)$ is the box function $b(t)$. The "generalized derivative" of a step is the ______ function.

**21** What is the Laplace transform of $y'''(t)$ when you are given $Y(s)$ and $y(0), y'(0), y''(0)$?

**22** The *Pontryagin maximum principle* says that the optimal control is "bang-bang"—it only takes on the extreme values permitted by the constraints. To go from rest at $x = 0$ to rest at $x = 1$ in minimum time, use maximum acceleration $A$ and deceleration $-B$. At what time $t$ do you change from the accelerator to the brake? (This is the fastest driving between two red lights.)

## 8.6 Convolution (Fourier and Laplace)

This section is about multiplication. **Convolution is a different way to multiply functions.** It is also a way to multiply vectors. The rule for vectors may look new, but actually you learned it in third grade. Let me start with ordinary multiplication of numbers, and build up to convolution of vectors and convolution of functions.

When 112 is multiplied by 2 1 3, watch how we collect nine small multiplications:

$$\begin{array}{ccccc} & & 1 & 1 & 2 \\ & & 2 & 1 & 3 \\ \hline & & 3 & 3 & 6 \\ & 1 & 1 & 2 & \\ 2 & 2 & 4 & & \\ \hline \mathbf{2} & \mathbf{3} & \mathbf{8} & \mathbf{5} & \mathbf{6} \end{array} \qquad \begin{array}{ccccc} & & a & b & c \\ & & 2 & 1 & 3 \\ \hline & & 3a & 3b & 3c \\ & a & b & c & \\ 2a & 2b & 2c & & \\ \hline \bullet & \bullet & \bullet & \bullet & \bullet \end{array}$$

We don't think about this pattern—it is so familiar. In our minds we are just multiplying 112 by 213 in small steps. The new idea is to think of $(1, 1, 2)$ as a vector and $(2, 1, 3)$ as another vector. **The convolution of those vectors is the vector $(2, 3, 8, 5, 6)$.**

I need a new symbol $*$ for the convolution of two vectors $\boldsymbol{c}$ and $\boldsymbol{d}$ :

**Convolution of vectors** $\quad \boldsymbol{c} * \boldsymbol{d} = (c_0, c_1, \ldots) * (d_0, d_1, \ldots) = (\boldsymbol{c_0 d_0}, \boldsymbol{c_0 d_1 + c_1 d_0}, \ldots)$

That line ends with an important hint about $\boldsymbol{c} * \boldsymbol{d}$, if we can see it. First, every $c_i$ multiplies every $d_j$. (Those are the nine small multiplications.) Then the nine products are collected in a special way. We put $c_0 d_1$ with $c_1 d_0$. **The next component of $\boldsymbol{c} * \boldsymbol{d}$ will be $\boldsymbol{c_0 d_2 + c_1 d_1 + c_2 d_0}$.**

In the third grade multiplication, we are collecting together all the products $c_i d_j$ that go in the 100*s* column. Those were $300 + 100 + 400$. To express this with algebra, the $n^{th}$ component of $\boldsymbol{c} * \boldsymbol{d}$ will be $\boldsymbol{c_0 d_n} + \boldsymbol{c_1 d_{n-1}} + \cdots + \boldsymbol{c_n d_0}$. **These are all the products $c_i d_j$ with $i + j = n$.**

**Convolution $\boldsymbol{c} * \boldsymbol{d} = \boldsymbol{d} * \boldsymbol{c}$**
$$(\boldsymbol{c} * \boldsymbol{d})_n = \sum_{i+j=n} c_i d_j = \sum_i c_i d_{n-i}. \qquad (1)$$

The summation symbol allows the vectors to be infinitely long. The key point is that small multiplications $c_i d_j$ go together when $\boldsymbol{i + j = n}$, which is the same as $\boldsymbol{j = n - i}$. Let me show that rule again, this time for $2 + x + 3x^2$ times $1 + x + 2x^2$. **We are collecting all the pieces that multiply each power $x^n$.**

$$\begin{array}{r} 1 + x + 2x^2 \\ 2 + x + 3x^2 \\ \hline 3x^2 + 3x^3 + 6x^4 \\ x + x^2 + 2x^3 \quad\;\; \\ 2 + 2x + 4x^2 \qquad\qquad\quad \\ \hline \mathbf{2} + \mathbf{3}x + \mathbf{8}x^2 + \mathbf{5}x^3 + \mathbf{6}x^4 \end{array}$$

When we multiply polynomials, we take the convolution of the vectors of coefficients.

$(2, 1, 3) * (1, 1, 2) = (\mathbf{2, 3, 8, 5, 6})$

We will connect convolution of coefficients to multiplication of Fourier series. First, allow me to show one more example that collects the small multiplications $c_i d_j$ in the same "convolution way." That example is a matrix-vector multiplication $C\boldsymbol{d}$. The matrix $C$ has the numbers $c_0, c_1, \ldots$ along its diagonals and $C$ **times** $\boldsymbol{d}$ **is exactly the convolution** $\boldsymbol{c} * \boldsymbol{d}$.

**$C\boldsymbol{d} = \boldsymbol{c} * \boldsymbol{d}$**
**Constant diagonals**
**Toeplitz matrix**
**Shift invariant**

$$\begin{bmatrix} c_0 & & \\ c_1 & c_0 & \\ c_2 & c_1 & c_0 \\ & c_2 & c_1 \\ & & c_2 \end{bmatrix} \begin{bmatrix} d_0 \\ d_1 \\ d_2 \end{bmatrix} = \begin{bmatrix} c_0 d_0 \\ c_1 d_0 + c_0 d_1 \\ c_2 d_0 + c_1 d_1 + c_2 d_0 \\ c_2 d_1 + c_1 d_2 \\ c_2 d_2 \end{bmatrix} \qquad (2)$$

These "convolution matrices" are the key to signal processing. In that highly active world, the matrix $C$ is a *filter*. The way to understand this filter is through its frequency response $c_0 + c_1 e^{-i\theta} + c_2 e^{-2i\theta}$.

We are ready to connect convolution with Fourier series and Laplace transforms.

## Multiplying $f(x)g(x)$ is Convolution of Coefficients

Convolution answers a question that we unavoidably ask. When $\sum c_k e^{ikx}$ multiplies $\sum d_l e^{ilx}$ (call those functions $f(x)$ and $g(x)$), *what are the Fourier coefficients of the function* $h(x) = f(x)g(x)$? The answer is certainly not $c_k d_k$. We have to multiply every coefficient $c_k$ times every coefficient $d_l$. All those small multiplications $c_k d_l$ produce the coefficients of $(\sum c_k e^{ikx})(\sum d_l e^{ilx})$. The logic of the convolution rule has two steps:

1. $c_k e^{ikx}$ times $d_l e^{ilx}$ equals $c_k d_l e^{inx}$ when $\boldsymbol{k + l = n}$.

2. The $e^{inx}$ term in $f(x)g(x)$ contains every product $c_k d_l$ in which $\boldsymbol{l = n - k}$.

**The $n^{\text{th}}$ Fourier coefficient of $(\sum c_k e^{ikx})(\sum d_l e^{ilx})$ is the $n^{\text{th}}$ component of $\boldsymbol{c} * \boldsymbol{d}$:**

**Multiply functions $f, g$**
**Convolve coefficients $\boldsymbol{c}, \boldsymbol{d}$**

$$\textbf{Coefficient of } fg = (\boldsymbol{c} * \boldsymbol{d})_n = \sum_{k=-\infty}^{\infty} c_k\, d_{n-k}. \qquad (3)$$

**Example 1** The "identity vector" in convolution is $\boldsymbol{\delta} = (\ldots, 0, 0, 1, 0, 0, \ldots)$. Then $\boldsymbol{\delta} * \boldsymbol{d} = \boldsymbol{d}$ for every vector $\boldsymbol{d}$. The "identity function" is $i(x) = 1$. Then $i(x)g(x) = g(x)$ for every function $g$. The Fourier coefficients of $i(x) = 1$ are exactly $\boldsymbol{\delta}$.

You see how **convolution in frequency space ($k$ - space)** leads to **multiplication in function space ($x$ - space)**. This is the central idea of the convolution rule.

**Example 2** The **autocorrelation** of a vector $\boldsymbol{c}$ is the convolution $\boldsymbol{c} * \boldsymbol{c}'$. That vector $\boldsymbol{c}'$ is the reverse of $\boldsymbol{c}$. The components of $\boldsymbol{c}'$ are the Fourier coefficients $\overline{c}_{-k}$ of $\overline{f(x)}$. So autocorrelation $\boldsymbol{c} * \boldsymbol{c}'$ gives the Fourier coefficients of the product $f(x)\overline{f(x)} = |f(x)|^2$:

$$f\overline{f} = (1 + e^{ix})(1 + e^{-ix}) = \mathbf{1}\, e^{-ix} + \mathbf{2} + \mathbf{1}\, e^{ix} \qquad \boldsymbol{c} * \boldsymbol{c}' = (0, 1, 1) * (1, 1, 0) = (\mathbf{1}, \mathbf{2}, \mathbf{1}).$$

The autocorrelation of the box vector (0,1,1) is the hat vector (**1,2,1**). $Box * box = hat$.

## Convolution of Functions

The reverse question is equally important and has to be answered. If $f(x)$ and $g(x)$ have Fourier coefficients $c_k$ and $d_k$, **what function has the Fourier coefficients $c_k d_k$** ? We are multiplying vectors in $k$-space. Then we have convolution $f * g$ of functions in $x$-space!

**Periodic Convolution** $$(f * g)(x)=\int_0^{2\pi} f(y)g(x-y)dy=\int_0^{2\pi} g(y)f(x-y)dy. \qquad (4)$$

Vector convolution is $(\boldsymbol{c} * \boldsymbol{d})_n = \sum c_i d_{n-i}$. The key is $\boldsymbol{i} + (\boldsymbol{n} - \boldsymbol{i}) = \boldsymbol{n}$. Convolution of functions has an integral instead of a sum (of course). Above all we notice that $\boldsymbol{y} + (\boldsymbol{x} - \boldsymbol{y}) = \boldsymbol{x}$. The pattern stays exactly the same when the functions are not periodic and the integrals go from $-\infty$ to $\infty$:

**Infinite Convolution** $$(f * g)(x) = \int_{-\infty}^{\infty} f(y)g(x-y)\,dy = \int_{-\infty}^{\infty} g(y)f(x-y)\,dy. \qquad (5)$$

For the Laplace transform, all functions are zero for $t < 0$. Change $x$ and $y$ to $t$ and $T$.

**One-sided Laplace** $$(f * g)(t) = \int_{\mathbf{0}}^{\boldsymbol{t}} f(T)g(t-T)\,dT \quad \text{because} \quad \begin{matrix} f(T) = 0 \text{ for } T < 0 \\ g(t-T) = 0 \text{ for } T > t \end{matrix}$$

## Solving Differential Equations by Convolution

I want to apply convolution to the main problem of this book—the solution of equations like $y' - ay = f(t)$ and $y'' + y = f(x)$. Those are easy problems and we know the answers. Simplicity is good, it keeps the main point clear. Convolution will offer us a new way to write the solutions $y(t)$ from Laplace and $y(x)$ from Fourier.

I will recall the old ways to solve the same equations. The next page has a summary of the outstanding examples in this book—*linear equations with constant coefficients.*

**Example 3** Solve the equation $y' - ay = f(t)$ by convolution, starting from $y(0) = 0$.

*Solution* Take the Laplace transform of both sides, and divide to find $Y(s)$:

$$sY(s) - aY(s) = F(s) \quad \text{gives} \quad Y(s) = \frac{F(s)}{s-a} = \boldsymbol{G(s)\,F(s)}. \qquad (6)$$

*The transform $F(s)$ of the driving function is multiplied by the "transfer function" $G(s)$.* In this problem $G(s) = 1/(s-a)$. Then $y(t)$ is the inverse transform of $Y(s) = G(s)F(s)$.

The key is convolution. **Multiplication in $s$ - space becomes convolution in $t$ - space.** This rule gives the solution $\boldsymbol{y = g * f}$ from $Y = GF$. Then we prove the rule.

**The inverse transform of the transfer function $G(s)$ is the impulse response $g(t)$.** For the equation $y' - ay = f(t)$, the transfer function is $G(s) = 1/(s-a)$ and its inverse transform is $g(t) = e^{at}$. Then the multiplication $Y(s) = G(s)F(s)$ becomes a convolution of the impulse response $e^{at}$ with the driving function $f(t)$:

| **Solution by convolution** | $y(t) = g(t) * f(t) = \displaystyle\int_{T=0}^{t} e^{a(t-T)} f(T)\,dT$ | (7) |
|---|---|---|

*Please recognize this solution.* We are integrating $e^{-at} f(t)$ for the fourth time ! The central problem of Chapter 1 was $y' - ay = f(t)$ (*or* $q(t)$). There we proposed three methods.

1. The **integrating factor** $e^{-at}$ multiplies $y' - ay = f(t)$. Integrate $(e^{-at}y)' = e^{-at}f$.
2. **Variation of parameters** in the null solution $y_n = Ce^{at}$ gives $y_p(t) = C(t)\,e^{at}$.
3. Every input $f(T)$ is multiplied by its **growth factor** $e^{a(t-T)}$. Combine the outputs.
4. (New) The solution $y(t)$ is the **convolution** of $f(t)$ with the impulse response $e^{at}$.

The impulse response is $g(t) = g * \delta$, when the input is the impulse $f(t) = \delta(t)$. The forced response is $y = g * f$, when the force is $f(t)$. **Always the convolution of the driving force $f(t)$ with the Green's function $g(t)$ produces the output $y(t)$.**

*Confession* I used Green's name partly because the letter $g$ appeared so conveniently. My deeper reason is to express a central idea that connects differential equations and matrix equations—the two themes of this book. *Convolution with the impulse response (the Green's function) is just like multiplication by the inverse matrix $A^{-1}$.*

Here is the message that comes from $AA^{-1} = I$. The vector $\boldsymbol{g}_j$ in column $j$ of $A^{-1}$ is the response to the delta vector $\boldsymbol{\delta}_j = (\cdot, 0, \mathbf{1}, 0, \cdot)$ in column $j$ of the identity matrix.

$$A\boldsymbol{g}_j = \boldsymbol{\delta}_j \quad \textbf{in linear algebra} \qquad \boldsymbol{g}' - a\boldsymbol{g} = \boldsymbol{\delta}(t) \quad \textbf{in differential equations}$$

I hope you find this helpful. The Green's function $g(t-T)$ gives the response at time $t$ to a unit impulse at time $T$. The total response at $t$ is the integral of impulses $f(T)$ times responses $g(t-T)$. Compare with the solution $\boldsymbol{v} = A^{-1}\boldsymbol{b}$ to a matrix equation $A\boldsymbol{v} = \boldsymbol{b}$.

The inverse matrix $A^{-1}$ gives the response at position $i$ to a unit impulse at position $j$. The solution $\boldsymbol{v} = A^{-1}\boldsymbol{b}$ is the sum over all $j$ of impulses $b_j$ times those responses.

For shift-invariant equations, the response at $t$ to an impulse at $T$ depends only on the elapsed time $t - T$. For shift-invariant matrices, the responses $(A^{-1})_{ij}$ depend only on $i - j$. The differential equation has **constant coefficients**. The Toeplitz matrix has **constant diagonals**. Here $A$ is a difference matrix and $A^{-1}$ is a sum matrix.

$$A\boldsymbol{v} = \begin{bmatrix} 1 & & \\ -1 & 1 & \\ 0 & -1 & 1 \end{bmatrix} \begin{bmatrix} v_1 \\ v_2 \\ v_3 \end{bmatrix} = \begin{bmatrix} b_1 \\ b_2 \\ b_3 \end{bmatrix} \qquad \boldsymbol{v} = A^{-1}\boldsymbol{b} = \begin{bmatrix} 1 & & \\ 1 & 1 & \\ 1 & 1 & 1 \end{bmatrix} \begin{bmatrix} b_1 \\ b_2 \\ b_3 \end{bmatrix}. \qquad (8)$$

**Example 4** (**Fourier**) Solve the equation $-y'' + y = f(x)$ for $-\infty < x < \infty$.

**Solution** This is a boundary value problem, with $y = 0$ at the endpoints $x = -\infty$ and $x = \infty$. Take the Fourier transform of every term, so the two derivatives in $y''$ become multiplications by $ik$:

$$-\boldsymbol{y'' + y = f(x)} \qquad -(ik)^2\hat{y} + \hat{y} = \hat{f}(k) \qquad \boldsymbol{\hat{y}(k) = \frac{\hat{f}(k)}{k^2+1} = \hat{g}(k)\,\hat{f}(k)}. \tag{9}$$

In $k$-space, the transform $\hat{f}(k)$ is multiplied by $\hat{g}(k) = 1/(k^2+1)$. In $x$-space, the right side $f(x)$ is convolved with the Green's function $g(x)$. *That Green's function $g(x)$ is the solution when the right side $f(x)$ is a delta function $\delta(x)$.*

To complete the solution we need $g(x)$. The transform approach would invert $\hat{g}(k) = 1/(k^2+1)$. The direct approach is to solve $-g'' + g = \delta(x)$. Remember that $\delta(x) = 0$ for $x > 0$ and $x < 0$:

$$\begin{array}{llll} \boldsymbol{x > 0} & -g'' + g = 0 \text{ gives } g = c_1e^x + c_2e^{-x} & & \text{Then } g(\infty) = 0 \text{ requires } c_1 = 0 \\ \boldsymbol{x < 0} & -g'' + g = 0 \text{ gives } g = C_1e^x + C_2e^{-x} & & \text{Then } g(-\infty) = 0 \text{ requires } C_2 = 0 \end{array}$$

The action is all at $x = 0$. There is no jump in the function $g(x)$, so that $C_1 = c_2$. The minus sign in $-g'' + g = \delta(x)$ produces a **drop of 1** in the slope $g'(x)$ at $x = 0$. Comparing the slopes $-c_2e^{-x}$ and $C_1e^x$ at $x = 0$ gives $C_1 + c_2 = 1$. The coefficients are $C_1 = c_2 = \frac{1}{2}$ and the Green's function $g(x)$ is found:

$$\boldsymbol{g(x)} = \begin{cases} \frac{1}{2}e^{-x} & \text{for } x > 0 \\ \frac{1}{2}e^{x} & \text{for } x < 0 \end{cases} \qquad \text{and convolution gives } y(x) = \int_{-\infty}^{\infty} f(X)g(x-X)\,dX.$$

Compare with this second order equation in time, when Fourier changes to Laplace. Now we have initial values at $t = 0$ instead of boundary values at $x = \pm\infty$.

**Example 5** Solve the equation $y'' + y = f(t)$ starting from $y(0) = y'(0) = 0$.

**Solution** Take the Laplace transform of both sides, and divide by $s^2 + 1$ to find $Y(s)$:

$$s^2Y(s) + Y(s) = F(s) \quad \text{gives} \quad Y(s) = \boldsymbol{\frac{F(s)}{s^2+1} = F(s)G(s)}. \tag{10}$$

The transfer function is $G(s) = 1/(s^2+1)$. That is the Laplace transform of the impulse response (the growth factor) $\boldsymbol{g(t) = \sin t}$. (Problem 8.5.2 confirms that $(\sin t)''$ does surprisingly produce $\delta(t)$. The slope is zero for $t < 0$, and $(\sin t)'$ jumps to $\cos 0 = 1$ at $t = 0$.) Multiplication $F(s)G(s)$ corresponds to convolution $f * g$:

$$\textbf{Laplace convolution} \qquad y(t) = f(t) * g(t) = \int_0^t f(T)\sin(t-T)\,dT. \tag{11}$$

This solves Example 5 quickly—the crucial step is to be able to invert $G(s)$ to find $g(t)$.

## Proof of the Convolution Rule

We need to prove that the Laplace transform of $f(t) * g(t)$ is $F(s)G(s)$. Convolution becomes multiplication. Similarly the Fourier transform of $f(x) * g(x)$ is $\widehat{f}(k)\widehat{g}(k)$.

An integral over $T$ produces $f * g$, and then an integral over $t$ gives its transform. The key is to reverse the order in that double integral. Integrate first with respect to $t$.

$$\int_{t=0}^{\infty}\left(\int_{T=0}^{\infty} f(T)g(t-T)\,dT\right)e^{-st}\,dt = \int_{T=0}^{\infty}\left(\int_{t=0}^{\infty} g(t-T)e^{-s(t-T)}\,dt\right)f(T)e^{-sT}\,dT.$$

It was safe to extend the integration to $T = \infty$, since $g(t-T) = 0$ for $T > t$. Also safe to insert $e^{sT}$ and $e^{-sT}$; their product is 1. The inner integral on the right is exactly the Laplace transform $G(s)$, when $t - T$ is replaced by $\tau$:

$$\int_{t=0}^{\infty} g(t-T)e^{-s(t-T)}\,dt = \int_{\tau=-T}^{\infty} g(\tau)e^{-s\tau}\,d\tau = \int_{\tau=0}^{\infty} g(\tau)e^{-s\tau}\,d\tau = G(s). \qquad (12)$$

Since the inner integral is $G(s)$, the double integral is $F(s)\,G(s)$ as desired:

$$\int_{T=0}^{\infty} G(s)f(T)\,e^{-sT}\,dT = F(s)\,G(s). \quad \text{The convolution rule is proved.}$$

The same rule holds for Fourier transforms, except the integrals have $-\infty < x < \infty$ and $-\infty < k < \infty$. With those limits we don't have or need the one-sided condition that $g(t) = 0$ for $t < 0$. The steps are the same and we reach the same conclusion. *The Fourier transform of* $f(x) * g(x)$ *is* $\widehat{f}(k)\,\widehat{g}(k)$.

## Point-Spread Functions and Deconvolution

I must not leave the impression that convolution is only useful in solving differential equations. The truth is, we solved those equations earlier. Our solutions now have the neat form $y = f * g$, but they were already found without convolutions. A better application is a telescope looking at the night sky, or a CT-scanner looking inside you.

A telescope produces a *blurred image*. When the actual star is a point source, we don't see that delta function. **The image of $\boldsymbol{\delta(x, y)}$ is a point-spread function $\boldsymbol{g(x, y)}$**: the response to an impulse, the spreading of a point. With diffraction you see an "Airy disk" at the center. The radius of this disk gives the limit of resolution for a telescope.

*When the star is shifted, the image is shifted.* The source $\delta(x - x_0, y - y_0)$ produces the image $g(x - x_0, y - y_0)$. It is bright at the location $x_0, y_0$ of the star, and $g$ gets dark quickly away from that point. The image of the whole sky is an integral of blurred points.

The true brightness of the night sky is given by a function $f(x, y)$. *The image we see is the convolution* $c = f * g$. But if we do know the blurring function $g(x, y)$, *deconvolution will bring back $f(x, y)$ from $f * g$.* In transform space, the scanner multiplies by $G$ and the post-processor divides by $G$. Here is deconvolution:

$$c = f * g \text{ transforms to } C = FG. \quad \text{The inverse transform of } F = \frac{C}{G} \text{ gives } f.$$

The manufacturer knows the point-spread function $g$ and its Fourier transform $G$. The telescope or the CT-scanner comes equipped with a code for deconvolution. Transform the blurred output $c$ to $C$, divide by $G$, and invert $F = C/G$ to find the true source function $f$.

Note that two-dimensional functions $f(x, y)$ have two-dimensional transforms $\widehat{f}(k, l)$. The Fourier basis functions of $x$ and $y$ are $e^{ikx}e^{ily}$ with two frequencies $k$ and $l$.

## Cyclic Convolution and the DFT

The Discrete Fourier Transform connects $\boldsymbol{c} = (c_0, \ldots, c_{N-1})$ to $\boldsymbol{f} = (f_0, \ldots, f_{N-1})$. The Fourier matrix gives $\boldsymbol{Fc} = \boldsymbol{f}$. Computations are fast, because all the vectors are $N$-dimensional and the FFT is available. A convolution rule will lead directly to fast multiplication and fast algorithms. This is convolution in practice.

The rule has to change from $\boldsymbol{c} * \boldsymbol{d} = (1, 1, 2) * (2, 1, 3) = (2, 3, 8, 5, 6)$. When the inputs $\boldsymbol{c}$ and $\boldsymbol{d}$ have $N$ components, *their cyclic convolution also has $N$ components. The new symbol in* $(1, 1, 2) \circledast (2, 1, 3) = (7, 9, 8)$ *indicates "cyclic" by a circle in* $\circledast$.

The key is that $\boldsymbol{w^3 = 1}$. Cyclic convolution folds $5w^3 + 6w^4$ back into $5 + 6w$.

$$(\mathbf{1} + \mathbf{1}w + \mathbf{2}w^2)(\mathbf{2} + \mathbf{1}w + \mathbf{3}w^2) = 2 + 3w + 8w^2 + 5w^3 + 6w^4 = \mathbf{7} + \mathbf{9}w + \mathbf{8}w^2.$$

In the same way, $(0, \mathbf{1}, 0) \circledast (0, 0, \mathbf{1}) = (\mathbf{1}, 0, 0)$ because $w$ times $w^2$ equals $w^3 = 1$. I will use this example to test the cyclic convolution rule.

**Cyclic convolution rule for the $N$-point transform**

> The $k$th component of $F(\boldsymbol{c} \circledast \boldsymbol{d})$ is $(F\boldsymbol{c})_k$ times $(F\boldsymbol{d})_k$. That word "times" means: Multiply $1, w, w^2$ from $F\boldsymbol{c}$ and $1, w^2, w^4$ from $F\boldsymbol{d}$ to get $1, w^3, w^6$, which is $1, 1, 1$.

$$\boldsymbol{F} = \begin{bmatrix} \mathbf{1} & \mathbf{1} & \mathbf{1} \\ \mathbf{1} & \boldsymbol{w} & \boldsymbol{w^2} \\ \mathbf{1} & \boldsymbol{w^2} & \boldsymbol{w^4} \end{bmatrix} \quad F\begin{bmatrix} 0 \\ 1 \\ 0 \end{bmatrix} = \begin{bmatrix} \mathbf{1} \\ \boldsymbol{w} \\ \boldsymbol{w^2} \end{bmatrix} \text{ times } F\begin{bmatrix} 0 \\ 0 \\ 1 \end{bmatrix} = \begin{bmatrix} \mathbf{1} \\ \boldsymbol{w^2} \\ \boldsymbol{w^4} \end{bmatrix} \text{ is } F\begin{bmatrix} 1 \\ 0 \\ 0 \end{bmatrix} = \begin{bmatrix} \mathbf{1} \\ \mathbf{1} \\ \mathbf{1} \end{bmatrix}$$

The convolution $\boldsymbol{c} \circledast \boldsymbol{d}$ has $N^2$ small multiplications. Component by component multiplication of two vectors only needs $N$. So the convolution rule gives a fast way to multiply two very long $N$-digit numbers (as in the prime factors that banks use for security). When you multiply the numbers, you are convolving those digits.

**Transform the numbers to $f$ and $g$. Multiply transforms by $f_k g_k$. Transform back.**

When the cost of these three discrete transforms is included, the FFT saves the day:

**Go to $k$-space, multiply, go back** $N^2$ multiplications are reduced to $N+3N\log N$. In MATLAB, component-by-component multiplication is indicated by $\boldsymbol{f}.*\boldsymbol{g}$ (point-star).

$$F(\boldsymbol{c} \circledast \boldsymbol{d}) = (F\boldsymbol{c}).*(F\boldsymbol{d}) \qquad \text{ifft}(\boldsymbol{c} \circledast \boldsymbol{d}) = N * \text{ifft}(\boldsymbol{c}).*\text{ifft}(\boldsymbol{d}) \tag{13}$$

Note that the fft command transforms $\boldsymbol{f}$ to $\boldsymbol{c}$ using $\overline{w} = e^{-2\pi i/N}$ and the matrix $\overline{F}$. The ifft command inverts that transform using $w = e^{2\pi i/N}$ and the Fourier matrix $F$. The factor $N$ appears in equation (13) because $F\overline{F} = NI$.

## Circulant Matrices

Multiplication by an infinite constant-diagonal matrix gives an infinite convolution. When row $n$ of $C_\infty$ multiplies $\boldsymbol{d}$, this adds up the small multiplications $c_i d_j$ with $i + j = n$:

**Infinite convolution**

$$C_\infty \boldsymbol{d} = \begin{bmatrix} \bullet & \bullet & \bullet & \bullet \\ c_0 & c_{-1} & c_{-2} & \bullet \\ c_1 & c_0 & c_{-1} & c_{-2} \\ c_2 & c_1 & c_0 & c_{-1} \\ \bullet & c_2 & c_1 & c_0 \end{bmatrix} \begin{bmatrix} \bullet \\ d_0 \\ d_1 \\ d_2 \\ \bullet \end{bmatrix} = \boldsymbol{c} * \boldsymbol{d}. \tag{14}$$

Similarly, **cyclic convolution comes from an $N$ by $N$ matrix**. The matrix is called a "circulant" because every diagonal wraps around (based on $w^N = 1$). All diagonals have $N$ equal entries. The diagonal with $c_1$ is highlighted for $N = 4$:

**Cyclic convolution**
**Circulant matrix**

$$C\boldsymbol{d} = \begin{bmatrix} c_0 & c_3 & c_2 & \boldsymbol{c_1} \\ \boldsymbol{c_1} & c_0 & c_3 & c_2 \\ c_2 & \boldsymbol{c_1} & c_0 & c_3 \\ c_3 & c_2 & \boldsymbol{c_1} & c_0 \end{bmatrix} \begin{bmatrix} d_0 \\ d_1 \\ d_2 \\ d_3 \end{bmatrix} = \boldsymbol{c} \circledast \boldsymbol{d}. \tag{15}$$

Notice how the top row produces $c_0d_0 + c_3d_1 + c_2d_2 + c_1d_3$. Those subscripts $0 + 0$ and $3 + 1$ and $2 + 2$ are all zero when $N = 4$. *In this cyclic world*, 2 *and* 2 *add to* 0. That comes from $w^2w^2 = w^4 = w^0$.

Circulant matrices are remarkable. If you multiply circulants $B$ and $C$ you get another circulant. That product $BC$ gives convolution with the vector $\boldsymbol{b} \circledast \boldsymbol{c}$. The amazing part is the eigenvalues from the DFT and eigenvectors from the Fourier matrix:

The eigenvalues of $C$ are the components of the discrete transform $F\boldsymbol{c}$
The eigenvectors of *every* $C$ are the columns of $F$ (also the columns of $\overline{F}$ and $F^{-1}$)

We can verify two eigenvalues $\lambda = c_0 + c_1 + c_2$ and $c_0 + c_1w + c_2w^2$ for this circulant:

$$\begin{bmatrix} c_0 & c_2 & c_1 \\ c_1 & c_0 & c_2 \\ c_2 & c_1 & c_0 \end{bmatrix} \begin{bmatrix} 1 \\ 1 \\ 1 \end{bmatrix} = \lambda \begin{bmatrix} 1 \\ 1 \\ 1 \end{bmatrix} \qquad \begin{bmatrix} c_0 & c_2 & c_1 \\ c_1 & c_0 & c_2 \\ c_2 & c_1 & c_0 \end{bmatrix} \begin{bmatrix} 1 \\ w^2 \\ w \end{bmatrix} = \lambda \begin{bmatrix} 1 \\ w^2 \\ w \end{bmatrix}. \tag{16}$$

The equation $FC = \Lambda F$ is the cyclic convolution rule $F(\boldsymbol{c} \circledast \boldsymbol{d}) = (F\boldsymbol{c}).*(F\boldsymbol{d})$.

## The End of the Book

The book is ending on a high note. Constant coefficient problems have taken a big step from $Ay'' + By' + Cy = 0$. Now we have transforms (Fourier and Laplace) and convolutions. The discrete problems bring constant diagonal matrices. Cyclic problems bring circulants. Time to stop!

I should really say, *stop and look back*. The book has emphasized linear problems, because these are the equations we can understand. It is true that life is not linear. If the input is multiplied by 10, the output might be multiplied by 8 or 12 and not 10. But in most real problems, the input is multiplied or divided by less than 1.1. Then a linear model replaces a curve by its tangent lines (this is the key to calculus). To understand applied mathematics, we need differential equations and linear algebra.

## ■ REVIEW OF THE KEY IDEAS ■

1. Convolution $(1,2,3) * (4,5,6)$ is the multiplication $123 \times 456$ without carrying.
2. $(\sum c_k e^{ikx})(\sum d_l e^{ilx})$ has $(c * d)_n = \sum c_k d_{n-k}$ as the coefficient of $e^{inx}$. **Multiply functions $\leftrightarrow$ convolve coefficients** as in $(1+2x+3x^2)(4+5x+6x^2)$.
3. Differential equations transform to $Y(s) = F(s)G(s)$. Then $y(t) = f(t) * g(t) =$ driving force $*$ impulse response. The impulse response $g(t)$ is the Green's function.
4. **Shift invariance**: Constant coefficient equations and constant diagonal matrices.
5. Circulants $C\boldsymbol{d}$ give cyclic convolution $\boldsymbol{c} \circledast \boldsymbol{d}$. Multiply components $(F\boldsymbol{c}).*(F\boldsymbol{d})$.

## Problem Set 8.6

**1** Find the convolution $\boldsymbol{v} * \boldsymbol{w}$ and also the cyclic convolution $\boldsymbol{v} \circledast \boldsymbol{w}$:

(a) $\boldsymbol{v} = (1,2)$ and $\boldsymbol{w} = (2,1)$ (b) $\boldsymbol{v} = (1,2,3)$ and $\boldsymbol{w} = (4,5,6)$.

**2** Compute the convolution $(1,3,1) * (2,2,3) = (a,b,c,d,e)$. To check your answer, add $a+b+c+d+e$. That total should be 35 since $1+3+1=5$ and $2+2+3=7$ and $5 \times 7 = 35$.

**3** Multiply $1+3x+x^2$ times $2+2x+3x^2$ to find $a+bx+cx^2+dx^3+ex^4$. Your multiplication was the same as the convolution $(1,3,1) * (2,2,3)$ in Problem 2. When $x = 1$, your multiplication shows why $1+3+1=5$ times $2+2+3=7$ agrees with $a+b+c+d+e = 35$.

**4** (Deconvolution) Which vector $\boldsymbol{v}$ would you convolve with $\boldsymbol{w} = (1,2,3)$ to get $\boldsymbol{v} * \boldsymbol{w} = (0,1,2,3,0)$? Which $\boldsymbol{v}$ gives $\boldsymbol{v} \circledast \boldsymbol{w} = (3,1,2)$?

**5** (a) For the periodic functions $f(x) = 4$ and $g(x) = 2\cos x$, show that $f * g$ is **zero** (the zero function)!

(b) In frequency space ($k$-space) you are multiplying the Fourier coefficients of 4 and $2\cos x$. Those coefficients are $c_0 = 4$ and $d_1 = d_{-1} = 1$. Therefore every product $c_k d_k$ is ____.

**6** For periodic functions $f = \sum c_k e^{ikx}$ and $g = \sum d_k e^{ikx}$, the Fourier coefficients of $f * g$ are $\boldsymbol{2\pi c_k d_k}$. Test this factor $2\pi$ when $f(x) = 1$ and $g(x) = 1$ by computing $f * g$ from its definition (4).

**7** Show by integration that the periodic convolution $\int_0^{2\pi} \cos x \cos(t - x)dx$ is $\pi \cos t$. In $k$-space you are squaring Fourier coefficients $c_1 = c_{-1} = \frac{1}{2}$ to get $\frac{1}{4}$ and $\frac{1}{4}$; these are the coefficients of $\frac{1}{2}\cos t$. The $2\pi$ in Problem 6 makes $\pi \cos t$ correct.

**8** Explain why $f * g$ is the same as $g * f$ (periodic or infinite convolution).

**9** What 3 by 3 circulant matrix $C$ produces cyclic convolution with the vector $\boldsymbol{c} = (1, 2, 3)$? Then $C\boldsymbol{d}$ equals $\boldsymbol{c} \circledast \boldsymbol{d}$ for every vector $\boldsymbol{d}$. Compute $\boldsymbol{c} \circledast \boldsymbol{d}$ for $\boldsymbol{d} = (0, 1, 0)$.

**10** What 2 by 2 circulant matrix $C$ produces cyclic convolution with $\boldsymbol{c} = (1, 1)$? Show in four ways that this $C$ is not invertible. Deconvolution is impossible.

(1) Find the determinant of $C$. (2) Find the eigenvalues of $C$.

(3) Find $\boldsymbol{d}$ so that $C\boldsymbol{d} = \boldsymbol{c} \circledast \boldsymbol{d}$ is zero. (4) $F\boldsymbol{c}$ has a zero component.

**11** (a) Change $b(x) * \delta(x - 1)$ to a multiplication $\widehat{b}\,\widehat{d}$. Transform the box function $b(x) = \{1 \text{ for } 0 \le x \le 1\}$ to $\widehat{b}(k) = \int_0^1 e^{-ikx}dx$. The shifted delta transforms to $\widehat{d}(k) = \int \delta(x - 1)e^{-ikx}dx$.

(b) Show that your result $\widehat{b}\,\widehat{d}$ is the transform of a shifted box function. Then convolution with $\delta(x - 1)$ shifts the box.

**12** Take the Laplace transform of these equations to find the transfer function $G(s)$:

(a) $Ay'' + By' + Cy = \delta(t)$ (b) $y' - 5y = \delta(t)$ (c) $2y(t) - y(t-1) = \delta(t)$

**13** Take the Laplace transform of $y'''' = \delta(t)$ to find $Y(s)$. From the Transform Table in Section 8.5 find $y(t)$. You will see $y''' = 1$ and $y'''' = 0$. But $y(t) = 0$ for negative $t$, so your $y'''$ is actually a unit step function and your $y''''$ is actually $\delta(t)$.

**14** Solve these equations by Laplace transform to find $Y(s)$. Invert that transform with the Table in Section 8.5 to recognize $y(t)$.

(a) $y' - 6y = e^{-t}$, $y(0) = 2$ (b) $y'' + 9y = 1$, $y(0) = y'(0) = 0$.

**15** Find the Laplace transform of the shifted step $H(t-3)$ that jumps from 0 to 1 at $t=3$. Solve $y'-ay=H(t-3)$ with $y(0)=0$ by finding the Laplace transform $Y(s)$ and then its inverse transform $y(t)$ : one part for $t<3$, second part for $t\geq 3$.

**16** Solve $y'=1$ with $y(0)=4$—a trivial question. Then solve this problem the slow way by finding $Y(s)$ and inverting that transform.

**17** The solution $y(t)$ is the convolution of the input $f(t)$ with what function $g(t)$ ?

(a) $y'-ay=f(t)$ with $y(0)=3$ (b) $y'-$ (integral of $y$) $=f(t)$.

**18** For $y'-ay=f(t)$ with $y(0)=3$, we could replace that initial value by adding $3\delta(t)$ to the forcing function $f(t)$. Explain that sentence.

**19** What is $\delta(t)*\delta(t)$ ? What is $\delta(t-1)*\delta(t-2)$ ? What is $\delta(t-1)$ *times* $\delta(t-2)$?

**20** By Laplace transform, solve $y'=y$ with $y(0)=1$ to find a very familiar $y(t)$.

**21** By Fourier transform as in (9), solve $-y''+y=$ box function $b(x)$ on $0\leq x\leq 1$.

**22** There is a big difference in the solutions to $y''+By'+Cy=f(x)$, between the cases $B^2<4C$ and $B^2>4C$. Solve $y''+y=\delta$ and $y''-y=\delta$ with $y(\pm\infty)=0$.

**23** *(Review)* Why do the constant $f(t)=1$ and the unit step $H(t)$ have the same Laplace transform $1/s$ ? Answer: Because the transform does not notice _____.

# MATRIX FACTORIZATIONS

1. $A = LU = \begin{pmatrix} \text{lower triangular } L \\ \text{1's on the diagonal} \end{pmatrix} \begin{pmatrix} \text{upper triangular } U \\ \text{pivots on the diagonal} \end{pmatrix}$

   **Requirements**: No row exchanges as Gaussian elimination reduces $A$ to $U$.

2. $A = LDU = \begin{pmatrix} \text{lower triangular } L \\ \text{1's on the diagonal} \end{pmatrix} \begin{pmatrix} \text{pivot matrix} \\ D \text{ is diagonal} \end{pmatrix} \begin{pmatrix} \text{upper triangular } U \\ \text{1's on the diagonal} \end{pmatrix}$

   **Requirements**: No row exchanges. The pivots in $D$ are divided out to leave 1's on the diagonal of $U$. If $A$ is symmetric then $U$ is $L^{\mathrm{T}}$ and $\boldsymbol{A} = \boldsymbol{LDL^{\mathrm{T}}}$.

3. $\boldsymbol{PA} = \boldsymbol{LU}$ (permutation matrix $P$ to avoid zeros in the pivot positions).

   **Requirements**: $A$ is invertible. Then $P, L, U$ are invertible. $P$ does all of the row exchanges in advance, to allow normal $LU$. Alternative: $A = L_1 P_1 U_1$.

4. $\boldsymbol{EA} = \boldsymbol{R}$ ($m$ by $m$ invertible $E$) (any matrix $A$) = rref($A$).

   **Requirements**: None! *The reduced row echelon form* $R$ has $r$ pivot rows and pivot columns. The only nonzero in a pivot column is the unit pivot. The last $m - r$ rows of $E$ are a basis for the left nullspace of $A$; they multiply $A$ to give zero rows in $R$. The first $r$ columns of $E^{-1}$ are a basis for the column space of $A$.

5. $\boldsymbol{S} = \boldsymbol{C^{\mathrm{T}}C}$ = (lower triangular) (upper triangular) with $\sqrt{D}$ on both diagonals

   **Requirements**: $S$ is symmetric and positive definite (all $n$ pivots in $D$ are positive). This *Cholesky factorization* $C$ = chol($S$) has $C^{\mathrm{T}} = L\sqrt{D}$, so $C^{\mathrm{T}}C = LDL^{\mathrm{T}}$.

6. $\boldsymbol{A} = \boldsymbol{QR}$ = (orthonormal columns in $Q$) (upper triangular $R$).

   **Requirements**: $A$ has independent columns. Those are *orthogonalized* in $Q$ by the Gram-Schmidt or Householder process. If $A$ is square then $Q^{-1} = Q^{\mathrm{T}}$.

7. $\boldsymbol{A} = \boldsymbol{V\Lambda V^{-1}}$ = (eigenvectors in $V$) (eigenvalues in $\Lambda$) (left eigenvectors in $V^{-1}$).

   **Requirements**: $A$ must have $n$ linearly independent eigenvectors.

8. $\boldsymbol{S} = \boldsymbol{Q\Lambda Q^{\mathrm{T}}}$ = (orthogonal matrix $Q$) (real eigenvalue matrix $\Lambda$) ($Q^{\mathrm{T}}$ is $Q^{-1}$).

   **Requirements**: $S$ is *real and symmetric*. This is the Spectral Theorem.

**9.** $A = MJM^{-1}$ = (generalized eigenvectors in $M$) (Jordan blocks in $J$) ($M^{-1}$).

**Requirements**: $A$ is any square matrix. This *Jordan form* $J$ has a block for each independent eigenvector of $A$. Every block has only one eigenvalue.

**10.** $$A = U\Sigma V^{\mathrm{T}} = \begin{pmatrix} \text{orthogonal} \\ U \text{ is } m \times n \end{pmatrix} \begin{pmatrix} m \times n \text{ singular value matrix} \\ \sigma_1, \ldots, \sigma_r \text{ on its diagonal} \end{pmatrix} \begin{pmatrix} \text{orthogonal} \\ V \text{ is } n \times n \end{pmatrix}.$$

**Requirements**: None. This *singular value decomposition* (**SVD**) has the eigenvectors of $AA^{\mathrm{T}}$ in $U$ and eigenvectors of $A^{\mathrm{T}}A$ in $V$; $\sigma_i = \sqrt{\lambda_i(A^{\mathrm{T}}A)} = \sqrt{\lambda_i(AA^{\mathrm{T}})}$.

**11.** $$A^{+} = V\Sigma^{+}U^{\mathrm{T}} = \begin{pmatrix} \text{orthogonal} \\ n \times n \end{pmatrix} \begin{pmatrix} n \times m \text{ pseudoinverse of } \Sigma \\ 1/\sigma_1, \ldots, 1/\sigma_r \text{ on diagonal} \end{pmatrix} \begin{pmatrix} \text{orthogonal} \\ m \times m \end{pmatrix}.$$

**Requirements**: None. The *pseudoinverse* $A^{+}$ has $A^{+}A$ = projection onto row space of $A$ and $AA^{+}$ = projection onto column space. The shortest least-squares solution to $A\boldsymbol{x} = \boldsymbol{b}$ is $\widehat{\boldsymbol{x}} = A^{+}\boldsymbol{b}$. This solves $A^{\mathrm{T}}A\widehat{\boldsymbol{x}} = A^{\mathrm{T}}\boldsymbol{b}$. When $A$ is invertible: $A^{+} = A^{-1}$.

**12.** $A = QH$ = (orthogonal matrix $Q$) (symmetric positive definite matrix $H$).

**Requirements**: $A$ is invertible. This *polar decomposition* has $H^2 = A^{\mathrm{T}}A$. The factor $H$ is semidefinite if $A$ is singular. The reverse polar decomposition $A = KQ$ has $K^2 = AA^{\mathrm{T}}$. Both have $Q = UV^{\mathrm{T}}$ from the SVD.

**13.** $A = U\Lambda U^{-1}$ = (unitary $U$) (eigenvalue matrix $\Lambda$) ($U^{-1}$ which is $U^{\mathrm{H}} = \overline{U}^{\mathrm{T}}$).

**Requirements**: $A$ is *normal*: $A^{\mathrm{H}}A = AA^{\mathrm{H}}$. Its orthonormal (and possibly complex) eigenvectors are the columns of $U$. Complex $\lambda$'s unless $A = A^{\mathrm{H}}$: Hermitian case.

**14.** $A = UTU^{-1}$ = (unitary $U$) (triangular $T$ with $\lambda$'s on diagonal) ($U^{-1} = U^{\mathrm{H}}$).

**Requirements**: *Schur triangularization* of any square $A$. There is a matrix $U$ with orthonormal columns that makes $U^{-1}AU$ triangular:

**15.** $$F_n = \begin{bmatrix} I & D \\ I & -D \end{bmatrix} \begin{bmatrix} F_{n/2} & \\ & F_{n/2} \end{bmatrix} \begin{bmatrix} \text{even-odd} \\ \text{permutation} \end{bmatrix} = \text{one step of the (recursive) } \textbf{FFT}.$$

**Requirements**: $F_n$ = Fourier matrix with entries $w^{jk}$ where $w^n = 1$: $F_n\overline{F}_n = nI$. $D$ has $1, w, \ldots, w^{n/2-1}$ on its diagonal. For $n = 2^{\ell}$ the *Fast Fourier Transform* will compute $F_n\boldsymbol{x}$ with only $\frac{1}{2}n\ell = \frac{1}{2}n\log_2 n$ multiplications from $\ell$ stages of $D$'s.

# Properties of Determinants

1 ***The determinant of the n by n identity matrix is* 1.**

2 ***The determinant changes sign when two rows are exchanged*** (sign reversal):

3 ***The determinant is a linear function of each row separately*** (all other rows stay fixed).

| | |
|---|---|
| **multiply row 1 by any number $t$** | $\begin{vmatrix} ta & tb \\ c & d \end{vmatrix} = t \begin{vmatrix} a & b \\ c & d \end{vmatrix}$ |
| **add row 1 of $A$ to row 1 of $A'$** | $\begin{vmatrix} a+a' & b+b' \\ c & d \end{vmatrix} = \begin{vmatrix} a & b \\ c & d \end{vmatrix} + \begin{vmatrix} a' & b' \\ c & d \end{vmatrix}.$ |

Pay special attention to rules 1–3. They completely determine the number det $A$.

4 ***If two rows of $A$ are equal, then* det $A = 0$.**

5 ***Subtracting a multiple of one row from another row leaves* det $A$ *unchanged.***

$\ell$ times row 1 from row 2 $\quad \begin{vmatrix} a & b \\ c-\ell a & d-\ell b \end{vmatrix} = \begin{vmatrix} a & b \\ c & d \end{vmatrix}.$

6 ***A matrix with a row of zeros has* det $A = 0$.**

7 ***If $A$ is triangular then* det $A = a_{11}a_{22}\cdots a_{nn}$ *= product of diagonal entries.***

8 ***If $A$ is singular then* det $A = 0$. *If $A$ is invertible then* det $A \neq 0$.**

***Proof*** Elimination goes from $A$ to $U$. If $A$ is singular then $U$ has a zero row. The rules give det $A$ = det $U$ = 0. If $A$ is invertible then $U$ has the pivots along its diagonal. The product of nonzero pivots (using rule 7) gives a nonzero determinant:

**Multiply pivots** $\quad$ **det $A = \pm$ det $U = \pm$ (product of the pivots).**

9 ***The determinant of $AB$ is* det $A$ *times* det $B$: $|AB| = |A||B|$.**

$A$ times $A^{-1}$ $\quad$ $\boldsymbol{AA^{-1} = I}$ so $\quad$ **(det $A$)(det $A^{-1}$) = det $I$ = 1.**

10 ***The transpose $A^{\mathrm{T}}$ has the same determinant as $A$.***

# Index

**D**

## T



## U

## Index of Symbols

# LINEAR ALGEBRA IN A NUTSHELL

((*The matrix* $A$ *is* $n$ *by* $n$))

| **Nonsingular** | **Singular** |
|---|---|
| $A$ is invertible | $A$ is not invertible |
| The columns are independent | The columns are dependent |
| The rows are independent | The rows are dependent |
| The determinant is not zero | The determinant is zero |
| $A\boldsymbol{x} = \mathbf{0}$ has one solution $\boldsymbol{x} = \mathbf{0}$ | $A\boldsymbol{x} = \mathbf{0}$ has infinitely many solutions |
| $A\boldsymbol{x} = \boldsymbol{b}$ has one solution $\boldsymbol{x} = A^{-1}\boldsymbol{b}$ | $A\boldsymbol{x} = \boldsymbol{b}$ has no solution or infinitely many |
| $A$ has $n$ (nonzero) pivots | $A$ has $r < n$ pivots |
| $A$ has full rank $r = n$ | $A$ has rank $r < n$ |
| The reduced row echelon form is $R = I$ | $R$ has at least one zero row |
| The column space is all of $\mathbf{R}^n$ | The column space has dimension $r < n$ |
| The row space is all of $\mathbf{R}^n$ | The row space has dimension $r < n$ |
| All eigenvalues are nonzero | Zero is an eigenvalue of $A$ |
| $A^{\mathrm{T}}A$ is symmetric positive definite | $A^{\mathrm{T}}A$ is only semidefinite |
| $A$ has $n$ (positive) singular values | $A$ has $r < n$ singular values |